PCR
PRIMER
A LABORATORY MANUAL

PCR PRIMER

A LABORATORY MANUAL

EDITED BY

CARL W. DIEFFENBACH
National Institute of Allergy
and Infectious Diseases

GABRIELA S. DVEKSLER
Uniformed Services University
of the Health Sciences

Cold Spring Harbor Laboratory Press 1995

PCR PRIMER
A LABORATORY MANUAL

Printed in the United States of America
Design by Emily Harste

Library of Congress Cataloging-in-Publication Data

PCR primer : a laboratory manual / edited by Carl W. Dieffenbach,
Gabriela S. Dveksler.
p. cm.
Includes bibliographical references and index.
ISBN 0-87969-447-5 (cloth : alk. paper). -- ISBN 0-87969-448-3
(comb : alk paper)
1. Polymerase chain reaction--Laboratory manuals.
I. Dieffenbach, Carl W. II. Dveksler, Gabriela S.
QP606.D46P359 1995 95-32277
574.87′3282--dc20 CIP

15 14 13 12 11 10

All Cold Spring Harbor Laboratory Press publications may be ordered directly from Cold Spring Harbor Laboratory Press, 500 Sunnyside Boulevard, Woodbury, New York 11797-2924. Phone: 1-800-843-4388 in Continental U.S. and Canada. All other locations: (516) 422-4100. E-mail: cshpress@cshl.org. FAX: (516) 422-4097. For a complete catalog of all Cold Spring Harbor Laboratory Press publications, visit our World Wide Web Site: http://www.cshlpress.com.

Contents

Preface

Technology by its very nature is a developing entity. Thus, to freeze it in time–to get a snapshot of it–is a contrary act. When the technology is the polymerase chain reaction, this is even more true, because this technology continues to develop and expand at a rapid pace. However, from the perspective of investigators in the laboratory trying to cope with the twists and turns of the growth of PCR, it seemed important to try to provide an organized record of these techniques for use in the laboratory. To that end, we published a series of "Manual Supplements" in *PCR Methods and Applications.* The supplements were meant as a workshop for developing the components of this book and to give the journal's readers an opportunity to try the protocols and comment on them. The original protocols were reorganized, revised, and updated by their authors for a more formal appearance in this book. Additional protocols and appendices were added and are new for *PCR Primer: A Laboratory Manual.*

This PCR manual attempts to follow the path of two previous laboratory manuals published by Cold Spring Harbor Laboratory: *Molecular Cloning* and *Antibodies.* We, like many other investigators, have greatly benefited from the background information and the easy-to-follow presentation of protocols in these two excellent manuals. When Judy Cuddihy and John Inglis at Cold Spring Harbor approached us to write this book, our first concern was to be able to live up to the standards established by the previous manuals. In all, we have tried to provide a practical picture of the range of PCR and amplification techniques in a form that enables researchers to decide which versions are applicable to their research problems and then to use them in their laboratories. The first sections of the book are dedicated to the more basic but fundamental ideas in PCR: template preparation, primer design, troubleshooting, and contamination control methods. The other sections of the book have more sophisticated methodologies that will be applicable to particular experimental needs but that rely for their success on the understanding of the parameters covered in the first sections. We have also included the emerging alternative amplification methods that will no doubt complement and sometimes replace PCR as the amplification method of choice. We hope that the detailed description of techniques in this manual will demystify the myriad forms of PCR and amplification and will enable you to get them up and running quickly and successfully in your laboratory.

Our thanks to all the contributors to this book who have been patient with our demands and timely in handing in their manuscripts. We thank the staff at Cold Spring Harbor Laboratory Press, especially Patricia Barker, Inez Sialiano, Susan Schaefer, and Maryliz Dickerson, for their efficiency. Our gratitude goes to our respective families (Ann, Sara, and Rebecca Dieffenbach and Pablo Gutman) for their patience while we were taking our weekends to work on this book. We are especially grateful to Pablo Gutman for editing parts of this book. A special thanks also goes to Judy Cuddihy at Cold Spring Harbor, whose experience with previous manuals, advice, and help gave us the energy needed to complete this project. We hope this manual is useful and enhances your PCR experiments.

Gabriela S. Dveksler
Carl W. Dieffenbach

List of Abbreviations

ACE, angiotensin I-converting enzyme
AFLP, amplified fragment length polymorphism
AMV RT, avian myeloblastosis virus reverse transcriptase
APOB, Apolipoprotein B
AP-PCR, arbitrarily primed PCR
ARMS, amplification refractory mutation system
ASPCR, allele-specific PCR
ASRA, allele-specific restriction analysis
BAC, bacterial artificial chromosome
BBSH, bead-based sandwich hybridization
BGH, bovine growth hormone
BNF, buffered neutral formalin
BSA, bovine serum albumin
CMV, cytomegalovirus
cPCR, competitive PCR
CPCR, capture PCR
cpm, counts per minute
CRS, competitive reference standard
dATP, deoxyadenosine triphosphate
dCTP, deoxycytidine triphosphate
DEPC, diethylpyrocarbonate
DGGE, denaturing gradient gel electrophoresis
dGTP, deoxyguanosine triphosphate
DHFR, dihydrofolate reductase
DIFF-PCR, differential PCR
DMS, dimethylsulfate
DMSO, dimethylsulfoxide
DN-PCR, double-nested PCR
dNTP, any of the four deoxynucleotide triphosphates: dATP, dCTP, dGTP, dTTP
DOS, degenerate oligonucleotide sequences
DTT, dithiothreitol
dTTP, deoxythymidine triphosphate
dUTP, deoxyuridine triphosphate
ELISA, enzyme-linked immunosorbent assay
ELOSA, enzyme-linked oligonucleotide sorbent assay
E-PCR, expression PCR
FAF-ELOSA, fluorescein-antifluorescein-based ELOSA
F-SSCP, fluorescence single-strand conformational polymorphism
G-LCR, gapped LCR
HMA, heteroduplex mobility assay
HN-PCR, heminested PCR
HTA, heteroduplex tracking assay
ICVPCR, internally controlled virion PCR

IMS, immunomagnetic separation
IP-10, 4′-aminomethyl-4,5-dimethylisopsoralen
JOE, 6-carboxy 4′,5′-dichloro-2′,7′-dimethyoxy fluorescein
J-PCR, junction PCR
KGM, potassium glutamate
LAPCR, long and accurate PCR
LAR, ligase amplification reaction
LCR, ligase chain reaction
LDR, ligase detection reaction
LIC-PCR, ligation-independent cloning PCR
LMPCR, ligation-mediated PCR
LOH, loss of heterozygosity
LS-PCR, low-stringency PCR
MAAP, multiple arbitrary amplicon profiling
MAMA, mismatch amplification mutation assay
MAPREC, mutant analysis by PCR and restriction enzyme cleavage
MARMS, multiple amplification refractory mutation system
MoMLV, Moloney murine leukemia virus
MSSCP, multiplex single-strand conformational polymorphism
NASBA, nucleic acid sequence-based amplification
PAC, P1-derived artificial chromosome
PAH, phenylalanine hydroxylase
PASA, PCR amplification of specific alleles
PBL, peripheral blood lymphocytes
PBMC, peripheral blood mononuclear cells
PCR-PIRA, PCR-primer introduced restriction analysis
PEG, polyethylene glycol
PET, paraffin-embedded tissue
Pfu, Pyrococcus furiosus
Pwo, Pyrococcus woesei
RACE, random amplification of cDNA ends
RAP-PCR, RNA arbitrarily primed PCR
RAPD, random amplified polymorphic DNA
RCR, repair chain reaction
RDA, representational difference analysis
RFLP, restriction fragment length polymorphism
RL/RT/PCR, RNA ligase, reverse transcription-PCR
RNA-PCR, PCR starting with RNA
rPCR, random amplification of whole DNA sequences
RPCR, recombination PCR
RS-PCR, restriction site PCR
RSO, restriction site oligonucleotide
RT, reverse transcriptase
RT-PCR, reverse transcriptase PCR
RT-RPCR, reverse transcriptase rapid PCR
SDA, strand displacement amplification
SDM, site-directed mutagenesis
2-SLA, two-stage linear amplification
3-SLA, three-stage linear amplification
SNuPE, single nucleotide primer extension
SOE, spliced overlap extension
SpCCM, solid-phase chemical cleavage
3SR, self-sustained sequence replication
SS, sucrose synthase
SSB, single-stranded DNA-binding protein
SSCP, single-strand conformational polymorphism
STR, short tandem repeat
STS, sequence tagged site
Taq, Thermus aquaticus
TAS, transcription-based amplification systems
Tbr, Thermus brockianus
TD PCR, touchdown PCR
TdT, terminal deoxynucleotidyl transferase
Tfl, Thermus flavus
Tli, Thermococcus litoralis
TRF, time-resolved fluorometry
Tth, Thermus thermophilus
UNG or UDG, uracil *N*-glycosylase or uracil DNA glycosylase
UP, universal product
UTL, untranslated leader
VNTR, variable number tandem repeat
X-PCR, xeno-competitive PCR
YAC, yeast artificial chromosome

PCR
PRIMER
A LABORATORY MANUAL

Introduction to PCR

The polymerase chain reaction (PCR), the repetitive bidirectional DNA synthesis via primer extension of a region of nucleic acid, is simple in design and can be applied in seemingly endless ways. Because PCR is so much more than just mixing reagents in a tube and running a machine, this section outlines the best ways to utilize the available laboratory space in which PCR procedures are set up and includes protocols on how to avoid PCR contamination. In addition to discussing where to perform clean PCR, this section also describes how to get started with a standard PCR protocol.

The components of a standard PCR protocol using *Taq* DNA polymerase are as follows:

10× Enzyme-specific reaction buffer:
- 10–50 mM Tris-HCl, pH 7.5–9.0
- 6–50 mM KCl or $(NH_4)_2SO_4$
- 1.5–5.0 mM $MgCl_2$ or $MgSO_4$

0.2 mM of each dATP, dGTP, dCTP, and dTTP
0.1–1.0 μM of each oligonucleotide primer
2.0–2.5 units of a thermostable DNA polymerase
Nucleic acid template, 10^2–10^5 copies
Distilled water to 100 μl

PCR amplification of a template requires two oligonucleotide primers, the four deoxynucleotide triphosphates (dNTPs), magnesium ions in molar excess of the dNTPs, and a thermostable DNA polymerase to perform DNA synthesis. The quantities of oligonucleotide primers, dNTPs, and magnesium may vary for each specific application. Because PCR tends to be an empirical technology, there are differences in recommended conditions from protocol to protocol throughout this book. These recommended conditions may need to be optimized for different DNA sequences and oligonucleotide primers.

"A Standard PCR Protocol: Rapid Isolation of DNA and PCR Assay for β-Globin" provides an example of a basic DNA-PCR protocol, whereas "Specificity, Efficiency, and Fidelity of the PCR" addresses the effects of varying the components of the reaction mix on product yield and enzyme fidelity. Different parameters can be adjusted if the PCR yield is suboptimal; these parameters, as well as the possible addition of co-solvents, are discussed in "Optimization and Troubleshooting in PCR."

A common and easy protocol to improve product yield and increase specificity in the amplification is the hot start technique. There are a number of recommended methods for this, including uracil-*N*-glycosylase and dUTP, a wax barrier, a wax bead impregnated with magnesium, or an anti-*Taq* DNA polymerase monoclonal antibody. With all of these methods, after heating the reaction to 92°C for the first time, all the reaction components mix, and DNA synthesis occurs only from accurately hybridized primers. The use of hot start is detailed in "Setting Up a PCR Laboratory" and "Enzymatic Control of Carryover Contamination in PCR."

The commercially available thermostable DNA polymerases have different pH optima as well as different salt requirements. The recommended buffers, reverse transcriptase, and exonucleolytic activities, as well as the manufacturers of some of the more popular thermostable DNA polymerases, are shown in Table 1 (pages 4, 5).

An additional variable to consider is the final volume of the reaction. PCR requires rapid changes of temperature, which are accomplished by the thermal cycler. As a general rule, reactions are usually between 20 and 100 μl. Large-volume samples will be inefficiently heated and cooled, while small-volume reactions render insufficient product for manipulation and analysis.

Three distinct events must occur during a PCR cycle: (1) denaturation of the template, (2) primer annealing, and (3) DNA synthesis by a thermostable polymerase.

1. DNA denaturation occurs when the reaction is heated to 92–96°C. The time required to denature the DNA depends on its complexity, the geometry of the tube, the thermal cycler, and the volume of the reaction. For DNA sequences that have a high G+C content, the addition of glycerol, longer denaturation times, and the use of nucleotide analogs have been reported to improve the yield of the PCR.

2. After denaturation, the oligonucleotide primers hybridize to their complementary single-stranded target sequences. The temperature of this step varies from 37°C to 65°C, depending on the homology of the primers for the target sequence as well as the base composi-

tion of the oligonucleotides. Primers are present at a significantly greater concentration than the target DNA, and are shorter in length; as a result, they hybridize to their complementary sequences at an annealing rate several orders of magnitude faster than the target DNA duplex can reanneal.

3. The last step is the extension of the oligonucleotide primer by a thermostable polymerase. Traditionally, this portion of the cycle is carried out at 72°C. The time required to copy the template fully depends on the length of the PCR product. Depending on the PCR thermal cycler being used, it is feasible in certain circumstances to use two-step PCR rather than the traditional three steps as discussed in "Specificity, Efficiency, and Fidelity of the PCR."

The most serious issue with the widespread use of PCR is the contamination of reactions with target nucleic acids. This can occur during steps prior to the actual amplification reaction. "Enzymatic Control of Carryover Contamination in PCR" and "Ultraviolet Irradiation of Surfaces to Reduce PCR Contamination" provide specific protocols to reduce contamination, and "Setting Up a PCR Laboratory" discusses how to integrate these methods into good laboratory practices and improved laboratory design.

One of the significant breakthroughs in PCR is the ability to amplify DNA segments of up to 45 kb efficiently. This occurs through the use of a combination of two thermostable DNA polymerases, one with proofreading activity and one lacking this function. As discussed in "Long-distance PCR," the successful amplification of very long PCR products requires the optimization of buffer conditions as well as of the temperatures and length of each part of the cycle.

The quantity and quality of the starting material for amplification—the template nucleic acid—is of central importance in PCR. The size of the PCR product that can be amplified is dependent on both the reaction conditions and the template quality.

Sample preparation is discussed in detail in Section 2. One point that is occasionally overlooked when setting up a PCR is the quantity or copy number of target sequences added. The quantity of template in a reaction should be measured by the number of copies of the target sequence present, not by weight. The presence of excess template can prevent successful amplification.

The appropriate handling of both the template nucleic acid and the resulting PCR products is required to obtain meaningful data from a PCR. For this reason, we have stressed the importance of methods to control both sample- and PCR product-derived contamination. Without these precautions, and the methods to validate and assess reaction sensitivity and specificity that are detailed in this section, PCR can become more of a bane than a boon to your research effort.

Table 1 Thermostable Polymerases, Buffers, Activities, and Availability

Enzyme and supplier[a]	Exo activity[b]	Buffer/pH (RT)[c]	Salt	Divalent cation[c]	Additional additives[d]
Taq DNA polymerase[B,L,M,N,P,T] *Thermus aquaticus*	5	10 mM Tris-HCl, pH 8.3	50 mM KCl	1.5–5.0 mM $MgCl_2$	BSA, NP-40, Tween 20
Stoffel fragment[P] (Carboxy-terminal 544 amino acids of *Taq* DNA polymerase)	NO	10 mM Tris-HCl, pH 8.3	10 mM KCl	2.0–10.0 mM $MgCl_2$	BSA, Tween 20
UlTma DNA polymerase[P] *Thermotoga maritima*	5	10 mM Tris-HCl, pH 8.8	10 mM KCl	1.5–5.0 mM $MgCl_2$	Tween 20
Tth DNA polymerase[B,E,P,T] *Thermus thermophilus*	5	10 mM Tris-HCl, pH 8.3	90 mM KCl	1.0–2.0 mM $MnCl_2$, reverse transcriptase activity 2.0–4.0 mM $MgCl_2$, DNA synthesis EGTA used to chelate the Mn^{++} ion	glycerol, Tween 20 has potent RT activity with Mn^{++}
Pfu DNA polymerase[S] *Pyrococcus furiosus* (native)	3	20 mM Tris-HCl, pH 8.2	10 mM KCl 6 mM $(NH_4)_2SO_4$	1.5–2.5 mM $MgCl_2$	BSA, Triton X-100
Pfu DNA polymerase[S] (recombinant and Exo$^-$ forms)	3 exo$^-$ = NO	20 mM Tris-HCl, pH 7.5	–	8.0 mM $MgCl_2$	BSA

Vent, DeepVent[N] *Thermococcus litoralis, Pyrococcus GB-D*	3 $Exo^- = NO$	20 mM Tris-HCl, pH 8.8	10 mM KCl 10 mM $(NH_4)_2SO_4$	1.5–5.0 mM $MgSO_4$	Triton X-100
Tli DNA polymerase[M] *Thermococcus litoralis*	5	10 mM Tris-HCl, pH 9.0	50 mM KCl	1.5–5.0 mM $MgCl_2$	Triton X-100
Hot Tub DNA polymerase[H] *Thermus ubiquitus*	NO	50 mM Tris-HCl, pH 9.0	20 mM $(NH_4)_2SO_4$	0.7–2.0 mM $MgCl_2$	
Tfl DNA polymerase[E,M] *Thermus flavus*	5	50 mM Tris-HCl, pH 9.0 or 20 mM Tris-acetate, pH 9.0	20 mM $(NH_4)_2SO_4$ 70 mM K-acetate[M]	1.5–5.0 mM $MgCl_2$ for DNA synthesis 1.5–5.0 mM $MnSO_4$ for reverse transcriptase activity	0.5% Tween 20, has potent RT activity with Mn^{++}
Pwo DNA polymerase[B] *Pyrococcus woesei*	3	10 mM Tris-HCl, pH 8.85	20 mM $(NH_4)_2SO_4$	1.5–4.0 mM $MgSO_4$	BSA, cannot use dUTP
Tbr DNA polymerase[A,F] *Thermus brockianus*	5	10 mM Tris-HCl, pH 8.8	50 mM KCl	1.5–5.0 mM $MgCl_2$	Triton X-100

[a]Name of enzyme and microorganism from which the enzyme was isolated or cloned. Included as a superscript are the suppliers. Complete information about suppliers is provided in the Appendix. [A]AmRESCO, [B]Boehringer Mannheim, [C]CLONTECH, [E]Epicentre Technologies, [F]Finnzymes OY, [H]Amersham, [L]Life Technologies, Inc., [N]New England Biolabs, [M]Promega Corp., [P]Perkin Elmer, Applied Biosystems Division, [S]Stratagene, [T]TaKaRa(PanVara).

[b]Exonuclease activity: 3′-5′ exo = 3, 5′-3′ exo = 5, no exonuclease activity = NO.

[c]The buffer, salt, and divalent cation conditions used for routine amplifications.

[d]Additional additives and comments about specific enzymes.

Setting Up a PCR Laboratory

Carl W. Dieffenbach,[1] Elizabeth A. Dragon,[2] and Gabriela S. Dveksler[3]

[1]Division of AIDS, NIAID, National Institutes of Health, Bethesda, Maryland 20852
[2]Roche Molecular Systems, Somerville, New Jersey 08876-1700
[3]Department of Pathology, Uniformed Services University of the Health Sciences, Bethesda, Maryland 20814

INTRODUCTION

Because of the nature of PCR, it is critical that the only DNA that enters the reaction is the template added by the investigator. Thus, PCR must be performed in a DNA-free, clean environment. The issue of contamination and the cleanliness required to perform contamination-free PCR have been compared to the good microbiological techniques used for handling pathogens (J. Sninsky, pers. comm.). The major difference here is that the "biohazard" infects the PCR, not the researcher. This chapter provides guidance for the establishment and maintenance of a clean environment for any PCR-based assay system, regardless of the number of samples being processed. These suggestions work best if implemented before a contamination problem occurs; if contamination is already a problem, strategies for handling this situation are also provided.

As the use of PCR grows in areas such as clinical diagnosis of genetic diseases (Wang et al. 1992) or the monitoring of viral burden in patients receiving antiretroviral therapy (Piatak et al. 1993), rational guidelines for installing a PCR facility and monitoring for contamination will need to be advanced. This discussion deals with the establishment of two types of laboratories—those performing contamination-sensitive PCR assays, such as measurement of quantities of target sequences in a sample, and those that are using PCR as a contamination-insensitive molecular biology tool, as in mutagenesis of a DNA clone.

The purpose of planning ahead in considering the design, location, and execution of PCR in discrete areas is to avoid contamination of the new PCR assays with old PCR products, molecular clones, or sample-to-sample contamination. To date, four approaches have been devised to prevent contamination. The first is the physical separation of the individual parts of the PCR into sample preparation, pre-PCR, and post-PCR locations (Kwok and Higuchi 1989). This approach should be a central part of any contamination control strategy and can be scaled to suit the needs of the investigator. The physical separation of parts of the PCR process requires some additional space, money, and supplies to equip and maintain a larger infrastructure. However, these components alone are not foolproof, because good laboratory practice is still required for the prevention of sample-to-sample contamination. The second method, the use of uracil DNA-glycosylase (UNG) and deoxyuridine triphosphate (dUTP) substituted for thymidine triphosphate (dTTP), is effective only against contamination with dUTP-labeled PCR products (Longo et al. 1990). The third method, the use of UV light, is effective against all types of contamination. However, this approach is limited because it cannot destroy all of the contamination; UV light only reduces the contamination by several logs, and it is less effective if the DNA fragment is less than 300 bp (Sarkar and Sommer 1990, 1991). The final method is the derivatization of single- and double-stranded DNA with chemical adducts, such as isopsoralen. These adducts prevent the contaminating DNA from serving as a substrate in the reaction (Cimino et al. 1991). The implementation and use of UNG and UV contamination control systems are described in subsequent chapters (Hartley and Rashtchian; Cone and Fairfax) in this section. With these caveats in mind, we suggest the following guidelines be considered when establishing a PCR laboratory.

ESTABLISHMENT OF A PCR LABORATORY

To perform PCR for the repetitive detection of a specific sequence, three distinct areas are required. The specific technical operations and reagents for each one are detailed below. There is new interest in using PCR for the quantitative detection of target sequences, such as human immunodeficiency virus (HIV) (Piatak et al. 1993; Mulder et al. 1994). As the interest in quantitation of specific RNAs and the importance of measurements of viral burden by RNA-PCR grow, there is an increasing need for contamination-free, reliable PCR.

Sample Preparation Area

This room is specific for sample preparation only. The following special precautions should be taken in the preparation and handling of the reagents to be used in nucleic acid extraction:

1. PCR products or DNA clones containing the sequence to be amplified cannot be handled in this room.

2. Tissue cultures, tissue specimens, and serum samples are all brought into the sample preparation room and processed for the extraction of DNA or RNA, depending on the application.

3. Tools used in sample processing should not be used for general molecular cloning or manipulation of the target sequence.

4. DNA samples should be manipulated with specialized barrier or positive-displacement pipettes, which prevent the carryover of aerosols created during pipetting.

5. Large volumes should be pipetted with individually wrapped, sterile, disposable pipettes.

6. Aerosols should be minimized by briefly centrifuging the tubes prior to opening; also, tubes should not be popped open, which creates an aerosol.

7. Lab coats and gloves should be worn at all times, and gloves should be changed frequently, particularly between each step of the purification process. Lab coats should be dedicated to the sample preparation area and washed frequently.

The method chosen for the purification of template can have a significant impact on the risk of contamination. In general, the simpler the method that gives reliable results, the better, because less sample manipulation will be required. Always use freshly prepared or properly stored unused reagents and buffers for nucleic acid extraction. Do not use reagents that have been exposed previously to other samples.

If your laboratory or institution does not have the space for a specific region for sample preparation, consider making arrangements with colleagues to borrow space and the necessary supplies for sample preparation. This arrangement should only be made with laboratories that have never performed molecular cloning with any of the sequences you are interested in amplifying. Although other laboratories are surely contaminated with DNA, this is irrelevant if your primer sets will not amplify their molecular clones. This ad hoc approach can work for both the sample preparation and pre-PCR areas.

Sample Preparation and RNA-PCR

The extra steps associated with RNA-PCR require additional sample handling and, therefore, there is increased chance of sample-to-sample contamination. To avoid this problem, the reverse transcription step can be performed in the sample preparation area. The use of UNG with RNA-PCR to prevent contamination has also been reported (Pang et al. 1992), and this method is described by Hartley and Rashtchian in a later chapter in this section. This approach is particularly valid if the reverse transcription is performed with random hexamer primers rather than with the specific antisense oligonucleotide primer. In this method, all of the RNA present in the sample is converted to cDNA with relatively equal efficiency. Because no single, specific product is being produced, it is less likely to become a source of contamination.

The reverse transcription reaction should be terminated by boiling the reaction, which kills the reverse transcriptase and denatures the RNA:DNA duplex. Alternatively, the RNA strand can be eliminated by RNase H treatment or base hydrolysis. For many applications, r*Tth* polymerase can be used as an alternative to traditional two-step reverse transcriptase and *Taq* polymerase (Myers and Gelfand 1991). The advantages of using r*Tth* are significant: (1) there is a single buffer system; (2) dUTP and UNG can be incorporated in the reverse transcription reaction; and (3) there is improved specificity of the reaction resulting from less mispriming because the cDNA synthesis reaction is performed at a higher temperature with r*Tth*. For laboratories performing large numbers of assays on RNA samples, the reduced handling and improved contamination control make the r*Tth* system an attractive alternative to the two-enzyme RNA-PCR systems.

Pre-PCR Area

An area devoted to the preparation of the individual reactions is essential. This area must be maintained clean and free of all sources of contamination from molecular cloning and sample preparation. Requirements for the pre-PCR area are reagents and equipment; specifically, positive-displacement pipettes that are dedicated to the pre-PCR area.

Each laboratory or department must make a decision as to whether or not synthesis of the primers and probes for its assays will be performed internally or prepared externally. If the synthesis capabilities exist internally, both the synthesis and the subsequent purification of the primers must take place in an area removed from post-PCR activities, sample preparation, and standard molecular biology. Any of the above-mentioned activities can lead to the inadvertent contamination of primers with DNA that will be impossible to remove and could cause spurious results. Again, the pipettors used for handling of primers should be dedicated to this purpose.

Handling of Reagents in the PCR Laboratory

Specific consideration should be given to the preparation and maintenance of clean PCR components.

1. All of the solutions used should be prepared free from contaminating nucleic acids and/or nucleases (both DNases and RNases). To ensure that the nucleic acid preparations routinely amplify properly, always use the highest-quality components for each solution. This should prevent problems due to the introduction of heavy metal ions, nucleases, or other unspecified contaminants. Gloves should be worn at all times when preparing reagents, handling samples, setting up reactions, and performing the subsequent detection of the amplified product.

2. The water used in all PCR reagents should be the highest quality—freshly distilled/deionized, filtered using a 0.22-micron filter, and autoclaved. We have found that USP-certified water that has been filtered and autoclaved is also sufficient for use in PCR. Contamination may also be avoided by using a new bottle for each set of experiments. Routine analysis of the water from the source tap should be performed to determine the conductivity of the water, as well as the possible contamination of the water supply with bacteria or fungi. Never assume that the supply is clean. Significant bacterial contamination has been detected even in house-distilled or -deionized water systems that employ UV sterilization. It is important to remember that bacteria, fungi, and algae grow in water storage systems (i.e., plastic water jugs); therefore, to minimize the chance of contamination of the water used for reagent preparation, always use freshly collected and processed water.

3. The addition of antimicrobials such as sodium azide is recommended for any reagents that will be stored from 20°C to 25°C. The inclusion of 0.025% sodium azide in the amplification reagents or sample preparation reagents does not inhibit the amplification reaction.

4. All reagents should be made up in large volumes. Test to determine if the reagent performs satisfactorily, and then aliquot the reagent into single-usage volumes for storage. Using aliquots of a proven reagent (stored under the appropriate conditions) establishes consistency from experiment to experiment.

5. Disposable, sterile bottles and tubes should be used for all reagents and sample preparation procedures. Glassware that is washed and autoclaved in a common-use laboratory kitchen is a potential source of contaminating nucleic acids. This is particularly true

when many molecular biology laboratories share a large centralized wash room.

6. Newly made reagents should be tested before they are used to prepare new specimens. If possible, try to keep aliquots of a few samples that can be used either to demonstrate a performance standard with a given set of reagents, or, if the system is prone to difficulties, to keep a representative problematic specimen to check reagent performance.

7. The pipettes used in sample preparation and pre-PCR should be carefully stored when not in use. Storage in airtight, self-sealing bags is an effective way to keep pipettes free from contamination.

Lab coats cannot leave the pre-PCR area. Also, the movement of personnel must be carefully thought out. When experiments are planned or in progress, researchers should not move at all from the "dirty" molecular biology rooms into the "clean" sample preparation and pre-PCR rooms. It is also best if a different investigator is responsible for the analysis of the results in the post-PCR area. If one individual is responsible for the entire assay, then that person should move unidirectionally from pre- to post-PCR.

Construction of PCR Mixtures in the Pre-PCR Area

The pre-PCR area should contain storage for the reagents needed for PCR.

1. Ready-to-use "master mix" solutions can be prepared, aliquoted, and stored at either –20°C or 4°C. (dNTPs at the lower concentrations used for PCR buffers are stable at –20°C for months.) These are useful if the laboratory is involved in the amplification of one or a few specific sequences. These master mix solutions have all but one of the necessary components for amplification to occur (i.e., no Mg^{++} or no enzyme). Either a 10X or 2X final concentration can be prepared and stored for easy, convenient use. This allows experiment-to-experiment consistency and removes a significant chance for the introduction of experimental error by miscalculation or a pipetting error when preparing complex reaction mixtures. The possibility for inadvertent contamination of the PCR mixture is eliminated if these reagents are aliquoted into single-use tubes using a clean pipettor, with or without wax sealing.

2. If your laboratory uses multiple primer sets so that construction of single-use reaction mixtures containing all the reagents is not cost-

effective, consider aliquoting and storing the individual components of the PCR in daily-use sizes. This method provides a degree of protection should an inadvertent contamination of one of the stock reagents occur. However, the master mix must be carefully constructed and aliquoted each time a PCR is performed. In this situation, you should consider the use of wax to seal in the reagents and to provide for a hot start.

3. As a rule, you should have a panel of negative, weak, and strong positive control samples to assay the efficiency and cleanliness of the sample preparation and pre-PCR processes. In addition, you may wish to validate the final sample preparation buffer by using a known weak positive to demonstrate that there is no inhibition of amplification.

4. The negative controls that are run alongside each set of samples should be constructed to assay for sample-to-sample contamination as well as for contamination with PCR products. The negative controls should include all reagents used except the input nucleic acid.

5. When positive controls are to be performed, the quantity of nucleic acid being manipulated should be minimized for two reasons. First, a limited number of copies of the target sequence (10^1–10^4) should be included in the reaction to serve as a valid control. Second, by minimizing the amount of DNA, there is less chance of contamination of other samples by DNA aerosol.

6. Because control reactions are essential, the properties of the control template should be considered. The control template should have identical amplification properties to the natural target. In the past, this has meant that many investigators used a cloned version of the natural template, making it very difficult to track contamination of PCR assays with the positive control. By constructing a control that is a different size from the natural target and yet maintains the amplification efficiency of the natural target, the problem with contamination with the positive control can be monitored.

Options for Contamination Control

A powerful enzymatic method for elimination of one form of contamination—the use of UNG—has been devised (Longo et al. 1990). This technique efficiently eliminates contamination arising from PCR products. An alternative method of contamination control is the use of UV light. This method does not completely eliminate the contamination problem, but reduces it by several orders of magnitude. However, UV can effectively control contamination arising from molecular clones that are present in the laboratory. Thus, UV serves as a prophylactic treatment of reagents if low-level contamination is a problem. Both of these approaches to contamination control should be considered. The integration of these procedures is described in this section.

If you are using PCR only in a contamination-insensitive manner, it is not critical to maintain sample preparation and pre-PCR areas. Often the substrate in the PCR consists of cloned sequences used at 10^4 to 10^6 copies per reaction. However, most laboratories often perform a mixture of contamination-sensitive and -insensitive PCR. Reagents found in all PCR procedures, such as nucleotides and *Taq* DNA polymerase, need to be aliquoted or purchased in sufficient quantities to ensure that there are adequate supplies for the contamination-sensitive reactions. Reagents can always be used from the pre-PCR area for contamination-insensitive reactions; however, once used in the standard molecular cloning lab, they must not be returned to the pre-PCR area. Reagents for PCR sequencing should also be stored separately from other PCR components.

Location of the PCR Machine

Where the PCR machine should reside seems like a trivial issue. However, if the machine is to be used for multiple applications, including ones that are contamination-insensitive, the machine should be located in a room where PCR products will be handled. This decision is a trade-off. You must consider the fact that standard hot start techniques (the addition of a key ingredient to each reaction at the start of the first cycle) cannot be performed on a PCR machine in a post-PCR area. The use of Ampliwax PCR Gems (Perkin-Elmer) to separate the reaction into two parts or the use of the monoclonal antibody to *Taq* polymerase (TaqStart antibody, CLONTECH) provides different systems for hot start without opening or manipulating the samples once they are placed in the machine. An alternative hot start method is the use of UNG as follows: After the standard UNG incubation, include a single 2-minute incubation at 50°C prior to the first denaturation cycle (Kinard and Spadoro, pers. comm.). This provides sufficient time for UNG to degrade all the mispriming events that have been initiated in the reaction.

Post-PCR Area

After the PCR assays have been performed, the samples need to be analyzed and the data interpreted. An area should be set aside specifically for the post-reaction manipulation of samples. It is imperative that all reagents, disposables, and equipment used in post-PCR activities be dedicated solely to this purpose. Never use equipment or reagents from this area of the laboratory in any pre-PCR activity.

The major source of contamination seen in PCR laboratories is the DNA obtained as product from previous PCR procedures; this arises from the microaerosols that are generated during the pipetting and manipulation of the PCR samples. Although these aerosols cause no problems if they are confined to the post-PCR area, they can cause havoc if they travel to the pre-PCR area on a pair of gloves, in a pipettor, on an investigator, or on a lab coat. If the laboratory has opted not to use the UNG and dUTP system to control PCR contamination, it is extremely critical that there be no mixing of reagents and personnel between the post-PCR and the pre-PCR areas. Therefore, the traffic flow in the laboratory should be unidirectional, always moving from "clean" to "dirty."

Within the post-PCR area, the tools for analysis should be dedicated to post-PCR use. If the laboratory is using the 96-well ELISA format for analysis, the specialized equipment can be shared with protein-based or cell culture-based assays but not with anything involved in pre-PCR. If analysis is to be performed by gel electrophoresis, then be sure that all the necessary equipment is located in standard molecular biology areas where contamination-sensitive PCR will not be performed.

From a laboratory management perspective, the division of labor for research projects is critical. However, unless a laboratory is involved in the high-volume throughput of hundreds of samples per week, it is not possible to have laboratory personnel dedicated solely to working in the pre-PCR area. Thus, investigators should plan to perform the clean work first; then, when that is completed, move on to perform the standard molecular biology from which PCR contamination can arise. The establishment of "clean days" and "dirty days" can be considered. On clean days, sample preparation and pre-PCR experiments are performed, and no one in the lab can perform studies that are prone to generate contaminating aerosols, such as plasmid DNA isolation, manipulation of phage, or analysis of PCR products.

The Perspective of the Small Laboratory

For a laboratory that is studying the structure, function, and expression of a particular gene, the questions of use of PCR, planning for a pre-PCR area, and maintaining contamination-free areas require careful thought. If the laboratory is considering establishing a PCR assay to detect a sequence that has been manipulated extensively within the laboratory, several critical changes need to occur. First, new equipment must be purchased, primarily micropipettors and positive-displacement devices that will be dedicated to a pre-PCR area. Next, sites where plasmid and phage clones of the sequence of interest have not been manipulated in the laboratory must be identified. Often a sample preparation/pre-PCR area can be as little as 3–4 linear feet of bench space, a single drawer, and less than one cubic foot of freezer (–20°C) space. Ideally, available space meeting these requirements should be on a separate floor from the home laboratory. If such space is not available during normal working hours, it is possible that the space is available early in the morning or in the evening. It is preferable to perform the sample preparation and pre-PCR procedures early in the day, before your colleagues unknowingly begin aerosolizing DNA all over the lab.

Anticipation of contamination, acknowledging that it can happen, and acting to prevent it are important first steps to trouble-free PCR. If contamination has not yet been seen in the PCR, there are two possibilities—it is not there or you have not looked hard enough.

Dr. John Sninsky has proposed that PCR contamination be considered as a form of infection. If standard sterile techniques that would be applied to tissue culture or microbiological manipulations are applied to PCR, then the risk of contamination is greatly reduced. Above all else, common sense should prevail.

ACKNOWLEDGMENTS Partial support for the work described in this chapter was provided by Uniformed Services University of the Health Sciences grant CO-74ET to G.S.D.

REFERENCES

Cimino, G.D., K.C. Metchate, J.W. Tessman, J.C. Hearst, and S.T. Issacs. 1991. Post-PCR sterilization: A method to control carryover contamination for the polymerase chain reaction. *Nucleic Acids Res.* **19:** 99–107.

Kwok, S. and R. Higuchi. 1989. Avoiding false positives with PCR. *Nature* **339:** 237–238.

Longo, M.C., M.S. Berninger, and J.L. Hartley. 1990. Use of uracil DNA glycosylase to control carryover contamination in polymerase chain reactions. *Gene* **93:** 125–128.

Mulder, J., N. McKinney, C. Christopherson, J. Sninsky, L. Greenfield, and S. Kwok. 1994. Rapid and simple PCR assay for quantitation of human immunodeficiency virus type 1 RNA in plasma: Application to acute retroviral infection. *J. Clin. Microbiol.* **32:** 292–300.

Myers, T.W. and D.H. Gelfand. 1991. Reverse transcription and DNA amplification by a *Thermus thermophilus* DNA polymerase. *Biochemistry* **30:** 7661–7666.

Pang, J., J. Modlin, and R. Yolken. 1992. Use of modified nucleotides and uracil-DNA glycosylase (UNG) for the control of contamination in the PCR-based amplification of RNA. *Mol. Cell. Probes* **6:** 251–256.

Piatak, M., Jr., M.S. Saag, L.C. Yang, S.J. Clark, J.C. Kappes, K.C. Luk, B.H. Hahn, G.M. Shaw, and J.D. Lifson. 1993. High levels of HIV-1 in plasma during all stages of infection determined by competitive PCR. *Science* **259:** 1749–1754.

Sarkar, G. and S.S. Sommer. 1990. Shedding light on PCR contamination. *Nature* **343:** 27.

——. 1991. Parameters affecting susceptibility of PCR contamination to UV inactivation. *BioTechniques* **10:** 590–594.

Wang, X., T. Chen, D. Kim, and S. Piomelli. 1992. Prevention of carryover contamination in the detection of beta S and beta C genes by polymerase chain reaction. *Am. J. Hematol.* **40:** 146–148.

A Standard PCR Protocol: Rapid Isolation of DNA and PCR Assay for β-Globin

Maryanne T. Vahey,[1] Michael T. Wong,[2] and Nelson L. Michael[1]

[1]Division of Retrovirology, Walter Reed Army Institute of Research, Rockville, Maryland 20850
[2]Department of Infectious Diseases, Wilford Hall Medical Center, Lackland Air Force Base, San Antonio, Texas 78236

INTRODUCTION

This chapter describes a standard PCR protocol starting from DNA. This protocol contains all the essentials of a basic PCR protocol: sample preparation, master mix for the PCR, appropriate positive and negative controls, and a reliable detection system. Each of these facets of PCR is dealt with in more detail throughout this book. This protocol describes a simple and effective method to extract the total DNA complement from peripheral blood mononuclear cells (PBMC). The copy number of β-globin DNA is determined by quantitative PCR and represents the number of cell equivalents in a specific DNA sample. This technique is useful for standardizing numbers of cell equivalents from sample to sample as well as from assay to assay.

REAGENTS

Proteinase K (GIBCO/BRL, cat. no. 5530UA): Resuspended at 10 mg/ml, made fresh, held on ice, and used within 10 minutes

DLB (detergent lysis buffer):
- 10 mM Tris-HCl, pH 7.5
- 2.5 mM $MgCl_2$
- 0.45% Triton X-100
- 0.45% Tween 20

PROTOCOL

Preparation of Cellular DNA

1. Aliquot 0.5–1.0 x 10^6 PBMC into sterile 1.5-ml microfuge tubes. This protocol assumes that some or all of the specimens are HIV-infected.

Note: HIV-infected cells are handled in a biohazard level 2+ (BL2+) facility in Class II Biosafety hoods.

2. Centrifuge tubes at 1500 rpm for 15 minutes at room temperature.
3. Carefully siphon off the supernatant using clean, disposable micropipette tips. Discard tips and supernatant into containers with a virucidal solution.
4. Resuspend the pellet and lyse the cells with 100 μl of DLB.
5. Add 1.2 μl of fresh proteinase K.
6. Cap tubes and spray their exteriors with 70% ethanol.
7. Remove the samples from the hood and place them in a 60°C water bath or heat block for 1 hour. If the water bath is used, the water level should only cover the volume of the lysate. DO NOT ALLOW THE TUBES TO SUBMERGE.
8. If the tubes contain biohazardous material, spray them down with 70% ethanol. It is now safe to remove the samples from the BL2+ facility, and the remainder of the work may be carried out on a BL2 bench top.
9. Transfer the tubes to a 95°C heat block for 15 minutes to neutralize the proteinase K.
10. Snap-cool the tubes in ice water and store them at –80°C until needed for the PCR.

REAGENTS

dNTPs, 25 mM (Promega, cat. no. U1240)

Taq DNA polymerase (Perkin-Elmer, cat. no. N801-0060)

Mineral oil (Perkin-Elmer, cat. no. 186-2302)
β-Globin primers, 20 μM
Sense: GAA GAG CCA AGG ACA GGT AC
Antisense: CAA CTT CAT CCA CGT TCA CC

Liquid hybridization probe:
AAG TCA GGG CAG AGC CAT CTA TTG CTT ACA

10x PCR buffer (Perkin-Elmer, cat. no. N808-0006)

0.5-ml sterile PCR tubes (Perkin Reaction Tubes, cat. no. N801-0180)
β-Globin DNA dilution series: 5 x 10^4, 1 x 10^4, 5 x 10^3, 1 x 10^3, 5 x 10^2, 1 x 10^2 cell equivalents

PROTOCOL

PCR Assay for β-Globin DNA

1. In a pre-PCR area, assemble and label the PCR (0.5 ml) tubes and place them in a rack. Tubes to be labeled are: the assay blank (H_2O), clinical specimen DNAs, and β-globin DNA standard dilution series. Add 10 μl of sterile H_2O to the tube labeled as the assay blank.

2. Add 8 μl of H_2O and 2 μl of clinical specimen DNA to the tubes labeled as clinical specimens.

3. Add 10 μl of each dilution, working from the low concentration to the high concentration, to tubes appropriately labeled for β-globin DNA standard curve dilution series.

4. Determine the number of reaction tubes to be run and add two. Make a master mix by multiplying this number by the following volumes:

Component	Volume (μl/rxn)	Final concentration
Water	68.7	
10x Buffer	10.0	$[Mg^{++}] = 1.64$ mM
β-Globin sense, 20 μM	5.0	1.0 μM
β-Globin antisense, 20 μM	5.0	1.0 μM
dNTPs, 25 mM each	0.8	200 μM each
Taq DNA polymerase, 5 units/μl	0.5 μl 90 μl/reaction + template = 100 μl	2.5 units/tube

5. Starting with the assay blank tube, continuing with the clinical specimen tubes, and finally with the β-globin DNA standard curve tubes, aliquot 90 μl of PCR master mix to each tube to bring the final volume of each reaction to 100 μl.

6. Add 50 μl of mineral oil to all tubes using a micropipettor and a disposable pipette tip. Change tips between each tube.

7. Vortex each tube and briefly centrifuge.

8. Place tubes into the preheated thermal cycler and begin the cycling according to the following schedule:

Denature	30 seconds at 95°C
Anneal	30 seconds at 50°C
Extend	3 minutes at 72°C
22 cycles	
Time delay	10 minutes at 72°C
Soak	10 minutes at 4°C (this can be left overnight)

9. Store samples at 4°C until used in the liquid hybridization assay, which should be performed no later than the next day, if possible.

ANALYSIS OF RESULTS

In this assay system, quantitation is achieved by direct comparison of unknowns with a standard curve. The β-globin control template consists of nucleotides −195 through +73 of the human β-globin gene listed in GenBank as HUMMBB5E. This sequence of approximately 278 bp is cloned into the *Sma* site of the vector pBluescript IIKS in the forward orientation (total length of the plasmid and insert is approxi-

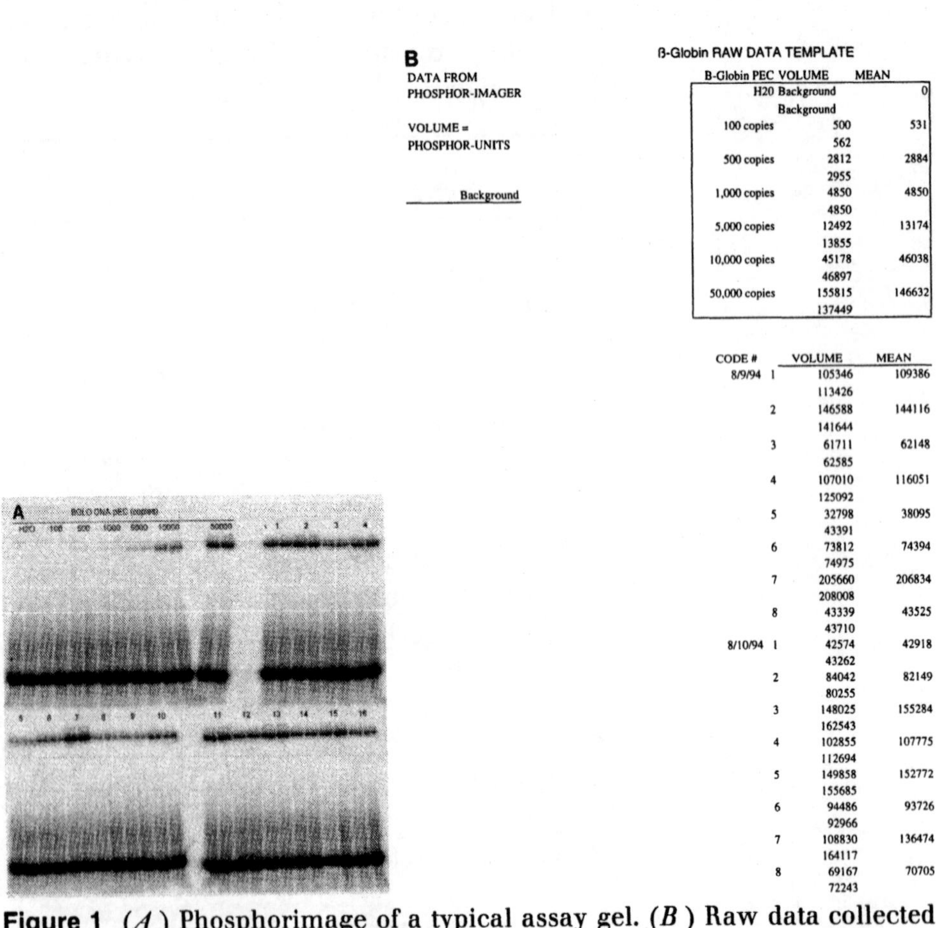

ß-Globin RAW DATA TEMPLATE

B-Globin PEC	VOLUME	MEAN
H20 Background		0
Background		
100 copies	500	531
	562	
500 copies	2812	2884
	2955	
1,000 copies	4850	4850
	4850	
5,000 copies	12492	13174
	13855	
10,000 copies	45178	46038
	46897	
50,000 copies	155815	146632
	137449	

CODE #		VOLUME	MEAN
8/9/94	1	105346	109386
		113426	
	2	146588	144116
		141644	
	3	61711	62148
		62585	
	4	107010	116051
		125092	
	5	32798	38095
		43391	
	6	73812	74394
		74975	
	7	205660	206834
		208008	
	8	43339	43525
		43710	
8/10/94	1	42574	42918
		43262	
	2	84042	82149
		80255	
	3	148025	155284
		162543	
	4	102855	107775
		112694	
	5	149858	152772
		155685	
	6	94486	93726
		92966	
	7	108830	136474
		164117	
	8	69167	70705
		72243	

Figure 1 (*A*) Phosphorimage of a typical assay gel. (*B*) Raw data collected from the phosphorimager. The label pEC refers to the cloned assay standard. The numbered lanes are patient samples.

mately 3403 bp). The β-globin sense primer is a 20-nucleotide oligonucleotide located in the human genome at position −195 to −176. The antisense β-globin primer is a 20-nucleotide oligonucleotide located at position +53 to +73 and with the sense primer generates a 278-bp PCR product. The probe for the β-globin PCR product is complementary to the sequence positions −77 to −48 within the PCR product generated by the primers. The assay as described here ideally contains 20,000 cell equivalents, and thus the actual number of cell equivalents can be standardized for each sample.

The β-globin standard PCR products from this assay are analyzed using the liquid hybridization assay described in detail in the chapter by Vahey and Wong in Section 5. Briefly, the liquid hybridization is performed by mixing 30 μl of PCR product with 5×10^4 cpm of a ^{32}P-labeled oligonucleotide probe in hybridization buffer. The entire reaction mixture is heated and incubated for 5 minutes at 94°C and for 10 minutes at 55°C and then stored at 4°C.

Four microliters of loading buffer is added to 10 μl of sample and is applied to duplicate wells of an acrylamide minigel. Following electrophoresis to resolve the bound versus free probe, the gel is removed from the glass plates and exposed to a storage phosphor plate with subsequent analysis in a phosphorimager. See Figures 1 and 2.

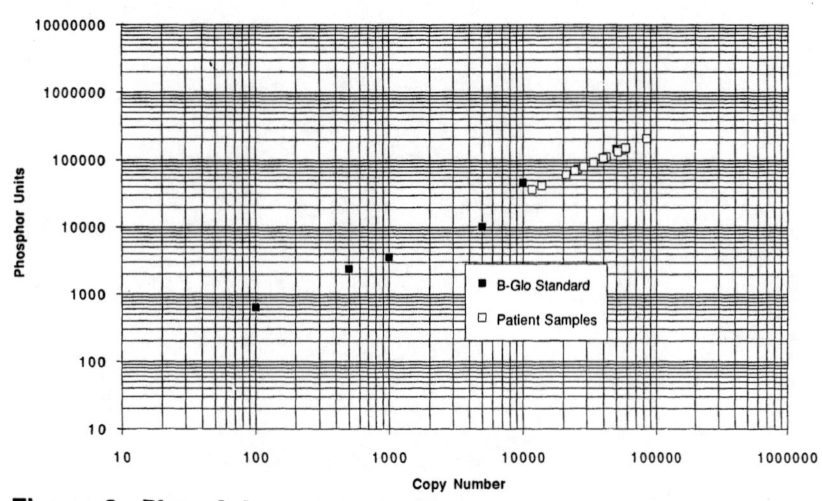

Figure 2 Plot of the standard curve and values of copy numbers (or cell equivalents) for the patient samples assayed.

Enzymatic Control of Carryover Contamination in PCR

James L. Hartley and Ayoub Rashtchian

Life Technologies, Inc., Gaithersburg, Maryland 20884-9980

INTRODUCTION

PCR produces an abundance of amplified DNA product from traces of input DNA. It was apparent early that due to its exquisite sensitivity, PCR is especially susceptible to contamination. In some applications (e.g., cloning a gene, preparing a probe), contamination is not a concern. However, if a primer pair is used many times, if the PCR is designed to be very sensitive, or if the presence or the absence of amplification of a target sequence has diagnostic implications, then possible contamination must be eliminated for the PCR results to be meaningful.

Contaminating DNA can originate from three sources: DNA from other test samples, DNA from experimental materials such as recombinant clones, or DNA generated by previous PCR amplification of the same target sequence. This last source of contamination, often called "carryover" contamination, has proven to be the most troublesome.

Early users of PCR noted that carryover contamination could be a significant problem because of the abundance of DNA generated by PCR and the ease with which such DNA can be reamplified (Gibbs and Chamberlain 1989; Kwok and Higuchi 1989). Detecting carryover contamination, e.g., by including negative control reactions, is essential. Prevention is clearly preferred, however, because correcting the problem can be costly, and testing of samples probably needs to cease until a thorough clean-up can be effected. This most likely means discarding all suspected reagents and cleaning, or even replacing, equipment. A last resort, one not always possible, is to change to a different primer pair, so as to amplify a different region of the target DNA.

This chapter focuses on the enzymatic elimination of PCR product carryover. This approach modifies the PCR so that the products of previous PCR amplifications are discriminated against. For this approach to work, a discriminating process of some kind must intervene after the last cycle of a first PCR or before the first cycle of a subsequent PCR. Because PCR uses DNA primers to detect a DNA target, this process must either act on PCR product DNA before primers and target DNA are added, or it must discriminate in favor of the true target and against possible PCR-derived DNA. Various ways of achieving this discrimination have been proposed. Physical methods are discussed in Dieffenbach et al. and in Cone and Fairfax, both this volume.

Several enzymatic ways of eliminating carryover contamination have been demonstrated. Pretreatment of PCR products with nucleases is based on the principle that oligonucleotide primers, being single-stranded, are resistant to restriction endonucleases, but carryover contaminants with known (and preferably multiple) cleavage sites should be cut efficiently and made unamplifiable (DeFilippes 1991). This was, in fact, observed, and different restriction enzymes have provided different degrees of decontamination. Surprisingly, DNase I could also be used successfully (Furrer et al. 1990). Target DNA must be added after inactivation of the nucleases.

In a second class of methods, the PCR primers are modified. These methods are based on the fact that for PCR to proceed, the primer DNA, which after PCR is found at the 5′ end of each DNA strand, must itself be copied at each cycle. If primers contain uracil bases (Longo et al. 1990), reamplification of PCR products may be inhibited with the enzyme uracil DNA glycosylase (UDG; also called uracil *N*-glycosylase or UNG). If primers contain a 3′ ribonucleotide, treatment of PCR products with ribonuclease or alkali releases primer sequences and inhibits reamplification (Rys and Persing 1993; Walder et al. 1993).

The most widely used decontamination method for diagnostic PCR is based on substituting PCR product DNA with deoxyuracil bases in place of thymines (Longo et al. 1990). A schematic of this method is shown in Figure 1. The DNA produced in such reactions is normal in most respects (e.g., it is cut by many restriction enzymes [Bodnar et al. 1983; Wang et al. 1992] and hybridizes to probes [Wang et al. 1992]), except that it contains tens or hundreds of deoxyuridines. Preincubation of all amplification reactions with the enzyme UDG results in removal of dU from carryover DNA (but does not affect DNA, dUTP, or RNA), creating tens or hundreds of abasic sites. DNA polymerases stall at these sites. Furthermore, such sites are heat-labile and break during temperature cycling. Either type of damage prevents amplification. If dUTP is used routinely in all PCR amplifications, then all PCR products contain uracil and are susceptible to UDG. The method is robust and, because it acts on complete reactions just prior to temperature cycling (i.e., all components including target DNA are present), no carryover PCR product, regardless of source, should escape destruction.

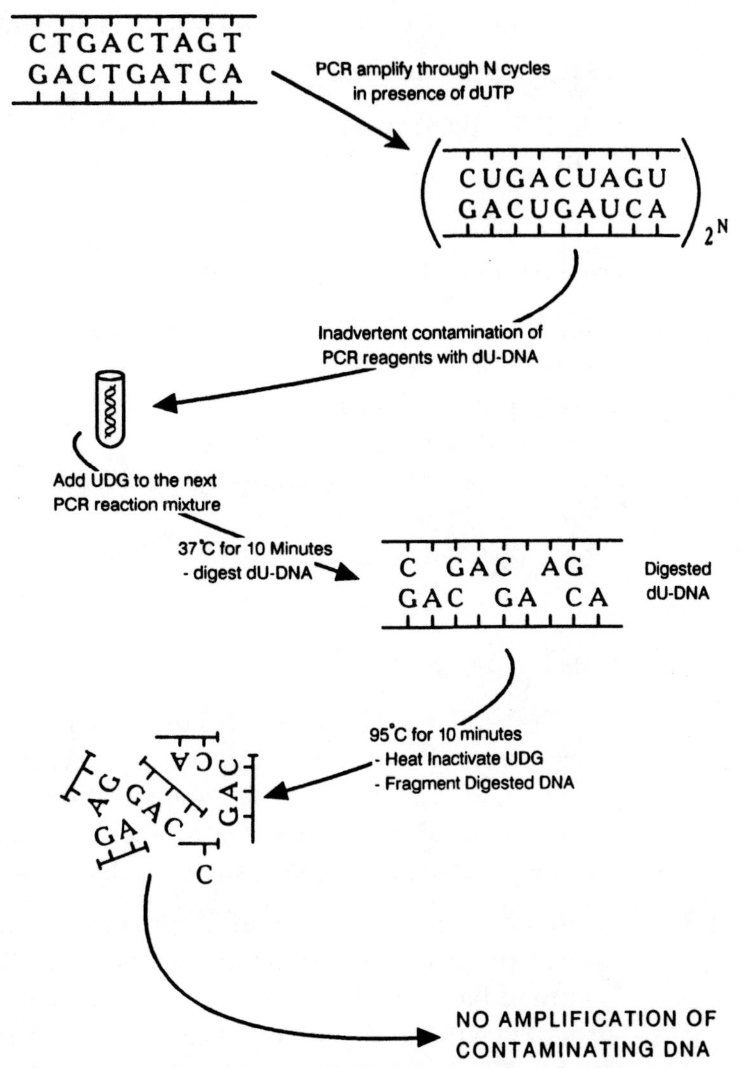

Figure 1 Schematic representation of UDG method for prevention of carryover contamination in PCR.

It is important to note that longer PCR products make UDG decontamination more efficient (but see discussion below for application to "long PCR"). In one study, contamination of product shorter than 100 bp could not be completely eliminated with UDG (Espy et al. 1993). Smaller DNAs may contain too few uracil bases to guarantee complete destruction by UDG.

REAGENTS

dUTP nucleotide mix: dATP, dCTP, dGTP, and dUTP mix at 10 mM each (see below for higher dUTP concentration if amplification yield is low)

Uracil DNA glycosylase, 1 unit/μl (Life Technologies, cat. no. 18054-015)

10X PCR buffer (100 mM Tris-HCl, pH 8.3, 500 mM KCl)

25 mM $MgCl_2$ solution

Taq DNA polymerase, 5 units/μl (Life Technologies, cat. no. 18038-018)

PROTOCOL

To a 0.5-ml microcentrifuge tube on ice add the following:

10X PCR buffer	5 μl
10 mM dUTP mix	1 μl
PCR primers (10 μM each)	1 μl
Uracil DNA glycosylase, 1 unit/μl	1 unit
Taq DNA polymerase, 5 units/μl	0.5 μl
25 mM $MgCl_2$	3 μl (see below)
Sterile distilled water	38.5 μl

Notes

- *Magnesium concentration.* For most PCR primer pairs, $MgCl_2$ at a concentration of 1.5 mM is satisfactory. If amplification is not adequate, adjustments to Mg concentration may be helpful. If nucleotide (e.g., dUTP) concentrations are increased, the Mg concentration must also be raised, since nucleotides chelate Mg ions.
- *dUTP concentration.* The fundamental requirement for UDG control of carryover amplification is that all PCR products contain uracil bases replacing thymine. This is accomplished by replacing dTTP with dUTP in reaction mixtures. For some amplifications this direct substitution (usually 0.2 mM) has no effect on amplification potential (Rys and Persing 1993; Kox et al. 1994); for some target/primer combinations, however, the concentration of dUTP must be raised to 0.6 mM or even 1 mM to reach comparable sensitivities (Wang et al. 1992; Hohlfeld et al. 1994). It is important that magnesium concentrations also are raised in such situations (see above). Recommendation: Start with dUTP equimolar with other dNTPs, test higher concentrations if necessary.
- *Decontamination.* To decontaminate PCR procedures, UDG is added to all assays and they are incubated prior to temperature cycling. Because the UDG of commerce is derived from *Escherichia coli,* this step was originally done at 37°C (Longo et al. 1990). Studies have verified decontamination at room temperature (Wang et al. 1992; De Wit et al. 1993). Subsequently, a short time at 50°C has been found to be effective (Hohlfeld et al. 1994; Kox et al. 1994). In addition, it has been claimed that this step increases PCR specificity (Mulder et al. 1994). The amount of UDG required ap-

pears to be quite variable: As little as 0.01 unit has been used (Kox et al. 1994). The balance is between cost of enzyme and completeness of decontamination. Recommendation: If UDG cost is an issue, test different amounts with different levels of expected contamination. Otherwise, one unit per assay is effective for at least 10^9 contaminant molecules (Longo et al. 1990).

- *Temperature cycling, gel electrophoresis, and hybridization.* Normal procedures are followed. Many clinical studies have tested the specificity of amplification by a variety of hybridization tests (see, e.g., Rys and Persing 1993; Mulder et al. 1994). Recommendation: Use normal protocols.
- *Residual UDG activity.* UDG from *E. coli* that has been through many PCR cycles recovers a small fraction of its catalytic activity when it is returned to lower temperatures (Thornton et al. 1992). Although this activity is easily detectable with appropriate assays, PCR products often undergo subsequent treatment (gel electrophoresis, denaturation, hybridization, etc.) that eliminates residual UDG activity. However, it is only prudent to store reactions with dU-containing DNA and residual UDG activity at –20°C when not in use. One study added an equal volume of chloroform to each reaction after cycling was complete to keep the UDG inactive (Hohlfeld et al. 1994). Recommendation: Store PCR products at –20°C.

DISCUSSION

The efficacy of the use of UDG for decontamination of PCR has been demonstrated by several groups. Longo et al. (1990) showed that intentional contamination with $>10^{10}$ molecules of PCR product did not yield product detectable by ethidium bromide staining when reamplification was attempted following a 10-minute UDG incubation. In the clinical laboratory setting, the UDG decontamination procedure has been used in the development of several diagnostic assays based on PCR, such as the detection of *Mycobacterium leprae* (De Wit et al. 1993), *Mycobacterium tuberculosis* (Nolte et al. 1993), human immunodeficiency virus (HIV) (Butcher and Spadoro 1992), and Lyme disease (Dodge et al. 1992). Commercial versions of these tests incorporate the UDG decontamination technology under the trademark AmpErase (Roche Diagnostic Systems).

In "long PCR," DNAs longer than 30 kb have been amplified using mixtures of DNA polymerases, one with and one without a 3′ exonuclease (proofreading) activity (Barnes 1994; Cheng et al. 1994). Addition of dUTP to these reactions has uniformly inhibited amplification of long products (>4 kb; D. Shuster and A. Rashtchian, unpubl.). This phenomenon appears to be related to inhibition of the tested proofreading polymerases (*Pfu*, Vent, and DeepVent) by dU-containing DNA. Therefore, UDG-based decontamination of long PCR may not be possible with currently available enzymes.

Carryover contamination is a significant source of false-positive results when primer pairs are used repeatedly in DNA amplification. Routine procedures for detecting carryover contamination are essential, but they must be complemented by prevention. Enzymatic control measures such as those described here carry a cost; e.g., modification of the standard PCR or special procedures and facilities. It is necessary to balance the costs of these measures with the benefit derived from controlling contamination and obtaining reliable results.

REFERENCES

Barnes, W.M. 1994. PCR amplification of up to 35-kb DNA with high fidelity and high yield from lambda bacteriophage templates. *Proc. Natl. Acad. Sci.* **91:** 2216–2220.

Bodnar, J.W., W. Zempsky, D. Warder, C. Bergson, and D.C. Ward. 1983. Effect of nucleotide analogs on the cleavage of DNA by the restriction enzymes *Alu*I, *Dde*I, *Hin*fI, *Rsa*I, and *Taq*I. *J. Biol. Chem.* **258:** 15206–15213.

Butcher, A. and J. Spadoro. 1992. Using PCR for detection of HIV-1 infection. *Clin. Immunol. News* **12:** 73–76.

Cheng, S., C. Fockler, W.M. Barnes, and R. Higuchi. 1994. Effective amplification of long targets from cloned inserts and human genomic DNA. *Proc. Natl. Acad. Sci.* **91:** 5695–5699.

DeFillipes, F.M. 1991. Decontaminating the polymerase chain reaction. *BioTechniques* **10:** 26–29.

De Wit, M.Y.L., J.T. Douglas, J. McFadden, and P.R. Klatser. 1993. Polymerase chain reaction for detection of *Mycobacterium leprae* in nasal swab specimens. *J. Clin. Microbiol.* **31:** 502–506.

Dodge, D.E., R. Nersesian, and R. Sun. 1992. Diagnosis of the Lyme disease spirochete *Borrelia burgdorferi*. *Clin. Immunol. News* **12:** 69–73.

Espy, M.J., T.F. Smith, and D.H. Persing. 1993. Dependence of polymerase chain reaction product inactivation protocols on amplicon length and sequence composition. *J. Clin. Microbiol.* **31:** 2361–2365.

Furrer, B., U. Candrian, P. Wieland, and J. Luthy. 1990. Improving PCR efficiency. *Nature* **346:** 324.

Gibbs, R.A. and J.S. Chamberlain. 1989. The polymerase chain reaction: A meeting report. *Genes Dev.* **3:** 1095–1098.

Hohlfeld, P., F. Daffos, J.-M. Costa, P. Thulliez, F. Forestier, and M. Vidaud. 1994. Prenatal diagnosis of congenital toxoplasmosis with a polymerase-chain-reaction test on amniotic fluid. *New Engl. J. Med.* **331:** 695–699.

Kox, L.F.F., D. Rhienthong, A. Miranda, N. Udomsantisuk, K. Ellis, J. van Leeuwen, S. van Heusden, S. Kuijper, and A.H.J. Kolk. 1994. A more reliable PCR for detection of *Mycobacterium tuberculosis* in clinical samples. *J. Clin. Microbiol.* **32:** 672–678.

Kwok, S. and R. Higuchi. 1989. Avoiding false positives with PCR. *Nature* **339:** 237–238.

Longo, M.C., M.S. Berninger, and J.L. Hartley. 1990. Use of uracil DNA glycosylase to control carryover contamination in polymerase chain reactions. *Gene* **93:** 125–128.

Mulder, J., N. McKinney, C. Christopherson, J. Sninsky, L. Greenfield, and S. Kwok. 1994. Rapid and simple PCR assay for quantitation of human immunodeficiency virus type 1 RNA in plasma: Application to acute retroviral infection. *J. Clin. Microbiol.* **32:** 292–300.

Nolte, F.S., B. Metchock, J.E. McGowan, A. Edwards, O. Okwumabua, C. Thurmond, P.S. Mitchell, B. Plikaytis, and T. Shinnick. 1993. Direct detection of *Mycobacterium tuberculosis* in sputum by polymerase chain reaction and DNA hybridization. *J. Clin. Microbiol.* **31:** 1777–1782.

Rys, R.N. and D.H. Persing. 1993. Preventing false positives: Quantitative evaluation of three protocols for inactivation of polymerase chain reaction amplification products. *J. Clin. Microbiol.* **31:** 2356–2360.

Thornton, C.G., J.L. Hartley, and A. Rashtchian. 1992. Utilizing uracil DNA glycosylase to control carryover contamination in PCR: Characterization of residual UDG activity following thermal cycling. *BioTechniques* **13:** 180–183.

Walder, R.Y., J.R. Hayes, and J.A. Walder. 1993. Use of

PCR primers containing a 3′-terminal ribose residue to prevent cross-contamination of amplified sequences. *Nucleic Acids Res.* **21:** 4339–4343.

Wang, X., T. Chen, D. Kim, and S. Piomelli. 1992. Prevention of carryover contamination in the detection of β^s and β^c genes by polymerase chain reaction. *Am. J. Hematol.* **40:** 146–148.

Ultraviolet Irradiation of Surfaces to Reduce PCR Contamination

Richard W. Cone[1] and Marilynn R. Fairfax[2]

[1]Division of Infectious Diseases, Department of Medicine, University Hospital, CH-8091 Zurich, Switzerland
[2]Department of Pathology, Wayne State University School of Medicine, Detroit, Michigan 48201

INTRODUCTION

False-positive PCRs arise from contamination with exogenous genomes, plasmids, or PCR products (Kwok and Higuchi 1989). Contaminated laboratory surfaces represent one of the many potential sources of exogenous DNA.

UV irradiation of dry DNA provides just one tool in the arsenal necessary to prevent PCR contamination. Although this type of decontamination was recommended previously as a way to "quickly damage any DNA left on exposed surfaces" (Kwok and Higuchi 1989), further work has revealed a slow time course and sequence dependence (Fairfax et al. 1991; Fox et al. 1991; Sarkar and Sommer 1991). UV irradiation has also been proposed for decontaminating DNA in reagent solutions (Isaacs et al. 1991; Meier et al. 1993; Sarkar and Sommer 1993), a procedure that has met with mixed reviews (Fox et al. 1991; Dwyer and Saksena 1992; Frothingham et al. 1992).

Most UV-induced DNA damage occurs via the formation of cyclobutane rings between neighboring pyrimidine bases, thymidine or cytidine. The cyclobutane rings form intrastrand pyrimidine dimers that inhibit polymerase-mediated chain elongation. Dimer formation is reversible, establishing a steady-state equilibrium that favors monomers over dimers. As such, <10% of the possible pyrimidine dimers actually exist in irradiated DNA at one time (Gordon and Haseltine 1982).

UV irradiation of laboratory surfaces has some important limitations. First, the surface must be perpendicular to the light source to achieve optimal light intensity. Skewed surfaces dilute the intensity, and three-dimensional objects, such as pipettors, cannot be effectively decontaminated by UV light because only a fraction of the surface actually faces the light source. This drawback is compounded by the fact that almost all laboratory surfaces, such as pipettors, centrifuges, door handles, and test tube racks, present potential sources of contamination (Cone et al. 1990). Second, other materials dried with the target DNA, such as irrelevant DNA and nucleotides, can shield the target, making inactivation less efficient (Frothingham et al. 1992). Third, very short PCR products may not contain adequate numbers of neighboring pyrimidines to make them susceptible targets. The UV sensitivity of an amplified region can be estimated by counting the number of dimerizable sites (neighboring pyrimidines: CT, TT, TC, CC) in each single strand of the sequence. Based on theoretical considerations (Gordon and Haseltine 1989) and limited experimental data (Fairfax et al. 1991; Meier et al. 1993), sequences with <10 dimerizable sites will be relatively UV-resistant.

This procedure describes a method for reducing DNA contamination on laboratory surfaces by using UV light to inactivate dried DNA (Fairfax et al. 1991). Different procedures have been proposed for UV inactivation of contaminating DNA in solutions (Isaacs et al. 1991; Meier et al. 1993; Sarkar and Sommer 1993). Although UV irradiation can be helpful, meticulous technique remains the most important method for preventing contamination. In particular, UV irradiation is not an effective replacement for the physical separation of sample preparation in a pre-PCR laboratory from PCR product analysis in a post-PCR laboratory. UV irradiation can, however, provide an additional margin of safety for keeping the PCR laboratory contamination-free.

SUPPLIES AND REAGENTS

UV light ballast UF-36-2 (American Ultraviolet)
Two UV lamps, model G36T6L (American Ultraviolet)
Markline timer switch (M.H. Rhodes)
Shortwave UV radiometer J-225 (American Ultraviolet)
Purified template DNA
35 x 10-mm tissue culture dishes (Corning)
10 mM Tris-HCl (pH 8.0)

PROTOCOL

UV Irradiation of Surfaces

Caution: UV irradiation is mutagenic and can cause visual loss or blindness. Wear UV-protective glasses and cover exposed skin when working with UV light.

INSTALLATION

1. Mount the ballast and two lamps approximately 1 meter over the work surface. The UV light source can be located at any distance from the surface, but as the distance increases, stronger lights will be necessary to achieve the same light intensity at the work surface. Installation of an in-line timer switch for automatic lamp shutoff can help to conserve the limited UV lamp life.

2. Document the UV light (254 nm) intensity at the work surface by measuring it with a UV meter. This measure of UV intensity establishes the baseline performance of the UV lamp installation. Lamp performance can then be checked by comparing future light intensity measurements with this one. We achieved effective decontamination with an intensity of 400 $\mu W/cm^2$ at the work surface using the above equipment.

Measuring DNA Inactivation

STANDARDS

1. Obtain a concentrated solution of purified template DNA, such as genomic DNA, plasmid DNA, or PCR products.

2. Establish the minimum amplifiable concentration by making duplicate tenfold dilutions of the DNA in 10 mM Tris-HCl (pH 8.0) and then amplifying an aliquot of each dilution.

3. Determine the most dilute specimen that was PCR positive in duplicate and call the DNA concentration in that dilution the minimum amplifiable concentration.

4. Prepare a concentrated DNA standard from the original DNA solution that is 10^6–10^8 times more concentrated than the minimum amplifiable concentration.

5. Prepare 12 test targets, each composed of 100 μl of the concentrated DNA standard spread in the center of a plastic petri dish and dried at room temperature.

EXPERIMENTAL PROTOCOL

1. Place all 12 uncovered petri dishes with dry DNA in the area to be decontaminated. When ready to begin this 8-hour experiment, remove 3 dishes from the area and cover them. Turn on the UV lights.

2. After each UV irradiation time point (2, 4, and 8 hours), remove and cover 3 more dishes. Resuspend the DNA by adding 100 μl of 10 mM Tris-HCl (pH 8.0) to each dish. Agitate thoroughly by pipetting repeatedly and swirling the dish for several minutes, and remove the liquid to a labeled tube.

DETERMINING UV SENSITIVITY

1. Quantitate the amount of amplifiable DNA in each sample by amplifying serial tenfold dilutions as described above.

2. Plot the results with time on the x-axis (0, 2, 4, and 8 hours) and the number of tenfold dilutions to achieve the minimum amplifiable concentration on the y-axis (Fairfax et al. 1991).

The data should reveal a time-dependent decrease in DNA concentration. For instance, if the minimum amplifiable concentration of the 0-hour time point was reached at a 10^{-7} dilution and the minimum amplifiable concentration of the 4-hour time point was reached at a 10^{-4} dilution, then 4 hours of irradiation would have resulted in a 1,000-fold reduction. Although inactivation of even tenfold could be considered useful, susceptible targets can routinely be inactivated by 10,000-fold or more.

Decontamination Procedure

Decontaminate the work space after use by turning on the UV lights. Turn off the UV lights before resuming work. The minimum duration of UV illumination required for effective DNA inactivation can be determined from the measurement of DNA inactivation procedure described above. Alternatively, the UV lights can remain on at all times when the work area is not in use.

TROUBLESHOOTING

If contamination persists, look for shadowed work space areas and nonperpendicular surfaces that escape effective irradiation, and seek other contamination sources, such as reagents, equipment, or surfaces outside of the immediate work area that could contact the operator during setup.

If contamination recurs after it was eliminated by UV irradiation, remeasure the UV intensity at the work surface as described above and compare it with the original intensity. Replace the UV lamps as necessary. UV lamps will still look blue even though their UV output has decreased.

Recurrent contamination may also indicate that separation of contaminating DNA from the PCR setup area is not adequate. It is essential to maintain strict isolation of specimens to prevent contamination from PCR products, plasmids containing target DNA, and very high levels of concentrated target DNA, such as purified viral or bacterial DNA. Therefore, UV irradiation is an adjunct to proper technique.

REFERENCES

Cone, R.W., A.C. Hobson, M.W. Huang, and M.R. Fairfax. 1990. Polymerase chain reaction decontamination: The wipe test. *Lancet* **336:** 686–687.

Dwyer, D.E. and N. Saksena. 1992. Failure of ultraviolet irradiation and autoclaving to eliminate PCR contamination [letter]. *Mol. Cell. Probes* **6:** 87–88.

Fairfax, M., M. Metcalf, L. Corey, and R.W. Cone. 1991. Slow inactivation of dry PCR templates by UV light. *PCR Methods Appl.* **1:** 142–143.

Fox, J.C., M. Ait-Khaled, A. Webster, and V.C. Emery. 1991. Eliminating PCR contamination: Is UV irradiation the answer? *J. Virol. Methods* **33:** 375–382.

Frothingham, R., R.B. Blitchington, D.H. Lee, R.C. Greene, and K.H. Wilson. 1992. UV absorption complicates PCR decontamination. *BioTechniques* **13:** 208–210.

Gordon, L.K. and W.A. Haseltine. 1982. Quantitation of cyclobutane pyrimidine dimer formation in double- and single-stranded DNA fragments of defined sequence. *Radiat. Res.* **89:** 99–112.

Isaacs, S.T., J.W. Tessman, K.C. Metchette, J.E. Hearst, and G.D. Cimino. 1991. Post-PCR sterilization: Development and application to an HIV-1 diagnostic assay. *Nucleic Acids Res.* **19:** 109–116.

Kwok, S. and R. Higuchi. 1989. Avoiding false positives with PCR. *Nature* **339:** 237–238.

Meier, A., D.H. Persing, M. Finken, and E.C. Bottger. 1993. Elimination of contaminating DNA within polymerase chain reaction reagents: Implications for a general approach to detection of uncultured pathogens. *J. Clin. Microbiol.* **31:** 646–652.

Sarkar, G. and S.S. Sommer. 1991. Parameters affecting the susceptibility of PCR contamination to UV inactivation. *BioTechniques* **10:** 589–594.

——. 1993. Removal of DNA contamination in polymerase chain reaction reagents by ultraviolet irradiation. *Methods Enzymol.* **218:** 381–388.

Specificity, Efficiency, and Fidelity of PCR

Rita S. Cha[1] and William G. Thilly[2]

[1]Dana-Farber Cancer Institute, Boston, Massachusetts 02115
[2]Center for Environmental Health Sciences and Division of Toxicology, Whitaker College of Health Sciences and Technology, Massachusetts Institute of Technology, Cambridge, Massachusetts 02139

INTRODUCTION

The efficacy of PCR is measured by its specificity, efficiency (i.e., yield), and fidelity. A highly specific PCR generates one and only one amplification product that is the intended target sequence. More efficient amplification generates more products with fewer cycles. A highly accurate (i.e., high-fidelity) PCR contains a negligible amount of DNA polymerase-induced errors in its product. An ideal PCR would have high specificity, yield, and fidelity. Studies indicate that each of these three parameters is influenced by numerous components of PCR, including the buffer conditions, the PCR cycling regime (i.e., temperature and duration of each step), and DNA polymerases. Unfortunately, adjusting conditions for maximum specificity may not be compatible with high yield; likewise, optimizing for the fidelity of PCR may result in reduced efficiency. Thus, when setting up a PCR, one should know which of the three parameters is the most important for its intended application, and optimize the PCR accordingly. For instance, for direct sequencing analysis of a homogeneous population of cells (either by sequencing or by restriction fragment length polymorphism [RFLP]), the yield and specificity of PCR is more important than the fidelity. On the other hand, for studies of individual DNA molecules or rare mutants in a heterogeneous population, fidelity of PCR is vital. This chapter discusses essential components of PCR and how each influences the specificity, efficiency, and fidelity of PCR.

SETTING UP PCR

Template

Virtually all forms of DNA and RNA are suitable substrates for PCR. These include genomic, plasmid, and phage DNA, previously amplified DNA, cDNA, and mRNA. Samples prepared via standard molecular methodologies (Sambrook et al. 1989) are sufficiently pure for PCR, and usually no extra purification steps are required. Shearing of genomic DNA during DNA extraction does not affect the efficiency of PCR (at least for the fragments that are less than about 2 kb). In general, the efficiency of PCR is greater for smaller-size template DNA (i.e., previously amplified fragment, plasmid, or phage DNA) than for high-molecular-weight (i.e., undigested eukaryotic genomic) DNA. Thus, mechanical shearing and/or rare restriction enzyme digestion of genomic DNA prior to PCR are suggested for increasing the yield (Coen 1991).

Typically, 0.1–1 μg of mammalian genomic DNA is utilized per PCR (Saiki et al. 1985; Scharf et al. 1986; Mullis and Faloona 1987; Keohavong et al. 1988b; Sambrook et al. 1989). For reproducible PCR, less than 10 μg of DNA is recommended. Assuming that a haploid mammalian genome (3×10^9 bp) weighs about 3.4×10^{-12} g, 1 μg of genomic DNA corresponds to approximately 3×10^5 copies of autosomal genes. For bacterial genomic DNA or a plasmid DNA, which represent a much less complex genome, as little as picogram (10^{-12} g) to nanogram (10^{-9} g) quantities are used per reaction (Sambrook et al. 1989; Coen 1991). Previously amplified DNA fragments have also been utilized as PCR templates. Purification of the amplified product is highly recommended if the initial PCR generated a number of unspecific bands, or if a different set of primers (i.e., internal primers) are to be utilized for the subsequent PCR. On the other hand, if the amplification reaction contains only the intended target product, and the purpose of the subsequent PCR is simply to increase the overall yield utilizing the same set of primers, no further purification is required. One could simply take out a small aliquot of the original PCR mixture and subject it to a second round of PCR. In addition to the purified form of DNA, PCR from cells has also been demonstrated. In this laboratory, direct amplification of the hypoxanthine-guanine phosphoribosyltransferase (HPRT) exon 3 fragment from 1×10^5 human cells (following proteinase treatment to open up the cells) has been routinely carried out (P. Keohavong, unpubl.).

Primer Design

For many applications of PCR, primers are designed to be exactly complementary to the template. However, for other applications, such as allele-specific PCR, the engineering of mutations or new restriction endonuclease sites into a specific region of the genome, and cloning of homologous genes where sequence information is lacking, base pair mismatches are introduced either intentionally or unavoidably (Coen 1991). In either case, an ideal set of primers should hybridize efficiently to the target sequence with negligible hybridization to other related sequences that are present in the sample. Primers are typically 15–30 bases long. Assuming that the nucleotide sequences of the genome are randomly distributed, the probability of finding a match using a set of 20-base-long primers is $(1/4)^{(20 + 20)} = 9 \times 10^{-26}$. Because there are 3×10^9 bp per haploid mammalian genome, it is highly unlikely that this set of primers will find another perfectly matched template in the genome. However, amplification of unspecific products in PCR using a set of 20-base-long primers is not uncommon. This is likely due to the fact that primers containing a number of mismatches still are amplified under most PCR conditions, and that the nucleotide sequences of the genome are, in fact, not randomly distributed. Researchers have been successful in eliminating unspecific PCR products by adding the final ingredient (usually the polymerase) when the reaction mixture is hot (hot start PCR; D'Aquila et al. 1991) or by using nested primers (Mullis and Faloona 1987). To optimize the specificity of the genes suspected to be duplicated in the genome, primer sequences should be selected from intronic regions of the gene, because they are divergent even in members of tandemly repeated gene families.

Reaction Mixture

The "standard" buffer for *Taq* polymerase-mediated PCR contains 50 mM KCl, 10 mM Tris-HCl (pH 8.3 at room temperature), and 1.5 mM $MgCl_2$ (Coen 1991). The "standard" buffers for other DNA polymerases including modified T7 or Sequenase (Keohavong et al. 1988b), T4 (Keohavong et al. 1988a), Klenow (Mullis and Faloona 1987), Vent (New England Biolabs 1990, 1991), and *Pfu* (see Stratagene catalog) are also available. Although the standard buffer works well for a wide range of templates and oligonucleotide primers, the "optimal" buffer for a particular PCR varies depending on the target and the primer sequences, and the concentrations of other components in the reaction (i.e., dNTP and primers). Therefore, these so-called "standard" conditions should be regarded as a point of departure to explore modifications and potential improvements. In particular, the concentration of Mg^{++} should be optimized whenever a new combination of target and primers is first used or when the concentration of dNTPs or primers is altered. dNTPs are the major source of phosphate groups in the reaction, and any change in their concentration affects the concentration of available Mg^{++}. The presence of divalent cations is critical, and it has been shown that magnesium ions are superior to manganese, and that calcium ions are ineffective (Chien et al. 1976). In addition to the standard components of the PCR buffer mentioned above, some researchers routinely use additional components such as gelatin, Triton X-100, or bovine serum albumin for stabilizing enzymes, and glycerol (Cha et al. 1992; Cheng et al. 1994; Varadaraj and Skinner 1994), dimethylsulfoxide (DMSO; Mullis and Faloona 1987; Cheng et al. 1994; Varadaraj and Skinner 1994), or formamide (Sarkar et al. 1990; Cheng et al. 1994; Varadaraj and Skinner 1994) for enhancing specificity. It has been proposed that these reagents enhance the specificity of PCR by lowering melting and strand separation temperatures (Cheng et al. 1994). This in turn facilitates denaturation of the template and increases the specificity of primer annealing.

Primers and dNTP

To maximize the efficiency of PCR, one must ensure that the reaction mixture contains nonlimiting amounts of primers and dNTPs. Typically, in a 100-μl reaction mixture, between 0.3 μM (1.8×10^{13} molecules) and 3 μM (1.8×10^{14} molecules) of each primer and between 37 μM (2.2×10^{15} molecules) and 1.5 mM (9×10^{16} molecules) of each dNTP are utilized. For a genomic DNA PCR containing 1 μg of template DNA (3×10^5 copies of autosomal genes), the molar ratio between the primers and the genomic target sequence is at least 10^8 to 1. Having such a large excess of primers ensures that once template DNA becomes denatured, it will anneal to primers rather than to itself. Because the maximum copy number of amplified target sequence is about 10^{12} copies (see Fig. 2 and Exponential Phase of PCR, below), each primer is always in at least 10-fold excess of the target sequence (assuming that primers are not consumed by generating unspecific amplification products). The ratio between the primer and template is also important with regard to the specificity of PCR. If the ratio is too high, PCR is more prone to generate unspecific amplification products, and primer-dimers are also formed. However, if the ratio is too low (i.e., <0.1% of the standard condition for the genomic DNA PCR), the efficiency of PCR is greatly compromised.

For primers, the fraction of free (i.e., unincorporated) primers is strictly dependent on how many target sequences are generated. The fraction of free dNTPs, however, depends not only on the number of target sequences generated, but also on the size of the target sequence. For example, generating 10^{12} copies of a 100-bp target sequence consumes $10^{12} \times 100 = 10^{14}$ dNTP molecules. On the other hand, generating 10^{12} copies of a 2-kb fragment consumes 2×10^{15} dNTP molecules and effectively decreases the concentration of free dNTP. This, in turn, has a deteriorative effect on the overall efficiency of PCR. Thus, for amplifying a large target sequence, a higher concentration of dNTP is recommended (Keohavong et al. 1988b).

PCR Cycle

A typical PCR cycle consists of three steps: (1) a denaturation step (1–2-minute incubation at ≥94°C); (2) a primer annealing (or hybridization) step (1–2-minute incubation at 50–55°C); and (3) an extension step (1–2-minute incubation at 72°C). It has been hypothesized that each of the three steps in the cycle requires a minimal amount of time to be effective, whereas too much time at each step can be both wasteful (time wise) and deleterious to the DNA polymerase (Coen 1991). On the other hand, at least for relatively short DNA fragments (i.e., 100–200 bp long), PCR consisting of two steps (e.g., a denaturation step: 94°C incubation step for 1 minute followed by a primer hybridization/extension step at 50–57°C for 1 minute) can generate as much product as a three-step PCR (Cha et al. 1992). This has been the case for at least four different sets of primers tested on two different genes (Cha et al. 1992). It is possible that due to the high processivity of *Taq*, primers that anneal to the template become fully extended during the short time period during which the reaction mixture reaches the optimal temperature for *Taq* polymerase (70°C–75°C) between the 50°C to 94°C transition.

This notion is also consistent with the results of "rapid PCR" (Wittwer and Garling 1991). In an attempt to increase the speed of temperature cycling (i.e., reduce ramp times), researchers have used capillary tubes as containers and air as the heat-transfer medium for PCR. Standard protocols for a 30-cycle amplification using microfuge tubes are usually 2–6 hours in length. Using a rapid cycler, the authors completed 35 cycles of three-step PCR in 15 minutes. In this rapid PCR, each cycle consisted of a ≤1-second denaturation step at 94°C, a ≤1-second annealing step at 45°C, and a 10-second elongation step at 72°C. In addition to improving cycle times, the rapid cycle PCR amplification was more specific than three-step PCR using a conventional thermal cycler. One possible limitation of the currently available rapid PCR technique is its small reaction volume (10 μl). Because of these volume constraints, only 50 ng of DNA was used as a PCR template. Since 50 ng of mammalian DNA represents about 1.5×10^4 copies, it would not be useful for detecting rare mutations in mammalian cells. Nevertheless, rapid PCR could be used effectively for the analysis of less complex genomes (i.e., bacteria, plasmid, or phage) and/or homogeneous populations. It should also be pointed out that rapid PCR generates as much product as conventional heat-block PCR (i.e., $1–5 \times 10^{12}$ copies) (P. Andre and W. Thilly, unpubl.).

Finally, as in the case of the "standard" PCR buffer, the "standard" three-step PCR regime should also be viewed as a point of departure from which further improvement can be made. In general, higher an-  nealing temperature and shorter time allowed for annealing and extension steps improve the specificity of PCR. Also, in amplifying large fragments (i.e., >1 kb), it is necessary to increase the duration of each step to get efficient amplification (Kwok et al. 1990; Coen 1991).

EXPONENTIAL PHASE OF PCR

To set up an informative and analytical PCR, one must understand the kinetics of specific product accumulation during PCR. A schematic representation of different products accumulating as a function of cycle is depicted in Figure 1. The desired blunt-ended duplex fragments appear for the first time during the third cycle of the PCR, and from this point on, this product accumulates exponentially according to the formula, $N_f = N_0 (1 + Y)^n$, where N_f is the final copy number of the double-stranded target sequence, N_0 is the initial copy number, Y is the efficiency of primer extension per cycle, and n is the number of PCR cycles under conditions of exponential amplification (Keohavong et al. 1988b). As depicted in Figure 2, in most cases, once the final copy number of the desired fragment (N_f) reaches about 10^{12}, its efficiency per cycle (Y) drops dramatically, and the product stops accumulating exponentially. This drop in efficiency likely reflects that enzymes become limiting in the reaction. Because products are accumulating exponentially, adding twice as much enzyme at this stage will only support one additional cycle of PCR. The exponential phase of a PCR refers to the early cycle period during which the products accumulate in a manner that is consistent with the equation above. Continuing PCR beyond this point often results in amplification of unspecific bands, the appearance of small deletion mutant bands, and, in certain instances, the disappearance of the specific product (G. Hu, unpubl.). One can overcome these undesired effects of "overamplification" and achieve additional amplification by taking a small aliquot of the reaction mixture that has already undergone 10^6–10^7 doublings and placing it in a fresh reaction mixture.

For many applications of PCR, especially the ones that are quantitative in nature, it is critical that amplification is carried out in the exponential phase of PCR (Fig. 2). Numerous laboratories have studied the efficiencies of different DNA polymerases that are utilized in PCR. As a result, we now have a fairly good idea regarding how efficient different DNA polymerases are in a typical PCR (Keohavong and Thilly 1988; Ling et al. 1991). By using the equation above, a knowledge of the initial copy number permits one to estimate how many cycles will be required for the final copy number to reach about 10^{12}. For example, for *Taq* PCR (assuming efficiency per cycle of 70%) starting with 1 μg of genomic DNA (e.g., 3×10^5 copies of mammalian genome), the equation becomes $10^{12} = 3 \times 10^5 (1 + 0.7)^n$. Solving for n gives 28.6, indicating that in this hypothetical case, the desired product will accumulate exponentially up to about cycle number 29. Thus, an analysis that is quantitative in nature must be carried out on the samples that are taken out at or before the 29th cycle. In fact, since

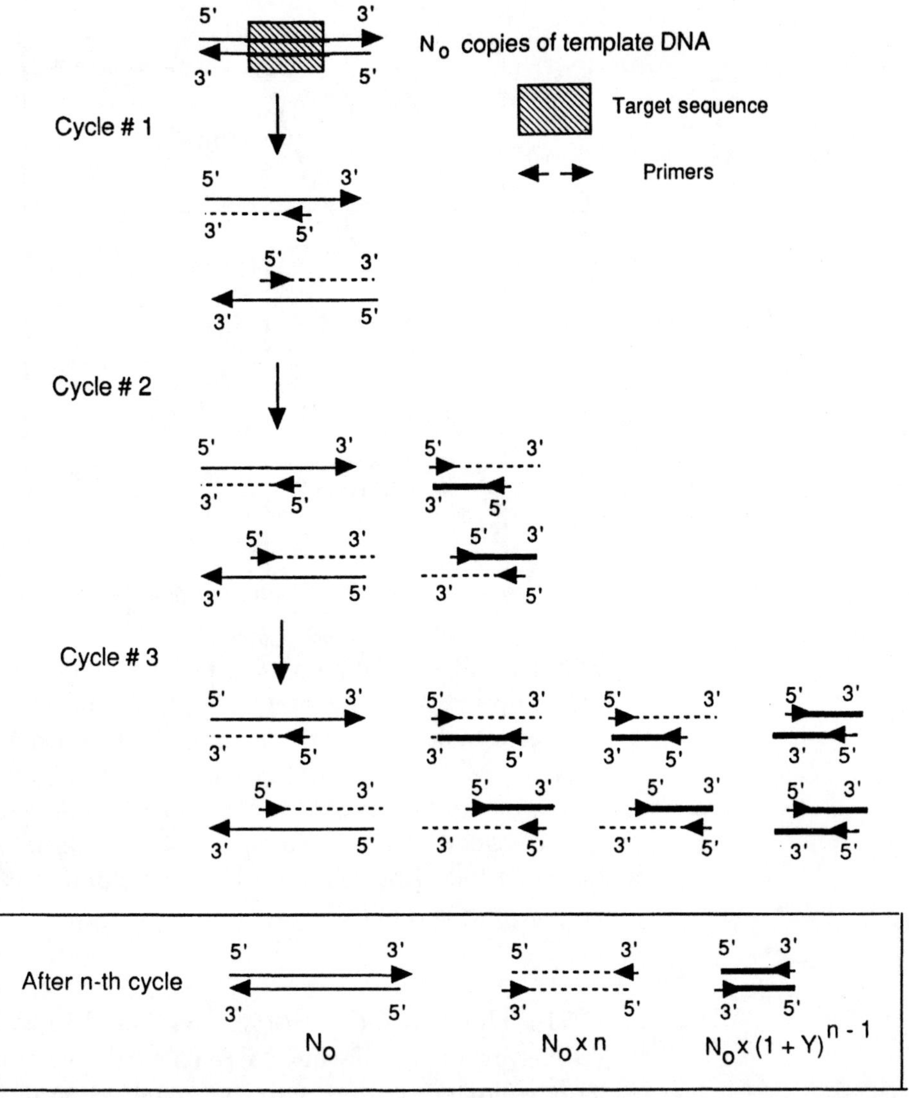

Figure 1 Schematic representation of PCR. N_0 copies of duplex template DNA are subjected to n cycles of PCR. During each cycle, duplex DNA is denatured by heating, which then allows primers (arrows) to anneal to the target sequence (hatched square). In the presence of DNA polymerase and dNTPs, primer extension takes place. The desired blunt-ended duplex product (thick bars with arrows) appears during the third cycle and accumulates exponentially during subsequent cycles. Following n cycles of exponential PCR, there will be N_0 $(1 + Y)^{n-1}$ copies of the duplex target sequence.

10^{12} copies of a particular sequence are sufficient for most applications in molecular biology, there is no apparent reason to carry out additional cycles.

The efficiency of the same polymerase can vary significantly, depending on the nature of the target sequence, the primer sequences, and the reaction conditions (Eckert and Kunkel 1991; Ling et al.

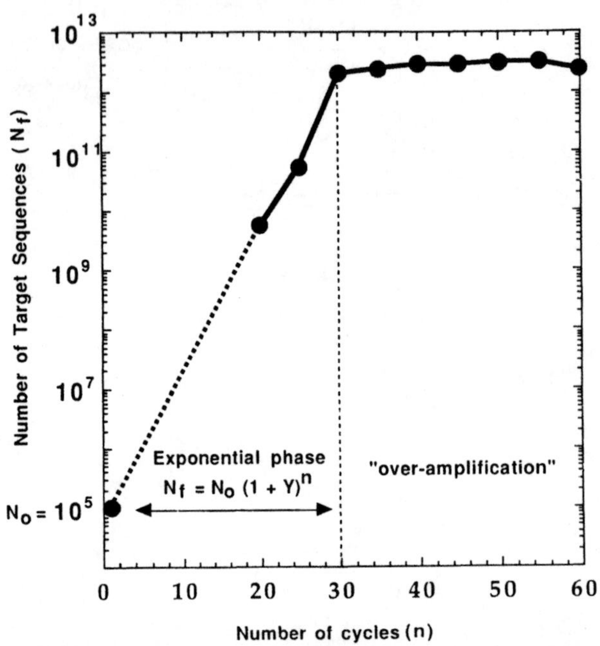

Figure 2 Accumulation of target sequence during PCR as a function of number of cycles. Approximately 10^5 (N_o) copies of rat Ha-*ras* gene exon 1 are subjected to 60 cycles of PCR under a standard *Taq* PCR condition (Chien et al. 1976). A 2.5-µl aliquot is taken at 20, 25, 30, 35, 40, 45, 50, 55, and 60 cycles (n) and analyzed on a polyacrylamide gel. The number of target sequences generated at each stage (N_f) is estimated based on the intensity of the band following ethidium bromide staining. *Taq* (2.5 units) is added following 30 cycles of PCR.

1991). Therefore, the efficiencies listed in Table 1 may not reflect the efficiency of a different PCR carried out under different conditions. The reported values can be used to make a reasonable estimate. Nevertheless, because each specific PCR amplification has a different efficiency, to carry out an accurate quantitative analysis, one needs to determine the efficiency of the particular PCR (see Fig. 2).

DNA POLYMERASES AND PCR

In vitro DNA replication has been accomplished by DNA polymerases from many different sources (Saiki et al. 1985, 1988; Mullis and Faloona 1987; Keohavong et al. 1988a; New England Biolabs 1990, 1991; see also Stratagene catalog). The initial PCR procedure described by Saiki et al. (1985) used the Klenow fragment of *Escherichia coli* DNA polymerase I. This enzyme was heat labile, and, as a result, fresh enzyme had to be added during each cycle following the denaturation and primer hybridization steps. Introduction of the thermostable *Taq* polymerase in PCR (Saiki et al. 1988) subsequently alleviated this tedium and made possible automation of the thermal cycling portion of the procedure. For PCR, thermostable DNA

Table 1 Summary of PCR Conditions

Enzyme	dNTP (mM)	pH	Mg (mM)	Efficiency per cycle (%)	Error rate (error/bp incorporated)	PCR-induced mutant fraction[a] (%)	No. of cycles required[b]	References
Pfu	0.1	8.4	1.5	60	7×10^{-7}	0.3	30	P. Andre (unpubl.)
T4	2.15	8.0	5	56	3×10^{-6}	2	32	Keohavong and Thilly (1989)
T7	3.5	8.0	2.5	90	4.4×10^{-5}	13	22	Keohavong and Thilly (1989)
Vent	0.5–1.5	8.5	7.5	70	4.5×10^{-5}	16	26	Ling et al. (1991)
Taq	0.5–1.5	8.0	5	36	7.2×10^{-5}	25	45	Ling et al. (1991)
Taq	16.6	8.8	10	88	2×10^{-4}	56	22	Dunning et al. (1988); Keohavong and Thilly (1989)
Klenow	1.5	7.9	10	80	1.3×10^{-4}	41	24	Mullis and Faloona (1987); Keohavong and Thilly (1989)

[a]Fraction of PCR-induced noise following 10^6-fold amplification of 200-bp target sequence given the error rate.
[b]Number of cycles required to obtain 10^6-fold amplification given the efficiency per cycle.

polymerases (e.g., *Taq*, Vent, and *Pfu*) are preferred over heat-labile polymerases (e.g., T4, T7, and Klenow) simply because they are much easier to handle and, most importantly, they are amenable to automation.

Studies have shown that different DNA polymerases have distinct characteristics that affect the efficacy of PCR. For example, *Taq* polymerase does not have the $3' \rightarrow 5'$ exonuclease proofreading function, and, as a result, it has a relatively high error rate in PCR (Table 1). On the other hand, its inability to edit mispaired $3'$ ends has been an asset for researchers who developed the allele-specific PCR based on the concept that primers containing mismatches at the $3'$ end were not extended as efficiently as the perfectly matched primers (Newton et al. 1989; Wu et al. 1989; Kwok et al. 1990; Cha et al. 1992; Bottema and Sommer 1993). This concept would not have worked for enzymes with exonuclease activities, because once the $3'$ mismatch was recognized by the polymerase, it would first be repaired and would then be extended, thus abolishing the specificity conferred by the $3'$ mismatches. As applications of PCR become increasingly sophisticated and specific, distinctive properties of polymerases should be utilized to meet specific needs.

Fidelity of in vitro DNA polymerization is perhaps one of the most intensively studied subjects in PCR. For many applications of PCR, where a relatively homogeneous DNA population is analyzed (i.e., direct sequencing or restriction endonuclease digestion), the polymerase-induced mutations during PCR are of little concern. In general, polymerase-induced mutations are randomly distributed over the amplified fragment, and an accurate consensus sequence is usually obtained. However, PCR is also used for studies of rare molecules in heterogeneous populations. Examples include the study of allelic polymorphisms in individual mRNA transcripts (Frohman et al. 1988; Lacy et al. 1989), the characterization of the allelic stages of single sperm cells (Li et al. 1990) or single DNA molecules (Jeffreys et al. 1990; Ruano et al. 1990), and the characterization of rare mutations in a tissue (Cha et al. 1992) or a population of cells in culture. For these applications, it is vital that the polymerase-induced mutant sequences do not mask the rare DNA sequences. Each polymerase-induced error, once introduced, is amplified exponentially along with the original wild-type sequences during subsequent cycles. This results in an overall increase in the fraction of polymerase-induced mutant sequences as a function of the number of amplification cycles. Analyses that use small amounts of template DNA are especially prone to PCR-induced artifacts. For example, if one were to carry out PCR with 10 copies of template DNA, any polymerase-induced mutation during the first few cycles would appear as a major mutant population in the final PCR products. Because the number of copies of template DNA is low and the error rate of *Taq* polymerase is about 10^{-4}, the probability of this event occurring is low (i.e., 10^{-3}). However, if such an event should occur, the particular mutation induced by the polymerase would comprise as much as 10% of the final PCR products. One can prevent this "jackpot" artifact by starting with a large amount of template DNA (i.e., $\geq 10^5$ copies). In this case, about 10 mutations are introduced on the average during the first cycle of the PCR; however, all of these mutations constitute only about $1/10^5$ of the final products.

Under low-fidelity conditions (i.e., *Taq* or Klenow PCR), this mutant fraction can become significant. For example, following one million-fold amplification by a DNA polymerase with an error rate of 10^{-4}, the PCR-induced error constitutes as much as 33% of the 200-bp-long amplified products.[1] Assuming that polymerase errors are uniformly distributed, the error frequency per base, on average, is 1.7×10^{-3} ($0.33 \times 200 = 0.0017$). This level of PCR-induced noise will certainly hinder attempts to characterize rare mutations in tissue cul-

[1]The fraction of PCR-induced mutants is calculated according to a formula $F(>1) = 1 - e^{-bfd}$, where b is the length of the target sequence, f is the error rate, and d is the number of doublings (Newton et al. 1989; Wu et al. 1989). Thus, following a 10^6-fold amplification (e.g., 20 doublings) of a 200-bp fragment at an error rate of 10^{-4}/bp incorporated will lead to an estimated PCR-induced mutant fraction of 33% ($1 - e^{-(200)(10^{-4})(20)} = 0.33$).

ture or in animals and humans, where the expected mutant frequency of a particular mutation could be as low as 10^{-7} or 10^{-8}.

The fidelity of PCR varies depending on reaction conditions and the nature of the target sequences. In the past, several groups have found conditions that permitted more accurate PCR by modifying reaction buffer conditions. For instance, Ling et al. (1991) were able to reduce the error rate of *Taq* PCR by a factor of 2.8 (from 2×10^{-4} to 7.2×10^{-5}) by modifying reaction conditions. One may assess the significance of this 2.8-fold improvement on *Taq* PCR fidelity by comparing the fractions of PCR-induced noise before and after the improvement. According to the formula, $F(>1) = 1 - e^{-bfd}$ (see footnote 1) (Eckert and Kunkel 1991), 56% of the PCR product amplified under the low-fidelity condition is *Taq* polymerase-induced noise (Table 1). On the other hand, only 25% of the PCR product generated under the high-fidelity condition is polymerase-induced noise. In this case, a 2.8-fold reduction in the *Taq* polymerase error rate reduced the overall PCR-induced mutant fraction by more than half (Table 1). Thus, it is indeed possible to improve the overall fidelity of PCR substantially by adjusting reaction conditions. Nevertheless, it must be pointed out that despite much effort to optimize the fidelity of *Taq*, T7, and Vent PCR by altering reaction conditions, their improved fidelity has never reached the level of *Pfu* or T4 polymerase (Keohavong and Thilly 1989; Ling et al. 1991), suggesting that some intrinsic properties of the polymerase also contribute to its overall error rate. Regarding the error rates of exo+ polymerases, one should realize that the measured error rate reflects the average value of a heterogeneous population of DNA polymerases; this heterogeneity presumably arises as a result of errors during transcription of the gene. It is possible that some of the transcription errors are introduced in the region of the gene that is critical for fidelity of the polymerase (i.e., the proofreading function), and thus increase the average error rate. If this is the case, one may be able to enhance the fidelity of exo+ polymerase PCR by devising a means to physically separate or biologically inactivate these rare exo– mutant polymerases (W. Thilly, unpubl.).

In addition to the error rate during PCR, the kinds of mutations that are introduced during PCR are also dependent on DNA polymerases. Whereas GC to AT transitions are the predominant mutations for T4 and T7 polymerases, AT to GC transitions are most frequently observed with *Taq* polymerase (Keohavong and Thilly 1989). *Taq* polymerase is also highly prone to generating deletion mutations if the template DNA has the potential to form secondary structures (Cariello et al. 1991). The Klenow fragment induces possible transitions and deletions of 2 and 4 bp. These observations again suggest that each polymerase has distinctive modes of operation regarding fidelity in in vitro replication.

The findings that different polymerases induce different types of mutations in PCR also have a very practical value in designing PCR-based experiments. For example, if one were to look for a rare allele that had undergone a GC to AT transition, it would be best to use *Taq* polymerase for PCR. Because *Taq* predominantly induces AT to GC transitions (Keohavong and Thilly 1989), using *Taq* will minimize false-positive cases that may arise as a result of a *Taq* polymerase-induced artifact. In another hypothetical case, assume that *Taq* PCR followed by sequencing analysis, either by cloning and sequencing or by denaturant gradient gel electrophoresis (DGGE)-type analysis followed by sequencing, reveals that in the population of cells analyzed, a rare AT to GC mutant allele exists at a frequency of 10^{-3}. However, this mutation is the type of mutation expected from *Taq* amplification, thus one is not sure if this is a true variant in the original sample or a PCR artifact. To distinguish between these two possibilities, the same analysis can be carried out again using a T7 or T4 polymerase, to see whether the AT to GC mutations appear again. If this AT to GC mutation appears following PCR mediated by two different enzymes with different mutational specificities, then it is fair to say that the mutation existed in the original sample.

Because of its thermostability, reliability, and durability, *Taq* DNA polymerase has been most widely used in PCR. However, as summarized in Table 1, the fidelity of *Taq* (2×10^{-4} error/bp per duplication) is the lowest among DNA polymerases whose fidelity has been measured. This, in turn, effectively prevents using *Taq* polymerase in a PCR where the fidelity is of concern. Recently, a number of additional thermostable enzymes have been isolated. Unlike *Taq*, which does not have the $3' \rightarrow 5'$ exonuclease proofreading function, these newly isolated enzymes (e.g., Vent and *Pfu*) do have the editing function, and, as expected, they are more accurate than *Taq* polymerase. Error rates for *Pfu* and Vent have been estimated to be 7×10^{-7} and 4.5×10^{-5} errors/bp per duplication, respectively (Table 1). The fraction of PCR-induced noise in a 200-bp target sequence following a 10^6-fold amplification in these cases would be 0.3% for *Pfu* and 16% for Vent polymerases (56% for *Taq* PCR; see Table 1).

ANALYZING FIDELITY OF PCR

For many applications of PCR where rare variants are involved, the fidelity of PCR is an important concern. A number of laboratories have studied the fidelity of PCR, and the error rates of commonly utilized DNA polymerases are known (Table 1). However, because the fidelity of a polymerase varies significantly depending on the reaction conditions and the nature of the target sequences, it needs be determined on a sequence-by-sequence and/or a reaction condition-by-reaction condition basis. There are at least three independent methods of measuring the fidelity of PCR: (1) the forward mutation assay, (2) the reversion mutation assay, and (3) the DGGE-type analysis.

Forward Mutation Assay

The forward mutation assay consists of cloning individual DNA molecules from the amplified population and determining the number of DNA sequence changes according to what fraction of the cloned population displays a particular phenotype (Loeb and Kunkel 1982; Eckert and Kunkel 1990, 1991). For example, one can assess the error rate during synthesis of the *lacZ* gene by the frequency of light blue and colorless (mutant) plaques among the total plaques scored. The nature of mutations can also be determined by DNA sequence analysis of a collection of the mutants.

Reversion Mutation Assay

The second method is a reversion mutation assay using a phage template DNA that contains specific mutations resulting in a measurable phenotype (i.e., *lacZ* $^-$, or colorless phenotype). In these assays, polymerase-induced errors are scored as DNA sequence changes that revert the mutant to a wild-type or pseudo-wild-type phenotype. This approach is especially useful for highly accurate polymerases (Eckert and Kunkel 1991).

Reversion assays are focused on a limited subset of errors occurring at only a few sites. As mentioned above, in general, polymerase-induced mutations are randomly distributed throughout a target sequence. However, a number of locations in the target sequence are more prone to polymerase-induced errors (Keohavong and Thilly 1989). Thus, error rates measured by the reversion assay may vary significantly depending on the nature of the initial mutations placed in the phage template.

DGGE-type Analysis

DGGE is a system that separates DNA fragments harboring small changes (i.e., single-base substitutions, small additions, or deletions) based on their sequences. In this case, DGGE is used to separate polymerase-induced mutant sequences from the correctly amplified sequences. By measuring the fraction of signals coming from the portion of the gel corresponding to the polymerase-induced mutant sequences (heteroduplex fraction), one can calculate the fidelity of the enzyme according to a formula, $f = HeF/(b \times d)$, where f is the error rate (errors per bp incorporated per duplication), HeF is the heteroduplex fraction, b is the length of the single-strand low-melting domain in which mutants can be detected, and d is the number of DNA duplications (Keohavong and Thilly 1989; Ling et al. 1991). Unlike the other two assays in which only the changes that result in phenotypic changes are scored as PCR-induced mutations, DGGE allows the visualization and detection of all the mutations introduced in the target sequence. This feature makes DGGE the most comprehensive and sensitive means of measuring PCR fidelity among the currently available techniques.

ACKNOWLEDGMENTS Work in the authors' laboratories is in part supported by grants from the National Institute of Environmental Health Sciences (grants NIEHS 5-P01-ES03926 and P01-ES05622) and the U.S. Department of Energy (grant DE-FG02-ER60448).

REFERENCES

Bottema, C.D.K. and S.S. Sommer. 1993. PCR amplification of specific alleles: Rapid detection of known mutation and polymorphisms. *Mutat. Res.* **288:** 93–102.

Cariello, N.F., W.G. Thilly, J.A. Swenberg, and T.R. Skopek. 1991. Deletion mutagenesis during polymerase chain reaction: Dependence on DNA polymerase. *Gene* **99:** 105–108.

Cha, R.S., H. Zarbl, P. Keohavong, and W.G. Thilly. 1992. Mismatch amplification mutation assay (MAMA): Application to the c-H-*ras* gene. *PCR Methods Appl.* **2:** 14–20.

Cheng, S., C. Fockler, W.M. Barnes, and R. Higuchi. 1994. Effective amplification of long targets from cloned inserts and human genomic DNA. *Proc. Natl. Acad. Sci.* **91:** 5695–5699.

Chien, A., D.B. Edgar, and J.M. Trela. 1976. Deoxyribonucleic acid polymerase from the extreme thermophile *Thermus aquaticus. J. Bacteriol.* **127:** 1550.

Coen, D.M. 1991. The polymerase chain reaction. In *Current protocols* (ed. F.M. Ausubel et al.), pp. 15.01–15.40. Wiley, New York.

D'Aquila, R.T., L.J. Bechtel, J.A. Videler, J.J. Eron, P. Gorczyaca, and J.C. Kaplan. 1991. Maximizing sensitivity and specificity of PCR by preamplification heating. *Nucleic Acids Res.* **19:** 3749.

Dunning, A.M., P. Talmud, and S.E. Humphries. 1988. Errors in the polymerase chain reaction. *Nucleic Acids Res.* **16:** 10393.

Eckert, K.A. and T.A. Kunkel. 1990. High fidelity DNA synthesis by the *Thermus aquaticus* DNA polymerase. *Nucleic Acids Res.* **18:** 3739–3744.

——. 1991. DNA polymerase fidelity and the polymerase chain reaction. *PCR Methods Appl.* **1:** 17–24.

Frohman, M.A., M.K. Dush, and G.R. Martin. 1988. Rapid production of full-length cDNAs from rare transcripts: Amplification using a single gene specific oligonucleotide primer. *Proc. Natl. Acad. Sci.* **85:** 8998–9002.

Jeffreys, A.J., R. Neumann, and V. Wilson. 1990. Repeat unit sequence variation in minisatellites: A novel source of DNA polymorphism for studying variation and mutation by single molecule analysis. *Cell* **60:** 473–485.

Keohavong, P. and W.G. Thilly. 1989. Fidelity of DNA polymerases in DNA amplification. *Proc. Natl. Acad. Sci.* **86:** 9253–9257.

Keohavong, P., A.G. Kat, N.F. Cariello, and W.G. Thilly. 1988a. Laboratory methods: DNA amplification in vitro using T4 DNA polymerase. *DNA* **7:** 63–70.

Keohavong, P., C.C. Wang, R.S. Cha, and W.G. Thilly. 1988b. Enzymatic amplification and characterization of large DNA fragments from genomic DNA. *Gene* **71:** 211–216.

Kwok, S., D.E. Kellogg, N. McKinney, D. Spasic, L. Goda, C. Levenson, and J.J. Sninsky. 1990. Effects of primer-template mismatches on the polymerase chain reaction: Human immunodeficiency virus type 1 model studies. *Nucleic Acids Res.* **18:** 999–1005.

Lacy, M.J., L.K. McNeil, M.E. Roth, and D.M. Kranz. 1989. T-cell receptor δ-chain diversity in peripheral lymphocytes. *Proc. Natl. Acad. Sci.* **86:** 1023–1026.

Li, H., X. Cui, and N. Arnheim. 1990. Direct electrophoretic detection of the allelic state of a single DNA molecule in human sperm by using the polymerase chain reaction. *Proc. Natl. Acad. Sci.* **87:** 4580–4584.

Ling, L.L., P. Keohavong, C. Dias, and W.G. Thilly. 1991. Optimization of the polymerase chain reaction with regard to fidelity: Modified T7, *Taq*, and Vent DNA polymerases. *PCR Methods Appl.* **1:** 63–69.

Loeb, L.A. and T.A. Kunkel. 1982. Fidelity of DNA synthesis. *Annu. Rev. Biochem.* **52:** 429–457.

Mullis, K.B. and F.A. Faloona. 1987. Specific synthesis

of DNA in vitro via a polymerase-catalysed chain reaction. *Methods Enzymol.* **155:** 335–350.

New England Biolabs. 1990, 1991. Vent™ DNA polymerase technical bulletin. Beverly, Massachusetts.

Newton, C.R., A. Graham, L.E. Heptinstall, S.J. Powell, C. Summers, N. Kalsheker, J.C. Smith, and A.F. Markham. 1989. Analysis of any point mutation in DNA. The amplification refractory mutation system. *Nucleic Acids Res.* **17:** 2503–3516.

Ruano, G., K.K. Kidd, and J.C. Stephens. 1990. Haplotype of multiple polymorphisms resolved by enzymatic amplification of single DNA molecules. *Proc. Natl. Acad. Sci.* **87:** 6296–6300.

Saiki, R.K., S. Scharf, F. Faloona, K.B. Mullis, G.T. Horn, H.A. Erlich, and N. Arnheim. 1985. Enzymatic amplification of β-globin genomic sequences and restriction site analysis for diagnosis of sickle cell anemia. *Science* **230:** 1350–1354.

Saiki, R.K., D.H. Gelfand, S. Stoffel, S.J. Scharf, R. Higuchi, G.T. Horn, K.B. Mullis, and H.A. Erlich. 1988. Primer-directed enzymatic amplification of DNA with thermostable DNA polymerase. *Science* **239:** 487–491.

Sambrook, J., E.F. Fritsch, and T. Maniatis. 1989. *Molecular cloning: A laboratory manual,* 2nd edition. Cold Spring Harbor Laboratory, Cold Spring Harbor, New York.

Sarkar, G., S. Kapelner, and S.S. Sommer. 1990. Formamide can dramatically improve the specificity of PCR. *Nucleic Acids Res.* **18:** 7465.

Scharf, S.J., G.T. Horn, and H.A. Erlich. 1986. Direct cloning and sequence analysis of enzymatically amplified genomic sequences. *Science* **233:** 1076–1078.

Varadaraj, K. and D.M. Skinner. 1994. Denaturants or cosolvents improve the specificity of PCR amplification of a G+C rich DNA using genetically engineered DNA polymerases. *Gene* **140:** 1–5.

Wittwer, C.T. and D.J. Garling. 1991. Rapid cycle DNA amplification: Time and temperature optimization. *BioTechniques* **10:** 76–83.

Wu, D.Y., L. Ugozzoli, B.J. Pal, and R.B. Wallace. 1989. Allele-specific enzymatic amplification of β-globin genomic DNA for diagnosis of sickle cell anemia. *Proc. Natl. Acad. Sci.* **86:** 2757–2760.

Optimization and Troubleshooting in PCR

Kenneth H. Roux

Department of Biological Science, Florida State University, Tallahassee, Florida 32306-3050

INTRODUCTION

The use of PCR to generate large amounts of a desired product can be a double-edged sword. Failure to amplify at optimum conditions can lead to the generation of multiple undefined and unwanted products, even to the exclusion of the desired product. At the other extreme, no product may be amplified. A typical response at this point is to vary one or more of the many parameters that are known to contribute to primer-template fidelity and primer extension. High on the list of optimization variables are Mg^{++} concentrations, buffer pH, and cycling conditions. With regard to the last variable, the annealing temperature is most important. The situation is further complicated by the fact that some of the variables are quite interdependent. For example, because dNTPs directly chelate a proportional number of Mg^{++} ions, an increase in the concentration of dNTPs decreases the concentration of free Mg^{++} available to influence polymerase function.

Touchdown PCR

Touchdown (TD) PCR represents a fundamentally different approach to PCR optimization (Don et al. 1991). Rather than using multiple reaction tubes, each with different reagent concentration and/or cycling parameters, a single tube or a small set of tubes is run under cycling conditions that inherently favor amplification of the desired amplicon, often to the exclusion of artifactual amplicons and primer-dimers. Multiple cycles are programmed so that the annealing segments in sequential cycles are run at incrementally lower temperatures (see below). As cycling progresses, the annealing-segment temperature, which was selected to be initially above the suspected T_m, gradually declines to, and falls below, this level. This strategy helps ensure that the first primer-template hybridization events involve only those reactants with the greatest complementarity, i.e., those yielding the target amplicon. Even though the annealing temperature may eventually drop down to the T_m of nonspecific hybridizations, the target amplicon will have already begun its geometric amplification and is thus in a position to outcompete any lagging (nonspecific) PCR products during the remaining cycles. Because the aim is to avoid low-T_m priming during the earlier cycles, it is imperative that the hot start modification (D'Aquila et al. 1991; Erlich et al. 1991; Mullis 1991) (see below) be used with TD PCR. TD PCR should be viewed not as a method of determining the optimum cycling conditions for a specific PCR, but as a potential one-step method for approaching optimal amplification. We have found that a variety of otherwise satisfactory single-amplicon-yielding reactions are rendered more robust (i.e., yield more product) when subjected to TD PCR (K.H. Roux and K.H. Hecker, unpubl.).

TD PCR is of particular value when the degree of identity between the primer and template is unknown (Roux 1994). This situation often arises when primers are designed on the basis of amino acid sequences, members of a multigene family are amplified, or evolutionary PCR is attempted; i.e., amplification of DNA from one species using primers with identity to a homologous segment of another species. In such cases, the mismatches between the primers and template may have lowered the T_m of the target amplicon enough to approach those of the spurious priming sites. Degenerate primers with multiple base variation or inosine residues are often used in such situations (Knoth et al. 1988; Lee et al. 1988; Patil and Dekker 1990; Batzer et al. 1991; Peterson et al. 1991), but the greater variety of sequences in the former case and the relaxed stringency in the latter case might tend to increase the chances of nonspecific priming. Moreover, in some cases the locations of potential base mismatches are unknown. Although TD PCR can be used with degenerate primers (Batzer et al. 1991), we have shown that nondegenerate primers displaying a significant degree of template-sequence mismatch can yield single-target amplicons of single-copy genes from genomic DNA under standard buffer conditions (Roux 1994). Even mismatches clustered near the 3′ end of the primer are tolerated.

PROTOCOL

Programming the Thermal Cycler for TD PCR

The goal in programming for TD PCR is to produce a series of cycles with progressively lower annealing temperatures. The annealing temperature range should span about 15°C and extend from at least a few degrees above to 10 or so degrees below the estimated T_m. For example, for a calculated primer-template T_m of 62°C with no degeneracy, program the thermal cycler to decrease the annealing temperature 1°C every second cycle (i.e., run 2 cycles per degree) from 65°C to 50°C, followed by 15 additional cycles at 50°C.

Some thermal cyclers (e.g., Perkin-Elmer model 9600 and MJ Research model PTC-100) readily accommodate TD PCR and are easily programmed to decrease the temperature of a segment automatically by a fixed amount per cycle (e.g., 0.5°C/cycle). For others, a long series of files must be linked or extensive strings of commands entered. In these latter cases, it may be more convenient to create a "generic" TD PCR program covering a broader temperature range (~20°C) than to reprogram every time the range needs to be modified by a few degrees. Another alternative to programming restrictions and inconvenience is to use fewer but more abrupt steps (e.g., seven 2°C steps or five 3°C steps); however, doing so may decrease the chances for discriminating between products with two closely spaced T_m values.

The continued presence of spurious bands following TD PCR indicates that the initial annealing temperature was too low, that there is a relatively small gap between the T_m values of the target and unwanted amplicons, and/or that the unwanted amplicons are being more efficiently amplified. Raising the number of cycles per 1°C descending step to 3 or 4 will give the target amplicon added competitive advantage before the initiation of the spurious amplification. A proportional number of cycles should be removed from the end of the program to prevent excess cycling and the concomitant degradation of the amplicon and generation of high-molecular-weight smears (Bell and DeMarini 1991).

Modifications of TD PCR for use with degenerate and mismatched primers include lowering the annealing temperature range (e.g., 50°C declining to 35°C) and running the last 15 cycles at 50°C.

Optimization Strategy

The example given is for TD PCR, but the same principles apply to conventional PCR.

1. Design optimal primer pairs that are closely matched in T_m. For additional discussion of primer design, see Dieffenbach et al. in Section 3 (this volume).

2. Calculate or estimate approximate T_m. Program the thermal cycler for TD PCR as described above.

3. Set up several standard hot start PCR mixes incorporating a range of Mg^{++} concentrations and including appropriate positive and negative controls. Use 10^4–10^5 copies of the template.

4. Amplify as above and analyze products.

   a. If weak or no product is detected:

      - Subject reaction tubes to 10 additional cycles at constant annealing temperature (i.e., 55°C) and recheck.
      - Reamplify 10-fold dilutions (1:10 to 1:1000) of initial TD PCR at fixed annealing temperature for 30 cycles.
      - Use more template and check for inhibitor in template preparation by spiking original PCR mix with dilutions of known positive (demonstrably amplifiable) template.
      - Add, extend, or increase the temperature of the initial template denaturation step prior to cycling (5 minutes at 95°C is standard).
      - Vary concentrations of buffer components (pH, *Taq* DNA polymerase, dNTPs, primers).
      - Add enhancers to PCR mix (see below).
      - Reamplify dilutions (1:10 to 1:1000) of the first reaction using nested primers.
      - Abandon this primer set, design new primers, and begin again. Depending on one's degree of impatience and tolerance for frustration, this step might supersede any of the above.

   b. If multiple products or a high-molecular-weight smear is observed:

      - Raise the maximum and minimum annealing temperatures (i.e., move the range upward) in the TD PCR program.
      - Remove some cycles from the bottom of the range and/or from the terminal constant temperature cycles.
      - Increase the number of cycles per degree annealing temperature by 1 cycle, i.e., to 3 cycles per degree. Doing so may necessitate removing some lower end and/or terminal cycles to prevent smearing due to excess cycling.
      - Vary concentrations of buffer components (pH, *Taq* DNA polymerase, dNTPs, primers).
      - Attempt band purification followed by reamplification. Target bands can be cut from gels and allowed to diffuse out or be liberated by freeze/thaw cycles or enzymatic gel digestion. Alternatively, a small plug of gel can be removed with a micropipette tip or, most simply, by stabbing the band directly in the gel with an autoclaved toothpick and inoculating a fresh reaction tube.

- Reamplify 1:10^4 and 1:10^5 dilutions of first reaction using nested primers.
- If all else fails, abandon primer set, design new primers, and begin again.

OTHER OPTIMIZATION STRATEGIES

Several other optimization strategies have been developed for standard PCR, although most are applicable to TD PCR as well. Each is discussed briefly below. Variables that affect PCR product specificity and yield are listed in Table 1.

Enhancing Agents

Various additives such as DMSO (1–10%), PEG-6000 (5–15%), glycerol (5–20%), nonionic detergents, formamide (1.25–10%), and bovine serum albumin (10–100 μg/ml) can also be incorporated into the reaction to increase specificity and yield (Pomp and Medrano 1991; Newton and Graham 1994). In fact, some reactions may amplify only in the presence of such additives (Pomp and Medrano 1991). Several optimization kits incorporating these and other enhancing agents and a variety of buffers are currently marketed (e.g, by Continental Laboratory Products, Invitrogen, Perkin-Elmer, and Stratagene). Additional discussion of PCR optimization and contamination-avoidance strategies can be found in Newton and Graham (1994).

Table 1 Conditions Favoring Enhanced Specificity

	Use hot start
	Use TD PCR (favors enhanced specificity and yield)
	Optimize primer design
↓	Mg^{++}
↓	dNTP (also favors higher fidelity)
	Optimize pH
↓	*Taq* DNA polymerase
↓	Cycle segment lengths
↓	Number of cycles
↑	Annealing temperature
↓	Inhibitors
↑	Ramp speed
↑	Chance that target temperature is achieved in each tube
	Add and optimize enhancer(s)
↓	Primer concentration
↓	Primer degeneracy
↑	Template denaturation efficiency

Adjusting conditions in the direction opposite that listed above usually favors increased sensitivity (i.e., more product) and the concomitant risk of nonspecific amplification. The aim is to strike a balance between these two opposing tendencies. ↑ and ↓ signify increase and decrease, respectively.

Matrix Analyses

The basic challenge is to devise an optimization protocol that is efficient in both time and cost. A full matrix analysis in which several values for each of the variables are tested in combination with each of the other variables can quickly become overwhelmingly cumbersome and costly. The size of the matrix can be significantly pared down by applying the Taguchi method (Taguchi 1986), in which several key variables are simultaneously altered (Cobb and Clarkson 1994). A more typical strategy is to run a simple matrix analysis focused on those parameters most likely to have the greatest impact on PCR primer hybridization and enzyme fidelity, e.g., Mg^{++} concentration and annealing temperature.

Mg^{++} Concentration

Mg^{++} concentration is the easiest parameter to manipulate because all concentration variations can be run simultaneously in separate tubes. Suppliers of *Taq* polymerase now provide the $MgCl_2$ solution separate from the standard reaction buffer to simplify its adjustment. A typical two-step optimization series might first include Mg^{++} at 0.5-mM increments from 0.5 to 5.0 mM and, after the range is narrowed, a second round covered by several 0.2- or 0.3-mM increments.

Annealing Temperature

Optimization of annealing temperature begins with calculation of the T_m values of the primer-template pairs by one of several methods, the simplest being $T_m = 4(G + C) + 2(A + T)$. A single-base mismatch lowers the T_m by about 5°C. More complex formulas can also be used (Sambrook et al. 1989; Sharrocks 1994), but in practice, because the T_m is variously affected by the individual buffer components and even the primer and template concentrations, any calculated T_m value should be regarded as an approximation. Several reactions run at temperature increments (2–5°C) straddling a point 5°C below the calculated T_m give a first approximation of the optimum annealing temperature for a given set of reaction conditions. It should be noted that some primers, for reasons that are not entirely apparent, are refractory to optimization (He et al. 1994). One possible explanation may be that unique characteristics of the target amplicon give a T_m above the temperature of the denaturation cycle segment (Sharrocks 1994). If permissible, it may be more time- and cost-efficient simply to design a second set of primers that hybridize to neighboring DNA.

Cycle Number, Reamplification, and Product Smearing

Increasing the number of cycles may enhance an anemic reaction, but this modification can also lead to the generation of spurious bands and to smears composed of high-molecular-weight products rich in single-stranded DNA (Bell and DeMarini 1991). Similar smearing can occur under normal conditions if the quantity of starting template is too great, as often occurs in attempts to reamplify from a previous PCR. A general rule of thumb is to use 1 μl of a 1:10^4 to 1:10^5 dilution of a PCR product if a gel band is detectable.

Nested PCR

Nested and semi-nested PCRs are often quite successful in reducing or eliminating unwanted products while at the same time dramatically increasing sensitivity (Mullis and Faloona 1987; Gibbs 1990; Mullis 1991; Zhang and Ehrlich 1994; Zimmerman et al. 1994). An initial set of primers straddling the DNA segment of interest is first amplified under standard conditions. Spurious products are frequently primed with one or both primers and contain irrelevant sequences internally. An aliquot of the reaction-product mix is then subjected to an additional round of amplification using primers complementary to the sequences internal to the first set of primers. Only the legitimate product should be amplified in this second round. This approach is often successful even if the desired product is initially below the level of detection by ethidium bromide staining and in the presence of visible spurious bands. Semi-nested PCR, in which a second primer is internal to only one end of the target segment, can be equally effective (Zhang and Ehrlich 1994). This variation is often required for gene walking or attempts at 5′ or 3′ RACE in which the template DNA sequence internal to only one of the primers is known.

A second form of artifact, known as jumping PCR, may not be eliminated by nested PCR. Incompletely extended products can occasionally rehybridize to an adjacent segment of DNA, perhaps to a similar gene element, to prime an unintended product (Huang and Jeang 1994). In such instances, the sequence internal to one or both primers is still present, but the amplicon size differs.

If nested PCR methods are employed, better results may be obtained if the first and second rounds of amplification are terminated after 20 or so cycles rather than the usual 30–35. This modification minimizes the chances of generating unwanted high-molecular-weight bands and smears (Bell and DeMarini 1991; Zhang and Ehrlich 1994). Such artifacts often contain considerable single-stranded DNA and appear to be the result of mispriming by DNA products amplified in earlier cycles. Nested PCR is extremely sensitive; as little as a single copy of a viral gene has been detected in a background of 10^6 genomes (Zimmerman et al. 1994).

Hot Start PCR

Even brief incubations of a PCR mix at temperatures significantly below the T_m can result in primer-dimer and nonspecific priming. Hot start PCR methods (D'Aquila et al. 1991; Erlich et al. 1991; Mullis 1991) can dramatically reduce these problems. The aim is to withhold at least one of the critical components from participating in the reaction until the temperature in the first cycle rises above the T_m of the reactants. For example, in smaller assays incorporating an oil overlay, one of the components common to all tubes (e.g., *Taq* DNA polymerase) can be initially withheld and added only after the temperature rises above 80°C during the first denaturing stage. Alternatively, a wax bead can be melted over the bulk of the reaction mix in each tube and allowed to solidify, and the withheld component can be pipetted on top of the wax cap. These beads can be made in the laboratory (Bassam and Caetano-Anolles 1993; Wainwright and Seifert 1993) or purchased (Ampliwax PCR Gems, Perkin-Elmer). During the temperature ramp into the first denaturation segment, the wax melts and the final component becomes incorporated and mixed by convection in each tube, a great convenience when dealing with large numbers of tubes. A recent hot start variation involves adding specific anti-*Taq* DNA polymerase antibody (TaqStart Antibody, CLONTECH) to the PCR tubes prior to the addition of *Taq* DNA polymerase. The antibody prevents polymerase activity from beginning until the temperature rises to dissociate and denature the blocking antibody. This modification is compatible with newer thermal cyclers and techniques that seek to avoid the extra handling and purification steps accompanying oil and wax addition and sample recovery.

TROUBLESHOOTING

- *Little or no detectable product.* You have adjusted the Mg^{++} concentration, buffer pH, and cycling parameters; added more cycles; and tried lower annealing temperatures and TD PCR. You still see no product on ethidium bromide-stained gels (acrylamide gels are considerably more sensitive than agarose gels), yet your positive controls indicate no reagent problems. What should be the next step? Lengthening the initial denaturation step and/or increasing the temperature increases the likelihood that the template DNA is fully denatured to provide the maximal number of priming sites. Standard conditions for this optional step are 5 minutes at 95°C. An in-tube thermocouple can be used to predetermine that the indicated temperature corresponds to the actual sample temperature. Amplification may have occurred, but it could have been inefficient. If so, the amplicons can be revealed by a probe of the dried gel or a blot. A secondary amplification using the same primers or, preferably, nested primers may be all that is needed to generate a specific product. Serial 10-fold dilutions ranging from 1:100 to 1:10,000 should be used.

Little or no product may indicate the presence of inhibitors in the DNA sample. Numerous inhibitors of PCR have been described. These include ionic detergents (e.g., SDS and Sarkosyl) (Weyant et al. 1990), phenol, heparin (Beutler et al. 1990), xylene cyanol, and bromphenol blue (Hoppe et al. 1992). Test for inhibitor in the template preparation by spiking the original PCR mix with dilutions of known positive (demonstrably amplifiable) template. Reextraction, ethanol precipitation, and/or centrifugal ultrafiltration may resolve the problem. Proteinase K carryover can serve to digest the *Taq* DNA polymerase, but it is readily denatured by a 5-minute incubation at 95°C.

ACKNOWLEDGMENTS I thank Rani Dhanarajan, Dan Garza, and Karl Hecker for their valuable comments.

REFERENCES

Bassam, B.J. and G. Caetano-Anolles. 1993. Automated "hot start" PCR using mineral oil and paraffin wax. *BioTechniques* **14:** 30–34.

Batzer, M.A., J.E. Carlton, and P.L. Deininger. 1991. Enhanced evolutionary PCR using oligonucleotides with inosine at the 3′-terminus. *Nucleic Acids Res.* **19:** 5081.

Bell, D.A. and D. DeMarini. 1991. Excessive cycling converts PCR products to random-length higher molecular weight fragments. *Nucleic Acids Res.* **19:** 5079.

Beutler, E., T. Gelbart, and W. Kuhl. 1990. Interference of heparin with the polymerase chain reaction. *BioTechniques* **9:** 166.

Cobb, B.D. and J.M. Clarkson. 1994. A simple procedure for optimizing the polymerase chain reaction (PCR) using modified Taguchi methods. *Nucleic Acids Res.* **22:** 3801–3805.

D'Aquila, R.T., L.J. Bechtel, J.A. Viteler, J.J. Eron, P. Gorczyca, and J.C. Kaplin. 1991. Maximizing sensitivity and specificity of PCR by preamplification heating. *Nucleic Acids Res.* **19:** 3749.

Don, R.H., P.T. Cox, B.J. Wainwright, K. Baker, and J.S. Mattick. 1991. "Touchdown" PCR to circumvent spurious priming during gene amplification. *Nucleic Acids Res.* **19:** 4008.

Erlich, H.A., D. Gelfand, and J.J. Sninsky. 1991. Recent advances in the polymerase chain reaction. *Science* **252:** 1643–1651.

Gibbs, R.A. 1990. DNA amplification by the polymerase chain reaction. *Anal. Chem.* **62:** 1202–1214.

He, Q., M. Marjamaki, H. Soini, J. Mertsola, and M.K. Viljanen. 1994. Primers are decisive for sensitivity of PCR. *BioTechniques* **17:** 82–87.

Hoppe, B.L., B.M. Conti-Tronconi, and R.M. Horton. 1992. Gel-loading dyes compatible with PCR. *BioTechniques* **12:** 679–680.

Huang, L.-M. and K.-T. Jeang. 1994. Long-range jumping of incompletely extended polymerase chain fragments generates unexpected products. *BioTechniques* **16:** 242–246.

Knoth, K., S. Roberds, C. Poteet, and M. Tamkun. 1988. Highly degenerate inosine-containing primers specifically amplify rare cDNA using the polymerase chain reaction. *Nucleic Acids Res.* **16:** 10932.

Lee, C., X. Wu, R.A. Gibbs, R.G. Cook, D.M. Muzny, and C.T. Caskey. 1988. Generation of cDNA probes directed by amino acid sequence: Cloning of urate oxidase. *Science* **239:** 1288–1291.

Mullis, K.B. 1991. The polymerase chain reaction in an anemic mode: How to avoid cold oligodeoxyribonuclear fusion. *PCR Methods Appl.* **1:** 1–4.

Mullis, K. and F.A. Faloona. 1987. Specific synthesis of DNA *in vitro* via a polymerase-catalyzed chain reaction. *Methods Enzymol.* **155:** 335–350.

Newton, C.R. and A. Graham. 1994. *PCR.* Bios Scientific, Oxford.

Patil, R.V. and E.E. Dekker. 1990. PCR amplification of an *Escherichia coli* gene using mixed primers containing deoxyinosine at ambiguous positions in degenerate amino acid codons. *Nucleic Acids Res.* **18:** 3080.

Peterson, M.G., J. Inostroza, M.E. Maxon, O. Flores, A. Adomon, D. Reinberg, and R. Tjian. 1991. Structure and functional properties of human general transcription factor IIE. *Nature* **354:** 369–373.

Pomp, D. and J.F. Medrano. 1991. Organic solvents as facilitators of polymerase chain reaction. *BioTechniques* **10:** 58–59.

Roux, K.H. 1994. Using mismatched primer-template pairs in touchdown PCR. *BioTechniques* **16:** 812–814.

Sambrook, J., E.F. Fritsch, and T. Maniatis. 1989. *Molecular cloning: A laboratory manual*, 2nd edition. Cold Spring Harbor Laboratory, Cold Spring Harbor, New York.

Sharrocks, A.D. 1994. The design of primers for PCR. In *PCR technology: Current innovations* (ed. H.G. Griffin and A.M. Griffin), pp. 5–11. CRC Press, Boca Raton, Florida.

Taguchi, G. 1986. *Introduction to quality engineering.* Asian Productivity Organisation, UNIPUB, New York.

Wainwright, L.A. and H.S. Seifert. 1993. Paraffin beads can replace mineral oil as an evaporation barrier in PCR. *BioTechniques* **14:** 34–36.

Weyant, R.S., P. Edmonds, and B. Swaminathan. 1990. Effect of ionic and nonionic detergents on the *Taq* polymerase. *BioTechniques* **9:** 308–309.

Zhang, X-Y. and M. Ehrlich. 1994. Detection and quantitation of low numbers of chromosomes containing *bcl-2* oncogene translocations using semi-nested PCR. *BioTechniques* **16:** 502–507.

Zimmermann, K., K. Pischinger, and J.W. Mannhalter. 1994. Nested primer PCR detection limits of HIV-1 in a background of increasing numbers of lysed cells. *BioTechniques* **17:** 18–20.

Long-distance PCR

Orit S. Foord and Elise A. Rose

Advanced Center for Genetic Technology, Applied Biosystems Division, The Perkin-Elmer Corporation, Foster City, California 94404

INTRODUCTION

The ability routinely and specifically to amplify and detect PCR products ranging in size from less than 1 kb to more than 50 kb, regardless of target template sequence or structure, would revolutionize a broad spectrum of biological research applications of the PCR (Rose 1991; Ohler and Rose 1992; Ohler et al. 1993; Foord and Rose 1994):

1. The ability to amplify fragments up to 20–50 kb would potentially enable the isolation of an entire gene from a cDNA, thereby obviating the time-consuming task of screening a genomic library for the target gene.
2. The generation of a 20-kb fragment would span approximately half of the DNA cloned into a cosmid, thus making it possible to access the entire insert from as few as two amplifications initiating from the left and right vector cloning sites.
3. Amplification of 50-kb targets would provide the same benefits to the analysis of bacterial artificial chromosome (BAC) and P1-derived artificial chromosome (PAC) clones, as well as small yeast artificial chromosomes (YACs) (Green and Olson 1990; Shizuya et al. 1992; Ioannou et al. 1994).
4. Large genomic fragments could be isolated from complex genomes, as well as from hybrid cell lines or from microdissected or flow-sorted chromosomal regions (Nelson et al. 1989).
5. Generation of large cDNAs from an uncloned pool of mRNA extension products would be feasible. (Applications 4 and 5 would be especially useful when studying tissues or species where appropriate libraries are unavailable.)
6. Long-distance PCR facilitates the amplification of eukaryotic genomic DNA segments containing introns of varying number and lengths, thus facilitating the delineation of intron/exon boundaries.

Applications with particular relevance to genome research include:

1. Analysis of the higher-molecular-weight restriction fragment length polymorphisms within a population.
2. Amplification of large regions containing genes with expandable triplet repeats, such as the Huntington's disease gene, making presymptomatic diagnosis by PCR technically feasible.
3. Isolation of uncharacterized DNA located outside of adjacent known sequences and generation of a wide size range of amplification products to expedite isolation of uncloned DNA (gaps) represented in physical maps and to maintain order and orientation of closely linked loci during analysis.
4. Enhancement of the utility of methodologies involving amplification of regions between specific interspersed repetitive elements (or between combinations of types of repeat units), thus facilitating fingerprinting of DNA fragments, as well as ordering and orienting contiguous DNA clones (Nelson et al. 1989).
5. Application to directed transposon-based mapping and sequencing, a potentially powerful strategy for rapidly and efficiently analyzing templates of 100 kb or more (Ohler and Rose 1992; Ohler et al. 1993; Foord and Rose 1994; E.A. Rose, in prep.).

Limitations in the size of PCR products that have been reliably generated in the past have been on the order of 3–4 kb (Erlich et al. 1991). Several reports have appeared describing amplification of larger PCR fragments (Jeffreys et al. 1988; Rychlik et al. 1990; Krishnan et al. 1991; Maga and Richardson 1991; Kainz et al. 1992; Ohler and Rose 1992; Ohler et al. 1993), but relatively little effort has been made to optimize conditions for the general use of long-distance amplification. Moreover, research involving long-distance PCR has used templates of known size and, in many cases, known sequence. The result has been retrofitting experimental design to obtain the expected results. Attempts to employ these methods for long-distance PCR in other primer-template systems have been generally unsuccessful.

For long-distance PCR to be of general use to the scientific community, a robust methodology must be available not only for the routine amplification of a range of specific high-molecular-weight DNA products but, simultaneously, for efficient extension through those genome regions that have been notoriously recalcitrant to PCR.

REAGENTS

For a 50-μl reaction volume, useful starting conditions include a reaction mixture containing the following:

2.5 mM $MgCl_2$
25 mM Tris-HCl, pH 8.9–9.0 at 20–25°C
1.0 mM total dNTPs

30 pmoles–50 μmoles of upstream primer
30 pmoles–50 μmoles of downstream primer (similar melting temperatures, and preferably >55°C)
0.01% gelatin
1x r*Tth* chelating buffer containing 5% glycerol (Perkin-Elmer)
2.5 units of r*Tth* DNA polymerase (Perkin-Elmer)
50 ng of template DNA
AmpliWax PCR Gem for hot start

PROTOCOL

Schedule for Thermal Cycling

Thermal cycling is performed in the GeneAmp PCR System 9600 under the following conditions:

10 cycles, each consisting of:
- 10 seconds at 95°C
- 30 seconds at 58°C (or higher, depending on primer melting temperatures)
- 3 minutes at 72°C

Followed by 20 cycles, each consisting of:
- 10 seconds at 95°C
- 30 seconds at 58°C (or higher, depending on primer melting temperatures)
- 3 minutes at 72°C + 30-second automatic increment every cycle

Note: Advances have been made over the past few years in extending the length over which one can reliably amplify specific DNA templates. A kit is now available (the GeneAmp XL PCR Kit; Perkin-Elmer) that uses a combination of r*Tth* DNA polymerase and Vent DNA polymerase in a fashion similar to that originally described by Barnes (1994) and provides a useful starting point for long-distance PCR (along with troubleshooting tips).

DISCUSSION

The following topics are important considerations in designing a long-distance PCR methodology: the components of the reaction mixture, methodology issues, thermal cycling conditions, and methods for detecting the amplified product.

Reaction Mixture Components

The components of the PCR mixture alone do not ensure successful DNA amplification. Thermal cycling parameters and DNA detection methodologies are equally important for achieving reproducible and specific PCR, especially when addressing the ability to generate high-molecular-weight PCR products. Several generally accepted "standard" PCR protocols must be altered when attempting to extend the PCR product size above 3–4 kb. A discussion of each of the reaction mixture components, as they relate to long-distance PCR, follows.

HIGH-QUALITY TEMPLATE

Degraded, nicked, and unpurified DNA from many sources has been used successfully in routine PCR. Amplification of higher-molecular-weight PCR products is not as efficient as the generation of smaller fragments (Rychlik et al. 1990; Krishnan et al. 1991; Rose 1991; Kainz et al. 1992; Ohler and Rose 1992; Ohler et al. 1993; Foord and Rose 1994). Therefore, it is important that the majority of DNA present in a long-distance PCR mixture serve as a high-quality template, because high-integrity DNA provides a greater concentration of intact initial template available for PCR. The protocols and precautions of genome researchers involved in the cloning and analysis of high-molecular-weight DNA are relevant when amplifying PCR products approaching 10–50 kb. Isolation procedures that yield the highest quality vector and genomic DNA are therefore encouraged. The analysis of genomic DNA on a low-percentage (e.g., 0.4%) agarose gel or pulsed-field gel electrophoresis is recommended prior to PCR. We have found that whole chromosomal DNA prepared in "pulsed-field quality" low-melting-point agarose (Rose et al. 1990), followed by agarose digestion, generates a superior yield of large PCR product as compared with DNA isolated in solution (Blin and Stafford 1976). Size analysis of these DNAs using pulsed-field electrophoresis demonstrates that the DNA prepared in low-melting-point agarose contains intact chromosomes (Mathew et al. 1988; Smith et al. 1988; Rose et al. 1990; Foord and Rose 1994), with DNA averaging 50 megabases or more in size, whereas the average size of DNA prepared in solution is 75–500 kb.

Template concentration directly affects reproducibility, specificity, and the yield of the amplification product. Excess template increases mispriming, and consequently, the extent of nonspecificity. If too little initial template is used, the amplification of the higher-molecular-weight, lower-copy-number target is less reproducible, with the PCR product often not detectable. Initial template concentration should be optimized to achieve a balance between specificity and yield given the expected size range of the PCR product(s) being generated. In long-distance PCR studies that involve the use of plasmid or lambda templates, various investigators have amplified their targets from 0.1 ng to 100 ng of starting template (Rychlik et al. 1990; Maga and Richardson 1991; Ponce and Micol 1991; Ohler and Rose 1992; Ohler et al. 1993). Studies involving the use of genomic DNA in long-distance PCR have employed as little as 25–100 ng of starting template (Jeffreys et al. 1988; Foord and Rose 1994).

PRIMERS

Optimal primer design provides a balance between specificity and efficiency (or yield) of amplification (Dieffenbach et al. 1993). Specificity reflects the frequency with which mispriming occurs, resulting in unrelated amplification products (Dieffenbach et al. 1993). Many of the generally accepted rules for primer design apply to long-distance PCR. For example, we and other workers have found that primer length is optimal at 18–25 nucleotides (Jeffreys et al. 1988; Rychlik et al. 1990; Krishnan et al. 1991; Maga and Richardson 1991; Ponce and Micol 1991; Ohler and Rose 1992; Ohler et al. 1993). When designing primers for amplification of regions from complex genomes, or regions known to be particularly G+C-rich, increased length provides additional stability. The addition of each incremental nucleotide to the primer confers approximately four times more specificity (Dieffenbach et al. 1993). As in standard PCR, complementarity at the 3′ ends of the primers as well as secondary structure should be avoided (Dieffenbach et al. 1993). This is important in long-distance PCR because any competition for reagents (such as occurs in the formation of primer-dimer) diminishes yield of the desired product. It is also often beneficial to consider the stability of primers at the 3′ ends (Kwok et al. 1990).

The concentration of primers used in long-distance PCR does not differ from standard PCR conditions. Primer concentration should be nonlimiting, but not sufficiently in excess to promote mispriming and primer oligomerization. Successful long PCR products have been generated using primer concentrations ranging from 10 pmoles to 200 nmoles of each primer per reaction (Jeffreys et al. 1988; Rychlik et al. 1990; Krishnan et al. 1991; Maga and Richardson 1991; Ponce and Micol 1991; Oher and Rose 1992; Ohler et al. 1993). Note, however, that high primer concentrations decrease the time necessary for sufficient annealing and therefore may *increase* specificity. We routinely use 30 pmoles–200 nmoles of each primer for a 50-μl PCR.

A number of software programs are available for primer design. For long-distance PCR, Rychlik and Rhoads (1989) suggest the use of the nearest-neighbor algorithm for selection of melting temperatures. We also have found that use of this calculation yields superior results. The melting temperatures of primers should be as high as possible to promote specificity without compromising yield of product. In addition, the nucleotide composition of primers should reflect the nucleotide composition of the region from which the desired target is to be amplified (e.g., if the target region is 50% G+C, then the primers should also be 50% G+C).

Although it is generally recommended that the melting temperatures of primer pairs be balanced (Innis and Gelfand 1990), differences in the melting temperatures of primer pairs are not of particular consequence in routine short-distance PCR. For the generation of long PCR products, however, we have found that the primer design should be optimized so that both primers have melting temperatures within 1–2°C of each other (Ohler and Rose 1992; Ohler et al. 1993).

Even very small divergence in primer pair melting temperatures promotes mispriming of the oligonucleotide with the higher melting temperature (Rychlik and Rhodes 1989; Ohler and Rose 1992).

ENZYMES

More than a dozen different thermostable enzymes have been used for PCR amplification. The utility of many of these enzymes in long-distance PCR has been evaluated in several reports (Jeffreys et al. 1988; Rychlik et al. 1990; Krishnan et al. 1991; Maga and Richardson 1991; Ponce and Micol 1991; Kainz et al. 1992; Ohler and Rose 1992; Ohler et al. 1993; Barnes 1994). DNA polymerases with 5′→ 3′ exonuclease ("nick translation") activity have yielded consistently superior performance in long-distance PCR as compared with those enzymes exhibiting 3′→ 5′ exonuclease ("proofreading") activity, both 3′→ 5′ and 5′→ 3′ exonuclease activities, or neither exonuclease activity. In our early studies, we examined enzymes with various combinations of 5′→ 3′ and 3′→ 5′ exonuclease activities, in the buffers recommended by their manufacturers (see Table 1) (Ohler and Rose 1992). Although some of the characteristics of these enzymes have yet to be defined, they clearly differ in a number of additional activities that are potentially important for long-distance PCR, including extension rate, processivity, fidelity, thermostability, and thermal activity profile.

In our hands, under the conditions utilized, r*Tth* DNA polymerase (the recombinant form of the DNA polymerase from *Thermus thermophilus*; Perkin-Elmer) at 0.5–2.5 units/50 μl reaction provided the most consistent successful performance in long-distance PCR (Ohler and Rose 1992). We and others have also found that Hot Tub DNA polymerase (isolated from *Thermus ubiquitus*; Amersham) can be used reliably to generate PCR products as large as 6–15.6 kb (Maga and Richardson 1991; Kainz et al. 1992; Ohler and Rose 1992). AmpliTaq DNA polymerase (the recombinant form of the DNA polymerase from *Thermus aquaticus*; Perkin-Elmer) and native *Taq* DNA polymerase (Perkin-Elmer) have also been used to generate long PCR products, but the results have not been as specific or reproducible as required (Kainz et al. 1992; Ohler and Rose 1992; Barnes 1994). In contrast, enzymes with 3′→ 5′ exonuclease activity, such as Vent DNA polymerase (isolated from *Thermus litoralis*; New England Biolabs) (Krishnan et al. 1991; Ohler and Rose 1992) or *Pfu* DNA polymerase (isolated from *Pyrococcus furiosus*; Stratagene) (Barnes 1994), failed to generate specific PCR products greater than 6 kb in length consistently. Most of our efforts have therefore focused on optimizing conditions for use of r*Tth* DNA polymerase in long-distance PCR (Ohler and Rose 1992; Ohler et al. 1993; Foord and Rose 1994).

Table 1 Exonuclease Activities of DNA Polymerases

DNA polymerase	5′→ 3′ Exonuclease	3′→ 5′ Exonuclease	References
AmpliTaq	+	–	1,2
r*Tth*	+	–	3,4
Hot Tub	+	–	5
Vent	–	+	6
Tma	+	+	7
AmpliTaq exo⁻ mutein	–	–	8
Stoffel fragment	–	–	9

+ indicates that enzyme possesses activity. – indicates that enzyme lacks activity.

References: [1]Lawyer et al. (1993); [2]Perkin-Elmer DNA polymerase package insert (1991); [3]Myers and Gelfand (1991); [4]Perkin-Elmer r*Tth* DNA polymerase package insert (1991); [5]Amersham DNA polymerase package insert (1991); [6]New England Biolabs DNA polymerase package insert (1991); [7]Perkin-Elmer DNA polymerase package insert (1993); [8]R. Abramson (pers. comm.); [9]Lawyer et al. (1993).

Recombinant *Tth* DNA polymerase has a half-life of 20 minutes at 95°C (sufficient to remain active over 30 or more cycles during which the enzyme is transiently exposed to these extremely high denaturation temperatures), optimal DNA polymerase activity in the same temperature range at which stringent primer annealing occurs (65°C–75°C), 5′→ 3′ exonuclease activity, and no detectable 3′→ 5′ exonuclease activity (Myers and Gelfand 1991; see also r*Tth* DNA polymerase package insert, Perkin-Elmer Cetus, 1991). (Lack of 3′→ 5′ exonuclease activity minimizes the likelihood that primers and single-stranded template are destroyed during PCR.) *Tth* DNA polymerase's processivity (i.e., the number of nucleotides replicated before the enzyme dissociates from the template) is 30–40 nucleotides, and its extension rate is approximately 60 nucleotides per second, in 100 mM KCl (r*Tth* DNA polymerase package insert, Perkin-Elmer Cetus, 1991).

The use of combinations of thermostable enzymes with different activities might be beneficial in amplifying even longer templates (Ohler and Rose 1992), and recent experiments reported by Barnes (1994) support this approach. These studies used lambda DNA as template; PCR products up to 35 kb in size were generated using an amino-terminal deletion mutant of *Taq* DNA polymerase (Klentaq1) in combination with *Pfu* DNA polymerase. A ratio of approximately 180:1 Klentaq1:*Pfu* (0.1875 unit *Pfu* plus 33.75 units of Klentaq1 in a total combined volume of 1.2 μl) provided optimal results in these studies. Other combinations of enzymes either lacking exonuclease activity or displaying 3′→ 5′ and/or 5′→ 3′ exonuclease activities have also proved useful in generating long PCR products; however, the use of these different enzyme combinations requires determination of different optimal relative concentrations.

The thermostable enzymes currently used for PCR amplification are most likely repair enzymes, which characteristically replicate short stretches of nucleotides with low processivity and low extension rate. This is in contrast to replication enzymes (Kornberg and Baker 1991), which display high processivity and rates of extension. The use of such replication enzymes alone might significantly enhance the ability to obtain high yield of very long PCR products (20–50 kb). The isolation of thermostable replication enzymes for use in PCR would clearly make this approach testable.

Finally, long-distance PCR might be improved by using alternative thermostable DNA polymerases (including chimeric enzymes and mutant enzymes with alterations to specific active sites) and combining such thermostable enzymes with thermostable pyrophosphatases and accessory proteins (e.g., thermostable helicases and single-strand DNA displacement proteins). Studies concerning the use of such proteins in long-distance PCR are currently in progress.

BUFFERS

Standard PCR protocols recommend a buffer of 10 mM Tris-HCl (pH 8.3–8.4 at 20–25°C) for both *Taq* DNA polymerase and r*Tth* DNA polymerase. This common Tris-HCl buffer system was intended to represent a starting point from which reaction conditions could be optimized for specific primer-template systems. However, it does not represent the optimal buffer system for long-distance PCR. We find that long-distance PCR with r*Tth* DNA polymerase works well with a buffer containing 5% glycerol (v/v), 25 mM Tris-HCl (pH 8.9 at room temperature), 100 mM KCl, 0.75 mM EGTA, 0.05% Tween 20, 2.5 mM $MgCl_2$, and 0.01% gelatin (Ohler and Rose 1992). The EGTA is present to chelate manganese and inhibit any reverse transcriptase activity that might otherwise be exhibited by this enzyme. As in standard PCR with r*Tth*, a magnesium-dependent DNA polymerase, optimal activity is observed in a concentration range of 1.5–2.5 mM $MgCl_2$ (Myers and Gelfand 1991). Total dNTPs are added to 1 mM (250 μM each), again in nonlimiting quantities sufficient for generation of the desired long products (Ohler and Rose 1992).

Given the large temperature dependence of pKa in Tris buffers, Ponce and Mikol (1991) proposed the use of a less temperature-sensitive Tricine buffer. These investigators used a 300 mM Tricine buffer, pH 8.4 at room temperature, to obtain specific 6.2-kb products with *Taq* DNA polymerase. The use of less temperature-sensitive buffers such as Tricine, or a combination of buffers with pKa values falling within the neutral pH range at optimal enzyme extension temperatures (more similar to in vivo conditions), might enhance the ability to obtain long PCR products.

Additional components useful in amplifying long templates include 5% glycerol and 0.01% gelatin. The presence of gelatin in excess of

0.05% inhibits long-distance PCR. Finally, we routinely siliconize our PCR tubes to prevent the enzyme and low-copy templates from sticking to the walls (Ohler and Rose 1992).

Methodology Issues

WAX-MEDIATED HOT STARTS

Hot starts enhance PCR specificity by eliminating the production of nonspecific products resulting from pre-PCR mispriming and primer oligomerization (primer-dimer) during the initial steps of PCR (Chou et al. 1991). AmpliWax PCR Gems (Perkin-Elmer) are an effective and convenient tool for implementing this technique. Uniformity and yield of PCR also improve using the hot start approach. The uniformity effect is most likely due to providing a vapor barrier of consistent mass in each reaction tube, and the enhanced yield from focusing all reagents on the production of the specific desired product. By permitting specific PCR to occur at temperatures that would otherwise result in the formation of nonspecific products, wax-mediated hot starts provide the additional benefit of decreasing the time required for PCR optimization.

REACTION VOLUME

Most PCR amplifications, including long-distance PCR, have been carried out in 100-μl volumes. Recently, there has been a trend toward the use of lower reaction volumes (e.g., 50 μl), which yield superior results (Ohler and Rose 1992). This is attributable to improved heat transfer, and such results are further enhanced by the use of products such as the thin-walled MicroAmp reaction tubes in the Perkin-Elmer GeneAmp PCR System 9600, an instrument designed for the tight fit of these tubes (Haff et al. 1991). This point is important because the use of tubes that do not fit properly into thermal cyclers, or that have not been designed for thermal cycling, results in inefficient heat transfer, yielding inferior results if used in long-distance PCR. Rapid cycling in glass capillary tubes might also provide excellent heat transfer. Heat transfer could be improved even further if PCR could be performed in submicroliter volumes in microfabricated instruments. Technical advances in this area may provide significant benefits in the amplification of long templates.

Thermal Cycling Conditions

One cycle in a PCR amplification reaction consists of three steps that can be carried out at two or three separate temperatures. A typical cycle consists of a denaturation step at 90–97°C, an annealing step at

40–75°C, and an extension step at 70–75°C. It is often helpful to precede the first denaturation step with an initial predenaturation at 95°C for 3 minutes. A common "standard" PCR protocol might involve 30 cycles, each consisting of denaturation for 1 minute at 94°C, annealing for 1 minute at 55°C, and extension for 1 minute at 72°C. For optimal PCR, however, the specific times and temperatures used for each step are chosen depending on the characteristics of the primer-template system being used. Two-temperature PCR can be employed because most thermostable DNA polymerases used in PCR are actively extending off the primers over the entire temperature range between the points commonly chosen for annealing and extension (enabling one to use a combined anneal/extend step). The use of two-temperature protocols would be expected to increase specificity in long PCR because it decreases the amount of time during which mispriming may occur.

Although denaturation occurs rapidly, longer denaturation times might be expected to produce superior results when amplifying long templates. At 95°C, *Taq* DNA polymerase still maintains 50% of its original activity after 40 minutes, and r*Tth* has a half-life of 20 minutes, both of which are more than sufficient to provide adequate activity throughout the 25–30 cycles of most PCRs during which the enzyme is transiently exposed to these high temperatures (Gelfand 1989).

Annealing temperatures are chosen on the basis of the primer melting temperatures, taking into consideration estimated template length and G+C content (see Primers section above). We routinely use the equations defined by Rychlik for the calculation of optimal temperature for primer-template annealing (Rychlik and Rhoads 1989; Rychlik et al. 1990). In general, the annealing temperature is set as high as possible to promote specificity. These calculations have proven to be the most reliable in defining optimal annealing conditions when template length and nucleotide composition are unknown.

Taq DNA polymerase extends at a rate of 0.25 nucleotides per second at 22°C, 1.5 nucleotides per second at 37°C, 24 nucleotides per second at 55°C, greater than 60 nucleotides per second at 70°C, and 150 nucleotides per second at 75–80°C (Gelfand 1989). Thus, at commonly chosen extension temperatures such as 70–72°C, *Taq* DNA polymerase would be expected to be extending at a rate of greater than 3.5 kb per minute. As a general rule, extension times of 1 minute per kilobase are more than sufficient to generate the expected PCR product (Gelfand 1989). The same rule applies when using r*Tth*, because this enzyme has an extension rate that is very similar to that of *Taq* DNA polymerase.

We have used this rationale in defining optimal conditions for long-distance PCR using a GeneAmp PCR System 9600 (Perkin-

Elmer). Although other types of DNA thermal cyclers may have different performance characteristics, the general guidelines provided above for designing long PCR thermal cycling conditions should still apply. It should be noted, however, that any features of an instrument's design that compromise its ability to regulate temperature or incubation times reliably and reproducibly, or that result in inferior heat transfer, would be expected to make the generation of long PCR products more difficult.

For a 12.2-kb template, using r*Tth* DNA polymerase and the type of reaction mixture described in the sections above, we have demonstrated success in generating specific PCR products by running 10 cycles, each consisting of a 1-minute denaturation at 94°C, a 1-minute annealing step at 55°C, and a 3-minute extension step at 72°C, followed by 20 cycles in which the extension step is automatically extended 30 seconds in each cycle. Automated hot starts are performed using AmpliWax PCR Gems, as described above (Ohler and Rose 1992; Ohler et al. 1993; Foord and Rose 1994).

The ability to program the thermal cycler to increase extension times automatically in the later cycles of the reaction is particularly useful. DNA polymerases extend primers discontinuously through a succession of reactions, moving on and off the DNA template. As the PCR product begins to accumulate in the later cycles, the ability to extend a significant proportion of primers over a long distance in a given unit of time might be limited by the relative decrease in enzyme molecules per template. Thus, increasing the extension time in each of the later PCR cycles could increase the likelihood of synthesizing long PCR products (Ohler and Rose 1992). Whether or not this rationale is correct, the use of autosegment extension does result in more consistent generation of long PCR products.

Product Detection

To date, the most common method for detecting PCR products is the use of ethidium bromide fluorescence with a standard UV lightbox. To achieve sufficient sensitivity for detecting low-copy-number PCR products of high molecular weight, it is generally necessary to use the entire reaction volume. To achieve the necessary sensitivity, as well as to overcome background problems resulting from nonspecific amplification, probing with radioactively labeled single-copy DNA or oligonucleotide probes has been used (Jeffreys et al. 1988; Strausbaugh et al. 1990). Problems with this approach are associated with the use of radioactivity, including the instability of probes, the use of biohazardous materials, the length of time required for detection, and difficulty in automation. Southern hybridization with oligonucleotide probes conjugated with enzymes or with biotin, followed by either colorimetric or chemiluminescent detection, overcomes many of the problems associated with the use of radioactivity. These methods still suffer from a lack of sufficient sensitivity, however, if one needs to detect sub-nanogram quantities of a PCR product. The same problems apply to the use of probes labeled with currently available fluorescein dyes.

The use of both primers labeled with biotin and primers end-labeled with fluorescent tags is not adequate for detecting low-abundance, high-molecular-weight DNA products (Ohler and Rose 1992). Detection of large amplification products is improved significantly if the number of signal-generating units incorporated per DNA molecule is increased (Ohler and Rose 1992; Ohler et al. 1993), as with use of a DNA probe containing multiple biotinylated dUTPs or direct incorporation of biotinylated dUTPs into the PCR product. Optimal sensitivity is obtained either by intercalation of a fluorescent agent into the PCR product (Ohler and Rose 1992; Ohler et al. 1993) or by direct incorporation of fluorescently labeled nucleotides during the PCR amplification (Foord and Rose 1994; E.A. Rose, in prep.). Such fluorescent PCR products may be detected at femtomolar concentrations.

We have detected PCR products as large as 9 kb using the inherent fluorescence of the DNA intercalating agent ethidium bromide (Ohler and Rose 1992). An Applied Biosystems model 362A Fluorescent Fragment Analyzer, which employs an argon-based laser system, enabled the rapid and sensitive detection and sizing of PCR products. A major asset of this system is the ability to use internal size standards to adjust for slight variations in migration of DNA containing intercalated dyes. Another advantage of using this system is that it allows us to analyze only 1–5 μl of a 50-μl PCR sample mixture, leaving the remaining PCR product available for other post-PCR procedures.

We have also used an alternative fluorescent intercalating reagent, the thiazole orange dimer, TOTO-1 (1,1′-(4,4′,7,7′-tetramethyl-4,7-diazaundecamethylene)-bis-4-[3-methyl-2,3-dihydro-(benzo-1,3-thiazole)-2-methylidene]-quinolinium tetraiodide) (Glazer and Rye 1992; Rye et al. 1992). The sensitivity of DNA detection achievable with TOTO-1, as compared with the use of conventional ethidium bromide staining of large PCR products, is several orders of magnitude greater. Again using a model 362A Fluorescent Fragment Analyzer, we were able to detect PCR products ranging in size from 1 kb to 12.2 kb, where the larger PCR products were present at femtomolar concentrations in sub-microliter volumes (Ohler et al. 1993).

A series of newly developed fluorescent dUTP analogs has also been used for detecting long PCR products generated at low copy number (E.A. Rose, in prep.). Similar to other fluorescent nucleotides, these compounds contain either rhodamine or fluorescein derivatives as fluorophores. The dUTP analogs used in these studies differ in structure, however, from previously described fluorescent nucleotides. This modification in synthesis results in an approximately 500- to 1000-fold greater incorporation efficiency of these derivatives, and the detection of PCR products is correspondingly enhanced. Analysis of PCR fragments labeled with these novel nucleotides, again using the Fluorescent Fragment Analyzer, enables us to detect attomole amounts of DNA using 1/100th–1/1000th of a 50-μl PCR mixture. This ability to analyze such small volumes with adequate sensitivity of detection will be critical in the future as further reductions in the scale of PCR are achieved. Use of these newly developed dUTP compounds has proven to be the most sensitive detection system available to us for the analysis of DNA templates containing approximately 50–80% A+T. For genome regions containing a higher G+C content, we are investigating the use of an analogous dCTP fluorescent derivative.

Although all of the methodologies discussed above use standard agarose gel electrophoresis to separate PCR products of different sizes, it should be noted that DNA fragments do not separate in a logarithmic fashion as their length approaches 50 kb. Even in the 20–50-kb range, longer- and lower-percentage agarose gels may be required to obtain adequate separation and accurate sizing. Pulsed-field gel electrophoresis is recommended for the resolution of DNA fragments greater than 50 kb in length. Enhanced detection of very large DNA fragments can still be achieved in an automated fashion, using an Applied Biosystems model 373A Fluorescent Fragment Analyzer with PCR products separated by field inversion gel electrophoresis (E.A. Rose, in prep.).

DEVELOPING LONG-DISTANCE PCR

In the near future it will be critical to gain more experience applying this technology to unknown templates, as opposed to well-defined model systems. Further improvements will be developed involving enzyme formulations, reaction components, thermal cycling conditions, and detection methodologies for longer PCR products. We believe that the ability to amplify targets in the 20–50-kb range and beyond routinely and specifically will have a revolutionary impact on many aspects of molecular biology research. Attention will then shift from development of the methods themselves to the truly exciting work involving applications of long-distance PCR to address a vast array of basic questions in the study of biology.

ACKNOWLEDGMENTS

E.A.R. is funded by National Institutes of Health grant R01-HG00565.

REFERENCES

Barnes, W.M. 1994. PCR amplification of up to 35-kb DNA with high fidelity and high yield from lambda bacteriophage templates. *Proc. Natl. Acad. Sci.* **91:** 2216–2220.

Blin, N. and D.W. Stafford. 1976. A general method for isolation of high molecular weight DNA from eukaryotes. *Nucleic Acids Res.* **3:** 2303.

Chou, Q., M. Russell, D. Birch, J. Raymond, and W. Bloch. 1991. Prevention of pre-PCR mis-priming and primer dimerization improves low copy number amplification. *Nucleic Acids Res.* **20:** 7.

Dieffenbach, C.W., T.M.J. Lowe, and G.S. Dveksler. 1993. General concepts for PCR design. *PCR Methods Appl.* **3:** 530–537.

Erlich, H., D. Gelfand, and J. Sininsky. 1991. Recent advances in the polymerase chain reaction. *Science* **252:** 1643–1650.

Foord, O. and E.A. Rose. 1994. Long-distance PCR. *PCR Methods Appl.* **3:** S149–S161.

Gelfand, D.H. 1989. Taq DNA polymerase. In *PCR technology: Principles and applications for DNA amplification* (ed. H.E. Erlich), pp. 17–22. Stockton Press, New York.

Glazer, A.N. and H.S. Rye. 1992. Stable dye-DNA intercalation complexes as reagents for high-sensitivity fluorescent detection. *Nature* **359:** 859–861.

Green, E.D. and M.V. Olson. 1990. Systematic screening of yeast artificial chromosome libraries by use of the polymerase chain reaction. *Proc. Natl. Acad. Sci.* **87:** 1213–1217.

Haff, L., J.G. Atwood, J. DiCesare, E. Katz, E. Picozza, J. Williams, and T. Woudenberg. 1991. A high performance system for automation of the polymerase chain reaction. *BioTechniques* **10:** 102–112.

Innis, M.A. and D.H. Gelfand. 1990. Optimization of PCRs. In *PCR protocols: A guide to methods and applications* (ed. M.A. Innis et al.), pp. 3–12. Academic Press, San Diego, California.

Ioannou, P., C. Amemiyaa, A. Garnes, P. Kroisel, C. Chen, M. Batzer, A. Carrano, H. Shizuya, and P.J. DeJong. 1994. A new bacteriophage P1-derived vector for the propagation of large human DNA fragments. *Nat. Genet.* **6:** 84–89.

Jeffreys, A.L., V. Wilson, R. Neumann, and J. Keyte. 1988. Amplification of human mini-satellites by the polymerase chain reaction: Towards DNA fingerprinting of single cells. *Nucleic Acids Res.* **16:** 10953–10971.

Kainz, P., A. Schmiedlechner, and B. Strack. 1992. In vitro amplification of DNA >10 kb. *Anal. Biochem.* **202:** 46–49.

Kornberg, A. and T. Baker. 1991. *DNA replication.* W.H. Freeman, New York.

Krishnan, B.R., D. Kersulyte, I. Brikun, C.M. Berg, and D.E. Berg. 1991. Direct and crossover PCR amplification to facilitate Tn*5supF*-based sequencing of lambda clones. *Nucleic Acids Res.* **19:** 6177–6182.

Kwok, S., C.E. Kellog, N. McKinney, D. Spasic, L. Goda, C. Levenson, and J.J. Sninsky. 1990. Effects of primer-template mismatches on the polymerase chain reaction: Human immunodeficiency virus 1 model studies. *Nucleic Acids Res.* **18:** 999–1005.

Maga, E.A. and T. Richardson. 1991. Amplification of a 9.0 kb fragment using PCR. *BioTechniques* **11:** 185–186.

Mathew, M.K., C.L. Smith, and C.R. Cantor. 1988. High-resolution separation and accurate size determination in pulsed field gel electrophoresis of DNA. I. DNA size and standards and the effect of agarose and temperature. *Biochemistry* **27:** 9204–9210.

Myers, T.W. and D.H. Gelfand. 1991. Reverse transcription and DNA amplification by a *Thermus thermophilus* DNA polymerase. *Biochemistry* **30:** 7661–7666.

Nelson, D.L., S.A. Ledbetter, L. Corbo, M.F. Victoria, R. Ramírez-Solis, T. Webster, D.H. Ledbetter, and C.T. Caskey. 1989. Alu polymerase chain reaction: A method for rapid isolation of human specific sequences from complex DNA sources. *Proc. Natl. Acad. Sci.* **86:** 6686–6690.

Ohler, L. and E.A. Rose. 1992. Optimization of long distance PCR using a transposon-based model system. *PCR Methods Appl.* **2:** 51–59.

Ohler, L., M. Zollo, E. Mansfield, and E.A. Rose. 1993. Use of a sensitive fluorescent intercalating dye to detect PCR products of low copy number and high molecular weight. *PCR Methods Appl.* **3:** 85–140.

Ponce, M.R. and J. Micol. 1991. PCR amplification of long DNA fragments. *Nucleic Acids Res.* **20:** 623.

Rose, E.A. 1991. Applications of the polymerase chain reaction to genome analysis. *FASEB J.* **5:** 46–51.

Rose, E.A., T. Glaser, C. Jones, C.L. Smith, W.H. Lewis, K.M. Call, M. Minden, E. Chamagne, L. Bonetta, H. Yeger, and D. Housman. 1990. Complete physical map of the WAGR region of 11p13 localizes a candidate Wilms' tumor gene. *Cell* **60:** 495–508.

Rychlik, W. and R.E. Rhoads. 1989. A computer program for the choosing of oligonucleotides for filter hybridization, sequencing and in vitro amplification of DNA. *Nucleic Acids Res.* **17:** 8543–8551.

Rychlik, W., W.J. Spencer, and R.E. Rhoads. 1990. Optimization of the annealing temperature for DNA amplification in vitro. *Nucleic Acids Res.* **18:** 6409–6412.

Rye, H.S., S. Yue, D.E. Wemmer, M.A. Quesada, R.P. Haugland, R.A. Mathies, and A.N. Glazer. 1992. Stable fluorescent complexes of double stranded DNA with bis-intercalating asymmetric cyanine dyes: Properties and applications. *Nucleic Acids Res.* **20:** 2803–2812.

Shizuya, H., B. Birren, U.J. Kim, V. Mancino, T. Slepak, Y. Tachiiri, and M.I. Simon. 1992. Cloning and stable maintenance of 300-kilobase-pair fragments of human DNA in *Escherichia coli* using an F-factor-based vector. *Proc. Natl. Acad. Sci.* **89:** 8794–8797.

Smith, C.L., S.R. Kico, and C.R. Cantor. 1988. Pulsed field gel electrophoresis and the technology of large DNA molecules. In *Genome analysis: A practical approach* (ed. K. Davies), pp. 41–47. IRL Press, Oxford, England.

Strausbaugh, L.D., M.T. Bourke, M.T. Sommer, M.E. Coon, and C.M. Berg. 1990. Probe mapping to facilitate transposon based DNA sequencing. *Proc. Natl. Acad. Sci.* **86:** 5908–5912.

Sample Preparation

Sample preparation, the purification of nucleic acids from specimens collected for analysis, is the critical first step for successful PCR. Because of the variety of PCR applications, the starting materials for nucleic acid extraction may include such diverse specimens as 17- to 20-million-year-old fossilized magnolia leaves (Golenberg et al. 1990), muscle recovered from the skins of extinct animals (Higuchi et al. 1984), human remains recovered from peat bogs (Lawlor et al. 1991), single human hairs (Higuchi et al. 1988), and paraffin-embedded biopsy specimens (Greer et al. 1991, and "PCR Amplification from Paraffin-embedded Tissues" in this section). Clearly, each specific sample type places unique constraints and requirements on the method of nucleic acid extraction. For example, substantially different purification procedures would be required to prepare samples suitable for long and accurate PCR (LAPCR) (Barnes 1994a,b; Chang et al. 1994; and "Long-distance PCR" in Section 1) than for the amplification of a region of mitochondrial DNA (Higuchi et al. 1988). The chapters in this section, as well as others throughout this manual, provide a comprehensive set of nucleic acid extraction methods from nearly all sources.

"Construction of a Subtractive cDNA Library Using Magnetic Beads and PCR" in Section 6 provides a protocol for RNA extraction from plant tissue. "RNA Purification" in this section describes the principles and methods for the extraction of RNA from animal cells and tissues. For optimal results, a fresh or rapidly frozen specimen is recommended, because of the ubiquitous distribution and stability of RNases. RNA extraction protocols require the use of chaotropic agents, detergents, and often, phenol and chloroform. As detailed in

"RNA Purification," additional purification steps are often required to produce RNA that is free of reverse transcriptase inhibitors, such as SDS and EDTA. Although RNA extracted from other sources, such as paraffin-embedded blocks, has been employed, the quality and reproducibility of PCR performed with this material has been extremely variable (Weizsacker 1991). For DNA extraction, the size of the desired amplicon determines the type of specimen and purification method recommended. As discussed in "Long-distance PCR" in Section 1 and in Barnes (1994a,b), high-molecular-weight genomic DNA prepared by traditional DNA extraction methods has proven to be a reliable PCR substrate. The critical factor in sample preparation for long-distance PCR is to minimize the amount of single-strand breaks and abasic sites. Because the DNA that serves as a template in the amplification reaction is single-stranded, minimizing nicks maximizes the length of the DNA substrate. Similarly, abasic sites cause the polymerase to stall and fall off the template. These sites are also prone to undergo a β-elimination reaction upon heating. Thus, both kinds of damage lead to shorter and less amplifiable substrates.

When synthesizing products in the 150-bp to 2000-bp range, the purification protocol described in "PCR Amplification from Paraffin-embedded Tissues" provides a method for preparing DNA. An application of this method to DNA from fresh or frozen lymphocytes is provided in "A Standard PCR Protocol: Rapid Isolation of DNA and PCR Assay for β-Globin" in Section 1. "Rapid Preparation of DNA for PCR Amplification with GeneReleaser" provides a protocol for the preparation of DNA from virtually any type of cell or tissue. This approach is different from traditional sample preparation procedures, where the DNA is purified free of degrading enzymes and polymerase inhibitors and then can be stored for a period of time prior to use. With the use of GeneReleaser in this procedure, the sample must be processed immediately prior to amplification. Storage and banking occur at the whole-cell stage. For some applications, such as for dried blood specimens, this works extremely well. In general, most DNA extraction methods avoid the use of organic solvents. The absence of chloroform and phenol extraction steps in the purification of the template eliminates the requirement for an ethanol precipitation.

Recommended purification procedures tend to minimize the number of steps involved in the process of template preparation. This not only reduces excessive shearing of DNA, it also minimizes the opportunities for sample-to-sample contamination. The prevention of contamination of PCR by either sample-to-sample contamination or PCR product carryover is discussed in Section 1.

Finally, there should be special considerations when choosing a fixative for in situ PCR applications. The nature of the fixative, fixation times, and requirements for protease digestion in the preparation of nucleic acid templates suitable for in situ PCR are analyzed in detail in Section 4, in "In Situ PCR."

REFERENCES

Barnes, W.M. 1994a. PCR amplification of up to 35-kb DNA with high fidelity and high yield from lambda-bacteriophage templates. *Proc. Natl. Acad. Sci.* **91:** 2216–2220.

——. 1994b. Tips and tricks for long and accurate PCR. *Trends Biochem. Sci.* **19:** 342.

Chang, S., C. Fockler, W.M. Barnes, and R. Higuchi. 1994. Effective amplification of long targets from cloned inserts and human genomic DNA. *Proc. Natl. Acad. Sci.* **91:** 5695–5700.

Golenberg, E.M., D.E. Giannassi, M.T. Clegg, C.J. Smiley, M. Durbin, D. Henderson, and G. Zurawski. 1990. Chloroplast DNA sequence from a miocene Magnolia species. *Nature* **344:** 656–658.

Greer, C.E., J.K. Lund, and M.M. Manos. 1991. PCR amplification from paraffin-embedded tissues: Recommendations on fixatives for long-term storage and prospective studies. *PCR Methods Appl.* **1:** 46–50.

Higuchi, R., C.H. von Beroldingen, G.F. Sensabaugh, and H.A. Erlich. 1988. DNA typing from single hairs. *Nature* **332:** 543–546.

Higuchi, R., B. Bowman, M. Freiburger, O.A. Ryder, and A.C. Wilson. 1984. DNA sequencs from the quagga, an extinct member of the horse family. *Nature* **312:** 282–284.

Lawlor, D.A., C.D. Dickel, W.W. Hauswirth, and P. Parham. 1991. Ancient HLA genes from 7,500-year-old archaeological remains. *Nature* **349:** 785–788.

Weizsacker, F.V. 1991. A simple and rapid method for detection of RNA in formalin-fixed, paraffin-embedded tissues by PCR amplification. *Biochem. Biophys. Res. Commun.* **174:** 176–180.

Rapid Preparation of DNA for PCR Amplification with GeneReleaser

Elliott P. Dawson,[1] James R. Harris,[1] and James R. Hudson[2]

[1]BioVentures, Inc., Murfreesboro, Tennessee 37129
[2]Research Genetics, Huntsville, Alabama 35801

INTRODUCTION

Many techniques and reagents have been described in the literature for the purpose of preparing nucleic acids for amplification. Traditional sample preparation methods used to obtain DNA generally rely on disrupting cells by mechanical or detergent action. In some cases, an enzymatic treatment may be necessary to digest proteins or break down other cellular constituents. The cellular material is usually then subjected to a solvent extraction, usually phenol/chloroform, and the nucleic acids are obtained in the aqueous phase following partitioning into two phases. The nucleic acids may be further purified by precipitation and redissolving in an appropriate buffer. DNA obtained by these techniques is of high purity and suitable for use in most, if not all, molecular biology techniques. However, these classic techniques are not very rapid, and they require several manipulations and transfers of materials and reagents that are not readily automated or suitable for processing large numbers of specimens. When a small amount of material is available for analysis, the classic methods of DNA preparation are particularly unsuitable.

Newer methods and reagents have become available in the last several years. This chapter reviews some of the applications of GeneReleaser, a second-generation reagent for the preparation of various specimens for PCR amplification. A second-generation PCR sample preparation reagent is one in which nucleic acids (DNA, RNA, or both) are obtained in a form suitable for amplification with a minimum of sample preparation and reagent manipulation and frequently omits, substitutes, or consolidates many of the reagents, transfers, and steps characteristic of classic nucleic acid preparation procedures.

These second-generation reagents and protocols generally overcome some of the issues of sample preparation with respect to speed, economy, efficiency, and throughput that classic nucleic acid preparation methods lack. However, the nucleic acids obtained by the newer methodologies may not be suitable for all molecular biology techniques, because the material is not as pure, or is produced in relatively small quantities.

GeneReleaser was developed in our laboratories for the rapid isolation of DNA from whole blood for amplification of HLA sequences used for typing prospective bone marrow donors. Since its introduction, GeneReleaser has been used for sample preparation with samples as diverse as whole blood spots (Babu et al. 1993; Nerurkar et al. 1993), large numbers of samples of whole blood (Hutchin et al. 1993; Dawson and Harris 1994), plant specimens (Dawson et al. 1994; Levy et al. 1994), veterinary specimens (Gwaltney et al. 1994; Stone et al. 1994), whole tissue (Baker and Palumbi 1994), and archival tissue of historic interest (Hunt et al. 1995).

When GeneReleaser is employed for sample preparation using the standard procedures that follow, the samples are simultaneously prepared and consumed. A fresh aliquot of the sample must be obtained or must have been stored in order to repeat a specific amplification. With modification and planning at the time of sample collection, this alteration in approach is easily accommodated. The multiple sample preparation procedure does much to overcome this disadvantage. In the end, selection of a specific procedure or reagent is best evaluated by the researcher's circumstances, the nature of the specimens under analysis, and the use for which the DNA is intended.

Specific protocols for the use of GeneReleaser with various specimens, and the PCR results from its use, are described with permission from BioVentures, Inc. Examples of PCR results obtained with DNA extracted by GeneReleaser treatment from whole blood, blood spots, paraffin-embedded tissue, *Mycobacterium tuberculosis*, fish, and plant materials are described.

REAGENTS

Thermostable DNA polymerases from a number of suppliers work well with GeneReleaser-prepared DNA. GeneReleaser (BioVentures) reagent is provided as a 1-ml suspension. GeneReleaser is composed of a 20% suspension of an activated hydrophilic proprietary polymer in aqueous 0.01 M Tris-HCl, 0.05 M KCl, 10% dimethyl sulfoxide (DMSO), pH 7.4–7.6. The hydrophilic polymer aids in cell disruption and serves as a nucleation site for precipitation and absorption of proteins. The polymer also has some chelating activity and affinity for lipids and polysaccharides and has very low affinity for nucleic acids. The polymer seems to remove or reduce uncharacterized inhibitors of

PCR. Kits contain 1.5-ml polypropylene tubes and disposable pestles that fit the tubes for sample grinding. Kits also include a 96-well polypropylene rack that is compatible with most microwave ovens.

PROTOCOLS FOR WHOLE BLOOD

Whole blood is often employed as a source of nucleic acids. Frequently, infectious agents are found in the blood and can be detected by PCR because of its extreme sensitivity. Collection of whole blood for use with GeneReleaser should be performed by standard venipuncture or by other acceptable techniques. GeneReleaser is compatible with the commonly used anticoagulants EDTA, heparin, and citrate; EDTA and heparin are most commonly used and are the preferred anticoagulants for use with GeneReleaser. However, when these anticoagulants are employed they must not exceed the final concentrations generally used for the collection of human whole blood (EDTA Na^+ or K^+ at 1.5 mg/ml or less; heparin 15–26 IU/ml or less). Blood collected in either EDTA or heparin and stored at room temperature, 4°C, or –20°C for periods of up to 3 years has been used successfully in our laboratory for PCR.

Treatment of Whole Blood

1. Place 1 μl of well-mixed whole blood into the bottom of each thermal cycling tube for each specimen.

2. Thoroughly resuspend the contents of the GeneReleaser tube by inverting 10–20 times.

3. Add 20 μl of the resuspended GeneReleaser to each tube.

4. Perform the thermal cycling program described below.

5. Perform the amplification reaction according to the optimized protocol.

Thermal Cycling Program for Use with GeneReleaser

Note: Reaction volumes and concentrations should be adjusted proportionally to allow for the 20-μl GeneReleaser volume. In some instances, a new magnesium titration may need to be performed to assure optimum amplification. We suggest the range of 1.5 mM–4.0 mM Mg^{++} in 0.5-mM increments for this titration.

Following the thermal cycler manufacturer's procedure and using the default transition rates between temperatures, enter and run the following program:

Overlay the specimens with mineral oil.

	Temperature	Time
1.	65°C	hold 30 seconds
2.	8°C	hold 30 seconds
3.	65°C	hold 90 seconds
4.	97°C	hold 180 seconds
5.	8°C	hold 60 seconds
6.	65°C	hold 180 seconds
7.	97°C	hold 60 seconds
8.	65°C	hold 60 seconds
9.	80°C	hold for up to 5 minutes

10. Add amplification reagents without vortexing or otherwise mixing the amplification tube contents. Vortexing or vigorously mixing the polymer-precipitated impurities will resuspend inhibitors into the aqueous phase and precipitate the thermostable DNA polymerase.

11. Begin the optimized amplification.

    Note: It is very important that the very first denaturing step of the first cycle be at 94°C for 2–5 minutes, depending on the brand of cycler, reaction volumes, etc.

Microwave Procedure for Use with GeneReleaser

We have found that microwave treatment of specimens affords more rapid sample preparation and facilitates the amplification of the more intractable specimens.

1. Mix 1 μl of specimen with 20 μl of GeneReleaser as described above.

   Note: It is extremely important not to use tubes any larger than these because the samples will be boiled away or tubes will rupture!!

2. Unlike the thermal cycle program, vortex the tubes containing specimen and GeneReleaser for about 10–30 seconds.

3. An oil overlay is optional.

4. Place the closed tubes in a polyethylene or polypropylene rack.

5. Place the rack in microwave oven and heat at maximum power setting for 5–7 minutes: 5 minutes if wattage is 900 or higher, 7 minutes if wattage is 500. The optimum is 4500 watt-minutes.

6. Remove the rack from the microwave oven and place the tubes on a preheated thermal cycler at 80–90°C.

7. Add the PCR master mix and begin the amplification cycles.

   Note: It is very important that the very first denaturing step of the first cycle be at 94°C for 2–5 minutes, depending on the brand of cycler, reaction volumes, etc.

Evaluation of Microwave Function

To evaluate the microwave, perform the following experiment:

1. Place 40 μl of deionized water in the same size and type of tube as used for PCR.

2. Overlay each tube with mineral oil.

3. Close the tubes, place in a microwave-safe rack, and heat on HIGH for 5 minutes.

4. If any caps pop or if the tubes distort in any manner, place a separate beaker in the microwave with 150 ml of ambient-temperature deionized water and repeat the above procedure. The beaker of water serves as a heat ballast.

5. If tubes open or distort after performing step 4, reduce the power by 10% increments and increase the time by 1-minute increments and repeat step 4 until the tubes no longer open or distort. Once standardized, the time and power levels can be used for all future GeneReleaser applications.

6. If a water ballast was used as described above, allow the tubes to remain in the microwave undisturbed for 2 minutes following the completion of the microwave cycles. This is to prevent any sudden boiling of the ballast after it has been heated.

Whole Blood Preparation for Multiple Analysis

In some instances, the researcher may wish to perform a single sample preparation for the purpose of performing multiple PCRs on a particular sample. The procedure below should be followed:

1. Collect whole blood by standard venipuncture or other collection techniques.

2. Transfer an aliquot of whole blood representing a population of 10^6 nucleated cells (~200 μl) to a 0.5-ml amplification tube.

3. Add an equal volume of 0.14 M NH_4Cl, 0.017 M Tris-HCl, pH 7.4, vortex for 5 seconds, and place on ice for 5 minutes.

4. Remove the tubes from the ice and centrifuge at 1000*g* for 1 minute.

5. Repeat steps 3 and 4 if the pellet is excessively tinted with hemoglobin (some color will remain).

6. Remove and discard the supernatant.

7. Wash the pellet once with 200 μl of 1X PCR buffer and centrifuge 1 minute at 1000*g*. To the remaining pellet, add 20 μl of GeneReleaser and 20 μl of TE (10 mM Tris-HCl, pH 8.0, 1 mM EDTA).

8. Vortex the cell pellet and the added GeneReleaser for about 30 seconds until thoroughly mixed. Overlay the sample with mineral oil.

9. Treat the material from step 8 by either the microwave or thermal cycle lysis procedures described above.

10. Centrifuge the tube for 30 seconds and transfer 20 μl of the aqueous phase to a fresh tube. Bring to a total volume of 200 μl by the addition of 180 μl of TE.

11. Use 5 μl of the material from step 10 as a source of DNA for PCR. DNA obtained by this procedure is stable for 60 days at 4°C or for more than 1 year at –20°C.

Amplification of HLA-DRB from Preserved Whole Blood

The use of GeneReleaser in the preparation of whole blood specimens for amplification of a 272-bp region of HLA-DRB by PCR was investigated. Ten samples of whole blood were collected (1.5 mg/ml Na_2EDTA as preservative). The blood was further aliquoted and stored at 25°C, 4°C, and –20°C. DNA was obtained from each specimen either by treatment with GeneReleaser using the thermal cycle lysis procedure or by a boiling lysis procedure (Higuchi 1989).

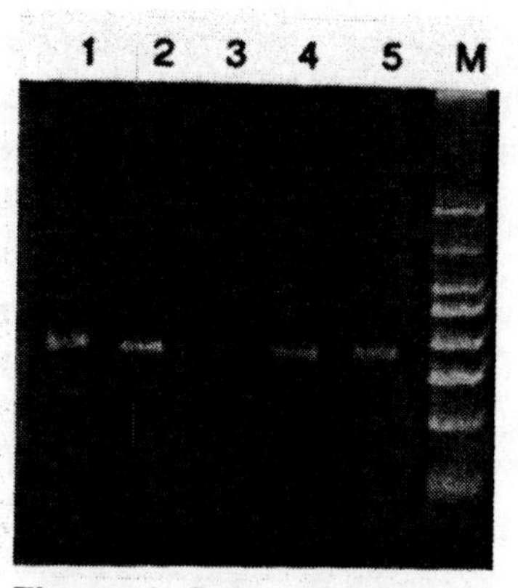

Figure 1 Representational amplification results for HLA-DRB (272-bp product). (Lane *1*) Positive control; (lanes *2,4,5*) GeneReleaser-treated whole blood (3 of 10) stored at 25°C, 4°C, −20°C, respectively. (Lane *3*) Sample (1 of 10) identical to that of lane *2* (stored at 25°C) treated according to the method of Higuchi (1989). (M) BioMarker Low (1000, 700, 500, 400, 300, 200, 100, and 50 bp dsDNA). The PCR was performed for 32 cycles.

Amplification for HLA-DRB alleles was performed using published primers and conditions (Vaughn 1991) in a final volume of 100 μl. The results are shown in Figure 1.

Samples of whole blood can be successfully amplified for DRB and other gene sequences after being stored for extended periods of time using GeneReleaser as a specimen preparation reagent. The small sample volume and the minimum number of manipulations, as well as the performance of the entire procedure in the amplification tube, vastly simplify the preparation of large numbers of samples of whole blood for PCR.

Blood Spots

Frequently, blood and other body fluids are collected on paper for analysis, especially for the screening of newborn infants for phenylketonuria (PKU). Blood is usually obtained by capillary collection onto filters (Schleicher and Schuell 903 Specimen Collection Paper). Our experience is that such filters make ideal collection and storage systems for obtaining specimens for DNA analysis from collaborators abroad or at distant locations. This is especially the case from regions of the world where collection, storage, and shipment of whole blood or other body fluids under refrigerated conditions is not feasible. In some instances, the amount of blood or other fluid obtainable may be very limited, and the filters provide an ideal means of collecting and preserving such scarce specimens. Specimen collection papers are also useful for studying the frequency, distribution, and incidence of occurrence of inherited disorders within a population, especially for retrospective studies (McCabe et al. 1987). Blood collection filters are also useful for collecting specimens from animals; for example, to determine the success of insertion of genes in transgenic mice. Infectious disease agents can also be identified from dried blood spots using the PCR. The typical circular application zones found on these filters are 12.5 mm in diameter and can absorb about 50 μl of whole blood in the entire zonal area of 123 mm^2. Consequently, a 2.5-mm^2 section of the filter zone represents about 1 μl of whole blood from a 50-μl application of whole blood.

Amplification of Regions from Human Growth Hormone

Portions (2 mm^2) of two blood spots were treated using the microwave protocol. The treated specimens were placed at 80°C on a Perkin-Elmer model 480 thermal cycler, and 80 μl of a master mix (primer set used produces a 542-bp product) containing all the components (except for template DNA) for amplification was added. Amplification was performed using the following cycling conditions:

Cycles	Denature	Anneal	Extend
1	94°C-3 min	none	none
30	94°C-1 min	60°C-1 min	72°C-1.5 min
1	none	none	72°C-8 min

Following amplification, 10 μl of the reaction products was analyzed by electrophoresis on 7.5% polyacrylamide gel and detected by ethidium bromide staining. The results are shown in Figure 2.

PROTOCOL FOR PARAFFIN-EMBEDDED TISSUES

Paraffin-embedded tissues and other fixed tissues represent archival specimens. Because PCR can amplify relatively short regions of DNA relative to intact chromosomal DNA and is somewhat tolerant of degraded DNA, fixed tissues are amenable to molecular genetic studies. Such tissues represent vast resources for prospective and retrospective molecular genetic or infectious disease studies. Most paraffin-embedded specimens encountered have been formalin-fixed. Most preparations in the literature require enzymatic digestion of the specimen prior to analysis (Sepp et al. 1994). In many instances, GeneReleaser can be used to prepare specimens without any prior enzymatic treatment. In some cases, however, digestion with proteinase K may be required. This is especially true if amplified products of greater than 700 bp are to be produced. The method that follows has been used successfully to amplify a number of paraffin-embedded tissues. The example shown is for a 1.4-kb product derived from the p53 gene. To achieve adequate sensitivity, 40–50 cycles of amplification should be performed. As much as 20–50% of the reaction mixture may need to be analyzed. (See also "PCR Amplification from Paraffin-embedded Tissues" in Section 2.)

Figure 2 Amplification of a 542-bp segment of human growth hormone gene from dried blood spots stored on filter disks. (Lanes *1,2*) Replicates of filter of 11/23/94. (Lane *3*) BioMarker Low (1000, 700, 500, 400, 300, 200, 100 and 50 bp). (Lanes *4,5*) Replicates of filter of 10/31/94. (Lanes *6,7*) Replicates of EDTA whole blood of 12/13/91.

Processing Paraffin-embedded Tissues

1. Deparaffinize slide-mounted tissue using a standard technique. A 100% acetone wash 2x, 5 minutes each, is required for all lipid-rich tissue, such as brain tissue.

2. Wet the slide by sequential washes in
   100% ethanol 2x, 5 minutes each
   70% ethanol 2x, 5 minutes each
   30% ethanol 2x, 5 minutes each
   100% deionized water 2x, 5 minutes each

3. Wash the tissue on the slide again in
   TE 1x, 5 minutes
   100% deionized water 1x, 5 minutes

4. Scrape tissue (~1 mm^2 to 9 mm^2 in size) from the slide into a 1.5-ml tube.

5. Add 20 μl of GeneReleaser to the tissue in the tube.

6. Using a disposable pestle, grind the tissue with the pestle.

7. Transfer the entire contents to a standard PCR tube.

8. Overlay the contents with mineral oil and treat using the microwave protocol. Add 80 μl of 1.25x PCR master mix.

9. Transfer the tubes to a thermal cycler and proceed with the amplification.

Amplification of Human p53 Gene Sequences from Paraffin-embedded Tissues

Three surgically obtained, unstained, formalin-fixed, paraffin-embedded tissues affixed to microscope slides were processed according to the procedure described above. Following microwave treatment, the tubes were placed at 80°C for 5 minutes. Eighty microliters of a 1.25X PCR master mix containing all the components, except template DNA, was added to each tube. The following cycling conditions were used:

Cycles	Denature	Anneal	Extend
1	94°C-3 min	none	none
5	94°C-1 min	58°C-45 sec	72°C-3 min
35	94°C-1 min	58°C-1 min	72°C-3.5 min
1	none	none	72°C-10 min

Following amplification, 10 μl of the reaction product was analyzed by electrophoresis on 7.5% polyacrylamide gel and detected by ethidium bromide staining. The results are shown in Figure 3.

PROTOCOLS FOR WHOLE TISSUES

Whole tissues provide an excellent source of genomic DNA. A simple protocol using GeneReleaser to obtain DNA from fresh tissues is described.

Whole Tissue Homogenization

1. Cut a 1-mm^3 section from either fresh or frozen tissue that has been rinsed with sterile water to remove any surface contamination.

2. Place the section into the bottom of the 1.5-ml tube provided with the tube and pestle set.

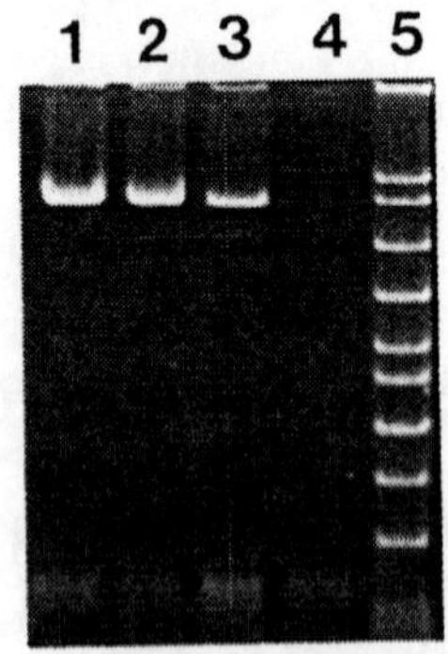

Figure 3 (Lanes *1,2,3*) 1.4-kb product from human tissue sections. (Lane *4*) Negative control. (Lane *5*) BioMarker Extended (2000, 1500, 1000, 700, 500, 400, 300, 200, 100, and 50 bp).

3. Add 25 μl of TE to the tube containing the sectioned tissue.

4. Mince the section of tissue by thrusting the pestle provided against the tissue and twisting the pestle to compress the tissue against the walls of the tube. Ten thrusts with the pestle are sufficient.

5. As each specimen is homogenized, maintain the tube containing the homogenate at 4°C.

6. Transfer 1 μl of the tissue homogenate obtained above into a 0.5-ml standard amplification tube.

   Note: Be sure that the tissue homogenate is well mixed prior to removing the aliquot.

7. Resuspend the GeneReleaser mixture either by vortexing 2–3 seconds or by 10 inversions.

8. Add 20 μl of the GeneReleaser suspension to the 1 μl of homogenate in the amplification tube.

9. Vortex the tube 2–3 seconds.

10. Overlay the mixture with mineral oil.

11. Place tubes in the 96-well rack, and microwave on HIGH for 5–7 minutes.

12. Transfer the tubes to a thermal cycler preheated to 80°C and allow them to equilibrate for 5 minutes.

13. Initiate amplification by the addition of 80 μl of a 1.25x master mix containing all components for the amplification.

14. Begin the optimized amplification.

    Note: It is very important that the very first denaturing step of the first cycle be at 94°C for 2–5 minutes, depending on the brand of cycler, reaction volumes, etc.

Amplification of Sequence from Fish Fins

Dorsal fins from three steelhead salmon were processed according to the protocol for whole tissue described above. Following microwaving, the tubes were placed at 80°C, and 80 μl of a 1.25x PCR master mix was added. Thermal cycling conditions for amplification were:

Cycles	Denature	Anneal	Extend
1	94°C-4 min	48°C-1 min	72°C-1 min
30	94°C-1 min	48°C-1 min	72°C-1 min
1	none	none	72°C-8 min

Following amplification, 5 µl of the reaction product was analyzed by electrophoresis on 4% agarose gel and detected by ethidium bromide staining. The results are shown in Figure 4.

Plant Materials

The application of molecular biological techniques to the selection, propagation, and modification of plants of agricultural interest has steadily grown. A number of PCR-based techniques have been developed and applied to these processes. In many instances, it is desirable and preferred to analyze a large number of specimen samples. Although hundreds of specimens can be analyzed and screened by PCR, in many instances, several thousand to several million assays are desired to assure, enhance, and accelerate plant selection and breeding programs. Because of their complex biochemistry and production of secondary metabolic products that inhibit polymerase activity, plant materials represent some of the most challenging tissues from which to obtain DNA suitable for use in PCR. Recently, several DNA isolation methods intended to overcome some of the problems associated with plant materials have been reported. The simplest is one in which fresh tobacco leaf or root tissues are placed directly into the PCR buffer to provide a source of DNA for PCR amplification (Berthomieu and Mayer 1991). Other methods (Tai and Tanksley 1990; Edwards et al. 1991; Langridge et al. 1991; Oard and Dronavalli 1992), although reliable, are still too complicated to match the ease of PCR

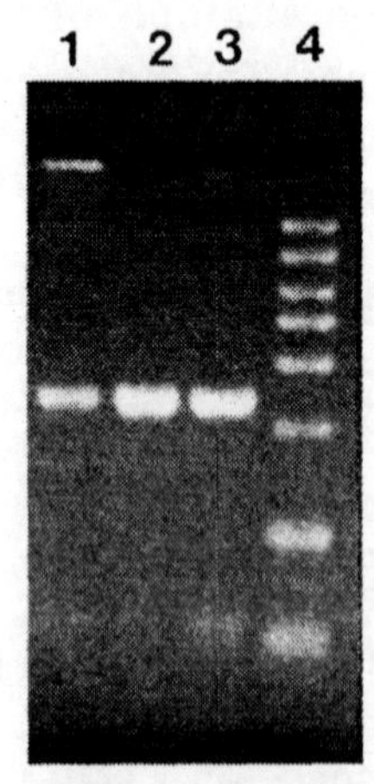

Figure 4 (Lanes *1–3*) 220-bp product from 3 different steelheads. (Lane *4*) BioMarker Low (1000, 700, 500, 400, 300, 200, 100, and 50 bp).

for the analysis of large numbers of samples in a relatively short period of time. Furthermore, most of the procedures are multistep processes requiring several reagent solutions and multiple manipulations or transfers. All of these manipulations increase the possibility of cross-contamination.

Plant Homogenization

1. Using 1.5-ml snap cap standard conical centrifuge tubes, punch a round leaf section by inserting the leaf between the base of the open cap and the tube and opening and closing the cap.

2. Using the disposable pestle, grind or mince the leaf material with 100 μl of sterile H_2O or 1X TE in the 1.5-ml tube. Repeat this with two replicate specimens using 50 μl and 25 μl of buffer for the homogenization.

3. The homogenized material may be used immediately or stored at −20°C until ready for use.

4. Transfer 1 μl of the tissue homogenates obtained above into a 0.5-ml standard amplification tube.

   Note: Be sure that the homogenate is well mixed prior to removing the aliquot.

5. Resuspend the GeneReleaser mixture either by vortexing 2–3 seconds or by 10 inversions.

6. Add 20 μl of the GeneReleaser suspension to the 1 μl of homogenate in the amplification tube.

7. Vortex the tube 2–3 seconds.

8. Overlay the mixture with mineral oil.

9. Place tubes in the 96-well rack, and microwave on HIGH for 5–7 minutes.

10. Transfer the tubes to a thermal cycler preheated to 80°C and allow them to equilibrate for 5 minutes.

11. Initiate amplification by the addition of 80 μl of a 1.25X master mix containing all components for the amplification.

12. Begin the optimized amplification.

    Note: It is very important that the very first denaturing step of the first cycle be at 94°C for 2–5 minutes, depending on the brand of cycler, reaction volumes, etc.

13. Assay the products using a standard method.

14. Select the volume of homogenization buffer that gave the best product yield of the three volumes described above for future homogenizations and amplifications from the same type of source material.

Amplification of Sequences from Transgenic Maize

Three varieties of maize were grown at ICI Seeds Research Group (Slater, Iowa). Two of the varieties (A and B) contained transgenic DNA sequences, and the third variety (BE81) was free of any transgenic DNA sequences. Maize leaf samples were collected from each of the three varieties by punching a portion of the leaf into a standard 1.5-ml snap cap centrifuge tube using the cap as the punch. Paired replicates of the maize leaf materials were processed according to the method of Edwards et al. (1991) and by treatment with GeneReleaser as described above. Amplification was performed using conditions previously optimized for the amplification of purified DNA from the transgenic maize varieties. The PCR products were assayed by standard agarose gel electrophoresis, visualized by ethidium bromide staining, and photographed. A set of water controls was processed in parallel with the samples. The results appear in Figure 5.

TROUBLESHOOTING

Most of the problems that we have encountered with the use of GeneReleaser have been with amplification conditions and not with release of nucleic acids from specimens. About 95% of these problems can be resolved by evaluating the amplification conditions, making

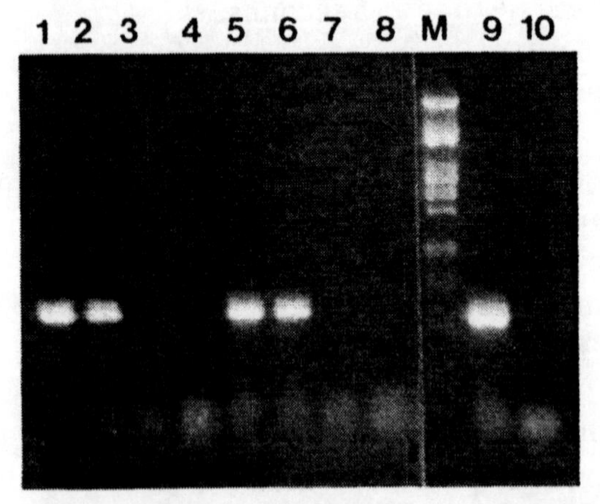

Figure 5 Results from method comparison of GeneReleaser (lanes *1–8*) and multistep miniprep (lanes *9–10*) (Edwards et al. 1991). Lanes *1*, *5*, and *9* are from transgenic maize A; lanes *2* and *6* are from transgenic maize B; lanes *3* and *7* are from control plant BE81; lanes *4*, *8*, and *10* are water controls.

minor modifications of cycling conditions, and following general guidelines for PCR (Saiki 1989). Occasionally, however, too much DNA or an inhibitor is present in the specimen. Often, diluting such specimens prior to PCR results in good amplification.

REFERENCES

Babu, P.G., C. Gnanmuthu, N.K. Saraswathi, R. Nerurkar, R. Vanagihara, and T.J. John. 1993. HTLV-I-associated myelopathy in South India (letter). *Aids Res. Hum. Retroviruses.* **9:** 499–500.

Baker, C.S. and S.R. Palumbi. 1994. Which whales are hunted? A molecular genetic approach to monitoring whaling. *Science* **265:** 1538–1539.

Berthomieu, P. and C. Mayer. 1991. Direct amplification of plant genomic DNA from leaf and root pieces using PCR. *Plant Mol. Biol.* **17:** 555.

Dawson, E.P. and J.R. Harris. 1994. Rapid sample preparation of DNA for amplification of HLA-DRB from whole blood. *J. NIH Res.* **6:** 76.

Dawson, E.P., K. Wang, S. Jiao, J.R. Harris, and J.R. Hudson. 1994. DNA archival storage and retrieval systems. In *Conservation of plant genes II: Utilization of ancient and modern DNA* (ed. R.P. Adams et al.), pp. 93–99. Missouri Botanical Garden, St. Louis.

Edwards, K., C. Johnstone, and C. Thomson. 1991. A simple and rapid method for the preparation of plant genomic DNA for PCR analysis. *Nucleic Acids Res.* **19:** 1349.

Gwaltney, S.M. and R.D. Oberst. 1994. Comparison of an improved polymerase chain reaction protocol and the indirect hemagglutination assay in the detection of *Eperythrozoon suis* infection. *J. Vet. Diagn.* **6:** 321–325.

Higuchi, R. 1989. Rapid efficient DNA extraction for PCR from cells or blood. *Amplifications* **2:** 1,3.

Hunt, D.M., K.S. Dulai, J.K. Bowmaker, and J.D. Mollon. 1995. The chemistry of John Dalton's color blindness. *Science* **267:** 984–988.

Hutchin, T., I. Hayworth, K. Higashi, N. Fischel-Ghodsian, M. Stoneking, N. Saha, C. Arnos, and G. Cortopassi. 1993. A molecular basis for human hypersensitivity to aminoglycoside antibiotics. *Nucleic Acids Res.* **21:** 4174–4179.

Langridge, U., M. Schwall, and P. Langridge. 1991. Squashes of plant tissue as substrate for PCR. *Nucleic Acids Res.* **19:** 6954.

Levy, L., I.-M. Lee, and A. Hadidi. 1994. Simple and rapid preparation of infected plant tissue extracts for PCR amplification of virus, viroid, and MLO nucleic acids. *J. Virol. Methods* **49:** 295–304.

McCabe, E.R.B., S. Huang, W.K. Seltzer, and M.L. Law. 1987. DNA microextraction from dried blood spots on filter paper blotters: Potential applications to newborn screening. *Hum. Genet.* **75:** 213–216.

Nerurkar, V.R., P.G. Babu, K.-J. Song, R.R. Melland, C. Gnanamuthu, N.K. Saraswathi, M. Chandy, M.S. Godec, T.J. John, and R. Yanagihara. 1993. Sequence analysis of human T cell lymphotropic virus type I strains from southern India: Gene amplification and direct sequencing from whole blood blotted on filter paper. *J. Gen. Virol.* **74:** 2799–2805.

Oard, J.H. and S. Dronavalli. 1992. Rapid isolation of rice and maize DNA analysis by random-primer PCR. *Plant Mol. Biol. Rep.* **10:** 236.

Saiki, R.K. 1989. The design and optimization of the PCR. In *PCR technology: The principles and applications for DNA amplification* (ed. H.A. Erlich), pp. 7–16. Stockton Press, New York, New York.

Sepp, R., I. Szabo, H. Uda, and H. Sakamoto. 1994. Rapid technique for DNA extraction from routinely processed archival tissue for use in PCR. *J. Clin. Pathol.* **47:** 318–323.

Stone, G.G., R.D. Oberst, M.P. Hays, S. McVey, and M.M. Chengappa. 1994. Detection of *Salmonella* serovars from clinical samples by enrichment broth cultivation-PCR procedure. *J. Clin. Microbiol.* **32:** 1742–1749.

Tai, T.H. and S.D. Tanksley. 1990. A rapid and inexpensive method for the isolation of total DNA from dehydrated plant tissue. *Plant Mol. Biol. Rep.* **8:** 297.

Vaughan, R.W. 1991. PCR-SSO typing for HLA-DRB alleles. *Eur. J. Immunogenet.* **18:** 69–80.

PCR Amplification from Paraffin-embedded Tissues: Sample Preparation and the Effects of Fixation

Catherine E. Greer,[1] Cosette M. Wheeler,[2] and M. Michele Manos[3]

[1]Department of Virology, Chiron Corporation, Emeryville, California 94608
[2]Department of Cell Biology and Center for Population Health, University of New Mexico, Albuquerque, New Mexico 87131
[3]Department of Molecular Microbiology and Immunology and Infectious Diseases, Johns Hopkins School of Public Health, Baltimore, Maryland 21205

INTRODUCTION

Overview

The exquisite sensitivity of PCR has afforded molecular studies of fixed paraffin-embedded tissue (PET) specimens, which comprise most archival clinical material. Combined with subsequent hybridization methods or DNA sequencing, PCR has provided the sensitive and specific detection of infectious agents (Shibata et al. 1988; Cao et al. 1989; Brandsma et al. 1990) and host genetic alterations (Burmer et al. 1989; Lyons et al. 1990; Thibodeau et al. 1993; van der Riet et al. 1994).

The preparation of PETs for use in PCR amplification is theoretically very simple. Preparation necessitates dissolving the paraffin from the tissue slice and subsequently treating the dried tissue to liberate DNA (Wright and Manos 1990). Complete DNA purification is possible, but often unnecessary. Such extensive purification procedures must be weighed against the increasing risk of sample contamination with each manipulation. (This is of extreme importance when PETs are being analyzed for the presence of an infectious agent.)

The success of any PCR-based study of fixed, paraffin-embedded material depends on several factors, including (1) the fixative used in the tissue processing; (2) the duration of the fixation; (3) the age of the paraffin block; and (4) the length of the DNA fragment to be amplified. Below, we review the results of studies conducted to evaluate the suitability of PETs as subsequent PCR targets. We also provide updated protocols and methodologic considerations for retrospective PCR-based studies.

Effects of Fixation

Although PCR DNA amplification is a powerful tool for retrospective studies, not all preservation or fixation methods render DNA that is suitable for subsequent amplification (Crisan et al. 1990; Ben-Ezra et al. 1991; Greer et al. 1991a,b). Previously, we reported extensive analyses of the effects of commonly used fixation methods on the efficiency of subsequent PCR amplification (Greer et al. 1991a,b). In those studies (see Table 1), the effect of fixation was measured by the ability of the DNA in a treated tissue to act as a template for the amplification of DNA fragments of increasing lengths. The effect of each fixation method tested is clearly reflected by the maximum product length obtained from each treated tissue. Of the specimens tested, those most successful in subsequent PCR amplifications are fixed in ethanol, acetone, or OmniFix, followed by 10% buffered neutral formalin (BNF). Another group of fixatives, including Zamboni's, Clarke's, paraformaldehyde, formalin-alcohol-acetic acid, and methacarn, compromise amplification efficiency. Tissues fixed in highly acidic solutions (Carnoy's, Zenker's, or Bouin's) are seriously compromised for amplification and are not considered desirable.

The length of time a sample is maintained in a fixative is also an important factor. Our previous studies indicate (Table 1) that after 24 hours of tissue fixation, the ability to amplify large PCR products decreases with all fixatives tested except ethanol, acetone, and OmniFix. The nonacidic fixatives afford the subsequent amplification of

Table 1 Effect of Fixation on Subsequent PCR

Fixative	Maximum length product after 24-hr fixation (bp)	Largest fragment generated (bp) and duration of fixation	
Acetone	1327	1327	8 days
Alcoholic formalin	989	989	24 hr
Bouin's	0	110	1 hr
10% BNF	1327	1327	24 hr
Carnoy's	268	989	4 hr
Clark's	989	1327	4 hr
Ethanol	1327	1327	30 days
Formalin-alcohol-acetic acid	989	1327	1 hr
Methacarn	989	1327	1 hr
Paraformaldehyde	989	1327	1 hr
OmniFix	1327	1327	72 hr
		989	30 days
Zamboni's	989	1327	4 hr
Zenker's	110	268	4 hr

Data are summarized from previous reports (Greer et al. 1991a,b).

fragments 536 bp or more in length. Although many clinical laboratories routinely fix tissues for 24 hours or less, some tissues may be treated for up to several days, thereby reducing the amplifiable fragment size. For example, biopsies are often placed in buffered formalin and shipped to reference laboratories for embedding and analysis. A more extreme situation exists when tissue sampling occurs in remote regions, thus requiring fixation and storage for extended periods of time prior to analysis. Consequently, the length of time a tissue is immersed in a fixative can be as critical as the type of fixative used. Both factors should be taken into consideration when planning either a retrospective or prospective study using PETs. When the DNA is compromised, a scheme involving smaller amplification products (less than 200 bp) is desirable.

Effects of Specimen Age

The approach used to assess the effects of fixation on subsequent DNA amplification was also applied to test the effects of specimen age. An unpublished study conducted at the University of New Mexico included 240 PET samples representing all invasive cervical carcinomas that were diagnosed at the University of New Mexico Hospital over a 20-year period. These specimens were tested for the ability to generate three sizes (268 bp, 536 bp, and 989 bp) of PCR fragments. All samples had been fixed in 10% BNF, and specimens were grouped according to age with at least 20 specimens for each time point. The results (Fig. 1) showed that after 16 years, 90% of the samples were sufficient for the amplification of the 268-bp β-globin fragment. Successful amplification of this fragment decreased to 45% for 20-year-old specimens. As the size of the amplified fragment was increased from 268 bp to 536 bp, a significant effect of specimen age was detected. Only 60% of 5-year-old specimens were sufficient for amplification of a 536-bp fragment. The most dramatic effect of specimen age was seen when the amplified fragment size was increased to 989 bp. Here a linear decrease in the number of successful amplifications was observed, so that by 5 years, there were no samples sufficient for the amplification of the 989-bp fragment. Clearly, in the case of samples older than 5 years, the smaller the fragment, the greater the likelihood of successful amplification.

Because of the many variables involved in PET fixation and processing, and the differing efficiencies of primer pairs, it is recommended that comprehensive pilot studies be conducted to assess DNA sufficiency of any collection of PETs. Results of these studies help determine the optimal amplification product lengths that can be obtained from the specimens chosen.

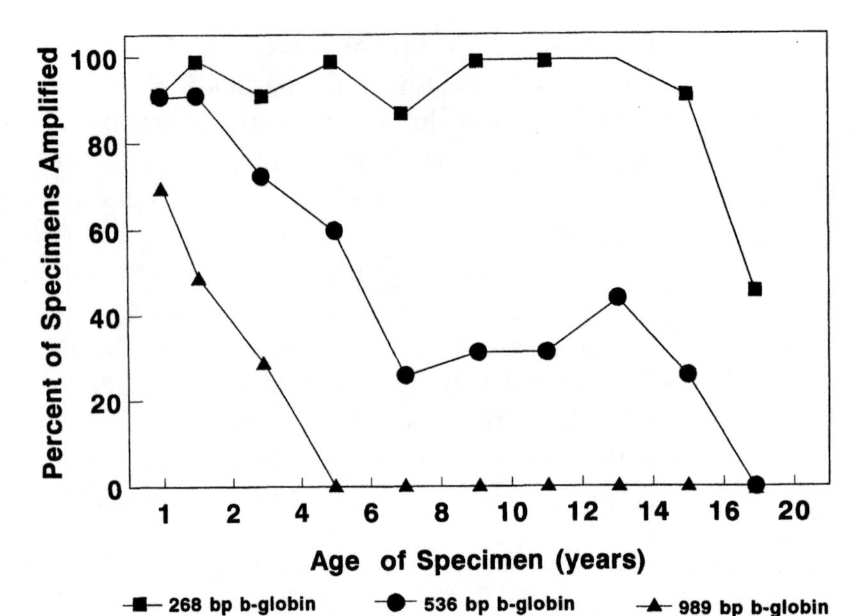

Figure 1 Effect of specimen age on maximum size of amplification product. Tissue blocks used in this study were 1–20 years old. Invasive cervical cancer biopsies that had been fixed in 10% BNF and paraffin-embedded were deparaffinized, subjected to proteinase K digestion, and amplified with various β-globin primer pairs. Each time point represents at least 20 PETs from the specimen age group.

REAGENTS AND MATERIALS

Tissue sections
10% bleach solution (freshly diluted)
Octane, xylene, or AmeriClear (Baxter)
100% ethanol
HPLC-grade acetone (optional)
Proteinase K (20 mg/ml stock solution)
Digestion buffer: 50 mM Tris-HCl (pH 8.5), 1 mM EDTA, 1% Laureth-12 ("Macol LA-12," PPG Company [formerly Mazer Chemicals]) or 0.5% Tween 20 or 1% Laureth-10
2-inch sterile gauze pads (Johnson & Johnson brand, Thomas Scientific, cat. no. 2904-C12)
1.5-ml microfuge tubes, flat top and tight fit (Sarstedt; siliconized, cat. no. 3207, non-siliconized, cat. no. 3210)
Cap-Locks (Intermountain Scientific, cat. no. C-3270-1)
Disposable plastic pipettes (Baxter, cat. no. P5212-205)
*Dry heat blocks at 55°C and 95°C
*Microfuge (PCR-product and plasmid clean)
*PCR-product clean test tube racks
*Vortex mixer
Rotating or rocking platform shaker
Quartz sand (Sigma #S-9887)

*These items must be dedicated for clean, pre-PCR use.

PROTOCOLS

Preparation of Tissue Sections

A fundamental safeguard to prevent PCR product contamination is to process samples in an area physically separated from post-PCR sample analysis. Considerable effort is required to prevent sample-to-sample contamination during the sectioning of paraffin blocks. The microtome, microtome blade, and any equipment used in the sectioning area must be rigorously cleaned between each paraffin block by squirting *freshly diluted* 10% bleach (Dychdala 1977; Hoffman et al. 1981) onto gauze pads and carefully cleaning the microtome. The bleach wash should be followed by an ethanol rinse to prevent corrosion of the microtome. The blade must be removed and carefully wiped clean of any debris with another clean, bleach-soaked gauze pad. Any tissue remaining on the blade may easily contaminate the next sample. Disposable blades provide the greatest protection from block-to-block contamination. Finally, it is essential to change gloves between the cleaning of the microtome and the sectioning of each new block.

Once the microtome has been well cleaned, the first section is taken to expose a PCR-"clean" surface. Replicate sections (5–20 μm) can be cut from each block and a single section placed in a sterile, 1.5-ml microfuge tube. The thickness of the section depends on the size of the tissue. For a small biopsy (2–3 mm), 10- to 20-μm sections may be required, whereas larger tissues (5 x 5 mm) can be sectioned at 5 μm. Although multiple sections can be placed in a single tube, fewer thick sections are more practical for processing.

If the targeted tissue is localized within a limited portion of the block, such as the area of tumor invasion, it is critically important to prepare adjacent, flanking "first and last" sections (5 μm) for hematoxylin and eosin (H & E) staining. First, this procedure ensures that in the case of a negative result, the tissue of interest was present. Second, it allows confirmation of the original histologic diagnosis. In some cases, the tissue may be microdissected within the block or on a mounted section. This is common, for example, when identifying tumor-specific mutations. Because such manipulations are vulnerable to minute amounts of contamination, this is not recommended for infectious disease studies.

Deparaffinizing Sections

1. Centrifuge tissue to the bottom of the 1.5-ml tube (about 5 seconds). This prevents the section from escaping when the tube is opened.

2. Open the tube by holding a clean gauze pad over the cap and gently prying off the top. Never "pop" or "flip" the tube open with

your thumb or touch the inside of the cap; this can cause sample-to-sample contamination.

3. Add 1 ml of octane (or AmeriClear or xylene) and gently vortex to loosen the paraffin from the bottom of the tube. Most paraffins dissolve quickly (2–5 minutes), but others may require gentle vortexing or continued mixing on a rotating platform shaker at room temperature (up to 30 minutes) to be dissolved. Deparaffinized tissue is opaque and "fluffy" in appearance, whereas undissolved paraffin is solid white and rigid.

4. Pellet the tissue and any remaining paraffin by centrifugation for 2–5 minutes at full speed in a microfuge.

5. *Carefully* remove the solvent with a single-use, fine-tipped pipette. Do not disturb the tissue because it can be easily dislodged. Do not remove any tissue while pipetting, but if this does occur, expel the tissue into the tube and repeat the centrifugation. (If Pasteur pipettes are used, each must be cotton-plugged to prevent contamination of the pipette bulb and thus other samples).

6. Repeat steps 3–5 if any paraffin remains.

7. Carefully add 0.5 ml of 100% ethanol to the tube and mix well.

8. Centrifuge for 2–5 minutes and carefully remove the ethanol.

9. A drop (10–30 μl) of HPLC-grade acetone can be added to speed the evaporation of the ethanol. The open tubes should be carefully placed in a sand-filled, 55°C heat block to dry the tissue.

   Note: The open tubes are vulnerable to contamination at this point. Therefore, do not allow contact between individual tubes, and eliminate air flow around the tubes. PCR product- or plasmid-containing tubes should never be used in this area and, specifically, this heat block. The sand in the block should be changed frequently to reduce the possibility of contamination.

10. The appearance of the tissue changes from opaque when wet to solid white when dry. Handle the tubes carefully, because static electricity can cause the dry tissue to pop out of the tube.

    Note: Never use speed-vacs or vacuum bottles, where sample-to-sample contamination readily occurs, to dry the tissue pellets.

A recent report suggested the use of microwave treatment of paraffin-embedded tissues suspended in buffer as an alternative method of deparaffinization (Banerjee et al. 1995). Although we have no experience with this method, it appears to merit further investigation. The article's authors stated that they had consistent amplification of fragments of 400–600 bp in length.

Proteinase K Digestion

1. Suspend the dry pellet in fresh digestion buffer (typically, 100 μl). The required volume varies with the amount of tissue present after deparaffinization. For example, a 3 x 5 mm x 5 μm tissue pellet should be digested in approximately 250 μl of buffer. Smaller pellets, 2 x 3 mm x 5 μm, should be digested in 50 μl of buffer. In general, the dried tissue should occupy about 25–30% of the volume of digestion buffer when the buffer is *first* added to the tube. The tissue must be completely immersed in the digestion buffer. If necessary, vortex and briefly centrifuge tubes before incubation.

2. Incubate at 55°C for 3 hours (small biopsies) to overnight (larger pieces of tissue). Intermittent mixing or continuous rocking of the tubes may help with larger pieces of tissues. Very large specimens may require a longer incubation (up to 48 hours) and additional proteinase K.

3. Just prior to proteinase K inactivation, briefly centrifuge the tubes to ensure that all liquid is at the bottom of the tube. Place tubes in a 95°C heat block for exactly 10 minutes. Heating longer can damage the DNA, and heating less than 8 minutes may not fully inactivate the proteinase K. Additional time is required for volumes greater than 0.5 ml.

   Note: For this incubation, cap locks are usually necessary to prevent caps from popping open. Alternatively, some brands of microfuge tubes (Sarstedt, Costar) can accommodate this high temperature step and do not require cap locks when heat-inactivating volumes less than 0.6 ml.

4. Just prior to amplification, pellet any remaining debris by microcentrifugation for approximately 10 seconds. Amplify 1 μl and 10 μl of the crude DNA preparation.

5. It is recommended that the samples be amplified promptly after digestion. However, the digested tissue can be stored for 1 month at –20°C or below. Some targets, such as multiple-copy genomic sequences, are amplifiable after several months.

Alternative Methods of Crude DNA Extraction

Recently, several reports have introduced simplified methods for PET sample preparation and subsequent amplification. Although these procedures may be more rapid than those outlined in this chapter, they may not be suitable in all situations. Methods such as sonication (Heller et al. 1991, 1992) and boiling (Kallio et al. 1991) do not extract longer DNA fragments as efficiently as proteinase K digestion (Forsthoefel et al. 1992; C.E. Greer et al., unpubl.) and may not liberate adequate copies of target DNA. This can be particularly problematic if the desired target is present in low copy numbers or if the concentration of background DNA is particularly high. Furthermore, these methods may promote the preferential amplification of small DNA fragments. These issues must be addressed and determined experimentally with pilot studies using several samples of the PET collection. Ultimately, the decision must be based on the requirements of an individual study.

Preparation from Cytology Slides

Although amplification of DNA extracted from cells fixed on microscope slides has not been widely investigated, some researchers have utilized these materials (Smits et al. 1992; L. Villa, pers. comm.).

1. Remove the coverslip by immersion in xylene for up to 2 days.

2. Remove H & E stains by successive incubations in ethanol solutions with increasing concentrations of water. Complete removal of the stains is required for successful amplification.

3. Remove the cells from the slide by the addition of a proteinase K digestion buffer and carefully scrape the cells into a 1.5-ml microfuge tube.

4. Digest the samples at 50°C for 60 hours.

After digestion, the proteinase K is heat-inactivated and samples are amplified directly. In some cases, further DNA purification (phenol/chloroform extraction, ethanol precipitation) may be required.

GENERAL CONSIDERATIONS

- Be conscious of cross-contamination and use techniques that reduce its occurrence.

- Never work directly over open vials.

- Be extremely careful opening any tubes.

- Always centrifuge tubes before opening.

- Use gauze pads to cover the lid when opening tubes to prevent aerosols.
- Gently pry or pull open tubes; never "pop" or "flip" caps open with your thumb.
- Do not create aerosols, particularly when pipetting solvents.
- Wear clean, dedicated lab coats (Kitcnin et al. 1990).
- Wear clean undergarments (Kitcnin et al. 1990).
- Wear gloves and change them frequently.

Appropriate Controls for Sample Processing and Amplification

Because of the numerous manipulations in this protocol, negative controls must be included to monitor and identify sample-to-sample contamination. With microbial DNA detection, a negative control specimen should be a PET that does not contain the sequences being amplified, but that does contain internal control sequences (such as β-globin). For example, we use BNF-fixed, paraffin-embedded appendix tissue as a negative control for our human papillomavirus studies. A negative control PET section should be the first sample processed and then interspersed (after every tenth sample) starting with the microtome sectioning. Controls must be carried through all phases of sample processing and amplification. Positive controls are also useful, but should only be included as the final sample, to reduce the possibility of contaminating other samples. These PET controls are included in addition to amplification controls such as no addition, and positive and negative purified DNA samples. (We and other investigators have suggested that such controls should be required by journal editors for papers submitted for publication.)

Recommendations for the Assessment of Sample DNA

As noted above, a preliminary study to assess the quality of the specimen DNA should be made when using archival materials. In addition, the use of a primer pair as an internal or coamplification control is critical when testing for infectious agents. Without this marker of DNA sufficiency (product size and quantity), the lack of detection of the infectious agent in a particular specimen is not informative. Therefore, we recommend the use of a single primer pair (of the appropriate size) as a coamplification control for routine sample analysis. As mentioned previously, these primers should amplify a fragment of greater length than the PCR fragment of interest.

Table 2 Sequences of Human β-Globin Amplification Primers

Primers	
PC03	5′ ACACAACTGTGTTCACTAGC 3′
PC04	CAACTTCATCCACGTTCACC
GH20	GAAGAGCCAAGGACAGGTAC
KM29	GGTTGGCCAATCTACTCCCAGG
RS40	ATTTTCCCACCCTTAGGCTG
RS42	GCTCACTCAGTGTGGCAAAG
RS80	TGGTAGCTGGATTGTAGCTG

Primer pairs	*Predicted product size*
PC03/PC04	110 bp
GH20/PC04	268 bp
RS42/KM29	536 bp
RS80/RS40	989 bp
KM29/RS80	1327 bp

The predicted DNA fragment sizes for each of the five primer pairs are listed (Greer et al. 1991b).

Listed in Table 2 are five oligonucleotide primer pairs (Greer et al. 1991b) located in the human β-globin gene that can be used to assess sample DNA sufficiency to produce amplification fragments of 110 bp, 268 bp, 536 bp, 989 bp, and 1327 bp. Amplification parameters are as follows:

1. Aliquots (1 and 10 μl) of prepared samples are amplified separately with each of the four primer pairs. Each 100-μl reaction should contain 1 μl or 10 μl of sample DNA; 100 nM of each primer; 200 μM of each dNTP (dATP, dGTP, dCTP, dTTP); 2.5 units of *Taq* DNA polymerase (AmpliTaq, Perkin-Elmer); 50 mM KCl, 4 mM $MgCl_2$, and 10 mM Tris-HCl (pH 8.5).

2. A 100-μl mineral oil overlay is added to prevent evaporation during thermal cycling (not required when using a Perkin-Elmer TC9600).

3. Cycling parameters are 1 minute at 95°C, 1 minute at 55°C, and 2 minutes at 72°C for 40 cycles (or the number of cycles you plan to use in your assay), followed by an additional 5 minutes at 72°C (DNA Thermal Cycler, Perkin-Elmer).

Aliquots of amplified DNA can be resolved on either 7% (w/v) polyacrylamide gels or 1% agarose gels, stained with ethidium bromide, and photographed under UV light. The resulting profiles give an indication of how efficiently increasing sizes of DNA fragments can be amplified. A control DNA fragment at least 100 bp longer than the fragment of interest should be readily amplified. If it is not, then either a scheme generating a smaller PCR product must be sought, or alternative specimens must be found for study.

TROUBLESHOOTING

No amplification? There can be many explanations:

- *Too much or too little crude DNA in amplification.* Amplify a range of sample volumes, usually 1 μl to 20 μl. If the digestion is too concentrated, a dilution of the digested sample may be required. If dilution is frequently required, an increase in the digestion volume is indicated. Similarly, if volumes greater than 20 μl are routinely required, a reduction in digestion volume is indicated. In some cases, use of higher volumes may inhibit PCR (due to inhibitors); then the DNA must be concentrated by further purification.

- *DNA may be too degraded.* This may be the result of several factors, including the fixation process and the age of the PET sample. To determine the maximum "amplifiable" fragment length, follow the recommendations above. In addition, determining the average size of the sample DNA directly, by agarose gel electrophoresis, is also informative.

- *Incomplete heat-inactivation of proteinase K may result in digestion of the* Taq *polymerase.* Repeat the heat-inactivation. Spin tubes prior to opening, making sure all liquid is in the bottom of the tube. Check the temperature of the heat source to be sure the sample temperature reaches 95°C. Large volumes of digestion buffer (> 0.5 ml) will require more time to reach 95°C.

- *Cycling intervals are insufficient.* The cycling parameters for PETs may require modification to accommodate the fragmented genomic DNA, particularly when amplifying DNA fragments greater than 400 bp. An increase in time from 1 minute (or less) to 2 minutes may be useful during the 72°C extension. In addition, an increase in the number of cycles (e.g., from 30 to 40) is also recommended to accommodate inefficient amplification during early cycles.

- *Too much nonspecific priming.* A hot start may remedy this (Chou et al. 1992). Either the use of AmpliWax PCR Gems (Perkin-Elmer) or the delayed addition of *Taq* may greatly reduce nonspecific bands while increasing the amount of specific product generated. In some cases, this can also be very helpful in increasing sensitivity.

- *PCR inhibitors remain in sample.* We have observed that further DNA purification is helpful in some cases. Proteinase K-digested material can subsequently be subjected to phenol/chloroform extraction and ethanol precipitation of the DNA. It is possible that the addition of this step may also function to liberate the DNA better from the highly cross-linked protein matrix obtained after fixation.

Complete DNA purification should only be implemented when absolutely necessary, because these additional manipulations provide additional opportunities for contamination. (Some laboratories doing genetic studies routinely prepare purified DNA from PETs to obtain sufficient material for multiple PCR analyses. In these cases, contamination is much less of an issue than in infectious disease work.)

DISCUSSION

The use of paraffin-embedded tissues in PCR-based studies has resulted in many exciting new insights in the areas of cancer research, genetic and infectious disease, and molecular epidemiology. However, this tool has some limitations due to the intrinsic properties of PETs. As discussed in this paper, tissue fixation and the age of embedded tissue are important factors affecting the size of target DNA that can successfully be amplified. Although each paraffin-embedded tissue has individual characteristics, a general rating can be made regarding the quality of DNA derived from PETs from a particular time period and institution. We have found considerable variability in the quality of DNA derived from different institutions using BNF. This variability may be due to modification of fixation procedures and/or the quality or the age of the chemicals used in the fixation process. We stress that pilot studies are essential.

Many researchers derive DNA sequence information from PCR products derived from PETs. Extensive studies have not addressed the accuracy of sequence information from PETs. Fortunately, most studies use sufficient amounts of input DNA such that artifacts are unlikely to affect results. However, in experiments where minute amounts of target are available for PCR, some concerns about the effects of PET DNA damage are warranted. We have shown that many fixatives, particularly those containing acid, cause a significant decrease in the length of genomic DNA that can be amplified. Acids may hydrolyze glycosidic bonds, thus generating abasic sites in DNA. Randall and colleagues (1987) extensively studied the kinetics of nucleotide insertion opposite abasic sites in DNA using *Drosophila* DNA polymerase-α and found that the specificity of nucleotide insertion was 6–11 times greater for A over G and 20–50 times greater for A over C and T. If *Taq* polymerase has its own preferences, this has implications for the analysis of point mutations from low-copy-number targets. Furthermore, observations made from studying ancient or highly degraded DNA demonstrated that *Taq* polymerase can "jump" to another template during PCR when the polymerase encounters strand scission or abasic sites (Pääbo et al. 1990). Such jumping can generate artifactual hybrids, for example, between alleles or microbial genomes, when amplifying from a few copies of input target.

DNA amplification methods have allowed archival PETs to become routinely useful clinical investigative material for molecular genetic studies. Because retrospective studies can often provide the most cost-effective and expedient approaches to epidemiologic questions, PCR-based analyses of PETs have contributed to our recent progress in many areas of biomedical research.

ACKNOWLEDGMENTS

We thank Luisa Villa, Ken Kinzler, and David Sidransky for helpful discussions, and Kent Thudium and Matthew McClelland for their comments on the manuscript.

REFERENCES

Banerjee, S.K., W.F. Makdisi, A.P. Weston, S.M. Mitchell, and D.R. Campbell. 1995. Microwave-based DNA extraction from paraffin-embedded tissue for PCR amplification. *BioTechniques* **18:** 768–773.

Ben-Ezra, J., D.A. Johnson, J. Rossi, N. Cook, and A. Wu. 1991. Effect of fixation on the amplification of nucleic acids from paraffin-embedded material by the polymerase chain reaction. *J. Histochem. Cytochem.* **39:** 351–354.

Brandsma, J., A.J. Lewis, A.L. Abramson, and M.M. Manos. 1990. Detection and typing of papillomavirus DNA in formalin-fixed paraffin-embedded tissue. *Arch. Otolaryngol.* **116:** 844–848.

Burmer, G.C., P.S. Rabinovitch, and L.A. Loeb. 1989. Analysis of c-Ki-*ras* mutations in human colon carcinoma by cell sorting, polymerase chain reaction, and DNA sequencing. *Cancer Res.* **49:** 2141–2146.

Cao, M., X. Xiao, B. Egbert, T.M. Darragh, and T.S.B. Yen. 1989. Rapid detection of cutaneous herpes simplex virus infection with the polymerase chain reaction. *J. Invest. Dermatol.* **82:** 391–392.

Chou, Q., M. Russell, D.E. Birch, J. Raymond, and W. Bloch. 1992. Prevention of pre-PCR mis-priming and primer dimerization improves low-copy-number amplifications. *Nucleic Acids Res.* **20:** 1717–1723.

Crisan, D., E.M. Cadoff, J.C. Mattson, and K.A. Hartle. 1990. Polymerase chain reaction: Amplification of DNA from fixed tissue. *Clin. Biochem.* **23:** 489–495.

Dychdala, G.R. 1977. Chlorine and chlorine compounds. In *Disinfection, sterilization, and preservation* (ed. S. Stanton), pp. 157–182. Lea and Febiger, Philadelphia.

Forsthoefel, K.F., A.C. Papp, P.J. Snyder, and T.W. Prior. 1992. Optimization of DNA extraction from formalin-fixed tissue and its clinical application in Duchenne muscular dystrophy. *Am. J. Clin. Pathol.* **98:** 98–104.

Greer, C.E., J.K. Lund, and M.M. Manos. 1991a. PCR amplification from paraffin-embedded tissues: Recommendations on fixatives for long-term storage and prospective studies. *PCR Methods Appl.* **1:** 46–50.

Greer, C.E., S.L. Peterson, N.B. Kiviat, and M.M. Manos. 1991b. PCR amplification from paraffin-embedded tissues: Effects of fixative and fixation time. *Am. J. Clin. Pathol.* **95:** 117–124.

Heller, M.J., R.A. Robinson, L.J. Burgart, C.J. TenEyck, and W.W. Wilke. 1992. DNA extraction by sonication: A comparison of fresh, frozen, and paraffin-embedded tissues for use in polymerase chain reaction assays. *Mod. Pathol.* **5:** 203–206.

Heller, M.J., L.J. Burgart, C.J. TenEyck, M.E. Anderson, T.C. Greiner, and R.A. Robinson. 1991. An efficient method for the extraction of DNA from formalin-fixed, paraffin-embedded tissue by sonication. *BioTechniques* **11:** 372–377.

Hoffman, P.N., J.E. Death, and D. Coates. 1981. The

stability of sodium hypochlorite solutions. In *Disinfectants: Their use and evaluation of effectiveness* (ed. C.H. Collins), pp. 77–83. Academic Press, London.

Kallio, P., S. Syrjanen, A. Tervahauta, and K. Syrjanen. 1991. A simple method for isolation of DNA from formalin-fixed paraffin-embedded samples for PCR. *J. Virol. Methods* **35:** 39–47.

Kitcnin, P.A., Z. Szotyori, C. Fromholc, and N. Almond. 1990. Avoidance of false positives. *Nature* **344:** 201.

Lyons, J., C.A. Landis, G. Harsh, L. Vallar, K. Grunewald, H. Feichtinger, Q.-Y. Duh, O.H. Clark, E. Kawasaki, H. Bourne, and F. McCormick. 1990. Two G protein oncogenes in human endocrine tumors. *Science* **249:** 655–659.

Pääbo, S., D.M. Irwin, and A.C. Wilson. 1990. DNA damage promotes jumping between templates during enzymatic amplification. *J. Biol. Chem.* **265:** 4718–4721.

Randall, S.K., R. Eritja, B.E. Kaplan, J. Petruska, and M.F. Goodman. 1987. Nucleotide insertion kinetics opposite abasic lesions in DNA. *J. Biol. Chem.* **262:** 6864–6870.

Shibata, D.K., N. Arnheim, and W.J. Martin. 1988. Detection of human papillomavirus in paraffin-embedded tissue using the polymerase chain reaction. *J. Exp. Med.* **167:** 225–230.

Smits, H.L., L.M. Tieben, S.P. Tjong-A-Hung, M.F. Jebbink, R.P. Minnaar, C.L. Jansen, and J.T. Schegget. 1992. Detection and typing of human papillomaviruses present in fixed and stained archival cervical smears by a consensus polymerase chain reaction and direct sequence analysis allow the identification of a broad spectrum of human papillomavirus types. *J. Gen. Virol.* **73:** 3263–3268.

Thibodeau, S.N., G. Bren, and D. Schaid. 1993. Microsatellite instability in cancer of the proximal colon. *Science* **260:** 816–819.

van der Riet, P., D. Karp, E. Farmer, Q. Wei, L. Grossman, K. Tokino, J.M. Ruppert, and D. Sidransky. 1994. Progression of basal cell carcinoma through loss of chromosome 9q and inactivation of a single p53 allele. *Cancer Research* **54:** 25–27.

Wright, D.K. and M.M. Manos. 1990. Sample preparation from paraffin-embedded tissues. In *PCR protocols: A guide to methods and applications* (ed. M.A. Innis et al.), pp. 153–158. Academic Press, Berkeley, California.

RNA Purification

Jeffrey Adamovicz and William C. Gause

Department of Microbiology and Immunology, Uniformed Services University of the Health Sciences, Bethesda, Maryland 20814

INTRODUCTION

RNA can be isolated from eukaryotic cells by many methods. The cell type and the method of isolation influence the likelihood of successful isolation of undegraded RNA. Isolation of total RNA from eukaryotic cells yields several different RNA species. Most cellular RNA species are ribosomal RNA (rRNA). When total RNA is subjected to gel electrophoresis, a characteristic banding pattern representing three prominent rRNA molecules (28S, 18S, and 5S) is observed. These three bands are the major components of the 60S and 40S ribosomal subunits and make up about 80% of the cell's total RNA. The second most abundant class of RNA includes transfer RNA (tRNA) and small nuclear RNA (snRNA). The majority of these RNAs are either homogeneous in size or of a defined sequence and may be isolated by techniques directed at their common physical characteristics (Adams 1992). Together, these species along with rRNA represent 95–99% of the cell's total RNA. The least abundant class of cellular RNA is messenger RNA (mRNA). This chapter describes methods for the isolation of RNA containing mRNA that can be readily reverse-transcribed to complementary DNA (cDNA).

mRNA molecules are heterogeneous in terms of size and sequence, but eukaryotic mRNA molecules possess a variable-length tract of polyadenylic acid residues at their 3′ terminus. This tail simplifies the isolation of mRNA on oligo(dT/dU)-cellulose matrices such as beads or chromatography columns (Berger and Kimmel 1987; Sambrook et al. 1989). Commercial kits such as FastTrack and Micro-FastTrack from Invitrogen are available that are designed to isolate mRNA from large sample sizes or small sample sizes, respectively. The advantage of these kits is that they do not require the previous isolation of total RNA but instead directly isolate mRNA on oligo(dT) pellets, which are then eluted by spin-column chromatography. The yield of mRNA obtained from these kits is reported to be approximately twice that of CsCl gradient centrifugation. Two additional, less costly protocols for the isolation of mRNA are included below. A disadvantage of poly(A)$^+$-selected mRNA is that purity usually varies between samples and, consequently, so does the ratio of mRNA to total RNA, thereby contributing another source of experimental variation. For the reverse transcriptase (RT)-PCR assay, mRNA is usually at a sufficient concentration in total RNA to be readily amplified. Therefore, the analysis of mRNA is usually accomplished by the isolation of total cellular RNA; the direct isolation of mRNA should only be attempted if the isolation of total RNA is not appropriate.

The protocols for the isolation of total RNA presented in this chapter are based on the separation of RNA from other cellular macromolecules through differences in solubility or sedimentation. The essential steps for RNA isolation are cell disruption with or without subcellular fractionation, RNase inactivation and deproteinization of the RNA, and physical separation of the RNA from other macromolecules. The physical separation of RNA involves extracting the nucleic acid by one of three primary methods: (1) phenol extraction with or without chloroform and/or a denaturant-like sodium dodecyl sulfate (SDS); (2) extraction with guanidinium salts; and (3) extractions using combinations of phenol, ionic detergents, and guanidinium salts.

The RNA may be extracted from whole tissues, cells, or cellular fractions. If the target RNA is scarce or difficult to precipitate, exogenous carrier RNA may be added during the extraction (O'Garra et al. 1992). The RNA commonly used as the carrier RNA is commercially available bacterial or yeast rRNA. Following extraction of RNA into an aqueous phase, it is precipitated with a salt in either ethanol or isopropanol. The salt makes the nucleic acid insoluble, and its composition is dependent on how the RNA is to be used (Wallace 1987).

After the RNA has been precipitated, it must be solubilized and quantified. The RNA can be quantified spectrophotometrically by measuring the A_{260} of an aliquot. The purity of the RNA sample is assessed by measuring the $A_{260/280}$ ratio. The 260/280 ratios should be 1.8 or above. It actually may be more important to ensure that the samples have similar ratios as opposed to a specific ratio (Yamaguchi et al. 1992). If the ratio is low (below 1.6), it is most likely due to contamination with protein, and the sample should be reextracted.

If degradation does occur, it is not detectable with spectrophotometry. A simple method to detect RNA degradation is to load 5–10 μg of RNA on a formaldehyde agarose gel and examine the appearance of the 28S/18S rRNA bands (Tsang et al. 1993). If these bands are extensively degraded, they appear as indistinct smears or are absent in part or completely. If degradation is detected, it is most likely caused by an endogenous RNase, but it may also be caused by an exogenous RNase.

Prior to the isolation of RNA, several measures should be taken in the laboratory to eliminate exogenous RNases. The use of sterile, disposable plasticware for all reagents greatly reduces the risk of degradation because it is essentially free of RNase. If glassware is used, it should be baked at 200°C for 4 hours. All solutions should be prepared with RNase-free water. The RNase in the water may be inactivated with diethylpyrocarbonate (DEPC) (Sambrook et al. 1989). RNases may also be avoided by obtaining high-quality distilled and deionized water. The most abundant source of exogenous RNase contamination is from the hands, skin, hair, or contaminated lab surfaces, and sample contamination can be controlled by wearing gloves and using laboratory paper to cover contaminated surfaces.

The most serious obstacle to isolating mRNA is the presence of endogenous RNases. These enzymes rapidly degrade RNA, and they are very abundant in certain tissues like the pancreas. Endogenous RNase activity is greatly reduced by the use of sufficient quantities of guanidinium salts, phenol, SDS, or other strong denaturants. These denaturants should be rapidly mixed with the cells and tissues using a powerful homogenizer such as the Polytron (Brinkmann). Although the RNA is relatively stable during extraction, it may be exposed to exogenous or endogenous RNase during subsequent isolation steps. The actions of RNase may be prevented by adding vanadyl-ribonucleoside complex after eluting the RNA to a final concentration of 10 mM. This RNase inhibitor interferes with in vitro translation assays and must be phenol-extracted prior to using RNA for this application. An alternative RNase inhibitor is derived from human placenta and marketed under numerous names. An example is RNAsin from Promega, which may be added at a final concentration of 500 units/ml to purified RNA. The use of SDS during the extraction can serve to inhibit RNase, but a LiCl precipitation step to recover RNA without precipitating the SDS must be included. Other inhibitors of RNase include reducing agents, such as β-mercaptoethanol. The addition of an RNase inhibitor should be considered when unexplained degradation occurs. The use of RNase inhibitors can also extend the length of time RNA can be stored.

Total cellular RNA can be isolated by several methods (Berger and Kimmel 1987; Sambrook et al. 1989). One of the most effective methods uses guanidinium salts; this was first described by Cox (1968) and later improved (Chirgwin et al. 1979; Favaloro et al. 1980). Guanidinium salts disrupt cells and denature protein, allowing RNA to remain soluble while simultaneously inhibiting RNase. A drawback of this technique is that it is not possible to obtain subcellular fractions because subcellular components are completely disrupted. The technique has been modified by the addition of a phenol-chloroform extraction step that allows the separation of RNA from DNA (Chomczynski and Sacchi 1987). At the acid pH in this protocol, the solubility of DNA is decreased and most of it remains in the organic phase, whereas the RNA partitions to the aqueous phase.

A modified protocol for the single-step isolation of total RNA with acid guanidinium-thiocyanate-phenol-chloroform is described below and has been used with considerable success in our laboratory. This method for the isolation of total RNA from eukaryotic cells is a rapid and effective means of isolating large amounts of RNA while reducing the risk of degradation by RNase. It is useful when a large number of samples must be processed simultaneously. The basic method has been modified by many investigators for specific applications (Puissant and Houdebine 1990; McCaustland et al. 1991; Murphy et al. 1993; Yang and Xu 1993; Mies 1994; Santos and Gouvea 1994).

If a higher yield of relatively pure and undegraded mRNA is required, the acid guanidinium technique can be combined with cesium chloride density gradient centrifugation (Aviv and Leder 1972; Chirgwin et al. 1979). This method is used to separate RNA from other macromolecules by differences in sedimentation. The second protocol describes modifications for the isolation of mRNA by this method. The disadvantage of this method is that it requires more time and can only be used for processing a few samples simultaneously. For most purposes, this labor-intensive approach is probably unnecessary, but it may be required for the isolation of RNA from "difficult" sources such as plant tissue. This method may be modified for the preparation of cellular fractions when the location of a particular RNA is known and an enrichment of the species is required. A case in which this procedure would be useful is the isolation of unprocessed mRNA from the nuclear fraction of mammalian cells (Berger and Kimmel 1987; Sambrook et al. 1989) or the isolation of cytoplasmic RNA without the contamination of nuclear RNA (Wilkinson 1988b). The separation of the nuclear fraction from cytoplasmic fractions can also be important if DNA contamination of the cytoplasmic fractions would interfere with analysis of cytoplasmic RNA. A less time-consuming method for the separation and simultaneous extraction of nuclear and cytoplasmic RNA has been described by Wilkinson (1988a).

A third protocol for isolating RNA is SDS-phenol extraction, which has been used for the isolation of RNA from virally infected or transfected mammalian cells (Stallcup and Washington 1983; Sambrook et al. 1989; Chattopadhhyay et al. 1993). This technique is also useful when isolating mRNA from cytoplasmic fractions of cells previously transfected with plasmid DNA (Liu et al. 1994). Additionally, the method is advantageous for use with subcellular fractions and produces high RNA yields. It is also advantageous because SDS is an effective inhibitor of RNase. A disadvantage of this method is that it usually yields RNA with low 260/280 absorbance ratios (Liu et al. 1994), due to protein contamination through incomplete deproteinization of RNA by SDS. An improved protocol that includes chloroform extraction is offered below.

In summary, the source and relative abundance of mRNA as well as the presence of RNase must be considered during the selection of an isolation method. The techniques discussed use solubility, buoyancy, or structural differences to separate RNA from other cellular macromolecules. Generally, we have found that the acid guanidinium-thiocyanate-phenol-chloroform technique is the most practical for the preparation of RNA to be reverse-transcribed to cDNA for PCR. The most important factor in isolating undegraded RNA is protection from RNase. After determining the method to reduce RNase activity, the isolation of cellular RNA should be a simple, quick task tailored to individual experimental requirements.

REAGENTS

RNAzol B or equivalent
ACS (American Chemical Society) grade chloroform
ACS grade isopropanol
ACS grade ethanol, 75%
ACS grade isoamyl alcohol

PROTOCOL

The Acid Guanidinium-Thiocyanate-Phenol-Chloroform Method

This is a modification of the acid guanidinium-thiocyanate-phenol-chloroform (RNAzol B) method from Tel-Test Inc. for the isolation of total RNA from whole spleen or cultured splenic cells (Chomczynski 1991; Svetić et al. 1991; Gause and Adamovicz 1994).

1. Sample preparation

   a. Determine the approximate weight of any tissue samples to be studied.

   b. Distribute RNAzol to polypropylene tubes based on 2 ml RNAzol per 100 mg of tissue or 0.2 ml per 1 x 10^6 cultured cells. The RNAzol should be kept cold and protected from light; *take care when handling the RNAzol because it is highly caustic.*

   c. Place the tissue sample in RNAzol; homogenize it thoroughly.

   d. Snap-freeze the sample in liquid nitrogen and store at −70°C for future extraction.

   Note: The homogenizer head should be washed sequentially between samples with distilled water, 75% ethanol, and RNAzol.

2. RNA extraction

   a. Thaw frozen homogenate for approximately 5 minutes in a 37°C water bath.

   b. Add 0.2 ml of a 24:1 mixture of chloroform/isoamyl alcohol for every 2 ml of homogenate and shake the samples vigorously for 15 seconds. The isoamyl alcohol serves as an antifoaming agent.

   c. Let the samples sit on ice for 5 minutes.

   d. Centrifuge the samples at 12,000*g* for 15 minutes in a refrigerated centrifuge (4°C).

   e. Carefully remove the aqueous (top) phase that contains the RNA and transfer it to another tube, which is then stored on ice.

   Note: Take care to avoid removing the white interface layer because this contains protein which can hamper your ability to quantify the RNA.

3. RNA precipitation

   a. To each sample, add a volume of cold isopropanol that is equal to the volume of the aqueous phase.

   b. Mix the tubes gently and store the samples for 15 minutes on ice.

   c. Centrifuge the samples at 12,000*g* for 15 minutes at 4°C. The RNA should form a whitish/yellow pellet at the bottom of the tube.

4. RNA washing

   a. Carefully decant the isopropanol.

   b. Wash the RNA pellet by adding one volume of cold 75% ethanol and resuspend the RNA pellet by shaking or pipetting.

   c. Centrifuge the sample at 12,000*g* for 8 minutes at 4°C.

   d. Carefully decant the ethanol and dry the pellets by air or under vacuum. Vacuum drying is more expedient, but take care not to overdry the pellet because this makes it harder to dissolve the RNA later.

5. RNA storage

   a. Dissolve the RNA pellets in 20–50 μl of distilled, deionized water. If the quality of the water is in question, or if it is known to contain divalent cations, then add 10 mM EDTA to the RNA. The EDTA scavenges cations, particularly Mg^{++}, which is known to catalyze random breaks in RNA.

b. Freeze and then thaw the samples once to increase solubility.

c. Store the samples for up to 1 year at –70°C. Place the samples in formamide at –20°C if long-term storage is required (Chomczynski 1992).

ANALYSIS AND TROUBLESHOOTING

One potential problem with this protocol is RNA cross-contamination from the homogenizer head if it is not properly cleaned between samples. The head should be cleaned as outlined above. Contamination of individual RNA samples with DNA can also be problematic. DNA contamination can be reduced by the addition of an RNA-selective preparatory ethanol precipitation step (Siebert and Chenchik 1993). If the RNA preparation is to be used in an RT-PCR assay, the importance of DNA contamination is reduced through the correct design of PCR primers that span introns. DNA contamination could be critical in experiments where the genomic structure of the gene of interest is not known or the gene lacks introns. In these situations, it is advisable to perform the ethanol RNA precipitation step described above or to treat the sample with RNase-free DNase.

Another situation that requires a modification of the basic protocol is the extraction of RNA from tissues containing a high amount of lipid, like adipose tissue. Two preparatory chloroform extraction steps can be added to remove the contaminating lipid (Louveau et al. 1991). Other published modifications of the basic protocol of Chomczynski and Sacchi (1987) include the use of LiCl, which is highly selective for the precipitation of RNA and is extremely useful for small numbers of cells (Meier 1988) or for large numbers of samples with low amounts of RNA (Birnboim 1988; Vauti and Siess 1993; Walther et al. 1993). Although these methods yield more RNA, they are incompatible for use with reverse transcription unless the lithium is subsequently removed by an ethanol wash step (Wallace 1987). The method described in the outlined protocol is applicable for most lymphoid tissues or cell cultures. The yield from 100 mg of murine spleen is approximately 200 μg of total RNA. Total RNA collected by this method is suitable for use in a variety of assays, including RT-PCR.

REAGENTS

Homogenization buffer (5 M guanidine isothiocyanate, 0.2 M Tris-acetate [pH 8.5], 0.62% sodium lauroyl sarcosine)
Molecular biology reagent grade β-mercaptoethanol
Freshly prepared 20% polyvinyl pyrrolidone
5.7 M Cesium chloride
ACS grade chloroform
ACS grade isoamyl alcohol
ACS grade ethanol 75%, 100%
1% SDS

2 M Sodium acetate (pH 4.2)
Sterile cheesecloth/nylon mesh
TE-saturated phenol (pH 7.6) with 10 mM Tris-HCl (pH 8.0) and 1 mM sodium EDTA (pH 8.0), with 0.04% 8-hydroxyquinolone

PROTOCOL

The Guanidine Isothiocyanate Cesium Chloride Density Gradient Separation Method for Preparation of RNA from Plant Tissue

Isolation of mammalian RNA by guanidinium thiocyanate cesium chloride is adequately described in Sambrook et al. (1989) and no significant changes to this method are needed. Published protocols modify this method for the isolation of RNA from difficult tissues, including plants with high levels of phenolic terpenoids, tannins, and other unidentified compounds from species like the cotton plant and pine tree. The protocol described below is adapted from previously published protocols of Baker et al. (1990) and John (1992) and includes the use of polyvinyl pyrrolidone (PVP). This compound complexes with polyphenolics and makes them insoluble.

1. Tissue preparation
   a. Snap-freeze the plant tissue (0.5–2 g) to be studied in liquid nitrogen and powder in a mortar.
   b. Add 0.7% β-mercaptoethanol and 1% PVP to the buffer just prior to adding the tissue sample.
   c. Transfer the powder to a silanized tube containing four volumes of buffer (w/v).
   d. Homogenize the mixture in a Polytron or equivalent machine for 1.5 minutes. If DNA in the preparation makes it viscous, the DNA should be sheared by drawing the homogenate through a 20-gauge needle.
   e. Filter the homogenate through cheesecloth, then place onto a 5.7 M cesium chloride layer in an appropriate ultracentrifuge tube, and centrifuge as described previously (Sambrook et al. 1989).

2. RNA extraction
   a. After completion of the centrifugation, carefully decant the CsCl pad and then cut off the top two-thirds of the gradient tube. The RNA pellet should remain attached to the tube's bottom.
   b. Dislodge the RNA pellet, place in a sterile 1.5-ml microfuge tube, and wash in 1 ml of cold 75% ethanol to remove any contaminating CsCl.

c. Centrifuge the samples at 12,000*g* for 15 minutes at 4°C.

d. Decant the ethanol and resuspend the RNA in 450 μl of TE-phenol. Then add 100 μl of SDS and 450 μl of a 24:1 mixture of chloroform/isoamyl alcohol and shake the samples vigorously for 15 seconds. Let the samples sit on ice for 5 minutes and ensure that the pellet is completely solubilized or the yield will be adversely affected.

e. Centrifuge the samples at 12,000*g* for 15 minutes at 4°C.

f. Carefully remove the aqueous (top) phase that contains the RNA and transfer it to another tube, which is then stored on ice.

3. RNA precipitation

a. To each sample, add 1/10 volume of cold (4°C) ammonium acetate followed by two volumes of cold (–20°C) absolute ethanol.

b. Mix the tubes gently and store the samples overnight at –20°C.

c. The following morning, centrifuge the samples at 12,000*g* for 10 minutes at 4°C. The RNA should form a whitish/yellow pellet at the bottom of the tube.

4. RNA washing

a. Carefully decant the ethanol.

b. Wash the RNA pellet by adding one volume of cold 75% ethanol, and resuspend the RNA pellet by shaking or pipetting.

c. Centrifuge the sample at 12,000*g* for 8 minutes at 4°C.

d. Carefully decant the ethanol and dry the pellets by air or under vacuum. Vacuum drying is more expedient, but take care not to overdry the pellet because this makes it harder to dissolve the RNA later.

5. RNA storage

a. Dissolve the RNA pellets in 20–50 μl of distilled, deionized water. If the quality of the water is in question, or if it is known to contain divalent cations, then add 10 mM EDTA to the RNA solution. The EDTA scavenges cations, particularly Mg^{++}, which is known to catalyze random breaks in RNA.

b. Freeze and then thaw the samples once to increase solubility.

c. Store the samples for up to 1 year at –70°C. Place the samples in formamide at –20°C if long-term storage is required (Chomczynski 1992).

ANALYSIS AND TROUBLESHOOTING

The mRNA isolated by this method is suitable for use in numerous assays, including RT-PCR. A microprocedure adapted for RT-PCR with low numbers of cultured cells is described in Rappolee et al. (1989). Another adaptation to remove proteoglycans, which can interfere with isopycnic gradient separations, involves a 0°C extraction step and is described in Groppe and Morse (1993). The substitution of cesium trifluroacetate for cesium chloride improves the yield and quality of RNA isolated from cartilage (Smale and Sasse 1992) and may be a good alternative for other "difficult" tissues.

REAGENTS

1% SDS
Water-saturated phenol (pH 7) with 0.04% 8-hydroxyquinolone
2 M Sodium acetate (pH 4.2)
ACS grade chloroform
ACS grade isoamyl alcohol
5 M Sodium chloride
ACS grade ethanol 75%, 100%

PROTOCOL

The SDS-Phenol-Chloroform Method

The SDS phenol-chloroform extraction method for use with mammalian cells or tissue is described here. This is a modified technique derived from previous protocols (Chattopadhhyay et al. 1993; Liu et al. 1994).

1. Sample preparation

   a. Determine the approximate weight of any tissue samples to be studied.

   b. Distribute a mixture of SDS/phenol (1:2, v/v) to polypropylene tubes based on 0.5 ml SDS and 1.0 ml phenol per 100 mg of tissue or 0.2 ml SDS and 0.4 ml phenol per 1 × 10^6 cultured cells. The SDS should be stored at room temperature and the phenol at 4°C and protected from light; *take care when handling the phenol because it is highly caustic.*

   c. Place the tissue sample in the tubes and homogenize for a minimum of 30 seconds.

   Note: Between samples, clean the homogenizer head sequentially with distilled water, 75% ethanol, and SDS/phenol (1:2).

2. RNA extraction

   a. To the lysate, add 1/10 volume of 4°C sodium acetate and vortex the samples vigorously for 15 seconds.

b. Centrifuge the samples at 12,000*g* for 10 minutes at 4°C.

c. Carefully remove the aqueous (top) phase that contains the RNA and transfer it to another tube, which is then stored on ice.

Note: Take care to avoid removing the white interface layer because this contains protein, which can hamper your ability to quantify the RNA.

d. To the aqueous phase, add one volume of a mixture containing five parts phenol and one part chloroform/isoamyl alcohol (24:1). Vortex this solution for 15 seconds and then repeat the centrifugation step.

e. Remove the aqueous phase and place it in another tube.

3. RNA precipitation

a. To each sample, add 1/25 volume of cold (4°C) 5 M NaCl followed by two volumes of cold (–20°C) absolute ethanol.

b. Mix the tubes gently and store the samples overnight at –20°C.

c. The following morning, centrifuge the samples at 12,000*g* for 10 minutes at 4°C. The RNA should form a whitish/yellow pellet at the bottom of the tube.

4. RNA washing

a. Carefully decant the ethanol.

b. Wash the RNA pellet by adding one volume of cold 75% ethanol, and resuspend the RNA pellet by shaking or pipetting.

c. Centrifuge the sample at 12,000*g* for 8 minutes at 4°C.

d. Carefully decant the ethanol and dry the pellets by air or under vacuum. Vacuum drying is more expedient, but take care not to overdry the pellet because this makes it harder to dissolve the RNA later.

5. RNA storage

a. Dissolve the RNA pellets in 20–50 µl of distilled, deionized water. If the quality of the water is in question, or if it is known to contain divalent cations, then add 10 mM EDTA to the RNA solution. The EDTA scavenges cations, particularly Mg^{++}, which is known to catalyze random breaks in RNA.

b. Freeze and then thaw the samples once to increase solubility.

c. Store the samples for up to 1 year at –70°C. Place the samples in formamide at –20°C if long-term storage is required (Chomczynski 1992).

ANALYSIS AND TROUBLESHOOTING

The extraction of RNA from certain tissue samples may require the addition of proteinase K. Proteinase K helps degrade RNase and other proteins and results in a higher RNA yield. If proteinase K is to be included, it should be added to the SDS in step 1 in the following proportions: 0.5 ml of 1% SDS with 10 mM EDTA (pH 8.0) and 1 ml of phenol with 10 μl of proteinase K (10 mg/ml stored at –20°C). Following homogenization, the proteinase K digestion should be completed in a 45°C water bath for 1 hour. The higher temperature is required to minimize nuclease activity during the digestion. Complete the remainder of the protocol as outlined.

A major contaminant that may present with this method is polysaccharide. This contamination should be detectable spectrophotometrically at A_{230}. A modification designed to remove this contamination from cultured plant cells is described in Shirzadegan et al. (1991). Other modifications of this method for use with yeast (Schmitt et al. 1990) or cultured cells have been described in Peppel and Baglioni (1990) and Salvatori et al. (1992). The total RNA yield for this method is 20 μg per 1×10^6 cells, but the yield varies depending on the cell type. The RNA is suitable for use in numerous assays, including RT-PCR.

REAGENTS

Method 1

10 mM Tris-HCl (pH 7.5)
1 mM EDTA
0.1% SDS
5 M Sodium chloride
Oligo(dT)-cellulose
ACS grade absolute alcohol

Method 2

HYBOND messenger affinity paper (Amersham)
RNAzol B or equivalent
0.5 M Sodium chloride
ACS grade ethanol 75%

PROTOCOLS

Two Methods for the Isolation of Poly(A)+ RNA

The first method is a modification of the method of Celano et al. (1993) and can be used with any of the primary RNA isolation methods to enhance the isolation of poly(A)+ mRNA. The second method is adapted from Sheardown (1992) for the in situ isolation of poly(A)+ mRNA from cultured cells or from cell suspensions. It is designed for direct application of RT-PCR to captured poly(A)+ mRNA.

Method 1

1. Sample preparation

   a. Suspend 1 mg of total RNA in 600 µl of cellulose buffer (10 mM Tris-HCl, 1 mM EDTA, and 0.1% SDS), and place this mixture in a 1.5-ml snap top tube.

   b. Heat the tube for 5 minutes in a 65°C water bath and quench the sample on ice for 5 minutes. This step relaxes RNA secondary structure and RNA aggregation.

   c. Add 60 µl of 5 M NaCl to the RNA mixture and then transfer the entire solution to a tube containing hydrated oligo(dT)-cellulose (60 mg/600 µl). Mix the RNA/cellulose thoroughly by inverting the tube several times, and remove air bubbles by flicking the bottom.

   d. Incubate the tube for 10 minutes in a 37°C water bath.

   e. Centrifuge the tube at 12,000*g* for 8 minutes at room temperature.

   f. Use a pipette to remove and discard the supernatant.

   g. Wash the pellet by adding 1 ml of cellulose buffer, mix by inversion, and repeat the centrifugation.

   h. Remove the supernatant and discard as before, then repeat the wash using 1 ml of RNase-free deionized and distilled water.

2. RNA elution

   a. Add 400 µl of RNase-free water to the cellulose pellet and thoroughly mix to resuspend the pellet.

   b. Elute the RNA by heating in a 65°C water bath for 5 minutes.

   c. Pellet the cellulose by centrifuging the tube at 12,000*g* for 8 minutes at room temperature.

   d. Remove the supernatant containing the mRNA to another tube and store on ice for RNA precipitation (below).

   e. Repeat the elution step two more times and collect the supernatant; store on ice.

3. RNA precipitation

   a. Precipitate the RNA by adding 1/25 volume of 5 M NaCl to the supernatant followed by 2 1/2 volumes of absolute alcohol.

   b. Mix this solution thoroughly by inversion and place at –20°C overnight.

c. The following morning, centrifuge the samples at 12,000*g* for 10 minutes at 4°C. The RNA may form a whitish pellet at the bottom of the tube or it may not be visible. If the pellet is not visible, add 200 µl of 75% ethanol and pipette (wash) the sides of the tube thoroughly.

d. Centrifuge the sample at 12,000*g* for 10 minutes at 4°C.

4. RNA washing

a. Carefully decant the alcohol.

b. Wash the RNA pellet by adding 400 µl of cold 75% ethanol and resuspend the RNA pellet by shaking or pipetting.

c. Centrifuge the sample at 12,000*g* for 8 minutes at 4°C.

d. Carefully decant the ethanol and dry the pellets by air or under vacuum. Vacuum drying is more expedient, but take care not to overdry the pellet because this makes it harder to dissolve the RNA later.

5. RNA storage

a. Dissolve the RNA pellets in 20–50 µl of distilled, deionized water. If the quality of the water is in question or if it is known to contain divalent cations, then add 10 mM EDTA to the RNA solution. The EDTA scavenges cations, particularly Mg^{++}, which is known to catalyze random breaks in RNA.

b. Freeze and then thaw the samples once to increase solubility.

c. Store the samples for up to 1 year at –70°C. Place the samples in formamide at –20°C if long-term storage is required (Chomczynski 1992).

Method 2

1. Sample preparation

a. Dilute cell preparations to a concentration of 1000 cells/5 µl.

b. Cut HYBOND paper into 2-mm^2 pieces, soak in 0.5 M NaCl for 30 minutes, and air-dry on paper towels or blotting paper.

c. Add 5 µl of cell suspension on top of a square of HYBOND paper and lyse the cells by adding 5 µl of RNAzol B.

d. Add 20 µl of 0.5 M NaCl in 1-µl drops to bind poly (A)$^+$ mRNA to poly(dU) on the paper.

2. RNA washing

   a. Allow all the liquid to soak through the HYBOND paper to the underlying absorbent paper.

   b. Place the HYBOND paper in a 1.5-ml microcentrifuge tube and add 1 ml of 0.5 M NaCl.

   c. Vortex the paper for 15 seconds and decant the NaCl; then repeat the wash. After decanting the NaCl, wash the paper in two successive changes of 75% ethanol. At this point the paper may be stored in 75% ethanol at –20°C, or the ethanol can be decanted and the paper dried by vacuum centrifugation.

   d. Thoroughly dry the paper, and then place it in a tube for reverse transcription.

ANALYSIS AND TROUBLESHOOTING

Isolation of mRNA by Method 1 should produce an enriched source for a desired mRNA species free of other forms of RNA. The amount of mRNA recovered should equal 1–2.5% of the input RNA. The purity of the mRNA can be confirmed by running 5 μg of RNA on a formaldehyde gel containing ethidium bromide. There should be a lack of ribosomal bands and a smear of mRNA from about the 20-kb marker down, with the greatest concentration from 5 kb to 10 kb. A sixfold enhancement of mRNA should occur as detected by Northern blotting of total RNA versus an equal amount of poly(A)$^+$ RNA for housekeeping gene expression with β-actin or HPRT probes.

The method of RNA precipitation described here is a change in the published protocol. The use of NaCl allows any SDS that is present to remain in solution while the RNA is precipitated (Wallace 1987). If the isolated RNA is to be used in an in vitro translation assay, substitute 2.0 M ammonium acetate for NaCl. Otherwise, the poly(A)$^+$ mRNA isolated in this way may be used for numerous assays or RT-PCR.

The isolation and use of mRNA by Method 2 is limited in that it can only be used with RT-PCR. To determine if the isolation was successful, RT can be performed using either random primers or poly(A) primers. The PCR can be completed by direct transfer of the paper to a tube containing PCR master mix. In this case, there should not be a final denaturing step during the RT reaction. If a denaturing step is included in the RT reaction, then the cDNA in solution should be used as a template for PCR.

REFERENCES

Adams, R.L.P. 1992. *The biochemistry of nucleic acids.* Chapman and Hall, London.

Aviv, H. and P. Leder. 1972. Purification of biologically active globin messenger RNA by chromatography on oligothymidylic acid-cellulose. *Proc. Natl. Acad. Sci.* **69:** 1408–1412.

Baker, S.S., C.L. Rugh, and J.C. Kamalay. 1990. RNA and DNA isolation from recalcitrant plant tissues.

BioTechniques **9:** 268–272.

Berger, S.L. and A.R. Kimmel. 1987. *Guide to molecular cloning techniques.* Academic Press, Orlando, Florida.

Birnboim, H.C. 1988. Rapid extraction of high molecular weight RNA from cultured cells and granulocytes for Northern analysis. *Nucleic Acids Res.* **16:** 1487–1497.

Celano, P., P.M. Vertino, and R.A.J. Casero. 1993. Isolation of polyadenylated RNA from cultured cells and intact tissues. *BioTechniques* **15:** 26–28.

Chattopadhhyay, N., R. Kher, and M. Godbole. 1993. Inexpensive SDS/phenol method for RNA extraction from tissues. *BioTechniques* **15:** 24–25.

Chirgwin, J.M., A.E. Przybyla, R.J. MacDonald, and W.J. Rutter. 1979. Isolation of biologically active ribonucleic acid from sources enriched in ribonuclease. *Biochemistry* **18:** 5294–5299.

Chomczynski, P. 1991. *RNAzol™ B isolation of RNA.* Tel-Test Bulletin 2. Tel-Test Inc., Friendswood, Texas.

——. 1992. Solubilization in formamide protects RNA from degradation. *Nucleic Acids Res.* **20:** 3791–3792.

Chomczynski, P. and N. Sacchi. 1987. Single-step method of RNA isolation by acid guanidinium thiocyanate-phenol-chloroform extraction. *Anal. Biochem.* **162:** 156–159.

Cox, R.A. 1968. The use of guanidinium chloride in the isolation of nucleic acids. *Methods Enzymol.* **126:** 120–129.

Favaloro, J., R. Treisman, and R. Kamen. 1980. Transcription maps of polyoma virus-specific RNA: Analysis by two-dimensional nuclease S1 gel mapping. *Methods Enzymol.* **65:** 718–749.

Gause, W. and J. Adamovicz. 1994. The use of the PCR to quantitate gene expression. *PCR Methods Appl.* **3:** S123–S135.

Groppe, J.C. and D.E. Morse. 1993. Isolation of full-length RNA templates for reverse transcription from tissues rich in RNase and proteoglycans. *Anal. Biochem.* **210:** 337–343.

John, M.E. 1992. An efficient method for isolation of RNA and DNA from plants containing polyphenolics. *Nucleic Acids Res.* **20:** 2381.

Liu, Z., D.B. Batt, and G.G. Carmichael. 1994. An improved rapid method of isolating RNA from cultured cells by SDS-acid phenol/chloroform extraction. *BioTechniques* **16:** 56–57.

Louveau, I., S. Chaudhuri, and T.D. Etherton. 1991. An improved method for isolating RNA from porcine adipose tissue. *Anal. Biochem.* **196:** 308–310.

McCaustland, K.A., S. Bi, M.A. Purdy, and D.W. Bradley. 1991. Application of two RNA extraction methods prior to amplification of hepatitis E virus nucleic acid by the polymerase chain reaction. *J. Virological Methods* **35:** 331–342.

Meier, R. 1988. A universal and efficient protocol for the isolation of RNA from tissues and cultured cells. *Nucleic Acids Res.* **16:** 2340.

Mies, C. 1994. A simple, rapid method for isolating RNA from paraffin-embedded tissues for reverse transcription-polymerase chain reaction (RT-PCR). *J. Histochem. Cytochem.* **42:** 811–813.

Murphy, E., S. Hieny, A. Sher, and A. O'Garra. 1993. Detection of in vivo expression of interleukin-10 using a semi-quantitative polymerase chain reaction method in *Schistosoma mansoni* infected mice. *J. Immunol. Methods* **162:** 211–223.

O'Garra, A., R. Chang, N. Go, R. Hastings, G. Houghton, and M. Howard. 1992. Ly-1 B (B-1) cells are the main source of B cell-derived interleukin 10. *Eur. J. Immunol.* **22:** 711–717.

Peppel, K. and C. Baglioni. 1990. A simple and fast method to extract RNA from tissue culture cells. *BioTechniques* **9:** 134–136.

Puissant, C. and L.M. Houdebine. 1990. An improvement of the single step method of RNA isolation by acid guanidinium thiocyanate-phenol-chloroform extraction. *BioTechniques* **8:** 148–149.

Rappolee, D.A., A. Wang, D. Mark, and Z. Werb. 1989. Novel method for studying mRNA phenotypes in single or small numbers of cells. *J. Cell. Biochem.* **39:** 1–11.

Salvatori, R., R.S. Bockman, and P.T.J. Guidon. 1992. A simple modification of the Peppel/Baglioni method for RNA isolation from cell culture. *BioTechniques* **13:** 510–511.

Sambrook, J., E.F. Fritsch, and T. Maniatis. 1989. *Molecular cloning: A laboratory manual,* 2nd edition, pp. 7.1–7.83. Cold Spring Harbor Laboratory Press, Cold Spring Harbor, New York.

Santos, N. and V. Gouvea. 1994. Improved method for purification of viral RNA from fecal specimens for rotavirus detection. *J. Virol. Methods* **46:** 11–21.

Schmitt, M.E., T.E. Brown, and B.L. Trumpower. 1990. A rapid and simple method for preparation of RNA from *Saccharomyces cerevisiae. Nucleic Acids Res.* **18:** 3091–3092.

Sheardown, S.A. 1992. A simple method for affinity purification and PCR amplification of poly(A)$^+$ mRNA. *Trends Genet.* **8:** 121.

Shirzadegan, M., P. Christie, and J.R. Seemann. 1991. An efficient method for isolation of RNA from tissue cultured plant cells. *Nucleic Acids Res.* **21:** 6055.

Siebert, P.D. and A. Chenchik. 1993. Modified acid guanidinium thiocyanate-phenol-chloroform RNA extraction method which greatly reduces DNA contamination. *Nucleic Acids Res.* **21:** 2019–2020.

Smale, G. and J. Sasse. 1992. RNA isolation from cartilage using density gradient centrifugation in cesium trifluoroacetate: An RNA preparation technique effective in the presence of high proteoglycan content. *Anal. Biochem.* **203:** 352–356.

Stallcup, M.R. and L.D. Washington. 1983. Region-specific initiation of mouse mammary tumor virus RNA synthesis by endogenous RNA polymerase II in preparations of cell nuclei. *J. Biol. Chem.* **258:** 2802–2807.

Svetić, A., F.D. Finkelman, Y.C. Jian, C.W. Dieffenbach, D.E. Scott, K.F. McCarthy, A.D. Steinberg, and W.C. Gause. 1991. Cytokine gene expression after *in vivo* primary immunization with goat antibody to mouse IgD antibody. *J. Immunol.* **147:** 2391.

Tsang, S.S., X. Yin, C. Guzzo-Arkuran, V.S. Jones, and A.J. Davison. 1993. Loss of resolution in gel electrophoresis of RNA: A problem associated with the presence of formaldehyde gradients. *BioTechniques* **14:** 380–381.

Vauti, F. and W. Siess. 1993. Simple method of RNA isolation from human leucocytic cell lines. *Nucleic Acids Res.* **21:** 4852–4853.

Wallace, D.M. 1987. Precipitation of nucleic acids. *Methods Enzymol.* **152:** 41–48.

Walther, W., U. Stein, and W. Uckert. 1993. Rapid method of total RNA mini-preparation from eucaryotic cells. *Nucleic Acids Res.* **21:** 1682.

Wilkinson, M. 1988a. A rapid and convenient method for isolation of nuclear, cytoplasmic and total cellular RNA. *Nucleic Acids Res.* **16:** 10934.

———. 1988b. RNA isolation: A mini-prep method. *Nucleic Acids Res.* **16:** 10933.

Yamaguchi, M., C.W. Dieffenbach, R. Connolly, D.F. Cruess, W. Baur, and J.B. Sharefkin. 1992. Effect of different laboratory techniques for guanidinium-phenol-chloroform RNA extraction on A_{260}/A_{280} and on accuracy of mRNA quantitation by reverse transcriptase-PCR. *PCR Methods Appl.* **1:** 286–290.

Yang, F. and X. Xu. 1993. A new method of RNA preparation for detection of hepatitis A virus in environmental samples by the polymerase chain reaction. *J. Virol. Methods* **43:** 77–84.

Primer Design

Thermostable polymerases with corresponding buffers and nucleoside triphosphates used in PCR are commercially available in ready-to-use form and at convenient concentrations. Yet, with all this prepared starting material, the amplification of certain templates can fail. Assuming that all of the reagents have been added in the proper concentrations, two critical reaction components are left to the researcher. The first is the nucleic acid template, which should be as pure as possible and should not contain DNA polymerase inhibitors (although when it comes to template purity, PCR is more permissive than many other molecular biology techniques). The second is the selection of oligonucleotide primers, often critical for the overall success of an amplification reaction.

The selection of a primer is distinctively challenging when considering applications such as multiplex PCR or nested PCR, or when designing primers based on amino acid sequences. The use of degenerate primers and base analogs, as well as considerations of primer length and codon usage in different species when designing a primer based on the amino acid sequence of a peptide, are discussed in "Design and Use of Mismatched and Degenerate Primers."

Computer-assisted primer design is more effective than manual or random selection. Some of the factors that affect the performance of primers used in PCR—melting temperatures and possible homology among primers—are well defined and can be easily encoded in computer software. The speed of computers allows calculations of all possible permutations of a primer's placement, length, and relation to other primers that meet conditions specified by the user. From the thousands of combinations tested, parameters can be adjusted so that only those primers suitable for the needs of a particular experiment are presented. Thus, the overall "quality" (as defined by the user in the program parameters) of the primers selected by using computer software is guaranteed to be better than those derived "manually."

As shown in the different protocols throughout this book, primers can be designed without previous knowledge of the template sequence. These include primers that have a random nucleotide sequence or homopolymers (i.e., oligo[dT]). Complete homology with the template is not required and, therefore, primers may contain promoter elements, restriction enzyme recognition sites, or a variety of modifications at their 5′ ends. These modifications to the primer do nothing to hinder the PCR, but they do aid the researcher in the future use of the amplicon.

The challenge inherent in primer design is perhaps most evident in PCR using more than one primer pair, as described in "Multiplex PCR." Because more than one primer set is added to a reaction, each primer sequence has to be compared to the rest of the primers in the reaction to avoid amplification failure due to primer-dimer formation. Some of the computer programs listed in the Appendix can compare multiple primer pairs to determine their acceptability for use in multiplex PCR. Some primer design programs are included in sophisticated and expensive software packages with multiple capabilities, whereas other programs are limited to primer design.

General Concepts for PCR Primer Design

Carl W. Dieffenbach,[1] Todd M.J. Lowe,[2] and Gabriela S. Dveksler[3]

[1]Division of AIDS, NIAID, National Institutes of Health, Bethesda, Maryland 20852
[2]Department of Molecular Biology, Washington University, St. Louis, Missouri 63130
[3]Department of Pathology, Uniformed Services University of the Health Sciences, Bethesda, Maryland 20814

INTRODUCTION

One critical parameter for successful amplification in a PCR is the correct design of the oligonucleotide primers. The selected primer sequences determine the size and location of the PCR product, as well as define the T_m of the amplified region, a physical parameter that has been shown to be important in product yield. Well-designed primers can help avoid the generation of background and nonspecific products as well as distinguish between cDNA or genomic templates in RNA-PCR (see Section 5). Primer design also greatly affects the yield of the product. When poorly designed primers are used, no or very little product is obtained, whereas correctly designed primers generate an amount of product close to the theoretical values of product accumulation in the exponential phase of the reaction. Optimization of reaction conditions, such as adjusting the magnesium concentration and using specific cosolvents such as dimethylsulfoxide (DMSO), formamide, and glycerol, may be required even with a good primer pair. However, primers that do not follow the basic rules of primer design may not benefit from changes in the reaction conditions. In this case, designing a new primer set might save both time and money.

Currently, there are a wide range of computer programs for performing primer selection; they vary significantly in selection criteria, comprehensiveness, interactive design, and user-friendliness (Rychlik and Rhoads 1989; Lowe et al. 1990; Pallansch et al. 1990; Lucas et al. 1991; O'Hara and Venezia 1991; Tamura et al. 1991; Griffais et al.

1991; Makarova et al. 1992; Montpetit et al. 1992; Osborne 1992). Specialty primer design software programs that offer enhanced user interfaces, additional features, and updated selection criteria (Rychlik and Rhoads 1989; Lowe et al. 1990) are also available, as are primer design options that have been added to larger, more general software packages. This chapter describes the basic rules for the design of a good primer set.

PARAMETERS FOR BASIC PCR DESIGN

The aim of primer design is to obtain a balance between two goals: specificity of amplification and efficiency of amplification. Specificity is the frequency with which a mispriming event occurs. Primers with mediocre-to-poor specificity tend to produce PCR products with extra unrelated and undesirable amplicons, as seen on an ethidium bromide-stained agarose gel. The efficiency of a primer is how close to the theoretical optimum of a twofold increase of product for each PCR cycle a primer pair can amplify a product.

Given a target DNA sequence, primer analysis software attempts to strike a balance between these two goals by using preselected default values for each of the primer design variables. These variables, listed below, have predictable effects on the specificity and efficiency of amplification. Depending on the experimental requirements, these "primer search parameters" can be adjusted to override the default values that are meant to be effective for only general PCR applications. For example, in medical diagnostic PCR applications, search parameters and reaction conditions are adjusted to increase specificity at the cost of some efficiency, because avoiding false-positive results is a higher priority than producing large quantities of amplified product. By carefully considering the following list of parameters when using primer design software, a more effective selection of primers will be achieved.

Primer Length

Specificity is generally controlled by the length of the primer and the annealing temperature of the PCR. Oligonucleotides between 18 and 24 bases tend to be very sequence-specific if the annealing temperature of the PCR is set within a few degrees of the primer T_m (defined as the dissociation temperature of the primer/template duplex). These oligonucleotides work very well for standard PCR of defined targets that do not have any sequence variation. The longer the primer, the smaller the fraction of primed templates there will be in the annealing step of the amplification. In exponential amplification, even a small inefficiency at each annealing step will propagate and result in a significant decrease in amplified product. In summary, to optimize

PCR, using primers of a minimal length that ensures melting temperatures of 54°C or higher provides the best chance for maintenance of specificity and efficiency.

Short oligonucleotides of 15 bases or less are useful only for a limited number of PCR protocols, such as the use of arbitrary or random short primers in the mapping of simple genomes and in the subtraction library protocol described by Liang and Pardee (1992) and Williams (1990). Depending on the organism's genome size, there is a bare minimum length, which will vary by a few nucleotides. In general, it is best to build in a margin of specificity for safety. For each additional nucleotide, a primer becomes four times more specific; thus, the minimum primer length used in most applications is 18 nucleotides. Clearly, if purified cDNA is being used, or if genomic DNA is not present, the length can be reduced because the risk of nonspecific primer-template interactions is greatly reduced. Yet, it is generally a good idea to design primers so that the synthesized oligonucleotides can be used in a variety of experimental conditions (18- to 24-mers); the marginal cost of oligonucleotides with 4–5 additional bases makes it worth the expense.

The upper limit on primer length is somewhat less critical and has more to do with reaction efficiency. For entropic reasons, the shorter the primer, the more quickly it anneals to target DNA and forms a stable double-stranded template to which DNA polymerase can bind. In general, oligonucleotide primers 28–35 bases long are necessary when amplifying sequences where a degree of heterogeneity is expected. This primer length has proven to be generally useful in two types of applications: (1) in amplifying closely related molecules such as isoforms of a protein or family of proteins within a species, as well as in the cloning of a gene from a different species than the one whose sequence is available to the researcher (Dveksler et al. 1993); and (2) in amplifying the sequences of viruses such as HIV-1 where sequence variation and the existence of a swarm are the hallmarks of the disease and, as a consequence, the possibility of having a set of primers with perfect complementarity to all the templates (in this example, all HIV-1 isolates) is not expected (Mack and Sninsky 1988; Ou et al. 1988).

In both of these cases, one first uses the primer design software to compare all available related sequences and to describe the DNA region with the least amount of sequence variability. These regions serve as starting places to select the primers. In some instances, the researcher already knows the function of the protein and its domains essential for performing that function. In these cases, comparing available sequences in the regions critical for the functional activity of the related proteins within the family aids in the definition of the sequences around which the design of new primers should be centered.

One example of this is the PCR cloning of an enzyme or receptor with a similar structure and function from a related species using the available structural data. With the amino acid sequence information and the help of codon usage tables for different species, both primers, or at least one of them, can be designed around the "conserved sequence." When selecting primers to amplify DNA from a different species, sequences at the 5′- or 3′-untranslated regions of the mRNA should be avoided because they may not necessarily have any degree of homology.

The placement of the 3′ end of the primer is critical for a successful PCR. If a conserved amino acid can be defined, the first 2 bases of the codon (or 3 bases in the case of an amino acid coded for by a single codon [methionine and tryptophan]) can serve as the 3′ end. Perfect base pairing between the primer's 3′ end and the template is optimal for obtaining good results, and minimal mismatches should exist within the last 5–6 nucleotides at the 3′ end of the primer. Attempts to compensate for the mismatches between the 3′ end of the primer and the template by lowering the annealing temperature of the reactions do not improve the results, and failure of the reaction is almost guaranteed. With this concept in mind, one should evaluate all possible strategies in the design of primers when the nucleotide sequence of the template to be amplified is not known with certainty. Cases like the one described above are encountered routinely when the researcher wishes to amplify a cDNA using information from a partial protein sequence (Dveksler et al. 1993). Several approaches, including the use of degenerate oligonucleotide primers that cover all possible combinations for the bases at the 3′ end of the primer in the pool as well as the use of inosine to replace the base corresponding to the third or variable position of certain amino acid codons, have been successful for cDNA cloning as well as for the detection of sequences with possible variations (Lin and Brown 1992). Much of this type of PCR study is empirical, and different primers may have to be synthesized to obtain the desired match.

Longer primers could also arise when extra sequence information, such as a T7 RNA polymerase-binding site, restriction sites, or a GC clamp, is added to primers (Loh et al. 1989; Sheffield et al. 1989; Kain et al. 1991). In general, the addition of unrelated sequence at the 5′ end of the primer does not alter the annealing of the sequence-specific portion of the primer. In some cases, when a significant number of bases that do not match the template sequence are added to the primer, four to five cycles of amplification can be performed at a lower annealing temperature; this is followed by the rest of the cycles at the annealing temperature calculated with the assumption that the sequence at the 5′ end of the primer is already incorporated into the template.

The addition of bases at the 5′ ends of the primers is frequently observed when the researcher needs to clone the PCR product. In these cases, the restriction enzyme sites of choice are those that do not cut within the DNA at sites other than the primer. To ensure subcloning of the whole amplified fragment of unknown sequence as a single piece, the addition of sites for enzymes that recognize 6 bases, or the addition of partially overlapping recognition sites for different enzymes, is recommended. An important consideration when adding restriction sites to a primer is that most enzymes require 2 or 3 nonspecific extra bases 5′ to their recognition sequence to cut efficiently, thus adding to the length of the nontemplate-specific portion of the primer (New England Biolabs catalog 1993/1994, pp. 180–181). Another drawback of long primer sequences is in the calculation of an accurate melting temperature necessary to establish the annealing temperature at which the PCR is to be performed. For primers shorter than 20 bases, an estimate of T_m can be calculated as $T_m = 4(G+C)+2(A+T)$ (Suggs et al. 1981), whereas for longer primers, the T_m requires the nearest-neighbor calculation, which takes into account thermodynamic parameters and is employed by most of the available computer programs for the design of PCR primers (Breslauer et al. 1986; Freier et al. 1986).

The Terminal Nucleotide in the PCR Primer

Kwok and colleagues have shown that the 3′-terminal position in the primers is essential for controlling mispriming (Kwok et al. 1990). For some of the applications described above, this chance of mispriming is useful. The other issue with the 3′ ends of the PCR primers is the prevention of homologies within a primer pair. Care must be taken that the primers are not complementary to each other, particularly at their 3′ ends. Complementarity between primers leads to the undesirable primer-dimer phenomenon, in which the obtained PCR product is the result of the amplification of the primers themselves. This sets up a competitive PCR situation between the primer-dimer product and the native template and is detrimental to the success of the amplification. In cases where multiple primer pairs are added in the same reaction (as in multiplex PCR), it is very important to double check for possible complementarity of all the primers added in the reaction. Generally, the computer programs do not allow primer pairs with 3′-end homologies; thus, when they are used in conjunction with the hot start technique, the chances of formation of primer-dimer products are greatly reduced (Chou et al. 1992).

Reasonable GC Content and T_m

PCR primers should maintain a reasonable GC content. Oligonucleotides 20 bases long with a 50% G+C content generally have T_m values in the range of 56–62°C, which provides a sufficient thermal window for efficient annealing. The GC content and T_m should be well matched within a primer pair. Poorly matched primer pairs can be less efficient and specific because loss of specificity arises with a lower T_m value; the primer with the higher T_m value has a greater chance of mispriming under these conditions. If too high a temperature is used, the primer of the pair with the lower T_m value may not function at all. This matching of GC content and T_m is critical when selecting a new pair of primers from a list of already-synthesized oligonucleotides within a sequence of interest for a new application. For this reason, we advocate the adoption of a standardized approach to primer design for the laboratory. By planning ahead, it is easier to mix and match selected primers, because they will all have similar physical characteristics.

PCR Product Length and Placement within the Target Sequence

All of the computer programs provide a place to select a range for the length of the PCR product. In general, the length of the PCR product has an impact on the efficiency of amplification (Rychlik et al. 1990). The length of a PCR product for a specific application is dependent in part on the template material. Clinical specimens prepared from fixed tissue samples tend to yield DNA that does not support the amplification of large products (Greer et al. 1991). It is relatively straightforward to obtain products greater than 3 kb from pure plasmid or high-molecular-weight DNA. For the purpose of detecting a DNA sequence, PCR products of 150–1000 bp are generally produced.

The specifics of the size of the desired products often depend on the application. If the purpose is to develop a clinical assay to detect a specific DNA fragment, a small DNA amplification product of 120–300 bp may be optimal. The product should be specific and efficient to produce and also contain enough information for use in a capture probe hybridization assay (Whetsell et al. 1992). Products in this size range can be produced using the two-step amplification cycling method, thereby shortening the length of the amplification procedure.

Other PCR approaches have different optimal product lengths. For example, to monitor gene expression by quantitative RNA-PCR, the product must be large enough so that a competitive template can be constructed and both the product and the competitor can be easily resolved on a gel. These products tend to run in the 250- to 750-bp range. Here the issue is to maximize the efficiency of both the reverse transcriptase step and the PCR.

In terms of placing the PCR primers within a cDNA sequence, two specific points should be kept in mind. First, try to keep the primers and product within the coding region of the mRNA, as this is the unique sequence responsible for the production of the protein, unlike the 3′-noncoding region, which shares homologies with many different mRNAs. Second, try to place the primers on different exons so that the RNA-specific PCR product is different in size from one arising from contaminating DNA.

If the purpose of the PCR is to clone a specific region of a gene or cDNA, then the size of the PCR product is preselected by the application. Here the computer program can provide information about selected primer sets that flank the desired area. In some instances, when the complete sequence is required for further experiments and the PCR product to be obtained is above the ideal length or the template is not of the best quality, overlapping PCR fragments can be amplified by designing the correct primers flanking unique restriction sites in the template sequence. The production of a fragment containing the entire sequence can then be obtained by cutting and pasting the amplified pieces. When approaching this kind of application, it is important to think ahead of time about the ideal method for cloning the PCR products and how the clone will be used in the future. For example, if utilizing restriction endonuclease sites at the end of the primers as described above, it is important to be sure that these enzymes do not cut within the amplified region. Again, the software programs can provide this information (Lowe et al. 1990).

A Simple Rule for Non-computer-based Selection

Occasionally, PCR primers must be selected from very defined regions at the 3′ and 5′ ends of a specific sequence. A simple method of primer design here is to choose regions that are deficient in a single nucleotide. Selecting primers in this way reduces the chance of extensive primer-primer homology. Here again, care must be taken to have a balanced primer pair in terms of length and base composition so that the T_m values of the primers are within 2–3°C of each other.

Nested PCR

In certain situations, there are unresolvable problems with the quantity and quality of the template to be amplified. Perhaps the actual quantity of target nucleic acid is very dilute relative to the rest of the material present, or there is a limit on the purity of the starting material. Both of these problems occur simultaneously in certain clinical applications (Albert and Fenyo 1990). In these circumstances, one approach to synthesizing a product reliably is to develop a nested PCR assay.

Generally, the sample is first amplified for 20–30 cycles using the outer primer set, then a very small aliquot of this reaction is amplified a second time for 15–25 cycles using the inner primer set. The inner set of PCR primers is positioned within the DNA so that the complementary sequence for the inner primer pair is present in the PCR product obtained in the first amplification reaction and available to form a template-primer complex. This has been shown to be more successful than diluting and reamplifying with the same primers (Albert and Fenyo 1990). The position of the inner primer set is often the determining factor in the overall structure of the nested approach and is a factor in determining the final product size. For example, in the nested PCR detection system, adapted from the original assay of Larder for the amino acid 215 mutation in the HIV-1 reverse transcriptase involved in zidovudine resistance, the 3′ end of one of the inner primers must match the mutations to produce a quality PCR product (Larder et al. 1991). A control inner primer set run in parallel detects the wild-type sequence. In general, the product of the inner set is small, 120–270 bp. The outer primer pair ideally should completely flank the inner product. When selecting nested primer sets, special care must be given to eliminating potential primer-dimers and matches between members of the inner and outer primer sets. Some of the software programs for primer design have the selection of nested primers as an option.

USING PRIMER DESIGN SOFTWARE

It is important to stress that the primer selection parameters described here are general and are not necessarily implemented in the same manner among the different primer selection programs. Thus, two programs using slightly different selection algorithms rarely, if ever, select the same primers, even if the basic parameters are set equivalently. These discrepancies are due to differences in the calculation methods and the order in which the selection criteria are applied. For example, calculating the temperature of primer-template annealing can be performed in one of several ways. The original formula of Suggs et al. (1981), $T_m = 2^oC \times (A+T) + 4^oC \times (G+C)$, is popular for its simplicity and roughly accurate prediction of oligonucleotide T_m. More recently, Rychlik et al. (1990) implemented T_m prediction based on nearest-neighbor thermodynamic parameters (Breslauer et al. 1986; Freier et al. 1986) that appear to be slightly more accurate. Other programs base primer annealing temperatures on formulas originally developed for DNA fragments over 100 nucleotides long (McConaughy et al. 1969). Thus, specifying a desired primer annealing temperature to be 60°C will produce different primers from the same target sequence. Further work by Rychlik et al. (1990) produced an empirically derived equation for the optimal annealing temperature of a primer pair that depends on nearest-neighbor calculations. Wu et al. (1991) have also empirically derived an equation based on primer length and GC content to determine optimal oligonucleotide annealing temperature. These examples illustrate how something as basic as primer T_m calculation can vary among the programs.

Second, different programs attack the task of primer selection very differently, applying selection criteria to reduce the number of possible primers the program must consider, while not eliminating potentially good candidates. For example, the program by Lowe et al. (1990) only considers primers that have a 3′-end CC, GG, CG, or GC dinucleotide, which may increase priming efficiency, but allows the user to specify a range of primer lengths. In contrast, the program by Rychlik and Rhoads (1989) does not impose this requirement, but checks primers of a single length specified by the user. Both of these approaches eliminate potentially good primers but still, in most cases, produce an adequate number of primers that meet all the conditions considered important by the authors.

In using the computer software described, keep in mind that the broader the selection parameters are made, the more cases the computer must consider, significantly affecting the time required for primer searches. This is one reason that search parameters should be kept as narrow and specific as possible when they are clearly dictated by experimental design. More restrictive search parameters usually result in faster searches and produce primers of greater quality. In programs that attempt more difficult selection tasks, such as choosing primers that are highly conserved across many species, or selection of degenerate primers from protein sequences, the basic criteria for primer selection often must be relaxed to find primers that satisfy more critical needs.

Using one of the available software programs in conjunction with the information presented here should result in a good primer set. The next task is the preparation of a good nucleic acid template.

REFERENCES

Albert, J. and E.M. Fenyo. 1990. Simple, sensitive and specific detection of human immunodeficiency virus type 1 in clinical specimens by polymerase chain reactions with nested primers. *J. Clin. Microbiol.* **28:** 1560–1564.

Breslauer, K.J., F. Ronald, H. Blocker, and L.A. Marky. 1986. Predicting DNA duplex stability from the base sequence. *Proc. Natl. Acad. Sci.* **83:** 3746–3750.

Chou, Q., M. Russell, D.E. Birch, J. Raymond, and W. Bloch. 1992. Prevention of pre-PCR mis-priming and primer dimerization improves low-copy number amplifications. *Nucleic Acids Res.* **20:** 1717–1723.

Dveksler, G.S., C.W. Dieffenbach, C.B. Cardellichio, K. McCuaig, M.N. Pensiero, G.-S. Jiang, N. Beauchemin, and K.V. Holmes. 1993. Several members of the mouse carcinoembryonic antigen-related glycoprotein family are functional receptors for the coronavirus mouse hepatitis virus-A59. *J. Virol.* **67:** 1–8.

Freier, S.M., R. Kierzek, J.A. Jaeger, N. Sugimoto, M.H. Caruthers, T. Neilson, and D.H. Turner. 1986. Improved free-energy parameters for predictions of RNA duplex stability. *Proc. Natl. Acad. Sci.* **83:** 9373–9377.

Greer, C.E., J.K. Lund, and M.M. Manos. 1991. PCR amplification from paraffin-embedded tissues: Recommendations on fixatives for long term storage and prospective studies. *PCR Methods Appl.* **1:** 46–50.

Griffais, R., P.M. Andre, and M. Thibon. 1991. K-tuple frequency in the human genome and polymerase chain reaction. *Nucleic Acids Res.* **19:** 3887–3891.

Kain, K.C., P.A. Orlandi, and D.E. Lanar. 1991. Universal promoter for gene expression without cloning: Expression PCR. *BioTechnology* **10:** 366–374.

Kwok, S., D.E. Kellog, N. McKinney, D. Spasic, L. Goda, C. Levenson, and J.J. Sninsky. 1990. Effects of primer-template mismatches on the polymerase chain reaction: Human immunodeficiency virus 1 model studies. *Nucleic Acids Res.* **18:** 999–1005.

Larder, B.A., P. Kellam, and S.D. Kemp. 1991. Zidovudine resistance predicted by direct detection of mutations in DNA from HIV-infected lymphocytes. *AIDS* **5:** 137–144.

Liang, P. and A. Pardee. 1992. Differential display of eukaryotic mRNAs by PCR. *Science* **257:** 967–971.

Lin, P.K.T. and D.M. Brown. 1992. Synthesis of oligodeoxyribonucleotides containing degenerate bases and their use as primers in the polymerase chain reaction. *Nucleic Acids Res.* **19:** 5149–5152.

Loh, E.Y., J.F. Elliot, S. Cwirla, L.L. Lanier, and M.M. Davis. 1989. Polymerase chain reaction with single-sided specificity: Analysis of T-cell receptor gamma chain. *Science* **243:** 217–220.

Lowe, T.M.J., J. Sharefkin, S.Q. Yang, and C.W. Dieffenbach. 1990. A computer program for selection of oligonucleotide primers for polymerase chain reaction. *Nucleic Acids Res.* **18:** 1757–1761.

Lucas, K., M. Busch, S. Mossinger, and J.A. Thompson. 1991. An improved microcomputer program for finding gene- or gene family-specific oligonucleotides suitable as primers for polymerase chain reactions or as probes. *CABIOS* **7:** 525–529.

Mack, D.H. and J.J. Sninsky. 1988 A sensitive method for the identification of uncharacterized viruses related to known virus groups: Hepadnavirus model system. *Proc. Natl. Acad. Sci.* **85:** 6977–6981.

Makarova, K.S., A.V. Mazin, Y.I. Wolf, and V.V. Soloviev. 1992. DIROM: An experimental design interactive system for directed mutagenesis and nucleic acid engineering. *CABIOS* **8:** 425–431.

McConaughy, B.L., C.L. Laird, and B.J. McCarthy. 1969. Nucleic acid reassociation in formamide. *Biochemistry* **8:** 3289–3295.

Montpetit, M.L., S. Cassol, T. Salas, and M.V. O'Shaughnessy. 1992. OLIGOSCAN: A computer program to assist in the design of PCR primers homologous to multiple DNA sequences. *J. Virol. Methods* **36:** 119–128.

O'Hara, P.J. and D. Venezia. 1991. PRIMGEN, a tool for designing primers from multiple alignments. *CABIOS* **7:** 533–534.

Osborne, B.I. 1992. HyperPCR: A Macintosh hypercard program for determination of optimal PCR annealing temperature. *CABIOS* **8:** 83.

Ou, C.-Y., S. Kwok, S.W. Mitchell, D.H. Mack, J.J. Sninsky, J.W. Krebs, P. Feorino, D. Warfield, and G. Schochetman. 1988. DNA amplification for direct detection of HIV-1 in DNA of peripheral blood mononuclear cells. *Science* **239:** 295–297.

Pallansch, L., H. Beswick, J. Talian, and P. Zelenka. 1990. Use of an RNA folding algorithm to choose regions for amplification by the polymerase chain reaction. *Anal. Biochem.* **185:** 57–62.

Rychlik, W. and R.E. Rhoads. 1989. A computer program for choosing optimal oligonucleotides for filter hybridization, sequencing and in vitro amplification of DNA. *Nucleic Acids Res.* **17:** 8543–8551.

Rychlik, W., W.J. Spencer, and R.E. Rhoads. 1990. Optimization of annealing temperature for DNA amplification in vitro. *Nucleic Acids Res.* **18:** 6409–6412.

Sheffield, V.C., D.R. Cox, L.S. Lerman, and R.M. Myers. 1989. Attachment of a 40-base pair G+C-rich sequence (GC-clamp) to genomic DNA fragments by polymerase chain reaction results in improved detection of single base changes. *Proc. Natl. Acad. Sci.* **86:** 232–236.

Suggs, S.V., T. Hirose, E.H. Myake, M.J. Kawashima, K.I. Johnson, and R.B. Wallace. 1981. Using purified genes. *ICN-UCLA Symp. Mol. Cell. Biol.* **23:** 683–693.

Tamura, T., S.R. Holbrook, and S.-H. Kim. 1991. A Macintosh computer program for designing DNA sequences that code for specific peptides and proteins. *BioTechniques* **10:** 782–784.

Whetsell, A., J. Drew, G. Milman, R. Hoff, E. Dragon, K. Alder, J. Hui, P. Otto, P. Gupta, H. Farzadegan, and S. Wolinsky. 1992. Comparison of three nonradioisotopic polymerase chain reaction-based methods for detection of human immunodeficiency virus type 1. *J. Clin. Microbiol.* **30:** 845–853.

Williams, J.G.K. 1990. DNA polymorphisms amplified by arbitrary primers are useful as genetic markers. *Nucleic Acids Res.* **18:** 6531–6535.

Wu, D.Y., W. Ugozzoli, B.K. Pal, J. Qian, and R.B. Wallace. 1991. The effect of temperature and oligonucleotide primer length on specificity and efficiency of amplification by the polymerase chain reaction. *DNA Cell Biol.* **10:** 233–238.

Design and Use of Mismatched and Degenerate Primers

Shirley Kwok, Sheng-Yung Chang, John J. Sninsky, and Alice Wang

Roche Molecular Systems, Inc., Alameda, California 94501

INTRODUCTION

Although simple in concept, PCR requires a myriad of complex interactions between template, primers, deoxynucleoside triphosphates, and DNA polymerase to accomplish targeted amplification successfully. Primer design in conjunction with changes in the concentrations of reaction components and thermal cycling parameters provides a versatility available in no other molecular technique. For example, PCR can be employed to accommodate mismatches in the primer-template duplex, thereby permitting amplification not only of related sequences, but also of uncharacterized sequences. It can also be designed to amplify a small number of mutant genes with single base alterations selectively in a vast background of normal genes.

The flexibility of PCR has greatly simplified molecular manipulations. Procedures for altering a particular template sequence, which in the past required several steps, can now be performed with only a few manipulations. Because 3′ and internal mismatches between primer-template duplexes are tolerated under appropriate conditions, point mutations and desired restriction endonuclease sites can be introduced directly into the primers. Likewise, nucleotide insertions and deletions can be similarly introduced into the amplified product via the primers.

Degenerate primers have made it possible to amplify related but distinct nucleic acid sequences as well as to amplify targets for which only amino acid sequences are available. The range of possible applications precludes providing detailed protocols for each. Instead, the critical parameters and a guide for representative use are summarized. Depending on the application, a subset of these recommendations or suggestions may suffice.

Mismatch Discrimination

Primers can be tailored to amplify selectively targets that vary by a single nucleotide (for review, see Ugozzoli and Wallace 1991). Genetic diseases commonly arise from single base-pair mutations. Sickle cell anemia, for example, results from A to T transversion in codon 6 of the human β-globin sequence (Marotta et al. 1977). Single point mutations at codons 12, 13, or 61 of the *ras* genes result in activation of the proto-oncogene that has been associated with a high frequency of human cancers (Bos 1989). Resistance to HIV antivirals such as zidovudine, ddI, ddC, and pyridinone reverse transcriptase inhibitors is conferred by multiple single-base mutations (Larder et al. 1991; Nunberg et al. 1991; St. Clair et al. 1991; Larder and Boucher 1991).

The discrimination of different PCR targets is based on the fact that *Taq* DNA polymerase lacks a 3′→ 5′ exonuclease activity (Lawyer et al. 1989) and that mismatched 3′ termini are extended at a lower rate than matched termini. Various mismatches are extended at different efficiencies (Creighton et al. 1992; Huang et al. 1992). Multiple acronyms have been used to describe sequence-specific amplification, including ASPCR (*a*llele-*s*pecific PCR) (Ugozzoli and Wallace 1991), ARMS (*a*mplification *r*efractory *m*utation *s*ystem) (Newton et al. 1989), MAMA (*m*ismatch *a*mplification *m*utation *a*ssay) (Cha et al. 1992), and PASA (*PCR a*mplification of *s*pecific *a*lleles) (Dutton and Sommer 1991). The appellation "allele-specific" is somewhat of a misnomer because rare somatic mutations can also be identified using this approach.

In designing primer pair systems to discriminate mutant from wild-type sequences, one needs first to examine the mutations involved. The placement of a mismatch at the 3′ terminus of a primer-template duplex is more detrimental to PCR than internal mismatches. However, not all 3′-terminal mismatches affect PCR equally. Kwok et al. (1990) demonstrated that even in a system designed to tolerate mismatches, 3′-terminal mismatches involving A:G, G:A, C:C, and G:G reduced product yield by 100-fold and A:A by 20-fold, whereas all other mismatches had little effect (Table 1). Because G:T mismatches are more stable than other mismatches (Kidd et al. 1983), G:T should be avoided when designing primers, although this mismatch has been used successfully (Li et al. 1990). On the other hand, one can still take advantage of this mismatch by targeting the complementary strand that forms an A:C mismatch. It is important to note that sequence context can significantly alter the properties of these mismatches, and as much as a 10-fold difference among sites has been reported (Mendelman et al. 1989; Huang et al. 1992).

Occasionally, a single-base mismatch at the 3′ terminus is insufficient to achieve the desired level of discrimination, particularly when the ratio of mutant to wild-type sequences is low. A 3′-terminal mis-

Table 1 Relative Amplification Efficiencies of 3′-Terminal Mismatches in the Presence of 800 μM dNTP

		Primer 3′ base			
		T	C	G	A
Corresponding	T	1.0	1.0	1.0	1.0
template	C	1.0	≤.01	1.0	1.0
base	G	1.0	1.0	1.0	≤.01
	A	1.0	1.0	≤.01	0.5

Product yields were normalized to be the perfect matches (1.0).
(Reprinted, with permission, from Kwok et al. 1990 [Oxford University Press].)

match coupled with an additional mismatch either 1, 2, or 3 bases from the 3′terminus can increase discrimination (Newton et al. 1989; Cha et al. 1992). Therefore, if necessary, the deliberate introduction of a second mutation 1–3 bases from the 3′terminus will destabilize the 3′end and provide even greater differentiation. To increase the ratio of mutant to wild-type sequences, an "enriched" PCR was employed by Kahn et al. (1991) to identify mutations at codon 12 of the Ki-*ras* gene. A restriction enzyme site was generated in the wild-type sequence using a mismatched primer in the first round of amplification. The wild-type sequences were eliminated by digesting the PCR products with the appropriate restriction enzyme, and mutant sequences were enriched with a second round of amplification using mutant-specific primers.

The use of pairs of sequence-specific primers for PCR facilitates direct haplotype determination (Lo et al. 1991; Sarkar and Sommer 1991). This procedure is particularly attractive for determining the haplotypes of individuals who are heterozygous at a number of polymorphic sites and/or in whom pedigree samples are not available. An example of this approach is illustrated in Figure 1. In this biallelic system, the polymorphic site at allele 1 is either T or C and the polymorphic site at allele 2 is either G or T. The four possible

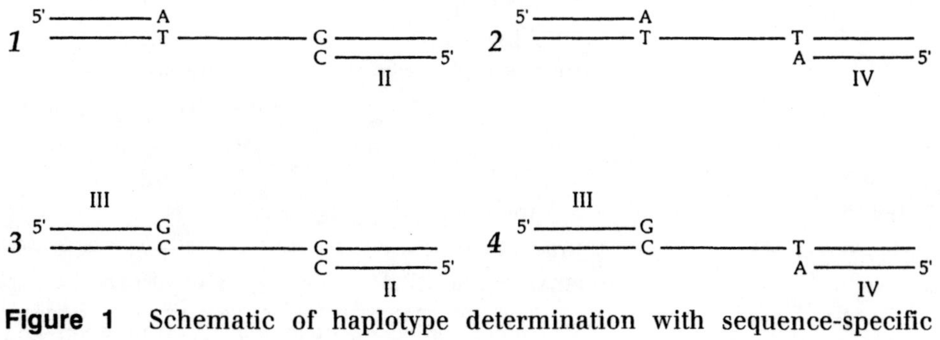

Figure 1 Schematic of haplotype determination with sequence-specific primers.

haplotypes (1, 2, 3, 4) can be differentially amplified by using four pairwise combinations of primers specific to each allele. Specifically, primer pairs I and II would amplify haplotype 1; I/III, haplotype 2; II/III, haplotype 3; and II/IV, haplotype 4. In addition, this procedure could be used to identify regions that contain as many as four *cis* mutations and would be available, for example, for the analysis of mutations in the HIV genome that confer zidovudine resistance.

In addition to 3′-terminal mismatches, other factors should also be considered. Shorter primer (< 20 bases, T_m < 55°C); lower dNTP levels (Ehlen and Dubeau 1989); higher annealing temperatures; lower primer, $MgCl_2$, and enzyme concentrations; and fewer cycles increase the stringency of the amplifications and can be used to skew amplifications to favor the target sequence (see Table 2). Tada et al. (1993) found that the Stoffel fragment, an amino-terminally truncated variant of *Taq* DNA polymerase, when used in combination with hot start (Chou et al. 1992), further enhanced discrimination. The detection of a small number of single-base mutant genes in a 10^4- and 10^5-fold excess of unaltered genes has been demonstrated successfully in the aforementioned studies.

Mismatch Tolerance

Although the effects of primer-template mismatches aid in the design of primers for sequence or allele-specific amplification, these effects can also be used to design primers to maximize detection of variant sequences. Most notably, retroviruses and RNA viruses replicate with less fidelity than many DNA viruses, and therefore viral genomes within and among patients may vary. False negatives due to mis-

Table 2 Flexibility of PCR Amplification

	Single nucleotide discrimination	Mismatch tolerance
[enzyme]	↓	↑
[$MgCl_2$]	↓	↑
[dNTPs]	↓	↑
[primers]	short/↓	long/↑
Nature of mismatch	purine:purine pyrimidine:pyrimidine> purine:pyrimidine avoid G:T	3′terminal T
Annealing temperature/time	↓/↑	↑/↓
extension step (time)	↓ or eliminate	↑

Parameters affecting mismatch discrimination and mismatch tolerance of PCR amplification. Brackets represent the concentration of the parameter.

matches at the 3′terminus can be minimized by designing primers with a 3′-terminal T. Primers that terminate in T were efficiently extended even when mismatched with T, G, or C (Kwok et al. 1990). Our unpublished studies likewise demonstrate that 3′-terminal dU primers also tolerate mismatches. However, as a rule, 3′-terminal mismatches should be avoided if at all possible to maximize efficient extension by *Taq* DNA polymerase.

For amplification of closely related, yet distinct, sequences, chimeric primers can be designed. This approach is particularly attractive when the number of base differences between two sequences in the primer-binding region is small. The primer used, then, is not homologous to either sequence but has a limited number of mismatches to both. By designing the primers so that the mismatches are in the middle or at the 5′terminus, amplification of both target sequences can occur without compromise. Kwok et al. (1988) have described this approach for amplification of HTLV-I and -II.

The use of primers that are at least 25 bases long ($T_m > 70°C$), high dNTP concentrations (800 μM), and annealing temperatures below the T_m of the primers will better accommodate mismatches (Table 2). The increase in specificity conferred by hot start PCR may not be suitable if accommodation of mismatches is desired.

Because single-base mismatches can significantly lower the melting temperature of a probe-target duplex, modified bases such as 5-bromodeoxyuracil and 5-methyldeoxycytosine have been used to increase duplex stability (Hoheisel et al. 1990). These bases have enabled the use of short oligonucleotides such as the *Not*I octadeoxynucleotide to probe genomic libraries (Hoheisel et al. 1990). The use of 2,6-diaminopurine, an adenine analog, may similarly enhance the stability of duplexes (Cheong et al. 1988). The stability of these nucleotide duplexes is influenced by base stacking and also by sequence context.

Recently, Nichols et al. (1994) described the use of a nondiscriminatory base analog, 1-(2′-deoxy-β-D-ribofuranosyl-3-nitropyrrole) or M, at ambiguous sites in DNA primers. Oligonucleotides that contained M at several sites were successfully used as primers for sequencing and PCR. The incorporation of M at the 3′terminus may improve mismatch tolerance.

Degenerate Primers

Some PCR applications require that primers not only tolerate mismatches, but also accommodate primer binding to an unknown sequence. Because of the degenerate nature of the genetic code, a given amino acid may be encoded by different triplets. The number of triplets that encode each amino acid is listed in Table 3. The degenerate

Table 3 Codon Degeneracy

Number of codons	Amino acids
1	Met, Trp
2	Phe, Tyr, His, Gln, Asn, Lys, Asp, Glu, Cys
3	Ile
4	Val, Pro, Thr, Ala, Gly
6	Leu, Ser, Arg

primers are designed as a pool of all the possible combinations of nucleotides that could code for a given amino acid sequence. Degenerate oligonucleotides can either be synthesized individually and then pooled or multiple bases can be programmed at one position in the DNA synthesizer. Because of the different stability of the blocked bases, care should be taken to use fresh reagents to ensure reproducible and equivalent synthesis of degenerate primers (Bartel and Szostak 1993). The degeneracy of a given primer pool can be determined by multiplying the number of possible nucleotides at each position. For example, the degree of degeneracy of the primer pool for "AlaAsnIleLysMet" is $4 \times 2 \times 3 \times 2 \times 1 = 64$, where the codon degeneracy for Ala=4, Asn=2, Ile=3, Lys=2, and Met=1. Mixed oligonucleotide primers derived from an amino acid sequence can be used to amplify a specific sequence from genomic DNA or cDNA that, in turn, can then be used as a probe to screen a genomic/cDNA library for the corresponding clones (Girgis et al. 1988; Lee et al. 1988). This is a powerful technique for obtaining a DNA probe when only a limited portion of a protein sequence is available for a sought-after gene. Degenerate primers based on the amino acid sequence of conserved regions were also used to search for members of a gene family (Wilks et al. 1989), homologous genes from different species (Kopin et al. 1990), or related viruses (Mack and Sninsky 1988; Manos et al. 1989; Shih et al. 1989).

When designing degenerate primers, the degeneracy of the genetic code for the selected amino acids of the region targeted for amplification must be examined. Obviously, selection of amino acids with the least degeneracy is desirable because it provides the greatest specificity. Several approaches can be considered for increasing the specificity of the amplification using degenerate primers.

1. The pools may be synthesized as subsets such that one pool may contain either a G or C at a particular position, whereas the other contains either an A or T at the same position.
2. The degeneracy of the mixed primer may be reduced by using the codon bias for translation (Wada et al. 1991).

3. The size of the degenerate primers can be as short as 4–6 amino acid codons in length. Six to nine base extensions that contain restriction enzyme sites can be added to the 5′terminus to facilitate cloning. Mack and Sninsky (1988) showed that for short primers, the 5′extensions facilitated amplification.
4. Degeneracy at the 3′end of the primer should be avoided, because single-base mismatches may obviate extension. The last base of the terminal codon can be omitted unless the amino acid is Met or Trp.
5. The inclusion of deoxyinosine at some ambiguous positions may reduce the complexity of the primer pool (Table 4) (Ohtsuka et al. 1985; Sakanari et al. 1989).

Inosine occurs naturally in the wobble position of the anticodon of some transfer RNAs and is known to form base pairs with A, C, and U in the translation process. Several experiments have suggested that deoxyinosine might be an "inert" base; its presence in an oligonucleotide sequence seems neither to disturb DNA duplex formation nor to destabilize the duplex (Martin et al. 1985; Ohtsuka et al. 1985). Independent of sequence effects, the order of stabilities of base-pairing is I:C> >I:A>I:T = I:G (Martin et al. 1985). The deoxyinosine-containing primers have been used successfully to amplify cDNA or genomic fragments for the generation of DNA probes from a peptide sequence (Schuchman et al. 1990; Aarts et al. 1991). Deoxyinosine was generally incorporated at base positions with three- to fourfold redundancy. This method may be useful for proteins with highly degenerate codons. Inosine usually directs the incorporation of C when in the template. Unfortunately, the studies to date do not discern whether inosine actually assists mismatched priming. Another approach, pioneered by Lin and Brown (1992), exploits the use of unconventional bases that are capable of base-pairing with either a purine or a pyrimidine. These unconventional bases are 6H, 8H-3,4-dihydropyrimido[4,5-c][1,2]oxazin-7-one (P) and 2-amino-6-methoxy-aminopurine (K), where P can base-pair with A and G, and K can base-pair with C and T. They demonstrated that oligonucleotides containing the degenerate bases P and K can be used as primers in PCR.

Table 4 Approaches in Designing Degenerate Primers

1. Pools can be synthesized as subsets
2. Use codon bias for translation
3. Use 4–6 amino acids (12–18 bases) in length
4. Avoid degenerate base at the 3′ terminus; omit the last base of the terminal codon unless the amino acid is Met or Trp
5. Consider using deoxyinosine for accommodating multiple components to reduce the degeneracy

Similarly, the universal nucleoside 1-(2′-deoxy-β-D-ribofuranosyl)-3-nitropyrrole (M) can also be used at ambiguous sites in DNA primers for amplification (Nichols et al. 1994).

It has been observed that the PCR thermal profile can dramatically alter the success rate of degenerate PCR (Compton 1990). The preferred degenerate PCR amplification profile starts with a non-stringent annealing temperature (35–45°C) for 2–5 cycles, followed by 25–40 cycles at a more stringent annealing temperature. The relaxed annealing conditions allow the short complementary primer portion to hybridize to the target. After the second cycle of amplification, the 5′extension becomes incorporated into the amplified product and serves as the template for subsequent rounds of amplification. By shifting to a more stringent annealing temperature, increased specificity can be achieved. Furthermore, a 4- or 5-minute ramp time between the annealing and extension temperatures may also result in greater specificity.

Mutagenesis Primers

With the advent of PCR, recombinant DNA procedures such as M13 phage site-directed mutagenesis and the construction of "fusion proteins" have been greatly simplified (Higuchi 1990; Scharf 1990). There are several advantages in generating a mutation with mismatched primers by PCR. First, PCR provides substantial flexibility in the types of mutations that can be introduced. For example, single-base substitutions, deletions, and insertions of different sizes can all be easily generated with one or two rounds of PCR. Second, the mutation(s) of interest located in the PCR product can be cloned into any desirable vector by a single cloning step. Third, almost all the clones harbor the expected mutation.

Conventional methods for the construction of sequences encoding a fusion protein require several manipulations. In addition, the sequence at the junction is generally not ideal because the restriction enzyme selected for the joining of the two sequences is usually not present in the native sequence. Therefore, it is difficult to construct a sequence encoding an authentic fusion protein. With PCR, the junction of two segments can easily be engineered into the primers, thereby eliminating extraneous bases.

The location of the mutated base(s) in the primers depends on the nature of the desired mutation. To introduce single-base substitutions and insertions at a location with a unique restriction enzyme site nearby, only two primers are needed (Fig. 2A). A mismatched mutagenesis primer (P1) with the desired mutation sequence and restriction enzyme site sequence (site X) is paired with another primer (P2) with the second restriction enzyme site (site Y) for PCR.

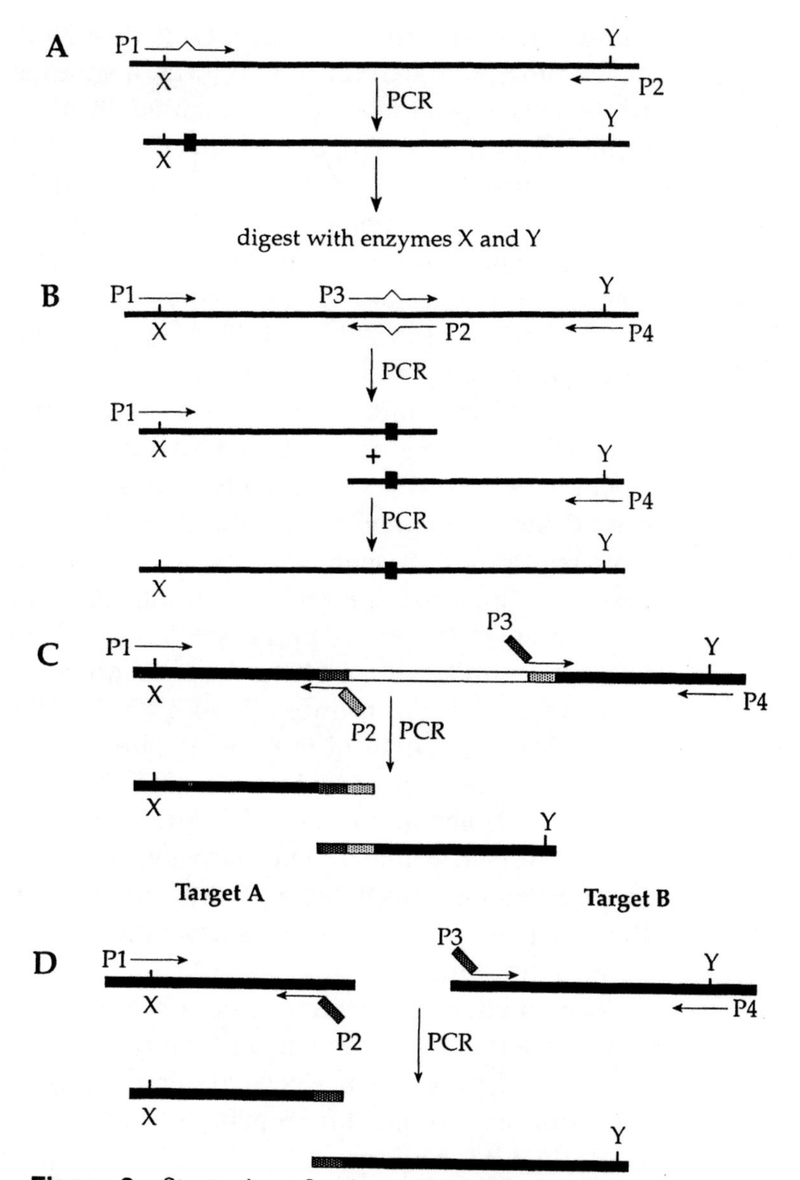

Figure 2 Strategies of using mismatched primers for generating mutations (*A–C*), and fusion of two fragments of DNA (*D*). Bends in arrows represent the mismatched sequence between primer and target. (*Closed box*) The mutation introduced by the mismatched primer. (X and Y) Two restriction enzyme sequences shown in the DNA sequence. The shaded areas with the same pattern have identical sequence.

The mismatch(es) should be placed in the middle of a 24- or 36-base oligonucleotide. The PCR product can be digested with enzymes X and Y and cloned into a suitable vector. For generating mutations in a sequence that does not contain a convenient restriction enzyme site nearby, two sets of primers are needed for two rounds of PCR (Fig. 2B). Primer 1 (P1) paired with mutagenesis primer P2 and mutagenesis primer 3 (complementary to primer P2) paired with primer 4 (P4) are used in two separate PCR assays in the first round of PCR. Following the first round of PCR, a small aliquot (1/20 to 1/50) of the two PCR products (or gel-purified PCR products) is mixed and amplified in the presence of primers 1 and 4 in the second round of PCR. Because the two PCR products from the first round of PCR have overlapping sequences at the "mutagenized end," one of the two partially matched duplexes formed after denaturation and annealing serves as the template in the second round of PCR. The PCR product from the second round of amplification is used for cloning.

The introduction of large deletions and the fusion of sequences can be accomplished in a similar manner. The mutagenesis primers for generating a large deletion should contain sequences flanking the deletion region (Fig. 2C). Both mutagenesis primers (P2 and P3) should have a region of more than 12 bases at the 3′terminus that matches the target sequence and a mismatched tail of more than 7 bases that matches the region flanking the other end of the deletion. The length of the overlapping sequences at the mutagenized end of two first-round PCR products is determined by the total number of bases in the tails of primers P2 and P3. For the joining of two target sequences, the length of the mismatched tail of mutagenesis primers should be more than 12 bases (Fig. 2D). Primers 1 and 4 illustrated in Figure 2B–D are in the regions with restriction enzyme sites for cloning of the PCR product. Alternatively, the restriction enzyme site for cloning can be introduced with the primer containing the sequence of the restriction site and a few extra bases to ensure enzyme digestion at the 5′terminus.

The length of the primer used in these experiments depends on the G/C content of the sequence and the type of desired mutation. In general, 12–14 bases of perfect complementarity (T_m ~40°C) at the 3′ terminus are sufficient for a primer to be annealed and extended, particularly if it is not especially AT-rich. The annealing temperature can also be controlled to achieve an efficient amplification.

If the nucleotide sequence of the mutated fragment is important for further analysis, the location of primer 2 in Figure 1A and primer pair P1/P4 in Figure 1B–D should be as close as possible to the mutagenesis primer(s) to minimize the number of nucleotides to be confirmed by sequencing. The conditions to minimize the misincorporation of *Taq* DNA polymerase should be observed. The parameters of the PCR reaction, such as the concentration of dNTP and $MgCl_2$, and pH, all have effects on the fidelity of the enzyme (Eckert and Kunkel 1990; Cha and Thilly, this volume). The amount of template and the number of cycles of amplification should be properly controlled to minimize amplification at the plateau phase. Other proofreading DNA polymerases, such as those for *Thermotoga maritima* (*UlTma*), *Thermococcus litoralis* (Vent), and *Pyrococcus furiosus* (*Pfu*), can be used as alternative enzymes for mutagenesis PCR. However, one should bear in mind that the proofreading activity of these enzymes also degrades primers, single-stranded PCR products, and template; therefore, their use may require careful optimization.

CONCLUSION

Although the published procedures noted were successful, very few comprehensive studies have been performed to explain fully the precise mechanisms involved. As a result, the amount of information available is perhaps less than what might have been expected given the large number of studies carried out. The procedures outlined above are intended to serve as guidelines for the design and use of mismatched and degenerate primers. Depending on the particular PCR application, a subset of these recommendations or suggestions may suffice. Familiarity with the accomplishments and limits of published studies, an understanding of the principles of enzymology and nucleic acid hybridization, and a willingness to reflect first on the goal before embarking on experimentation will increase an investigator's likelihood of success. PCR—like any good tool—can be productive with minimal familiarity. However, understanding the intricacies of how it works and the breadth of potential alterations permits greater exploration of its limits and its application to new areas.

ACKNOWLEDGMENTS

We thank Shari Kurita for graphics, Oxford University Press for granting us permission to publish Table 1, and JoAnn Williams for manuscript preparation. We thank David Gelfand for reviewing the manuscript. It is impossible to cite or acknowledge the large number of individuals both at Roche Molecular Systems and elsewhere who have contributed to the studies described in this manuscript. We thank all of them for their efforts.

REFERENCES

Aarts, J.M., J.G. Hontelez, P. Fischer, R. Verkerk, A. van Kamen, and P. Zabel. 1991. Acid phosphatase-1, a tightly linked molecular marker for root-knot nematode resistance in tomato: From protein to gene, using PCR and degenerate primers containing deoxyinosine. *Plant Mol. Biol.* **16:** 647–661.

Bartel, D.P. and J.W. Szostak. 1993. Isolation of new ribozymes from a large pool of random sequences. *Science* **261:** 1411–1418.

Bos, J.L. 1989. Ras oncogens in human cancer: A review. *Cancer Res.* **49:** 4682–4689.

Cha, R.S., H. Zarbl, P. Keohavong, and W.G. Thilly. 1992. Mismatch amplification mutation assay (MAMA): Application to the c-H-*ras* gene. *PCR Methods Appl.* **2:** 14–20.

Cheong, C., I. Tinoco, and A. Chollet. 1988. Thermodynamic studies of base pairing involving 2,6-diaminopurine. *Nucleic Acids Res.* **16:** 5115–5122.

Chou, Q., M. Russell, D.E. Birch, J. Raymond, and W. Bloch. 1992. Prevention of pre-PCR mis-priming and primer dimerization improves low-copy-number amplifications. *Nucleic Acids Res.* **20:** 1717–1723.

Compton, T. 1990. Degenerate primers for DNA amplification. In *PCR protocols: A guide to methods and applications* (ed. M.A. Innis et al.), pp. 39–45. Academic Press, San Diego, California.

Creighton, S., M. Huang, H. Cai, N. Arnheim, and M.F.

Goodman. 1992. Base mispair extension kinetics: Binding of avian myeloblastosis reverse transcriptase to matched and mismatched base pair termini. *J. Biol. Chem.* **267:** 2633–2639.

Dutton, C. and S.S. Sommer. 1991. Simultaneous detection of multiple single-base alleles at a polymorphic site. *BioTechniques* **11:** 700–702.

Eckert, K.A. and T.A. Kunkel. 1990. High fidelity DNA synthesis by *Thermus aquaticus* DNA polymerase. *Nucleic Acids Res.* **18:** 3739–3744.

Ehlen, T. and L. Dubeau. 1989. Detection of *ras* point mutations by polymerase chain reaction using mutation-specific, inosine-containing oligonucleotide primers. *Biochem. Biophys. Res. Commun.* **160:** 441–447.

Girgis, S.I., M. Sleviazaki, P. Denny, G.J.M. Ferrier, and S. Legon. 1988. Generation of DNA probes for peptides with highly degenerate codons using mixed primer PCR. *Nucleic Acids Res.* **16:** 10371.

Higuchi, R. 1990. Recombinant PCR. In *PCR protocols: A guide to methods and applications* (ed. M.A. Innis et al.), pp. 177–1831. Academic Press, San Diego, California.

Hoheisel, J.D., A.G. Craig, and H. Lehrach. 1990. Effect of 5-bromo- and 5-methyldeoxycytosine on duplex stability and discrimination of the NotI octadeoxynucleotide. *J. Biol. Chem.* **265:** 16656–1660.

Huang, M., N. Arnheim, and M.F. Goodman. 1992. Extension of base mispairs by Taq DNA polymerase: Implications for single nucleotide discrimination in PCR. *Nucleic Acids Res.* **20:** 4567–4573.

Kahn, S.M., W. Jiang, T.A. Culbertson, I.B. Weistein, G.M. Williams, N. Tomita, and Z. Ronai. 1991. Rapid and sensitive nonradioactive detection of mutant K-*ras* genes via "enriched" PCR amplification. *Oncogene* **6:** 1079–1083.

Kidd, V.J., R.B. Wallace, K. Itakura, and S.L.C. Woo. 1983. α_1-Antitrypsin deficiency detection by direct analysis of the mutation in the gene. *Nature* **304:** 230–234.

Kopin, A.S., M.B. Wheeler, and A.B. Leiter. 1990. Secretin: Structure of the precursor and tissue distribution of the mRNA. *Proc. Natl. Acad. Sci.* **87:** 2299–2303.

Kwok, S., D. Kellogg, G. Ehrlich, B. Poiesz, B. Bhagavati, and J.J. Sninsky. 1988. Characterization of a sequence of human T cell leukemia virus type I from a patient with chronic progressive myelopathy. *J. Inf. Dis.* **158:** 1193–1197.

Kwok, S., D.E. Kellogg, N. McKinney, D. Spasic, L. Goda, C. Levenson, and J.J. Sninsky. 1990. Effects of primer-template mismatches on the polymerase chain reaction: Human immunodeficiency virus type 1 model studies. *Nucleic Acids Res.* **18:** 999–1005.

Larder, B.A. and C.A.B. Boucher. 1993. PCR detection of human immunodeficiency virus drug resistance mutations. In *Diagnostic molecular biology* (ed. D.H. Persing et al.), pp. 527–533. American Society for Microbiology, Washington, D.C.

Larder, B.A., P. Kellan, and S.D. Kemp. 1991. Zidovudine resistance predicted by direct detection of mutations in DNA from HIV-infected lymphocytes. *AIDS* **5:** 137–144.

Lawyer, F.C., S. Stoffel, R.K. Saiki, K. Myambo, R. Drummond, and D.H. Gelfand. 1989. Isolation, characterization, and expression in *Escherichia coli* of the DNA polymerase gene from *Thermus aquaticus. J. Biol. Chem.* **264:** 6427–6436.

Lee, C.C., X. Wu, R.A. Gibbs, R.G. Cook, D.M. Muzny, and C.T. Caskey. 1988. Generation of cDNA probes directed by amino acid sequence: Cloning of urate oxidase. *Science* **239:** 1288–1291.

Li, H., X. Cui, and N. Arnheim. 1990. Direct electrophoretic detection of the allelic state of single DNA molecules in human sperm by using the polymerase chain reaction. *Proc. Natl. Acad. Sci.* **87:** 4580–4584.

Lin, P.K.T. and D.M. Brown. 1992. Synthesis of oligodeoxyribonucleotides containing degenerate bases and their use as primers in the polymerase chain reaction. *Nucleic Acids Res.* **20:** 5149–5152.

Lo, Y.-M.D., P. Patel, C.R. Newton, A.F. Markham, K.A. Fleming, and J.S. Wainscoat. 1991. Direct haplotype determination by double ARMS: Specificity, sensitivity and genetic applications. *Nucleic Acids Res.* **19:** 3561–3567.

Mack, D. and J.J. Sninsky. 1988. A sensitive method for the identification of uncharacterized viruses related to known virus groups: Hepadnavirus model system. *Proc. Natl. Acad. Sci.* **85:** 6977–6981.

Manos, M.M., Y. Ting, D.K. Wright, A.J. Lewis, T.R. Broker, and S.M. Wolinsky. 1989. Use of polymerase chain reaction amplification for the detection of genital human papillomaviruses. *Cancer Cells* **7:** 209–214.

Marotta, C.A., J.T. Wilson, B.J. Forget, and S.M. Weissman. 1977. Human β globin messenger RNA. *J. Biol. Chem.* **252:** 5040–5053.

Martin, F.H., M.M. Castro, F. Aboulela, and I. Tinoco, Jr. 1985. Base pairing involving deoxyinosine: Implication for probe design. *Nucleic Acids Res.* **13:** 8927–8938.

Mendelman, L.V., M.S. Boosalis, J. Petruska, and M.F. Goodman. 1989. Nearest neighbor influences on DNA polymerase insertion fidelity. *J. Biochem.*

264: 14415–14423.

Newton, C.R., A. Graham, L.E. Heptinstall, S.J. Powell, C. Summers, N. Kalshekar, J.C. Smith, and A.F. Markham. 1989. Analysis of any point mutation in DNA. The amplification refractory mutation systems (ARMS). *Nucleic Acids Res.* **17:** 2503–2516.

Nichols, R., P.C. Andrews. P. Zhang, and D.E. Bergstrom. 1994. A universal nucleoside for use at ambiguous sites in DNA primers. *Nature* **369:** 492–493.

Nunberg, J.H., W.A. Schleif, E.J. Boots, J.A. O'Brien, J.C. Quintero, J.M. Hoffman, E.A. Emini, and M.E. Goldman. 1991. Viral resistance to human immunodeficiency virus type 1-specific pyridinone reverse transcriptase inhibitors. *J. Virol.* **65:** 4887–4892.

Ohtsuka, E., S. Matsuka, M. Ikehara, Y. Takahashi, and K. Matsubara. 1985. An alternative approach to deoxyoligonucleotides as hybridization probes by insertion of deoxyinosine at ambiguous codon positions. *J. Biol. Chem.* **260:** 2605–2608.

Sakanari, J.A., C.E. Staunton, A.E. Eakin, and C.S. Craik. 1989. Serine proteases from nematode and protozoan parasites: Isolation of sequence homologs using generic molecular probes. *Proc. Natl. Acad. Sci.* **86:** 4863–4867.

Sarkar, G. and S.S. Sommer. 1991. Haplotyping by double PCR amplification of specific alleles. *BioTechniques* **10:** 436–440.

Scharf, S.J. 1990. Cloning with PCR. In *PCR protocols: A guide to methods and applications* (ed. M.A. Innis et al.), pp. 84–91. Academic Press, San Diego, California.

Schuchman, E.H., C.E. Jackson, and R.J. Desnick. 1990. Human arylsufatase B: MOPAC cloning, nucleotide sequence of a full-length cDNA, and regions of amino acid identity with arylsufatases A and C. *Genomics* **6:** 149–158.

Shih, A., R. Misra, and M.G. Rush. 1989. Detection of multiple, novel reverse transcriptase coding sequences in human nucleic acids: Relation to primate retroviruses. *J. Virol.* **63:** 64–75.

St. Clair, M.H., J.L. Martin, G. Tudor-Williams, M.C. Bach, C.L. Vavro, D.M. King, P. Kellan, S.D. Kemp, and B.A. Larder. 1991. Resistance to ddI and sensitivity to AZT induced by a mutation in HIV-1 reverse transcriptase. *Science* **253:** 1557–1559.

Tada, M., M. Omata, S. Kawai, H. Saisho, M. Ohto, R.K. Saiki, and J.J. Sninsky. 1993. Detection of *ras* gene mutations in pancreatic juice and peripheral blood of patients with pancreatic adenocarcinoma. *Cancer Res.* **53:** 1–3.

Ugozzoli, L. and R.B. Wallace. 1991. Allele-specific polymerase chain reaction. *Methods* **2:** 42–48.

Wada, K., Y. Wada, H. Doi, F. Ishibashi, G. Takashi, and T. Ikemura. 1991. Codon usage tabulated from the GenBank genetic sequence data. *Nucleic Acids Res.* **19:** 1981.

Wilks, F.F., R.R. Kurban, C.M. Hovens, and S.J. Ralph. 1989. The application of the polymerase chain reaction to cloning members of the protein tyrosine kinase family. *Gene* **85:** 67–74.

Multiplex PCR

Mary C. Edwards and Richard A. Gibbs

Institute for Molecular Genetics, Baylor College of Medicine, Houston, Texas 77030

INTRODUCTION

Ever since it was shown that PCR could simultaneously amplify multiple loci in the human dystrophin gene (Chamberlain et al. 1988), multiplex PCR has been firmly established as a general technique. A short list of multiplex PCR applications now includes pathogen identification, gender screening, linkage analysis, forensic studies, template quantitation, and genetic disease diagnosis. Multiplex PCR can be a two-amplicon system or it can amplify 13 or more separate regions of DNA. It may be the end point of analysis, or it may be preliminary to further analyses such as sequencing or hybridization. The steps for developing a multiplex PCR and the benefits of having multiple fragments amplified simultaneously, however, are similar in each system.

This chapter focuses on multiplex systems in which each primer pair targets a single locus, unlike RAPD (Williams et al. 1990) and alumorph (Zietkievicz et al. 1992) PCR, which amplify multiple loci with a single primer or primer set.

ADVANTAGES OF MULTIPLEX PCR

Internal Controls

Potential problems in PCR include false negatives due to reaction failure or false positives due to contamination. False negatives are often revealed in multiplex amplification because each amplicon provides an internal control for the other amplified fragments. For example, multiple exons may be amplified in assays that survey for gene deletions. Unless the entire region scanned by the multiplex PCR is deleted, amplification of some fragment(s) indicates that the reaction has not failed (Fig. 1A). Furthermore, because major deletions are usually contiguous (Chamberlain et al. 1992), results that suggest noncontiguous deletions based on the absence of bands usually reflect

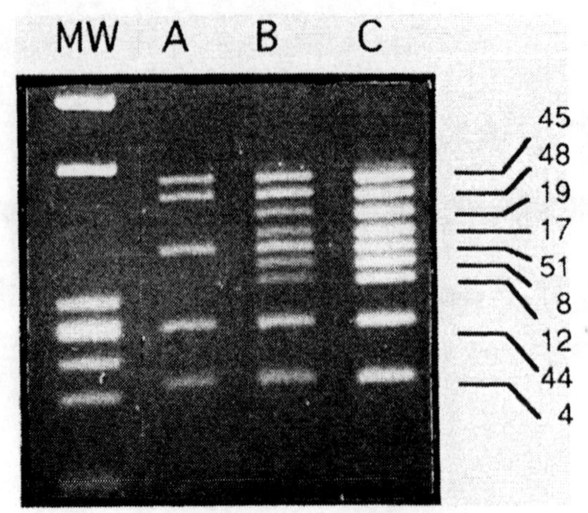

Figure 1 Multiplex amplification of nine exons in the human dystrophin gene (Chamberlain 1989). The normal pattern of amplification, indicating the presence of these exons in the patient, is shown in lane *C*. Lane *A* is missing bands corresponding to exons 8, 12, 17, and 19, suggesting that the patient has a deletion in the region of exons 8–19. The presence of some amplified bands serves to delineate the extent of the deletion and to indicate that the reaction has not failed. Lane *B* shows bands with approximately half the intensity for the exons that were not amplified in patient A, suggesting that this patient is a carrier for a deletion in that region. This demonstrates potential for a simple method of quantifying templates by multiplex PCR. The molecular-weight marker is ΦX174 DNA, digested with *Hae*III.

artifactual failure of some fragments to amplify. Complete PCR failure can be distinguished from an informative no-amplification result by adding a control amplicon external to the target sequence to the reaction (Ballabio et al. 1990; Levinson et al. 1992). In addition to monitoring PCR failure and artifacts, internal control amplicons can be designed to verify the presence of a target template. In multiplex assays where closely related templates such as pathogen strains are distinguished by amplifying differing sequences, primers for a sequence common to all templates provide a positive control for amplification (Bej et al. 1990, 1991; Kaltenboek et al. 1992; Wilton and Cousins 1992; Way et al. 1993).

Indication of Template Quality

The quality of the template may be determined more effectively in multiplex than in single locus PCR. Degraded templates give weaker signals for long PCR products than for short ones (Chamberlain et al. 1992). A loss in amplification efficiency due to PCR inhibitors in the template samples can be indicated by reduced amplification of an abundant control sequence in addition to the amplification of rarer target sequences in an otherwise standardized reaction (van der Vliet et al. 1993).

Indication of Template Quantity

The exponential amplification and internal standards of multiplex PCR can be used to assess the amount of a particular template in a sample. To quantitate templates accurately by multiplex PCR, the amount of reference template, the number of reaction cycles, and the minimum inhibition of the theoretical doubling of product for each cycle must be determined (Ferre 1992). In the simplest method of quantitation, the gene multiplexes for major deletions detect carriers or duplications in probands when the band intensity of abnormal amplicons is compared with that of normal, homozygous fragments in the multiplex (Fig. 1B) (Chamberlain et al. 1989; Gibbs et al. 1990; Abbs and Bobrow 1992; Ioannou et al. 1992). Cycling conditions for carrier testing must be determined carefully, because the variation in amplicon intensities will be masked if the reaction is allowed to cycle until the signal is saturated (Chamberlain et al. 1991). The diagnosis is most accurate when at least two other fragments are used in the comparison (Abbs and Bobrow 1992) and the analysis is performed by densitometry, fluorescent scanning on an automated DNA sequencer, or analysis of charge-coupled device camera images (M. Metzger, pers. comm.). Furthermore, preliminary studies suggest that signal intensities of fluorescent multiplex PCR products may reflect relative amounts of mixed, disproportionate DNAs in forensic samples (Klimpton et al. 1993).

The majority of multiplex quantitation assays compare the signal intensity of a reference sequence to the signal from another sequence in the same reaction, either directly or by extrapolating the result to standard curves (Ferre 1992). There are numerous coamplification assays based on this principle in the literature of competitive PCR with RNA or DNA standards.

Efficiency

The expense of reagents and the preparation time are less in multiplex PCR than in systems where several tubes of uniplex PCRs are used. A multiplex reaction is ideal for conserving costly polymerase and templates in short supply. For maximum efficiency of preparation time, the reactions can be prepared in bulk, randomly tested for quality, and stored frozen without enzyme or template until use (Chamberlain et al. 1988).

DESIGN AND DEVELOPMENT

Producing some multiplex PCR systems may be as simple as combining two sets of primers for which reaction conditions have been determined separately. However, other multiplex PCRs must be developed with careful consideration for the regions to be amplified, the relative sizes of the fragments, the dynamics of the primers, and the optimization of PCR technique to accommodate multiple fragments.

The following steps are involved in multiplex PCR. These steps are discussed in detail below.

1. Choose loci for analysis.
   a. Determine the PCR and detection system to be used.
   b. Distribute amplicons (localized at mutation hot spots, linked to genes, chromosomally unlinked, grouped close exons in a single amplicon, etc.).
   c. Design internal control fragment(s) (other exons, external sequences, host sequence, sequence conserved in all target templates, etc.).
2. Position primers in regions of well-characterized sequence, in relation to amplicon sizes.
3. Design primers with similar physical properties and reaction kinetics.
4. Develop PCR conditions separately for each primer set.
5. Add primer sets sequentially, altering conditions as necessary. Reduce nonspecific amplification (hot start, ionic detergents, short extension times, hottest annealing, change primer pairs as required). Vary relative concentrations of primer sets for equal amplification. Change buffer systems if necessary.
6. Adjust reaction components and cycling conditions for multiplex amplification. Mg^{++}, dNTP, and polymerase requirements may increase. Ideal extension times may be longer.

Selection of Multiplex Loci

The regions selected for multiplex amplification may be determined by the nature of the analysis; for example, deletion assays amplify exons, forensic assays distinguish individual variation at highly polymorphic markers, and microbial assays may exploit strain- or species-specific variation. The type of analysis to be done may lend itself to a PCR technique that was originally applied to uniplex amplification but that can be adapted for multiplex purposes, such as allele-specific PCR (Fortina et al. 1992b; Uggozoli and Wallace 1992), restriction site-generating PCR (Ng et al. 1991; Cremonesi et al. 1992; Gasparini et al. 1992), amplification refractory mutation system (ARMS) (Beinvenu et al. 1992; Ferrie et al. 1992; Fortina et al. 1992a;

Jawaheer et al. 1993), color complementation assay (Chehab and Kan 1989), and nested PCR (Levinson et al. 1992; Repp et al. 1993; Zazzi et al. 1993). For deletion assays of genes with many exons, the amplicons can be distributed to scan a wide region or a concentrated region at deletion hot spots. Close exons can be amplified by a pair of primers spanning both (Gibbs et al. 1990). Multiplex amplicons that distinguish similar templates, such as virus types, bacteria strains, or gene alleles, are ideally located in regions that are not extremely variable, where any given amplicon might contain the sequence of several genotypes (Repp et al. 1993). Unlike the exon-specific multiplexes in which multiple signals provide built-in positive controls, other multiplexes, such as those identifying pathogens, rely on a differential signal to distinguish templates and so benefit from an additional primer set to provide an internal control.

Positioning of Primers

Detailed sequence information for primer sites at the selected loci is important, because nonspecific amplification may occur at other sites with similar sequences (Chamberlain et al. 1991), or reduced amplification may occur at primer-template mismatched sites (Zazzi et al. 1993). Primers for exon-amplifying multiplexes are ideally placed in intronic sequences adjacent to the exons. This provides some margin for adjustment of fragment length or amplification quality as well as possible information about alterations affecting splice sites (Chamberlain et al. 1988). If the multiplex product is to be resolved electrophoretically, fragment sizes should be selected so that they may be separated easily from each other. The range of band sizes should not be so wide that all fragments cannot be resolved well on the same gel. However, with the use of fluorescently labeled primers, product ranges may overlap and yet be distinguished by color (Chehab and Kan 1989; Edwards et al. 1991; Ziegle et al. 1992; Klimpton et al. 1993). Fluorescently labeled multiplex primers aid diagnostics by representing product amounts more accurately than ethidium bromide stain and by reducing reaction time and nonspecific amplification with fewer PCR cycles needed to obtain a signal (Chamberlain et al. 1991).

Developing Primers and Reaction Conditions

Primer sequences should be designed so that their predicted hybridization kinetics are similar to those of other primers in the multiplex reaction. A G/C content of 40–60% and a length of 23–28 nucleotides are suggested as general guidelines for specific annealing at moderate temperatures (Gibbs et al. 1989). Primer annealing temperatures and concentrations may be calculated to some extent, but conditions will almost certainly have to be refined empirically in multiplex PCR. Conditions for each set of primers should be developed individually and modified if necessary as primer sets are added. Primer pairs that work separately but not when combined may be improved by a prior ethanol precipitation in 0.3 M sodium acetate (Beggs et al. 1990). The possibility of nonspecific priming and other artifacts is increased with each additional primer. Thus, primer pairs that give a "clean" signal alone but produce artifact bands in multiplex PCR may benefit from hot start PCR (Zazzi et al. 1993), the addition of organics, annealing at the highest possible temperature, or, if all else fails, reselection of the primer sequence (Chamberlain et al. 1991).

If equimolar primer concentrations do not yield uniform amplification signals for all fragments, the concentration of some primer pairs can be reduced in relation to others. This is particularly important in samples where one target is more abundant than others (Sunzeri et al. 1991). An inverse relation between the required oligonucleotide concentration in multiplex PCR and its A/T content, but not its length or melting temperature, has been suggested (Vandenvelde et al. 1990; Zazzi et al. 1993).

When all primer pairs are not compatible, it may be necessary to subgroup them in smaller multiplexes (Richards et al. 1991b). However, an earlier maximum of 18 primers in a multiplex (Chamberlain et al. 1989) has been surpassed recently by a dystrophin gene multiplex containing 26 primers (M. Morsey et al., in prep.) by further optimizing reaction conditions.

Titration of Reaction Components

It may be necessary to adjust concentrations of various reaction components to achieve a robust multiplex PCR. Mg^{++} and dNTP requirements generally increase with the number of amplicons in multiplex PCR, but the concentrations must be optimized because each primer pair may have different requirements (Chamberlain et al. 1991). Likewise, polymerase requirements generally increase with the number of primer pairs in the multiplex PCR (Chamberlain et al. 1991). Buffer systems may affect amplification dramatically. For example, the buffer recommended by the Cetus Corporation allows complete amplification of normal DNA in one Duchenne muscular dystrophy (DMD) 9-exon multiplex PCR (Beggs et al. 1990) but not in another (Chamberlain et al. 1989). Dimethylsulfoxide was found to be a beneficial ingredient (Vandenvelde et al. 1990; Uggozoli and Wallace 1992) or an inhibitor (Zazzi et al. 1993) in different multiplex systems. Other additives that minimize nonspecific binding in multiplex PCR are Tween 20 and Triton X-100 (Levinson et al. 1992), β-mercaptoethanol (Chamberlain et al. 1988; Gibbs et al. 1990; Runnebaum et al. 1991), and tetramethylammonium chloride (Uggozoli and Wallace 1992).

Adapting Thermal Cycling Conditions

Thermal cycling parameters are also determined largely by the sequence of the primer sets. Generally, extension times should be increased with the number of loci amplified in the reaction (Chamberlain et al. 1991). Doubling this time parameter in trial experiments is reasonable. However, long extension and annealing times could provide an opportunity for nonspecific amplification (Chamberlain et al. 1991).

Competition and Interference

An aspect of PCR that may be exacerbated in multiplex reactions is competition for resources and resulting artifacts. Differences in the yields of unequally amplified fragments are enhanced with each cycle (Ferrie et al. 1992). Sets of amplicons of varying lengths but similar sequence may show preferential amplification of the shortest, particularly if they share a common primer. This may be due to limited processivity or suppressed amplification of the outer, longer amplicon by the inner, shorter one when primers anneal on the same strand (Repp et al. 1993). This effect can be circumvented by initiating PCR with the long amplicon primers and by adding the primer for the shorter amplicon some cycles later (Bourque et al. 1993), or by using a low concentration of the short amplicon primer (Repp et al. 1993). In multiplex PCR of homologous amplicons with very similar lengths and no shared primers, no competition was reported (Manam and Nichols 1991). Suppressed amplification of one amplicon by another has been noted in a multiplex PCR in which sequence and primers were not shared (Bej et al. 1990), but coamplification was resolved by initiating the limited amplicon several cycles before the other.

Primer-template mismatches have been noted to be at a disadvantage relative to perfect matches in multiplex PCR, presumably due to competition for binding to the polymerase (Zazzi et al. 1993). When an ambiguous negative result for the presence of multiple HIV-1 sequences was suggested at a mismatched site, a uniplex of the failed amplicon generated a product (Zazzi et al. 1993).

Multiple sets of primers increase the possibility of primer complementarity at the 3′ends, leading to primer-dimers. These artifacts deplete the reaction of dNTPs and primers and outcompete the multiplex amplicons for polymerase (Vandenvelde et al. 1990; Bourque et al. 1993). This effect can be reduced by titrating primer concentrations and cycling conditions.

Post-PCR Analysis

More extensive analysis than gel electrophoresis of multiplex products is requisite for some systems, particularly those that identify point mutations or other small alterations. Additionally, the complexity of some multiplex reactions makes verification of specific PCR products by such methods as probing and sequencing desirable, even if limited to the development phase. Many of the techniques for product analysis developed for uniplex PCR can be applied directly to multiplex PCR.

A second multiplex reaction can be generated by using the product of the first as a template when high specificity is required (Zazzi et al. 1993). Alternatively, the second reaction may be based on the results of the first. In an example of the latter, carrier diagnosis is made by multiplexed simple tandem repeats (STRs) if deletions are not found by DMD exon multiplex PCR (Clemens et al. 1991; Schwartz et al. 1992). Uniplex reactions may be generated from multiplex reactions (Gibbs et al. 1990; Killimann et al. 1992; Levinson et al. 1992), or vice versa (Repp et al. 1993).

The product of a multiplex PCR may be sequenced to reveal new mutations or small alterations where major deletions are not present. This may be done directly from the multiplex reaction product (Manam and Nichols 1991; Lo et al. 1992), or the product may require further preparation prior to sequencing. The introduction of biotinylated and universal-tailed primers in nested PCR following multiplex PCR allows solid-phase sequencing of exons and flanking intronic sequence for small alterations (Gibbs et al. 1990; Killimann et al. 1992). Other multiplex reactions have been subcloned prior to sequencing in the development phase (Lohmann et al. 1992; Repp et al. 1993).

Treatment with restriction enzymes (Ng et al. 1991; Cremonesi et al. 1992; Gasparini et al. 1992; Levinson et al. 1992; Picci et al. 1992; Jawaheer et al. 1993; O'Keefe and Dobrovic 1993), hybridization of product with a probe (Serre et al. 1991; Sunzeri et al. 1991; Chehab and Wall 1992; Wattel et al. 1992; Golmolka et al. 1993; Shuber et al. 1993; van der Vliet et al. 1993; Vesy et al. 1993), and single-strand conformational polymorphism (SSCP) analysis (Runnebaum et al. 1991; Lo et al. 1992; Nigro et al. 1992) are also used routinely to analyze multiplex PCR products. The choice of the particular method is guided by the sequence of the amplified fragments. For example, mutations that abolish restriction sites may be detected by enzymatic digestion, whereas broad screening for sequence variants may be accomplished by SSCP. Finally, PCR multiplex products can be of use as a molecular-weight ladder because the lengths of the amplified fragments are known.

APPLICATIONS OF MULTIPLEX PCR

Gene Deletion and Mutation Detection

Major deletion multiplexes of X-linked human disease genes are designed to give positive or negative indication of the presence of an exon. The hypoxanthine phosphoribosyltransferase (HPRT) gene multiplex for Lesch–Nyhan syndrome (Gibbs et al. 1990) and the α-galactosidase A gene multiplex for Fabry disease (Kornreich and Desnick 1993) each amplify all coding regions. However, dystrophin gene multiplexes for Duchenne/Becker muscular dystrophy (DMD/BMD) (Chamberlain et al. 1989; Beggs et al. 1990; Abbs and Bobrow 1992; Covone et al. 1992) concentrate on hot-spot regions of deletion. The steroid sulfatase gene multiplex detects what are frequently whole-gene deletions (Ballabio et al. 1990). These reactions can provide a template for the detection of finer alterations if major deletions are not identified in a sample (Gibbs et al. 1990; Killimann et al. 1992; Nigro et al. 1992).

Mutations and small deletions in genes are detected by multiplex assays either directly by PCR or by subsequent analysis of PCR products. Results may be determined immediately by gel electrophoresis for RB1 gene exons amplified to reveal small deletions causing retinoblastoma (Lohmann et al. 1992) and for multiplex ARMS reactions of common population-specific β-thalassemia (Beinvenu et al. 1992; Fortina et al. 1992a) and cystic fibrosis (Ferrie et al. 1992; Fortina et al. 1992b) mutations. Several mutation types may be examined simultaneously, as in a multiplex reaction that detects a point mutation, a 4-base deletion, and complete deletion of the α-globin genes (Chehab and Kan 1989).

Other mutation-amplifying multiplexes rely on post-PCR manipulation of the reaction product for diagnosis. SSCP detects human p53 tumor suppressor gene mutations associated with breast cancer (Runnebaum et al. 1991). Cycle sequencing reveals activation mutations in mouse *ras* oncogenes (Manam and Nichols 1991). Techniques to identify the numerous, often population-specific, mutations in the cystic fibrosis transmembrane conductance regulator (CFTR) gene have spawned several multiplex systems: hybridization of exons to mutation-specific oligonucleotides (Serre et al. 1991; Chehab and Wall 1992; Shuber et al. 1993), restriction enzyme digestion of natural restriction sites at amplified mutations (Picci et al. 1992) or at primer-created restriction sites (Ng et al. 1991; Cremonesi et al. 1992; Gasparini et al. 1992), and use of allele-specific primers (Fortina et al. 1992b).

Genotyping by multiplex PCR employs similar techniques. ABO blood group alleles are distinguished by allele-specific primers (Uggozoli and Wallace 1992) or by enzymatic digestion of amplified product (O'Keefe and Dobrovic 1993). HLA–DR4 variants, associated with autoimmune diseases, are typed by multiplex ARMS (Jawaheer et al. 1993).

Multiplex PCR of sequence tagged sites has aided the physical mapping of breakpoints and loci on chromosome 16 using somatic cell hybrids (Richards et al. 1991a) and of the X chromosome in deletion patients (Worley et al. 1992). Other mapping applications are discussed below.

Polymorphic Repetitive DNA

Repetitive DNA polymorphisms are multiplexed for mapping, disease linkage, gender determination, and DNA typing/identification. Short tandem repeats (STRs) of 1–6 bp are convenient for multiplexing because they are numerous, highly polymorphic (Beckmann and Weber 1991), and may be coamplified without overlapping size ranges (Edwards et al. 1992). Multiplexes of relatively close repeats are employed for disease linkage, but chromosomally unlinked repeats are used for the identification of individuals (Edwards et al. 1991).

Multiplex PCR is an ideal technique for DNA typing because the probability of identical alleles in two individuals decreases with the number of polymorphic loci examined. Reactions have been developed with potential applications in paternity testing, forensic identification, and population genetics (Edwards et al. 1991, 1992; Klimpton et al. 1993).

Multiplexed polymorphic repeats determine whether family members have inherited an identical chromosome to the proband. Generally, diagnosis by STR markers is performed when assays to locate the mutation directly, such as the gene multiplex PCRs described above, are not informative. These assays lend themselves to multiplexing because examination of more than one marker linked to the disease gene reduces the possibility of missing recombination events and because occasional new mutations in an STR marker might suggest a mistyped or mislabeled individual unless other markers are examined (Mulley et al. 1991). $(CA)_n$ repeats have been multiplexed for the diagnosis of myotonic dystrophy (Mulley et al. 1991), DMD (Clemens et al. 1991; Schwartz et al. 1992), cystic fibrosis (Estivill et al. 1991; Morral and Estivill 1992), and Prader-Willi/Angelman syndromes (Mutirangura 1993). STRs multiplexed for examining potential associations of *V*β*6* human T-cell receptors with disease led to the identification of new gene family members (Golmolka et al. 1993).

Genetic mapping with multiplex-amplified STRs can augment physical mapping. Comparison of recombination frequencies among markers indicates their relative positions. Multiplex PCR aided the ordering of tightly linked chromosome 9 repeats (Furlong et al. 1993), and numerous X chromosome $(CA)_n$ repeats have been mapped (Huang et al. 1992; Worley et al. 1992). Such multiplexes may also assist the physical mapping of yeast artificial chromosomes (Edwards et al. 1991).

Repetitive DNA loci are incorporated in gender-determining multiplex assays. Embryos of families with X-linked disease can be sexed by coamplifying a Y-specific repetitive DNA locus with a gene sequence on both X and Y chromosomes (Levinson et al. 1992); multiplex results are confirmed by subsequent analyses. The Y-specific STR is amplified with an X-specific STR in a multiplex for forensic specimens (Pfitzinger et al. 1993).

Microbe Detection and Characterization

PCR analysis of bacteria is advantageous, because the culturing of some pathogens is lengthy or not possible. Bacterial multiplexes indicate a particular pathogen among others, or distinguish species or strains of the same genus. An amplicon of sequence conserved among several groups is often included in the reaction to indicate the presence of phylogenetically or epidemiologically similar, or environmentally associated, bacteria and to signal a functioning PCR. Multiplex assays with this format distinguish species of *Legionella* (Bej et al. 1990), *Mycobacterium* (Wilton and Cousins 1992), *Salmonella* (Way et al. 1993), *Escherichia coli*, and *Shigella* (Bej et al. 1991) and major groups of *Chlamydia* (Kaltenboek et al. 1992) from other genus members or associated bacteria. An assay for *Mycobacterium leprae* coamplifies human and pathogen DNA (van der Vliet et al. 1993). Multiplex assays differentiate forms of the insecticidal protein crystal-producing *Bacillus thuringiensis* (Bourque et al. 1993), Shiga-like toxin-producing *E. coli* (Gannon et al. 1992), and yeast (Pearson and McKee 1992).

Viral DNA is amplified by multiplex PCR to screen tissue samples or to examine associations of infection with disease. A fragment from the host genomic DNA is generally coamplified in these assays (Vandenvelde et al. 1990; Sunzeri et al. 1991; Zazzi et al. 1993). Human papillomavirus (HPV) associations with carcinomas or lesions (Soler et al. 1991; Toh et al. 1992), and adenovirus 12 with celiac disease (Vesy et al. 1993), have been examined. Multiplex assays detect or screen for HPV (Vandenvelde et al. 1990), human immunodeficiency virus type 1 (HIV-1) and human T-cell leukemic viruses (Sunzeri et al. 1991), human T lymphotrophic virus types I and II (Wattel et al. 1992), hepatitis B virus (Repp et al. 1993), parvovirus B19 (Sevall 1990), and hog cholera viruses (Wirz et al. 1993). HIV-1 infection can be detected by nested multiplexes of conserved regions (Zazzi et al. 1993).

CONCLUSIONS

The properties of multiplex PCR, including internal controls, indications of template quantity and quality, and less expense of time and reagents, make the technique a useful general tool and preferable in many instances to simultaneous uniplex PCR. Multiplex PCR methods exhibit great flexibility in experimental design and in overcoming limiting primer kinetics and fragment competition. A number of uniplex PCR techniques have been adapted to multiplex amplification for the diagnosis of genetic and infectious disease; for identification of individuals, populations, and pathogens; and to aid in defining the organization of the human genome.

Given the usefulness already demonstrated by multiplex PCR, future applications should be numerous. Maximizing the number of regions that may be concurrently amplified would have practical applications in genetic disease diagnosis at loci without apparent mutation hot spots or with many new mutations, and loci at which population-specific mutations have necessitated the development of multiplex reactions oriented to single populations. Multiplex amplification should be ideal whenever two or more sequences are examined by PCR for associations such as genetic linkage, environmental associations, and host/parasite and disease/infection relationships. The general refining of the PCR technique may be aided by multiplex amplification in that each additional primer set in the reaction often provides a challenge for optimizing the technique. Multiplex quantitation systems also provide information on the PCR technique by revealing influences on the exponential generation of product.

REFERENCES

Abbs, S. and M. Bobrow. 1992. Analysis of quantitative PCR for the diagnosis of deletion and duplication carriers in the dystrophin gene. *J. Med. Genet.* **29:** 191–196.

Ballabio, A., J.E. Ranier, J.S. Chamberlain, M. Zollo, and C.T. Caskey. 1990. Screening for steroid sulfatase (STS) gene deletions by multiplex DNA amplification. *Hum. Genet.* **84:** 571–573.

Beckmann, J.S. and J.L. Weber. 1991. Survey of human and rat microsatellites. *Genomics* **12:** 627–631.

Beggs, A.H., M. Koenig, F.M. Boyce, and L.M. Kunkel. 1990. Detection of 98% DMD/BMD gene deletions by PCR. *Hum. Genet.* **86:** 45–48.

Beinvenu, T., P. Sebillon, D. Labie, J.C. Kaplan, and C. Beldjord. 1992. Rapid and direct detection of the most frequent Mediterranean beta-thalassemic mutations by multiplex allele-specific enzymatic amplification. *Hum. Biol.* **64:** 107–113.

Bej, A.K., S.C. McCarty, and R.M. Atlas. 1991. Detection of coliform bacteria and *Escherichia coli* by multiplex polymerase chain reaction: Comparison with defined substrate and plating methods for water quality monitoring. *Appl. Environ. Microbiol.* **57:** 1473–1479.

Bej, A.K., M.H. Mahbubani, R. Miller, J.L. DiCesare, L. Haff, and R.M. Atlas. 1990. Multiplex PCR amplification and immobilized capture probes for detection of bacterial pathogens and indicators in water. *Mol. Cell. Probes* **4:** 353–365.

Bourque, S.N., J.R. Valero, J. Mercier, M.C. Lavoie, and R.C. Lavesque. 1993. Multiple polymerase chain reaction for detection and differentiation of the microbial insecticide *Bacillus thuringiensis. Appl. Environ. Microbiol.* **59:** 523–527.

Chamberlain, J.S., R.A. Gibbs, J.E. Ranier, and C.T. Caskey. 1991. Detection of gene deletions using multiplex polymerase chain reactions. In *Methods in molecular biology: Protocols in human molecular genetics* (ed. C. Mathew), vol. 9, pp. 299–312. Humana Press, Clifton, New Jersey.

Chamberlain, J.S., R.A. Gibbs, J.E. Ranier, P.N. Nguyen, and C.T. Caskey. 1988. Deletion screening of the Duchenne muscular dystrophy locus via multiplex DNA amplification. *Nucleic Acids Res.* **16:** 11141–11156.

—— 1989. Multiple PCR for the diagnosis of Duchenne muscular dystrophy. In *PCR protocols: A guide to methods and applications* (ed. D.H. Gelfand et al.), pp. 272–281. Academic Press, San Diego, California.

Chamberlain, J.S. and 33 coauthors. 1992. Diagnosis of Duchenne and Becker muscular dystrophies by polymerase chain reaction: A multicenter study. *J. Am. Med. Assoc.* **267:** 2609–2615.

Chehab, F.F. and Y.W. Kan. 1989. Detection of specific DNA sequences by fluorescence amplification: A color complementation assay. *Proc. Natl. Acad. Sci.* **86:** 9178–9182.

Chehab, F.F. and J. Wall. 1992. Detection of multiple cystic fibrosis mutations by reverse dot blot hybridization: A technology for carrier screening. *Hum. Genet.* **89:** 163–168.

Clemens, P.R., R.G. Fenwick, J.S. Chamberlain, R.A. Gibbs, M. de Andrade, R. Chakraborty, and C.T. Caskey. 1991. Carrier detection and prenatal diagnosis in Duchenne and Becker muscular dystrophy families, using dinucleotide repeat polymorphisms. *Am. J. Hum. Genet.* **49:** 951–960.

Covone, A.E., F. Caroli, and G. Romeo. 1992. Screening Duchenne and Becker muscular dystrophy patients for deletions in 30 exons of the dystrophin gene by three-multiplex PCR. *Am. J. Hum. Genet.* **51:** 675–677.

Cremonesi, L., E. Belloni, C. Magnani, M. Seia, and M. Ferrari. 1992. Multiplex PCR for rapid detection of three mutations in the cystic fibrosis gene. *PCR Methods Appl.* **1:** 297–298.

Edwards, A., A. Civitello, H.A. Hammond, and C.T. Caskey. 1991. DNA typing and genetic mapping with trimeric and tetrameric tandem repeats. *Am. J. Hum. Genet.* **49:** 746–756.

Edwards, A., H.A. Hammond, L. Jin, C.T. Caskey, and R. Chakroborty. 1992. Genetic variation at five trimeric and tetrameric tandem repeat loci in four human population groups. *Genomics* **12:** 241–253.

Estivill, X., N. Morral, T. Cassals, and V. Nunes. 1991. Prenatal diagnosis of cystic fibrosis by multiplex PCR of mutation and microsatellite alleles. *Lancet* **338:** 458.

Ferre, F. 1992. Quantitative or semi-quantitative PCR: Reality vs. myth. *PCR Methods Appl.* **2:** 1–9.

Ferrie, R.M., M.J. Schwartz, N.H. Robertson, S. Vaudin, M. Super, G. Malone, and S. Little. 1992. Development, multiplexing, and application of ARMS tests for common mutations in the CFTR gene. *Am. J. Hum. Genet.* **51:** 251–262.

Fortina, P., G. Dotti, R. Conant, G. Monokian, T. Parrella, W. Hitchcock, E. Rappaport, E. Schwartz, and S. Surrey. 1992a. Detection of the most common mutations causing beta-thalassemia in Mediterraneans using a multiplex amplification refractory mutation system (MARMS). *PCR Methods Appl.* **2:** 163–166.

Fortina, P., R. Conant, G. Monokian, G. Dotti, T. Parrella, W. Hitchcock, J. Kant, T. Scanlin, E. Rappaport, and E. Schwartz. 1992b. Non-radioactive detection of the most common mutations in the cystic fibrosis transmembrane conductance regulator gene by multiplex polymerase chain reaction. *Hum. Genet.* **90:** 375–378.

Furlong, R.A., R.G. Goudie, N.P. Carter, J.E.W. Lyall, N.A. Affara, and M.A. Ferguson-Smith. 1993. Analysis of four microsatellite markers on the long arm of chromosome 9 by meiotic recombination in flow-sorted single sperm. *Am. J. Hum. Genet.* **52:** 1191–1199.

Gannon, V.P., R.K. King, J.Y. Kim, and E.J. Thomas. 1992. Rapid and sensitive method for the detection of Shiga-like toxin-producing *Escherichia coli* in ground beef using the polymerase chain reaction. *Appl. Environ. Microbiol.* **58:** 3809–3815.

Gasparini, P., A. Bonizzato, M. Dognini, and P.F. Pignatti. 1992. Restriction site generating-polymerase chain reaction (RG-PCR) for the probeless detection of hidden genetic variation: Application to the study of some common cystic fibrosis mutations. *Mol. Cell. Probes* **6:** 1–7.

Gibbs, R.A., J.S. Chamberlain, and C.T. Caskey. 1989. Diagnosis of new mutation diseases using polymerase chain reaction. In *The polymerase chain reaction: Principles and applications* (ed. H. Erlich), pp. 171–192. Stockton Press, New York.

Gibbs, R.A., P.N. Nguyen, A. Edwards, A.B. Civitello, and C.T. Caskey. 1990. Multiple DNA deletion detection and exon sequencing of the hypoxanthine phosphoribosyltransferase gene in Lesch-Nyhan families. *Genomics* **7:** 235–244.

Golmolka, M., C. Epplen, J. Buitkamp, and J.T. Epplen. 1993. Novel members and germline polymorphisms in the human T-cell receptor Vb6 family. *Immunogenetics* **37:** 257–265.

Huang, T.H.-M., R.W. Cottingham Jr., D.H. Ledbetter, and H.Y. Zoghbi. 1992. Genetic mapping of four dinucleotide repeat loci, DXS435, DXS45, DXS454, DXS424, on the X chromosome using multiplex polymerase chain reaction. *Genomics* **13:** 375–380.

Ioannou, P., G. Christopoulos, K. Panayides, M. Kleanthous, and L. Middleton. 1992. Detection of Duchenne and Becker muscular dystrophy carriers by quantitative multiplex polymerase chain reaction analysis. *Neurology* **42:** 1783–1790.

Jawaheer, D., W.E. Ollier, and W. Thomson. 1993. Multiple ARMS-RFLP: A single and rapid method of HLA-DR4 subtyping. *Eur. J. Immunogenet.* **20:** 175–187.

Kaltenboek, B., K.G. Kansoulas, and J. Storz. 1992. Two-step polymerase chain reactions and restriction endonuclease analyses detect and differentiate *ompA* DNA of the *Chlamydia* spp. *J. Clin. Microbiol.* **30:** 1098–1104.

Killimann, M.W., A. Pizzuti, M. Grompe, and C.T. Caskey. 1992. Point mutations and polymorphisms in the human dystrophin gene identified in genomic DNA sequences amplified by multiplex PCR. *Hum. Genet.* **89:** 253–258.

Klimpton, C.P., P. Gill, A. Walton, A. Urquhart, E.S. Millican, and M. Adams. 1993. Automated DNA profiling employing multiplex amplification of short tandem repeat loci. *PCR Methods Appl.* **3:** 13–21.

Kornreich, R. and R.J. Desnick. 1993. Fabry disease: Detection of gene rearrangements in the human alpha-galactosidase A gene by multiplex PCR amplification. *Hum. Mutat.* **2:** 108–111.

Levinson, G., R.A. Fields, G.L. Harton, F.T. Palmer, A. Maddelena, E.F. Fugger, and J.D. Schulman. 1992. Reliable gender screening for human preimplantation embryos, using multiple DNA target-sequences. *Hum. Reprod.* **7:** 1304–1313.

Lo, K.W., C.H. Mok, G. Chung, D.P. Huang, F. Wong, M. Chan, J.C. Lee, and K.W. Tsao. 1992. Presence of p53 mutation in human cervical carcinomas associated with HPV-33 infection. *Anticancer Res.* **12:** 1989–1994.

Lohmann, D., B. Horsthemke, G. Gillessen-Kaesbach, F.H. Stefani, and H. Hofler. 1992. Detection of small RB1 gene deletions in retinoblastoma by multiplex PCR and high-resolution gel electrophoresis. *Hum. Genet.* **89:** 49–53.

Manam, S.S. and W.W. Nichols. 1991. Multiple polymerase chain reaction amplification and direct sequencing of homologous sequences: Point mutation of the *ras* gene. *Anal. Biochem.* **6:** 552–526.

Morral, N. and X. Estivill. 1992. Multiple PCR amplification of three microsatellites within the CFTR gene. *Genomics* **51:** 1362–1364.

Mulley, J.C., A.K. Gedeon, S.J. White, E.A. Haan, and R.I. Richards. 1991. Predictive diagnosis of myotonic dystrophy with flanking microsatellite markers. *J. Med. Genet.* **28:** 448–452.

Mutirangura, A. 1993. Multiplex PCR of three dinucleotide repeats in the Prader-Willi/Angelman critical region (15q11-q13): Molecular diagnosis and mechanism of uniparental disomy. *Hum. Mol. Genet.* **2:** 143–151.

Ng, I.S.L., R. Pace, M.V. Richard, K. Kobayashi, B. Kerem, L.-C. Tsui, and A.L. Beaudet. 1991. Method for analysis of multiple cystic fibrosis mutations. *Hum. Genet.* **87:** 613–617.

Nigro, N., L. Politano, G. Nigro, S.C. Romano, A.M. Molinari, and G.A. Puca. 1992. Detection of a nonsense mutation in the dystrophin gene by multiple SSCP. *Hum. Mol. Genet.* **1:** 517–520.

O'Keefe, D.S. and A. Dobrovic. 1993. A rapid and reliable method for genotyping the ABO blood group. *Hum. Mutat.* **2:** 67–70.

Pearson, B.M. and R.A. McKee. 1992. Rapid identification of *Saccharomyces cerevisiae*, *Zygosaccharomyces bailli* and *Zygosaccharomyces rouxii*. *Int. J. Food Microbiol.* **16:** 63–67.

Pfitzinger, H., B. Ludes, and P. Mangin. 1993. Sex determination of forensic samples: Coamplification and simultaneous detection of a Y-specific and an X-specific DNA sequence. *Int. J. Leg. Med.* **105:** 213–216.

Picci, L., F. Anglani, M. Scarpa, and F. Zachello. 1992. Screening for cystic fibrosis gene mutations by multiplex DNA amplification. *Hum. Genet.* **88:** 527–531.

Repp, R., S. Rhiel, K.H. Heermann, S. Schaefer, C. Keller, P. Ndumbe, F. Lambert, and W.H. Gerlich. 1993. Genotyping by multiplex polymerase chain reaction for detection of endemic hepatitis B virus transmission. *J. Clin. Microbiol.* **31:** 1095–1102.

Richards, I.R., K. Holman, S. Lane, G.R. Sutherland, and D.F. Callen. 1991a. Human chromosome 16 physical map: Mapping of somatic cell hybrids using multiplex PCR deletion analysis of sequence tagged sites. *Genomics* **10:** 1047–1052.

Richards, I.R., K. Holman, Y. Shen, H. Kozman, H. Harley, D. Brook, and D. Shaw. 1991b. Human glandular Kallikrein genes: Genetic and physical mapping of the KLK1 locus using a highly polymorphic microsatellite PCR marker. *Genomics* **11:** 77–82.

Runnebaum, I.B., M. Nagarajan, M. Bowman, D. Soto, and S. Sukumar. 1991. Mutations in p53 as potential markers for human breast cancer. *Proc. Natl. Acad. Sci.* **88:** 10657–10661.

Schwartz, J.S., J. Tarleton, B. Popovich, W.K. Seltzer, and E.P. Hoffman. 1992. Fluorescent multiple linkage analysis and carrier detection for Duchenne/Becker's muscular dystrophy. *Am. J. Hum. Genet.*

51: 721–729.

Serre, J.L., A. Taillandier, E. Mornet, B. Simon-Bouy, J. Boue, and A. Boue. 1991. Nearly 80% of cystic fibrosis heterozygotes and 64% of couples at risk may be detected through a unique screening of four mutations by ASO reverse dot blot. *Genomics* **11:** 1149–1151.

Sevall, J.S. 1990. Detection of parvovirus B19 by dot-blot and polymerase chain reaction. *Mol. Cell. Probes* **4:** 237–246.

Shuber, A.P., J. Skoletsky, R. Stern, and B.L. Handelin. 1993. Efficient 12-mutation testing in the CFTR gene: A general model for complex mutation analysis. *Hum. Mol. Genet.* **2:** 153–158.

Soler, C., P. Allibert, Y. Chardonnet, P. Cros, B. Matrand, and J. Thivolet. 1991. Detection of human papilloma virus types 6, 11, 16, and 18 in mucosal and cutaneous lesions by the multiplex polymerase chain reaction. *J. Virol. Methods* **35:** 143–157.

Sunzeri, F.J., T.-H. Lee, R.G. Brownlee, and M.P. Busch. 1991. Rapid simultaneous detection of multiple retroviral DNA sequences using the polymerase chain reaction and capillary DNA chromatography. *Blood* **77:** 879–886.

Toh, Y., H. Kuwano, S. Tanaka, K. Baba, H. Matsuda, K. Sugimachi, and R. Mori. 1992. Detection of human papillomavirus DNA in esophageal carcinoma in Japan by polymerase chain reaction. *Cancer* **70:** 2234–2238.

Uggozoli, L. and B. Wallace. 1992. Application of an allele-specific polymerase chain reaction to the direct determination of ABO blood group genotypes. *Genomics* **12:** 670–674.

Vandenvelde, C., M. Verstraete, and D. Van Beers. 1990. Fast multiple polymerase chain reaction on boiled clinical samples for rapid viral diagnosis. *J. Virol. Methods* **30:** 215–227.

van der Vliet, G.M., C.J. Hermans, and P.R. Klatser. 1993. Simple colorimetric microtiter plate hybridization assay for detection of amplified *Mycobacterium leprae* DNA. *J. Clin. Microbiol.* **31:** 665–670.

Vesy, C.J., J.K. Greenson, A.C. Papp, P.J. Snyder, S.J. Qualman, and T.W. Prior. 1993. Evaluation of celiac disease biopsies for adenovirus 12 DNA using a multiplex polymerase chain reaction. *Mod. Pathol.* **6:** 61–64.

Wattel, E., M. Mariotti, F. Agis, E. Gordien, O. Prou, A.M. Courouce, P. Rouger, S. Wain-Hobson, I.S. Chen, and J.J. Lefrere. 1992. Human T lymphotrophic virus (HTLV) type I and II DNA amplification in HTLV-I/II-seropositive blood donors of the French West Indies. *J. Infect. Dis.* **165:** 369–372.

Way, J.S., K.L. Josephson, S.D. Pillai, M. Abbaszadegan, C.P. Gerba, and I.L. Pepper. 1993. Specific detection of *Salmonella* spp. by multiplex polymerase chain reaction. *Appl. Environ. Microbiol.* **59:** 1473–1479.

Williams, J.G.K., A.R. Kubelik, K.J. Livak, J.A. Rafalaski, and S.V. Tingey. 1990. DNA polymorphisms amplified by arbitrary primers are useful as genetic markers. *Nucleic Acids Res.* **18:** 6531–6535.

Wilton, S. and D. Cousins. 1992. Detection and identification of multiple mycobacterial pathogens by DNA amplification in a single tube. *PCR Methods Appl.* **1:** 269–273.

Wirz, B., J.D. Traschin, H.K. Muller, and D.B. Mitchell. 1993. Detection of hog cholera virus and differentiation from other pestiviruses by polymerase chain reaction. *J. Clin. Microbiol.* **31:** 1148–1154.

Worley, K.C., J.A. Towbin, X.M. Zhu, D.F. Barker, A. Ballabio, J. Chamberlain, L.G. Biesecker, S.L. Blethen, P. Brosnan, J.E. Fox, W.B. Rizzo, G. Romeo, N. Sakuragawa, W.K. Seltzer, S. Yamaguchi, and E.R.B. McCabe. 1992. Identification of new markers in Xp21 between DXS28 (C7) and DMD. *Genomics* **13:** 957–961.

Zazzi, M., L. Romano, A. Brasini, and P.E. Valensin. 1993. Simultaneous amplification of multiple HIV-1 DNA sequences from clinical specimens by using nested-primer polymerase chain reaction. *AIDS Res. Hum. Retroviruses* **9:** 315–320.

Ziegle, J.S., Y. Su, K.P. Corcoran, L. Nie, P.E. Maynard, L.B. Hoff, L.J. McBride, M.N. Kronick, and S.R. Diehl. 1992. Application of automated DNA sizing technology for genotyping microsatellite loci. *Genomics* **14:** 1026–1031.

Zietkievicz, E., M. Labuda, D. Sinnett, F.H. Glorieux, and D. Labuda. 1992. Linkage mapping by simultanenous screening of multiple polymorphic loci using *Alu* oligonucleotide-directed PCR. *Proc. Natl. Acad. Sci.* **89:** 8448–8451.

Detection of PCR Products: Quantitation and Analysis

PCR technology is, in essence, a two-step process: the amplification of a segment of nucleic acid and the detection of the resulting DNA fragment. With the rapid development of PCR technology, there has been a concomitant increase in the number of methods for PCR product detection. To be of value, a detection system must accurately and reproducibly reflect the nature and quantity of the starting nucleic acid.

The simplest and most widely used detection method employs agarose gel electrophoresis to resolve and detect a product of the length predicted by the placement of the oligonucleotide primers. The appearance of a discrete band of the correct size is usually indicative of a successful reaction; however, the presence of a band does not provide additional information, nor does it prove that the observed product is, in fact, the expected one. A simple method for product monitoring is the selection of PCR primers flanking diagnostic restriction enzyme sites (Rappolee et al. 1988). Following PCR, a portion of the reaction is subjected to digestion with the appropriate restriction enzyme and electrophoresed to detect the DNA fragments of the predicted sizes.

An alternative method of detecting the PCR product involves the use of an oligonucleotide probe that hybridizes to the template and is located between the two primers employed in the PCR. This probe can be used to detect the PCR product after transfer to a solid support (Southern blotting) or by solution hybridization with the product, followed by gel electrophoresis to separate the free from the bound probe. These two oligonucleotide hybridization methods can be quantitated through the use of scintillation counting. The application of this technology to quantitative measurements of RNA copy number is detailed in Section 5 of this manual. These methods represent the most widely used research laboratory approaches for PCR quantitation.

Oligonucleotide probes have also proven essential for detecting the presence or absence of specific sequence variations. The power of PCR when combined with a detection system using allele-specific oligonucleotide probes was shown by Saiki et al. (1986). This technology is available in kit form, for example, for the detection and typing of HLA-DQα (Gyllensten and Allen 1991).

Other approaches to the detection of PCR products include systems using a nonradioactive ELISA-based format. These methods are useful in clinical research and diagnosis. The detection system described in "Immunological Detection of PCR Products" is an example of such an approach. ELISA-based formats are adaptable to numerous targets and can provide a readout compatible with quantitative PCR. Numerous other ELISA-based detection methods have been devised and, for certain molecular targets and organisms, marketed in kit form. A list of companies that market specific PCR quantitation kits is provided in *The Lab Manual Source Book* published by Cold Spring Harbor Laboratory Press.

Newer-generation detection systems that are useful for high throughput and quantitation have also been developed. These technologies are purchased as defined systems. "Quantitative PCR Using the AmpliSensor Assay" describes one such complete system for the detection and quantitation of PCR. The technology of the system, which is based on fluorescence energy transfer, provides a means of measuring the accumulation of the specific PCR product at each cycle. When the specific detection oligonucleotide, which consists of a primer-template structure that is configured for fluorescence energy transfer, is denatured, there is a loss of fluorescent signal. Thus, the decrease in fluorescence intensity correlates with the accumulation of PCR product. The choice of a system like this requires a significant capital expenditure and commits the laboratory to a specific technology for the foreseeable future. With the introduction of the TaqMan LS-50B system, Perkin-Elmer has devised an alternative fluorescence energy transfer-based detection system. In the TaqMan system, the fluorescent reporter and a quencher molecule are physically attached to the 5′ and 3′ ends of the detection oligonucleotide probe, respectively. The signal is generated during the detection part of PCR when the probe hybridizes to its target sequence, followed by the processive degradation of the detection oligonucleotide by the 5′-exonuclease activity of *Taq* DNA polymerase. This releases the reporter molecule from the physical proximity of the quencher. The fundamental difference between these two fluorescence energy transfer-based systems is that TaqMan produces an increasing signal with accumulating product, whereas AmpliSensor measures the loss of signal with accumulating product.

The electrochemoluminescence technology described in "One-tube Quantitative HIV-1 RNA NASBA" in Section 10 represents another state-of-the-art detection system. The system, combined with a set of quantified specific, externally added amplification standards and their cognate detection oligonucleotides, is adaptable to PCR as well. For both fluorescence energy transfer- and electrochemoluminescence-based detection systems, issues related to overall cost and contamination control when using such complex systems remain to be studied.

Where specific PCR protocols end and unique detection methods begin is sometimes difficult to determine. The chapters "Single-strand Conformational Polymorphism" (SSCP), "Sensitive and Fast Mutation Detection by Solid-phase Chemical Cleavage," and "Genetic Subtyping of Human Immunodeficiency Virus Using a Heteroduplex Mobility Assay" provide good methods for identifying point mutations and specific sequence heterogeneity and represent powerful tools for discovering new sequence variants or studying sequence variation in populations. These methods rely on the resolving power of acrylamide gels to detect, in the case of SSCP, the sequence-dependent variation in single-stranded DNA folding, and in the case of the heteroduplex mobility assay, the extent of homology of an unknown sample with a known set of standards. These methods are relatively easier to perform when compared with related methods such as denaturing gradient gel electrophoresis (Myers et al. 1987) and the GC clamp (Sheffield et al. 1989). However, the SSCP methodology may miss a fraction of the sequence variants.

As described above, well-defined, single-nucleotide changes are ideally monitored using allele-specific oligonucleotide probes. For screening and mapping the approximate position of the base pair change, the chapter "Sensitive and Fast Mutation Detection by Solid-phase Chemical Cleavage" provides a superior method for detecting all of the sequence changes within a region.

Some chapters in this section represent examples of the true power of PCR. One Achilles' heel of PCR has been the perception that the analysis of sequence changes or expression was limited to defined genes. Using the technology described in "DNA Fingerprinting Using Arbitrarily Primed PCR" and "RNA Fingerprinting Using Arbitrarily Primed PCR," it is now possible to detect, clone, and sequence differences that exist between cells or tissues at the nucleic acid level.

The chapter in this section entitled "In Situ PCR" represents a unique detection system. This approach results in the determination of the number of cells in a population that are positive for a specific target amplicon. This has proven to be an extremely useful technology, although it is sometimes difficult to implement. This method may move rapidly into general use, as several companies are now making reagents and supplies specifically tailored to in situ PCR.

REFERENCES

Gyllensten, U. and M. Allen. 1991. PCR-based HLA class II typing. *PCR Methods Appl.* **1:** 91–98.

Myers, R.M., T. Maniatis, and L.S. Lerman. 1987. Detection and localization of single base changes by denaturing gradient gel electrophoresis. *Methods Enzymol.* **155:** 501–527.

Rappolee, D.A., D. Mark, M.J. Banda, and Z. Werb. 1988. Wound macrophages express TGF-α and other growth factors in vivo: Analysis by mRNA phenotyping. *Science* **241:** 708–712.

Saiki, R.K., T.L. Bugawan, G.T. Horn, K.B. Mullis, and H.A. Erlich. 1986. Analysis of enzymatically amplified β-globin and HLA-DQα DNA with allele-specific oligonucleotide probes. *Nature* **324:** 163–166.

Sheffield, V.C., D.R. Cox, L.S. Lerman, and R.M. Myers. 1989. Attachment of a 40-base-pair G + C-rich sequence (GC-clamp) to genomic DNA fragments by the polymerase chain reaction results in improved detection of single-base changes. *Proc. Natl. Acad. Sci.* **86:** 232–236.

Immunological Detection of PCR Products

James G. Lazar

Digene Diagnostics Inc., Silver Spring, Maryland 20904

INTRODUCTION

To use PCR to its fullest potential, the identification and measurement of PCR products, or amplicons, must be improved compared to traditional methods. Amplicons must be identified by their specific nucleotide sequences to prevent false or ambiguous results due to primer-dimer formation, nonspecific amplification, or target sequence variation. Quantitation of amplicon levels is an important first step for the estimation of the relative number of initial target molecules.

Until several years ago, PCR was usually performed on clean model systems or on specimens that had been highly purified. Gel-based detection methods worked well for these applications when sample throughput, time, and per-sample costs were not important issues. However, now that PCR is a "mainstream" technique, moving from the realm of the molecular biology laboratory to a wide variety of fields, which are testing large numbers of samples and a wide variety of sample types, new detection methods are needed that meet the requirements of these users.

Advanced detection methods must provide significant advantages over traditional methods to justify the time, effort, and cost involved in converting to a new technology. Implementation of a new detection method in a laboratory typically requires significant investments in time to learn the new technology and to validate the new method's performance against the method currently used. Moreover, advanced detection methods may require significant investment in capital equipment and the purchase of expensive consumables and reagents.

To overcome these drawbacks, an ideal detection system should be designed to achieve the following goals:

1. Ten or fewer input target copies should be detectable after one round of amplification and the detection should be sequence-specific.

2. Primer-dimers and other nonspecific amplification products should not be detected.

3. All detection methods should be capable of incorporating the dUTP/UNG decontamination procedure to ensure against carry-over contamination (Hartley and Rashtchian 1993, this volume).

4. The detection method should be easy to use, should require little or no specialized training, and should minimize tedious and laborious procedures.

5. No purification of the PCR products should be required.

6. There should be as little hands-on time as possible, and the method should be amenable to automation with currently available equipment.

7. Throughput should be flexible so that small or large batch sizes can be run without wasting reagents or consumables.

8. The total cost of an advanced detection method should be competitive with traditional detection methods, notwithstanding performance considerations.

9. The detection system should use common equipment and reagents (except sequence-specific components such as primers and probes), and a universal procedure for detecting all PCR products.

One-time costs for adopting an advanced detection system are usually limited to capital equipment expenditures such as readers, shakers, and incubators—equipment that may already be available in many labs. Recurring costs include reagents, disposables, and labor. With advanced detection methods, however, cost savings may be realized from increased productivity, reduced hands-on time, higher sample throughput, fewer repeat amplifications due to ambiguous results, and reduced cost of amplification reagents if converting from nested-primer methods. By incorporating a universal detection system, the equipment and number of different reagents that laboratories must purchase and store are minimized, and multiple targets can be detected in the same assay, a feature that is especially important to small-volume users.

Oligonucleotide-based Detection Methods

Numerous new methods for the detection of PCR products are found in the literature (Hayashi et al. 1989; Lundeberg et al. 1990; Landgraf et al. 1991; Bush et al. 1992; Conway et al. 1992; Kolk et al. 1992; He et al. 1993; Rasmussen et al. 1994). Two common detection methods use specific oligonucleotide sequences to either capture or detect the PCR amplicons (Syvänen et al. 1988; Inouye and Hondo 1990; Keller et al. 1990). In both formats, the PCR mix contains one biotinylated primer and one unmodified primer. During amplification, the biotin-labeled primer is incorporated into the amplicon. In the oligonucleotide probe format (Fig. 1), the amplicon is denatured and hybridized to an enzyme-labeled oligonucleotide probe (Levenson and Chang 1990). The reaction mixture is simultaneously captured onto a streptavidin-coated plate, excess labeled oligonucleotide is washed away, and the enzyme label generates a signal with an appropriate substrate.

In the oligonucleotide capture format (Fig. 2), the amplicon is denatured and hybridized to a sequence-specific capture oligonucleotide that is prebound to a solid phase. Sequences that are not captured are washed away, and captured sequences are detected through the biotin label with enzyme-labeled streptavidin.

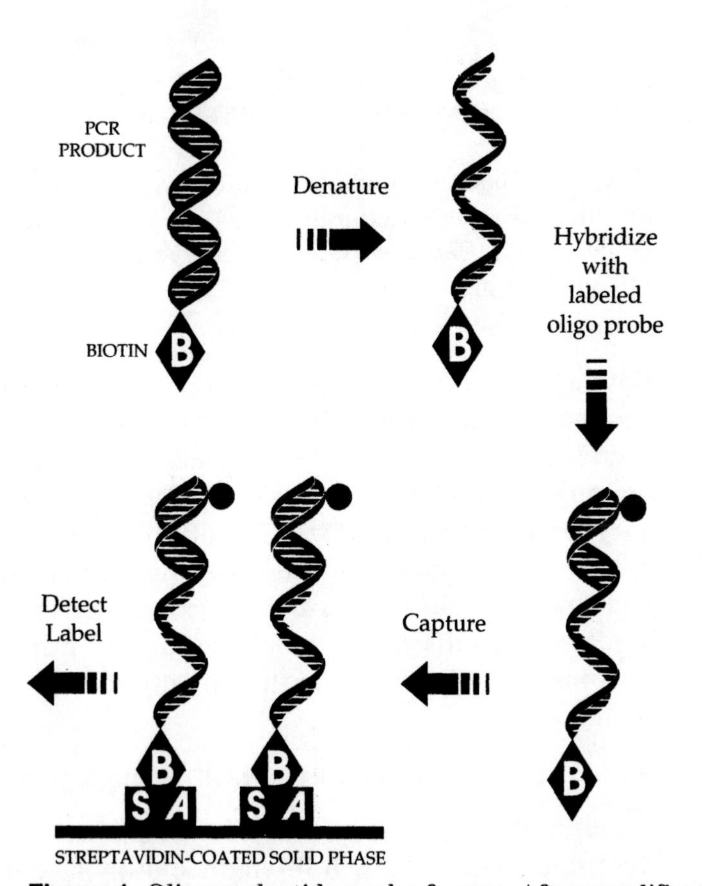

Figure 1 Oligonucleotide probe format. After amplification, the PCR product is denatured and hybridized to a labeled oligonucleotide. Hybrids are captured onto a streptavidin-coated plate, and the label generates a detectable signal.

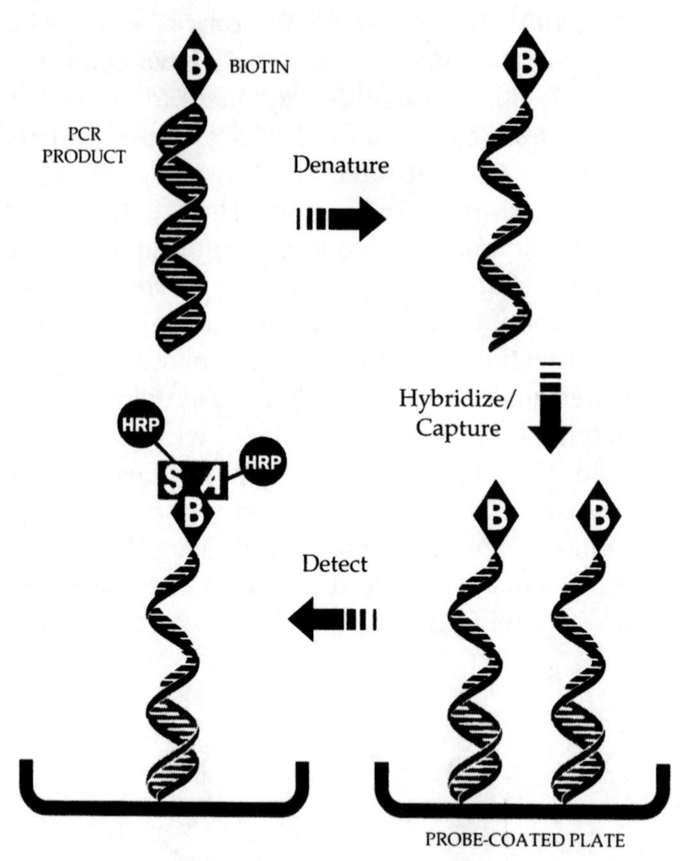

Figure 2 Oligonucleotide capture format. After amplification, PCR products are denatured and captured by an oligonucleotide probe bound to a plate (solid phase). Captured nucleic acids are detected by a label incorporated into the PCR product.

Both of these formats are straightforward and relatively simple to design and construct in either a blot or microplate format. Both formats are limited in sensitivity by the fact that only a few enzyme labels are bound per amplicon. Moreover, optimal hybridization conditions may vary significantly among different oligonucleotide probes due to sequence variations within the hybridization region. As a consequence, genotypically similar strains of an organism may be captured or detected with dissimilar efficiency due to minor sequence variation, thus making relational quantitation difficult.

Immunological Detection: The SHARP Signal System

An elegant ELISA method for the detection of PCR products has been described that utilizes full-length biotinylated RNA probes, solution hybridization, and a monoclonal antibody specific for RNA:DNA hybrids (Coutlée et al. 1989; Bobo et al. 1990). The SHARP Signal System (Digene Diagnostics) is a capture ELISA assay that also uses an antibody to RNA:DNA hybrids (Fig. 3) (Lazar 1993). With the SHARP Signal System, PCR is performed using one biotinylated and one unmodified primer. After amplification, a portion of the reaction mixture is denatured and then hybridized in solution to a complementary unlabeled RNA probe. Then the RNA:DNA hybrids are captured onto a streptavidin-coated microplate with an alkaline phosphatase-conjugated antibody specific for RNA:DNA hybrids. After washing, the detection signal is generated with a colorimetric substrate and read on a conventional microplate reader at 405 nm.

This detection system is sensitive because multiple antibodies, and hence multiple enzyme labels, can bind to each captured hybrid. Specificity is also enhanced because the formation of RNA:DNA hybrids is favored at the hybridization temperature of 65°C, and, because the RNA probes are complementary to the whole length of amplicon, small sequence variations that occur between different strains of the same species have a negligible effect on hybridization or detection efficiency (Young and Anderson 1985).

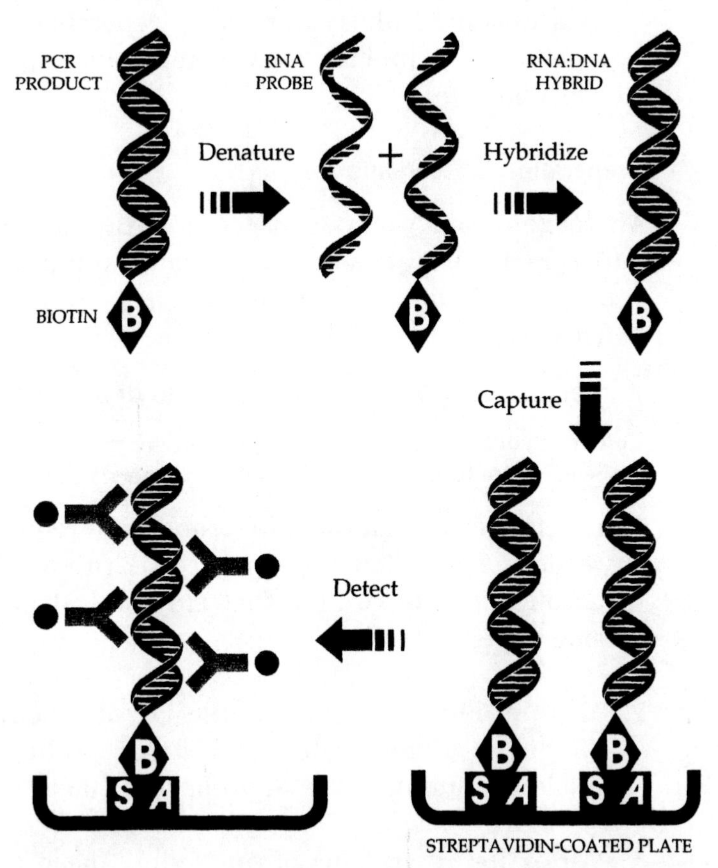

Figure 3 The SHARP Signal System. PCR products are denatured and hybridized to unlabeled RNA probes. RNA:DNA hybrids are captured onto a streptavidin-coated plate and detected with an enzyme-conjugated antibody specific for RNA:DNA hybrids.

REAGENTS

These are purchased in kit form as the SHARP Signal System.

RNA probe (supplied as part of a target-specific probe/primer set or by the user)

Sample diluent (a buffer containing carrier DNA and indicator dye)

Denaturation reagent (concentrated base [1.25 N NaOH] for sample denaturation)

Probe diluent (a neutralizing buffer containing buffer salts and hybridization accelerator)

Capture plate (streptavidin-coated microtiter plate for capture of biotinylated RNA:DNA hybrids)

Detection reagent (alkaline phosphatase-conjugated antibody directed against RNA:DNA hybrids and RNase in buffer)

Wash buffer (an optimized dry pack wash that is reconstituted before use)

Substrate concentrate (stabilized 50X *p*-nitrophenyl phosphate [PNPP])

Substrate diluent (diethanolamine buffer)

Materials required but not supplied are: PCR products for testing, 65°C water bath, 37°C incubator, vortex mixer, microplate shaker, hybridization tubes and rack, repeating positive displacement pipettor, colorimetric microplate reader, plate washer or 500-ml wash bottles

PROTOCOL

Preparation of Reagents

All reagents are ready to use except the target-specific probe, wash buffer, and substrate solution. Variable amounts of the probe mix and substrate solution may be prepared as required depending on the batch size.

TARGET-SPECIFIC PROBE (Prepare prior to denaturation incubation)

Note: Vial contents can become trapped in the vial lid. Centrifuge the target-specific probe briefly to bring liquid to the bottom of the vial. Tap tube gently to mix.

1. Determine the amount of probe mix required (25 μl/well). It is recommended that extra probe mix (up to 10 μl/well) be made to account for the volume that may be lost in pipette tips or on the side of the vial.

2. Make a 50-fold dilution of the target-specific probe in the probe diluent to yield a 500 pmoles/liter probe solution. The recommended dilution is 4 μl of probe into 200 μl of probe diluent.

3. Vortex for 30 seconds at maximum speed to mix thoroughly. Extreme care should be taken at this step. The probe diluent is viscous. Care should be taken to ensure thorough mixing when preparing the probe mix, because incomplete mixing may result in reduced signal.

WASH BUFFER (Prepare during capture incubation or earlier)

1. Add the contents of each wash buffer pack to 3 liters of distilled or deionized water and mix until dissolved. Fill a 500-ml squeeze bottle with the wash buffer, or use as directed in an automatic plate washer.

2. Seal the container to prevent contamination or evaporation.

3. Prepared wash buffer is stable for 6 months at 2–30°C. If necessary, equilibrate to 20–30°C before using. It is recommended that the wash apparatus be cleaned with bleach and rinsed thoroughly with deionized water once a month to prevent possible contamination from alkaline phosphatase present in bacteria and molds.

SUBSTRATE SOLUTION (Prepare during detection incubation)

1. Determine the amount of substrate solution required (100 μl/well). Extra substrate solution (up to 25 μl/well) should be made to account for the volume that may be lost in pipette tips or on the side of the container.

2. Aliquot the required amount of the substrate diluent and allow it to warm to 20–25°C.

3. Make a 50-fold dilution of the substrate in the substrate diluent. Mix well.

Test Procedure

Sample denaturation

1. Remove the PCR products (samples) and required kit reagents from the refrigerator and allow them to reach 20–25°C for at least 15–30 minutes. Mix thoroughly.

2. Place the Negative Assay Control (NAC), target-specific Positive Assay Control (PAC) (from the Probe/Primer Set), and samples to be tested in a microtube rack.

3. Label the hybridization tubes and place them in an appropriate rack or holder.

4. The PAC, NAC, and samples may be tested singly or, for greater accuracy, in duplicate. The instructions that follow are for single test units. They should be multiplied by the appropriate factor if multiple tests will be run for each sample.

5. Prepare the probe mix (see above).

6. Pipette 50 μl of the sample diluent into each hybridization tube using a standard or repeat pipettor.

7. Pipette 25 μl of the denaturation reagent into each hybridization tube using a standard or repeat pipettor.

8. Pipette 5 μl of the control or sample into each hybridization tube, using a new pipette tip for each transfer.

9. Cap the tubes and vortex each for 5 seconds at maximum speed, or shake vigorously (1200 ± 100 rpm) on a rotary shaker for 30 seconds. All samples should be purple. Incubate tubes for 10 ± 2 minutes at 20–25°C.

Sample hybridization

10. Pipette 25 μl of the probe mix into each hybridization tube. Cap the tubes and shake vigorously on a rotary shaker for 30 seconds at 1200 ± 100 rpm or vortex each tube for 5 seconds at maximum speed.

    Note: The probe mix is viscous. Care should be taken to ensure thorough mixing and that the required amount is completely dispensed into each tube. A reliable dispensing method is to use a repeat pipettor, or other positive-displacement pipetting device, and to dispense the probe mix into the bottom of the tube.

    The controls and specimens should turn yellow after mixing. Tubes that remain purple have not received the proper amount of the probe or have not been mixed properly. Tubes that remain purple after repeat mixing should be recorded, and those specimens should be retested.

11. Incubate hybridization tubes for 30 ± 5 minutes in a 65 ± 2°C water bath or thermal cycler.

Plate hybrid capture

12. Remove the hybridization tubes from the water bath and transfer the contents (100 μl) of each hybridization tube to the corresponding capture well, using a standard pipette with a new tip for each transfer.

13. Place a cover on the capture plates and shake on a microplate rotary shaker at 1200 ± 100 rpm at 20–25°C for 30 ± 5 minutes. Prepare the wash buffer during this incubation (see Preparation of Reagents).

Detection

14. Remove the capture plate from the microplate rotary shaker and decant the contents by inverting the plate over a sink and shaking vigorously. Tap excess liquid out of the inverted plate onto clean absorbent paper. Flush the sink with water after emptying the wells.

15. Pipette 100 µl of the detection reagent into each well.

16. Cover the capture plate with a plate cover, and shake at 1200 ± 100 rpm on a microplate rotary shaker for 30 ± 3 minutes at 20–25°C. Prepare the substrate solution during this incubation (see Preparation of Reagents).

17. Decant the detection reagent from the wells by inverting the plate over a sink and shaking vigorously.

18. Wash the capture plate five times with a vigorous stream of reconstituted wash buffer from the squeeze bottle. Between each washing, decant excess wash buffer by inverting the plate over a sink and shaking vigorously. Alternatively, an automated plate washer can be used.

19. Wash the plate one time with deionized water to remove bubbles. Decant excess liquid by inverting the plate over a sink and shaking vigorously. Tap excess liquid out of the inverted plate onto clean absorbent paper.

20. Pipette 100 µl of the substrate solution into each well, including the substrate blank wells. Make the addition in the same order that the plate will be read by the plate reader.

21. Cover plate with plate cover, and incubate at 37 ± 2°C for 1 hour. The linear range and sensitivity can be enhanced by increasing the substrate incubation time at 37 ± 2°C up to 24 hours.

22. Read absorbance at 405–410 nm. If the plate reader does not automatically perform a blank correction, calculate the mean of the substrate blanks and subtract the result from each of the specimen absorbance values.

23. Record pertinent assay information on the data sheet provided.

The SHARP Signal assay procedure is summarized in Figure 4. In most applications, overall time to results is less than 4 hours with

Prepare probe mix.
↓
Pipette 50 μl Sample Diluent into Hybridization Tubes.
↓
Pipette 25 μl Denaturation Reagent into each tube.
↓
Pipette 5 μl Sample or Control into each tube.
Shake at 1000 rpm for 30 seconds.
↓
Incubate at room temperature for 10 minutes.
↓
Pipette 25 μl Probe Mix into each tube.
↓
Shake for 30 seconds at room temperature.
↓
Incubate in a 65 °C waterbath for 30 minutes.
↓
Transfer from Hybridization Tube to Capture Plate well.
↓
Shake plate at 25°C for 30 minutes.
(Prepare Wash Buffer)
↓
Decant and blot Capture Plate.
Pipette 100 μl Detection Reagent into each well.
↓
Cover plate and shake at 25°C for 30 minutes.
↓
Decant and wash plate 5 times with 1X Wash Buffer. Wash plate 1X with deionized water.
↓
Pipette 100 μl Substrate into each well. Cover plate and incubate at 37°C for one hour.
↓
Read absorbance at 405 nm

Figure 4 Summary of the SHARP Signal System assay.

hands-on time of approximately 1 hour. To test large numbers of samples efficiently, a batch testing protocol has been developed.

MATERIALS

For the simultaneous testing of large numbers of samples using the SHARP Signal System, the Micronic tube rack (Matrix Technologies) is recommended. The tube rack holds 96 tubes in an 8 X 12 format, is compatible with multichannel pipettors, and accommodates both 1.2-ml and 1.4-ml tubes. Furthermore, the tube rack is open at the bottom, making it suitable for water bath incubation. After the samples are denatured and hybridized, the use of this rack makes it easier to transfer the sample to the capture microplates using a multichannel pipettor.

Micronic reusable tube rack with cover (Matrix Technologies, cat. no. 225-00)
1.2-ml bulk polypropylene tubes (USA Scientific Plastics, cat. no. 1412-000)
Plate sealers (Dynatech Laboratories, cat. no. 001-010-5701)

PROTOCOL

Batch Testing of Samples

1. Label tubes (1.2 ml or 1.4 ml) and place in the Micronic tube rack.

2. Using a repeat pipettor, dispense 50 μl of the sample diluent and then 25 μl of the denaturation reagent into each tube.

3. Pipette 5 μl of the control or sample into each tube using a new pipette tip for each transfer.

4. Seal the tubes with a plate sealer and cover with the tube rack cover.

5. Place the tube rack on a rotary shaker and shake vigorously for 30 seconds at 1200 ± 100 rpm. All samples should turn purple. Incubate for 10 ± 2 minutes at 20–25°C.

6. *Slowly* remove the plate sealer and discard.

7. Using a repeat pipettor, pipette 25 μl of the probe mix into each tube. Apply a new plate sealer to the tubes and replace the cover. Place the rack on a rotary shaker and shake vigorously for 30 seconds at 1200 ± 100 rpm. Ensure that the samples have turned yellow.

8. Place the rack in a 65 ± 2°C water bath for 30 ± 5 minutes.

9. Remove the rack from the water bath. *Slowly* remove the plate sealer and discard. Using a multichannel pipettor, transfer the contents (100 μl) from each tube to the corresponding capture well in the microplate.

10. Proceed with the SHARP Signal System Assay as described above.

In one laboratory, the SHARP Signal System has been automated with standard ELISA robotic equipment and is currently being used with excellent results for rapid PCR detection of human papillomavirus (HPV) in cervical specimens (Terry et al. 1994).

DISCUSSION

The SHARP Signal System is quite sensitive because multiple antibodies, and hence multiple enzyme labels, can react with each captured hybrid. In a model system, the assay has been shown to be at least 100 times more sensitive than ethidium bromide staining of PCR products in agarose gels, and it can detect approximately 10 pg of

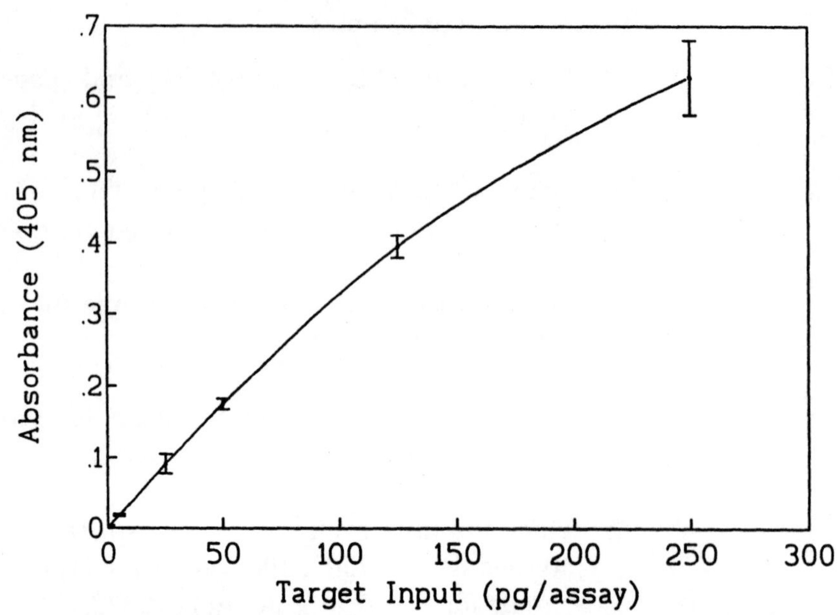

Figure 5 Sensitivity of the SHARP Signal System with 95% confidence intervals. PCR product from the L1 region of HPV-16 was quantitated, diluted, and tested in the SHARP Signal System.

biotinylated PCR product per well (Fig. 5). The detection assay is extremely flexible and easy to adapt to any target sequence because the RNA probes are unlabeled and can be easily produced by transcription of plasmid DNA containing a T7, SP6, or T3 RNA polymerase promoter. Alternatively, a promoter can be incorporated into a PCR primer and, after amplification with target DNA, the resulting PCR product can be used as the transcription template (Tsai and Dreher 1993; Urrutia et al. 1993).

The performance of the SHARP Signal System has been evaluated in a model system with several target analytes (Lazar et al. 1993b). In one study, plasmid DNA containing the 5′ noncoding region of the hepatitis C virus (HCV) was serially diluted in buffer containing sheared herring sperm DNA. Aliquots containing 1 μg of sheared herring sperm DNA and 0, 10, 10^2, 10^3, 10^4, and 10^5 copies of the HCV sequence were PCR amplified for 35 cycles with primers directed against the HCV 5′ noncoding region (the sense primer was biotinylated on the 5′ end). A 5-μl aliquot of each amplification reaction was analyzed by ethidium bromide gel analysis and by the SHARP Signal System. The results are summarized in Figure 6. The data show that 10 input copies are detectable and that the signal generated is proportional to the number of input copies.

In a recent study using clinical specimens, the SHARP Signal System was compared to a nested-primer method for the detection of HCV in serum (Lazar et al. 1993a). Specimens were processed by

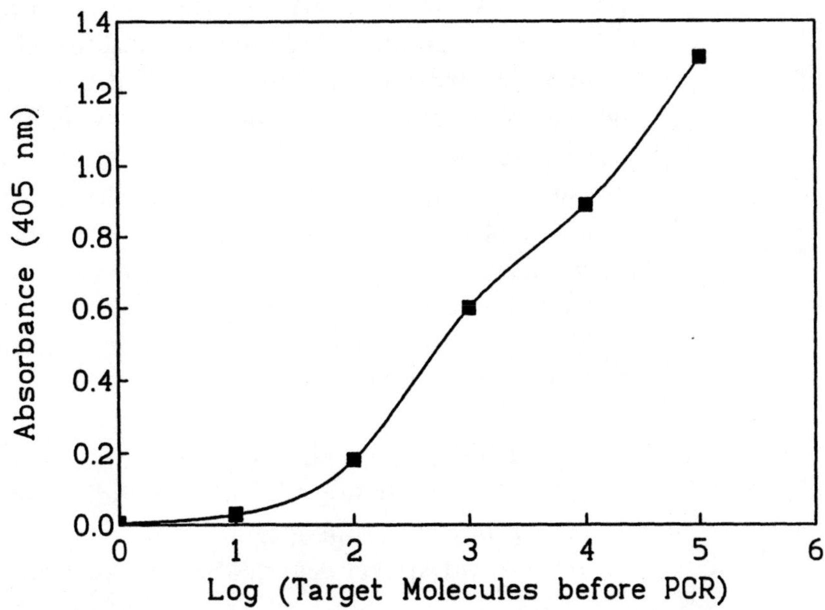

Figure 6 SHARP Signal System. Detection of amplified plasmid DNA containing the HCV 5′ noncoding region.

standard methods to preserve the RNA and to remove PCR inhibitors. cDNA copies of the RNA were prepared by a reverse transcriptase reaction using an antisense primer and the GeneAmp RNA PCR kit (Perkin-Elmer). One portion of the samples was then amplified with a single primer set containing one primer labeled with a 5′ biotin, and another portion of the samples was amplified using the nested-primer method. The samples amplified with the single primer set were analyzed by the SHARP Signal System after one round of amplification. The samples amplified with the nested-primer technique were analyzed by ethidium bromide gel analysis after the second round of amplification. Of the 99 clinical specimens tested, 47 were positive and 50 were negative by both methods. The remaining two samples were nested-primer positive, SHARP Signal System negative. Upon reanalysis of these samples, one sample was determined to be a nested-primer false positive. Upon repeat testing, the other sample was again negative by the SHARP Signal method, but was confirmed from clinical data to be a true positive. Overall, the SHARP Signal System exhibited 98% (97/99) agreement with the nested-primer method, and an overall sensitivity of 98% (47/48) and specificity of 100% (51/51) with respect to the adjudicated results.

The SHARP Signal System Assay for PCR Products provides sufficient reagents to perform 192 tests in an 8-well strip, 96-well microplate format. RNA probes, primers, and detection controls for the amplification and detection of specific infectious disease targets such as human immunodeficiency virus (HIV), HCV, hepatitis B virus

(HBV), HPV, *Mycobacterium tuberculosis* (Mtb), and cytomegalovirus (CMV) are currently available, and additional targets are under development. Alternatively, a custom RNA probe and primer set for specific targets of interest can be ordered from Digene Diagnostics. The methodology of the SHARP Signal System allows maximum flexibility because the same reagents, except for the RNA probe, are used for all assays and the detection procedure is universal for all targets. Thus, in a single assay on a single capture plate, a user can test simultaneously for several analytes.

The use of unlabeled RNA probes may be superior to conventional DNA probes for several reasons:

1. RNA:DNA hybrids are more stable than DNA:DNA hybrid structures, thus allowing the use of higher temperatures to increase the stringency of probe:target association and to favor probe:target hybridization over reannealing of the target or probe.

2. A posthybridization RNase digestion increases the specificity of detection and reduces background by removing free or nonspecifically bound RNA probe.

3. Strand-specific single-stranded probes are more sensitive and efficient in detecting target sequences because of a lack of competition from probe reassociation.

The flexibility of the system is further increased because the assay uses easily produced, unlabeled RNA probes, allowing users to develop their own probe/primer set as outlined by the protocol in Figure 7. A PCR product containing the target sequence of interest can be cloned directly into a plasmid containing a T7 RNA polymerase promoter (Buchman et al. 1992). After this plasmid DNA has been purified and linearized with an appropriate restriction enzyme, RNA is transcribed from the plasmid DNA, grown, purified by lithium chloride precipitation, and is then ready to use. Biotinylated and unmodified primers may be ordered from commercial suppliers; however, the strand of amplification product containing the biotinylated primer must be the opposite sense of the RNA probe or detection will not occur.

For even greater flexibility, standard cloning techniques can be used to prepare much longer RNA probes. For instance, the Digene HBV probe is a genomic, full-length RNA probe of approximately 3200 bases. By choosing the appropriate primer sets, many different genes can be amplified and then detected with the same probe in the same assay.

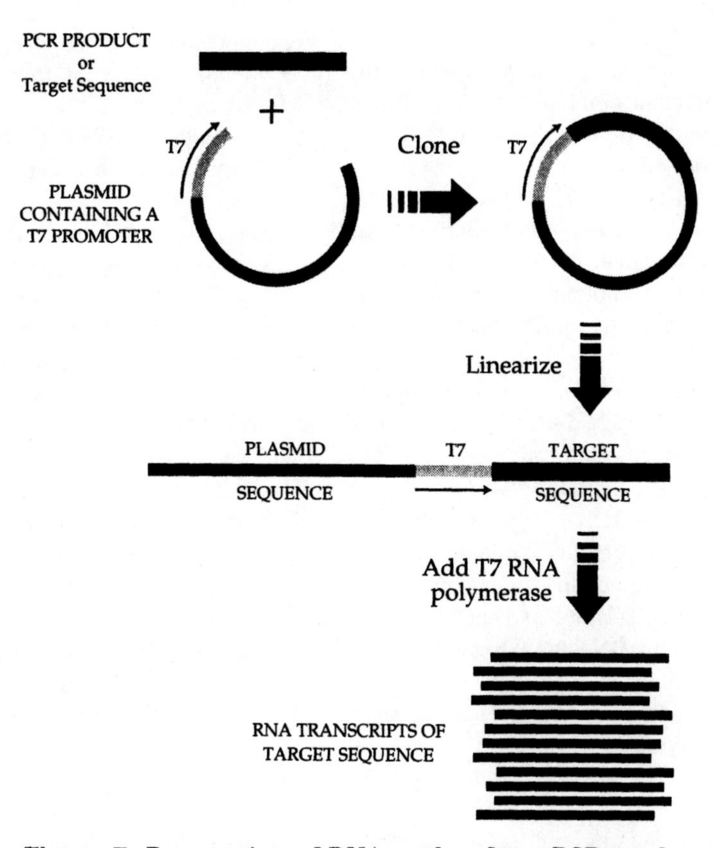

Figure 7 Preparation of RNA probes from PCR products or other DNA fragments. PCR products or other DNA fragments are cloned into a plasmid containing a T7 RNA polymerase promoter. After propagation and purification of the plasmid DNA, the DNA is linearized and RNA transcripts are produced.

CONCLUSIONS

Traditional, gel-based methods for the detection of PCR products lack sensitivity and specificity, or require tedious blotting techniques to achieve adequate results. To overcome these drawbacks, many techniques and assay formats have been developed that are significant improvements over traditional gel-based assays. Advanced detection assays developed in-house and commercially available assays are now being used that provide rapid, sensitive, and specific detection of PCR products in standard ELISA formats. Before choosing a new detection method, laboratories should examine their current and expected future needs and expectations with respect to assay format, per-sample cost, assay flexibility, sample throughput, capital investment, and technical support. The Digene SHARP Signal System is a rapid, sensitive, cost-effective, universal detection system for PCR products that comes close to achieving the goals of an ideal advanced detection system.

REFERENCES

Bobo, L., F. Coutlée, R.H. Yolken, T. Quinn, and R.P. Viscidi. 1990. Diagnosis of *Chlamydia trachomatis* cervical infection by detection of amplified DNA with an enzyme immunoassay. *J. Clin. Microbiol.*

28: 1968–1973.

Buchman, G.W., D.M. Schuster, and A. Rashtchian. 1992. Rapid and efficient cloning of PCR products using the CloneAmp™ System. *Focus* (Life Technologies, Inc.) **14:** 41–45.

Bush, C.E., L.J. Di Michele, W.R. Peterson, D.G. Sherman, and J.H. Godsey. 1992. Solid-phase time-resolved fluorescence detection of human immunodeficiency virus polymerase chain reaction amplification products. *Anal. Biochem.* **202:** 146–151.

Conway, B., L.J. Bechtel, K.A. Adler, R.T. D'Aquila, J.C. Kaplan, and M.S. Hirsch. 1992. Comparison of spot-blot and microtitre plate methods for the detection of HIV-1 PCR products. *Mol. Cell. Probes* **6:** 245–249.

Coutlée, F., L. Bobo, K. Mayur, R.H. Yolken, and R.P. Viscidi. 1989. Immunodetection of DNA with biotinylated RNA probes: A study of reactivity of a monoclonal antibody to DNA-RNA hybrids. *Anal. Biochem.* **181:** 96–105.

Hartley, J.L. and A. Rashtchian. 1993. Dealing with contamination: Enzymatic control of carryover contamination in PCR. *PCR Methods Appl.* **3:** S10–S14.

Hayashi, K., M. Orita, Y. Suzuki, and T. Sekiya. 1989. Use of labeled primers in polymerase chain reaction (LP-PCR) for a rapid detection of the product. *Nucleic Acids Res.* **17:** 3605.

He, Y., F. Coutlée, P. Saint-Antoine, C. Olivier, H. Voyer, and A. Kessous-Elbaz. 1993. Detection of polymerase chain reaction-amplified human immunodeficiency virus type 1 proviral DNA with a digoxigenin-labeled RNA probe and an enzyme-linked immunoassay. *J. Clin. Microbiol.* **31:** 1040–1047.

Inouye, S. and R. Hondo. 1990. Microplate hybridization of amplified viral DNA segment. *J. Clin. Microbiol.* **28:** 1469–1472.

Keller, G.H., D.P. Huang, J.W.K. Shih, and M.M. Manak. 1990. Detection of hepatitis B virus DNA in serum by polymerase chain reaction amplification and microtiter sandwich hybridization. *J. Clin. Microbiol.* **28:** 1411–1416.

Kolk, A.H.J., A.R.J. Schuitema, S. Kuijper, J. van Leeuwen, P.W.M. Hermans, J.D.A. van Embden, and R.A. Hartskeerl. 1992. Detection of *Mycobacterium tuberculosis* in clinical samples by using polymerase chain reaction and a nonradioactive detection system. *J. Clin. Microbiol.* **30:** 2567–2575.

Landgraf, A., B. Reckmann, and A. Pingoud. 1991. Direct analysis of polymerase chain reaction products using enzyme-linked immunosorbent assay techniques. *Anal. Biochem.* **198:** 86–91.

Lazar, J.G. 1993. A rapid and specific method for the detection of PCR products. American Biotechnology Laboratory, September, 14–16, Shelton, Connecticut.

Lazar, J.G., J.A. Tropp, and A.T. Lörincz. 1993a. *Comparison of the Digene SHARP Signal System and a nested-primer method for the detection of hepatitis C virus in clinical specimens using PCR.* Presented at ASM Conference on Molecular Diagnostics and Therapeutics, Jackson, Wyoming, September 1993.

Lazar, J.G., J.A. Tumulty, H. Salim, and S.S. Challberg. 1993b. *Sensitive and specific detection of PCR products: Application of the Digene SHARP Signal System.* Presented at the 93rd Annual Meeting of the American Society for Microbiology, Atlanta, Georgia, May 1993.

Levenson, C. and C.A. Chang. 1990. Nonisotopically labeled probes and primers. In *PCR protocols: A guide to methods and applications* (ed. M.A. Innis et al.), pp. 99–112. Academic Press, San Diego, California.

Lundeberg, J., J. Wahlberg, M. Holmberg, U. Pettersson, and M. Uhlén. 1990. Rapid colorimetric detection of *in vitro* amplified DNA sequences. *DNA Cell Biol.* **9:** 287–292.

Rasmussen, S.R., H.B. Rasmussen, M.R. Larsen, R. Hoff-Jørgensen, and R.J. Cano. 1994. Combined polymerase chain reaction-hybridization microplate assay used to detect bovine leukemia virus and *Salmonella. Clin. Chem.* **40:** 200–205.

Syvänen, A.C., M. Bengtström, J. Tenhunen, and H. Söderlund. 1988. Quantification of polymerase chain reaction products by affinity-based hybrid collection. *Nucleic Acids Res.* **16:** 11327–11338.

Terry, G., L. Ho, A. Szarewski, and J. Cuzick. 1994. Semiautomated detection of human papillomavirus DNA of high and low oncogenic potential in cervical smears. *Clin. Chem.* **40:** 1890–1892.

Tsai, C.H. and T.W. Dreher. 1993. *In vitro* transcription of RNAs with defined 3′ termini from PCR-generated templates. *BioTechniques* **14:** 58–61.

Urrutia, R., M.A. McNiven, and B. Kachar. 1993. Synthesis of RNA probes by the direct in vitro transcription of PCR-generated DNA templates. *J. Biochem. Biophys. Methods* **26:** 113–120.

Young, B.D. and M.L.M. Anderson. 1985. Quantitative analysis of solution hybridisation. In *Nucleic acid hybridisation: A practical approach* (ed. B.D. Hames and S.J. Higgins), pp. 47–71. IRL Press, Oxford.

Quantitative PCR Using the AmpliSensor Assay

Chang Ning John Wang, Kai Y. Wu, and Hwa-Tang Wang

AcuGen Systems, Biotronics Corporation, Lowell, Massachusetts 01851

INTRODUCTION

The AmpliSensor assay is a real-time, quantitative tool for PCR-based detection. The assay is based on the principle that fluorescence energy transfer can be used to detect duplex formation between complementary nucleic acid strands. If the two complementary strands are labeled with donor and acceptor fluorophores, respectively, fluorescence energy transfer between the fluorophores is facilitated when the strands are base-paired, or eliminated when base-pairing is disrupted. In this way, the extent of energy transfer can be used to measure the amount of duplex formation between the fluorophore-labeled oligonucleotide duplex, thereby relating the extent of duplex formation mediated by the DNA polymerase (see Fig. 1).

The AmpliSensor assay is a two-step reaction. In the first step, asymmetric PCR is carried out to its late-log phase before one of the target strands is significantly overproduced. In the second step, an AmpliSensor primer complementary to the overproduced target strand is added to prime a semi-nested reaction in concert with the excess primer. As the semi-nested amplification proceeds, the AmpliSensor duplex starts to dissociate as the AmpliSensor sequence is duplicated. As a result, the fluorophores configured for energy transfer are disengaged from each other, causing the energy transfer process preestablished in all of the AmpliSensor primers to disrupt for those primers involved in the amplification process. The measured fluorescence intensity is in proportion to the amount of AmpliSensor duplex left at the end of each amplification cycle. The decrease in the fluorescence intensity correlates proportionately to the initial target dosage and the extent of amplification.

The signal-generating mechanism of the AmpliSensor assay is tightly linked to the priming event. However, unlike typical PCR, where single-stranded primers are constantly subject to the pos-

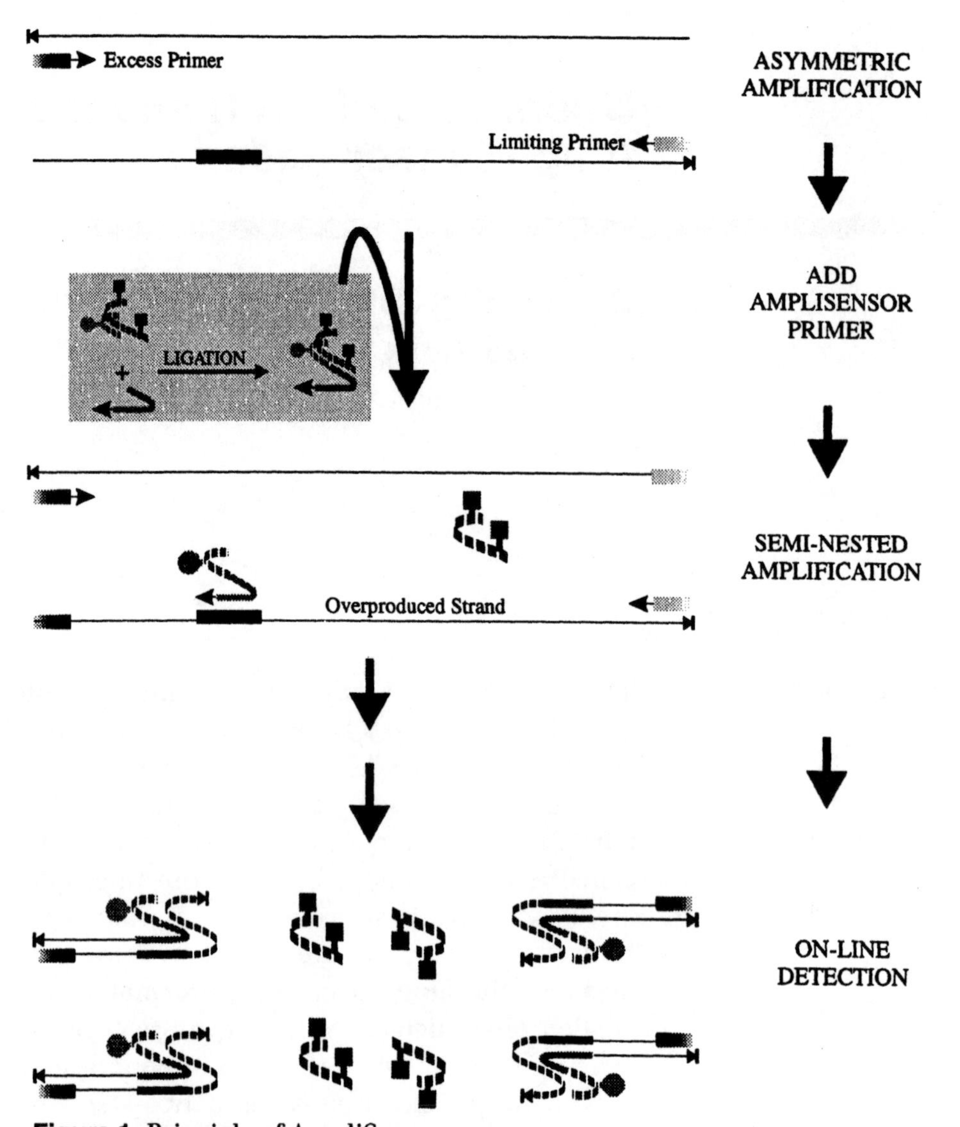

Figure 1 Principle of AmpliSensor assay.

sibility of nonspecific priming, the AmpliSensor primer is sequestered in a double-stranded form and stabilized against random priming. Consequently, the fluorescence signal correlates to the overall energy transfer efficiency in a predictable, sequence-specific manner, and the amplified product can be monitored directly.

In principle, one can adopt the AmpliSensor assay to monitor multiplex PCR for the degree of individual amplification as long as the signal of each AmpliSensor reaction can be correspondingly differentiated. For absolute quantitation, a housekeeping sequence can be monitored and normalized along with the target for its extent of amplification to ensure adequate sampling. A set of AmpliSensor primers specific to an internal mimic control can also be included in the amplification reaction to correct the difference in overall amplification efficiency between reactions. As a simple, homogeneous assay for PCR quantitation, the AmpliSensor technology is target-specific, and it is amenable to the normalization of signal loading and sample loading. Thus, it can reliably count the amplified product while it accounts for the efficiency difference among reactions.

The AmpliSensor protocol described below represents the use of a complete kit including a specifically designed computer interface and a programmable instrument to carry out the reaction. The power of this method lies in its ability to adapt any well-defined PCR procedure and quantify its product in a cycle-dependent manner.

REAGENTS AND MATERIALS

Universal AmpliSensor Kit (Biotronics, cat. no. 10-1688-10)
Taq DNA polymerase (Perkin-Elmer, cat. no. N801-0060)
96-well thermoplate temperature cycler (Biotronics, cat. no. 90-1030-T)
AG-9600 Analyzer or Minilyzer fluorescence thermoplate reader (Biotronics, cat. no. 90-1000-A)
A.S.A.P. data acquisition software (Biotronics, cat. no. 90-1000-A)
Electronic multi-dispenser (Biotronics, cat. no. 90-1040-P)
96-well thermoplate (Biotronics, cat. no. 95-1000-30)
Thermoplate cover (Biotronics, cat. no. 95-1000-35)
Mineral/dispensing oil (Biotronics cat. no. 95-1000-20)

10x dNTP mix
0.1 mM each of the four deoxyribonucleotides

10x PCR buffer
50 mM Tris-HCl (pH 8.7)
50 mM KCl
5 mM NH_4Cl
1 mM dithiothreitol
0.1% Triton X-100

10x Excess primer (20–40 ng/μl)

10x Limiting primer (7–10 ng/μl)

TE buffer
10 mM Tris-HCl (pH 8.0)
1 mM EDTA (pH 8.0)

40 mM $MgCl_2$
100 mM EDTA

Desktop computer (486 DX recommended)
Vortex
Pipette tips
Microcentrifuge tubes
Deionized or distilled water

PROTOCOLS

Ligation

1. Add the following reagents at room temperature to a 0.5-ml microcentrifuge tube:

5X coupling reagent	10 μl
5X Universal AmpliSensor	10 μl
Oligonucleotide primer (25 nucleotides long)	0.5 μg
Add H_2O to 50 μl	

   Both the 5X coupling reagent and 5X Universal AmpliSensor are components of the Universal AmpliSensor kit. The coupling reagent consists of proportionate T4 ligase, T4 polynucleotide kinase, and ATP in phosphate buffer. The Universal AmpliSensor is a quasi-stable signal duplex of two oligonucleotides each labeled with an energy donor and acceptor fluorophore, respectively. The two strands of the Universal AmpliSensor are unequal in length with the long strand 5′ overhanging the short one by 7 nucleotides (5′-GCGTCCC-3′). For effective ligation, the primer should encompass at its 5′ end a "hook" sequence, that is, 5′-GGGACGC-3′, complementing the overhang of the Universal AmpliSensor. The amount of primer used per reaction reflects the desirable 1:1 molar ratio of a 25-mer to the Universal AmpliSensor. For primers of different lengths, adjust the volume of AmpliSensor according to the following formula:

   $$V = (\text{Total volume}/5) \times (25/\text{Primer length})$$

   Thus, for a 28-mer in a 100-μl reaction, 18 μl of 5X Universal AmpliSensor is required. The optimum primer length including the "hook" sequence is 25–30 nucleotides.

2. Incubate the tube for 90 minutes at 37°C. Terminate the reaction by adding 50 μl of deionized H_2O, and heat for 5 minutes to 90°C to denature the enzymes. Place the ligated product in a freezer and label it as 5X AmpliSensor primer stock. The AmpliSensor-ligated primer can be batch-prepared in advance. However, once ligated, the primer should not be stored for more than 2 weeks.

Ligation Efficiency Estimation

1. Add the following reagents at room temperature to a 0.5-μl microcentrifuge tube:

10X PCR buffer	10 μl
10X dNTP	10 μl
40 mM $MgCl_2$	10 μl
Add H_2O to 50 μl	

This mix is intended for eight extension assays, which can be scaled up if necessary.

2. Aliquot 5 µl of the extension mix into eight wells of a thermoplate. Designate two of the wells for Sample, two for Negative, two for Standard, and two for Blank. Overlay 10 µl of mineral oil to each reaction.

3. Mix 2 µl of the 5X AmpliSensor primer stock with 8 µl of the 1X PCR buffer and label it as Sample solution. Mix 2 µl of the 5X Universal AmpliSensor with 98 µl of the 1X PCR buffer and label it as Reference solution. The preparation of the reference solution is intended for 20 reactions and can be stored in a freezer for later use.

4. Aliquot 5 µl of the Sample solution to the Sample wells, and dispense 5 µl of the Reference solution to both the Negative and Standard wells. To the Blank, add 5 µl of H_2O. The Negative serves as a control of baseline signal, whereas the Standard is a reference of maximum energy transfer.

5. Subject the reaction mixes in the thermoplate for 20 seconds to 94°C, for 20 seconds to 55°C, for 30 seconds to 72°C, and then for 30 seconds to 20°C to equilibrate the signal.

6. Read the fluorescence under the coupling mode using an AG-9600 Analyzer or Minilyzer. Upon finishing, save the data as base reading.

7. Add 1 unit of *Taq* DNA polymerase to both the Sample and Standard wells, and subject the reactions to three additional thermal cycles. Each cycle consists of 20 seconds at 94°C, 20 seconds at 55°C, and 30 seconds at 72°C. At the end of cycling, cool the reactions down to 20°C for 30 seconds.

   The polymerase will fill in the 7-nucleotide gap of the overhang for those AmpliSensor molecules that have not been ligated to the primer. The gap-filling stabilizes the quasi-stable AmpliSensor and thus leads to an enhancement of the overall fluorescence intensity. To estimate the coupling efficiency, the Sample is compared to the Standard for its relative fluorescence enhancement reading after the extension reaction.

8. Repeat the fluorescence reading under the same mode. When the data are saved, the coupling efficiency is displayed on the screen as a percentage of 100 for each sample.

9. Activate the report window to review the coupling status. A "+" status is assigned to samples of ligation efficiency greater than 70.

10. Proceed to asymmetric PCR for samples assigned with "+" status. Repeat ligation for samples of "–" status.

In general, the raw reading of the Standard after the extension should almost be doubled, otherwise the extension reaction should be deemed incomplete. The most common cause of low ligation efficiency is overestimation of primer concentration. Assuming that the 5′ end of the primer is not defective, the efficiency can be improved by elevating the primer ratio.

Asymmetric Preamplification

1. To make the PCR master mix, add the following to a 1.5-ml microcentrifuge tube:

10x PCR buffer	110 μl
40 mM $MgCl_2$	110 μl
10x dNTP mix	110 μl
10x Excess primer	110 μl
10x Limiting primer	110 μl
Add H_2O to 880 μl	

The PCR master mix is intended for 100 reactions, allowing 10% extra volumes. It can be scaled up or down according to the experimental need. If reagent delivery is done by the AmpliSensor Analyzer, the user is prompted for the desired sample and reagent volumes of each reaction. The volume of the master mix required for the entire assay is displayed on-screen in a message box so that the mix can be prepared for automated dispensing accordingly. The excess primer is preferably a few nucleotides longer than the limiting primer. The optimum length is 25 nucleotides for the excess primer and 20 nucleotides for the limiting primer. The asymmetric primer ratio is critical to the ultimate sensitivity and resolution of the assay, due to the fact that the amplitude of fluorescence signal change is determined by the over-priming rate of the excess primer. No less than 0.75 ng of limiting primer and 2–4 ng of the excess primer should be used for each microliter of the reaction.

2. Add 50 units of *Taq* DNA polymerase, mix thoroughly, and then aliquot 8 μl into each well of the thermoplate, except the well designated as Apex, to which add 10 μl of 1x PCR buffer instead. Overlay 10 μl of mineral oil to each reaction.

If the master mix is dispensed by the AmpliSensor Analyzer, 7 μl of mineral oil will be automatically overlaid.

3. To each aliquot, add either 2 μl of sample or 2 μl of the positive standards of known target dosage. Add 2 μl of TE buffer to the Negative well.

   To assure intra-assay consistency, run a duplicate for each reaction. A minimum of four positive standards are required for quantitative analysis by the A.S.A.P. (*A*mpli*S*ensor *A*nalysis *P*rogram; Windows-based software dedicated to the operation of the AG-9600 Analyzer or Minilyzer and the subsequent data processing and analysis) program. For threshold analysis, one should designate a positive standard of appropriate target copies as the cutoff threshold. The Negative serves as a no-template control to ensure that no nonspecific amplification occurs and the master mix is free of source contamination. The Apex represents a constant source of signal to which all the sample readings can be normalized.

4. Subject the reactions in the thermoplate to 25 thermal cycles. Each cycle consists of 94°C for 20 seconds, 55°C for 20 seconds, and 72°C for 30 seconds. At the end of the cycling, chase the reaction for 30 seconds at 72°C and cool it down to 4°C for 2 minutes.

   In general, for detection of 10–1000 copies of the target, the asymmetric preamplification should run for 25 cycles prior to moving into the semi-nested phase. For targets greater than 1000 copies per assay, the preamplification should run for 20 cycles instead. Select an annealing temperature between 50°C and 65°C according to the stringency requirement.

Semi-nested Amplification and On-line Detection

1. Add the following reagents to a 0.5-ml microcentrifuge tube to make the AmpliSensor mix:

5x AmpliSensor primer stock	66 μl
10x PCR buffer	33 μl
40 mM $MgCl_2$	33 μl
Add H_2O to 330 μl	

   The AmpliSensor mix is intended for 100 reactions, which can be scaled up or down according to the experimental need. If the AmpliSensor mix is delivered by the AmpliSensor Analyzer, the user is prompted for the desired volume of the AmpliSensor mix for each reaction. The total volume required is displayed in a message box so that the mix can be prepared for automated dispensing accordingly.

2. Aliquot 5 μl of the AmpliSensor mix to each reaction except for the Blank.

   For maximum sensitivity, use 5 μl of AmpliSensor mix per 10-μl reaction. The quantitative resolution and detection sensitivity can be modulated by adjusting the amount of AmpliSensor mix per reaction. The Analyzer is programmed to overlay an additional 5 μl of mineral oil after dispensing the mix.

3. Subject the reactions in the thermoplate to a thermal cycle of 94°C for 20 seconds, 60°C for 20 seconds, and 72°C for 30 seconds. Cool the reaction for 30 seconds at 20°C to establish a signal equilibrium, and then acquire the fluorescence reading under the standard mode using an AG-9600 Analyzer or Minilyzer. Save the data as a base reading.

   The initial base readings serve as a reference ceiling, to which all the corresponding readings acquired at later cycles are compared for their progressive diminution as a result of energy-transfer disruption.

4. Resume the thermal cycling for a defined number of cycles and chase the reaction for 30 seconds at 72°C at the end. Repeat the fluorescence reading under the same mode. Save the data as an assay-cycle reading.

   Each active file can store up to nine assay-cycle readings, and a serial number is assigned to each assay cycle. A detection index is derived for each reading according to the algorithm detailed below in Data Processing and Interpretation. The magnitude of the detection index is directly proportional to the amount of amplification product accumulated at the current cycle.

5. Activate the graph subwindow to review the current data in either bar chart or graph format. Otherwise, activate the report subwindow to reveal the quantitative data or the diagnostic status of each sample.

   Data analysis can be performed by either cutoff or quantitative mode checked off from a decision-making dialog box anytime during the data acquisition. The quantitative analysis yields quantitative information, and the cutoff analysis assigns the diagnostic status to each sample. The user may choose either linear regression or polynomial curve fitting to establish the standard curve for quantitative analysis. Linear regression is particularly suitable for small-range, high-resolution quantitation.

6. Click the "print" button to print the data. The data can be retrieved in either tabulated, graph, or matrix format.

DATA PROCESSING AND INTERPRETATION

Prior to normalization, raw readings from all the cycles must be corrected for the background noise by subtracting the blank value. Dividing the base-cycle sample reading by the reading of the Apex yields the sample quotient r_i for sample *i*. The normalized sample reading $N_{i,j}$ of cycle *j* is derived by dividing the assay-cycle sample reading $R_{i,j}$ by r_i and subtracting the value δ. The value δ reflects the background signal derived from the added AmpliSensor molecule as if they are 100% dissociated. The normalized sample readings are then compared to the normalized negative reading to yield the detection indexes $I_{i,j}$, which manifest the extent of sample signal deviated from that of the negative control.

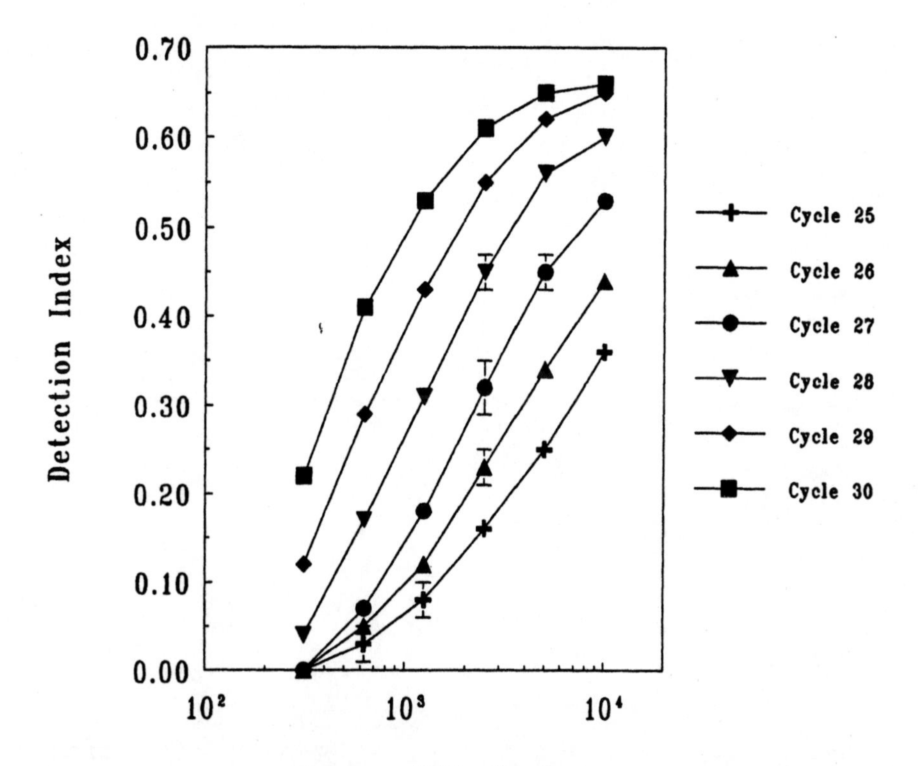

Figure 2 Cycle-dependent accumulation of an amplified 165-bp IGF-1 fragment.

The following equations express the relationship of these terms:

$I_{i,j} = 1 - (N_{i,j} / N_{neg,j})$, where
$N_{i,j} = (R_{i,j} / r_i) - \delta$
$N_{neg,j} = (R_{neg,j} / r_i) - \delta$
$r_i = R_{i,base} / R_{apex,base}$ and
$R_{i,j} = \textit{Raw reading} - \textit{Blank value}$

The detection index varies in a cycle-dependent manner and can be used to extrapolate the amount of specific PCR product accumulated at the assay cycle.

ANTICIPATED RESULT In contrast to end-point analysis such as a gel-based assay, the AmpliSensor assay can resolve a twofold target dosage difference in a kinetic manner. The best-resolution window is progressively exposed toward higher sensitivity as the amplification cycle proceeds (Fig. 2).

DNA Fingerprinting Using Arbitrarily Primed PCR

Michael McClelland and John Welsh

California Institute of Biological Research, La Jolla, California 92037

INTRODUCTION

Fingerprinting of genomes with arbitrary primers (Welsh and McClelland 1990; Williams et al. 1990) is a versatile method for detecting polymorphisms for genetic mapping, phylogenetics, and population biology. The method generates a fingerprint using arbitrarily selected primers under conditions where the primer will initiate synthesis on DNA, even when the match with the template is imperfect. Some of these priming events occur on opposite strands. The most efficient of these pairs of priming events compete with each other during amplification to produce a fingerprint of a few to over 100 prominent PCR products. Such products have often been referred to as random amplified polymorphic DNAs (RAPDs) (Williams et al. 1990). More than 1000 publications have used this strategy.

Over the past few years, a multitude of variants and improvements of this method have been developed. In this chapter, we highlight some of these protocols and justify our current preferred protocol for DNA fingerprinting. Details of these protocols can be found in Bowditch et al. (1993); Clark and Lanigan (1993); Welsh and McClelland (1993); Williams et al. (1993); and Yu et al. (1993). The next chapter discusses RNA fingerprinting using arbitrarily primed PCR (AP-PCR) (Liang and Pardee 1992; Welsh et al. 1992).

Much has been made in the literature of differences in primer length or in the gel system used for resolving PCR-based fingerprints. The fundamental principle used does not change, but these differences do allow a number of subtle variations in the design of a fingerprinting experiment. When designing such an experiment, it is important to decide how many PCR products are wanted per lane and whether to use agarose, nondenaturing acrylamide, or denaturing acrylamide. One choice is to generate relatively few PCR products and to resolve these on an agarose gel. The more prominent of these products can then be scored quite reliably. Such fingerprints can be generated using 10-base primers and AmpliTaq (Perkin-Elmer) in the manner first described by Williams et al. (1990, 1993). Alternatively, a larger number of fragments can be generated and resolved on an acrylamide gel. There are a number of ways to achieve this:

1. Select, in a preliminary screen, 10-mers that yield a large number of fragments using AmpliTaq holoenzyme.

2. Use longer primers, such as 18-mers (Welsh and McClelland 1990).

3. Use primers that are biased toward more common sequences in the genome by performing a statistical analysis.

4. Use very short primers in very large amounts to generate extremely complex fingerprints (Caetano-Anolles et al. 1992).

The protocol presented here uses arbitrary 10-mers in combination with the AmpliTaq Stoffel fragment (or KlenTaq). This enzyme generates almost twice as many easily visible PCR products as does AmpliTaq using the same primer and template DNA. Furthermore, this enzyme increases the number of arbitrary 10-mers that give productive fingerprints from about 75% to over 90% (Sobral and Honeycutt 1993).

At least for primers of 10 bases or longer, the more complex the PCR fingerprint, the more reproducible it seems to be. However, too many products make the pattern difficult to interpret. Thus, reliable fingerprints can be obtained by screening arbitrary primers for moderately complex patterns and then using these on the population under study.

If a simple pattern is desired, then it is important to be aware of the "context effect." This situation arises for very simple, but not for complex, patterns. The fingerprint is the result of a competition between many PCR products, and the fewer the winners, the greater the effect of their presence and absence on the probability of other PCR products being amplified. For example, if two similar but nonidentical genomes give very simple fingerprints with the same primer and differ by one or more of the major PCR products, then one must be concerned that the difference(s) may affect the probability of amplification of other PCR products. One symptom of such a context effect in a mapping population would be nonparental products that appear due to the absence of a prominent polymorphism from one parent in some offspring. This phenomenon may be a good reason to avoid using simple fingerprint patterns that contain fewer than 10 prominent PCR products.

For the purposes of fingerprinting, it is often best to use arbitrary primers in pairwise combinations (Welsh and McClelland 1991a). A few primers can be used in a very large number of pairwise combinations. For example, 20 primers can be used in 380 (19 x 20/2) different combinations. One potential disadvantage of using pairwise combinations of primers is that some of the products will have the same primer at both ends, and these products may be shared between fingerprints using that primer.

One advantage of using primers in pairwise combinations is that the products can be directly sequenced using conventional PCR sequencing kits. However, these sequences are not always of the highest quality, perhaps because no part of a fingerprinting gel is entirely free of other products. Sometimes it is best to clone the products first, as described in Welsh and McClelland (1991a).

Using 10-mers in Pairwise Combinations

For use in this protocol, the DNAs should be of similar quality and of at least 10 kb average length. Surprisingly, it is mainly consistency in DNA quality that is most important; thus, for example, one can use crude lysates of bacterial DNA if all the strains are treated in the same way (Welsh and McClelland 1990, 1993).

REAGENTS AND EQUIPMENT

Thermal cycler (Perkin-Elmer 9600 model)
PCR tubes (MicroAmp, Perkin-Elmer)
2x AP-PCR mixture
- 20 mM Tris-HCl, pH 8.3
- 20 mM KCl
- 10 mM $MgCl_2$
- 0.2 mM each dNTP
- 0.1 unit/μl *Taq* DNA polymerase Stoffel fragment (Perkin-Elmer)
- 0.1 μCi/μl [α-^{32}P]dCTP

Two arbitrary 10-mer primers (0.4 μM each, Genosys)

PROTOCOL

All procedures can be set up on ice or at room temperature, as convenient. Reactions are prepared at two DNA concentrations. It is wise to initially titrate the DNA over two orders of magnitude to find the concentrations that give the most robust fingerprints. For mammalian DNA, the best fingerprints are usually obtained in the range of 5–50 ng per 20-μl reaction volume. The optimal concentration of the primer must be determined empirically. For most 10-mers, the optimum is around 0.4 μM. Note, however, that we have not been able to fingerprint bacterial genomes using pairs of primers except at high template concentrations (>1 μg). Perhaps for simple genomes the primer-dimers compete effectively. We recommend that only one primer be used at 0.4 μM for small genomes.

1. Prepare DNA at 2X final concentration and distribute 10 μl to tubes.

2. Add 10 μl of 2X AP-PCR mixture. The volume proportions of DNA to the reaction mixture can, of course, be changed, as can the total final reaction volume so long as the final concentrations of reaction components are kept the same. Many laboratories run 10-μl reactions.

3. Set the thermal cycling profile for 10-mers as follows: 1 minute at 94°C, 1 minute at 35°C, 2 minutes at 72°C, for 40 cycles.

4. Dilute the products 1:4 in 80% formamide containing 10 mM EDTA and tracking dye, heat for 15 minutes to 65°C, and electrophorese 2 μl through a denaturing sequencing-type polyacrylamide gel or an MDE gel (Baker). Electrophorese sequencing gels in 1.0X TBE at 50 watts at 65°C for 3 hours. Electrophorese MDE gels in 0.5X TBE at 7 watts at room temperature for 18 hours. The gel may be dried and exposed to X-ray film without hindering the ability to clone fragments from the gel. Alternatively, the radioactive label can be omitted entirely and native agarose gels followed by ethidium bromide staining can be used.

5. If desired, bands can be excised from the gel with a razor blade, eluted, reamplified, and cloned (see the next chapter on RNA fingerprinting).

TROUBLESHOOTING

The issue of reproducibility is of much concern, because some failures in PCR-based fingerprinting have been published and may discourage other investigators from using the method. One concern is that the patterns may vary from day to day or from lab to lab (see, e.g., Meunier and Grimont 1993). The other concern is unreliability within the same experiment on the same day.

The problem of intraexperiment variability has been overstated. First, almost all of these problems are due to inadequately prepared DNA. The easiest way to find out if the DNA is of sufficient quality is to perform twofold serial dilutions of the DNA over a wide range from about 200 ng to 200 pg. If the DNA does not produce reliable fingerprints over a number of twofold dilutions with a number of different primers, then the DNA quality is suspect. Second, primers that give moderately complex patterns should be used, as described earlier. High-quality fingerprints have been obtained from genomes in every kingdom and over a wide spectrum of G+C contents. Thus, a failure of this kind must be attributed to inadequate DNA or reagents. We have reached the somewhat surprising conclusion that when DNA quality is adequate, the reliability of the fingerprints seems to derive primarily from fingerprint complexity.

One cannot know if the difference between two genomic fingerprints is real if the experiment is not controlled for DNA quality and quantity. Thus, initially, every experiment must include fingerprinting for *at least two* concentrations of genomic DNA for each individual. Any differences between individuals that do not occur at both genomic DNA concentrations should be rejected. Too many concentration-dependent differences should lead to concern about the DNA or reagents.

Day-to-day variation and inter-laboratory variation constitute a more genuine concern. Such variation occurs because all PCR-based fingerprinting varieties are sensitive to the buffer conditions, enzyme quality, and the primer preparation. Although this is easy to control in a particular experiment, it is harder to control between experiments. The ratio of intensities among products within a single fingerprint lane may vary from day to day. However, as long as the fingerprint pattern is complex, the variation does not extend to variability in the presence or absence of bands from day to day, and the ratio of intensities of a particular product compared between lanes does not change. These differences between experiments and between experimenters are generally rather subtle, so they can be accommodated by the simple expedient of fingerprinting DNA from reference strains on each gel.

What Gel to Use?

Some researchers prefer agarose gels and simpler fingerprints because they believe agarose gels are easier to set up and because agarose gels can be stained with ethidium bromide, thereby avoiding radioactivity. When beginning a project on fingerprinting, it may be wise to run products on an agarose gel to ensure that the amplification has been robust. It is disappointing to run an acrylamide gel and later find that an error was made in the experiment. However, after the technology is mastered, the use of a protocol that generates a relatively complex fingerprinting pattern and the extra work of running a denaturing acrylamide gel are rewarded by much more and better-resolved data than can be obtained using agarose. When such data are scored, there is a greater chance that PCR products of a particular size are homologs because there is almost single-base resolution of fragments. In addition, the use of denaturing gels eliminates the problem of uneven amplification of the two strands that, on a native gel, yields a double-stranded product and a fainter variable product for the single strand that is in excess.

We have also separated *denatured* DNA fingerprints on a *native* acrylamide gel. This strategy (McClelland et al. 1994) combines single-strand conformation polymorphisms (SSCP) (Hayashi 1991) with PCR-based fingerprinting. An example is shown in Figure 1. SSCP gels allow increased confidence in scoring the usual presence/absence of polymorphisms that result from arbitrarily

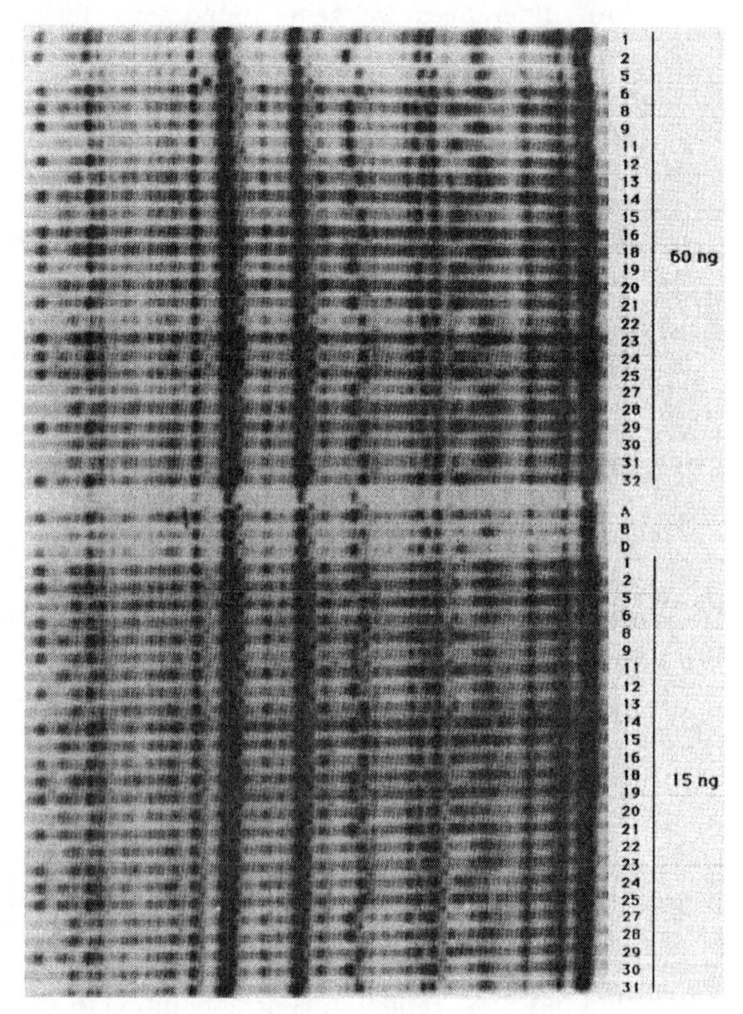

Figure 1 AP-PCR-SSCP for genetic mapping. A total of 15 ng and 60 ng of genomic DNA from mouse strains in the C57BL/6J (B) versus DBA (D) recombinant inbred mapping population was fingerprinted using the pair of primers S2 (5′-CCTCTGACTG) and S1 (5′-GAGGTCCACA) (Operon) and the conditions described in the protocol. The fingerprints were resolved on 5% MDE + 5% glycerol, dried, then autoradiographed. HydroLink-MDE was purchased from Baker. The numbers above each lane indicate the recombinant inbred lines. "A" indicates another mouse strain A/J. Only the top half of the gel is shown. Fragments visualized range from about 400 bases to 800 bases. Numerous presence/absence and length polymorphisms are visible.

primed PCR because each strand of these polymorphisms occurs at different parts of the gel, allowing two opportunities for scoring. Furthermore, polymorphisms that could not be scored by other gel systems, i.e., those that carry internal sequence polymorphisms, were also detected. It is worthwhile to use SSCP gels to increase throughput and reliability of scoring when mapping by PCR fingerprinting. However, because each strand usually occurs in a different part of the gel, polymorphisms are often counted twice. The resulting nonindependence of characters means SSCP is probably not appropriate for population biology and phylogenetics.

Motif Sequences Encoded in Primers

Completely arbitrary priming lies at one end of a spectrum of possible targeting strategies for fingerprinting. The other end of the spectrum uses primers derived from known perfect or near-perfect dispersed repeats, e.g., Alu-PCR (Nelson et al. 1989), tDNA-intergenic length polymorphisms (Welsh and McClelland 1991b), or REP-PCR (Versalovic et al. 1991). In this spectrum lie a cornucopia of other repeats, such as purine-pyrimidine motifs, that have been successfully used to produce PCR fingerprints (Welsh et al. 1991). These microsatellite repeats are useful because primers directed toward them reveal more polymorphisms than seen in the average arbitrarily primed fingerprint, making them particularly useful for detecting polymorphisms between closely related individuals (Welsh et al. 1991; Wu et al. 1994). Primer pairs directed toward tRNA genes are also useful because the tRNA gene clusters evolve more slowly than most of the rest of the genome, which is under less stringent selection pressure. The patterns produced by tDNA-intergenic length polymorphism primers can be used to compare genomes at a higher taxonomic level than is possible with AP-PCR (Welsh and McClelland 1991b). Other primers biased to rarer conserved motifs, such as those that occur in some promoters or in gene families, have also been tried (Birkenmeier et al. 1992). Finally, statistics can be applied to the known sequences in the DNA database for a particular species or related species. These data can be used to develop primers that carry sequences that are rare or common in a genome. Such primers influence the number of PCR products in a fingerprint.

One might expect that primers directed against sequences such as purine-pyrimidine repeats or other dispersed repeats may generate significantly more reliable fingerprints than completely arbitrary primers because the interaction of the primer with the template will be better. In some cases, such as primers directed against tRNA motifs, we have confirmed a considerable bias by sequencing the products (McClelland et al. 1992; Welsh and McClelland 1992). We also found that microsatellite repeat primers often amplify the expected microsatellite repeats. However, these microsatellite targets probably represent the minority of PCR products on the gel, despite the fact that microsatellites are quite prevalent in most genomes. The other products are arbitrarily primed at nonmotif locations (Welsh et al. 1991). Primers directed toward rarer motifs have a correspondingly lower success targeting the expected motif.

In those cases where the motif is poorly conserved or where the motif is rare, it might seem logical to attempt to increase the stringency of the motif-PCR. However, such attempts have generally failed because the fingerprints become unreliable at higher stringency. Increasing the stringency for a poorly matched or rare motif still does not allow motif sites to dominate versus the best arbitrary events. Eventually, as the stringency is raised higher, the whole fingerprint becomes unreliable. Nevertheless, there is nothing to be lost by using a primer with a rare motif sequence at low stringency. At a minimum, a good arbitrary fingerprint can be expected.

ACKNOWLEDGMENTS We thank Rhonda Honeycutt for helpful discussions and comments on the manuscript. This work was supported in part by National Institutes of Health grants AI-34829, NS-33377, and CA-68822 to M.M. and AI-32644 to J.W.

REFERENCES

Birkenmeier, E.H., U. Schneider, and S.J. Thurston. 1992. Fingerprinting genomes by use of PCR primers that encode protein motifs or contain sequences that regulate gene expression. *Mamm. Genome* **3:** 537–545.

Bowditch, B.M., D.G. Albright, J.G. Williams, and M.J. Braun. 1993. Use of randomly amplified polymorphic DNA markers in comparative genome studies. *Methods Enzymol.* **224:** 294–309.

Caetano-Anolles, G., B.J. Bassam, and P.M. Gresshoff. 1992. Primer-template interactions during DNA amplification fingerprinting with single arbitrary oligonucleotides. *Mol. Gen. Genet.* **235:** 157–165.

Clark, A.G. and C.M.S. Lanigan. 1993. Prospects for estimating nucleotide divergence with RAPDs. *Mol. Biol. Evol.* **10:** 1096–1111.

Hayashi, K. 1991. PCR-SSCP: A simple and sensitive method for detection of mutations in the genomic DNA. *PCR Methods Appl.* **1:** 34–38.

Liang, P. and A. Pardee. 1992. Differential display of eukaryotic messenger RNA by means of the polymerase chain reaction. *Science* **257:** 967–971.

McClelland, M., C. Peterson, and J. Welsh. 1992. Length polymorphisms in tRNA intergenic spacers detected using the polymerase chain reaction can distinguish streptococcal strains and species. *J. Clin. Microbiol.* **30:** 1499–1504.

McClelland, M., H. Arensdorf, R. Cheng, and J. Welsh. 1994. Arbitrarily primed PCR fingerprints resolved on SSCP gels. *Nucleic. Acids. Res.* **22:** 1770–1771.

Meunier, J.R. and P.A.D. Grimont. 1993. Factors affecting reproducibility of amplified polymorphic DNA fingerprint. *Res. Microbiol.* **144:** 373–379.

Nelson, D.L., S.A. Ledbetter, L. Corbo, M.F. Victoria, R. Ramirez-Solis, T.D. Webster, D.H. Ledbetter, and C.T. Caskey. 1989. Alu polymerase chain reaction: A method for rapid isolation of human specific sequences. *Proc. Natl. Acad. Sci.* **86:** 6686–6690.

Sobral, B.W.S and R.J. Honeycutt. 1993. High output genetic mapping of polyploids using PCR-generated markers. *Theor. Appl. Genet.* **86:** 105–112.

Versalovic, J., K. Thearith, and J.R. Lupski. 1991. Distribution of repetitive DNA sequences in eubacteria and application to fingerprinting of bacterial genomes. *Nucleic Acids Res.* **19:** 6823–6831.

Welsh, J. and M. McClelland. 1990. Fingerprinting genomes using PCR with arbitrary primers. *Nucleic Acids Res.* **18:** 7213–7218.

——. 1991a. Genomic fingerprinting with AP-PCR using pairwise combinations of primers: Application to genetic mapping of the mouse. *Nucleic Acids Res.* **19:** 5275–5279.

——. 1991b. Genomic fingerprints produced by PCR with consensus tRNA gene primers. *Nucleic Acids Res.* **19:** 861–866.

——. 1992. PCR-amplified length polymorphisms in tRNA intergenic spacers for categorizing staphylococci. *Mol. Microbiol.* **6:** 1673–1680.

——. 1993. The characterization of pathogenic microorganisms by genomic fingerprinting using arbitrarily primed polymerase chain reaction (AP-PCR). In *Diagnostic molecular microbiology* (ed. D.H. Persing et al.), pp. 595–602. ASM Press, Washington D.C.

Welsh, J., C. Petersen, and M. McClelland. 1991. Polymorphisms generated by arbitrarily primed PCR in the mouse: Application to strain identification and genetic mapping. *Nucleic Acids Res.* **19:** 303–306.

Welsh, J., K. Chada, S.S. Dalal, D. Ralph, R. Cheng, and M. McClelland. 1992. Arbitrarily primed PCR fingerprinting of RNA. *Nucleic Acids Res.* **20:** 4965–4970.

Williams, J.G., M.K. Hanafey, J.A. Rafalski, and S.V. Tingey. 1993. Genetic analysis using random amplified polymorphic DNA markers. *Methods Enzymol.* **218:** 704–740.

Williams, J.G., A.R. Kubelik, K.J. Livak, J.A. Rafalski, and S.V. Tingey. 1990. DNA polymorphisms amplified by arbitrary primers are useful as genetic markers. *Nucleic Acids Res.* **18:** 6531–6535.

Wu, K.-S., R. Jones, L. Danneberger, and P.A. Scolnik. 1994. Detection of microsatellite polymorphisms without cloning. *Nucleic Acids Res.* **22:** 3257–3258.

Yu, K.F., A. Van Denze, and K.P. Pauls. 1993. Random amplified polymorphic DNA (RAPD) analysis. In *Methods in plant molecular biology and technology* (ed. B.R. Glick and J.E. Thompson). CRC Press, Boca Raton, Florida.

RNA Fingerprinting Using Arbitrarily Primed PCR

Michael McClelland and John Welsh

California Institute of Biological Research, La Jolla, California 92037

INTRODUCTION

One great strength of the family of arbitrarily primed PCR (AP-PCR) methods is their simplicity (Welsh and McClelland 1990; Williams et al. 1990). The extension of AP-PCR fingerprinting to RNA (Liang and Pardee 1992; Welsh et al. 1992) has resulted in a tool with exciting potential for detecting differential gene expression. It is now possible to obtain a partly abundance-normalized sample of cDNAs produced in a single tube in a few hours. Fragments of differentially expressed genes can be cloned directly from PCR-amplified products isolated from the gel.

The method can provide a complex phenotype reflecting changes in the abundances of hundreds of RNAs under various conditions. Comparison of RNA fingerprints from different treatment groups allows one to draw inferences regarding gene regulation. Hypotheses regarding signal transduction can be tested, and new hypotheses can be generated using this information. In this chapter, we justify our current preferred protocol for RNA fingerprinting. The previous chapter discusses DNA fingerprinting.

The first published variation of AP-PCR applied to RNA uses a 3′ anchor primer such as oligo(dT)CA for reverse transcription and an arbitrary primer for priming the second-strand cDNA (Liang and Pardee 1992). At these priming sites, the 3′ seven or eight nucleotides of the primer usually match well with the template, but the nucleotides toward the 5′ end of the primer also influence which sequences amplify. The term "differential display" has been coined for this variant and could be applied to all variants of RNA fingerprinting using arbitrary primers.

In one of the many alternatives for RNA fingerprinting with arbitrary primers, first-strand cDNA synthesis is primed using oligo(dT). The first-strand cDNA can then be fingerprinted using a single arbitrary primer and two initial low-stringency cycles, incorporating the arbitrary primer at both ends for subsequent PCR amplification (McClelland and Welsh 1990).

In another variant of RNA fingerprinting using AP-PCR, synthesis of first-strand cDNA is initiated from an arbitrary primer at those sites in the RNA that best match the primer. Second-strand synthesis is achieved by arbitrary priming using a thermostable DNA polymerase, at sites where the primer finds the best matches. Poorer matches at one end of the amplified sequence can be compensated for by very good matches at the other end. These two steps result in a collection of molecules that are flanked at their 3′ and 5′ ends by the exact sequence (and complement) of the arbitrary primer. These serve as templates for high-stringency PCR amplification and result in fingerprints similar in appearance to those generated from genomic DNA. This variation on the AP-PCR strategy (Welsh and McClelland 1990) has been called RNA arbitrarily primed PCR (RAP-PCR) (Welsh et al. 1992; for review, see McClelland et al. 1995). An example is given in Figure 1. Because both primers are internal to the transcript, open reading frames are found in about 30% of products using this variation of RNA fingerprinting.

The ratio of the intensities of RAP-PCR products between samples correlates with the ratio of abundances of the corresponding RNA (Welsh et al. 1992; Wong et al. 1993). Although the intensities of different bands within the same fingerprint vary independently, the intensity of a band between fingerprints appears to be proportional to the concentration of its corresponding template sequence (Welsh et al. 1992; Wong et al. 1993). Extraordinary as this may seem at first glance, it is not difficult to rationalize. Each fingerprint is dominated by many products that do not vary in quantity or intensity between samples, and each differentially expressed product constitutes only a tiny fraction of the total mass of DNA synthesized. In a carefully performed experiment, the total mass of DNA synthesized from sample to sample is constant. These factors ensure that the amplification efficiency for any particular product is identical between samples. Furthermore, the ratios of intensities for a particular product reflect the single remaining variable, the template RNA ratios. The semiquantitative nature of AP-PCR has also been observed in DNA fingerprinting at heterozygous loci in the F_1 progeny of diploids (Welsh et al. 1991). This property of AP-PCR has allowed the ploidy of chromosomes in tumor cells to be determined by comparison with fingerprints from the normal diploid genome of the same individual (Peinado et al. 1992).

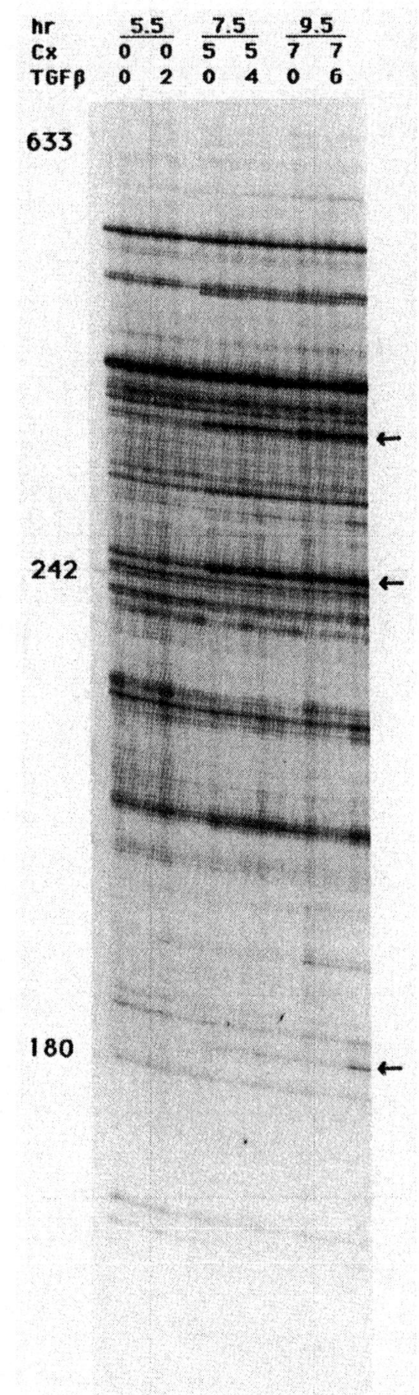

Figure 1 RAP-PCR fingerprint. Lung epithelial cells were released from confluence and treated with TGF-β or CX or both for the times indicated and then harvested. RNA was prepared as described in the text. The primers used were kinase A1 (5′-GAGGGTGCCTT) carrying a Ser/Thr kinase motif (manufactured by Genosys) and S2 (5′-CCTCTGACTG) (Operon Technologies). Each RNA was fingerprinted at three RNA concentrations, 80 ng, 40 ng, and 20 ng per 20 μl of reverse transcription reaction. Separation was on a 5% polyacrylamide-50% urea sequencing gel with electrophoresis in 1X TBE at 50 watts for 3 hours. The gel was dried and autoradiography was performed with X-Omat film (Kodak) for 12 hours. Some differentially amplified cDNAs are indicated by arrows. The image shows only a portion of the data. Single-stranded DNA markers are indicated in bases.

REAGENTS AND EQUIPMENT

Thermal Cycler (Perkin-Elmer 9600 model)
PCR tubes (MicroAmp, Perkin-Elmer)
Arbitrary primer, e.g., M13 sequencing or reverse sequencing primer
2X DNase I treatment mixture: 20 mM Tris-HCl, pH 8.0, 20 mM $MgCl_2$, 200 units/ml RNase-free DNase I (Boehringer Mannheim)

2x First-strand reaction mixture: 100 mM Tris-HCl, pH 8.3, 100 mM KCl, 8 mM $MgCl_2$, 20 mM DTT, 0.4 mM each dNTP, 20 μM primer, and 1 unit/μl MuLVRT (Stratagene)

2x Second-strand reaction mixture: 10 mM Tris-HCl, pH 8.3, 25 mM KCl, 2 mM $MgCl_2$, 0.1 μCi/μl [α-^{32}P]dCTP, and 0.2 units/μl *Taq* DNA polymerase (AmpliTaq; Perkin-Elmer). Addition of a second primer at this step (at about 20 μM) is optional.

PROTOCOL

RAP-PCR Fingerprinting Using a Single 18-Base Primer

1. Prepare total RNA by guanidinium thiocyanate-cesium chloride centrifugation (Chirgwin et al. 1979) or guanidinium thiocyanate-acid phenol-chloroform extraction (Sambrook et al. 1989). Dissolve the final pellet in 100 μl of water.

2. Add 100 μl of 2x DNase I treatment mixture and incubate for 30 minutes at 37°C. Extract with phenol-chloroform and precipitate with ethanol. Measure RNA concentration spectrophotometrically.

3. Prepare treated RNA at two concentrations of about 20 ng/μl and 4 ng/μl by dilution in water. Template concentrations that yield reproducible fingerprints depend on RNA quality and, to some extent, on the choice and quality of the primers and must therefore be determined empirically.

4. Assemble reactions on ice. Add 10 μl of 2x first-strand reaction mixture to 10 μl of RNA at each concentration. Ramp the reaction over 5 minutes to 37°C in a 96-well format thermal cycler (Perkin-Elmer) and hold at that temperature for an additional 10 minutes followed by 2 minutes at 94°C to inactivate the polymerase. Finally, cool to 4°C.

5. Add 20 μl of 2x second-strand reaction mixture to each 20 μl of the first-strand synthesis reaction. Cycle through one low-stringency step (1 minute at 94°C, 5 minutes at 40°C, 5 minutes at 72°C) followed by 35 high-stringency steps (1 minute at 94°C, 1 minute at 60°C, 1 minute at 72°C).

6. Dilute 4 μl of each reaction in 18 μl of 95% formamide, heat for 2 minutes to 94°C, then load 1.5 μl on a 5% polyacrylamide-50% urea sequencing gel and electrophorese in 1x TBE at 50 watts for 3 hours or until the xylene cyanol reaches three-quarters of the way to the bottom of the gel. Dry the gel and perform autoradiography with X-Omat film (Kodak) for 12 hours. Fragments visualized generally range from about 50 bases to 1000 bases.

Some primers work better than others. It is usually a good idea to screen several primers and use those that give the most qualitatively robust patterns for further work. Note that *Thermus thermophilus* (*Tth*) polymerase has reverse transcriptase activity and can substitute for both the RT and the DNA polymerase when a buffer containing Mn^{++} is used.

Our protocol for RNA fingerprinting with 10-base primers is similar to that described for longer primers. However, *Taq* Stoffel fragment is used instead of *Taq* polymerase (Sobral and Honeycutt 1993). This necessitates a dilution scheme that adjusts buffer components such that they are compatible with *Taq* Stoffel fragment.

REAGENTS

2x First-strand reaction mixture:

- 100 mM Tris-HCl, pH 8.3
- 100 mM KCl
- 8 mM $MgCl_2$
- 20 mM DTT
- 0.4 mM each dNTP
- 4 μM primer
- 2 units/μl MuLVRT (Stratagene)

Except for the primer concentration, this is the same as the first-strand reaction mixture for longer primers.

2x AP-PCR mixture:

- 20 mM Tris-HCl, pH 8.3
- 20 mM KCl
- 8 mM $MgCl_2$
- 0.4 mM each dNTP
- 0.1 μCi/μl [α-^{32}P]dCTP
- 0.4 units/μl *Taq* DNA polymerase (Stoffel fragment: Perkin-Elmer)
- 8 μM secondary primer

PROTOCOL

RNA Fingerprinting Using a Pair of 10-Base Primers Applied Sequentially

In this protocol, enough first-strand cDNA is prepared to allow for several second-strand reactions using different primers to be performed. The protocol has been designed using the spreadsheet Microsoft Excel, allowing automatic changes in the number of RNA preparations to be fingerprinted, RNA concentrations, and the number of first and second primers to be used, while maintaining the proper buffer components (see Fig. 2). Here we describe the case where one first strand-primer is used on six different RNA preparations at three concentrations, followed by four different secondary primers.

1. Prepare RNA as described above.

2. First-strand synthesis with a single primer is achieved as follows: Combine 5 µl total RNA at 200, 100, and 50 ng/µl with 5 µl of first-strand reaction mixture on ice, ramp over 5 minutes to 37°C, hold for 30 minutes at 37°C, heat for 5 minutes to 94°C to inactivate the reverse transcriptase, and cool to 4°C until the next step.

3. Dilute the reaction mixture with water to four times its initial volume. In this case, the 10-µl reaction is diluted to 40 µl by the addition of 30 µl water. This dilutes the buffer components to 12.5 mM Tris-HCl, pH 8.3, 12.5 mM KCl, 1 mM $MgCl_2$, 50 µM dNTP, and 1 µM initial primer. This concentration of primer is sufficient to sustain PCR amplification in subsequent steps.

4. Combine 10 µl of the diluted first-strand reaction with 10 µl of 2X AP-PCR mixture, which contains the secondary primer. At this point, the reaction components are:

   16.25 mM Tris-HCl, pH 8.3
   16.25 mM KCl
   4.5 mM $MgCl_2$

Primers-Specify:

Primers are as follows:		Primer concentration:
First strand primers:	A	100 µM
Second strand primers:	B, C, D, E	100 µM

RNA:

RNA stock concentration:	200 ng/µl
RNA: Diluted by	1, 2, 4 fold

Experiment Parameters-Specify:

Number of first strand primers:	1	
Number of second strand primers, overall:	4	Should be 4 or more for
Number of RNAs:	6	convenient pipetting volumes.
Number of concentrations of RNA:	3	

First strand synthesis Reaction Mixture without primer.

Set up reactions on ice.		No. reactions:
	Per Reaction:	18.00
10 x RT (50,50,4)	1 µl	18.00 µl
100 mM DTT	1	18.00
5 mM dNTP	0.4	7.20
water	1.55	27.90
MuLV-RT (75 U/µl)	0.25	4.50
	4.20	75.60

10 x RT (50,50,4) is 500 mM Tris pH 8.3, 500 mM KCl, 40 mM MgCl2.

Reaction Mixture with First Primers:

Divide the Reaction Mix. into	1		aliquots:	75.60 µl
Add 100 µM primers:	A			14.40
final concentration will be:	16.00	µM		90.00

Combine: 5.00 µl with an equal volume of each RNA.

Final reverse transcription reaction has a volume of : 10.00

Synthesis: 5 min ramp to 37°C, Hold 60 min. Heat to 94°C for 2 min to kill RT.

Bring final volume to: 40.00 µl by adding: 30.00 µl Water

Buffer concentrations at this point:

Tris	12.50	mM
KCl	12.50	mM
Mg	1.00	mM
Primer	4.00	µM
dNTP	0.05	mM

Figure 2 (*See facing page for legend.*)

0.23 mM dNTP
0.5 μM first primer
4 μM second primer
0.2 units/μl *Taq* DNA polymerase Stoffel fragment

Cycle the reactions through 45 cycles of the following temperature profile: 30 seconds at 94°C, 1 minute at 35°C, and 2 minutes at 72°C.

5. Analyze the reactions on sequencing gels as described above.

There are at least two good reasons to use this latter protocol. First, the number of visible products in the fingerprints is increased relative to using an 18-mer or one 10-mer primer. Furthermore, those products that contain the first and second primer are likely to be oriented with the first primer at the 3′ end and the second primer at the 5′

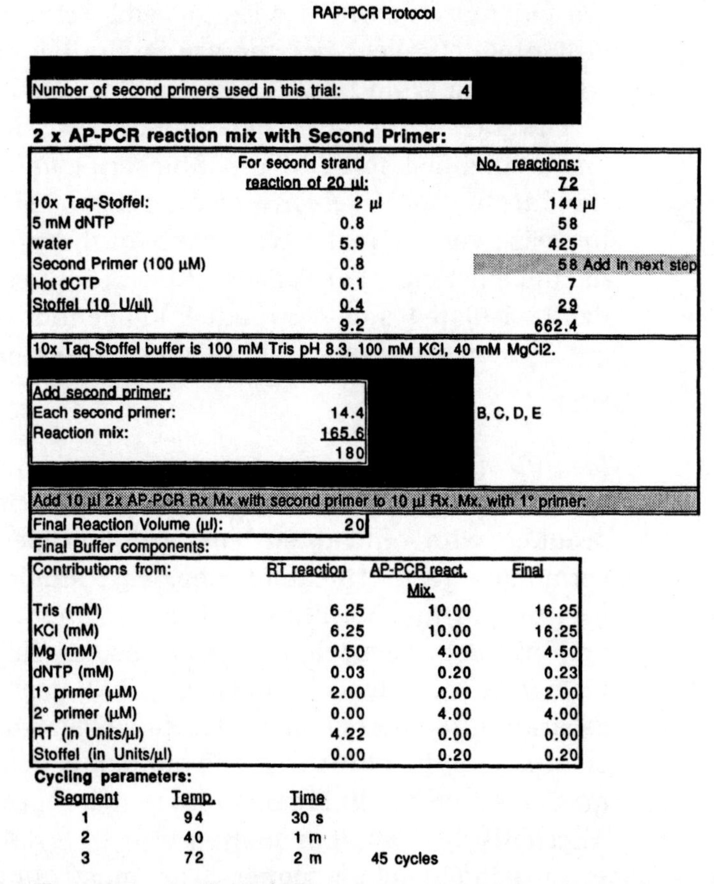

RAP-PCR Protocol

Number of second primers used in this trial: 4

2 x AP-PCR reaction mix with Second Primer:

	For second strand reaction of 20 μl:	No. reactions: 72	
10x Taq-Stoffel:	2 μl	144 μl	
5 mM dNTP	0.8	58	
water	5.9	425	
Second Primer (100 μM)	0.8	58	Add in next step
Hot dCTP	0.1	7	
Stoffel (10 U/μl)	0.4	29	
	9.2	662.4	

10x Taq-Stoffel buffer is 100 mM Tris pH 8.3, 100 mM KCl, 40 mM MgCl2.

Add second primer:		
Each second primer:	14.4	B, C, D, E
Reaction mix:	165.6	
	180	

Add 10 μl 2x AP-PCR Rx Mx with second primer to 10 μl Rx. Mx. with 1° primer:

Final Reaction Volume (μl): 20

Final Buffer components:

Contributions from:	RT reaction	AP-PCR react. Mix.	Final
Tris (mM)	6.25	10.00	16.25
KCl (mM)	6.25	10.00	16.25
Mg (mM)	0.50	4.00	4.50
dNTP (mM)	0.03	0.20	0.23
1° primer (μM)	2.00	0.00	2.00
2° primer (μM)	0.00	4.00	4.00
RT (in Units/μl)	4.22	0.00	0.00
Stoffel (in Units/μl)	0.00	0.20	0.20

Cycling parameters:

Segment	Temp.	Time	
1	94	30 s	
2	40	1 m	
3	72	2 m	45 cycles

Figure 2 An RAP-PCR protocol for the use of pairwise sequential 10-mer primers. This protocol, written in Microsoft Excel, allows the input of many variables. The example shown is set up for six RNA samples at four RNA concentrations, to be sampled by one initial primer and four subsequent primers. Cycling is performed according to the text.

end, thereby providing information about the sense strand of the original RNA. More than 50% of products in reactions using pairwise combinations of primers contain both primers and are distinct from those generated with either primer individually (Welsh and McClelland 1991).

ANALYSIS OF RESULTS AND TROUBLESHOOTING

- *Preparation of clones.* Differentially amplified RAP-PCR products are cut from the gel and eluted into 50 μl of TE for 2 hours at 65°C. A 5-μl aliquot of the RAP-PCR products is reamplified using the same oligonucleotide primers used to generate the fingerprint (Welsh et al. 1991). *Taq* polymerase often adds an extra base to the 3′ end of PCR products. Which base is added depends on the last base of the double-stranded sequence. For example, if the last base is an A, then an extra A is added. If the last base is a T, then it is removed and replaced with an A to produce an A:A mismatch. This phenomenon makes PCR products hard to blunt-end clone. We use the simple expedient of reamplifying polymorphic products using *Pfu* polymerase, which does not add extra bases to the 3′ end of PCR products. Vent polymerase is similar to *Pfu* in this property (Garrity and Wold 1992).

  The RAP-PCR products are cloned (e.g., into the pCR-Script vector, a modified form of the Bluescript vector, Stratagene) using standard protocols. *Eco*RV or *Srf* I are used during the ligation to linearize vectors that have closed on themselves, thereby improving the efficiency. For each RAP-PCR product, six or more independently isolated single-stranded phagemids are sequenced using the Sequenase reagent kit v2 (USB/Amersham) and [α-^{35}S]dATP (NEN).

- *Identification of the correct clone.* A common difficulty with cloning AP-PCR or RAP-PCR products is contamination of the isolated PCR product with unwanted comigrating species. In principle, a secondary gel purification step, e.g., single-strand conformation polymorphism (SSCP), could be included after the first PCR amplification to enrich further for the correct product. Recently, we have also had success purifying RAP products using SSCP gel separation (Suzuki et al. 1991). However, after reamplification and cloning into the pBlueScript plasmid, it is generally sufficient to sequence six clones to identify one that is represented twice or more. Alternatively, complementation tests (c-tests) can be performed on several independent clones. The most abundant product of the secondary PCR amplification is more likely than low-level background contamination to c-test positively. We have shown that confirmation of the correct product can be achieved by Southern transferring the RAP-PCR gel, or another gel produced by the same

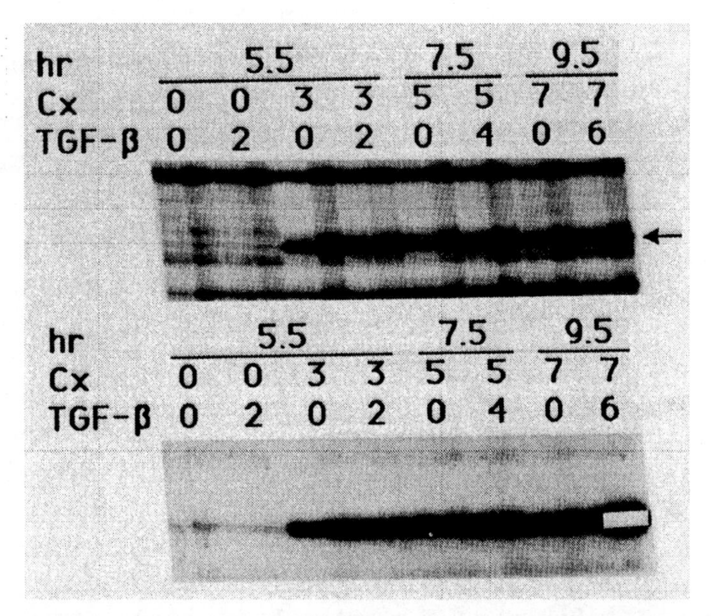

Figure 3 A product corresponding to a CX-inducible gene was excised from an RAP-PCR gel, reamplified, and cloned. The original RNA fingerprint was transferred to a nylon filter, and the filter was probed with the clone to verify that the correct band had been cloned. (*Top*) A close-up of the appropriate region from the RAP-PCR gel. (*Bottom*) The same region transferred to a membrane and probed with a clone from the differentially amplified band. The clone hybridized to the correct region and had the expected differential distribution between lanes, indicating that the clone represented the correct differentially amplified band.

primer, and using the putative clone as a probe (see Fig. 3) (Peinado et al. 1992; Wong and McClelland 1992). These Southern blots are much simpler than typical Southern blots; they have exposure times of only a few minutes because of the low complexity of the material on the gel. We have tested over 20 products in this way. Only two clones that were prevalent in the sequencing proved to be incorrect when tested against Southern blots of the RAP-PCR gels.

• *Southern blots.* DNA from RAP-PCR fingerprints is transferred to a membrane overnight (Duralon-UV, Stratagene) by capillary action using 20x SSC buffer and UV cross-linked. Blots are prehybridized with 5x SSC, 0.5% blocking reagent (Boehringer Mannheim), and 0.8% SDS for 4 hours at 65°C. Hybridization to blots is performed using conventional methods (Sambrook et al. 1989) and conditions recommended by the vendors. Radiolabeled probes can be synthesized by PCR from cloned RAP-PCR products using the T7-T3 oligonucleotide primer set (Stratagene) and [α-^{32}P]dCTP.

When clones are then used on Northern blots, a minority give no signal because the message is too rare, but they do give positive Southern blots, confirming probe quality. However, the subset that give positive Northern blots generally also display the expected differential gene expression.

DISCUSSION

Abundance Normalization

The effect of RNA abundance on the efficiency of sampling by RNA fingerprinting is an important issue. Experiments that used 10-base-long primers perfectly matched at both ends for a rare message did not lead to the targeted cDNA dominating the fingerprint (Liang and Pardee 1992). Instead, other products still competed effectively to produce a fingerprint indicating the lack of abundance normalization, a problem that is typical of most AP-PCR methods. Therefore, two RNAs with similar arbitrary priming and amplification efficiencies that differ by 100-fold in abundance have fingerprint products displaying a 100-fold ratio of intensities.

How can this problem be overcome? One obvious possibility is to increase the stringency of the arbitrary priming steps, thereby sampling only those transcripts with perfect or near-perfect matches. By increasing the number of bases that match in the arbitrary priming steps, one can increase the normalization. In an RNA sample of 10^8-base complexity, a perfect match of about 8.5 bases at each of the two initial priming events would theoretically produce an average of only one PCR product. This would be a normalized sampling because the match would entirely control the probability of occurrence. This strategy, however, would be a very slow method to yield information on gene expression and also appears to be difficult to execute. Increasing the stringency in the first arbitrary steps has little effect on the fingerprint until a stringency is reached where the fingerprint becomes unreliable. In order that many RNA molecules be sampled simultaneously, mismatches must be tolerated. Whereas the melting temperatures for primer-template interactions drop precipitously in response to increasing mismatches, the slope of the melting curve tends to become smaller because cooperative melting is compromised. Then, a small increase in stringency (temperature) has only a modest effect on selectivity. The selectivity imparted by a single additional base pair can be easily overcome by the abundance factor. Even when a better match is thermodynamically favored, poor matches can initiate significant synthesis over the course of a reaction. Thus, it is difficult to normalize for abundance completely under low-stringency conditions (i.e., conditions that permit mismatches). To circumvent this phenomenon and improve normalization, we have used a nested strategy.

Nested RAP-PCR is a method designed to normalize the fingerprint partially with respect to mRNA abundance (McClelland et al. 1993; Ralph et al. 1993). The strategy is very similar to standard nested PCR methods, except that we do not know a priori the internal sequences of the amplified products. In this method, the fingerprinting protocol is applied to the RNA, as described above. Then, a small aliquot of the first RAP-PCR fingerprint is further amplified at *high stringency* using a second nested primer having one, two, or three additional *arbitrarily chosen* nucleotides at the 3′ end of the first primer sequence. Because sampling subsequent to the first low-stringency fingerprinting is at high stringency, and the polymerase has a strong preference for a perfectly matched 3′ primer end, the arguments pertaining to the shape of the melting curve discussed above do not apply (McClelland et al. 1993; Ralph et al. 1993).

A Hypothetical Set of Circularly Permuted Nested Primers

```
A1    AACCCCACCA GAGAGACC
A2     ACCCCACCAG AGAGACCA
A3      CCCCACCAGA GAGACCAA
A4       CCCACCAGAG AGACCAAC
A5        CCACCAGAGA GACCAACC
A6         CACCAGAGAG ACCAACCC
A7          ACCAGAGAGA CCAACCCC
A8           CCAGAGAGAC CAACCCCA
A9            CAGAGAGACC AACCCCAC
A10            AGAGAGACCA ACCCCACC
A11             GAGAGACCAA CCCCACCA
A12              AGAGACCAAC CCCACCAG
A13               GAGACCAACC CCACCAGA
A14                AGACCAACCC CACCAGAG
A15                 GACCAACCCC ACCAGAGA
A16                  ACCAACCCCA CCAGAGAG
A17                   CCAACCCCAC CAGAGAGA
A18                    CAACCCCACC AGAGAGAC
A1                      AACCCCACCA GAGAGACC etc.
```

This particular set of primers excludes thymine because this is the least discriminatory base when present at the 3′ end of the primer (Kwok et al. 1990).

This strategy partially abundance-normalizes the sampling that occurs during RNA fingerprinting. Consider again the two messages that have equally good matches and equally good amplification efficiency but differ by 100-fold in abundance. The products derived from them differ by 100-fold after RAP-PCR. Thus, RAP-PCR fingerprinting produces a background of products that are not visible on the gel and that include products derived from low-abundance messages. A secondary round of amplification using a primer identical to the first except for an additional nucleotide at the 3′ end of the molecule can be expected to amplify selectively those molecules in the background that, by chance, share this additional nucleotide. The additional nucleotide occurs in 1/16 of the background molecules, accounting for both ends. There are many more molecules of low abundance in the RNA population than messages of high abundance, so most products produced by the high-stringency nesting step should derive from the low-abundance, high-complexity class.

Each additional nucleotide at the 3′ end of the initial primer contributes, in principle, a factor of 1/16 to the selectivity. Two extra bases at each end lead to 256-fold selection, and three bases at each end lead to 4096-fold selection. Four bases at each end generally do not yield reliable products. In practice, the selectivity is probably somewhat less than simple statistics might suggest because, whereas *Taq* polymerase is severely biased against extending a mismatch at the last nucleotide, it is more tolerant of mismatches at the second or third positions. Additional selectivity might therefore be achieved by successive nesting experiments. Nonetheless, our initial experiments are consistent with the interpretation that considerable additional selectivity is achieved by this nested priming strategy.

If nesting is successful using total RNA as a starting material, hnRNA sequences will inevitably be sampled, even if a primer biased toward the poly(A) tail is used. So far, most of the approximately 50 differentially expressed genes we have isolated using *un*nested RAP-PCR from various systems have given strong, sharp signals in Northern analysis, which indicates that the patterns derive predominately from mRNA. However, in nested primer experiments, where complexity becomes progressively more important than abundance, we anticipate that a greater proportion of the patterns will derive from hnRNA, which still contains introns. If this is considered a problem for a particular experimental design, then poly(A) selection or priming with oligo(dT) or oligo(dT)NN-clamped primers (Liang and Pardee 1992) might be preferred.

Comparison of RNA Fingerprinting Protocols

What are the advantages and disadvantages of the RAP-PCR protocol (Welsh et al. 1992) compared to that of Liang and Pardee (1992)? In contrast to an arbitrary selection step in both directions (Welsh and McClelland 1990; Welsh et al. 1992), the protocol of Liang and Pardee uses an oligo(dT)CA, or similar oligo(dT)-XM, primer for reverse transcription followed by an arbitrary primer in the other direction. The method samples 3′ ends that are mostly noncoding. In contrast, arbitrary priming from both ends can sample open reading frames (ORFs) in about 30% of mRNA products (Welsh et al. 1992; Ralph et al. 1993; Wong et al. 1993) because the authentic reading frame often occurs in arbitrarily selected stretches of a few hundred bases in a typical mRNA sequence. Thus, our strategy has the advantage that conserved ORFs between species or protein family members can occasionally be observed in database searches without further cloning (see, e.g., Ralph et al. 1993).

One variant of the protocol we use can be designed to prime arbitrarily only once after denaturation before switching to high-stringency annealing (e.g., 60°C for an 18-base primer); any contaminating dsDNA is usually primed only once and is therefore not efficiently amplified during PCR (Welsh et al. 1992). The protocols employing 10-mers or an anchored oligo(dT)-NN primer use 35°C annealing steps throughout and can sample contaminating dsDNA in the classic manner of RAPD (Williams et al. 1990, 1993). Finally, the oligo(dT)-XM primers are as highly promiscuous as other primers, if not more so, and must result in many products that are sampled from hnRNA or inside mRNAs and other products in which the oligo(dT)-CA primer occurs at both ends. To ensure sampling of only mRNA and not hnRNA or residual genomic DNA, the RNA must first be poly(A)-selected.

Each of these approaches has strengths and weaknesses. Although the protocol using an anchored oligo(dT)-NN primer is excellent, arbitrary priming from both directions, followed by nesting, can have significant advantages, not least of which is the sampling of ORFs.

Further Improvements

The RAP-PCR method is amenable to various possible improvements and, given the previously demonstrated utility of RAP-PCR, it is worth developing some of these possibilities. For example, in principle, *Tth* DNA polymerase could be used as both the reverse transcriptase and the DNA polymerase (Myers and Gelfand 1991), reducing the number of experimental steps. Conditions exist that are compatible with both properties of the enzyme (Young et al. 1993). Alternatively, a viral reverse transcriptase and *Taq* DNA polymerase can be mixed from the beginning. These enzymes also have different optima, but it should be possible to find a compatible buffer system.

Sampling Genes That Carry Conserved Sequence Motifs

One possibility for further adaptation of RAP-PCR is to encode conserved motif sequences in the primers. Then one could search for differentially expressed genes using motifs directed toward particular gene families of relevance to a particular biological phenomenon. Thus, one could attempt to bias in favor of zinc fingers, as we did with the primer ZF-1 (Ralph et al. 1993). Such efforts are an extension of the work with motifs that have been tried for fingerprinting total genomic DNA, such as promoter and amino acid motifs (Birkenmeier et al. 1992) or hypervariable purine-pyrimidine repeats (Welsh et al. 1991). Although the latter strategy does generate some products derived from RY repeat regions, the use of protein motifs is more difficult. RY repeats, such as GT repeats (Stallings et al. 1991), are much more common than protein sequence motifs. Rarer motifs must compete with a larger number of arbitrary priming events. It should be reiterated that increasing the stringency during the initial AP-PCR generally reduces the reliability of the fingerprint. Furthermore, the consequence of wobble bases in the translation of a conserved amino acid motif is that a primer that has no redundant positions will match perfectly with only a small fraction of the nucleic acid sequences encoding the motif. Nevertheless, this strategy is well worth exploring, and it is very likely that a series of papers using primers that carry motifs will appear soon. Indeed, Stone and Wharton (1994) present such an experiment.

Comparison with Other Methods

Genetic and reverse genetic methods for cloning interesting genes derive their power from the ability of the investigator to discern and follow a phenotype. Methods that rely on a biological assay of gene function have been very useful, particularly in the area of cancer research. However, appropriate bioassays for most genes are difficult to devise. When no biological assay can be found, or when genetics is inconvenient or impossible, other methods must be employed. Methods for detecting and cloning differentially expressed genes that do not rely on a biological assay of phenotype include subtractive hybridization and differential screening strategies, as well as RAP-PCR fingerprinting, which provides a molecular, rather than a biological, phenotype.

In a typical *differential screening* experiment, radioactive probes are made from cDNA from two RNA sources and used to screen a cDNA library prepared from one of the two. Occasionally, clones from the library hybridize to one or the other but not both probes. Unfortunately, low-abundance messages do not yield sufficient probe mass to allow favorable hybridization kinetics and detection levels. This problem is only partly alleviated by screening a library with a subtracted cDNA probe (Watson et al. 1990). Nevertheless, some significant discoveries have been made using this approach (see, e.g., Kuo et al. 1986; An et al. 1992). Improvements that allow the use of RNA from limited sources have been described (see, e.g., Brunet et al. 1990; Smith and Grindley 1992). Unlike RAP-PCR, which can be used to screen many RNA samples simultaneously, differential screening is designed for use in a sequential series of pairwise comparisons.

Subtractive hybridization (for review, see Sargent 1987) is technically challenging and has been used successfully by only a few laboratories. This method suffers from drawbacks similar to those of differential screening. The difficulties with subtractive hybridization methods derive mainly from the fact that rare messages have unfavorable hybridization kinetics, and many interesting genes are of this class. Abundant genes hybridize faster and more completely than low-abundance genes, making them more amenable to subtractive methods. A second problem is that subtraction by exhaustive hybridization of driver and target is employed to avoid sequences that are not differentially expressed. This has the undesirable effect of obscuring significant differences in gene expression that do not fall into the "all-or-nothing" category. Another shortcoming of the method is that, unlike RAP-PCR, subtractive hybridization is most conveniently used for pairwise comparisons. Additional comparisons require Northern analysis or similar assays.

Subtractive hybridization and differential screening have proven useful for uncovering differentially expressed genes between pairs of RNA populations. Nevertheless, these methods are rather cumbersome. It would be difficult to use these tools to provide a large number of genes with associated biological information from a large number of pairwise comparisons. RAP-PCR can be a powerful tool in this regard.

Theoretically, *representational difference analysis* (RDA) (Lisitsyn et al. 1993), which uses an iterative PCR-based subtractive approach to enrich differences in genomic DNAs, could be adapted to isolate genes that are differentially expressed between RNA samples. At present, such an application has not been published. However, a possible combination between RDA and RAP-PCR can be envisioned.

One advantage of PCR-based RNA fingerprinting compared to subtractive hybridization and differential screening is that the method yields results in a matter of a few days rather than after many steps that take weeks, a feature that RAP-PCR shares with two-dimensional gel electrophoresis of proteins (Rasmussen et al. 1992; Celis et al. 1993). However, the main advantage of this technique compared to subtractive hybridization, differential screening, and two-dimensional protein electrophoresis is that the method can compare more than two RNA samples simultaneously. Thus, the effect of a number of conditions can be assayed for any differentially amplified products that are observed. In a well-designed experiment, one can select genes that fall into a relatively narrow category for further study. This has become very important because the advent of this technique means that finding differentially expressed genes is no longer rate-limiting.

The Size of an RAP-PCR Experiment and the Issue of Complete Sampling

How thoroughly can we explore differential gene expression using RNA fingerprinting? This is perhaps the most important question when comparing RNA fingerprinting with other methods, such as subtractive hybridization. The number of fingerprinting reactions necessary to achieve various levels of coverage of the RNA complexity can be calculated. Neglecting for the moment problems of abundance normalization, we can determine the probability, P(0), of *missing* a message as $P(0) = e^{-\mu}$. In order to have a 95% chance of detecting a *particular* message, $P(0) = 0.05$, and $\mu = 3$, three times the total complexity (i.e., number of different kinds of messages) must be sampled. Assuming two RNA concentrations (providing information on reproducibility), 50 bands per lane, and 100 lanes per gel, $50 \times 100/2 = 2{,}500$ messages can be displayed. Assuming a complexity of 15,000 messages and a factor of 3 to account for multiple sampling of the same message, the entire complexity of the cell could, in principle, be represented in $(15{,}000/2{,}500) \times 3 = 18$ gels at the 95% confidence level. In an experiment involving more than one RNA, say n different RNA sources, $n \times 18$ gels must be run to achieve this level of coverage of all RNAs. This calculation ignores the effect of incomplete abundance normalization: Rarer messages are represented less frequently among the visible products. Thus, $n \times 18$ gels is a considerable *underestimate* of the effort needed to sample 95% of the RNAs. Indeed, the method may be incapable of sampling the rarest RNAs and is therefore not appropriate for finding a single particular RNA. Nevertheless, if there are several important molecular markers associated with a phenomenon, the chances of encountering some of them are excellent. Thus, RAP-PCR is ideally suited to situations where the number of genes differentially expressed is fairly high (i.e., >1 per 1000 transcripts).

Much of the technological development associated with sequencing, such as fluorescence-tagged primers, automated gel reading, and capillary electrophoresis, is readily adaptable to RAP-PCR. Therefore, a reasonable fraction of the genes expressed in many situations can be surveyed given existing technology, and this capability will be greatly enhanced by further technological developments.

A systematic way to divide up the pool of mRNAs (and a new acronym) has been suggested to survey the whole mRNA population with a series of other arbitrary primers in pairwise combinations (Bauer et al. 1993). However, such calculations assume that the fingerprints are entirely normalized for abundance. Also, although it is intuitively appealing to divide the RNA population into a number of separate pools using different anchored primers, surprisingly, there is no increase in the efficiency of surveying the RNA population. For example, 95% coverage using 12 anchored pools still requires exactly the same number of gel lanes as does using pairwise combinations of arbitrary primers.

Using RAP-PCR to Estimate the Proportion of Genes That Accompany a Particular Physiological Change

The essence of RAP-PCR is its ability to define genes in terms of regulatory sets. For example, consider only three RNA preparations: **A**, **B**, and **C**. These represent samples extracted from a given cell type treated in three different ways. Information can be obtained regarding the level of gene expression in 27 possible categories (3^3). Most genes fall into the first of these categories:

	A	B	C		A	B	C
1	0	0	0	2	↑	0	0
3	↓	0	0	4	0	↑	0
5	0	↓	0	6	0	0	↑
7	0	0	↓	8	↑	↑	0
9	↑	↓	0	10	↑	0	↑
11	↑	0	↓	12	↓	↑	0
13	↓	↓	0	14	↓	0	↑
15	↓	0	↓	16	0	↑	↓
17	0	↑	↓	18	0	↓	↑
19	0	↓	↓	20	↑	↑	↑
21	↓	↓	↓	22	↓	↑	↑
23	↑	↓	↑	24	↑	↑	↓
25	↑	↓	↓	26	↓	↑	↓
27	↓	↓	↑				

(0 = no change; ↑ = increased expression; ↓ = decreased expression).

When the quantitative level of gene expression is also considered, then the number of categories increases further. In addition, the treatments of the cells to generate **A, B,** and **C** can be used in seven possible combinations, increasing even further the information about each sampled gene:

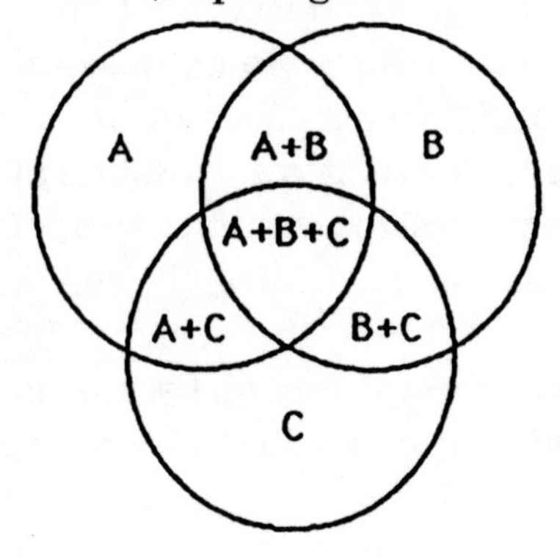

One tremendously useful resource that can be developed very easily using RAP-PCR is a collection of genes that fit a wide variety of regulatory categories. This could be achieved by comparing, in parallel, as many RNA preparations from an isogenic source as is practical. Concentrating on surveying only two RNA samples does not take advantage of the power of RAP-PCR. For this reason we would argue that, in general, the more treatments compared, the more is gained from using the RAP-PCR strategy.

The behavior of each transcript sampled in various experimentally manipulated regulatory scenarios can be determined. In a sense, what is obtained is a "byte" of information for each anonymous gene sampled, each "bit" of which consists of the transcript abundance in a particular sample. One goal could be to detect and isolate fragments of genes that fit every observable byte or regulatory category. This information should be of interest because the promoters of such genes will also carry *in cis* functions associated with these regulatory categories. This could be of great utility in the study of promoter function at the level of sequence specificity. The promoters of such genes or targeted insertions into the genes could also be very useful for expressing heterologous genes in a particular subset of situations, as may be necessary in some gene therapies, for instance.

In principle, there can be 6561 (i.e., 3^8) possible bytes of ternary bits for eight RNA samples (encoding increases, decreases, and no change) or more possible bytes if further information on the level of gene expression is included in each bit, such as separating increases in gene expression above and below a fourfold induction threshold. It is probable that some of the regulatory categories (i.e., bytes) will be very rare and only rarely sampled but, with these exceptions, one could clone examples from each category.

Even before such genes are cloned, we can obtain a very interesting set of data from the fingerprints. When the overlap in gene expression between tissues was estimated in the 1960s and 1970s (see, e.g., Davidson and Britten 1973), this information became a classic addition to textbooks despite the rather one-dimensional data that could be obtained. The reason these data were so interesting was that there was really no practical alternative to determine the scale of the similarities and differences in gene expression between tissues. RAP-PCR fingerprinting of RNA samples can be analyzed in a way that is orders of magnitude better than was possible at that time; the estimate of overlapping gene expression can be extrapolated from the distributions of many thousands of arbitrarily selected examples collected very quickly and easily. Thus, the scale of the intersection between RNA preparations can be estimated by simultaneous pairwise comparisons and examples of genes in different regulatory categories can be cloned if desired.

One obvious reservation about such data is the issue of abundance normalization. If nesting is not used, the observed products are biased toward highly expressed genes, which is a virtue if they are to be used, for example, as biomarkers, or if the corresponding promoter is to be used to express heterologous sequences in a particular expression distribution. The issue of abundance normalization can be addressed indirectly, as needed, by comparing the distributions of genes in progressively greater nestings of the initial fingerprint. If there is a systematic difference in the tissue distribution for genes at each level of nesting, this can be attributed to differences in the properties of more common versus rarer messages. Such an observation would not be difficult to quantitate (e.g., by RNase protection), and the trend would be illuminating, just as were the measurements by hybridization kinetics.

In principle, using a 96-well plate and 96-well gel format, one can perform PCR on two concentrations of eight RNA samples using four primers or primer pairs. Nesting each of these four primers to a depth of three bases yields at least 12 fingerprints in three further gels even before the possibilities for alternative nestings with different 3′ ends are factored in. Given 50 bands per lane, this single set of experiments yields 600 genes sampled. Such information can be gathered for many thousands of anonymous genes, a fraction of which may be selected for cloning, using the byte criteria outlined above. Furthermore, the number of products that vary can be compared to the number that do not vary between different RNA samples. One can then extrapolate the overall level of the response that distinguishes the two RNA samples (within the limitations of abundance normalization discussed earlier). As an example, we present a synopsis of our first use of this idea below.

An Estimate of the Proportion of Genes That Respond to TGF-β and Cycloheximide

We have fingerprinted lung epithelial cells in the G_1 stage of the cell cycle under the following conditions:

1. Transforming growth factor β (TGF-β) treatment, which halts progression of these cells into S phase

2. Cycloheximide (CX), which halts protein synthesis

3. Cycloheximide plus TGF-β, which reveals primary response genes to TGF-β

4. Without treatment

In a series of experiments with different primer pairs, we have surveyed at least 616 genes, of which 28 genes respond to TGF-β or to CX. An example of such a fingerprint is shown in Figure 1. We have distinguished between those genes that are secondary-response genes to TGF-β and those that are primary-response genes (the latter are also regulated by TGF-β even when CX is present). In addition, we observed an unexpected class of genes that are induced by CX but are then repressed by TGF-β.

We have also estimated the scale of each of these responses. For example, we determined that in the first hours of G_1, 0.8% (or less) of all transcripts sampled are up-regulated by TGF-β as a primary response. The abundances of 3% (or less) of all transcripts sampled are increased by CX, of which about one-third are down-regulated by TGF-β treatment. Extrapolation from these percentages to the rest of the RNA population is illuminating, despite a systematic bias due to abundance-based sampling. There is no reason to reject the possibility that all RNA abundance categories, including rarer RNAs that are less likely to be sampled in this protocol, may be affected to a similar degree as the sampled RNAs. Indeed, assuming that there are 15,000 expressed genes in the cell, then fragments from more than 3% of all the genes (>616/15,000) that are up-regulated by TGF-β and CX have been identified using only six primers or primer pairs. Cloning such genes constitutes a step in the direction of describing the complete response to these agents and is an essential prelude to determining the physiological role of each of these genes.

In a sense, such calculations of the scale of differential gene expression are an extension of the hybridization kinetics work of the 1970s. However, as discussed above, the complex molecular phenotype generated by RAP-PCR reflects changes in abundances of hundreds of individual RNAs. Many different RNA samples can be compared simultaneously, and genes from assorted regulatory categories can be cloned directly. Hypotheses regarding the interactions of signal transduction pathways can be tested and new hypotheses can be generated, for example, by comparing the effect of different combinations of hormones or drugs on differential gene expression.

ACKNOWLEDGMENTS

We thank Rhonda Honeycutt for helpful comments on the manuscript. This work was supported in part by National Institutes of Health grants AI-34829, NS-33377, and CA-68822 to M.M. and AI-32644 to J.W.

REFERENCES

An, G., T.H. Huang, J. Tesfaigzi, J. Garcia Heras, D.H. Ledbetter, D. M. Carlson, and R. Wu. 1992. An unusual expression of a squamous cell marker, small proline-rich protein gene, in tracheobronchial epithelium: Differential regulation and gene mapping. *Am. J. Respir. Cell Mol. Biol.* **7:** 104–111.

Bauer, D., H. Muller, J. Reich, H. Riedel, V. Ahrenkiel, P. Warthoe, and M. Strauss. 1993. Identification of differentially expressed mRNA species by an improved display technique (DDRT-PCR). *Nucleic Acids Res.* **21:** 4272–4280.

Birkenmeier, E.H., U. Schneider, and S.J. Thurston. 1992. Fingerprinting genomes by use of PCR primers that encode protein motifs or contain sequences that regulate gene expression. *Mamm. Genome* **3:** 537–545.

Brunet, J.F., E. Shapiro, S.A. Foster, E.R. Kandel, and Y. Iino. 1990. Identification of a peptide specific for *Aplysia* sensory neurons by PCR-based differential screening. *Science* **252:** 856–859.

Celis, J.E., H.H. Rasmussen, P. Madsen, H. Leffers, B. Honore, K. Dejgaard, P. Gromov, H.J. Hoffmann, M. Nielson, A. Vassilev, O. Vintermyr, J. Hao, A. Celis, B. Basse, J.B. Lauridsen, G.P. Ratz, A.H. Andersen, E. Walbum, I. Kjaegaard, M. Puype, J. Van Damme, J. Vanderkerckhove, et al. 1993. The human keratinocyte two-dimensional gel protein database (update 1993). *Electrophoresis* **14:** 1091–1198.

Chirgwin, J., A. Przybyla, R. MacDonald, and W.J. Rutter. 1979. Isolation of biologically active ribonucleic acid from sources enriched in ribonuclease. *Biochemistry* **18:** 5294–5299.

Davidson, E.H. and R.J. Britten. 1973. Organization, transcription and regulation in the animal genome. *Q. Rev. Biol.* **48:** 565–613.

Garrity, P.A. and B.J. Wold. 1992. Effects of different DNA polymerases in ligation-mediated PCR: Enhanced genomic sequencing and in vivo footprinting. *Proc. Natl. Acad. Sci.* **89:** 1021–1025.

Kuo, C.H., K. Yamagata, R.K. Moyzis, M.W. Bitensky, and N. Miki. 1986. Multiple opsin mRNA species in bovine retina. *Brain Res.* **387:** 251–260.

Kwok, S., D.E. Kellogg, N. McKinney, D. Spasic, L. Goda, C. Levenson, and J.J. Sninsky. 1990. Effects of primer-template mismatches on the polymerase chain reaction: Human immunodeficiency virus type 1 model studies. *Nucleic Acids Res.* **18:** 999–1005.

Liang, P. and A. Pardee. 1992. Differential display of eukaryotic messenger RNA by means of the polymerase chain reaction. *Science* **257:** 967–971.

Lisitsyn N., N. Lisitsyn, and M. Wigler. 1993. Cloning the differences between two complex genomes. *Science* **259:** 946–951.

McClelland M. and J. Welsh. 1990. Arbitrarily primed polymerase chain reaction method for fingerprinting. Patent application filed October 1990.

McClelland M., F. Mathieu-Daude, and J. Welsh. 1995. Differential display: RNA fingerprinting by arbitrarily primed PCR. *Trends Genet.* (in press).

McClelland, M., K. Chada, J. Welsh, and D. Ralph. 1993. Arbitrary primed PCR fingerprinting of RNA applied to mapping differentially expressed genes. In *Symposium on DNA fingerprinting: State of the science,* November, 1992 (ed. S.D. Pena et al.). Birkhauser Verlag, Basel, Switzerland.

Myers, T.W. and D.H. Gelfand. 1991. Reverse transcription and DNA amplification by a *Thermus thermophilus* DNA polymerase. *Biochemistry* **30:** 7661–7666.

Peinado, M.A., S. Malkhosyan, A. Velazquez, and M. Perucho. 1992. Isolation and characterization of allelic losses and gains in colorectal tumors by arbitrarily primed polymerase chain reaction. *Proc. Natl. Acad. Sci.* **89:** 10065–10069.

Ralph, D., J. Welsh, and M. McClelland. 1993. RNA fingerprinting using arbitrarily primed PCR identifies differentially regulated RNAs in Mink lung (Mv1Lu) cells growth arrested by TGF-β. *Proc. Natl. Acad. Sci.* **90:** 10710–10714.

Rasmussen, H.H., J. van Damme, M. Puype, B. Gesser, J.E. Celis, and J. Vandekerckhove. 1992. Microsequences of 145 proteins recorded in the two-dimensional gel protein database of normal human epidermal keratinocytes. *Electrophoresis* **13:** 960–969.

Sambrook, J., E.F. Fritsch, and T. Maniatis. 1989. *Molecular cloning: A laboratory manual,* 2nd edition. Cold Spring Harbor Laboratory Press, Cold Spring Harbor, New York.

Sargent, T.D. 1987. Isolation of differentially expressed genes. *Methods Enzymol.* **152:** 423–432.

Smith, D.E. and T. Grindley. 1992. Differential screening of a PCR-generated mouse embryo cDNA library: Glucose transporters are differentially expressed in early post implantation mouse embryos. *Development* **116:** 555–561.

Sobral, B.W.S and R.J. Honeycutt. 1993. High output genetic mapping of polyploids using PCR-generated markers. *Theor. Appl. Genet.* **86:** 105–112.

Stallings, R.L., A.F. Ford, D. Nelson, D.C. Torney, C.E. Hildebrand, and R.K. Moyzis. 1991. Evolution and distribution of (GT)n repetitive sequences in mammalian genomes. *Genomics* **10:** 807–815.

Stone, B. and W. Wharton. 1994. Targeted RNA fingerprinting: The cloning of differentially-expressed cDNA fragments enriched for members of the zinc finger gene family. *Nucleic Acids Res.* **22:** 2612–2618.

Suzuki, Y., T. Sekiya, and K. Hayashi. 1991. Allele-specific polymerase chain reaction: A method for amplification and sequence determination of a single component among a mixture of sequence variants. *Anal. Biochem.* **192:** 82–84.

Watson, J.B, E.F. Battenberg, K.K. Wong, F.E. Bloom, and J.G. Sutcliffe. 1990. Subtractive complementary DNA cloning of RC3, a rodent cortex-enriched mRNA encoding a novel 78 residue protein. *J. Neurosci. Res.* **26:** 397–408.

Welsh, J. and M. McClelland. 1990. Fingerprinting genomes using PCR with arbitrary primers. *Nucleic Acids Res.* **18:** 7213–7218.

——. 1991. Genomic fingerprinting with AP-PCR using pairwise combinations of primers: Application to genetic mapping of the mouse. *Nucleic Acids Res.* **19:** 5275–5279.

Welsh, J., C. Petersen, and M. McClelland. 1991. Polymorphisms generated by arbitrarily primed PCR in the mouse: Application to strain identification and genetic mapping. *Nucleic Acids Res.* **19:** 303–306.

Welsh, J., K. Chada, S.S. Dalal, D. Ralph, R. Cheng, and M. McClelland. 1992. Arbitrarily primed PCR fingerprinting of RNA. *Nucleic Acids Res.* **20:** 4965–4970.

Williams, J.G., M.K. Hanafey, J.A. Rafalski, and S.V. Tingey. 1993. Genetic analysis using random amplified polymorphic DNA markers. *Methods Enzymol.* **218:** 704–740.

Williams, J.G., A.R. Kubelik, K.J. Livak, J.A. Rafalski, and S.V. Tingey. 1990. DNA polymorphisms amplified by arbitrary primers are useful as genetic markers. *Nucleic Acids Res.* **18:** 6531–6535.

Wong, K.K. and M. McClelland. 1992. A *Bln*I restriction map of *Salmonella typhimurium*. *J. Bacteriol.* **174:** 1656–1661.

Wong, K.K., C.H. Mok, J. Welsh, M. McClelland, S.-W. Tsao, and R.S. Berkowitz. 1993. Identification of differentially expressed RNA in human ovarian carcinoma cells by arbitrarily primed PCR fingerprinting of total RNAs. *Int. J. Oncol.* **3:** 13–17.

Young, K.K., R.M. Resnick, and T.W. Myers. 1993. Detection of hepatitis C virus RNA by a combined reverse transcription-polymerase chain reaction assay. *J. Clin. Microbiol.* **31:** 882–886.

In Situ PCR

Gerard J. Nuovo

Department of Pathology, SUNY, Stony Brook, New York 11794-8691

INTRODUCTION

Since the discovery of the structure of DNA in 1953, a great deal of effort has been devoted to detecting specific target sequences. The first commonly used technique, filter hybridization, offered an excellent sensitivity of 1 copy per 100 cells. Restriction endonuclease digestion in conjunction with electrophoresis and varying stringencies (Southern blot hybridization) allowed for high specificity. However, the prerequisite nucleic acid extraction precluded localization of the target to its specific cell of origin. In many instances, this is critical information. For example, although HIV-1 DNA can be routinely detected from lymph nodes in seropositive, asymptomatic patients, it is very important to note which of the many different cell types that reside in this site are reservoirs of the virus prior to the development of AIDS. Furthermore, simple, although important, data such as the percentage of a given cell type that contains the target of interest cannot be obtained by Southern blot hybridization due to the obligatory destruction of tissue.

About 10 years ago, direct cellular localization of a DNA or RNA target was routinely achieved by in situ hybridization. Over the last 6 years, there have been dramatic improvements in the sensitivity of this method, especially as it relates to nonisotopic probes (Crum et al. 1988; Nuovo 1989; Nuovo and Richart 1989). However, despite these improvements, the relatively high detection threshold of in situ hybridization of about 10 copies per cell limits its usefulness (Nuovo 1993, 1994). Although this is not an issue in, for example, productive viral DNA infections where a given cell may contain thousands of copies, common situations such as latent viral infection or point mutations are not detectable by standard in situ hybridization. In addition, mRNA and viral RNAs are often present in copy numbers below the usual in situ hybridization threshold and thus may escape detection. Detection of these low-copy events, of course, has become routine over the last 8 years with the widespread use of PCR. With the hot start maneuver, one can routinely detect with PCR one copy in a background of 1 μg of total cellular DNA (Erlich et al. 1991; Chou et al. 1992; Nuovo et al. 1991). However, as with filter hybridization, the prerequisite nucleic acid extraction for PCR precludes the determination of which specific cell type contains the target of interest.

The last 4 years have shown a dramatic increase in the technology whereby the high sensitivity of PCR is combined with the cell-localizing ability of in situ hybridization. Although in situ PCR is still in its relative infancy, many groups have published protocols and data using in situ PCR techniques. This chapter presents protocols for in situ detection of PCR-amplified DNA and cDNA, and some of its applications, especially as it relates to HIV-1 and AIDS.

REAGENTS

10% buffered formalin (Polyscientific, cat. no. S182)
DEPC water (Research Genetics, cat. no. 750023)
Silane-coated slides (Oncor, cat. no. S1308)
Pepsin, proteinase K in situ reagents* (Enzo Diagnostics, cat. no. 32895)
RNase-free DNase (Boehringer Mannheim, cat. no. 776785)
RT-PCR kit** (Perkin-Elmer, cat. no. N808-0017)
In situ PCR cycler (Perkin-Elmer, cat. no. N804-0001)
Digoxigenin dUTP (Boehringer Mannheim, cat. no. 1093088)
Antidigoxigenin conjugate (Boehringer Mannheim, cat. no. 1093274)

*Many reagents, including the protease, buffers, washes, chromagen, probes, and counterstains, are part of comprehensive in situ hybridization kits. These are now marketed by several biotechnology companies.
**This kit includes the buffers, nucleotides, $MgCl_2$, *Taq* DNA polymerase, and control primers needed for RT-PCR.

PROTOCOL

This section provides an abbreviated protocol for doing in situ PCR, specifically RT in situ PCR. Detailed information about the various steps are provided in the remainder of this paper.

Preparation of Tissue Sections and Cell Samples

1. Fix the tissue samples in 10% buffered formalin, preferably from 8 to 15 hours, and then embed in paraffin.

2. For cell cultures, wash the cells once with PBS directly in the culture plate. Add 10% buffered formalin, let stand overnight, and then scrape the cells off the plate with a rubber policeman. Wash

the cells in diethylpyrocarbonate (DEPC) water twice, and centrifuge at 2000 rpm for 3 minutes. Resuspend the cells in 5 ml of DEPC water, and spot 50 μl on a slide.

3. Place several 4-μm paraffin tissue sections or three cell suspensions on silane-coated slides. The silane coating is essential for cell adherence during the procedure.

4. Remove the paraffin from the tissue samples by placing the slides in fresh xylene for 5 minutes and then in 100% ethanol for 5 minutes; air-dry.

Protease Digestion

1. Pepsin, trypsin, or proteinase K may be used. Pepsin or trypsin is preferred because these reagents are less stable than proteinase K and their activity is inhibited by increasing the pH during the wash step. The pepsin (2 mg/ml) is prepared by adding 20 mg of pepsin, 9.5 ml of water, and 0.5 ml of 2 N HCl.

   Note: For in situ PCR (with direct incorporation of the reporter molecule during PCR), the proper protease digestion time is highly dependent on the time of fixation in formalin. Table 1 gives the proper protease digestion time for a given formalin fixation time.

2. Inactivate the protease by washing the slide in DEPC water for 1 minute and then in 100% ethanol for 1 minute; air-dry. This simple wash step is sufficient to remove/inactivate the protease; heat-inactivation is not required.

Table 1 Effect of Protease Digestion Time on the Primer-independent Signal during In Situ PCR as a Function of the Time of Fixation in 10% Buffered Formalin

Fixation	Protease[a] digestion time (min)								
time	0	5	10	15	30	45	60	75	90
4 hr	0	1+	3+	2+	overdigested[b]				
6 hr	0	0	1+	3+	2+	overdigested[b]			
8 hr	0	0	0	0	1+	3+	–	–	–
15 hr	0	0	0	0	0	1+	1+	2+	3+
48 hr	0	0	0	0	0	0	1+	2+	3+
1 week	0	0	0	0	0	0	1+	1+	2–3+[c]

The signal was scored as follows: 0, 1+ (<25% of cells positive), 2+ (25-50% cells positive), and 3+ (>50% of cells positive). Signal measurements were made without knowledge of the reaction conditions.

[a]Protease is pepsin (2 mg/ml) at room temperature.

[b]Overdigested refers to loss of tissue morphology with a concomitant loss of the in situ PCR signal.

[c]The 2+ signal was with pepsin; the 3+ signal with proteinase K digestion (1 mg/ml).

DNase Digestion (for RT In Situ PCR)

It is important to DNase-digest two of the three tissue sections/cell suspensions. The *positive control* is the section that is not treated with DNase. An intense signal in at least 50% of cells demonstrates that protease digestion time is optimal; the signal represents DNA repair, genomic amplification, and mispriming. The *negative control* is the cells/tissue section that is DNase-digested and *not* treated with RT. The absence of signal demonstrates that the DNase digestion has rendered the native DNA template unavailable for DNA synthesis. The *test* is the section digested by DNase digestion that is then treated with RT.

1. To two of the three sections add: 1 μl of 10X buffer (the 10X buffer is made by adding 35 μl of 3 M sodium acetate, 5 μl 1 M $MgSO_4$, and 60 μl DEPC water), 1 μl of RNase-free DNase (Boehringer Mannheim, 10 units/μl), and 8 μl of DEPC water.

2. Cover the solution with the inside of the autoclaved plastic from the polypropylene bags to prevent drying. Place the slides in a humidity chamber at 37°C.

3. After overnight digestion, remove the coverslip, wash for 1 minute in DEPC water, then 100% ethanol, and air-dry.

RT Step

1. Place on one of the two sections treated with DNase the following solution: 2 μl of $MgCl_2$ (stock 25 mM), 1 μl of the RT buffer, 1 μl each of dATP, dCTP, dGTP, dTTP (stock of each nucleotide at 10 mM), 1.5 μl of DEPC water, 0.5 μl of 3′ primer (stock solution 20 μM), 0.5 μl of RNase inhibitor, and 0.5 μl of RT. These reagents are from the RT-PCR kit (Perkin-Elmer).

2. To prevent drying, anchor the coverslip with one small drop of nail polish, place the slide in an aluminum "boat" on the block of the thermal cycler, and incubate at 42°C for 30 minutes; cover with sterile mineral oil.

3. Remove the coverslip. Remove the oil with a 5-minute wash in xylene, followed by a 5-minute wash in 100% ethanol. Then air-dry.

PCR Step

1. For each slide prepare 25 μl of solution: 2.5 μl of PCR buffer, 4.5 μl of $MgCl_2$ (25 mM stock solution), 4.0 μl of dNTP solution (200 μM

final concentration of each nucleotide), 1.0 μl of 2% bovine serum albumin (BSA), 0.4 μl of digoxigenin dUTP solution (1 mM stock solution), 1 μl each of primer 1 and primer 2 (each stock 20 μM), 11 μl of water, and 0.5 μl of *Taq* DNA polymerase. The solutions are from the GeneAmp kit (Perkin-Elmer).

2. Add the solution in step 1 to the slide and cover with one large coverslip. Anchor the coverslip with two small drops of nail polish.

3. Ramp the PCR to 80°C, add preheated mineral oil, abort the file, go to 94°C for 3 minutes, and cycle at 55°C for 2 minutes, at 94°C for 1 minute for 15 cycles (in situ PCR), or 35 cycles (PCR in situ hybridization).

4. Remove the coverslip and polish. Wash for 5 minutes in xylene, 5 minutes in 100% ethanol, and then air-dry.

Detection Step

1. Wash the slides for 3 minutes in a 1X SSC solution that contains 0.2% BSA at 52°C.

2. Remove the excess wash solution. Add to each slide 100 μl of a 0.1 M Tris-HCl (pH 7.5) and 0.1 M NaCl solution with the anti-digoxigenin antibody (1:200 dilution) (Boehringer Mannheim). Incubate at 37°C for 30 minutes.

3. Wash the slides for 1 minute in the detection solution (0.1 M Tris-HCl, pH 9.5, and 0.1 M NaCl). Then incubate the slides in the detection solution (0.1 M Tris-HCl, pH 9.5, and 0.1 M NaCl) to which is added NBT/BCIP chromagen (Boehringer Mannheim). Incubate at 37°C for 5–15 minutes. Check the slides under a microscope and stop the reaction when the signal is strong.

4. Wash the slides in water for 1 minute, counterstain with nuclear fast red for 5 minutes, wash in water for 1 minute, then in 100% ethanol for 1 minute, and then in xylene for 1 minute. Mount coverslip using Permount. View under a microscope.

TROUBLESHOOTING

- If the positive control has >50% positive cells and the negative control shows no positive cells, then the RT/DNased section should show target-specific cDNA amplification.

- If the negative control shows positive cells, then increase the protease digestion time.

- If both the negative and positive controls do not show positive cells, then check the tissue morphology. If it is poorly preserved, decrease the protease digestion time. If it is well preserved, increase the protease digestion time.

DISCUSSION

Basics of In Situ PCR

TISSUE PREPARATION

The choice of fixative is essential for in situ PCR. Certain fixatives such as acetone and ethanol function by denaturing proteins and, in this way, render degradative enzymes inoperative. The most commonly used fixative in anatomic pathology is 10% buffered formalin, which cross-links proteins and nucleic acids and thus has a mode of action very different from acetone and ethanol. For some tissues, such as bone marrow and lymph node biopsies, formalin with picric acid (e.g., Bouin's solution) or with a heavy metal such as mercury (e.g., Zenker's solution) is sometimes employed because these solutions tend to give better nuclear detail with microscopic sections. Finally, immunohistochemistry at times requires frozen, unfixed tissues for preservation of the antigenic determinant, although many antigens are well preserved after formalin fixation and paraffin embedding.

Successful in situ hybridization and PCR can be done with unfixed, frozen tissue or after fixation in acetone, ethanol, or buffered formalin. Fixatives that include a heavy metal or picric acid do not allow PCR because of the rapid and extensive degradation of the DNA (Greer et al. 1989). Tissues fixed for more than 8 hours in solutions that contain either a heavy metal or picric acid do not permit a signal with in situ hybridization, although shorter-term fixation can yield intense signals (Nuovo and Silverstein 1988).

We performed a study of the utility of different fixatives with in situ PCR. The basis of these experiments was to use a cell line with a known target and determine which fixatives allowed a detection rate of 100% (Nuovo et al. 1993). The data from these experiments are presented in Table 2. Note that only buffered formalin fixation followed by protease digestion allowed a 100% detection rate of the *bcl-2* target with in situ PCR. It was surprising that acetone or ethanol fixation did not allow 100% detection rates under the conditions employed in the study. Possible explanations are that these fixations did not permit adequate amplification in the nuclei of most cells or that amplification occurred but that the PCR product migrated out of the cell into the amplifying solution. To test these hypotheses, the DNA from the cell and the amplifying solutions was retrieved after in situ PCR, extracted, and then tested by Southern blot hybridization for the 502-bp target from the *bcl-2* gene. These experiments demonstrated that the amplifying solution of the cells fixed in acetone or ethanol, but not

Table 2 Effect of Fixation Chemistry and Duration on the Detection of Amplified *bcl-2* DNA in Peripheral Blood Mononuclear Cells

	No protease digestion (% positive cells) (fixative)			
Fixation time	formalin	acetone	95% ethanol	Bouin's
5 min	5	2	14	0
15 hr	0	15	31	0
39 hr	n.d.	0	9	n.d.
	With protease digestion (% positive cells)			
Fixation time/ protease conditions	formalin	acetone	95% ethanol	Bouin's
5 min/2 mg/ml, 12 min	0	0	0	0
5 min/20 μg/ml, 1 min	1	0	0	—
5 min/20 μg/ml, 2 min	35	0	0	—
5 min/20 μg/ml, 5 min	0	0	0	—
15 hr/2 mg/ml, 12 min	100	0	0	0

(n.d.) Not done.

formalin, contained the *bcl-2* PCR product. These results led to the important topic of the migration barrier with in situ PCR. It was concluded that ethanol and acetone fixation do allow intranuclear DNA synthesis, but that, in many cells, the PCR product migrates out of the cell. However, formalin fixation appears to create a migration barrier that, under certain specified conditions, severely limits the movement of the PCR product from its site of synthesis in the cell. It is reasonable to assume that this relates to the protein-DNA latticework that occurs secondary to the cross-links created after formalin fixation.

Recently, there has been a concerted effort to find alternatives to formalin in the pathology laboratory because of the possible carcinogenic effects of this fixative. Fixatives that do not contain formalin include Histochoice and Streck's solution. These fixatives do permit in situ PCR with cell suspensions and tissue sections without the need for a protease digestion step. However, in our experience, these fixatives do not allow the precise subcellular localization of the cross-linking fixatives or a 100% detection rate (Nuovo 1994).

PROTEASE DIGESTION

The data in Table 2 underscore that 100% detection of the target was achieved only when prolonged formalin fixation was followed by protease digestion. Our attempts to obtain 100% detection rates with very brief (5 minute) fixation times with buffered formalin and no, or minimal, protease digestion were not successful. We expanded this study to tissues fixed in buffered formalin for times that ranged from 4 hours to 1 week (Nuovo et al. 1994). These data are presented in Table 1. Note that the signal evident with in situ PCR for formalin-fixed tissues is strongly dependent on the length of time for both formalin fixation and digestion with protease. It is very important to stress that these data apply to direct incorporation of the reporter molecule into the PCR product, which is called in situ PCR in this chapter. This is to be differentiated from PCR in situ hybridization, where the PCR product is detected by a labeled internal oligoprobe after a hybridization step. The optimal protease digestion time does not vary much with PCR in situ hybridization, where it is from 20 to 30 minutes for tissues that have been fixed for 4 hours to several days (Nuovo 1994).

Several proteases have been extensively tested for in situ hybridization. These include proteinase K, pepsin, trypsin, trypsinogen, and pronase. In our experience, all of these proteases allow successful in situ amplification of DNA and cDNA. We prefer trypsin or pepsin at a low pH, because a simple change of pH to 7 is enough to reduce the activity of the protease. Also, proteinase K tends to cause overdigestion more commonly. This can be recognized as the absence of signal and concomitant poor morphology (Fig. 1).

For those doing in situ PCR or, more commonly, RT in situ PCR, a very important point is evident from Table 1. The optimal protease time is dependent on the time of formalin fixation and can be defined by a 3+ signal (an intense signal in over 50% of cells). If the signal is not 3+ for a given protease digestion time, then it can be assumed that the time is suboptimal and that the digestion times should be increased until a 3+ signal is obtained. Inadequate protease digestion is

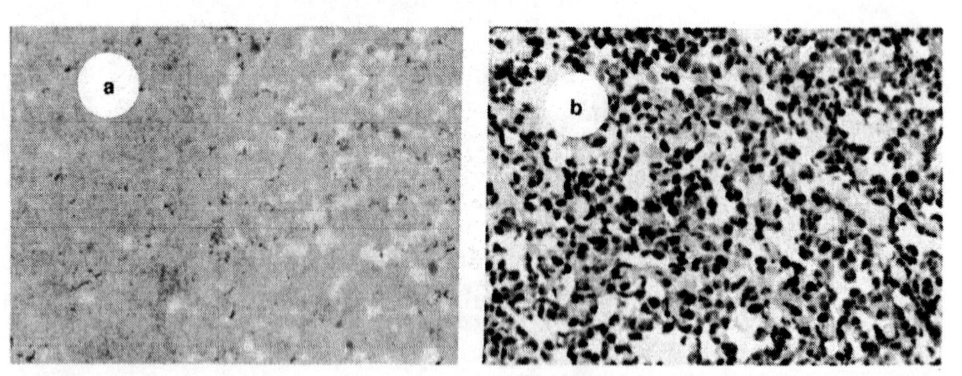

Figure 1 Overdigestion of tissue with protease. This tonsillar tissue was fixed in buffered formalin for 4 hr. No signal is evident with in situ PCR using *bcl-2* primers if the protease digestion time (pepsin 2 mg/ml) was 30 min (*a*); note the poor morphologic detail. If the digestion time was decreased to 10 min, an intense signal and good morphologic detail are evident (*b*).

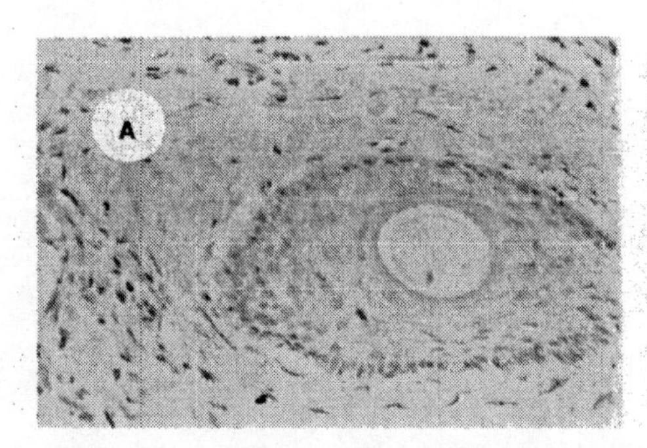

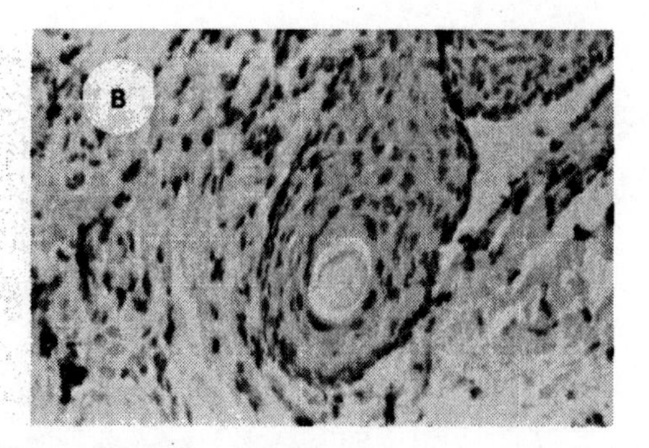

Figure 2 Relationship of protease digestion and formalin fixation times for successful in situ PCR. This skin biopsy was fixed in buffered formalin for 8 hr. No signal is evident with in situ PCR if the pepsin digestion time was 15 min (*A*). An intense signal is evident if the digestion time is increased to 45 min (*B*).

the most common cause of unsuccessful in situ PCR. This important principle is illustrated in Figure 2.

AMPLIFYING SOLUTION

There are some important differences in the composition of the amplifying solution when comparing in situ PCR to solution-phase PCR. The optimal $MgCl_2$ concentration for solution-phase PCR is usually from 1 to 4 mM, with 1.5 mM being a typical starting point. We varied the $MgCl_2$ concentration from 0 to 9 mM for a variety of primers and targets. In each case, an optimal signal for in situ PCR was obtained with 4.5 mM of $MgCl_2$ (Fig. 3) (Nuovo et al. 1993; Nuovo 1994). This relatively high and constant concentration of magnesium probably reflects its partial sequestration on the glass slide and cellular proteins.

The composition of the buffer (PCR buffer II, GeneAmp kit, Perkin-Elmer) and the concentration of the nucleotides (200 μM) and primers (1 μM) with in situ PCR are equivalent to those used for solution-phase PCR. The other major difference in the composition of the ampli-fying solution with regard to in situ PCR is the concentration of the DNA (*Taq*) polymerase (Perkin-Elmer). A usual concentration for the *Taq* DNA polymerase with solution-phase PCR is 1.5 units per 50 μl. In our experience, this concentration does not produce an intense signal with in situ PCR. Rather, 15 units per 50 μl are initially required. It is feasible that this higher concentration reflects partial adsorption of the polymerase onto the glass slide and plastic coverslip. To test this hypothesis, 1.5 μl of 2% BSA was added to the amplifying solution; BSA is commonly used in immunohistochemistry to inhibit nonspecific binding of proteins to the slide or tissue section. The addition of BSA allowed an intense hybridization signal after in situ PCR with a *Taq* concentration of 1.5 units per 50 μl (Nuovo et al. 1993; Nuovo 1994).

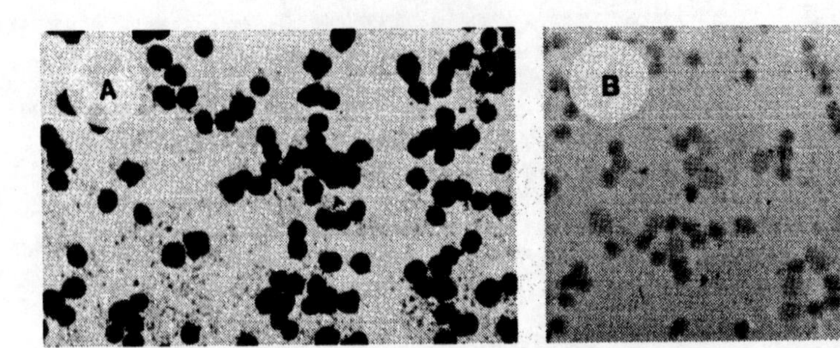

Figure 3 Importance of $MgCl_2$ concentration for successful in situ PCR. At a $MgCl_2$ concentration of 4.5 mM all peripheral blood mononuclear cells fixed for 15 hr in buffered formalin and digested have an intense signal using in situ PCR and *bcl-2* primers (*A*). No signal is evident if $MgCl_2$ is omitted (*B*). Similar poor results were obtained with magnesium concentrations of 1.5, 6.0, and 9.0 mM.

NONSPECIFIC PATHWAYS DURING IN SITU PCR

Studies have shown that primer-independent DNA synthesis occurs if paraffin-embedded fixed tissues are used for in situ PCR. This signal results from the formation of DNA nicks during the 65°C paraffin-embedding process. Because of this background, direct incorporation of the reporter group into the product is not recommended for formalin-fixed, paraffin-embedded tissues for DNA targets.

SPECIFIC PROTOCOLS

This section describes the protocols used for the in situ detection of PCR-amplified DNA and cDNA in intact cells.

PCR In Situ Hybridization

1. Place three paraffin-embedded tissue sections on silane slides. Remove the paraffin and digest with protease for 20–30 minutes.

2. Place PCR reagent over one section; use the others for negative and positive controls.

3. Place the slide on a thermal cycler. Do PCR for 30 cycles.

4. Wash in xylene, then ethanol.

5. Do in situ hybridization with an internal oligoprobe.

6. Wash in 1X SSC with 0.2% BSA at 50°C for 10 minutes, then detect the label on the probe (see Fig. 4).

Place at least three tissue sections or cytospins on a glass slide, because this allows the test to be performed with the negative control

1. Place 3 paraffin embedded tissue sections on silane slides, remove paraffin, protease digest for 20-30 min.

2. Place PCR reagent over 1 section; use others for - and + control. Place slide on thermal cycler.

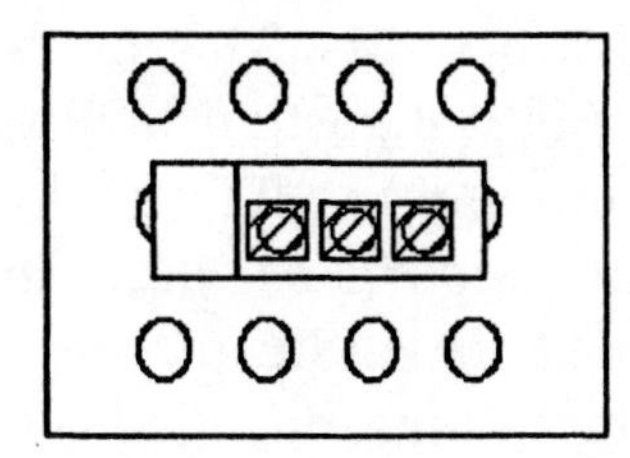

3. Do PCR for 30 cycles, wash in xylene and ETOH, then do in situ hybridization with internal oligoprobe.

4. Wash in 1XSSC with 0.2% BSA at 50C for 10 minutes, then detect label on probe.

Figure 4 Graphic representation of the PCR in situ hybridization protocol.

(e.g., no *Taq* DNA polymerase) or with the positive control (e.g., a target present in every cell) on the same glass slide. The hot start maneuver is essential for reliably detecting one target copy per cell using a single primer pair (Nuovo et al. 1991; Nuovo 1994).

In Situ PCR

With in situ PCR, the major procedural differences are the inclusion of a reporter molecule in the amplification step and, of course, the omission of the in situ hybridization step. A variety of reporter molecules may be used. Digoxigenin- and biotin-labeled nucleotides allow rapid and simple colorimetric detection of the amplified product by the use of the appropriate alkaline phosphatase-labeled conjugate (antidigoxigenen and streptavidin, respectively). In situ PCR is much quicker and simpler than PCR in situ hybridization. However, in our experience, it can only be employed for DNA targets if one is using frozen fixed tissues or cytospins that have not been heated. The power and widespread utility of in situ PCR are much more evident for RNA, where pretreatment overnight in a RNase-free DNase solution allows target-specific incorporation of the reporter molecule in the cDNA synthesized after the RT step.

RT In Situ PCR

The two steps of RT in situ PCR that differ from in situ PCR are the overnight digestion in RNase-free DNase done after the protease digestion and the RT step prior to in situ PCR. We have tried from 4 to 15 hours of DNase digestion and have shown that at least 7 hours are needed to eradicate completely the nonspecific signal after optimal protease digestion (Nuovo 1994). The RT step is done at 42°C for 30 minutes and the solution is made according to the manufacturer's recommendations (Perkin-Elmer, RT-PCR kit).

An essential point for successful RT in situ PCR is evident from Tables 1 and 2. Specifically, there is a strong relationship between the protease digestion and formalin fixation times. Adequate protease digestion is defined as an intense signal in over 50% of the cells and no signal after DNase digestion. It is important to realize that, with inadequate protease digestion, one usually sees a signal with DNase digestion that is stronger than that for the section not pretreated with DNase. Assuming that the nonspecific signal represents repair of DNA gaps, this DNase enhancement with suboptimal protease may represent the enlargement of these putative gaps under conditions where the DNase cannot completely destroy the DNA template. The key

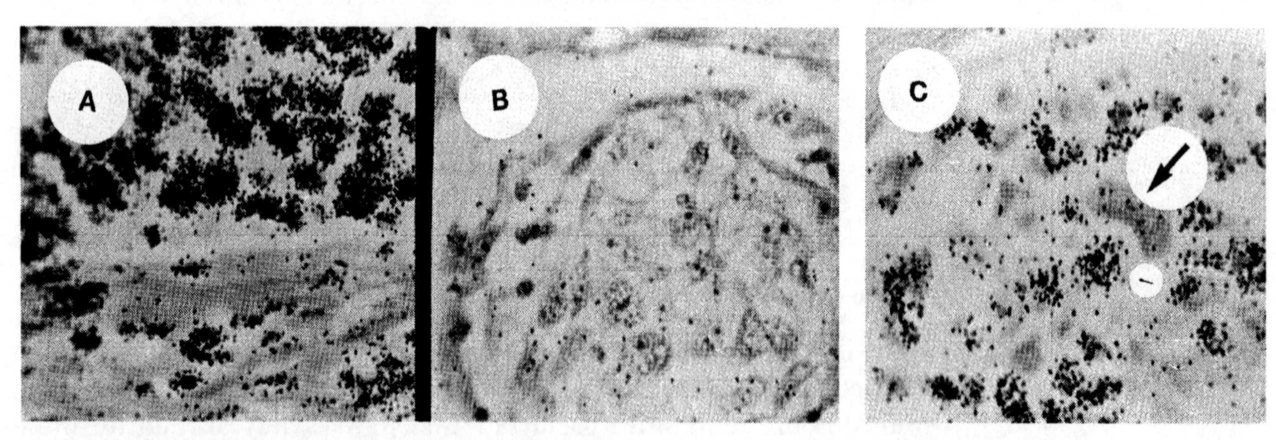

Figure 5 The importance of the positive and negative controls with RT in situ PCR. This cervical cancer was analyzed for MMP-92 expression using RT in situ PCR and direct incorporation of ^{3}H-dCTP. Note the intense nuclear signal with the positive control (no DNase, *A*) and the absence of signal with the negative control (DNase, no RT, *B*). These reactions were done on the same slide as the test and demonstrated that the protease and DNase digestions were adequate. On the same glass slide the RT in situ PCR for MMP-9 was done. Many cells contained this mRNA; note the cytoplasmic localization of the signal (*C*; small arrow, cytoplasm; large arrow, nucleus).

concept with regard to RT in situ PCR is that one must perform the positive control (no DNase) and negative control (DNase, no RT) on the same slide as the RT test. A successful run is defined by a strong signal in the positive control and no signal in the negative control (Fig. 5); note that ^{3}H is the reporter molecule. If a signal is seen with the negative control, then the most likely reason is inadequate protease digestion. Thus, the test should be repeated with increasing protease times. Another possible problem is no signal with the positive control and poor morphology. This signifies overdigestion with protease and should prompt retesting at lower protease digestion times (Fig. 1).

The final important point about RT in situ PCR concerns the localization of the signal. The signal should localize to different subcellular compartment(s) for the RNA than for the DNA in the positive control, where the signal should be pan-nuclear. This is illustrated in Figure 6. Note the dramatically different patterns evident with various

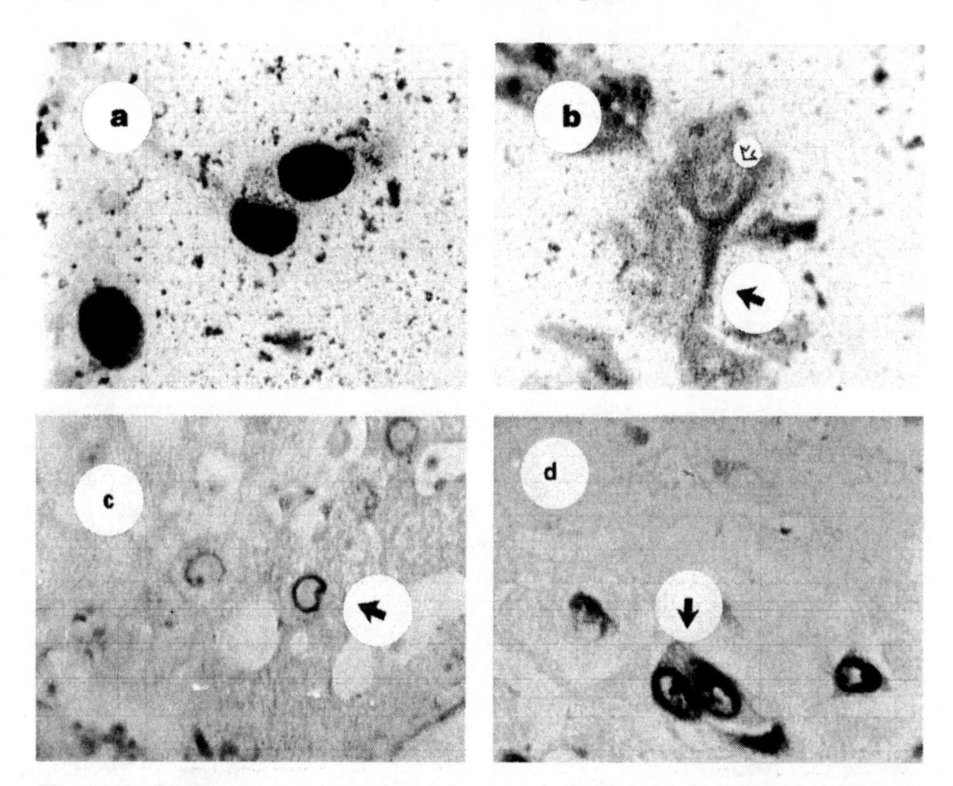

Figure 6 Different localizations of the signal with RT in situ PCR. The signal with the positive control, representing DNA repair, mispriming, and genomic DNA synthesis should localize to the nucleus, as is evident with this HT1080 cell line (*a*). The RNA-based signal should not be pan-nuclear. Variable patterns are seen. For example, a cytoplasmic signal is evident for MMP-92 in this cell line (*b*; open arrow, negative nucleus; large arrow, cytoplasmic signal). The signal for PCR-amplified hepatitis C RNA localizes to the junction of the nucleus and cytoplasm in the hepatocyte (*c*). A perinuclear and cytoplasmic signal is seen for the β chain of fibrinogen mRNA in trophoblasts (*d*; arrow, cytoplasm).

RNAs, suggesting different pathways from the nucleus to the cytoplasm.

SUMMARY

Many groups have now published data based on the in situ detection of PCR-amplified DNA and cDNA. As with standard in situ hybridization or PCR, variables that can affect the results with in situ PCR include the type of fixative and time of fixation, length of protease digestion, and the composition of the amplifying solution and oligoprobe cocktail. Investigators new to the field of in situ PCR should first try direct incorporation of the reporter molecule into paraffin-embedded tissue sections. Although nonspecific DNA synthesis is generated under these conditions, one can develop the confidence of synthesizing DNA inside the nucleus and appreciate the importance of the length of protease digestion time to successful RT in situ PCR. It is clear that the in situ detection of PCR-amplified DNA and cDNA will have a very strong impact on many diverse fields, such as oncogenesis, embryology, RNA trafficking, and detection of viral diseases, as it already has on our understanding of the pathogenesis of HIV-1 infection.

ACKNOWLEDGMENTS

The author thanks John Atwood, Phyllis MacConnell, Kim Rhatigan, Angella Forde, and Michele Margiotta for invaluable assistance.

REFERENCES

Chou, Q., M. Russell, D.E. Birch, J. Raymond, and W. Bloch. 1992. Prevention of pre-PCR mis-priming and primer dimerization improves low copy number amplifications. *Nucleic Acids Res.* **20:** 1717–1723.

Crum, C.P., G.J. Nuovo, D. Friedman, and S.J. Silverstein. 1988. A comparison of biotin and isotope labeled ribonucleic acid probes for in situ detection of HPV 16 ribonucleic acid in genital precancers. *Lab. Invest.* **58:** 354–359.

Erlich, H.A., D. Gelfand, and J.J. Sninsky. 1991. Recent advances in the polymerase chain reaction. *Science* **252:** 1643–1650.

Greer, C.E., S.L. Peterson, N.B. Kiviat, and M.M. Manos. 1991. PCR amplification from paraffin-embedded tissues: Effects of fixative and fixative times. *Am. J. Clin. Pathol.* **95:** 117–124.

Nuovo, G.J. 1989. A comparison of slot blot, Southern blot and in situ hybridization analysis for human papillomavirus DNA in genital tract lesions. *Obstet. Gynecol.* **74:** 673–677.

——. 1993. *Cytopathology of the cervix and vagina: An integrated approach.* Williams and Wilkins, Baltimore, Maryland.

——. 1994. *PCR in situ hybridization: Protocols and applications*, 2nd edition. Raven Press, New York.

Nuovo, G.J. and R.M. Richart. 1989. A comparison of different methodologies (biotin based and 35S based) for the detection of human papillomavirus DNA. *Lab. Invest.* **61:** 471–476.

Nuovo, G.J. and S.J. Silverstein. 1988. Comparison of formalin, buffered formalin, and Bouin's fixation on the detection of human papillomavirus DNA from genital lesions. *Lab. Invest.* **59:** 720–724.

Nuovo, G.J., F. Gallery, and P. MacConnell. 1994. Analysis of specific DNA synthesis during in situ PCR. *PCR Methods Appl.* **4:** 89–96.

Nuovo, G.J., F. Gallery, R. Hom, P. MacConnell, and W. Bloch. 1993. Importance of different variables for optimizing *in situ* detection of PCR-amplification. *PCR Methods Appl.* **2:** 305–312.

Nuovo, G.J., F. Gallery, P. MacConnell, J. Becker, and W. Bloch. 1991. An improved technique for the detection of DNA by *in situ* hybridization after PCR-amplification. *Am. J. Pathol.* **139:** 1239–1244.

Single-strand Conformational Polymorphism

Kazunobu Fujita and Jonathan Silver

Laboratory of Molecular Microbiology, NIAID, National Institutes of Health, Bethesda, Maryland 20892

INTRODUCTION

PCR-single-strand conformational polymorphism (SSCP) is a simple and powerful technique for identifying sequence changes in amplified DNA (Orita et al. 1989). It is based on the empirical observation that mobility of single-strand DNA in nondenaturing polyacrylamide gels is very sensitive to primary sequence, probably because slight sequence changes have major effects on conformation. In the absence of a robust theory for predicting mobility based on conformation and conformation based on sequence, SSCP remains largely empirical.

The major advantage of PCR-SSCP is its simplicity: Standard PCR is performed with trace amounts of [^{32}P]dCTP or labeled primers, and the product is analyzed on a nondenaturing polyacrylamide sequencing-type gel. The main limitation of PCR-SSCP is that it may miss certain mutations. Although 70% to >95% of mutations in various model systems have been detected with PCR-SSCP (Fan et al. 1993; Glavač and Dean 1993; Hayashi and Yandell 1993; Sheffield et al. 1993), achieving such high detection rates may require running gels under several conditions. SSCP is a powerful and convenient screening method, but it is not a definitive procedure for identifying all mutations.

We present a simple protocol for PCR-SSCP and discuss common problems and their solutions.

REAGENTS AND EQUIPMENT

Sequencing gel apparatus (we use an apparatus from Life Technologies with 31-cm x 40-cm glass plates, but any comparably sized apparatus can be substituted)

Gel drying apparatus

X-ray cassette with intensifying screen(s) to accommodate sequencing-size gels
3MM paper (Whatman)
30% acrylamide-0.8% bisacrylamide stock solution
10% ammonium persulfate in dH_2O (we prepare a fresh solution every 2–3 weeks and store it at 4°C)
5× TBE (1× is 89 mM Tris, 89 mM boric acid, 4 mM EDTA)

Caution: Acrylamide is a potent neurotoxin. Wear gloves and a mask when weighing acrylamide powders. To minimize the exposure to acrylamide, add dH_2O directly to a new bottle of acrylamide to give a final concentration of 30% acrylamide or purchase a ready-made mixture of acrylamide-bisacrylamide (e.g., ProtoGEL from National Diagnostics).

PROTOCOL

Amplification and Labeling of DNA by PCR

1. Use amplification protocol. Conditions that optimize yield depend on the specific amplicon and primers used. To amplify segments of HIV-1 from infected tissue culture cells, we set up 10-μl PCR mixtures containing 1 μg of genomic DNA, 5 pmoles of each primer, 0.5 units of *Taq* polymerase, each deoxynucleoside triphosphate at 0.2 mM, 10 mM Tris-HCl, pH 8.3, 50 mM KCl, 1.5 mM $MgCl_2$, and 0.01% gelatin.

2. Label DNA by adding 0.5 μCi of [^{32}P]dCTP (3000 Ci/mmole) to the PCR mixture. Alternatively, use primers that are 5′-end-labeled with ^{32}P. End-labeling can be achieved by mixing 1 μl of 10 μM 5′-OH primer, 5 μl of [γ-^{32}P]ATP (5000 Ci/mmole), 11 μl of dH_2O, 2 μl of manufacturer's 10× buffer, and 1 μl (10 units) of T4 polynucleotide kinase, incubating for 15 minutes at 37°C, followed by 10 minutes at 65°C to inactivate the kinase, and G-25 spun chromatography to remove unincorporated ATP.

Gel Preparation

1. Clean the sequencing gel glass plates with detergent solution, and rinse with water and then with ethanol.

2. Treat one plate with a silicon solution (e.g., UNELKO) to prevent sticking.

3. Assemble the plates with 0.4-mm spacers.

4. Mix in a 250-ml Ehrlenmeyer flask:

30% acrylamide-0.8% bisacrylamide	17 ml
5X TBE	20 ml
dH_2O	62 ml
TEMED (*N,N,N′,N′*-tetramethylenediamine)	30 μl

Mix well, then add 0.8 ml of 10% ammonium persulfate.

5. Cast the gel, paying special attention to avoid bubbles (Sambrook et al. 1989). After the gel polymerizes, cool it in a cold room for at least 30 minutes prior to sample loading (cooling is necessary only if the gel is to be run in the cold).

Sample Preparation

1. Mix equal volumes (~3 μl) of PCR product and sample buffer (95% formamide, 20 mM EDTA, pH 8.0, 0.05% xylene cyanol, and 0.05% bromophenol blue).

2. Heat to 85°C for 5 minutes to denature the DNA; then cool on ice.

Electrophoresis

1. Load 2 μl of denatured DNA per well.

2. Electrophorese at 4°C at a constant voltage of 500 V for 12–15 hours or at 1100 V for 3–5 hours in 1X TBE electrophoresis buffer.

Detection

1. After electrophoresis, separate the glass plates. The gel should stick to the non-silicon-treated plate.

2. Transfer the gel onto Whatmann 3MM paper, vacuum-dry, and autoradiograph.

Other Detection Methods

1. Silver staining is nearly as sensitive as radioisotopic labeling and provides a permanent record (Ainsworth et al. 1991; Mohabeer et al. 1991; Oto et al. 1993).

2. Fluorescent dyes attached to oligonucleotides, in conjunction with sensitive fluorescence detectors, offer advantages in terms of automated data acquisition. In addition, corrections can be made for lane-to-lane variation in mobility by running a marker DNA labeled with a different fluorophore in all lanes (Makino et al. 1992; Ellison et al. 1993).

TROUBLESHOOTING

- *No polymorphism is detected.* Several modifications of gel conditions have been reported to increase the sensitivity to detect single-base changes. For example, ranges of concentrations of acrylamide (usually from 4% to 12%) and of cross-linker bisacrylamide (usually from 2% to 3.4% of the concentration of acrylamide) have been reported to be beneficial in particular circumstances (Savov et al. 1992; Glavač and Dean 1993; Hayashi and Yandell 1993), as have additives such as 5–10% glycerol, 5% urea or formamide, and 10% dimethylsulfoxide or sucrose (Orita et al. 1989; Glavač and Dean 1993). Alterations in the gel-running temperature from 4°C to 37°C and changes in buffer concentration (possibly leading to changes in the gel-running temperature) may also help. Purine-rich strands may be more sensitive to base changes than pyrimidine-rich strands (Glavač and Dean 1993). Smaller fragments (<300 bp) are, in general, more likely to reveal single-base changes (Hayashi 1991; Hayashi and Yandell 1993; Sheffield et al. 1993), although fragment size and sequence context (the sequence of adjacent DNA) can have unpredictable effects on mobility shifts associated with particular base changes (Fan et al. 1993). RNA may be more sensitive to base change-induced mobility shifts than DNA, and RNA can be made from amplified DNA by incorporating RNA polymerase promoters at the 5′ ends of amplifying primers (Sarkar et al. 1992a). Estimates of the proportion of mutations detectable by SSCP are affected by the number of SSCP conditions tried (temperature, glycerol concentration, acrylamide and bisacrylamide concentrations, buffer concentration, etc.) but, in general, are in the 70% to >95% range for studies using two or three SSCP conditions (Fan et al. 1993; Glavač and Dean 1993; Hayashi and Yandell 1993; Sheffield et al. 1993).

  The power of SSCP to detect mutations may be increased by performing SSCP analysis on dideoxy sequencing ladders derived from test fragments rather than on the fragments themselves (Sarkar et al. 1992b). If there is a mutation, all dideoxy-terminated fragments greater than a certain size (corresponding to the position of the mutation) have non-wild-type sequence, and thus there are multiple chances to detect a mobility shift for each mutation.

- *The migration pattern of single-strand DNA is not reproducible or a complex pattern of bands is observed.* Free oligonucleotides present in a sample analyzed by SSCP can anneal to PCR product strands and generate shadow bands with altered mobility. Even at concentrations as low as 150–6 nM, free oligonucleotides have been reported to lead to mobility shifts and subspecies (Cai and Touitou 1993). This problem can be avoided by diluting PCR products (~10^{-2} or greater) or by removing oligonucleotides prior to SSCP analysis. For PCR products greater than 200 bp, oligonucleotides are easily and quickly removed by passage over Sephacryl S-300 spun columns (Pharmacia).

A high concentration of PCR products often leads to reannealing of complementary single strands, which can complicate the SSCP pattern. This can often be avoided by diluting PCR products prior to denaturation for SSCP; however, the use of dilute products necessitates sensitive detection methods such as autoradiography or silver staining. Alternative methods that have been reported to prevent reannealing include adding 33 mM methyl mercury (II) hydroxide to the denaturing sample buffer (Weghorst and Buzard 1993) or adding a "stacking" gel containing 75% formamide (Yap and McGee 1993) to keep complementary strands denatured until they separate in the gel.

Temperature variation during electrophoresis can cause artifacts. Temperature control with a special electrophoresis apparatus can solve this problem, but this apparatus adds cost and complexity.

The complexity of bands detected with SSCP can often be reduced and different species identified by incorporating a radiolabel at the 5′ end of each of the two PCR primers separately. It is also sometimes helpful to run a lane of undenatured PCR product to identify duplex DNA species.

DISCUSSION

SSCP has been used most extensively to screen for inherited mutations (Dean et al. 1990; Michaud et al. 1992; Leren et al. 1993) or to detect somatic mutations in cancer cells (Condie et al. 1993; Jacquemier et al. 1994). Because of its sensitivity to single-base changes, SSCP has been used to search for polymorphisms in cloned or amplified DNA that can then be used as genetic markers. Extensive genetic maps using SSCP markers have been constructed in the mouse (Beier 1993; Hunter et al. 1993). SSCP can be used to purify different alleles amplified from a heterozygous individual to facilitate sequencing (Suzuki et al. 1991). SSCP can also be used to quantitate input DNA in a PCR, by adding known amounts of a sequence variant that is presumed to amplify as efficiently as the PCR target but that migrates differently in SSCP (competitive PCR) (Yap and McGee 1992). In microbiology, SSCP has been used to classify virus strains (Lin et al. 1993).

SSCP can also be used to identify mutations that are selected for in various bacteriological or viral systems (Morohoshi et al. 1991; Fujita et al. 1992). For example, we used SSCP to identify changes in human immunodeficiency virus (HIV-1) associated with adaptation to growth in T-cell lines (Fujita et al. 1992). An advantage of SSCP in this setting is that non-wild-type SSCP bands correspond to mutations affecting a significant proportion of viral genomes, reflecting biological selection. We amplified overlapping segments of about 300 bp spanning the HIV envelope gene and found SSCP changes in one group of overlapping fragments. These fragments were cloned and sequenced to identify common sequence changes. We found that SSCP was also useful for identifying desired clones, because these clones contained inserts whose SSCP patterns matched the non-wild-type SSCP species seen on analysis of total proviral DNA. Variant species identified by SSCP were reinserted into a wild-type infectious viral clone and tested to see if they, in fact, conferred enhanced ability to grow in T-cell lines. When SSCP is used to identify sequence changes associated with a selected phenotype, it is very important to have a confirmatory test of the biological significance of identified sequence changes. Several completely in vitro systems have been described for "molecular evolution" under selective pressure (Robertson and Joyce 1990; Bartel et al. 1991); SSCP analysis might be useful for identifying strongly selected molecular variants in these systems.

REFERENCES

Ainsworth, P.J., L.C. Surh, and M.B. Coulter-Mackie. 1991. Diagnostic single strand conformational polymorphism, (SSCP): A simplified non-radioisotopic method as applied to a Tay-Sachs B1 variant. *Nucleic Acids Res.* **19:** 405–406.

Bartel, D.P., M.L. Zapp, M.R. Green, and J.W. Szostak. 1991. HIV-1 rev regulation involves recognition of non-Watson-Crick base pairs in viral RNA. *Cell* **67:** 529–536.

Beier, D.R. 1993. Single-strand conformation polymorphism (SSCP) analysis as a tool for genetic mapping. *Mamm. Genome* **4:** 627–631.

Cai, Q.-Q. and I. Touitou. 1993. Excess PCR primers may dramatically affect SSCP efficiency. *Nucleic Acids Res.* **21:** 3909–3910.

Condie, A., R. Eeles, A.L. Borresen, C. Coles, C. Cooper, and J. Prosser. 1993. Detection of point mutations in the p53 gene: Comparison of single-strand conformation polymorphism, constant denaturant gel electrophoresis, and hydroxylamine and osmium tetroxide techniques. *Hum. Mutat.* **2:** 58–66.

Dean, M., M.B. White, J. Amos, B. Gerrard, C. Stewart, K.-T. Khaw, and M. Leppert. 1990. Multiple mutations in highly conserved residues are found in mildly affected cystic fibrosis patients. *Cell* **61:** 863–870.

Ellison, J., M. Dean, and D. Goldman. 1993. Efficacy of fluorescence-based PCR-SSCP for detection of point mutations. *BioTechniques* **15:** 684–691.

Fan, E., D.B. Levin, B.W. Glickman, and D.M. Logan. 1993. Limitations in the use of SSCP analysis. *Mutat. Res.* **288:** 85–92.

Fujita, K., J. Silver, and K. Peden. 1992. Changes in both gp120 and gp41 can account for increased growth potential and expanded host range of human immunodeficiency virus type 1. *J. Virol.* **66:** 4445–4451.

Glavač, D. and M. Dean. 1993. Optimization of the single-strand conformation polymorphism (SSCP) technique for detection of point mutations. *Hum. Mutat.* **2:** 404–414.

Hayashi, K. 1991. PCR-SSCP: A simple and sensitive method for detection of mutations in genomic DNA. *PCR Methods Appl.* **1:** 34–38.

Hayashi K. and D. W. Yandell. 1993. How sensitive is PCR-SSCP? *Hum. Mutat.* **2:** 338–346.

Hunter, K.W., M.L. Watson, J. Rochelle, S. Ontiveros, D. Munroe, M.F. Seldin, and D.E. Housman. 1993. Single-strand conformational polymorphism (SSCP) mapping of the mouse genome: Integration of the SSCP, microsatellite, and gene maps of mouse chromosome 1. *Genomics* **18:** 510–519.

Jacquemier, J., J. Molès, F. Penault-Llorca, J. Adélaide, M. Torrente, P. Viens, D. Birnbaum, and C. Theillet. 1994. p53 immunohistochemical analysis in breast cancer with four monoclonal antibodies: Comparison of staining and PCR-SSCP results. *Br. J. Cancer* **69:** 846–852.

Leren, T.P., K. Solberg, O.K. Rødningen, L. Ose, S.

Tonstad, and K. Berg. 1993. Evaluation of running conditions for SSCP analysis: Application of SSCP for detection of point mutations in the LDL receptor gene. *PCR Methods Appl.* **3:** 159–162.

Lin, J.-C., B.K. De, and S.-C. Lin. 1993. Rapid and sensitive genotyping of Epstein-Barr virus using single-strand conformation polymorphism analysis of polymerase chain reaction products. *J. Virol. Methods* **43:** 233–246.

Makino, R., H. Yazyu, Y. Kishimoto, T. Sekiya, and K. Hayashi. 1992. F-SSCP: Fluorescence-based polymerase chain reaction-single-strand conformation polymorphism (PCR-SSCP) analysis. *PCR Methods Appl.* **2:** 10–13.

Michaud, J., L.C. Brody, G. Steel, G. Fontaine, L.S. Martin, D. Valle, and G. Mitchell. 1992. Strand-separating conformational polymorphism analysis: Efficacy of detection of point mutations in the human ornithine δ-aminotransferase gene. *Genomics* **13:** 389–394.

Mohabeer, A.J., A.L. Hiti, and W. J. Martin. 1991. Non-radioactive single strand conformation polymorphism (SSCP) using the Pharmacia "PhastSystem." *Nucleic Acids Res.* **19:** 3154.

Morohoshi, F., K. Hayashi, and N. Munakata. 1991. Molecular analysis of *Bacillus subtilis* ada mutants deficient in the adaptive response to simple alkylating agents. *J. Bacteriology* **173:** 7834–7840.

Orita, M., Y. Suzuki, T. Sekiya, and K. Hayashi. 1989. Rapid and sensitive detection of point mutations and DNA polymorphisms using the polymerase chain reaction. *Genomics* **5:** 874–879.

Oto, M., S. Miyake, and Y. Yuasa. 1993. Optimization of nonradioisotopic single strand conformation polymorphism analysis with a conventional minislab gel electrophoresis apparatus. *Anal. Biochem.* **213:** 19–22.

Robertson, D.L. and G.F. Joyce. 1990. Selection in vitro of an RNA enzyme that specifically cleaves single-stranded DNA. *Nature* **344:** 467–468.

Sambrook, J., E.F. Fritsch, and T. Maniatis. 1989. *Molecular cloning: A laboratory manual*, 2nd edition, pp. 13–52. Cold Spring Harbor Laboratory Press, Cold Spring Harbor, New York.

Sarkar, G., H.-S. Yoon, and S.S. Sommer. 1992a. Screening for mutations by RNA single-strand conformation polymorphism (rSSCP): Comparison with DNA-SSCP. *Nucleic Acids Res.* **20:** 871–878.

——. 1992b. Dideoxy fingerprinting (ddF): A rapid and efficient screen for the presence of mutations. *Genomics* **13:** 441–443.

Savov, A., D. Angelicheva, A. Jordanova, A. Eigel, and L. Kalaydjieva. 1992. High percentage acrylamide gels improve resolution in SSCP analysis. *Nucleic Acids Res.* **20:** 6741–6742.

Sheffield, V.C., J.S. Beck, A.E. Kwitek, D.W. Sandstrom, and E.M. Stone. 1993. The sensitivity of single-strand conformation polymorphism analysis for the detection of single base substitutions. *Genomics* **16:** 325–332.

Suzuki, Y., T. Sekiya, and K. Hayashi. 1991. Allele-specific polymerase chain reaction: A method for amplification and sequence determination of a single component among a mixture of sequence variants. *Anal. Biochem.* **192:** 82–84.

Weghorst, C.M. and G.S. Buzard. 1993. Enhanced single-strand conformation polymorphism (SSCP) detection of point mutations utilizing methylmercury hydroxide. *BioTechniques* **15:** 397–400.

Yap, E.P.H. and J.O. McGee. 1992. Nonisotopic SSCP and competitive PCR for DNA quantification: p53 in breast cancer cells. *Nucleic Acids Res.* **20:** 145.

——. 1993. Nonisotopic discontinuous phase single strand conformation polymorphism (DP-SSCP): Genetic profiling of D-loop of human mitochondrial (mt) DNA. *Nucleic Acids Res.* **21:** 4155.

Genetic Subtyping of Human Immunodeficiency Virus Using a Heteroduplex Mobility Assay

Eric L. Delwart,[1] Belinda Herring,[2] Allen G. Rodrigo,[3] and James I. Mullins[2]

[1]Aaron Diamond AIDS Research Center, New York University School of Medicine, New York, New York 10016
[2]Departments of Microbiology and Medicine, University of Washington, Seattle, Washington 98195
[3]Kingett Mitchell and Associates, Takapuna, Auckland, New Zealand

INTRODUCTION

Different strains of the same "species" of microorganism often display distinctive properties. In a growing number of cases, the linkage of phenotypic traits with genetic markers is allowing complicated biological assays to be replaced by genetic typing. The detection of such genetic variation has been revolutionized by PCR, which allows fragments of even the most complex genomes to be isolated in an essentially pure form in a matter of hours. Differences between gene segments can then be determined by direct sequencing of the PCR product. To expedite and extend genetic screening assays to greater numbers of samples, multiple nonsequencing methods have been developed that are simpler, typically require less complex apparatus, and are of lower cost than DNA sequencing. For example, heteroduplex analysis has been used in the field of medical genetics (Paw et al. 1990; Rommens et al. 1990; Cai et al. 1991; Clay et al. 1991; Farrar et al. 1991; Gyllensten and Allen 1991; Sorrentino et al. 1992) and for the detection of genetic polymorphisms in human populations (Ruano and Kidd 1992; Ruano et al. 1994). These methods have involved the use of nondenaturing as well as slightly denaturing gel electrophoresis conditions (Borresen et al. 1991; White et al. 1992). Vinyl polymer gels have also been used for the diagnosis of mutations within proto-oncogenes (Keen et al. 1991; Perrey and Carrel 1992; Soto and Sukumar 1992).

Nonsequencing genetic analysis methods can also facilitate the study of complex and rapidly evolving genetic systems such as RNA viruses. Recently described heteroduplex mobility assays (Delwart et al. 1993, 1994) used to classify HIV-1 strains into genetic subtypes are presented here. These assays also should be applicable to the analysis of other highly variable microorganisms.

Heteroduplexes are formed by simply denaturing and reannealing (usually by heating and cooling) partially complementary DNA strands. Sequence variation can then be detected by noting a reduced electrophoretic mobility of DNA heteroduplexes following electrophoresis through a polyacrylamide gel. The structural distortions of the DNA double helix caused by mismatched nucleotides or "gaps" (resulting from insertions or deletions in aligned regions) reduce the mobility of the DNA through the pores of the gel. The effects of these distortions on heteroduplex mobility are far less pronounced in larger-pore, agarose gels.

The structure of heteroduplexes has been analyzed using chemical and enzymatic probes (Bhattacharyya and Lilley 1989). The resulting model for an unpaired gap structure is of a nonflexible kink in the double helix to accommodate the likely extrahelical nucleotides. In contrast, mispaired nucleotides are thought to result in a more flexible, bubble-like distortion of the double helix in which every nucleotide is equally accessible to chemical modification and little overall perturbation of the helix structure is detected (Bhattacharyya and Lilley 1989). The stronger effects on mobility of unpaired versus mispaired nucleotides and of increasing gap size were also noticed using the HIV-1 envelope (*env*) DNA sequences (Delwart et al. 1993, 1994, 1995). Multiple pairwise heteroduplexes formed using previously sequenced variants indicated that a semiquantitative relationship existed between their electrophoretic mobility and the level of sequence divergence between the reannealed DNA strands (Delwart et al. 1993, 1994). Using regions of significant base pair mismatch and length variation (Fig. 1), this relationship can be used to classify HIV-1 strains rapidly into envelope sequence subtypes in what we refer to as the heteroduplex mobility assay (HMA) (Delwart et al. 1993).

A variation of HMA involves labeling the products of one PCR (e.g., using radioactive or fluorescent molecules) and reannealing them to a 100-fold mass excess of unlabeled DNA from another PCR, such that the labeled "probe" fragments are entirely driven into heteroduplexes formed with the unlabeled PCR product (the "driver"). Following polyacrylamide gel electrophoresis and detection, only heteroduplexes between the two sets of fragments should be detected. The mobility of the labeled heteroduplexes could therefore reveal the genetic relationships between complex mixtures of sequences (Delwart et al. 1993, 1994, 1995) as well as allow the detection of specific

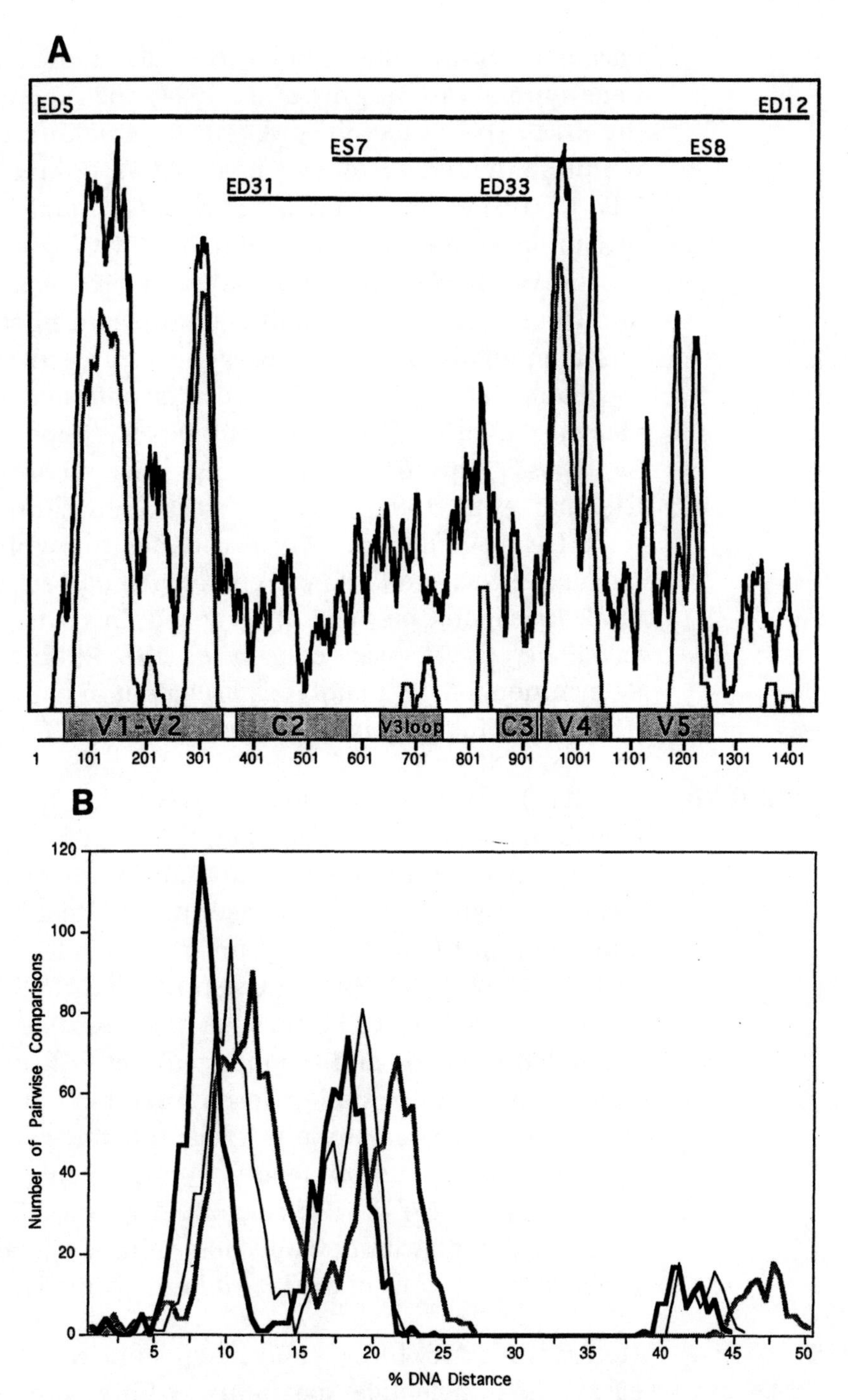

Figure 1 (*A*) Distribution of unpaired bases (*gray line*) and mismatched plus unpaired bases (*black line*) in aligned HIV sequences across the SU coding sequence. The number of differences was counted between individual pairs of sequences and tabulated along the y axis. Fragments amplified by the second-round primer sets used in this study are shown. (*B*) Degree of divergence between individual pairs of sequence sample using fragments delimited by the ED5–ED12 (*thick line*), ES7–ES8 (*thin line*), and ED31–ED33 (*gray line*) primers. Plots were generated using viruses from within the same individual, within the same subtype, between different subtypes, and between the main and O groups.

variants within these mixtures, as indicated by fast-migrating heteroduplexes (Delwart et al. 1994, 1995). This heteroduplex tracking assay (HTA) has been useful for tracking the evolution of HIV-1 within individuals (Delwart et al. 1994), in investigations of suspected HIV-1 transmission (Delwart et al. 1995), and in detecting the source of sample contamination (Sabino et al. 1994).

The World Health Organization (WHO) Network on HIV Isolation and Characterization recently conducted a pilot study comparing different methods of HIV-1 *env* gene sequence subtype determination (Osmanov et al. 1994). The results obtained using HMA, RNase A cleavage of RNA/DNA heteroduplexes (Lopez-Galindez et al. 1991; Sanchez-Palomino et al. 1993), and anchored-primer PCR (McCutchan et al. 1991, 1992a,b) were then compared with the results from DNA sequencing. All 54 isolates for which DNA sequence information was available were subtyped correctly by HMA (Bachmann et al. 1994), and because of its speed, low cost, ease of use, and high specificity, HMA was chosen as the initial screening method to determine the HIV-1 subtypes prevalent in countries in which vaccine trials are being prepared (Osmanov et al. 1994).

HIV-1 SUBTYPING KIT

An HIV-1 *env* gene subtyping kit, based on the heteroduplex mobility assays presented here, has been developed in collaboration with the WHO Network on HIV Isolation and Characterization and is available freely through the National Institutes of Health (NIH) AIDS Research and Reference Reagent Program in the United States and the Medical Research Council (MRC) of Great Britain AIDS Program in Europe. It currently consists of 23 reference plasmids from eight HIV-1 subtypes, a detailed protocol, and the four pairs of PCR primers described here. These protocols are subject to improvement, particularly with respect to the genomic regions analyzed and the gel conditions used. The rapid communication of technical improvements and the submission to the NIH and MRC AIDS reagent programs of subclones of new HIV subtypes for distribution for heteroduplex analysis will greatly improve later versions of the kit and are encouraged.

REAGENTS

Heteroduplex mobility analysis-specific reagent. (The asterisk indicates the nucleotide positions within the HIV-1-HXB2 genome [GenBank accession no. K03455]. All primers are diluted to 5 pmoles/μl.)

PCR primers

First round (2.0 kb from first exon *rev* to transmembrane region of gp41 in *env*)

ED3: 5′-TTAGGCATCTCCTATGGCAGGAAGAAGCGG; positions 5956–5985*

ED14: 5′-TCTTGCCTGGAGCTGTTTGATGCCCCAGAC; positions 7960–7931*

Second round (1.2 kb V1–V5 region of gp120)

ED5: 5′-ATGGGATCAAAGCCTAAAGCCATGTG; positions 6556–6581*
ED12: 5′-AGTGCTTCCTGCTGCTCCCAAGAACCCAAG; positions 7822–7792*

Alternative second round oligonucleotide primers (0.7 kb V3–V5 of gp120). Lowercase bases are complementary to the universal M13 primer (ES7) and reverse M13 primer (ES8). These primers allow direct sequencing of the PCR product and are not essential for HMA.

ES7: 5′-tgtaaaacgacggccagtCTGTTAAATGGCAGTCTAGC; positions 7001–7020*
ES8: 5′-caggaaacagctatgaccCACTTCTCCAATTGTCCCTCA; positions 7647–7667*

Alternative second round (0.5 kb C2–C3 of gp120)

ED31: 5′-CCTCAGCCATTACACAGGCCTGTCCAAAG; positions 6816–6844*
ED33: 5′-TTACAGTAGAAAAATTCCCCTC; positions 7359–7380*

PCR reagents
10X PCR buffer
  0.5 M KCl
  100 mM Tris-HCl, pH 8.3
  10% DMSO
  10% glycerol
10 mM $MgCl_2$
10X dNTP mixture: 2.0 mM each: dATP, dGTP, dCTP, dTTP
Positive DNA control: pNL4-3: diluted to approximately 10 copies per μl.
10X Annealing buffer
  1 M NaCl
  0.1 M Tris-HCl, pH 7.8
  20 mM EDTA
Gel electrophoresis equipment to run both horizontal agarose gels to check PCR efficacy, and 5% vertical polyacrylamide gels to analyze the heteroduplexes. Other gel systems, including fluorescence-based automated gene scanners, can also be used (data not shown).

Reference plasmids

The kit contains 23 plasmids carrying different HIV-1 *env* sequences. HMA with eight of these clones is shown in Figure 2.

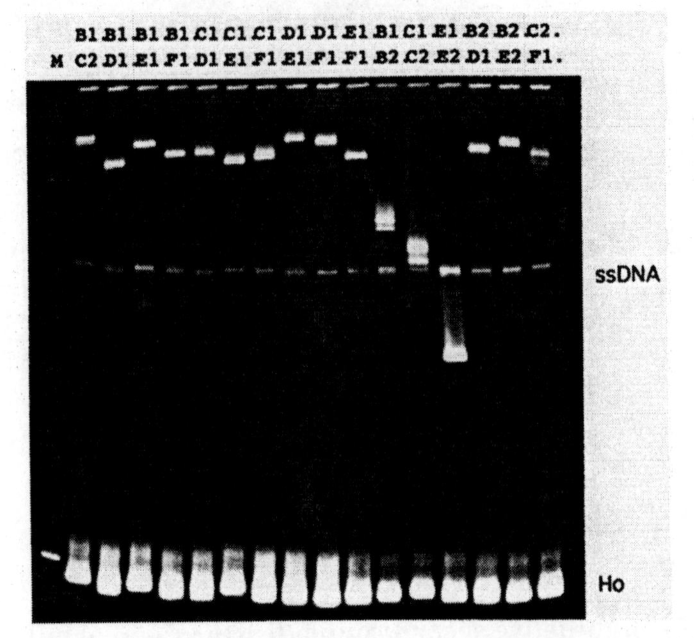

Figure 2 Heteroduplexes formed between pairs of DNA fragments PCR-amplified from subtype references. 1.2-kb ED12–ED5 *env* gene fragments were separated in a 5% polyacrylamide gel at 70 mA for 1000 V-hr. The subtypes of the two fragments compared are noted above each lane. Numbers next to the letters refer to different reference strains used in Figs. 2, 3, and 4, listed below along with the clone name and GenBank accession number (when available): (A1) RW20, U08794; (A2) IC144, no sequence; (A3) SF170, M66533; (B1) BR20, U08691; (B2) TH14, U08801; (B3) SF162, M65024; (C1) MA959, U08453; (C2) ZM18, L22954; (C3) IND868, U07103; (C4) BR25, U09133; (D1) UG21, U08804; (D2) UG38, U08806; (D3) UG46, U08809; (E1) TH22, U09131; (E2) TH06, U08810; (E3) CAR7, no sequence; (F1) BZ162, L22084; (F2) BZ163, L22085. (M) Molecular-weight marker. The 1.35-kb *Hae* III fragment from ΦX174 is visible. ssDNA refers to single-stranded DNA; Ho refers to the position of homoduplex migration.

PROTOCOLS

DNA Purification

The starting material for HMA is PCR-amplified DNA fragments. Substrates for PCR can include purified DNA or detergent-lysed cells from a variety of sources. Red-cell-depleted whole blood and fresh or frozen white blood cells can also be used (Mercier et al. 1990). However, keep in mind that amplification directly from cellular material is less sensitive than with purified DNA because of the presence of inhibitors, and thus is more noticeably dependent on the initial proviral DNA load in each sample. DNA purification protocols described by Dawson et al. and Vahey et al. (both this volume) provide reasonable starting substrates for this assay.

1. For the first-round PCR, combine:

5 μl of 10X PCR buffer
6.25 μl of 10 mM $MgCl_2$
29 μl of H_2O
5 μl of 10X dNTP mix
2 μl each of ED3 and ED14 primers (5 pmoles/μl)
0.5 μl of *Taq* DNA polymerase (5 units/μl)

2. Then add infected cell DNA (0.1–2.0 μg of DNA).

3. Cycle as follows: 3 cycles for 1 minute at 94°C, 1 minute at 55°C, and 1 minute at 72°C. Then perform 32 cycles for 15 seconds at 94°C, 45 seconds at 55°C, 1 minute at 72°C, and a final extension for 5 minutes at 72°C.

4. Store at –20°C until use.

5. For the second-round PCR, combine:

10 μl of 10X PCR buffer
12.5 μl of 10 mM $MgCl_2$
57 μl of H_2O
10 μl of 10X dNTP mix
4 μl each of ED5 and ED12 primers (5 pmoles/μl)
0.5 μl of *Taq* DNA polymerase (5 units/μl)

6. Add 2 μl of the first-round amplification, or 1 μl if amplifying an HMA reference plasmid at 10 ng/μl.

7. Cycle as in step 3.

8. Check for efficiency of amplification on a 1% agarose gel.

Heteroduplex Formation

The first step in the process of analysis of heteroduplexes is to assess the genetic diversity of each PCR-amplified sample. This analysis provides a baseline heteroduplex pattern (because of quasi-species diversity in vivo) with which to compare deliberately formed heteroduplexes. It is useful to run this sample on the same gel as the heteroduplexes formed between the reference plasmid-derived DNA and the unknown, so that bands present as a result of intra-quasi-species heteroduplexes can be identified (Figs. 3 and 4, below U lanes). PCR fragments generated from reference strains in plasmids do not result in detectable heteroduplexes, because, except for errors introduced during PCR, all duplexes should be perfectly complementary.

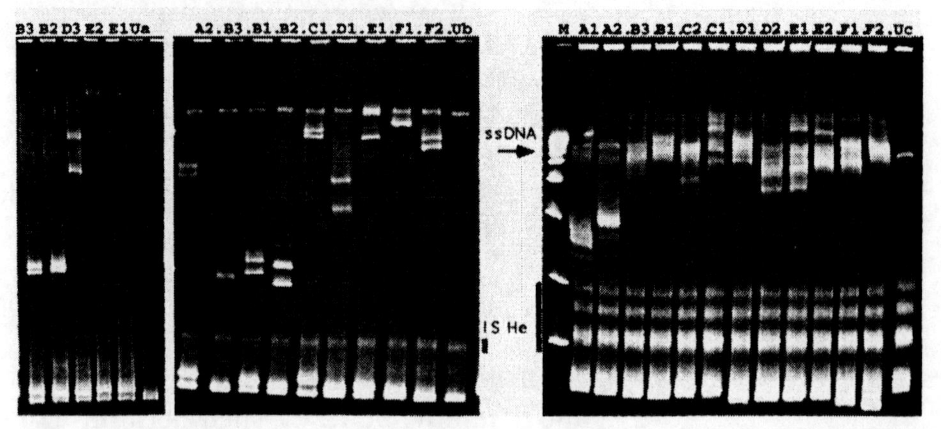

Figure 3 Heteroduplexes formed between DNA fragments amplified from infected blood samples and reference plasmids. 1.2-kb ED12–ED5 fragments were separated in a 5% polyacrylamide gel at 200 V for 6 hr. Reference fragments are noted above each lane and described in Fig. 2. The subtype determined for each unknown was B (Ua), B (Ub), and A (Uc). ssDNA and Ho are defined in Fig. 2. IS He refers to the position of intra-sample heteroduplexes, indicated by black bars next to the gels. The ssDNA is indicated with an arrow. Nonspecific amplification products in sample Ub are found just above the ssDNA indicator.

1. For unknown sample quasi-species heteroduplex formation, mix in a 500-μl Eppendorf tube:

   5 μl of the second-round PCR product
   5 μl of H_2O
   1.1 μl of 10x heteroduplex annealing buffer

2. Heat for 2 minutes at 94°C in a thermal cycler or water bath. Then snap-cool by rapidly moving the tubes to an ice-water bath.

3. For heteroduplexes between unknowns and reference standards, mix:

   5 μl of the second-round PCR product
   5 μl of the second-round PCR from the reference strain
   1.1 μl of 10x heteroduplex annealing buffer

4. Heat for 2 minutes at 94°C and snap-cool as above.

Polyacrylamide Gel Electrophoresis

1. Examine the heteroduplexes after electrophoresis on a 5% polyacrylamide gel (using a 30% acrylamide, 0.8% bisacrylamide stock) with 1x TBE buffer.

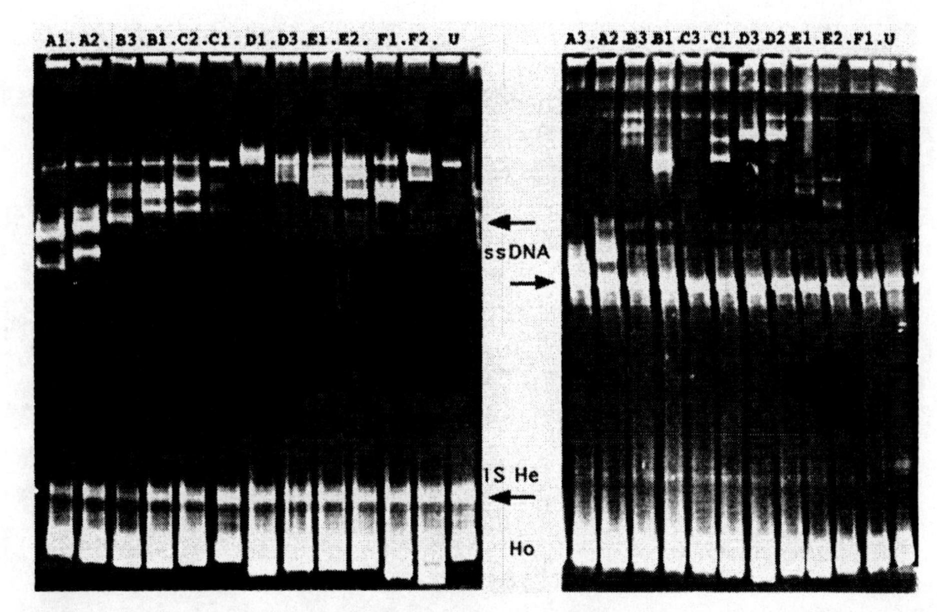

Figure 4 Heteroduplexes formed between different size DNA fragments from an HIV-1 subtype A unknown (U) and reference sequences. (*Left*) 1.2-kb ED12–ED5 region, 200 V electrophoresis conditions. (*Right*) 0.7-kb ES7–ES8 region, 250 V conditions. Heteroduplexes in the A3 and A2 lanes are within the ssDNA region. Reference fragments are noted above each lane and described in Fig. 2. ssDNA, IS He, and Ho are defined in Figs. 2 and 3. Nonspecific amplification products are seen above the ssDNA indicators.

2. Separate ED5–ED12 fragments (1.2 kb) at a constant 200 V for 6 hours or at a 70-mA constant current for 1000 V-hours. Separate the ES7–ES8 fragments (0.7 kb) at 250 V for 3 hours and the ED31–ED33 fragments (0.5 kb) at 250 V for 2 hours.

3. Add urea (10% w/v) to the gel mixtures to enhance the mobility shifts of the ED31–ED33 fragments if needed.

STRATEGY FOR SUBTYPING

Nested PCR is used to generate either 1.2-, 0.7-, or 0.5-kb *env* gene fragments from uncharacterized strains of HIV-1. The same size fragments are also amplified from a series of plasmids containing HIV-1 *env* genes from different subtypes that are used as references. Heteroduplexes formed between the unknown sample and the most closely related reference sequences exhibit the fastest electrophoretic mobilities and thus indicate the likely subtype of that strain (Figs. 2–5). For subtype assignment, heteroduplexes formed with a set of references from one subtype should have markedly faster mobilities than with other subtypes. Unknowns should initially be compared with each of the 23 reference strains available, both to obtain clear results and to gain familiarity with the assay. Subsequently, when

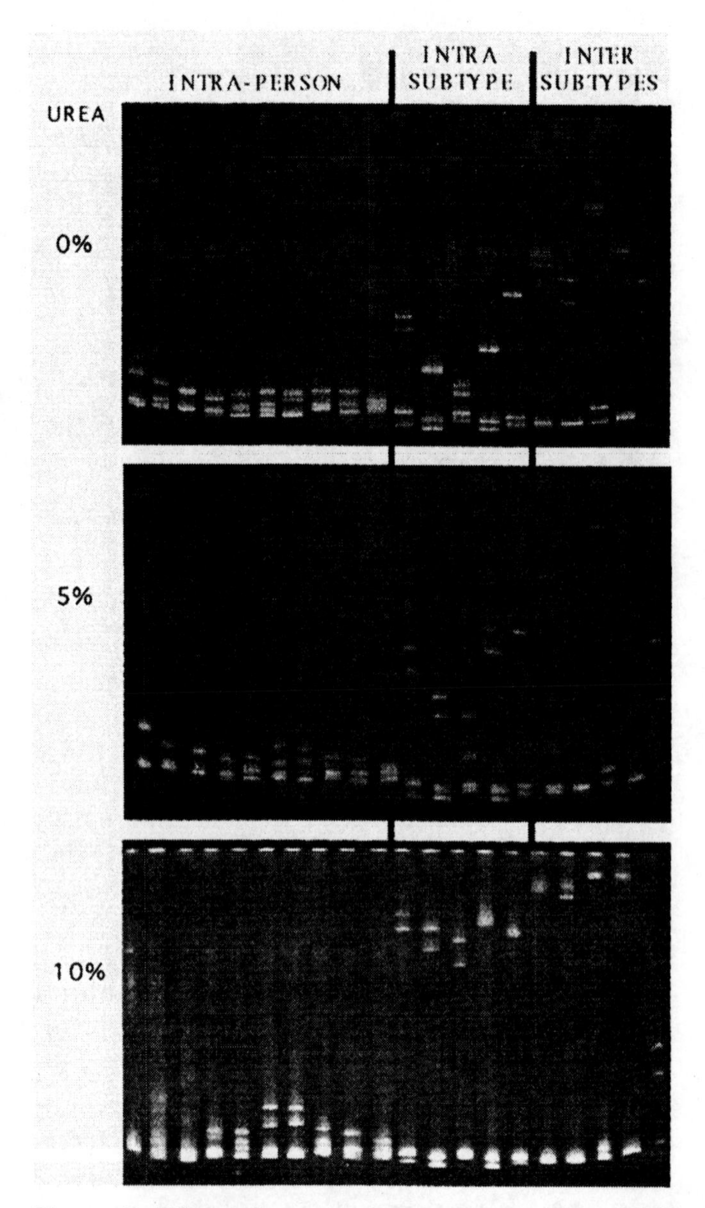

Figure 5 Effect of urea concentration on the mobility of fragments derived from the 0.5-kb ED31–ED33 region. Plasmids harboring *env* genes from the same person (intraperson), same subtype (intrasubtype), and different subtypes (intersubtypes) were used as the source of the DNA fragment pairs. Urea concentrations in the gel are indicated along the left side of the figure.

analyzing strains from a geographic region suspected of having a single or a few principal subtypes (e.g., subtype B in North America, subtypes B and E in Thailand), it is possible to reduce the number of reference strain comparisons required to make a definitive subtype assignment. When the expected subtype is not known and in geographic regions where multiple subtypes are present, such as in Uganda, it is necessary to compare each unknown with a panel of at least two or more references from each available subtype. Comparison to only a single reference sequence from a given subtype may result in ambiguous and, in rare cases, erroneous assignments.

There is great flexibility in the choice of reference strains for each subtype. Within a given geographic region, resident HIV strains of the same subtype are more likely to be closely related to each other than to geographically more distant strains of the same subtype. The use of locally derived references may therefore result in heteroduplexes that migrate faster with local unknowns, thereby permitting more rapid and confident assignment of subtype relationships. Use of closely related reference strains also allows confident subtype assignments using fewer pairwise comparisons.

Subtypes may also be assigned using smaller regions of the *env* region amplified with other second-round primers (ES7–ES8 [Fig. 4] and ED31–ED33 [Fig. 5]), and are particularly useful when ambiguous results are obtained with the larger fragment. The effect of gaps versus mismatches on the mobility of heteroduplexes is relatively stronger in shorter fragments than in longer fragments, particularly in the lower range of sequence divergence, such as that found within an individual (Delwart et al. 1993, 1994, 1995). However, very few or no gaps are found in the ED31- to ED33-amplified regions (Fig. 1A).

Some samples may be difficult to subtype with confidence by HMA. This generally indicates the detection of a genetic outlier within a known subtype (relative to the reference strains used), as shown in Figure 4, or the detection of a new subtype. It is also possible, however, that the difficulty stems from a large deletion or insertion in the V1–V2 or V4–V5 region, or, less likely, another region of *env* (Fig. 1A). Such deletions have not affected subtype assignments to date (see a discussion of very large deletions, below), involving >200 strains (Delwart et al. 1993, 1995; Bachmann et al. 1994; Osmanov et al. 1994). Nonetheless, the next step in characterizing an unusual strain would be to try another fragment of *env* or to determine a portion of the DNA sequence. As shown in Figure 4, even though the fastest-migrating ED5–ED12 heteroduplexes were found with subtype A references, their mobilities overlapped those displayed with other subtype references. Use of ES7–ES8, however, permitted unambiguous assignment of the unknown as belonging to subtype A.

There should be no limitation in performing HMA with other segments of the viral genome. The degree of variation presently required for good discrimination of heteroduplexes is within the range of 5–25%; hence, the degree of mismatch expected to be encountered should guide these decisions. Because the degree of DNA mismatch variation is similar in all three regions analyzed to date (Fig. 1B), the greater discrimination afforded by the larger fragments is largely the result of length variation and the associated adjacent mismatched nucleotides.

TROUBLESHOOTING: HETERODUPLEX PATTERN ANALYSIS

The Problem of Inherent Quasi-species Complexity

When amplifying viral sequences from a plasmid or a single provirus or RNA template, no heteroduplexes can be formed, because all DNA strands are perfectly complementary (except as a result of polymerase error during amplification) and only homoduplexes are seen on the polyacrylamide gel. When such fragments are reannealed together, a simple pattern of two single or overlapping heteroduplex bands is observed (Fig. 2) (Delwart et al. 1993, 1994). When amplifying sequences from a viral quasi-species, heteroduplexes can form between different, simultaneously amplified variants within the mixture (Delwart et al. 1993, 1994). In the latter situation, multiple heteroduplexes are seen when the PCR is analyzed on polyacrylamide gels (Figs. 3 and 4, U lanes). Heteroduplexes can take the form of sharp bands or a smear-like pattern. The complexity of the heteroduplex pattern and, by extension, the genetic diversity in a single sample can vary widely. For example, soon after HIV infection and often following virus isolation in vitro, quasi-species display a very low level of heterogeneity that is reflected by the presence primarily of homoduplexes in polyacrylamide gels (Delwart et al. 1994). Conversely, uncultured proviral DNA samples from individuals infected with HIV for more than 5 years typically display a high level of viral sequence complexity (Delwart et al. 1994). Complex quasi-species are seen as a combination of both homoduplexes and as heteroduplexes with reduced mobility. Quasi-species can consist of multiple variants in the absence of a clearly dominant sequence and, in these cases, homoduplexes are not necessarily visible and no particularly bright heteroduplex bands are observed on the polyacrylamide gel (i.e., only a smear is evident).

When locating the position of the major heteroduplexes formed between a reference (e.g., clone-derived) sequence and an uncharacterized strain, it is important to identify heteroduplexes formed by the unknown strain's quasi-species. For that reason, the PCR of the unknown sample is heated and cooled by itself and analyzed (preferably on the same gel as the heteroduplexes formed with the reference sequences). Heteroduplexes resulting from inherent quasi-species complexity can then be identified and disregarded when looking for interstrain heteroduplexes (Figs. 3 and 4).

If a single variant or a collection of highly related variants is amplified from the unknown sample, a single homoduplex band is seen in the gel. When such products are reannealed with a reference sequence, two fast-migrating homoduplexes (usually with indistinguishable or similar mobilities, e.g., see homoduplex bands at the bottom of Fig. 3) are seen. In addition, usually two (but occasionally comigrating) sharp heteroduplex bands are seen. In contrast, when a complex quasi-species is reannealed with a reference sequence, the heteroduplexes formed between the reference and the multiple variants can take the form of a series of bands or a diffuse smear. Occasionally, then, difficulties in identifying the interstrain heteroduplexes may be encountered. We have found it useful to dilute the genomic DNA serially prior to nested PCR to generate products derived from a less-complex mixture or a single variant. Assigning subtypes with less-complex quasi-species is easier because of the simpler pattern of heteroduplexes formed with reference sequences. Alternatively, one sample used in forming the heteroduplexes can be radiolabeled and used in a heteroduplex tracking assay.

Appearance of Single-stranded DNA

Discrete single or multiple bands migrating with a mobility of ~40% that of homoduplexes are sometimes seen when examining PCR products (including those formed by references alone) (Figs. 2–4). These bands correspond to collapsed single-stranded DNA (ssDNA) fragments that failed to reanneal with a complementary strand (Jensen and Straus 1993). Their uniform positioning makes them useful for visual comparison of heteroduplex mobilities.

Very Large Deletions

The detection of PCR products with reduced sizes in agarose gels is generally indicative of the presence of a subpopulation of viral genomes with large internal deletions (not of the type that characterizes normal variation in the V1–V2 and V4–V5 region [Fig. 1A]) that extend into normally conserved regions of the coding sequence and hence are derived from obviously defective genomes. Because short DNA fragments are preferentially amplified, the majority of the amplified DNA may in some cases consist of the smaller fragments (Edmonson and Mullins 1992). The occasional presence of these short amplification products results in heteroduplexes formed between fragments differing greatly in size that migrate near the top of the gel. Detection of these heteroduplexes formed within individual patients' quasi-species is thus another indication of deleted genomes. To prevent amplification of the smaller products, nested PCR is repeated using serially decreasing amounts of input genomic DNA until the shorter proviruses are diluted out and only the correct-size DNA fragment is amplified (Edmonson and Mullins 1992). This dilution procedure is usually successful, as deleted proviral genomes typically make up only a minority of target proviruses (data not shown).

Nonspecific Amplification Products

Another potential source of apparently very slowly migrating heteroduplexes is fragments formed by the amplification of nonviral DNA. Such nonspecific amplification products can be identified by their uniform presence in all lanes containing these products (Figs. 3 and 4).

GEL ELECTROPHORESIS: GENERAL CONSIDERATIONS

Heteroduplex mobility can be greatly slowed at higher temperatures or in the presence of denaturing agents in the gel matrix, particularly when the reannealed DNA strands are from very divergent strains. Temperature increases result in increases in duplex melting and thus slow heteroduplex mobility further. In Figure 5, the decrease in mobility with increasing concentration of the denaturant urea is greater for more divergent heteroduplexes than for the more related sequences. To compare data acquired across experiments, it is important to reproduce the electrophoresis conditions as closely as possible. For this purpose, the gel units, plates, acrylamide concentration, voltage/current, and buffer conditions should be carefully adhered to in each experiment.

PHYLOGENETIC INFERENCES

Given specific electrophoretic conditions, an equation can be derived to estimate the genetic distance between two DNA fragments based on their heteroduplex mobility (Delwart et al. 1993). The estimated genetic distances can then be used to derive fairly reliable phylogenetic relationships between multiple sequences without analyzing all of the possible pairwise heteroduplexes. We typically evaluate 25–33% of the $(N \times [N-1])/2$ possible comparisons (where N = number of sequences being compared).

Relative mobilities of heteroduplexes are typically estimated from photographs or video-captured images of ethidium-bromide-stained gels. The distance between the loading well and the midpoint between the two heteroduplexes is measured and divided by a value corresponding to the distance between the loading well and the midpoint between the two homoduplexes (the latter often migrate with the same mobility), as given below. Often, when the complexity of the unknown quasi-species is high, more than two heteroduplexes are formed with the reference sequence. In such instances, the approximate midpoint between the most prominent heteroduplexes has been used to estimate mobilities.

To determine the DNA distances, the sequences of each reference sequences, bounded by the second-round primers, are first aligned using one of the available computer programs (e.g., GENALIGN, IntelliGenetics), including manual refinement (e.g., MASE [Faulkner and Jurka 1988]). DNA distances are then calculated by counting mismatches after the removal of unpaired sites (gaps) introduced to maintain alignment (e.g., DNADIST from the PHYLIP software package [Felsenstein 1989]). This method was chosen to provide a comparison of HMA data to the currently most commonly used methods of HIV sequence analysis for the investigation of phylogenetic origins (Myers et al. 1993). These methods ignore gaps because there is no generally accepted means of weighting them. Weighting factors are normally available as a user-definable option, but they have not been thoroughly investigated to date. Despite this caveat, it is possible to determine a generally reliable relationship between heteroduplex mobility and DNA distance.

To estimate DNA distances from heteroduplex mobility data, standard curves are generated for the relevant electrophoresis conditions by reannealing pairs of DNA fragments of known sequence. Relative mobilities are then plotted against genetic distances, and the curve is approximated, for example, by an exponential function.

Using the 1.2-kb ED5–ED12 fragments and the exact electrophoresis conditions described above, constant 200-V conditions result in the following relationship: DNA distance = –ln[(mobility – 0.106)/0.94]/ 7.86. Using constant 70 mA (for 1000 V-hr) results in the following relationship: DNA distance = –ln[(mobility – 0.045)/1.14]13.55.

Once a calibration curve has been obtained, a matrix of distances between pairs of samples may be constructed. If comparisons between every pair of samples (including references) have been made, the matrix of genetic distances will be complete and can be used in any distance-based phylogenetic method, for example, the neighbor-joining (Saitou and Nei 1987; Fitch and Margoliash 1967) methods. If, however, only a fraction of all pairwise comparisons is made, then the matrix of genetic distances will be incomplete and one is limited, at present, to the use of the Fitch-Margoliash least-squares method. This method has been implemented in the computer program FITCH, part of J. Felsenstein's PHYLIP software (Felsenstein 1989). (PHYLIP may be obtained by "anonymous ftp" at genetics.washington.edu. Information files and programs are located in the /pub/phylip subdirectory. Users are required to register with J. Felsenstein [e-mail: joe@genetics.washington.edu]. PHYLIP is available for Macintosh, IBM-compatible, and UNIX computers.) With FITCH, the user has the option of including missing values in the distance matrix. This is done by using the subreplicate option. The documentation file included with the software describes how this may be done. FITCH writes the phylogenetic tree in parenthetical notation and places it in a file called "Treefile." This tree description can be used in a number of different phylogenetic programs including PAUP 3.1.1 (Swofford 1993) and MacClade (Maddison and Maddison 1992).

ACKNOWLEDGMENTS We thank Dr. Eugene G. Shpaer for his help in developing the computational analyses presented here. This work was supported by U.S. Public Health Service grant R01-AI-32885 and by a grant from the World Health Organization.

REFERENCES

Bachmann, M.H., E.L. Delwart, E.G. Shpaer, P. Lingenfelter, R. Singal, J.I. Mullins, and the WHO Network on HIV isolation and characterization. 1994. Rapid genetic characterization of HIV-1 from four WHO-sponsored vaccine evaluation sites using a heteroduplex mobility assay. *AIDS Res. Hum. Retroviruses* **10:** 1343–1351.

Bhattacharyya, A. and D.M.J. Lilley. 1989. The contrasting structures of mismatched DNA sequences containing looped-out bases (bulges) and multiple mismatches (bubbles). *Nucleic Acids Res.* **17:** 6821–6840.

Borresen, A.-L., E. Hovig, B. Smith-Sorensen, D. Malkin, S. Lystad, T.I. Andersen, J.M. Nesland, K.J. Isselbacher, and S.H. Friend. 1991. Constant denaturant gel electrophoresis as a rapid screening technique for p53 mutations. *Proc. Natl. Acad. Sci.* **88:** 8405–8409.

Cai, S.-P., B. Eng, Y.W. Kan, and D.H.K. Chui. 1991. A rapid and simple electrophoretic method for the detection of mutations involving small insertion and deletion: Application to ß-thalassemia. *Hum. Genet.* **87:** 728–730.

Clay, T.M., J.L. Bidwell, M.R. Howard, and B.A. Bradley. 1991. PCR-fingerprinting for selection of HLA matched unrelated marrow donors. *Lancet* **337:** 1049–1052.

Delwart, E.L., M.P. Busch, M.L. Kalish, J.W. Mosley, and J.I. Mullins. 1995. Rapid molecular epidemiology of HIV transmission. *AIDS Res. Hum. Retroviruses* (in press).

Delwart, E.L., H.W. Sheppard, B.D. Walker, J. Goudsmit, and J.I. Mullins. 1994. HIV-1 evolution *in vivo* tracked by DNA heteroduplex mobility assays. *J. Virol.* **68:** 6672–6683.

Delwart, E.L., E.G. Shpaer, F.E. McCutchan, J. Louwagie, M. Grez, H. Rübsamen-Waigmann, and J.J. Mullins. 1993. Genetic relationships determined by a DNA heteroduplex mobility assay: Analysis of HIV-1 *env* genes. *Science* **262:** 1257–1261.

Edmonson, P.L. and J.I. Mullins. 1992. Efficient amplification of half-genome sized fragments of human immunodeficiency virus from infected tissue samples. *Nucleic Acids Res.* **20:** 4933.

Farrar, G.J., P. Kenna, S.A. Jordan, R. Kumar-Singh, M.M. Humphries, E.M. Sharp, D.M. Sheils, and P. Humphries. 1991. A three-base-pair deletion in the peripherin-RDS gene in one form of retinitis pigmentosa. *Nature* **354:** 478–480.

Faulkner, D.V. and J. Jurka. 1988. Multiple aligned sequence editor (MASE). *Trends Biochem. Sci.* **13:** 321–322.

Felsenstein, J. 1989. PHYLIP–Phylogeny inference package. *Cladistics* **5:** 164.

Fitch. W.M. and E. Margoliash. 1967. Construction of phylogenetic trees. *Science* **155:** 279–284.

Gyllensten, U. and M. Allen. 1991. PCR-based HLA class II typing. *PCR Methods Appl.* **1:** 91–98.

Jensen, M.A. and N. Straus. 1993. Effect of PCR conditions on the formation of heteroduplex and single-stranded DNA products in the amplification of bacterial ribosomal DNA spacer regions. *PCR Methods Appl.* **3:** 186-194.

Keen, J., D. Lester, C. Inglehearn, A. Curtis, and S. Bhattacharyya. 1991. Rapid detection of single-base mismatches as heteroduplexes on Hydrolink gels. *Trends Genet.* **7:** 5.

Lopez-Galindez, C., J.M. Rojas, R. Najera, D.D. Richman, and M. Perucho. 1991. Characterization of genetic variation and 3′-azido-3′-deoxythymidine-resistance mutations of human immunodeficiency virus by the RNase A mismatch cleavage method. *Proc. Natl. Acad. Sci.* **88:** 4280–4284.

Maddison, W.P. and D.R. Maddison. 1992. *MacClade 3.01,* 3.01 edition. Sinauer and Associates, Sunderland, Massachusetts.

McCutchan, F.E., E. Sanders-Buell, C.W. Oster, R.R. Redfield, S.K. Hira, P.L. Perine, B.L. Ungar, and D.S. Burke. 1991. Genetic comparison of human immunodeficiency virus (HIV-1) isolates by polymerase chain reaction. *J. Acquired Immune Defic. Syndr.* **4:** 1241–1250.

McCutchan, F.E., B.L. Ungar, P. Hegerich, C.R. Roberts, A.K. Fowler, S.K. Hira, P.L. Perine, and D.S. Burke. 1992a. Genetic analysis of HIV-1 isolates from Zambia and an expanded phylogenetic tree for HIV-1. *J. Acquired Immune Defic. Syndr.* **5:** 441–449.

McCutchan, F.E., P.A. Hegerich, T.P. Brennan, P. Phanuphak, P. Singharaj, A. Jugsudee, P.W. Berman, A.M. Gray, A.K. Fowler, and D.S. Burke. 1992b. Genetic variants of HIV-1 in Thailand. *AIDS Res. Hum. Retroviruses* **8:** 1887–1895.

Mercier, B., C. Gaucher, O. Feugaes, and C. Mazurier. 1990. Direct PCR from whole blood, without DNA extraction. *Nucleic Acids Res.* **18:** 5908.

Myers, G., B. Korber, S. Wain-Hobson, R.F. Smith, and G.N. Pavlakis. 1993. Human retroviruses and AIDS 1993: A compilation and analysis of nucleic acid and amino acid sequences. Los Alamos National Laboratory, New Mexico.

Osmanov, S., L. Belsey, W. Heyward, J. Esparza, J. Bradac, B. Galvao-Castro, P. Van de Perre, E. Karita, S. Sempala, B. Tugume, B. Biryanwaho, C. Wasi, H. Rübsamen-Waigmann, H. von Briesen, U. Esser, M. Grez, H. Holmes, F.E. McCutchan, J. Louwagie, P. Hegerich, C. Lopez-Galindez, J.I. Mullins, E.L. Delwart, M.H. Bachmann, J. Goudsmit et al. 1994. HIV-1 variation in WHO-sponsored vaccine evaluation sites: Genetic screening, sequence analyses and preliminary biological characterization of representative viral strains. *AIDS Res. Hum. Retroviruses* **10:** 1327–1343.

Paw, B.H., P.T. Tieu, M.M. Kaback, J. Lim, and E. Neufeld. 1990. Frequency of three Hez A mutant alleles among Jewish and non-Jewish carriers identified in a Tay-Sachs screening program. *Am. J. Hum. Genet.* **47:** 698–705.

Perrey, D.J. and R.W. Carrel. 1992. Hydrolink gels: A rapid and simple approach to the detection of DNA mutations in thromboembolic disease. *J. Clin. Pathol.* **45:** 158–160.

Rommens, J., B.-S. Kerem, W. Greer, P. Chang, L.-C. Tsui, and P. Ray. 1990. Rapid nonradioactive detection of the major cystic fibrosis mutation. *Am. J. Hum. Genet.* **46:** 395–396.

Ruano, G. and K.K. Kidd. 1992. Modeling of heteroduplex formation during PCR from mixtures of DNA templates. *PCR Methods Appl.* **2:** 112–116.

Ruano, R., A.S. Deinard, S. Tishkoff, and K.K. Kidd. 1994. Detection of DNA sequence variation via deliberate heteroduplex formation from genomic DNAs amplified en masse in "population tubes." *PCR Methods Appl.* **3:** 225–231.

Sabino, E.C., E. Delwart, T.-H. Lee, A. Mayer, J.I. Mullins, and M.P. Busch. 1994. Identification of low-level contamination of blood as basis for detection of human immunodeficiency virus (HIV) DNA in anti-HIV-negative specimens. *J. Acquired Immune Defic. Syndr.* **7:** 853–859.

Saitou, N. and M. Nei. 1987. The neighbour-joining method: A new method for reconstructing phylogenetic trees. *Mol. Biol. Evol.* **4:** 406–425.

Sanchez-Palomino, S., J.M. Rojas, M.A. Martinez, E.M. Fenyo, R. Najera, E. Domingo, and C. Lopez-Galindez. 1993. Dilute passage promotes expression of genetic and phenotypic variants of human immunodeficiency virus type 1 in cell culture. *J. Virol.* **67:** 2938–2943.

Sorrentino, R., I. Cascino, and R. Tosi. 1992. Subgrouping of DR4 alleles by DNA heteroduplex analysis. *Hum. Immunol.* **33:** 18–23.

Soto, D. and S. Sukumar. 1992. Improved detection of mutations in the p53 gene as single-stranded conformational polymorphs and double-stranded heteroduplex DNA. *PCR Methods Appl.* **2:** 96–98.

Swofford, D.L. 1993. *PAUP 3.1.1* (Phylogenetic analysis using parsimony), 3.3.1 edition. Illinois Natural History Survey, Champaign, Illinois.

White, M.B., M. Carvalho, D. Derse, S.J. O'Brien, and M. Dean. 1992. Detecting single base substitutions as heteroduplex polymorphisms. *Genomics* **12:** 301–306.

Sensitive and Fast Mutation Detection by Solid-phase Chemical Cleavage

Lise Lotte Hansen,[1] Just Justesen,[2] and Torben A. Kruse[1]

Departments of [1]Human Genetics and [2]Molecular Biology, Aarhus University, Denmark

INTRODUCTION

Solid-phase chemical cleavage (SpCCM) is a fast and sensitive method for the detection of mutations. DNA fragments up to 2 kb can be analyzed in one operation, the position (within 10–15 bp) of the mutation can be determined, and the mismatched nucleotide(s) in the labeled DNA fragment will be known. Close to 100% of all mutations are found with this procedure (Cotton 1989). All reactions, from the first amplification of the test DNA to the search for mutations, can be carried out in microtiter plates (or microtubes in strips of eight). SpCCM is suitable for automation using a workstation to carry out the reactions and a fluorescent detection-based DNA sequencing system to analyze the cleaved fragments.

DNA from patients and control DNA are amplified by PCR using a biotinylated sense primer (or antisense primer). The control DNA is uniformly labeled with a radioactive isotope during PCR amplification. A heteroduplex between patient and control DNA is formed and is linked to streptavidin-coated magnetic beads through biotin on either the sense or antisense DNA strand. Using a magnet to attract the beads carrying the heteroduplex to one side of the wall of the tube, the solvent can then be removed, thus omitting the tedious precipitations, avoiding the loss of DNA, and significantly reducing the handling of the toxic chemicals. The heteroduplex is subjected to chemical modifications by osmium tetroxide or hydroxylamine, which modify mismatched thymines and cytosines, respectively (Cotton et al. 1988). Additionally, thymines and cytosines that are adjacent to mispaired bases may also react (Cotton and Campbell 1989). This is followed by cleavage of the backbone ribose at the modified bases. The DNA fragments are separated by polyacrylamide gel electrophoresis and visualized by autoradiography. The SpCCM procedure is illustrated in Figure 1.

REAGENTS

Osmium tetroxide (Aldrich or Fluka, cat. no. 75631)
Hydroxylamine (Aldrich or Fluka, cat. no. 55460)
Piperidine (Aldrich, cat. no. 10,409-4)
Pyridine (Sigma, cat. no. P-4036)
Diethylamine (Aldrich, cat. no. 11,000-0)
Taq DNA polymerase (Boehringer Mannheim)
10X *Taq* DNA polymerase buffer (Boehringer Mannheim)
dNTPs (Boehringer Mannheim)
Streptavidin-coated magnetic beads (2 ml) (DYNAL, cat. no. 112.05)
Magnet for microtubes (DYNAL MPC-9600, cat. no. 120.06)
Reaction tubes in strips of 8 (Advanced Biotechnologies, cat. no. AB-0266 Lot. 05)
QIAquick gel extraction kit (QIAGEN, cat. no. 28704)
Thermal cycler (Hybaid Omnigene)

PROTOCOLS

Preparation of Reagents

1. Make the bind and wash buffer (B&W) as follows: 10 mM Tris-HCl, pH 7.5, 1 mM EDTA, and 2.0 M NaCl. Autoclave and store at room temperature.

2. Make the annealing buffer as follows: 1.2 M NaCl, 12 mM Tris-HCl, pH 7.5, and 14 mM $MgCl_2$. Autoclave and store at room temperature.

3. Make the hydroxylamine solution, using a fume hood.

   a. Dissolve 1.39 grams of solid hydroxylamine in 1.6 ml of H_2O in a glass test tube by shaking under hot tap water.

   b. Add 1 ml of diethylamine dropwise and adjust the pH to approximately 6 by further adding up to 750 μl of diethylamine. To measure the pH accurately, add two drops of hydroxylamine solution to 2 ml of water. Never insert the electrode directly into an undiluted hydroxylamine solution.

   c. The solution can be stored for 7–10 days at 4°C.

4. Make the osmium tetroxide 10X buffer as follows:

   100 mM Tris-HCl, pH 7.7, 10 mM EDTA, and 15% pyridine

5. Make the osmium tetroxide solution, using a fume hood. In a fume hood, break an ampoule containing 0.5 grams of osmium tetroxide and place it in a glass bottle with 12.5 ml of H_2O. Make sure that

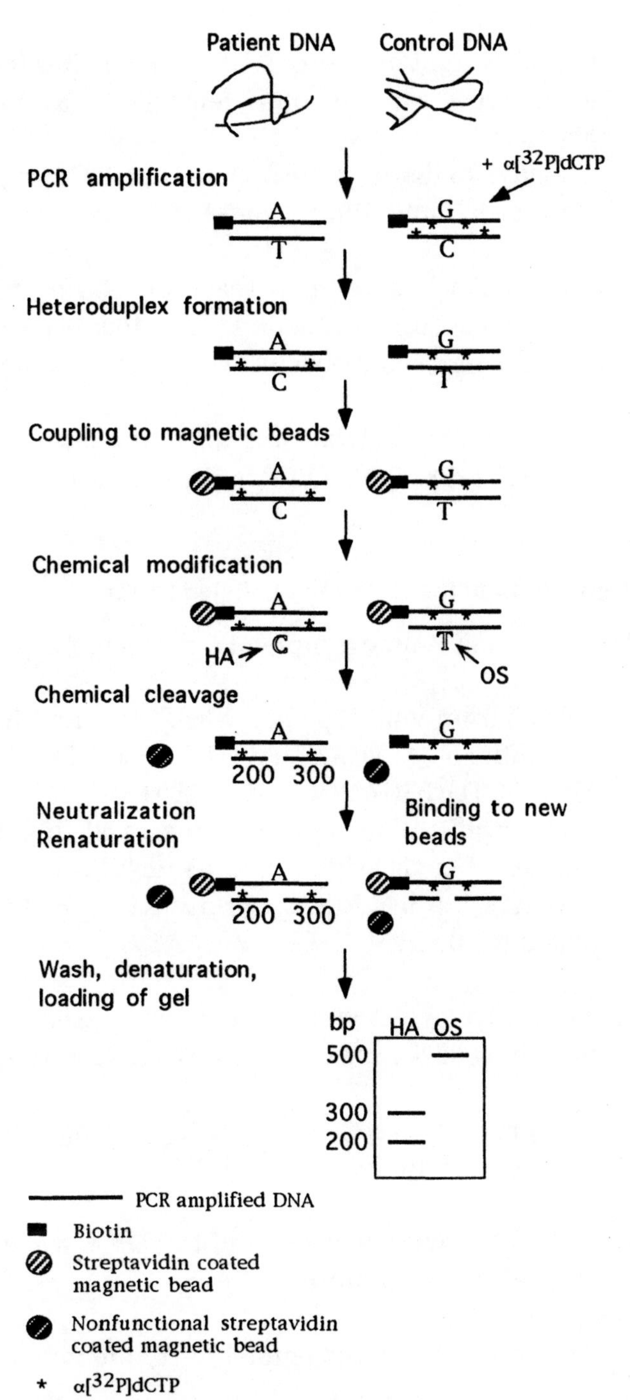

Figure 1 Schematic presentation of SpCCM. Patient and control DNA are PCR-amplified, one primer is labeled with biotin, and [α-^{32}P]dCTP is incorporated into the control DNA during amplification. Heteroduplexes are made and attached to streptavidin-coated magnetic beads. Chemical modification and cleavage are performed. The products are reannealed and attached to new streptavidin-coated magnetic beads, washed, and denatured. Nonbound DNA fragments are loaded onto a denaturing polyacrylamide gel. Modifications with hydroxylamine (HA) and osmium tetroxide (OS) take place in separate reaction tubes but are illustrated on one DNA strand. Modified nucleotides are illustrated with outline letters.

the lid is tight and well sealed by leaving the bottle in a container with screw cap. Osmium tetroxide may be inactivated if left in plastic.

Leave to dissolve for 2–3 days at 4°C wrapped in aluminum foil. Store at 4°C and dilute 1 in 5 before use.

A yellow color occurs upon reaction with pyridine (the buffer) and it indicates a successful reaction. The stock solution should be replaced when it turns green/greyish (usually after ~2 months).

Caution: Hydroxylamine and osmium tetroxide are health hazards and should be handled with gloves in a fume hood.

Preparation of the Uniformly Labeled Probe

The following is an example of a protocol for preparing the probe.

1. Exon 5 from wild-type p53 was PCR-amplified using primers:
   5A-biotin: 5′-*TTCAACTCTGTCTCCTTCCTCTTCC-3′;
   5B: 5′-CTGGGGACCCTGGGCAACC-3′ (* is biotin).
   PCR amplification conditions were: 50–100 ng of genomic DNA, 20 pmoles of each primer, 250 μM dNTP, 50 μCi [α-^{32}P]dCTP, 10 mM Tris-HCl, 1.5 mM $MgCl_2$, 50 mM KCl, pH 8.3. The volume was adjusted to 100 μl.

2. Thirty-five cycles were performed consisting of 1 minute at 94°C, 1 minute at 65°C, 1 minute at 72°C, followed by 10 minutes at 72°C.

3. The products were separated in a 2% agarose gel and stained with ethidium bromide.

4. The bands were excised and the DNA was eluted from the gel via QIAquick spin columns.

5. The probe was resuspended in H_2O to give 10,000 cpm/μl.

If optimal labeling of the DNA fragment is not obtained, a PCR amplification without the isotope can be made, 8–10 μl of the product can be analyzed on an agarose gel, and 2 μl can be used for a new PCR amplification with the isotope included.

Preparation of the Test DNA

1. Genomic DNA prepared from a breast cancer cell line HMT3522 (passage 45) and from the same cell line with a spontaneous muta-

tion in p53 (passage 379) was mixed: (normal [45]: mutated [379]) 100:0, 99:1, 95:5, 90:10, 0:100.

2. A total of 50 ng of each mixture and of the pure DNAs was PCR-amplified using the same conditions as for the probe, but without the isotope.

3. The product was analyzed on a 2% agarose gel.

The product can be purified via QIAquick spin columns if too many bands appear on the analyzing gel. Otherwise it is not necessary to purify the product.

Solid-phase Chemical Cleavage

Both DNA strands should be analyzed to be certain to detect all new mutations. One reaction is performed with biotin only on the sense primer, and the other reaction is performed with biotin on the antisense primer (Fig. 2).

Heteroduplex Formation

1. Mix 5 µl of the probe (50,000 cpm), 5 µl of the test DNA, and 10 µl of the 2x annealing buffer.

2. Incubate for 5 minutes at 100°C, followed by 60 minutes at 42°C. Centrifuge briefly.

3. Add 20 µl of Dynabeads (prewashed according to the manufacturer's instructions) to each annealing. Leave at room temperature with a gentle shaking to keep the beads in suspension for 15–30 minutes, depending on the length of the DNA fragments.

4. Place the tubes in the magnet for 30 seconds; then remove the supernatant while the tubes are in the magnet. A small amount of radioactive unbound DNA fragments will be present in the supernatant. Check the pellet with a Geiger counter to be sure that the majority of the radioactive label is still attached to the beads.

5. Remove the tube and wash the beads once with one volume of 2x B&W buffer.

6. Place the tubes in the magnet, and remove the B&W buffer.

7. Resuspend the beads in 26 µl of H_2O.

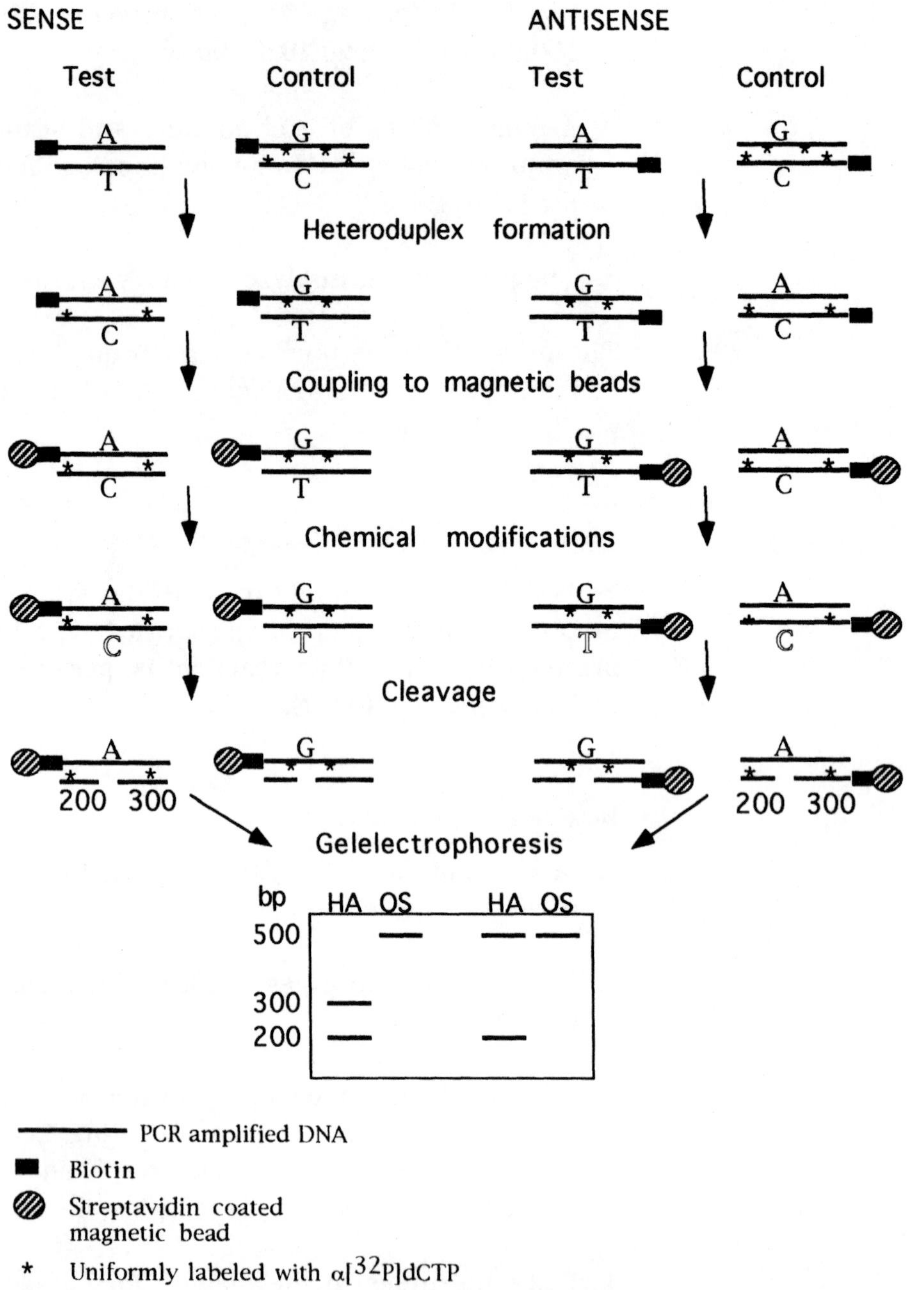

PCR amplified DNA

Biotin

Streptavidin coated magnetic bead

* Uniformly labeled with $\alpha[^{32}P]dCTP$

Figure 2 Illustration of the importance of performing two SpCCM experiments, one with biotinylated sense primer and one with biotinylated antisense primer, for the initial PCR amplification. The result obtained with a biotinylated sense primer is identical to chemical modification and cleavage of the DNA strands not attached to magnetic beads. There are two possible positions for the mutation, 200 or 300 bp from either end of the fragment. When compared with the experiment with biotin on the antisense DNA strands, the exact position of the mutation is revealed. Only the 200-bp fragment is present on the autoradiograph, which means that the 300-bp fragment is attached to the magnetic bead through the biotinylated primer. (HA) Hydroxylamine, (OS) osmium tetroxide. Outline letters illustrate modified nucleotides.

Chemical Cleavage Reaction

1. Distribute 6 μl of resuspended beads from each annealing to four tubes—two for the hydroxylamine and two for the osmium tetroxide reaction.

2. Add 20 μl of hydroxylamine solution to each of the hydroxylamine reaction tubes and incubate at 37°C for 10 and 30 minutes, respectively.

3. Add 2.5 μl of osmium tetroxide buffer and 15 μl of osmium tetroxide solution (1:5 in H_2O) to each of the osmium tetroxide reaction tubes, and incubate at 37°C for 1 and 5 minutes, respectively.

4. Place the tubes in the magnet for 30 seconds; then remove the supernatant while the tubes are in the magnet. Check the supernatant with the monitor. All counts should be in the pellet.

5. Remove the tube and wash the beads once with one volume of 2x B&W buffer.

6. Place the tubes in the magnet, and remove the B&W buffer.

7. Resuspend the beads in 50 μl of 10% piperidine (freshly diluted) and incubate for 30 minutes at 90°C.

8. Place the reactions on ice a few minutes, centrifuge briefly (omit if microtiter plates and paraffin oil are used), and add 25 μl of 1 M H_3PO_4 and 8 μl of 10x *Taq* DNA polymerase buffer. Incubate the samples for 5 minutes at 100°C and cool slowly to 30°C.

9. Add 5 μl of prewashed magnetic beads to each sample. Leave at room temperature with a gentle shaking for 15–30 minutes.

10. Place the tubes in the magnet for 30 seconds; then remove the supernatant while the tubes are in the magnet. It is important to check that the DNA fragments are attached to the beads and that there are no counts in the supernatant.

11. Remove the tube and wash the beads once with one volume of 2x B&W buffer.

12. Place the tubes in the magnet, and remove the B&W buffer.

13. Resuspend the beads in 5 μl of formamide dye (Sambrook et al. 1989), incubate for 3 minutes at 90°C, place the tubes in the

precooled magnet on ice, and load the supernatant onto a standard sequencing gel.

14. The beads can be washed and stored in one volume of the B&W buffer at 4°C.

Note: It is important to keep the Dynabeads in solution, either by gentle shaking or tapping of the tube. Microtiter plates made of polycarbonate are NOT resistant to piperidine. Use microtubes in strips of eight or microtiter plates made of polypropylene or polyethylene (Advanced Biotechnologies or Nunc). It may be necessary to incubate a small part of the microtiter plate or tube with 10% piperidine at 90°C in an Eppendorf tube to analyze the piperidine resistance.

Sensitivity of SpCCM

We used the following procedure to determine sensitivity:

1. A breast cell line HMT-3522 (passage 45) (Nielsen and Briand 1989) as control DNA was used to analyze the SpCCM sensitivity. This cell line has lost one allele on chromosome 17p, including the tumor suppressor gene p53.

2. The same cell line at passage 376 in which a mutation (H179N) has occurred in the p53 gene was used as test DNA (Moyret et al. 1994). DNA from 45 and 376 was mixed so that 376 comprised 0%, 1%, 5%, 10%, and 100%, respectively, of the total DNA amount.

3. These mixes were PCR-amplified using primers 5A-biotin and 5B. These primers are situated in the introns flanking exon 5.

4. A heteroduplex was formed between radioactively labeled PCR-amplified DNA from 45 and each of the amplified mixes.

5. The sense strands of the heteroduplexes were attached to prewashed magnetic beads (DYNAL) and subjected to solid-phase chemical cleavage. The mutation was detected (after 2 days of exposure of the X-ray film) in the presence of 5% DNA from the mutated cell line 376 (Fig. 3).

The p53 exon 5 is very GC-rich and has proved difficult to handle in the search for mutations by denaturing gradient gel electrophoresis (DGGE) because secondary structures are formed (Beck et al. 1993). We repeated the experiment mentioned above and performed traditional chemical cleavage. We could only detect the mutation in the presence of 100% of the mutated DNA that had formed heteroduplexes with the wild-type labeled DNA. A dense smear was seen at the top covering approximately one-third of the gel (Fig. 3). This is probably

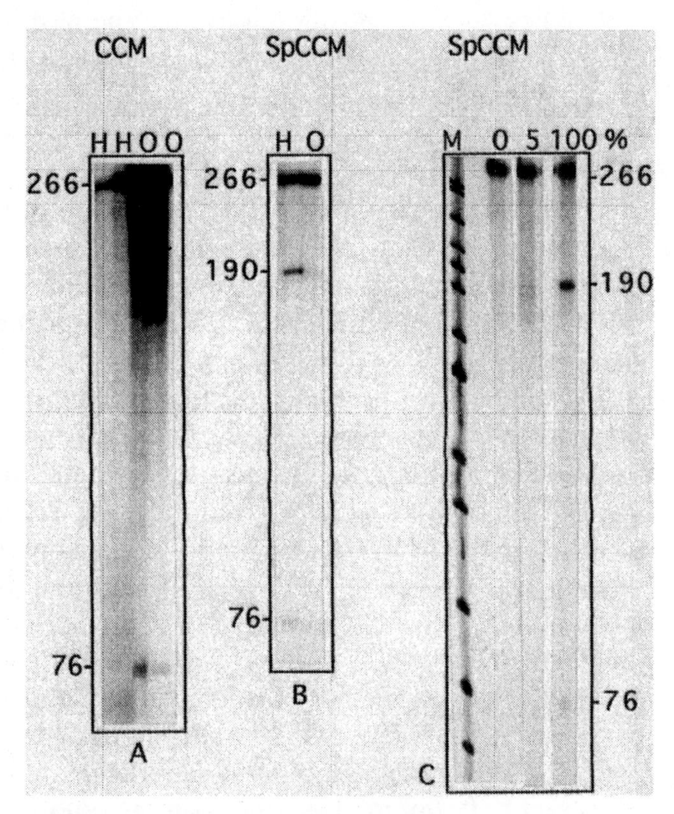

Figure 3 CCM and SpCCM performed on p53 exon 5, identifying the mutation H179D, which is a C→A substitution. (*A*) Heteroduplexes were formed between DNA from the breast cell line HMT-3522 passage 376 labeled with [α-^{32}P]dCTP and DNA from HMT-3522 passage 45 and subjected to CCM. The mutation, a mismatched T in the labeled DNA strand, was detected with osmium tetroxide (O). (*B*) SpCCM was performed on a heteroduplex consisting of HMT-3522 passage 45 labeled with [α-^{32}P]dCTP and HMT-3522 passage 376. The mutation, a mismatched C in the labeled DNA strand, was detected with hydroxylamine (H). (*C*) HM-3522 passages 45 and 376 were mixed, so that 376 comprised 0%, 5%, and 100%. The mixes were PCR-amplified and heteroduplexes were formed with HMT-3522 labeled with [α-^{32}P]dCTP and subjected to SpCCM. The marker (M) is pBR322 digested with *Msp*I.

due to the formation of secondary structures. We tried to improve the denaturing conditions by adding NP-40 and boiling the samples for 5 minutes immediately before loading the gels. None of these approaches improved the results. The smear was not seen in any of the solid-phase chemical cleavage reactions, probably because all unspecific, unbound DNA is washed away before the chemical reactions. The formation of secondary structures may be prevented by coupling one DNA strand to the magnetic bead.

ANALYSIS OF RESULTS

SpCCM reveals the position and the type of mismatched nucleotide(s) in the control DNA. DNA sequencing is the only other method with these characteristics. A mismatched thymine reacts with osmium tetroxide, and the length of the cleaved fragments shows the position of this thymine. If a cytosine is mismatched, a reaction takes place with hydroxylamine. A signal for both osmium tetroxide and hydroxylamine at the same fragment length indicates that more than one base is mismatched, and that we have either a deletion or an insertion. The detected nucleoside is situated in the radioactively labeled fragment.

To determine the exact position of a mutation within a patient sample, two experiments have to be performed, one in which the sense DNA strand is bound to the magnetic beads and one in which the antisense DNA strand is bound (Fig. 2). If more than one mismatch is present in the analyzed DNA fragment, it is not possible to determine the exact position of the mutations if the mismatched nucleotides are of the same type, e.g., G/C or A/T with the pyrimidines on the same DNA strand. If both a mismatched C and a T are present on the same DNA strand, a signal appears in both the hydroxylamine and the osmium tetroxide lanes, thus facilitating the interpretation of the position of the mismatches.

[γ-^{33}P]ATP nonbiotinylated primer end-labeling is another approach for obtaining good resolution between large fragments. If two identical pyrimidines are present on the same strand, only the distance from the end-labeled primer to the first mismatched pyrimidine is determined. Analysis should be carried out in two steps. One reaction is with PCR amplification of both test and control DNA with a biotin-labeled sense primer and [γ-^{33}P]ATP-labeled antisense primer. The other reaction is with reversed primer labeling.

TROUBLESHOOTING

- *No signal appears on the X-ray film.* The formation of radioactively labeled homoduplexes instead of heteroduplexes results in no cleavage. If the specific activity of the probe is too low, the nonradioactive DNA in the annealing reaction will be in relatively lower concentration, thereby preventing the formation of a sufficient percentage of heteroduplexes.

- *There are too many unspecific bands in each lane.* The explanation for this result is that the hydroxylamine or osmium tetroxide solutions are too old. Hydroxylamine should be replaced every 7–10 days, and osmium tetroxide develops a gray/green color when replacement is required. The bands are sequencing tracks of Cs (hydroxylamine) and Ts (osmium tetroxide).

- *There are too many well-defined bands in each lane.* The probe may consist of more than the desired DNA fragment. One must be more careful with the purification. Another explanation for this result is that there may be more than one mismatch between the probe and test DNA. The length of the fragments should add up to the full length of the uncleaved fragment.

DISCUSSION

If multiple DNA fragments are to be analyzed by SpCCM, we recommend performing a pilot experiment. Then the optimal reaction times for hydroxylamine and osmium tetroxide can be measured, thereby omitting half the reactions, as only one reaction for each chemical has to be made. We have found that incubating with piperidine for 15 minutes at 60°C resulted in sufficient cleavage when we used the p53 exon 5 sample, and restoration of the biotin-streptavidin linkage was much more efficient than after 30 minutes of piperidine treatment at 90°C.

It is possible to reduce the amount of magnetic beads from 20 to 10 μl per annealing reaction. The signal obtained after separation of the cleaved products by electrophoresis might be weak, but it is sufficient after 2 days of exposure.

The resolution of large DNA fragments in the polyacrylamide gel seems to be a limiting factor. If the mutation is situated close to one end of the fragment, it may be difficult to see the difference in the size of the uncleaved and cleaved fragments. Gradient gels may solve that problem.

Applications

SpCCM is very suitable for automation. A workstation can be used for all reactions, from the first PCR amplification of the patient DNA through the chemical modification and cleavage. The cleaved fragments can be analyzed on an automated DNA sequencing machine, if either one PCR amplification primer is labeled with fluorescent dye or if fluorescent dUTPs are incorporated during PCR amplification of the control DNA. One report on mutation detection by fluorescence with conventional CCM using the ABI 373 sequencing system (Haris et al. 1994) shows that it is possible to analyze three different DNA fragments per lane (multiplex) using three different dyes. Automation of fluorescent SpCCM will give us the opportunity to analyze a large number of DNA fragments and will reduce the handling of the hazardous chemicals.

SpCCM is a sensitive mutation detection method. Even when the mutated DNA comprises as little as 5% of the DNA analyzed, it can be detected after an overnight exposure. This is of great importance in the search for mutations in tumor DNA because this tissue is often contaminated with normal cells.

Cloned fragments can easily be analyzed for sequence variations, a factor that is important in expression studies. The recombinant bacteria, phage, or yeast can be placed in microtiter wells, denatured, and PCR-amplified followed by SpCCM.

ACKNOWLEDGMENTS We thank T.B. Christensen for excellent technical assistance. We thank Dr. P. Briand and Dr. M.W. Madsen for kindly donating the human breast epithelial cell line HMT-3522 (passage 45 and 376). Dr. P. Guldberg is thanked for providing us with primers and primer sequences for p53 exon 5. L.L.H. was supported by the Danish Cancer Society.

REFERENCES

Beck J.S., A.E. Kwitek, P.H. Cogen, A.K. Metzger, G.M. Duyk, and V.C. Sheffield. 1993. A denaturing gradient gel electrophoresis assay for sensitive detection of p53 mutations. *Hum. Genet.* **91:** 25–30.

Cotton, R.G. 1989. Detection of single base changes in nucleic acids. *Biochem. J.* **263:** 1–10.

Cotton, R.G. and R.D. Campbell. 1989. Chemical reactivity of matched cytosine and thymine bases near mismatched and unmatched bases in a heteroduplex between DNA strands with multiple differences. *Nucleic Acids Res.* **17:** 4223–4233.

Cotton, R.G.H., N.R. Rodrigues, and R.D. Campbell. 1988. Reactivity of cytosine and thymine in single-base-pair mismatches with hydroxylamine and osmium tetroxide and its application to the study of mutations. *Proc. Natl. Acad. Sci.* **85:** 4397–4401.

Haris, I.I., P.M. Green, D.R. Bentley, and F. Giannelli. 1994. Mutation detection by fluorescent chemical cleavage: Application to hemophilia B. *PCR Methods Appl.* **3:** 268–271.

Moyret, C., M.W. Madsen, J. Cooke, P. Briand, and C. Theillet. 1994. Gradual selection of a cellular clone presenting a mutation at codon 179 of the p53 gene during establishment of the immortalized human breast epithelial cell line HMT-3522. *Exp. Cell Res.* **215:** 380–385.

Nielsen, K.V. and P. Briand. 1989. Cytogenetic analysis of in vitro karyotype evolution in a cell line established from nonmalignant human mammary epithelium. *Cancer Genet. Cytogenet.* **39:** 103–118.

Sambrook, J., E.F. Fritsch, and T. Maniatis. 1989. *Molecular cloning: A laboratory manual,* 2nd edition. Cold Spring Harbor Laboratory Press, Cold Spring Harbor, New York.

PCR Starting from RNA

The amplification and quantitation of specific RNA species can be performed by several methods, including RNA-PCR/RT-PCR, QC-PCR (Piatak et al. 1993), NASBA (van Gemen et al. 1994), and bDNA (Urdea 1993). These methods detect or measure a defined RNA species, ranging from mRNAs for gene products to levels of viral RNA in plasma. NASBA and bDNA are discussed in Section 10. All of the PCR-based methodologies require the use of reverse transcriptase to synthesize cDNA from the RNA template. Methods for the preparation of quality RNA templates from plants, animal cells, and tissues are given in "Construction of a Subtractive cDNA Library Using Magnetic Beads and PCR" in Section 6 and in "RNA Purification" in Section 2.

Prior to initiating a study utilizing RNA-PCR, it is worthwhile to consider if this is the best experimental approach to take, as opposed to the more traditional methods of studying gene expression—Northern blot or RNase protection analysis. The following questions are helpful in making a decision.

Is the availability of cells or tissues for mRNA extraction and analysis limited? If limited quantities of mRNA are available, RNA-PCR may be the method of choice. For example, in the study of the effect of fluid shear on endothelin mRNA expression in human endothelial cells by Diamond and coworkers (1990), the average yield of RNA per monolayer was 8 μg. By pooling three monolayers per time point, the average yield was 25 μg. Two Northern blots could have been produced with this amount of material. Instead, RNA-PCR was performed for 10 different growth factors, cytokines, and controls. This amount of data could not have been obtained by probing the same Northern blots over and over. The methods for using RNA-PCR to obtain this type of relatively quantitative gene expression data are presented in "Use of the PCR to Quantitate Relative Differences in Gene Expression."

Is quantitative information regarding the level of mRNA expression necessary? RNA-PCR is a good method for screening cells and tissues for the expression of an mRNA. If quantitative information is required and mRNA amounts are limited, then RNA-PCR could be employed using a synthetic mRNA control amplified with the same set of primers as the mRNA for quantitation. This method uses an internal control to establish a standard curve and determine mRNA copy number. An example of this method is described in "Quantitative Liquid Hybridization PCR Method Employing Storage Phosphor Technology." However, if the amount of mRNA is not limited, then RNase protection assays are far superior for the purpose of accurate mRNA determination (Simeone et al. 1990).

Is the mRNA expressed at very low levels? Low-level expression of an mRNA species may arise if a small portion of the cells in a tissue express the gene of interest or if all cells produce the gene at subdetection levels. In either case, RNA-PCR is capable of finding the mRNA species. When studying gene expression in this range, it is important to keep in mind whether or not this expression is biologically relevant. If identification of the cells or gene of interest is important, then in situ methods, such as those described in "In Situ PCR" in Section 4, should be considered.

The individual steps in the RNA-PCR process are outlined in Figure 1. For RNA-PCR, choices must be made regarding the types of reverse transcriptase and the different possibilities for initiating or priming the cDNA reaction.

Choice of Reverse Transcriptase

A variety of reverse transcriptases are available for the synthesis of cDNA prior to RT-PCR. These enzymes include Moloney murine leukemia virus (MMLV) RT, avian myeloblastosis virus (AMV) RT, and the thermostable reverse transcriptases from *Thermus thermophilus* and *Thermus flavus.* There are also two RNase H^- mutants of MMLV RT, which are marketed as SuperScript and SuperScript II. All of these enzymes have been used successfully in RT-PCR. Because RT-PCR depends on the reverse transcription of RNA into cDNA, maximum conversion of RNA into cDNA is of critical importance to the success of RT-PCR. Depending on the purpose and design of the RT-PCR, the requirement for synthesis of full-length cDNA varies. For cloning applications, it is generally better if full-length cDNA is used. This is also true if the target region to be amplified is near the 5′ end of the mRNA. The RNase H^- derivatives of MMLV RT

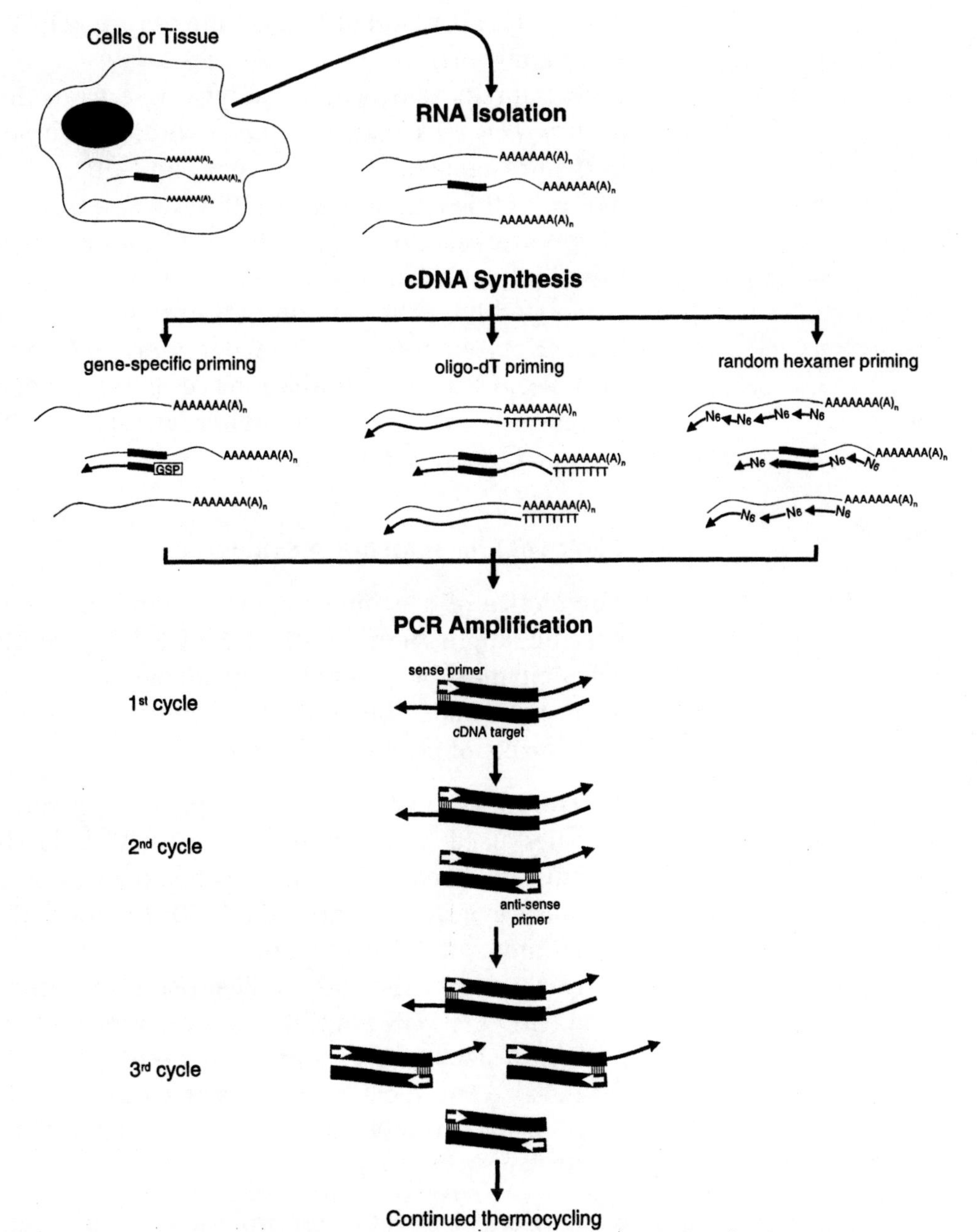

Figure 1 Schematic representation of various methods for the amplification of RNA by PCR. The abbreviation GSP is for gene-specific primer.

can convert a greater proportion of the RNA into cDNA and can synthesize longer cDNAs than other enzymes (Kotewicz et al. 1988; Gerard et al. 1992). These enzymes also operate at a higher temperature (50°C) than their wild-type counterparts and AMV RT. This property allows the synthesis of longer cDNAs from mRNA templates with secondary structure that are difficult to copy at lower temperatures. Gerard et al. (1992) have compared the cDNA yield from a 7.5-kb mRNA using different enzymes and have demonstrated that the use of

SuperScript II resulted in 2.5 times more cDNA when compared with other enzymes.

Tth polymerase, a DNA polymerase from the thermophilic eubacterium *Thermus thermophilus*, exhibits reverse transcriptase activity in the presence of Mn^{++} (Meyers and Gelfand 1991). The thermophilic nature of this enzyme allows the reverse transcription of RNA at high temperatures, which alleviates secondary structures present in the RNA template. More recently, Bicine buffer has been used to develop a one-step RT-PCR method that uses the *Tth* polymerase as a reverse transcriptase as well as DNA polymerase for amplification (Meyers et al. 1994). One of the limitations of using a thermostable reverse transcriptase is that it is not suitable for use with oligo(dT) or short random primers (see below).

Choice of Primer for cDNA Synthesis

The choice of a primer for cDNA synthesis is largely dictated by the specific application of the RT-PCR. A first-strand cDNA synthesis reaction may be primed using three different methods (see Fig. 1). The relative specificity of each primer for RNA influences the amount and variety of cDNA synthesized.

1. The most nonspecific of the primers, random hexamers, are typically used when a particular mRNA is difficult to copy in its entirety, because of the presence of sequences that cause the RT to abort synthesis (Compton 1990; Lee and Caskey 1990). With this method, all RNAs in a population serve as templates for first-strand cDNA synthesis, and the PCR primers confer the needed specificity during the PCR amplification reaction. Generally, 96% of all cDNA synthesized using random hexamers is from rRNA. To maximize the size of the cDNA synthesized using random hexamers, the ratio of primers to RNA may need to be determined empirically for each RNA preparation.

2. A method specific for mRNA is to use oligo(dT) as the primer. When the primer is hybridized to 3′ poly(A) tails, which are found in the vast majority of eukaryotic mRNAs (Frohman et al. 1988), only the mRNA is transcribed. Because poly(A)$^+$ RNA constitutes about 1–4% of a total RNA population, the amount and complexity of the resulting cDNA are considerably less than when random hexamers are used. Because of its high specificity, oligo(dT) priming generally does not require optimization of the primer/RNA ratio.

3. The most specific of the priming methods is to use an oligonucleotide containing sequence information that is complementary

to the target RNA. If the PCR amplification reaction uses two specific primers, first-strand synthesis can be primed with the amplification primer that hybridizes nearest to the 3′ terminus of the mRNA. The advantage of using a specific primer is that only the desired cDNA is produced, resulting in a more specific PCR amplification.

When measuring gene expression by RNA-PCR, it is critical that the minute amounts of contaminating DNA do not interfere and add to the signal derived from the mRNA. Two methods have been developed to circumvent this problem. A procedure for DNasing the RNA samples is provided by Rashtchian (1994). The second approach, when possible, is to place the PCR primers on different exons within the gene, eliminating the colinearity between the gene and the mRNA. As diagrammed in Figure 2, the PCR product from DNA is significantly larger and may not produce a product.

Many systems have been devised for the quantitation of mRNA by PCR, and three are presented here. The first two, "Use of the PCR to Quantitate Relative Differences in Gene Expression" and "Quantitative Liquid Hybridization PCR Method Employing Storage Phosphor Technology," detail methods that are applicable to the analysis of gene expression. The third, "Use of the SNuPE Assay to Quantitate Allele-specific Sequences Differing by a Single Nucleotide," is quite useful in defining the relative levels of expression of closely related genes. Whatever method is finally chosen for the analysis of gene expression, make sure that it will provide the necessary data.

Additional chapters are included in this section because they employ mRNA as the starting material. "Trapping Internal and 3′-Terminal Exons" describes a PCR-based method for defining the exon

PRIMER PLACEMENT

- **PCR is so sensitive that minute quantities of genomic DNA can give rise to products during amplification. Two approaches to this problem are:**
  - **Place PCR primers on different exons within a gene**
  - **For detection of mRNAs from unspliced genes, DNAse treat the RNA and run control reactions without adding reverse transcriptase**

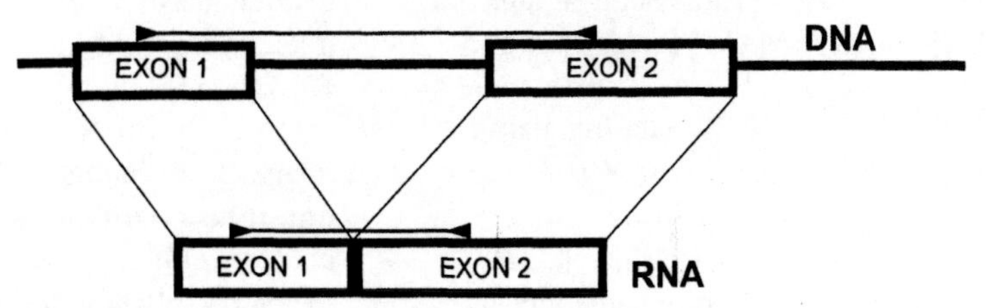

Figure 2 Scheme for primer placement.

borders of coding regions of incompletely defined genes. This technology is quite useful in genome mapping. The final chapter of this section describes a PCR-based method for rapidly attaching T7 RNA polymerase and eukaryotic translational control signals to specific DNA fragments. "Expression-PCR" is of great utility in producing specific control RNAs for use as RNA-PCR standards. Additionally, this method can be used for the rapid production of proteins from cloned DNA sequences for the study of site-directed mutations or as antigens for immunization.

The methods presented in this section demonstrate the real versatility and power of PCR. No other method yet developed can be applied to both the measurement of gene expression and the identification of yet-undefined sequences that are differentially regulated.

ACKNOWLEDGMENT The editors thank Dr. Ayoub Rashtchian, Life Technologies, Inc., for coauthoring this Section Introduction.

REFERENCES

Compton, T. 1990. Degenerate primers for DNA amplification. In *PCR protocols: A guide to methods and applications* (ed. M.A. Innis et al.), p. 39–45. Academic Press, San Diego, California.

Diamond, S.L., L.V. McIntire, J.B. Sharefkin, C.W. Dieffenbach, K. Fraisier-Scott, and S.G. Eskin. 1990. Tissue plasminogen activator messenger RNA levels increase in cultured human endothelial cells exposed to laminar shear stress. *J. Cell. Physiol.* **143:** 364–371.

Frohman, M.A., M.K. Dush, and G.R. Martin. 1988. Rapid production of full-length cDNAs from rare transcripts: Amplification using a single gene-specific oligonucleotide primer. *Proc. Natl. Acad. Sci.* **85:** 89–98.

Gerard, G.F., B.J. Schmidt, M.L. Kotewicz, and J.H. Campbell. 1992. cDNA synthesis by Moloney murine leukemia virus RNase H-minus reverse transcriptase possessing full DNA polymerase activity. *Focus* **14:** 91.

Kotewicz, M.L., C.M. Sampson, J.M. D'Alessio, and G.F. Gerard. 1988. Isolation of a cloned Moloney murine leukemia virus reverse transcriptase lacking ribonuclease H activity. *Nucleic Acids Res.* **16:** 265–277.

Lee, C.C. and T. Caskey. 1990. cDNA cloning using degenerate primers. In *PCR protocols: A guide to methods and applications* (ed. M.A. Innis et al.), p. 46–53. Academic Press, San Diego, California.

Liang, P. and A. Pardee. 1992. Differential display of eukaryotic messenger RNA by means of the polymerase chain reaction. *Science* **257:** 967–971.

Meyers, T.W. and D.H. Gelfand. 1991. Reverse transcription and DNA amplification by a *Thermus thermophilus* DNA polymerase. *Biochemistry* **30:** 7661–7666.

Meyers, T.W., C.L. Sigua, and D.H. Gelfand. 1994. High temperature reverse transcription and PCR by a *Thermus thermophilus* DNA polymerase. In *Proceedings of the 1994 Miami Bio/Technology Winter Symposium. Advances in gene technology: Molecular biology and human disease*, p. 87.

Piatak, M., Jr., K.C. Luk, B. Williams, and J.D. Lifson. Quantitative competitive polymerase chain reaction for accurate quantitation of HIV DNA and RNA species. *BioTechniques* **14:** 70–81.

Rashtchian, A. 1994. Amplification of RNA. *PCR Methods Appl.* **4:** S83–S91.

Simeone, A., D. Acampora, L. Arcioni, P.W. Andrews, E. Boncinelli, and F. Mavillio. 1990. Sequential activation of HOX2 homeobox genes by retinoic acid in human embryonal carcinoma cell. *Nature* **346:** 763–766.

Urdea, M. 1993. Synthesis and characterization of branched DNA (bDNA) for direct and quantitative detection of CMV, HBV, HCV and HIV. *Clin. Chem.* **39:** 725–736.

van Gemen, B., R. van Beuningen, A. Nabbe, D. van Strijp, S. Jurriaans, P. Lens, and T. Kievits. 1994. A one-tube quantitative HIV-1 RNA NASBA nucleic acid amplification assay using electrochemiluminescent (ecl) labelled probes. *J. Virol. Methods* **49:** 157–168.

Use of PCR to Quantitate Relative Differences in Gene Expression

William C. Gause and Jeffrey Adamovicz

Department of Microbiology and Immunology, Uniformed Services University of the Health Sciences, Bethesda, Maryland 20814

INTRODUCTION

Several techniques are currently available to measure changes in gene expression. These include the Northern blot, the RNase protection assay, in situ hybridization, and the reverse transcriptase polymerase chain reaction (RT-PCR). For many purposes, the Northern blot or the more sensitive RNase protection assay is sufficient for detecting quantitative differences between samples. However, if the sample quantity is low or the target message is rare such that these techniques are no longer practical, the more sensitive quantitative RT-PCR can be used. In cases where comparisons have been possible, results from the RT-PCR assay are quite comparable to results from Northern blot analysis (Murphy et al. 1990), slot blot analysis (Noonan et al. 1990), and in situ hybridization (Park and Mayo 1991). In the RT-PCR method, RNA is initially reverse-transcribed to cDNA and then the desired target cDNA species is amplified using specific primers. Fewer than 10 copies of target RNA are required for this procedure, and it has even been successful when the RNA was isolated from a single cell (Razin et al. 1991).

Because of this high sensitivity, RT-PCR is being used increasingly to quantitate small but physiologically relevant changes in gene expression that would otherwise be undetectable. For example, in a recent study involving the analysis of interleukin-4 (IL-4) gene expression after immunization with antigen, studies with blocking anti-IL-4 antibodies had previously demonstrated the importance of elevated IL-4 in the response, but neither Northern blot analysis nor in situ hybridization could detect corresponding elevations in IL-4 gene expression. A quantitative RT-PCR was developed to analyze IL-4 cytokine gene expression. The quantitative RT-PCR indeed showed that there was a 150-fold increase in IL-4 message in spleens from immunized mice (Svetić et al. 1991). Later studies using an ELISPOT assay, which measures protein secretion by individual cells, confirmed that the marked increase in IL-4 message was correlated with marked increases in IL-4 secretion (Morris et al. 1994). The advantage of the RT-PCR gene expression assay in this system is that tissue and cells taken directly from the animal can be measured. In contrast, protein assays usually require in vitro cell culture, in some cases with mitogens, that often results in the production of cytokines not originally produced by the cells in vivo.

PROTOCOL

RNA Isolation

This is a modification of the RNAzol B method from Tel-Test, Inc.

TISSUE PREPARATION

1. Determine the approximate weight of any tissue samples to be studied.

2. Distribute RNAzol to polypropylene tubes on the basis of 2 ml of RNAzol per 100 mg of tissue. The RNAzol should be kept cold and protected from light; *take care when handling the RNAzol because it is highly caustic.*

3. Place the tissue sample in RNAzol and homogenize it thoroughly.

4. Snap-freeze the sample in liquid nitrogen and store at –70°C for future extraction.

RNA EXTRACTION

1. Thaw frozen homogenate for approximately 5 minutes in a 37°C water bath.

2. Add 0.2 ml of a 24:1 mixture of chloroform/isoamyl alcohol for every 2 ml of homogenate and shake the samples vigorously for 15 seconds. Let the samples sit on ice for 5 minutes.

3. Centrifuge the samples at 12,000*g* for 15 minutes at 4°C.

4. Carefully remove the aqueous (top) phase that contains the RNA and transfer it to another tube; store on ice. Care must be taken to avoid the white interphase layer because this contains protein that will hamper the ability to quantify the RNA.

RNA PRECIPITATION

1. To each sample, add a volume of cold isopropanol that is equal to the volume of the aqueous phase.

2. Mix the tubes gently and store the samples on ice.

3. Centrifuge the samples at 12,000*g* for 15 minutes at 4°C. The RNA should form a whitish/yellow pellet at the bottom of the tube.

RNA WASHING

1. Carefully decant the isopropanol.

2. Wash the RNA pellet by adding one volume of cold 75% ethanol, and resuspend the RNA pellet by shaking or pipetting.

3. Centrifuge the sample at 12,000*g* for 8 minutes at 4°C.

4. Carefully decant the ethanol and dry the pellets by air or under vacuum. Vacuum drying is more expedient, but take care not to overdry the pellet.

RNA QUANTIFICATION

1. Solubilize the RNA pellets in 20–50 μl of distilled, deionized water. It is usually helpful to freeze and then thaw the samples to increase solubility.

2. Quantify the product spectrophotometrically by measuring the absorbance at 260 nm of an aliquot. The 260/280 ratio should be 1.8 or above. If the ratio is low, the sample should be reextracted.

Reverse Transcription

The reverse transcriptase reaction described below is based on a starting concentration of 3.6 μg of total RNA.

1. Prepare a master mix based on the number of samples you have with components in the following concentrations:

Deoxynucleotide triphosphates (Pharmacia) (2.5 mM *each* dNTPs)	2.5 μl
0.1 M Dithiothreitol (GIBCO/BRL)	2.0 μl
RNasin (40,000 units/ml) (Promega)	0.5 μl
Random primers (N)6 (20–40 units/ml) (Boehringer Mannheim)	2.0 μl
RNA (diluted with distilled/deionized H_2O)	11.8 μl
Volume	18.8 μl

2. Heat the above mixture for 5 minutes to 70°C and then quench on ice.

3. Centrifuge the samples briefly.

4. To each sample, add 6.2 μl of the following mix:

Reverse transcriptase buffer (GIBCO/BRL)	5.0 μl
Reverse transcriptase 200 units (GIBCO/BRL)	1.2 μl
Final volume	25.0 μl

5. Incubate the mixture at 37°C for 60 minutes followed by denaturing at 90°C for 5 minutes and quenching on ice for 5 minutes.

6. Store the samples at −70°C.

POLYMERASE CHAIN REACTION

The reaction described below is based on using 2.5 μl of reverse transcriptase reaction product (cDNA).

1. Prepare a master mix. Add components in the following concentrations, based on the number of samples.

Deoxynucleotide triphosphates (Pharmacia) (2.5 mM *each* dNTPs)	4.0 μl
Taq DNA polymerase buffer A	5.0 μl
$MgCl_2$ (25 mM)	3.0 μl
Taq DNA polymerase (5 units)	0.2 μl
Sense oligonucleotide primer (0.2 μg/μl)	2.0 μl
Antisense oligonucleotide primer (0.2 μg/μl)	2.0 μl
cDNA (from RT reaction)	2.5 μl
Distilled, deionized H_2O	31.3 μl
Final volume	50.0 μl

2. After mixing and spinning the PCR mixture down, overlay 50 μl of mineral oil.

3. Perform PCR cycling according to the following conditions:

   Denature samples for 5 minutes at 94°C.
   Cycles: denaturing for 45 seconds at 94°C
   annealing for 1 minute at 53°C
   extension for 2 minutes at 72°C
   Final extension for 7 minutes at 72°C.
   Soak at 4°C.

4. The samples can be stored for several months at 4°C.

Southern Blotting

This technique allows PCR products to be separated and quantitated. Briefly, a PCR product in solution is electrophoresed through an agarose gel that separates primers and nonspecific amplification products from the amplified cDNA of interest. Next, the product is denatured in an alkaline solution to remove secondary structure and then pH-neutralized and salt-saturated. The product is then transferred by capillary action to a solid support (nylon membrane). The PCR product is covalently linked to the nylon with UV light. The blot is then prehybridized to prevent nonspecific probe binding using blocking agents like Denhardt's and salmon sperm DNA. Following this, the blot is hybridized with radioactively end-labeled oligonucleotide probes specific for the product of interest. The blot is then washed to remove unbound probe, and the remaining signal is quantified. The specific steps are outlined below.

PREPARE AGAROSE GEL

For the purpose of transferring a PCR product to nylon membranes a 300-ml gel works best.

1. Prepare a 1% agarose TBE gel with ethidium bromide as follows.

2. Heat/dissolve 3 g of of agarose in 270 ml of distilled water.

3. Add 30 ml of 10X TBE and 30 μl (10 μg/μl) of ethidium bromide.

4. Cool to 50°C and pour.

PREPARE PCR SAMPLES

For our detection system, PhosphorImaging, we find that a 9-μl aliquot works best. Other detection methods that are not as sensitive may require a larger aliquot of the PCR product for transfer.

1. Carefully remove 9 μl of the PCR product and add to another tube, taking care to not transfer any of the mineral oil.

2. Add 1 μl of loading buffer to the samples and heat for 5 minutes to 65°C.

3. Quench the samples on ice and spin them down.

4. Load the gel. After loading the gel, add 1X TBE buffer and run the gel at 120 volts for 30–40 minutes.

BLOTTING

1. Denature the gel with a 1X solution of 1.5 M NaCl, 0.5 M NaOH, pH 13. Equilibrate the gel for 25 minutes in this solution, decant, and replace with fresh solution and repeat.

2. Rinse the gel twice with distilled water.

3. Neutralize the gel with 1.5 M NaCl, 1 M Tris-HCl, pH 7.5. Equilibrate the gel for 15 minutes in this solution, decant, and replace with fresh solution and repeat.

4. Rinse the gel twice with distilled water.

5. Saturate the gel in a 20X solution of SSPE (3 M NaCl, 0.2 M $NaH_2PO_4 \cdot H_2O$, 0.02 M EDTA, 0.213 M NaOH), pH 7.4. Equilibrate the gel for 30 minutes in this solution.

6. While the gel is soaking in 20X SSPE, prepare the following.

   a. Wet a Nytran (maximum strength plus-positively charged nylon membrane 0.2 μm) from Schleicher and Schuell in distilled water.

   b. Soak two pieces of thin blotting paper (Schleicher and Schuell, cat no. GB002) in 10X SSPE.

   c. Soak one piece of thick blotting paper (Schleicher and Schuell, cat no. GB004) in 5X SSPE.

   d. Set these items up to allow at least a 30-minute soak.

7. Remove the gel and invert it so that the bottom is facing up. Place the Nytran so that all the samples are covered, and exclude all air bubbles. Cut away and discard any excess agarose.

8. Place the GB002 blotting papers (presoaked in 10X SSPE) on top of the gel, followed by the GB004 paper (presoaked in 5X SSPE). Add 2–3 inches of dry GB004 blotting paper, and top with a metal or hard plastic tray followed by two 1-liter bottles of water.

9. The transfer is completed overnight but may sit for up to 48 hours.

PREHYBRIDIZATION

After the transfer to Nytran is completed, the DNA must be attached to the membrane. This is accomplished with UV cross-linking.

1. Place the blot in a UV cross-linker with the side that was in contact with the gel facing up. Expose it to 1200 Joules of UV energy.

2. Mark the blot with an indelible marker as to the approximate band position and orientation.

3. Add salmon sperm DNA (S.S.DNA) at a concentration of 50 μg/ml. Before being added to the prehybridization solution, the S.S.DNA must be denatured. This is done by heating the required amount of S.S.DNA (about 25–50 μg/ml of solution) for 5 minutes to 95°C and then quenching on ice.

4. Prewarm the prehybridization solution to 42°C.

5. Place the blot in a sealable food storage bag or other suitable container. Add the S.S.DNA to the prehybridization solution and then add the mixture to the blot and seal the bag or container. If a bag is used, special care must be taken to exclude air bubbles.

6. Place the bag or container in a 42°C water bath or oven and shake for 5 hours.

HYBRIDIZATION

This procedure is described for use with end-labeled oligonucleotide probes. Proper procedures for handling radioactive materials should be followed.

1. For probe preparation, prepare fresh 1X probe buffer:

1 M Tris-HCl, pH 7.6	5 μl
2 M $MgCl_2$	0.5 μl
0.5 M Dithiothreitol	1 μl
Distilled water	3.5 μl
Total	10.0 μl

2. Prepare the probe reaction mix:

Probe buffer	2.5 μl
Probe (0.2 μg/μl)	2.0 μl
T4 kinase (Pharmacia)(9700 units/ml)	1.0 μl
Distilled water	9.5 μl
[^{32}P]ATP (10 μCi/μl)	10.0 μl
Total	25.0 μl

3. Incubate for 40 minutes at 37°C.

4. After the incubation is complete, separate the nonincorporated label with a G-25 Sephadex spin column (5 Prime→3 Prime).

5. Determine the specific activity of the probe via liquid scintillation counting.

6. Store the hybridization solution at –20°C. A recipe for 100 ml is listed below:

20X SSPE	30 ml (final conc = 6X)
10% SDS	10 ml (final conc = 1X)
Distilled water	60 ml

7. Preheat the solution to 49°C.

8. Open the bag or container with the blot and the pour off the prehybridization solution.

9. Add 15 X 10^6 cpm of the probe to 10 ml of the hybridization solution, mix, and then add to the blot.

10. Reseal the bag or container and incubate at 49°C, while shaking, overnight.

Washing

LOW-STRINGENCY WASH

1. Remove the hybridization solution and discard properly.

2. Preheat the low-stringency wash described below to 49°C:

20X SSPE	180 ml (final conc = 6X)
10% SDS	6 ml (final conc = 0.1%)
Distilled water	414 ml
Total	600 ml

3. Wash the blot in enough wash solution to cover the blot thoroughly. Shake the blot while washing for 15 minutes at 49°C.

4. Remove the membrane and blot it dry with absorbent paper. At this point you may wish to check the blot with a portable radiation detection device. This will allow you to determine the effectiveness of the high-stringency wash.

HIGH-STRINGENCY WASH

1. Preheat the high-stringency wash described below to 49°C:

6X Wash solution	200 ml (final conc = 2X)
Distilled water	400 ml
Total	600 ml

2. Wash the blot as above, except the time required is only about 30 seconds. If additional washing is required, repeat the high-stringency step, or use a 1X wash. When the blot has been correctly washed, the background on blank areas of the membrane should be 200 cpm or less. Quantify the blot by a method of your choice.

Quantitation

After the blot is washed, the specific signal remaining must be quantitated. A commonly used method is exposure of the blot to film for a period of days or weeks followed by densitometry of the resulting signal. Although this method is sensitive, it is time-consuming. We have utilized a PhosphorImager (Molecular Dynamics) technique that is more sensitive and faster than conventional densitometry. This technique allows the capture of radioactive energy from the probe bound to the blot by exposure to a Phosphor storage screen. This signal is then transferred via the PhosphorImager and ImageQuant software to the computer screen, where the background signal and specific signal can be quantitated. The exposure time for the blot is dependent on the signal strength but usually is only 4–12 hours. The scanning and quantitation of the resulting signal are accomplished in minutes.

TROUBLESHOOTING

- *Could genomic DNA or other PCR products contaminate the reaction?* Contamination of PCR products can be problematic for interpretation of the results. Therefore, a negative control should be included during RT-PCR amplification and blotting. At the start of the RT reaction, a tube with no added RNA should be included. Instead of RNA, add an equivalent volume of water or RT buffer. This sample should be processed with the others and, following the RT reaction, an aliquot should be added to the PCR product. An aliquot from the PCR product should be Southern blotted as a control for any contaminating DNA or primers. Ideally, there should be no signal in the lane for the negative control. If there is competing genomic contamination represented as a second band, the PCR primers should be redesigned.

- *What if there are no bands on the Southern blot?* A positive control for the Southern blot should be included. RNA from a source known to express the gene of interest at levels comparable to the test samples is ideal. This control will allow determination of the specificity of the PCR. The control RNA sample should first be reverse-transcribed and then amplified to plateau level (30–40 cycles) and run on an agarose gel with ethidium bromide and size markers to determine if the expected size product is observed. Other contaminating bands may also be observed, but ideally there should only be one band. Next, the optimal number of PCR cycles should be determined as described previously (Svetić et al. 1991). The positive control for the Southern blot is then amplified at the empirically determined number of cycles simultaneously with the experimental group. The signal from the experimental group should match the positive control in relative position on the blot.

Lack of signal from the positive control and the other samples indicates a problem with the probe or the blotting procedure. If lanes other than the positive control are labeled, it probably indicates an error in the PCR for the positive control. If the positive control hybridizes very weakly or appears blotchy, it probably indicates a problem with denaturing, neutralizing, or saturating the gel; failure to transfer or to cross-link the blot properly may also be a problem.

If denaturing, neutralizing, or saturation is suspected to be the problem, the pH of stock solutions should be rechecked. If the transfer is suspected, the blot should be carefully repeated and the gel remnant examined under UV light for remaining traces of nucleic acid. The UV cross-linker should also be inspected to ensure that it is functioning properly. A final consideration is the stringency of the wash, which can be determined grossly with a hand-held radiation detector by monitoring the remaining signal through empirical experiments with either time or salt concentration as a variable. The absence of any signal from the experimental group in the presence of a strong signal from the positive control indicates a lack of expression of the gene of interest.

DISCUSSION

The Plateau Effect

The sensitivity of RT-PCR is a result of a chain reaction in which the products from one cycle of amplification serve as substrates for the next, resulting in an exponential increase in product. Theoretically, the amount of product doubles during each cycle of the PCR, but in actuality, beyond a certain number of cycles the efficiency of amplification decreases with increasing cycle number, resulting in the plateau effect as shown in Figure 1.

A number of factors may cause the plateau effect, including (1) the degradation of nucleotides or primers; (2) the inactivation of the DNA polymerase enzyme (*Taq* has a half-life of only 40 minutes at 95°C); (3) the reassociation of single-stranded PCR fragments before primers can anneal or be extended; (4) substrate excess, where there is more DNA than the amount of enzyme available to replicate it in the allotted polymerization time; (5) competition by nonspecific amplifi-

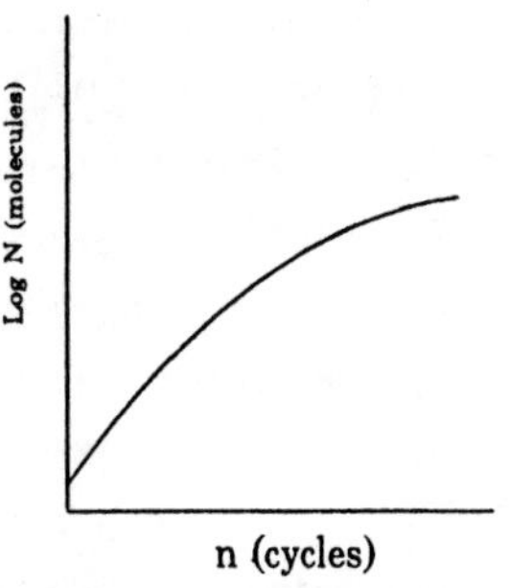

Figure 1 Relationship between product amplification and the number of cycles during PCR. A linear relationship occurs initially, followed by the plateau effect where increases in cycle number result in proportionally smaller increases in product. Although amplification efficiency (E) varies between target cDNAs even at early cycles, it inevitably decreases at higher cycle numbers.

cation products; and (6) the accumulation of inhibitors of polymerase activity, such as pyrophosphates.

The number of PCR cycles at which the plateau effect occurs varies greatly with the particular DNA sequence being amplified. Length, GC content, and the presence of any secondary structure in the sequence to be amplified are all important, as is the initial total quantity and concentration of the target DNA. As a result, the number of cycles at which the plateau effect occurs must be individually and empirically determined for each target sequence. This is particularly important in a quantitative PCR, and in the past, has led to considerable confusion. Amplifying various samples of target cDNA at a high number of cycles, e.g., 35, is useful for determining the presence or absence of a given target (Elhers and Smith 1991; Yamamura et al. 1991) but should not be used to compare differences in target quantities of less than 100-fold. This is because differences observed in the amount of detectable product after the plateau effect has been reached (an exception may be competitive PCR as we discuss later) are frequently artifacts and, upon further study, may show little relationship to the quantity of starting target material.

Noncompetitive RT-PCR

Noncompetitive RT-PCR relies on the observation, now well established, that prior to the onset of the plateau effect there is a linear relationship between the quantity of input RNA and final product during PCR amplification. The procedure we developed and used, as described in the protocol in this chapter, is widely used by a number of laboratories (Gendelman et al. 1990; Singer-Sam et al. 1990; Svetič et al. 1991, 1993; Wynn et al. 1993; Graziosi et al. 1994; Lu et al. 1994). To determine the number of cycles at which this linear relationship occurs, the initial sample of RNA should express high levels of target mRNA. Exogenous target should not be added to the RNA sample; instead, the sample should have endogenous target gene expression that is high, but no more than two to three times the highest levels that would be expected in an actual experiment. Five to ten serial dilutions of this sample, commonly 1:2, are made to span close to a 1000-fold concentration range. Complete sets of these dilution series are then amplified at one of several different cycle numbers. We have found with many cytokines that the cycle number at which a linear relationship is detected between input RNA and final product occurs between 18 and 25 cycles. Typically, the total RNA is diluted in buffer.

Although a linear relationship is detected under these circumstances (Svetić et al. 1991), the possibility exists that the RT-PCR technique is nonetheless nonquantitative, because decreases in the efficiency of priming may occur as mRNA specific for a given cytokine is diluted with mRNA specific for other proteins. We have directly tested this by diluting total RNA from tissue expressing high amounts of IL-4 mRNA in total RNA from tissue that expresses negligible amounts of IL-4 mRNA, so that the quantity and concentration of RNA remained the same but the amount of RNA from tissue that expressed high IL-4 mRNA levels decreased by 50% with each successive dilution. As shown in Figure 2, a linear relationship between input RNA and final RT-PCR product was maintained throughout the dilution range. This indicates that differences in the relative concentrations of target mRNA can be quantitated when total RNA concentration remains constant.

Often the low cycle number required to obtain the linear relationship shown in Figure 2 does not produce sufficient product for detection on UV transilluminators following ethidium bromide staining. To increase the sensitivity of detection, the product can be Southern-blotted and probed with a suitable end-labeled oligonucleotide that does not correspond to either of the original primers used for amplification, as was done in Figure 1. This approach adds an extra level of specificity to the procedure: If the primers cause any spurious nonspecific DNA amplification, often identified as unpredicted bands on an agarose gel, it will not be bound by the specific probe.

Using this procedure, one can obtain three levels of specificity: (1) amplification of the product with specific primers; (2) correspondence of the actual product size to the original estimated product size; and (3) hybridization of the product with an internal probe not corresponding to either primer. In addition, the low cycle number reduces artifacts such as nonspecific amplifications that could pose major obstacles at higher cycle numbers. Another technique for detecting the product involves the incorporation of labeled nucleotides into PCR products that are then resolved by gel electro-

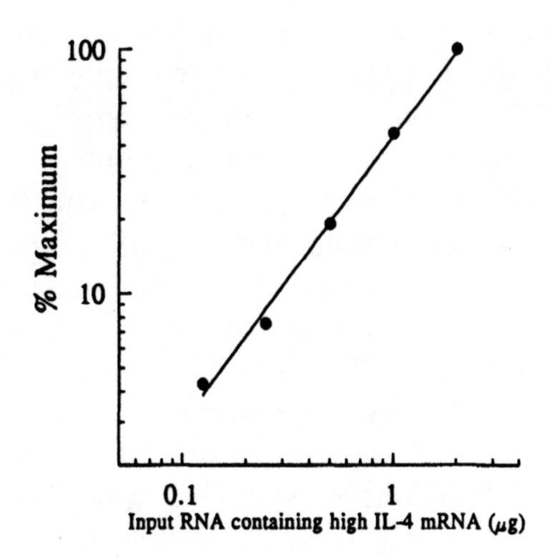

Figure 2 Densitometric analysis (*above*) and autoradiograph (*below*) of IL-4 RT-PCR signal as a function of the amount of input RNA.

phoresis. However, this approach is often associated with trace amounts of unincorporated label that can produce a "trail" of label throughout the lane of an electrophoretic gel. One can also use labeled primers at the beginning of the assay or as a final PCR step with a new internal labeled primer annealed to one strand of the PCR product and extended using *Taq* DNA polymerase, but both approaches often result in considerable background radioactivity and rarely give signals as well-defined and quantifiable as does the Southern blot (Ferre 1992; W.C. Gause, unpubl.).

Currently, an increasing number of investigators are using radioimaging and direct quantitation systems for the measurement of radiolabeled product. Autoradiography with densitometric quantitation is more cumbersome, particularly because it has a dynamic range of only 2–3 orders of magnitude, whereas systems such as AMBIS or phosphorimagers have ranges of 5 orders of magnitude or greater. Recently, nonradioactively labeled probes have been used successfully for Southern blot analysis of PCR products and subsequent quantitation of changes in cytokine gene expression with a video densitometer (Wynn et al. 1993). An ELISA-based quantitative RT-PCR assay has also been developed that uses an ELISA reader to measure amplified products (Alard et al. 1993). For safety reasons, these approaches using nonradioactive materials will probably become increasingly popular.

It is important to include an internal standard in the quantitative RT-PCR assay. DNA standards have been used for the quantification of RNA targets, but in these instances, variation in RT efficiency, an important source of variability, is not taken into account. Two types of internal standards are commonly used: an exogenous fragment added to the amplification reaction or an endogenous sequence or gene transcript that is normally present in the sample.

If an exogenous fragment is used, it can be an mRNA standard added to the target sample and amplified simultaneously with the endogenous target in a single PCR. The purpose of the exogenous internal standard is to allow the detection of differences in the amplification efficiency (E in Fig. 1) between reaction tubes within an assay or between assays. A known amount of internal standard is added to each sample and, after amplification, is quantitated. If differences in the amount of standard product are detected between reaction tubes, then the product derived from the target sequence is normalized to the standard in its respective tube. Although E has been considered a serious problem in the past (Gilliland et al. 1990b; Kellogg et al. 1990), it was partly a result of inferior thermal cyclers that exhibited temperature cycling variations from one well to another. With the advent of better machines, this has become less of a problem. For several currently available thermal cycling machines, the amount of variation from well to well has been reported by the manufacturers to be less than 10%. We have confirmed this in our laboratory by simultaneously amplifying target mRNA in all the wells of a thermal cycler at the number of cycles where the relationship between input RNA and final signal was linear (W.C. Gause, unpubl.). However, addition of an exogenous sequence may not be necessary where samples are compared within the same assay, but it is still useful for comparing samples in different experiments. The exogenous standard can also be used in noncompetitive assays to quantitate the absolute level of target or cDNA in the original sample, as was done by Wang et al. (1989). Although apparently successful, absolute quantitation of the PCR product relies on the assumption that the value of E is the same for both the target and standard mRNAs. This has to be determined empirically, because even slight differences in PCR product or primer sequence may affect the amplification efficiency. In many cases, however, quantitation of relative differences between samples in an experiment is most important and the inclusion of external standards to attempt to measure absolute levels is unnecessary.

An important alternative to the exogenous internal standard is the use of an endogenous sequence as the internal standard. Typically, a "housekeeping" gene is used, such as β-actin, HPRT, or GAPDH (Gendelman et al. 1990; Svetić et al. 1991; Peterson et al. 1994). The endogenous standard permits the detection of relative differences in the integrity of individual RNA samples. Even if the same quantity of RNA is used for each preparation, the final quantity of product may be greatly affected by differences in RNA integrity and the presence of inhibitors of reverse transcriptase. This is a particular problem, because the degree of RNA degradation can vary significantly between samples within a given experiment and, as mentioned earlier, RNA degradation is not detectable spectrophotometrically.

There are essentially two approaches for detecting the variation in RNA degradation between samples. One approach involves electrophoresing 5–10 μg of total RNA on an agarose gel and using ethidium bromide staining to determine if the ribosomal bands are intact. This procedure requires a large quantity of total RNA and, as a result, is often not practical. The second approach involves amplification of "housekeeping mRNA" from the same RNA preparation used for target amplification. Coamplification of the target and the housekeeping mRNA sequences is done in the same tube or in different tubes. Frequently, coamplification in the same tube is not practical because the housekeeping gene is expressed at such high levels in the RNA mixture that its amplification inhibits the amplification of the target mRNA (Becker-Andre and Hahlbrock 1989; Murphy et al. 1990; W.C. Gause, unpubl.). In these cases, it is usually necessary to amplify in separate tubes, although one can also wait until later stages of the amplification before adding the primers that amplify the endogenous standard primers (Kinoshita et al. 1992). Either way, with endogenous standards, variations in initial target RNA resulting from RNA degradation can be controlled by simply normalizing the final target product quantity to the amount of housekeeping gene product. For example, if two target PCR products A and B are quantitated at values of 50 and 100, respectively, and the corresponding housekeeping gene products are at 25 and 50, the final normalized values of A and B are both 2, demonstrating no difference between A and B regarding expression of the particular target gene measured.

An exogenous internal standard cannot be used to detect differences in mRNA integrity. Although useful for measuring differences in E during PCR, it is not a substitute for an endogenous standard. Endogenous standards also correct for differences in RNA purity resulting from DNA or protein contamination. Such contamination may result in misleading spectrophotometric readings and, consequently, incorrect concentration determinations.

In summary, the quantitative RT-PCR technique requires the use of an endogenous standard. Other endogenous standards besides typical housekeeping genes are also used. For example, if measurements are made of target gene expression from a particular cell subpopulation, a gene constitutively expressed by that subpopulation may be used. For example, T-cell receptor (TCR) Cα or Cβ has been used as an endogenous internal standard in the quantitation of changes in TCR Vβ expression (Choi et al. 1989; Paliard et al. 1991; Reiner et al. 1993).

Competitive RT-PCR

Competitive RT-PCR was developed to quantitate absolute values of target RNA. The procedure relies on the use of an external standard that "mimics" or closely imitates the target RNA species with respect to primer binding and other variables affecting PCR amplification. This is an important difference from the exogenous internal standard used in the noncompetitive assay, where the standard primer set can amplify a completely different target than the experimental target mRNA. In the competitive assay, the standard is similar enough to the target that competition with target occurs and, ideally, the experimental and standard target sequences amplify with the same efficiency (E in Fig. 1) but can be distinguished from each other following agarose gel electrophoresis. These standard fragments fall into two categories: homologous and heterologous competitor fragments (Becker-Andre and Hahlbrock 1989; Gilliland et al. 1990b; Uberla et al. 1991; Siebert and Larrick 1993). Homologous fragments differ only slightly from the target sequence with the addition of a unique restriction site or the presence of an additional sequence such as an intron that increases the molecular weight of the standard. Recently, several rapid PCR-based techniques have been used to develop homologous standards. One potential problem with this sort of standard is the formation of heteroduplexes between the standard and target sequences during the amplification process—an artifact that could interfere with quantitation. In contrast, heterologous competitor fragments differ from the target except for the flanking primer-template regions, which are identical. Thus, heteroduplex formation cannot occur and slight differences in the target and primer sequence can be easily created.

The competitive RT-PCR assay is usually performed by titrating a known quantity of the standard mRNA target against a constant amount of the experimental mRNA target. The concentration at which the product from the standard target equals the product produced from the experimental target is taken to be the starting concentration of the experimental mRNA. Given the differences between PCR mimics and the experimental target, particularly with heterologous standards, it is essential to determine empirically whether the target and mimic sequences amplify with similar efficiencies. To do this, one can plot the log of the product against the cycle number for the experimental and the standard target. The similarity of the slopes of the linear portion of the resulting two curves is indicative of the similarity of the amplification efficiency. Successful PCR mimics for a large number of genes have been developed. Many of these are commercially available from companies, including CLONTECH.

It has been suggested that the nature of competitive PCR makes it possible to obtain useful data after the reaction has reached the plateau phase. This can be a considerable advantage, because quantitation can be performed by simple agarose gel electrophoresis of ethidium bromide-stained PCR product. However, other workers caution that even the competitive PCR technique may not work well when the product is measured well after the plateau phase, particularly when the sequences of the target and standard molecules are different except for the primer sequences; i.e., heterologous (Pannetier et al. 1993).

A major source of variability usually not controlled for in competitive RT-PCR assays involves RNA purity and RNA integrity. In assays where DNA standards instead of RNA standards are used (Gilliland et al. 1990a; Li et al. 1991; Uberla et al. 1991), there is also no control for reverse transcriptase yield and uniformity. Because the nature of the RNA preparation is a major source of variation in any assay of gene expression, the reliability of this approach without additional endogenous internal standards is suspect. However, as mentioned earlier, because most investigators are interested in quantitating relative differences between control and treatment groups, the necessity for exogenous mimic RNA standards to attempt to determine absolute levels may be limited.

In summary, RT-PCR is a powerful technique for the quantitation of changes in gene expression. Several different approaches are widely used, including noncompetitive and competitive assays. With both assays, it is paramount to include endogenous standards, and if one plans to compare samples between assays or if the amplification efficiency varies significantly from well to well of the thermal cycler being used, it is advisable to also include external RNA standards. One should also standardize each of the procedures involved in the RT-PCR technique as much as possible, optimizing each step including (1) RNA isolation, (2) the reverse transcriptase reaction, (3) PCR amplification, (4) Southern blotting and hybridization, and (5) the quantitation of product.

The technique we have described in the Protocol section is used for quantitating differences in cytokine gene expression. As different laboratories become experienced with the quantitative RT-PCR technique, we expect it to be used increasingly in gene expression studies directed toward understanding immunity, development, differentiation, transformation, and tumorigenesis.

REFERENCES

Alard, P., O. Lantz, M. Sebagh, C.F. Calvo, D. Weill, G. Chavanel, A. Senik, and B. Charpentier. 1993. A versatile ELISA-PCR assay for mRNA quantitation from a few cells. *BioTechniques* **15:** 730–737.

Becker-Andre, M. and K. Hahlbrock. 1989. Absolute mRNA quantification using the polymerase chain reaction (PCR). A novel approach by a PCR aided transcript titration assay (PATTY). *Nucleic Acids Res.* **17:** 9437–9446.

Choi, Y.W., B. Kotzin, L. Herron, J. Callahan, P. Marrack, and J. Kappler. 1989. Interaction of *Staphylococcus aureus* toxin "superantigens" with human T cells. *Proc. Natl. Acad. Sci.* **86:** 8941–8945.

Elhers, S. and K.A. Smith. 1991. Differentiation of T cell lymphokine gene expression: The in vitro acquisition of T cell memory. *J. Exp. Med.* **173:** 25–36.

Ferre, F. 1992. Quantitative or semi-quantitative PCR: Reality versus myth. *PCR Methods Appl.* **2:** 1–9.

Gendelman, H.E., R.M. Friedman, S. Joe, L.M. Baca, J.A. Turpin, G. Dveksler, M.S. Meltzer, and C.W. Dieffenbach. 1990. A selective defect of interferon alpha production in human immunodeficiency virus-infected monocytes. *J. Exp. Med.* **172:** 1433–1442.

Gilliland, G., S. Perrin, and H.F. Bunn. 1990a. Competitive PCR for quantitation of mRNA. In *PCR protocols: A guide to methods and applications* (ed. M.A. Innis et al.), pp. 60–60. Academic Press, San Diego, California.

Gilliland, G., S. Perrin, K. Blanchard, and H.F. Bunn. 1990b. Analysis of cytokine mRNA and DNA: Detection and quantitation by competitive polymerase chain reaction. *Proc. Natl. Acad. Sci.* **87:** 2725–2729.

Graziosi, C., G. Pantaleo, K.R. Gantt, J.P. Fortin, J.F. Demarest, O.J. Cohen, R.P. Sekaly, and A.S. Fauci. 1994. Lack of evidence for the dichotomy of Th1 and Th2 predominance in HIV-infected individuals. *Science* **265:** 248–252.

Kellogg, D.E., J.J. Sninsky, and S. Kwok. 1990. Quantitation of HIV-1 proviral DNA relative to cellular DNA by the polymerase chain reaction. *Anal. Biochem.* **189:** 202–208.

Kinoshita, T., J. Imamura, H. Nagai, and K. Shimotohno. 1992. Quantification of gene expression over a wide range by the polymerase chain reaction. *Anal. Biochem.* **206:** 231–235.

Li, B., P.K. Swhajpal, A. Khanna, H. Vlassara, A. Cerami, K.H. Tenzel, and M. Suthanthiran. 1991. Differential regulation of transforming growth factor beta and interleukin 2 genes in human T cells: Demonstration by usage of novel competitor DNA constructs in the quantitative polymerase chain reaction. *J. Exp. Med.* **174:** 1259–1262.

Lu, P., X. Zhou, S.J. Chen, M. Moorman, S.C. Morris, F.D. Finkelman, P. Linsley, J.F. Urban, and W.C. Gause. 1994. CTLA-4 ligands are required to an in vivo interleukin 4 response to a gastrointestinal nematode parasite. *J. Exp. Med.* **180:** 693–698.

Morris, S.C., K.B. Madden, J.J. Adamovicz, W.C. Gause, B.R. Hubbard, M.K. Gately, and F.D. Finkelman. 1994. Effects of IL-12 on in vivo cytokine gene expression and Ig isotype selection. *J. Immunol.* **152:** 1047–1056.

Murphy, L.D., C.E. Herzog, J.B. Rudick, A.T. Fojo, and S.E. Bates. 1990. Use of the polymerase chain reaction in the quantitation of mdr-1 gene expression. *Biochemistry* **29:** 10351–10356.

Noonan, K.E., C. Beck, T.A. Holzmayer, J.E. Chin, J.S. Wunder, I.L. Andrulis, A.F. Gazdar, C.L. Willman, B. Griffith, D.D. Von Hoff, and I.B. Roninson. 1990. Quantitative analysis of MDR1 (multidrug resistance) gene expression in human tumors by polymerase chain reaction. *Proc. Natl. Acad. Sci.* **87:** 7160–7164.

Paliard, X., S.G. West, J.A. Lafferty, J.R. Clements, J.W. Kappler, P. Marrack, and B.L. Kotzin. 1991. Evidence for the effects of a superantigen in rheumatoid arthritis. *Science* **253:** 325–329.

Pannetier, C., S. Delassus, S. Darche, C. Sancier, and P. Kourilsky. 1993. Construction of recombinant RNA templates for use as internal standards in quantitative RT-PCR. *BioTechniques* **14:** 70–80.

Park, O. and K.E. Mayo. 1991. Transient expression of progesterone receptor messenger RNA in ovarian granulosa cells after the preovulatory luteinizing hormone surge. *Mol. Endocrinol.* **5:** 967–978.

Peterson, V.M., J.J. Adamovicz, T.B. Elliott, M.M. Moore, G.S. Madonna, W.E. Jackson, D. Ledney, and W.C. Gause. 1994. Gene expression of hematoregulatory cytokines is elevated endogenously after sublethal gamma irradiation and is differentially enhanced by therapeutic administration of biologic response modifiers. *J. Immunol.* **153:** 2320–2330.

Razin, E., K.B. Leslie, and J.W. Schrader. 1991. Connective tissue mast cell in contact with fibroblasts express IL-3 mRNA: Analysis of single cells by polymerase chain reaction. *J. Immunol.* **146:** 981–987.

Reiner, S.L., Z.E. Wang, F. Hatam, P. Scott, and R.M. Locksley. 1993. Th1 and Th2 cell antigen receptors in experimental leishmaniasis. *Science* **259:** 1457–1459.

Siebert, P.D. and J.W. Larrick. 1993. PCR MIMICs: Competitive DNA fragments for use as internal standards in quantitative PCR. *BioTechniques* **14:** 244–249.

Singer-Sam, J., M.O. Robinson, A.R. Bellve, M.I. Simon, and A.D. Riggs. 1990. Measurement by quantitative PCR of changes in HPRT, PGK-1, PGK-2, APRT, MTase, and Zfy gene transcripts during mouse spermatogenesis. *Nucleic Acids Res.* **18:** 1255–1259.

Svetić, A., F.D. Finkelman, Y.C. Jian, C.W. Dieffenbach, D.E. Scott, K.F. McCarthy, A.D. Steinberg, and W.C. Gause. 1991. Cytokine gene expression after in vivo primary immunization with goat antibody to mouse IgD antibody. *J. Immunol.* **147:** 2391–2397.

Svetić, A., K.B. Madden, X.D. Zhou, P. Lu, I.M. Katona, F.D. Finkelman, J.F. Urban, and W.C. Gause. 1993. A primary intestinal helminthic infection rapidly induces a gut-associated elevation of Th2-associated cytokines and IL-3. *J. Immunol.* **150:** 3434–3441.

Uberla, K., C. Platzer, T. Diamantstein, and T. Blankenstein. 1991. Generation of competitor DNA fragments for quantitative PCR. *PCR Methods Appl.* **1:** 136–139.

Wang, M., M.V. Doyle, and D.F. Mark. 1989. Quantitation of mRNA by the polymerase chain reaction. *Proc. Natl. Acad. Sci.* **86:** 9717–9721.

Wynn, T.A., I. Eltoum, A.W. Cheever, F.A. Lewis, W.C. Gause, and A. Sher. 1993. Analysis of cytokine gene mRNA expression during primary granuloma formation induced by eggs of *Schistosoma mansoni. J. Immunol.* **151:** 1430–1440.

Yamamura, M., K. Uyemura, R.J. Deans, K. Weinberg, T.H. Rea, B.R. Bloom, and R.L. Modlin. 1991. Defining protective responses to pathogens: Cytokine profiles in leprosy lesions. *Science* **254:** 277–279.

Quantitative Liquid Hybridization PCR Method Employing Storage Phosphor Technology

Maryanne T. Vahey[1] and Michael T. Wong[2]

[1]Division of Retrovirology, Walter Reed Army Institute of Research, Rockville, Maryland 20850

[2]Department of Infectious Diseases, Wilford Hall Medical Center, Lackland Air Force Base, San Antonio, Texas 78236

INTRODUCTION

The accurate and sensitive determination of viral burden is fundamental to the characterization of human immunodeficiency virus type-1 (HIV-1) disease progression and the assessment of interventions to prevent and control the disease (Michael et al. 1992; Piatek et al. 1993). The challenges to developing a quantitative assay for viral burden in HIV-infected persons are many-faceted.

The early stages of HIV infection are characterized by low levels of proviral DNA in circulating peripheral blood mononuclear cells (PBMCs). Thus, the assay must be capable of detecting as few as one copy of the viral genome in a background of 100,000 normal cells (Schnittman et al. 1989, 1991; Simmonds et al. 1990). In contrast, the course of HIV infection exhibits very high levels of viral burden in the plasma and serum compartments both during the initial seroconversion and in the final stages of AIDS, necessitating the ability for an assay to detect accurately in the range of 1×10^9 copies of viral RNA in 1.0 ml of acellular fluid (Coombs et al. 1989; Daar et al. 1991; Clark et al. 1992). Thus, the ideal assay requires a linear working range of seven to eight logs.

In any clinical application, conservation of clinical material by an assay procedure is of paramount importance. Efficient and representative extraction procedures must be designed to work with a minimum of cells and/or acellular fluid. Because patients may be followed over extensive periods of time, the need for an inexhaustible, unvarying gold standard for the assay is essential. Furthermore, because HIV exhibits a natural heterogeneity and patients are infected with viral swarms (Genesca et al. 1990), the assay must address the requirement for specificity and, ideally, for universal application. Finally, the requirement to assess the viral load in large numbers of clinical specimens from clinical trials and surveillance surveys makes it essential that data collection and manipulation be automated and that the turnaround time for results be as brief as possible.

In this chapter, we present a quantitative PCR assay with the essential features of sensitivity, accuracy, reproducibility, specificity, linear range, conservation of clinical materials, high throughput, and short turnaround time. These features make this assay amenable to the assessment of viral load in HIV-infected specimens. The method employs liquid hybridization (LH) in ^{32}P-labeled oligonucleotide probes that are specific to internal regions of the amplified product. The assay is standardized using the external coamplification of cloned template, which consists of cognate regions of the HIV genome cloned into a plasmid vector with a capacity to generate cRNA.

The size of the amplified product in the control template is identical to that expected in the clinical specimen. The isotopic signal is detected by the use of storage phosphor technology as marketed by the Molecular Dynamics Corporation. This technique is a patented process that, when compared to conventional exposures of X-ray film, has the advantages of dramatically reduced exposure times of minutes and not days, a linear sensitivity range that is 100 times greater than film, and a dynamic range that is 400 times greater than film (Johnson et al. 1990). In addition, all data generated, including the gel image and band intensity, are archived on a disc storage system in an Excel environment.

The assay system is presented as a series of protocols. The first two provide detailed procedures for the extraction of RNA from serum, the preparation of the PCR template, and the execution of the PCR assays. The controls included with these reactions are an RT-negative for each clinical specimen, a reagent blank, and the standard curve. This assay is based on the interpolation of a standard curve and the signal generated from an RT-PCR using viral RNA prepared from 250–500 μl of serum. In the third protocol, the techniques of product detection are presented. The details of assay standardization and performance are addressed in the final section.

REAGENTS

Tri-reagent (Molecular Research Center, cat no. TR.118.200)
4 M Sodium acetate solution (Sigma, cat. no. S2889)
1x PBS (GIBCO/BRL, cat. no. 630-5250PE)
Isopropanol (Fisher, cat. no. A405-7)
Chloroform (Sigma, cat. no. C 6038)
DTT (Boehringer Mannheim, cat. no. 100-032)
Glycogen (Boehringer Mannheim, cat. no. 901393)

RNA suspension buffer:
  25 μl RNasin 40 units/μl (Promega, cat. no. N2514)
  10 μl 0.1 M DTT
  965 μl sterile water
Carrier tRNA, 7.5 kb poly(A) (GIBCO/BRL, cat. no. 15621-014)
Bovine serum albumin (BSA-Sigma, cat. no. B2518).

BL2+ laboratory procedures are to be followed in parts of this protocol. A suitable sample preparation hood must be used while handling the samples, and the centrifugation should be done within the hood as well.

PROTOCOL

Preparation of HIV RNA from Serum

This protocol describes the methods for the pelleting of HIV virions from serum and the subsequent preparation of viral RNA.

1. Use 1.0 ml of human serum collected and stored without heparin as the starting material for each sample in this protocol. Remove the serum vial(s) from the freezer and thaw by manual agitation in a 37°C water bath. Once the vials are thawed completely, place on ice.

2. To the 1.0-ml volume of serum, add 1.0 ml of 1X PBS containing a final concentration of 5 μg/ml of BSA.

3. Centrifuge the mixture to pellet the virus in a microcentrifuge at 12,000*g* for 30 minutes at 4°C in a laminar-flow containment hood.

4. Remove the supernatant from the tube using a micropipettor and a disposable tip, taking care not to disturb the pellet of virus.

5. Add 800 μl of Tri-reagent to the pellet and mix well. Incubate for 5 minutes at room temperature. During the 5-minute incubation, the samples may be removed from BL2+ and taken to a BL2 laboratory after washing the outside of the tubes with 70% ethanol.

6. Add 160 μl of chloroform to the tube, mix well, and incubate for 3 minutes at room temperature.

7. Centrifuge the mixture at 12,000*g* for 15 minutes at 4°C.

8. Remove the entire volume of the upper phase (volume may vary) to a fresh tube.

9. Add 160 μl of chloroform and repeat steps 7 and 8.

10. Add 2.5 µl of 1 ng/µl carrier tRNA and 5.0 µl of 2 µg/ml glycogen. Vortex for 15 seconds. Add 400 µl of isopropanol, vortex, and hold for 30 minutes at –20°C.

11. Centrifuge at 12,000*g* for 15 minutes at 4°C.

12. Remove the supernatant, then wash the pellet with 1 ml of ice-cold 75% ethanol. Centrifuge at 12,000*g* for 15 minutes at 4°C.

13. Decant the supernatant and remove the excess liquid using a micropipettor and disposable tip, taking care to avoid the pellet. Air-dry the samples for 15 minutes.

14. Resuspend the pellet in 50 µl of RNA suspension buffer and store at –70°C until needed.

REAGENTS

Antisense GAG primer, 2.5 µM: CAT CCA TCC TAT TTG TTC CTG AAG G
Mineral oil (Perkin-Elmer, cat. no. 186-2302)
dNTPs, 25 mM (Promega, cat. no. U1240)
GAG primers, 20 µM:
GAG sense: CAA TGA GGA AGC TGC AGA ATG GGA TAG
GAG antisense: CAT CCA TCC TAT TTG TTC CTG AAG G
GAG RNA plasmid external control dilution series (100,000; 50,000; 10,000; 5,000; 1,000; 500; 100; 10; 5 copies)
RT (reverse transcriptase enzyme), 200 units/µl (GIBCO/BRL, Superscript, cat. no. 8053A)
5x First-strand buffer (GIBCO/BRL, cat. no. 8053A)
0.1 M DTT (GIBCO/BRL, cat. no. 8053A)
Taq DNA polymerase (Perkin-Elmer, cat. no. N801-0060)
10x PCR buffer (Perkin-Elmer, cat. no. N808-0006)

All reverse transcriptase reactions are to be assembled in a clean pre-PCR area, such as a laminar-flow hood.

PROTOCOL

RNA-PCR for GAG Product

This protocol is in two parts. Part one is used to prepare HIV GAG cDNA template from the RNA that has been extracted from the serum. Part two details the PCR amplification of the GAG cDNA.

1. Assemble and label the necessary 0.5-ml tubes for the assay, including contamination controls and the GAG RNA standard curve. To the assay blanks, add 10 µl of H_2O, and to the GAG standard curve, add 2 µl of H_2O. Tubes for samples are in pairs labeled as RT^- and RT^+.

2. To the tubes appropriately labeled as RNA clinical specimens (RT^+/RT^-), add 10 µl of the corresponding RNA clinical specimens.

3. Aliquot 8 µl of GAG RNA (see below) standard dilutions to the appropriate corresponding tubes, working from the low to the high dilutions. New tips should be used with each addition of template.

4. Add 4.0 µl of GAG antisense primer at 2.5 µM to each tube, changing tips after each addition.

5. Add 50 µl of mineral oil to all samples using a micropipettor and a disposable tip, again changing tips between each sample.

6. Vortex briefly, and then centrifuge the tubes at 12,000*g* for 5 seconds. Place the tubes into the preheated thermal cycler set for 10 minutes at 70°C. Then switch the temperature to 4°C and incubate for 2–3 minutes.

7. Make separate master mixes for reverse transcriptase reactions (RT^+) and control reactions without the enzyme (RT^- mock) by determining the number of reactions and adding two. The RT^+ tubes will include an assay blank, clinical specimens, and GAG RNA dilutions. The RT^- tubes will include an assay blank and clinical specimens only. Multiply these numbers by the respective volumes below:

For RT^+ reactions add:		For RT^- reactions add:	
5X First-strand buffer	4.0 µl	5X First-strand buffer	4.0 µl
0.1 M DTT	2.0 µl	0.1 M DTT	2.0 µl
RNasin (40 units/µl)	0.5 µl	RNasin (40 units/µl)	0.5 µl
dNTP mix (25 mM)	0.4 µl	dNTP mix (25 mM)	0.4 µl
RT enzyme (200 units/µl)	1.0 µl	H_2O	1.0 µl

Note: When there are a large number of samples to be run, it is sometimes necessary to run the RT^- reactions as a separate experiment. When running the RT^- reactions separately from the RT^+, it is not necessary to run the RT^- set with the GAG RNA dilution series.

8. Add 8 µl of the appropriate master mix to each tube.

9. Vortex the tubes briefly and centrifuge the tubes to separate the aqueous and oil phases.

10. Place the tubes in the thermal cyler for 15 minutes at 45°C, followed by 10 minutes at 95°C to heat-kill the RT enzyme and denature the cDNA from the RNA template. Soak for approximately 2–3 minutes at 4°C. The samples are now ready; proceed to the PCR amplification for GAG cDNA.

PROTOCOL

PCR Amplification of GAG cDNA

1. Determine the number of reaction tubes to be run and add two. (The number of tubes to be run includes all of the tubes from the RT assay.) Make a master mix by multiplying this number by the following volumes:

Component	Volume (μl/rxn)	Final Concentration
PCR-approved water	56.7	
10X Buffer	10.0	$[Mg^{++}] = 1.64$ mM
GAG sense, 20 μM	5.0	1.0 μM
CAA TGA GGA AGC TGC AGA ATG GGA TAG		
GAG antisense, 20 μM	5.0	1.0 μM
CAT CCA TCC TAT TTG TTC CTG AAG G		
dNTPs, 25 mM each	0.8	200 μM each
Taq DNA polymerase, 5 units/μl	0.5 μl	2.5 units/tube
	78 μl/reaction + template = 100 μl	

2. Starting from the assay blank tube, continuing with the clinical specimen tubes, and finally with the GAG RNA standard curve tubes, aliquot 78 μl of PCR master mix to each of the RT reaction tubes (22 μl) to bring the final volume of each reaction to 100 μl.

3. Vortex each tube briefly and centrifuge to separate the phases. Place tubes into the preheated thermal cycler at 95°C and begin cycling:

   Denature 95°C for 30 seconds
   Anneal 55°C for 30 seconds
   Extend 72°C for 3 minutes
   28 cycles

   Time delay: 10 minutes at 72°C

   Soak: 4°C (this can be left overnight)

4. Store samples at 4°C until use in liquid hybridization.

PCR Product Detection

This section describes the procedures for the preparation, use, and detection of a ^{32}P-labeled oligonucleotide probe.

REAGENTS

Polynucleotide kinase, T4 (Boehringer Mannheim, cat. no. 174645)
20 μM Oligonucleotide DNA probe:
GAG probe: ATG AGA GAA CCA AGG GGA AGT GAC ATA GCA

10× Kinase buffer
- 50 mM Tris-HCl, pH 7.6
- 10 mM $MgCl_2$
- 5 mM DTT
- 0.1 mM EDTA, pH 8.0

$TE_{30}N_{300}$ buffer
- 300 mM NaCl
- 10 mM Tris-HCl, pH 8.0
- 30 mM EDTA, pH 8.0

[γ-^{32}P]ATP (Amersham, cat. no. PB10168, 3000 Ci/mmole)

Select-D, G25 Sephadex columns (5 Prime→3 Prime, cat. no. 5301-233431)

PROTOCOL

Isotopic Labeling of Oligonucleotide Probes

1. To the appropriately labeled 1.5-ml microfuge tube, add the following reagents:

H_2O	11.5 μl
Oligonucleotide, 20 μM	0.5 μl
10× Kinase buffer	2.0 μl
Polynucleotide kinase	1.0 μl (8 units)
[γ-^{32}P]ATP	5.0 μl (add last)

2. Place the tube in water bath and incubate for 30 minutes at 37°C.

3. Purify the labeled oligonucleotide using a Sephadex G25 spun column.

4. Using a scintillation counter, determine the cpm per microliter incorporated into the purified probe.

5. Store the sample at –20°C in a Plexiglas container.

Liquid Hybridization of PCR Products with Isotopically Labeled Oligonucleotide Probes and Minigel Analysis of Liquid-hybridized PCR Products

These steps describe the procedures for the specific hybridization of PCR products by radiolabeled oligonucleotide probes complementary to the internal sequence of the PCR product. This is then followed by a method for the rapid separation, using acrylamide minigels, of hybridized PCR products from unincorporated reactants, and the preparation of the gels for subsequent phosphorimage analysis for the quantitation of the incorporated label in the PCR-specific band.

REAGENTS

1.1X OH buffer
667 µl 1 M NaCl
888 µl 0.5 M EDTA, pH 8.0
8.445 ml dH_2O

5X DNA loading buffer
5.0 ml 10X TBE
0.5 ml 1% w/v bromphenol blue
3.0 ml 100% glycerol
1.5 ml sterile distilled water

1X TBE (1:10 dilution of 10X TBE)
for 1 liter of 10X TBE:
108 g Tris base
55 g boric acid
40 ml 0.5 M EDTA, pH 8.0

40% Acrylamide/bis-acrylamide, 19:1 (Sigma, cat. no. A2917)

Multiflex gel loading tips, round (VWR Scientific, cat no. 53503-178)

TEMED (*N,N,N′,N′*-tetramethylenediamine) (GIBCO/BRL, cat. no. 5524UB

10% APS (ammonium persulfate, Bio-Rad Laboratories, cat. no. 161-0200)

PROTOCOL

1. Remove 30 µl of the PCR product and place it into an appropriately labeled 0.5-ml microfuge tube. Each tube will need 1 µl of probe, diluted to 5.0×10^4 cpm in H_2O.

2. Prepare the probe master mix so that each detection tube receives 5×10^4 cpm of probe in 1X–1.1X OH buffer.

3. Aliquot 10 µl of probe master mix into each labeled specimen tube. Mix and centrifuge to recover all of the volume.

4. Place the samples into a heat block and incubate:

   for 5 minutes at 94°C
   for 10 minutes at 55°C

   Store at 4°C until ready to load on gel.

5. Add 4 µl of 5X DNA loading buffer. Vortex and centrifuge at 12,000*g* for 5 seconds.

6. To the assembled minigels (see Table 1), load 10 ml of the sample to duplicate wells.

7. Run the gels for 30 minutes at 200 volts. The dye front will be near the bottom of the gel. Then process the gels for analysis.

Table 1 Minigel Assembly Method

1. Assemble minigel forms.
2. Make 10% polyacrylamide gel by adding the following reagents to a fresh, labeled 50-ml tube:

40% acrylamide/bis-acrylamide	5 ml
10x TBE	2 ml
dH_2O	13 ml
10% fresh APS	125 µl

   This will make enough for two gels for the Serva Dual Mini Slab apparatus.
3. Mix gently by inverting the tube several times and add 20 µl TEMED. Mix again.
4. Pour the gels using a pipet-aid with a disposable pipet to deliver the gel to each gel sandwich, taking care not to introduce bubbles. Insert the comb and allow the gels to polymerize (at least 15 minutes). It is helpful to use Kimwipes to wick away excess acrylamide solution.
5. Fill the gel rig with 1x TBE to the mark on the side of the lower buffer tank.
6. Remove the gasket and comb and rinse the bottom of the gel and wells with 1x TBE using a syringe with a multiflex gel loading tip.
7. Dry the back of the large plate and number the wells using a lab marker.
8. Carefully place the gel sandwich into the rig by inserting it at an angle to avoid introducing any bubbles under the gel.
9. Assemble the gel apparatus and fill the upper buffer tank with 1x TBE.
10. Pre-run the gels for 5 minutes at 200 volts. The minigels are now ready for loading.

8. Remove the gels from between the glass plates, wrap each in plastic wrap, and place in a cassette for analysis.

9. If using phosphor image technology, expose for 1 hour. The phosphor image provides a method of defining the exact cpm of each hybridized band. This information is readily exported as a data file to programs such as Excel. An example of a data set and a phosphor image is shown in Figure 1.

OVERVIEW

Assay Standardization

Standardization of the quantitative assay system is based on the use of cloned template material consisting of cognate sequences of genome that are cloned into a multipurpose cloning vehicle and transformed into a suitable bacterial host. This is a highly advantageous approach for use in high throughput clinical trials because (1) it minimizes the number of reactions that need to be done to generate the standard

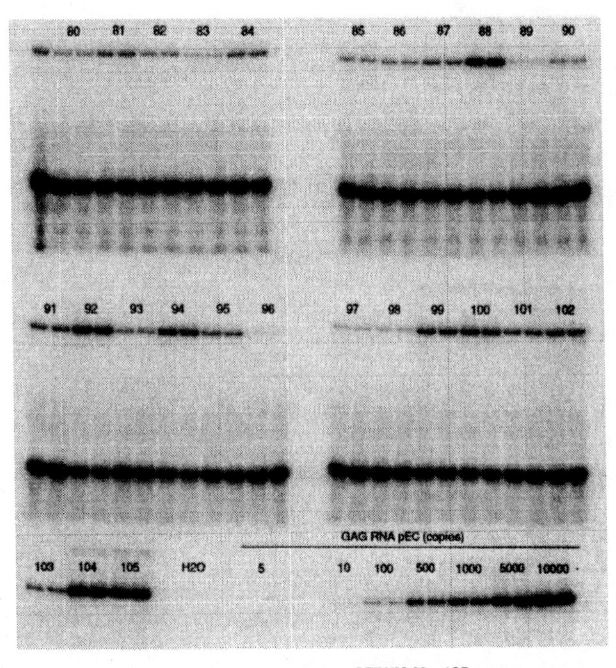

CALC 80-105

SERUM - Calculated Patient Copy Number

Slope(m)	Y-intr (b)	Copies(X)	Values(Y)	LOG Copies(X)	LOG Values (Y)
0.702441	2.963274	5	1578	0.70	3.20
		10	5241	1.00	3.72
		100	31681	2.00	4.50
		500	118159	2.70	5.07
		1000	167841	3.00	5.22
		5000	330812	3.70	5.52
		10000	329025	4.00	5.52

DATE	SAMPLE	Copies (X) (calculated)	Values (Y) (Known)	Copies (X) (Calculated)	Values (Y) (known)
3/31/88	80	326.04	53543	2.51	4.73
6/15/92	81	664.44	88284	2.82	4.95
5/11/93	82	335.43	54621	2.53	4.74
12/17/87	83	175.78	34692	2.24	4.54
8/1/89	84	636.24	85635	2.80	4.93
1/7/92	85	275.33	47547	2.44	4.68
1/14/88	86	288.95	49188	2.46	4.69
2/6/90	87	530.38	75359	2.72	4.88
2/2/91	88	2332.22	213272	3.37	5.33
1/2/88	89	115.75	25868	2.06	4.41
10/25/88	90	274.34	47427	2.44	4.68
1/21/88	91	934.23	112161	2.97	5.05
3/20/90	92	3930.89	307745	3.59	5.49
1/17/88	93	569.14	79187	2.76	4.90
2/7/89	94	2693.14	235954	3.43	5.37
5/8/90	95	889.48	108360	2.95	5.03
1/28/88	96	25.29	8886	1.40	3.95
8/7/90	97	114.21	25627	2.06	4.41
3/5/92	98	123.95	27143	2.09	4.43
4/21/88	99	1685.06	169738	3.23	5.23
8/15/89	100	2262.07	208745	3.35	5.32
5/6/88	101	840.03	104092	2.92	5.02
6/12/90	102	997.68	117459	3.00	5.07
4/12/93	103	544.65	76778	2.74	4.89
3/10/92	104	5380.08	383647	3.73	5.58
5/10/93	105	5732.46	401131	3.76	5.60

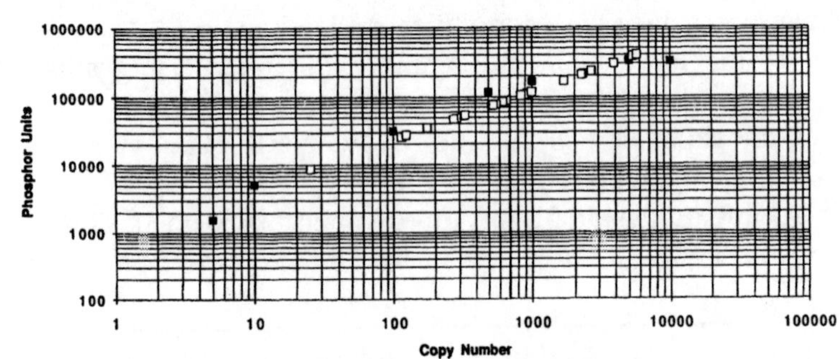

GAG RNA ACELLULAR FLUID

Sample #	Date	RT+	RNA Prep Correction Factor (5 or 10)	Starting Volume Correction Factor (1 or 2)	L.H. Correction Factor (3.33)	Diluted Factor Clinical	Final
80	3/31/88	326	1,630.22	3,260.45	10,857.29	NA	10,857
81	6/15/92	664	3,322.21	6,644.41	22,125.90	NA	22,126
82	5/11/93	335	1,677.14	3,354.28	11,169.76	NA	11,170
83	12/17/87	176	878.90	1,757.81	5,853.49	NA	5,853
84	8/1/89	636	3,181.20	6,362.41	21,186.81	NA	21,187
85	1/7/92	275	1,376.64	2,753.28	9,168.42	NA	9,168
86	1/14/88	289	1,444.77	2,889.53	9,622.15	NA	9,622
87	2/6/90	530	2,651.92	5,303.83	17,661.76	NA	17,662
88	2/2/91	2,332	11,661.12	23,322.25	77,663.09	NA	77,663
89	1/2/88	116	578.73	1,157.45	3,854.31	NA	3,854
90	10/25/88	274	1,371.68	2,743.37	9,135.41	NA	9,135
91	1/21/88	934	4,671.17	9,342.34	31,109.98	NA	31,110
92	3/20/90	3,931	19,654.47	39,308.94	130,898.78	NA	130,899
93	1/17/88	569	2,845.72	5,691.44	18,952.50	NA	18,953
94	2/7/89	2,693	13,465.72	26,931.44	89,681.71	NA	89,682
95	5/8/90	889	4,447.39	8,894.79	29,619.65	NA	29,620
96	1/28/88	25	126.43	252.87	842.04	NA	842
97	8/7/90	114	571.05	1,142.10	3,803.20	NA	3,803
98	3/5/92	124	619.75	1,239.49	4,127.50	NA	4,128
99	4/21/88	1,685	8,425.29	16,850.59	56,112.46	NA	56,112
100	8/15/89	2,262	11,310.35	22,620.70	75,326.93	NA	75,327
101	5/6/88	840	4,200.17	8,400.34	27,973.14	NA	27,973
102	6/12/90	998	4,988.40	9,976.79	33,222.72	NA	33,223
103	4/12/93	545	2,723.26	5,446.52	18,136.90	NA	18,137
104	3/10/92	5,380	26,900.38	53,800.75	179,156.50	NA	179,157
105	5/10/93	5,732	28,662.30	57,324.61	190,890.94	NA	190,891

RNA Prep Correction Factor:
Will be 5X if RNA was resuspended in 50uL
Will be 10X if RNA was resuspended in 100uL

Starting Volume Correction Factor:
Will be 1X if 1.0 ml of material was used
Will be 2X if 0.5 ml of material was used.

L.H. Correction Factor is A Constant & is Equal to 3.33

Clinical Dilution Factor (usually by 3)

Assay Dilution Factor (usually by 10)

* = Starting volumes less than 0.5 ml

Figure 1 Representative data set for the detection of HIV-genomic RNA in plasma. (*Upper left*) The phosphor image of the minigels showing a single band, high-molecular-weight PCR product band and a front of unincorporated material near the bottom of the gel (Note: the designation pEC refers to plasmid external control and is the shorthand used in the laboratory in reference to the cloned, template-derived standard curve). (*Upper right*) The calculation sheet for the interpolation of the standard curve and (*middle*) a graph of the standard curve, both of which are manipulated in Excel (the "copies calculated" values are the "raw copy numbers" appearing on the data reporting sheet [*bottom*]). The mean of two wells per sample is used in the determination.

curve, (2) it provides an inexhaustible supply of control material, and (3) the control material is invariant from reaction to reaction over the course of a trial, which may last for years.

The amplified control sequences generate the same size PCR product as expected in the clinical specimens, making the efficiencies of the PCR procedures equivalent. In the case of the RNA control system, cRNA is derived from the plasmid containing the cognate sequence using the T7 RNA polymerase. This allows the control template to be worked side by side with the clinical RNA template in the reverse transcriptase reaction and the subsequent PCR procedures.

The HIV GAG control template consists of the HIV MN sequence of approximately 1267 bp corresponding to nucleotides 495–1767 of the MN strain, which are cloned into the *Sma* site of pGEM-3Z (total length of the plasmid and insert is ~4009 bp) in the forward orientation and which can be treated with *Xba* for linearization and subsequent synthesis of cRNA using T7 RNA polymerase (total length of the linearized product is ~1300 bp). The GAG sense primer is a 27-nucleotide oligomer located in the HIV MN sequence at position 1412–1438. The GAG antisense primer is a 25-nucleotide oligomer located in the HIV MN sequence at position 1523–1547 and, with the GAG sense primer, generates a 136-bp PCR product. The probe for the HIV GAG PCR is a 30-nucleotide oligomer that hybridizes to the HIV MN sequence at position 1479–1506 to detect the 136-bp PCR product of the HIV GAG PCR.

The β-globin control template (used in the standardization of cellular assays) consists of nucleotides −195 through +73 of the human β-globin gene, listed in the GenBank as HUMMBB5E. This sequence of approximately 278 bp is cloned into the *Sma* site of the vector pBluescript IIKS in the forward orientation (total length of the plasmid and insert is ~3403 bp). The β-globin sense primer is a 20-nucleotide oligomer located in the human genome at position −195 to −176. The antisense β-globin primer is a 20-nucleotide oligomer located at position +53 to +73 and, with the sense primer, generates a 278-bp PCR product. The probe of the β-globin PCR product is complementary to the sequence positions −77 to −48 within the PCR product generated by the β-globin primers.

The protocols presented in this section detail the use of these cloned templates to standardize the PCR assay. Continuity of the values for the points in the standard curve, generated by formulations of the cloned template, allows direct comparison and assessment of assay performance. From lot to lot of cloned template, an empirically determined acceptable level of variance may be defined for the assay's performance. The tolerance for this assay system has been set at 1% for lot-to-lot variance on the standard curves for β-globin and HIV GAG DNA and RNA.

Preparation of Linearized Cloned Template for T7 Synthesis of cRNA

Double-strand plasmid DNA must be converted from a circular to a linear form to provide an appropriate substrate for the conversion of DNA to cRNA using RNA polymerases. This allows the standard curve for the RNA assay to employ RNA template in order to best represent the RNA starting material from patient isolates as processed in the LH-PCR assay.

REAGENTS

10 mg/ml BSA (bovine serum albumin, New England Biolabs, cat. no. 145S)
75% Ethanol
Plasmid (cloned template) DNA for GAG RNA
*Xba*I restriction enzyme (New England Biolabs, cat. no. 145S)
10x Restriction digest buffer (New England Biolabs, cat. no. 145S)
λ*Hind* III ladder (GIBCO/BRL, cat. no. 15207-012)
Phenol (GIBCO/BRL, cat. no. 550UA)
Chloroform (Baxter Scientific, cat. no. 4445)
Spectrophotometer
5x Oligonucleotide loading buffer
  5 ml 10x TBE
  2 ml 1% xylene cyanol FF
  3 ml 100% glycerol
1x TBE (1:10 dilution of 10x TBE)
TE
  1 ml 1 M Tris-HCl, pH 8.0 (0.01 M)
  200 μl 0.5 M EDTA, pH 8.0 (0.001 M)
  98.8 ml sterile H_2O
TEN
  2 ml 5 M NaCl (0.1 M)
  1 ml 1 M Tris-HCl, pH 8.0 (0.01 M)
  200 μl 0.5 M EDTA, pH 8.0 (0.03 M)
  96.8 ml sterile H_2O

Procedures are to be carried out on the bench top in a BL2 laboratory.

PROTOCOL

1. Turn water bath on to 37°C. Obtain the reagents from the freezer and allow them to thaw on ice.

2. Label a 1.5-ml microfuge tube and add the following reagents:

Plasmid (cloned template for HIV GAG)	50 μg
10x Restriction digest buffer 2	10 μl
BSA	1 μl
*Xba*I	100 units

Bring the volume to 100 μl with H_2O.

Note: Enzymes are supplied in glycerol; never add more enzyme than 10% of the total reaction volume.

3. Place the tube in a water bath and incubate for 1 hour at 37°C.

4. Meanwhile, set up a minigel form and make a 0.8% agarose gel by weighing 0.4 g of agarose and placing it in a Pyrex bottle, then adding 50 ml of 1X TBE.

   a. Loosely cap the bottle and heat in a microwave oven for 1–3 minutes, making sure that the agarose goes into solution.

   b. Carefully remove the bottle from the microwave using a protective glove.

   c. Gently swirl the solution and visually check to see if the agarose has completely dissolved. If it has not, reheat the bottle. Allow the solution to cool for 3 or 4 minutes.

   d. Add 2.5 μl of ethidium bromide solution, gently swirl the solution, pour it into the gel form, insert the well comb, and allow the gel to harden (~30 minutes).

5. After 1 hour, remove the linearized plasmid from the water bath. Label two 1.5-ml tubes, one for the linearized plasmid and one for the original circular plasmid.

6. Place a 1-μl aliquot of linearized plasmid into one of the 1.5-ml tubes, add 4 μl of H_2O and 1 μl of 5X oligonucleotide loading buffer.

7. Aliquot 200 ng of the original circular plasmid into the other 1.5-ml tube, add H_2O to bring the volume to 5 μl, then add 1 μl of 5X oligonucleotide loading buffer.

8. Once the agarose gel has hardened, remove the comb from the gel and place the gel into the gel apparatus. Fill the reservoir with enough 1X TBE to cover the gel.

9. Using a micropipettor and a disposable tip, load the entire 6 μl of each sample into one well each, and load 5 μl of λ*Hin*d III standards into an adjacent well.

10. Run the gel at 50 volts for 30–60 minutes.

11. After 30 minutes, carefully remove the gel from the apparatus and take to the UV light box. *(Use safety precautions when using the UV light box.)* Visualize the shift in bands at this point. The band from the linearized (cut) plasmid should shift up, as compared to the band from the circular (uncut) plasmid. After visual-

izing the shift in bands, continue to run the gel, setting a timer for 30 minutes. During this 30 minutes, proceed with the cleanup of the linearized DNA template.

12. Purify the linearized template by adding:

    200 µl of TEN to dilute the salt
    150 µl of phenol
    150 µl of chloroform

13. Vortex briefly to mix the contents of the tube.

14. Centrifuge the tube at 12,000*g* for 5 minutes at room temperature in a microcentrifuge. Visually inspect the tube to be sure the phases have separated. If they have not, centrifuge again.

15. Carefully remove the upper aqueous phase using a micropipettor and a disposable tip. Place this phase into a fresh, labeled 1.5-ml microfuge tube. Discard the interface and organic phase into an appropriate waste container.

16. Add 300 µl of chloroform to the collected aqueous phase. Vortex briefly to mix the contents of the tube.

17. Centrifuge the tube at 12,000*g* for 5 minutes at room temperature in a microcentrifuge. Visually inspect the tube to be sure the phases have separated. If they have not, centrifuge again.

19. Carefully remove the upper aqueous phase using a micropipettor and a disposable tip. Place this phase into a fresh, labeled 1.5-ml microfuge tube. Discard the interface and organic phase into an appropriate waste container.

20. Add 300 µl of cold 100% ethanol to the tube and mix the solutions well by inverting the tube several times.

21. Place in –20°C freezer for 1 hour.

22. Meanwhile, slide the gel out of the gel tray onto the box. Visualize and photograph the gel.

23. After 1 hour, remove the sample from the –20°C freezer and centrifuge the sample at 12,000*g* for 10 minutes at 4°C.

24. Carefully remove the supernatant using a micropipettor and a disposable tip and discard. Take care not to disturb the pellet.

25. Add 500 µl of cold 70% ethanol and centrifuge at 12,000*g* for 2 minutes at 4°C in a microfuge.

26. Decant the supernatant and remove the excess liquid using a micropipettor and disposable tip, taking care not to disturb the pellet. Air-dry the samples for 15 minutes.

27. Resuspend the pellet in 50 µl TE. Use gentle pipetting to aid in this process.

28. Store the purified, linearized plasmid DNA at 4°C.

Synthesis of cRNA Using T7 RNA Polymerase

This protocol describes a procedure for the generation of cRNA template from linearized DNA for use in the generation of the standard curve for the RNA-PCR assay.

REAGENTS

5x Transcription buffer (Promega, cat. no. P2075):
  200 mM Tris-HCl, pH 7.5
  30 mM $MgCl_2$
  10 mM spermidine
  50 mM NaCl
Linearized DNA template
0.1 M DTT (Promega, cat. no. P2075)
40 units/µl RNasin (Promega, cat. no. N2514)
2.5 mM rNTPs (Promega, cat. no. P1221, 10 mM each)
T7 RNA polymerase (Promega, cat. no. P2075)
DNase I, 10,000 units (Boehringer Mannheim, cat. no. 776785)
RNA Marker 1
Chloroform (Baxter Scientific, cat. no. 4443)
TE buffer
Formamide loading buffer:
  10 ml formamide
  200 µl 0.5 M EDTA, pH 8.0
  20 mg bromphenol blue
Phenol:chloroform:isoamyl alcohol (25:24:1)

This protocol is carried out on the bench top in a BL2 laboratory.
Be sure to use rNTPs, not dNTPs, when synthesizing cRNA.

Note: cRNA needs to be handled with extreme care to prevent contamination with RNases.

PROTOCOL

1. Turn water baths on to 40°C and 37°C. Collect reagents needed for the reaction and allow them to thaw on ice.

2. Label a 1.5-ml microfuge tube and place into a rack at room temperature.

3. Add the following, *in the order given,* to the labeled 1.5-ml microfuge tube held at room temperature:

5X Transcription buffer (T7 polymerase buffer)	20.0 μl
100 mM DTT	10.0 μl
RNasin (40 units/μl)	4.0 μl
2.5 mM rNTPs	20.0 μl
Linearized DNA template in TE (~1 μg)	2.0 μl
T7 RNA polymerase (1–2 μl)	20.0 units

4. Bring the reaction volume to 100 μl by adding the appropriate volume of water.

5. Place the reaction tube in a water bath and incubate for 120 minutes at 40°C.

6. Remove the tube and add 2.5 μl of DNase I (RNase-free).

7. Place the reaction tube in a water bath and incubate for 15 minutes at 37°C.

8. Remove the tube from the water bath.

9. Add 100 μl of phenol:chloroform:isoamyl alcohol to the sample.

10. Vortex briefly to mix the contents of the tube.

11. Centrifuge the tube at 12,000*g* for 15 seconds in a microcentrifuge. Visually inspect the tube to be sure the phases have separated. If they have not, centrifuge again.

12. Carefully remove the upper aqueous phase using a micropipettor and a disposable tip. Place this phase into a fresh, labeled 1.5-ml microfuge tube. Discard the interface and organic phase into an appropriate waste container.

13. Add 100 μl of chloroform to the collected aqueous phase.

14. Vortex briefly to mix the contents of the tube.

15. Centrifuge the tube at 12,000*g* for 15 seconds in a microcentrifuge. Visually inspect the tube to be sure the phases have separated. If they have not, centrifuge again.

16. Carefully remove the upper aqueous phase using a micropipettor and a disposable tip. Place this phase into a fresh, labeled 1.5-ml microfuge tube. Discard the interface and organic phase into an appropriate waste container.

17. Add 200 μl of cold 100% ethanol to the tube and mix the solutions well by inverting the tube several times.

18. Place in –20°C freezer for 1 hour.

19. Centrifuge the sample at 12,000*g* for 10 minutes at 4°C.

20. Carefully remove the supernatant using a micropipettor and a disposable tip and discard. Take care not to disturb the pellet.

21. Add 500 μl of cold 70% ethanol to the pellet and centrifuge at 12,000*g* for 2 minutes at 4°C in a microfuge.

22. Carefully remove the supernatant using a micropipettor and a disposable tip and discard. Take care not to disturb the pellet.

23. Remove the excess liquid using a micropipettor and a disposable tip, taking care not to disturb the pellet. Air-dry the samples for 15 minutes.

24. Dissolve the pellet in 50 μl of TE, pH 8.0, and place on ice.

25. The purified cRNA can be stored in the –80°C freezer.

26. To check the purity and concentration of cRNA, read the optical density of a 1:40 dilution using the following methods:

    a. Label two 1.5-ml microfuge tubes, one for blanking and one for the purified cRNA.

    b. Add 300 μl of water to the blanking tube.

    c. Add 292.5 μl of water to the purified cRNA tube. Then add 7.5 μl of the stock purified cRNA, and vortex briefly to mix.

    d. Place the entire volume of each tube into separate cuvettes.

    e. Blank the spectrophotometer at 260 nm with the water.

    f. Read the OD at 260 nm of the cRNA(s) and record the number. An OD at 280 nm may be read as well to give an estimate of the purity of the preparation (the ratio of 260/280 should be 2.0 for RNA).

    g. Calculate the concentration in μg/ml by multiplying the OD by 40 (the dilution of 1:40) and by 40 (an OD of 1 corresponds to ~40 μg/ml of RNA).

27. To check the purity and concentration visually, set up a 6% polyacrylamide/7 M urea gel (GEL-MIX 6, GIBCO/BRL) and characterize 300 ng of cRNA.

    Note: Turn on heat block to 95°C.

    a. Remove GEL-MIX 6 from the 4°C refrigerator and allow the solution to warm to room temperature.

    b. Set up the gel form. For pouring one gel, remove 10 ml of GEL-MIX 6 solution and place into a labeled tube.

    c. Add 60 μl of 10% (w/v) APS per gel to the tube containing the GEL-MIX 6 aliquot. Secure the tube cap.

    d. Mix the solutions well by gently inverting the tube several times.

    e. Pipette the gel mixture into the gel form using a disposable 10-ml pipet. Insert the comb and allow the gel to polymerize for 45–60 minutes.

    f. Set up the gel apparatus and pre-run the gel at 200 volts for 15 minutes.

    g. Label two 0.5-ml or two 1.5-ml microcentrifuge tubes. The tubes labeled here must be of appropriate size to fit into the heat block. One tube is for the RNA Marker 1 (Boehringer Mannheim), and the other is for purified cRNA.

    h. Aliquot 300 ng of the sample to a labeled tube, and the appropriate volume of RNA Marker 1 (Boehringer Mannheim) corresponding to 5 μg, to the other labeled tube.

       Note: The RNA Marker 1 generally comes in at 100 μg/100 μl, therefore, use 5 μl of the marker. Also note that the sample is in μg/ml from step 26. (Calculation: Divide [300 ng] by [concentration in ng/μl] = volume required.)

    i. Dry the samples in a SpeedVac for 5 minutes or less. Do not over-dry!

    j. Resuspend the sample and the marker in 5 μl of formamide loading buffer, vortex, and place the tubes in a 95°C heat block for 5 minutes to denature the RNA.

    k. Remove the tubes from the heat block, vortex, centrifuge briefly, and place directly on ice.

    l. Load each sample in separate wells in the gel.

    m. Run the gel at 200 volts for 30 minutes. The dye front will be near the bottom of the gel.

    n. Carefully remove the gel from the apparatus and place into the stain solution, gently rocking for 10 minutes at room temperature. (The stain solution is 100 ml of 1X TBE and 5 µl of 5 µg/ml ethidium bromide solution. This is made just prior to removing the gel.)

o. Carefully remove the gel from the stain solution and place on a UV light box to visualize the products. The expected product RNA is 1300 nucleotides in length and runs between the 1.6-kb and 1.0-kb bands on RNA Marker 1.

28. Photograph the gel.

Formulation of Cloned Template for PCR Assay Standardization

This protocol describes the procedures for determining the optical density of standard cloned template stocks and the calculations required to assemble the standard curve material.

REAGENTS

10 mg/ml carrier tRNA (Boehringer Mannheim, cat. no. 109495)
40 units/µl RNasin (Promega, cat. no. N2514)
0.1 M DTT (Boehringer Mannheim, cat. no. 100-032)
Purified cloned DNA plasmid or cRNA template prepared from cloned plasmid DNA
Diluent I: 4 µg/ml carrier tRNA (at 10.0 mg/ml) in sterile H_2O (used for initial dilutions and working dilutions for DNA)
Diluent II: 5 mM DTT (at 0.1 M) 1 unit/µl RNasin (at 40 units/µl), 4 µg/ml carrier tRNA (at 10.0 mg/ml) in sterile H_2O (used for working dilutions for RNA)

PROTOCOL

1. Turn the spectrophotometer on to warm up (this will take 10–15 minutes) and obtain the cloned template material from the cold box.

2. In a BL2 laminar-flow hood, label 1.5-ml microfuge tubes: one for blanking the instrument and one for each purified DNA cloned template for which the concentration of DNA is to be determined. To check the purity, read the optical density of a 1:40 dilution using the following methods:

   a. Add 300 µl of water to the instrument blanking tube.

   b. Add 292.5 µl of water to the purified DNA or RNA "standard" tubes. Then add 7.5 µl from the stock cloned template, and vortex briefly to mix.

   c. Place the entire volume of each tube into separate cuvettes.

   d. Blank the spectrophotometer at 260 nm with the water.

e. Read OD at 260 nm for all samples and record the number. An OD at 280 nm may be read to give an estimate of the purity of the preparations (the 260/280 ratio should be 1.8 for DNA).

f. Calculate the concentration in μg/ml by multiplying the OD by 50 (an OD of 1 corresponds to ~50 μg/ml of double-strand DNA).

3. To obtain stocks of each cloned template dilution it is necessary to perform several calculations:

   a. First, determine the mass weight in g/copy of the desired template. This is derived from the following formula: Mass wt (g/mole of nt) x Length (nt) x $1/6.0223 \times 10^{23}$ mole/copy = conc. in g/copy. The mass wt: RNA = 341 g/mole of nt. DNA = 648 g/mole of nt.

   b. Length of the PCR product of cloned templates used in the assay system:
   DNA globin template: 3403 nucleotides (construct is a β-globin DNA sequence of 268 bp cloned into pBluescript II KS at *Sma*I)
   DNA GAG template: 4009 nucleotides (construct is a HIV GAG sequence of 1266 bp cloned into pGEM-3Z at *Xba*I)
   RNA GAG template: 1300 nucleotides (this material is RNA obtained by T7 synthesis of a HIV GAG sequence cloned into pGEM-3Z at *Xba*I)

   c. Therefore, the mass weight in g/copy for each cloned template is as follows:
   β-globin: 3.66×10^{-18} g/copy
   GAG DNA: 4.31×10^{-18} g/copy
   GAG RNA: 7.36×10^{-19} g/copy

   d. To determine the highest concentration of standard material required of the assay and expressed in copy numbers: Multiply the desired target dilution (the highest copy value for each template) x 10 (gel loading correction factor: 1/3 of PCR product is used in the liquid hybridization; 1/3 of the liquid hybridization reaction is loaded on the gel).

   e. The dilutions of each template to be aliquoted as stock tubes are:
   β-globin DNA: 100; 500; 1,000; 5,000; 10,000; 50,000
   GAG DNA: 5; 10; 50; 100; 500; 1,000; 5,000
   GAG RNA: 5; 10; 100; 500; 1,000; 5,000; 10,000; 50,000; 100,000

   f. Therefore, the highest concentration for each template is:
   β-globin: 500,000 copies

GAG DNA: 50,000 copies
GAG RNA: 1,000,000 copies

g. To determine the formulation of the highest concentration: (Mass weight of template) (Highest concentration in copies)/ Template input volume of the PCR (0.01 ml for DNA; 0.008 ml for RNA) = Target concentration the highest concentration Thus:

β-globin: (3.66×10^{-18}) (500,000)/.01 = 1.83×10^{-10} g/ml or (183.1×10^{-6} μg/ml)

GAG DNA: (4.3×10^{-18}) (50,000)/.01 = 2.15×10^{-11} g/ml or (21.6×10^{-6} μg/ml)

GAG RNA: (7.36×10^{-19}) (1,000,000)/.008 = 9.2×10^{-11} g/ml or (92.0×10^{-6} μg/ml)

h. Calculate the dilution factor required to arrive at the number of copies of the highest concentration from the OD determination of stock template as follows.

i. Use the OD reading from step 2f expressed as μg/ml and divided by the number of copies desired in the highest concentration. For example, for formulating the RNA standard dilution: If the OD reading of the starting material is 921.1 μg/ml, then 921.1 μg/ml/92 $\times 10^{-6}$ μg/ml = 1.0012×10^{7}/1000 = 1.0012×10^{4}/1000 = 1.0012×10^{1} or 10. Therefore, the dilution factor is 1:10.

4. Determine diluent volumes. DNA will be diluted in Diluent I throughout the entire series, whereas RNA will be diluted in Diluent I for the stock solutions and in Diluent II for the working dilutions.

   a. Subsequent dilutions may be required to obtain the other stock dilution values. Most require 1:2 or 1:5 dilutions of the previous stock dilution.

   b. Diluent volumes should be placed into the tubes first. Change tips or pipettes with each addition of the volume from the previous dilution to the next tube. Each tube should be vortexed briefly prior to removing the volume to be placed into the next tube.

   c. In a laminar-flow hood, perform serial dilutions to obtain stocks of each plasmid external control dilution desired.

5. Dilutions for the formulation of the GAG DNA standards:

   Make 10 ml of Diluent I. This is 10 ml of sterile water and 4 μl of the carrier tRNA at 10.0 mg/ml.

Label nine 6-ml FALCON tubes and make dilutions as follows (final copy numbers in the tube are given on the right):

Tubes	Diluent	Copies
1. 1 μl of purified DNA stock	1 ml	
2. 1 μl from tube 1	1 ml	
3. no dilution		5,000
4. 200 μl from tube 3	800 μl	1,000
5. 500 μl from tube 4	500 μl	500
6. 200 μl from tube 5	800 μl	100
7. 500 μl from tube 6	500 μl	50
8. 200 μl from tube 7	800 μl	10
9. 500 μl from tube 8	500 μl	5

Note: The amount from tube 2 to tube 3 is dependent on the dilution factor and will be different each time. For example, if the dilution factor is 1:10, divide 1 ml (or 1000 μl) by 10 = 100.

6. Dilutions for the formulation of the β-globin standards:

Make 10 ml of Diluent I. This is 10 ml of sterile water and 4 μl of the carrier tRNA at 10.0 mg/ml.

Label eight 6-ml FALCON tubes and make dilutions as follows (final copy numbers in the tube are given on the right):

Tubes	Diluent	Copies
1. 1 μl of purified DNA stock	1 ml	
2. 1 μl from tube 1	1 ml	
3. no dilution		50,000
4. 200 μl from tube 3	800 μl	10,000
5. 500 μl from tube 4	500 μl	5,000
6. 200 μl from tube 5	800 μl	1,000
7. 500 μl from tube 6	500 μl	500
8. 200 μl from tube 7	800 μl	100

Note: The amount from tube 2 to tube 3 is dependent on the dilution factor and will be different each time.

7. Dilutions for the formulation of the GAG RNA standards:

Make 2.5 ml of Diluent I. This is 2.5 ml of sterile water and 1 μl of the carrier tRNA at 10.0 mg/ml.

Make 7 ml of Diluent II:

Sterile water	6472.2 μl
5 μM DTT (at 0.1 M)	350 μl
4 μg/ml carrier tRNA (at 10.0 mg/ml)	2.8 μl
1 unit/μl RNasin (at 40 units/μl)	175 μl

Label eleven 6-ml FALCON tubes and make dilutions as follows (final copy numbers in the tube are given on the right):

Tubes	Diluent	Copies
1. 1 μl of purified cRNA stock	1 ml	
2. 1 μl from tube 1	1 ml	
3. no dilution		100,000
4. 500 μl from tube 3	500 μl	50,000
5. 200 μl from tube 4	800 μl	10,000
6. 500 μl from tube 5	500 μl	5,000
7. 200 μl from tube 6	800 μl	1,000
8. 500 μl from tube 7	500 μl	500
9. 200 μl from tube 8	800 μl	100
10. 100 μl from tube 9	900 μl	10
11. 500 μl from tube 10	500 μl	5

Note: Tubes 1 and 2 are diluted in Diluent I. The rest are diluted in Diluent II. The amount from tube 2 to tube 3 is dependent on the dilution factor and will be different each time.

DISCUSSION

The assay system detailed in this chapter is an accurate and dependable method for the assessment of HIV in clinical specimens. The optimal performance of the assay depends on strict adherence to the standard operating procedures. In addition, over the 6 years that the assay has been used to quantitate HIV in clinical specimens, a body of information has been accumulated on the issue of assay performance and troubleshooting.

Prevention and detection of contamination is the most critical challenge to the routine execution of PCR on clinical samples. To minimize contamination and ideally to eliminate it, several steps can be taken.

1. Sample lysis and nucleic acid extraction and purification are performed in a room and a hood completely separated from the areas in which the PCR is carried out. The synthesis of primers and probes and the cloning and preparation of cloned template are also segregated away from the PCR areas.

2. Each of the technical staff has his/her own box of key reagents, which have been aliquoted in a clean laminar-flow hood and evaluated lot by lot for suitability for use in the PCR.

3. The PCR assay is set up in a clean laminar-flow hood dedicated to the template under study.

4. Staff members each acquire their own preference for ordering the setup of PCR procedures, but the general advice is to work from

the least concentrated target to the most and to take extreme caution in handling the sides of tubes, changing the pipette tips (which contain anti-aerosol filters) when indicated, and changing gloves often.

5. All aliquots of reagents used in the assay, including the sterile water, are maintained in small volumes and are never shared between workers.

6. The assay "reagent or water" blank is the determinant for the use or the rejection of the entire data set.

7. In the case of the RT-PCR assay, the execution of a mock RT tube or an RT^- is useful in that signal in this tube would indicate contamination with DNA template. To date, no RT^- GAG-specific PCR signals have been seen in our system. This indicates that HIV DNA contamination in serum is not a significant concern.

The use of positive controls in addition to the standards is redundant in this assay because the performance of the control template is indicative of that parameter in a quantitative manner.

There are factors that affect assay performance and reagent integrity. It has proven helpful to date each lot of reagent and to indicate the number of freeze-thaw cycles the standard curve template and the nucleotides have been through. As a rule, after about five freeze-thaws these reagents begin to deteriorate.

The appearance of no PCR product band is in most cases due to:

1. The omission of a key component of the reaction mix.

2. The use of the wrong primer.

3. The use of the wrong probe.

These problems are best avoided by the use of an assay checklist by the technical staff in which each reagent is checked off as it is used in the assembly of the reaction.

The appearance of spurious bands in the PCR products indicates most often that:

1. The assay reagents are deteriorating.

2. The clinical specimen template is deteriorating.

3. The wrong cycle program was run.

These problems can be avoided by a careful accounting of the date and condition of the assay reagents. Every effort is made to process clinical specimens immediately after nucleic acids are extracted. Storage of RNA template in alcohol at –80°C is acceptable, but long storage times and freeze-thaw cycles greatly reduce the integrity of the RNA template.

This assay system has been evaluated on cloned template and on blinded panels of clinical materials. It has yet to yield a false positive or to fail to detect an HIV signal in an infected specimen. It is capable of routinely attaining variances of less than 1% on repeated samples. The assay system demands an accomplished technical staff, a dedicated laboratory facility, and constant surveillance of assay performance. However, these measures are fundamental to the confident reporting of PCR data and their subsequent interpretation.

ACKNOWLEDGMENTS

We express our appreciation to Dr. Donald S. Burke for the hours of discussions on the critical need for a reliable quantitative assay for viral load in HIV-infected persons. We acknowledge the outstanding talents of the many technical staff who worked on the perfection of this assay system, especially John Cooley, Michael Nuzzo, Sandra Barrick, Deborah Dayhoff, and Cheryl Lewis. We are grateful to our colleagues who have employed this assay system as part of their research, especially Dr. Douglas Mayers, Dr. Nelson Michael, Dr. Deborah Birx, Dr. Robert Redfield, Dr. Andrew Artenstein, Dr. Kenneth Wagner, and Dr. Eric Kozlow. Most especially, we express our appreciation for the dedication of the clinical staff and the courage of the patients who comprise the Military Medical Consortium for Applied Retroviral Research. Funding for this study was provided by the U.S. Army Medical Research and Development Command. The opinions or assertions contained herein are the private views of the authors and are not to be construed as official or as reflecting the views of the Department of the Army, the Department of the Air Force, or the Department of Defense.

REFERENCES

Clark, S.J., M.S. Saag, W.D. Decker, S. Campbell-Hill, J.L. Roberson, P.J. Veldkamp, J.C. Kappes, B.H. Hahn, and G.M. Shaw. 1992. High titers of cytopathic virus in plasma of patients with symptomatic primary HIV-1 infection. *N. Engl. J. Med.* **324:** 954–960.

Coombs, R.W., A.C. Collier, J.P. Allain, B. Nikora, M. Leuther, G.F. Gyerset, and L. Corey. 1989. Plasma viremia in human immunodeficiency virus infection. *N. Engl. J. Med.* **321:** 1626–1631.

Daar, E.S., T. Moudgil, R.D. Meyer, and D.D. Ho. 1991. Transient high levels of viremia in patients with primary human immunodeficiency virus type 1 infection. *N. Engl. J. Med.* **324:** 961–964.

Genesca, J., R.Y.-H. Wang, H.J. Alter, and J.W.-K. Smith. 1990. Clinical correlation and genetic polymorphism of human immunodeficiency virus proviral DNA obtained after polymerase chain reaction amplification. *J. Infect. Dis.* **162:** 1025–1030.

Ho, D.D., T. Moudgil, and M. Alam. 1989. Quantitation of human immunodeficiency virus type 1 in the blood of infected persons. *N. Engl. J. Med.* **321:** 1621–1625.

Johnson, R.F., S.C. Pickett, and D.L. Barker. 1990. Autoradiography using storage phosphor technology. *Electrophoresis* **11:** 355–360.

Michael, N.L., M. Vahey, D.S. Burke, and R.R. Redfield. 1992. Viral DNA and mRNA expression correlate with the storage of human immunodeficiency virus (HIV) type 1 infection in humans: Evidence for viral replication in all stages of HIV disease. *J. Virol.* **66:** 310–316.

Piatek, M., M.S. Saag, L.C. Yang, S.J. Clark, J.C. Kappes, K.-C. Luk, B.H. Hahn, G.M. Shaw, and J.D. Lifson. 1993. High levels of HIV-1 in plasma during all stages of infection determined by competitive PCR. *Science* **259:** 1749–1754.

Schnittman, S.M., J.J. Greenhouse, H.C. Lane, P.F. Pierce, and A.S. Fauci. 1991. Frequent detection of HIV-1-specific mRNAs in infected individuals suggests ongoing active viral expression in all stages of disease. *AIDS Res. Hum. Retroviruses* **7:** 361–367.

Schnittman, S.M., M.C. Psallidopoulis, H.C. Lane, L. Thompson, M. Baseler, F. Massari, C.H. Fox, N.P. Salzman, and A.S. Fauci. 1989. The reservoir for HIV-1 in human peripheral blood is a T cell that maintains expression of CD4. *Science* **245:** 305–308.

Simmonds, P., P. Balfe, J.F. Peutherer, C.A. Ludlam, J.O. Bishop, and A.J.L. Brown. 1990. Human immunodeficiency virus infected individuals contain provirus in small number of peripheral mononuclear cells and at low copy numbers. *J. Virol.* **64:** 864–872.

Use of the SNuPE Assay to Quantitate Allele-specific Sequences Differing by a Single Nucleotide

Judith Singer-Sam

Department of Biology, Beckman Research Institute of the City of Hope, Duarte, California 91010

INTRODUCTION

The single-nucleotide primer extension assay (SNuPE) provides a means of identifying and quantifying allelic variants in DNA and RNA (Kuppuswamy et al. 1991; Singer-Sam et al. 1992; Singer-Sam and Riggs 1993). The assay is performed after amplification of a segment containing the allelic difference, which may be as small as a single nucleotide. An oligonucleotide primer that binds adjacent to the nucleotide difference is extended by a single base in the presence of a [^{32}P]dNTP specific for one of the two alleles; the ratio of incorporation in the two reactions is proportional to the relative amount of each allelic variant. The method can be used over a wide range, allowing detection of a given allele in up to 1000-fold excess of the other, depending on which nucleotides differ. With the use of an appropriate internal standard during amplification, the method can also be adapted for the absolute quantitation of specific sequences (Buzin et al. 1994).

Figure 1 illustrates the method as used for the quantitation of the ratio of *Pgk-1a/Pgk-1b* transcripts. RNA containing both allelic transcripts, which differ by a C/A transversion, is amplified by RT-PCR. The amplified product is gel-purified, and then mixed with *Taq* DNA polymerase, the SNuPE primer (shown by the arrow), and either [^{32}P]dCTP or [^{32}P]dATP. The mixture is heat-denatured, and the SNuPE primer is extended by one nucleotide. The amount of radioactivity in each extended primer is determined after denaturing gel electrophoresis and used to calculate the relative amount of each allelic transcript.

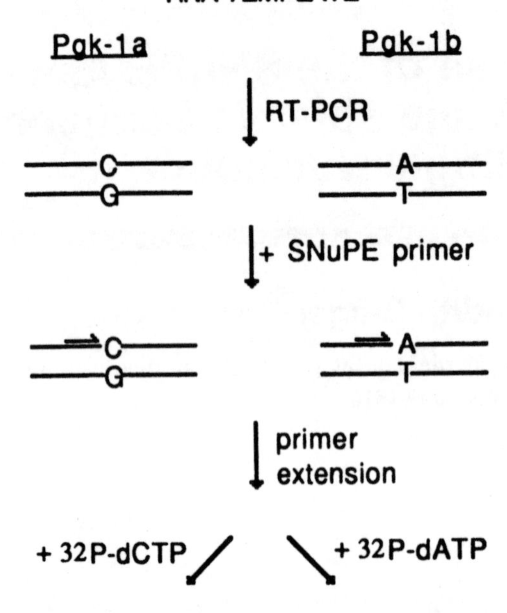

Figure 1 Schematic outline of the RT-PCR SNuPE assay (Singer-Sam et al. 1992). *Pgk-1a* and *Pgk-1b* are the two alleles of the *Pgk-1* gene.

REAGENTS

10X *Taq* buffer (100 mM Tris-HCl, pH 8.3, 500 mM KCl, 20 mM $MgCl_2$, 0.01% gelatin) (optimum pH and $MgCl_2$ concentration should be determined empirically for each primer)

TE (10 mM Tris-HCl, 1 mM EDTA, pH 8.0)

Gelase, 1 unit/μl (Epicentre)

AmpliTaq polymerase, 5 units/μl (Perkin-Elmer)

SNuPE primer, 18-mer or longer, with 3′ end just 5′ to allelic base difference (20 μM)

[^{32}P]dNTPs corresponding to bases which differ, 3000 Ci/mmole, diluted to 2 μCi/μl in H_2O just prior to incubation

20 mg/ml glycogen (Sigma)

Equilibrated phenol

Chloroform/isoamyl alcohol (24:1)

7.5 M Ammonium acetate

Low-melting-point agarose, ultrapure (BRL)

15% polyacrylamide-7 M urea gel in 0.1 M TBE, pH 8.3 (Maniatis et al. 1982)

Sequencing gel-loading buffer (Maniatis et al. 1982)

Ethanol

PROTOCOL

Purification of Amplified Product

1. Amplify the RNA or DNA sequence containing the allelic difference; a strong major band should be seen on an ethidium bromide-stained gel after amplification.

2. Extract amplified products with an equal volume of 1:1 phenol and chloroform/isoamyl alcohol, and precipitate with ethanol after addition of 1/2 volume of 7.5 M ammonium acetate (Maniatis et al. 1982). Add 20–40 μg of glycogen, if needed, to obtain visible pellet.

3. Resuspend the pellet in TE, run samples on low-melting agarose gel, and extract with GELase, following the instructions of the manufacturer.

4. Precipitate with 1/2 volume of 7.5 M ammonium acetate and 2 volumes of ethanol (add glycogen, 20–40 μg, as a carrier if needed), and resuspend in 10–30 μl of TE. Alternatively, Prep-a-Gene may be used (Bio-Rad Laboratories) and the DNA may be recovered in TE buffer.

5. Determine the concentration of each sample by a dilution series on an ethidum bromide-stained gel.

Quantitative SNuPE Assay

Refer to Kuppuswamy et al. (1991) and Singer-Sam et al. (1992). For each sample, two reactions are run, one with the [^{32}P]dNTP for each allele.

1. To each tube on ice, add in the following order: 10 ng of template DNA; a master mix containing *Taq* buffer, the SNuPE primer, and *Taq* DNA polymerase; and 2 μCi of [^{32}P]dNTP. The reaction mix contains, in a final volume of 10 μl: 1x *Taq* buffer, 1 μM SNuPE primer, and 0.75 units of *Taq* DNA polymerase.

2. Incubate the samples in a thermal cycler for one round of denaturation, annealing, and synthesis (example: 1 minute at 95°C, 2 minutes at 42°C, and 1 minute at 72°C). Place on ice.

3. As controls for background and maximum incorporation, incubate 10 ng of amplified product from each allele with the "incorrect" and appropriate [^{32}P]dNTP, respectively. Initially, controls should also include various ratios of the two alleles, mixed prior to amplification, to establish a standard curve.

4. After addition of the sequencing gel-loading buffer (10 μl), incubate each sample for 1–2 minutes at 90°C, and place on ice. Load 10 μl of each sample on a 1-mm-thick 15% polyacrylamide-urea gel (18.5 cm long x 16.5 cm wide) and electrophorese at 30 mA/gel until the bromphenol blue reaches the bottom (~1 hour).

5. Wrap the gel in plastic wrap and expose it to Kodak XAR-5 film for 2 hours to overnight at 4°C.

6. Cut out the portion of the gel between the bromphenol blue and xylene cyanol markers. Wrap the gel slice in plastic wrap.

7. Determine the amount of radioactivity above lane background in each (n+1) band (n=length of SNuPE primer) using a radioisotope scanning system. Occasionally the base 3′ to the incorporated [^{32}P]dNTP is identical to it, and one sees (n+2) products. In that case, the contribution of both (n+1) and (n+2) products should be included.

8. For each sample, the ratio of cpm incorporated with each of the two [^{32}P]dNTPs is calculated. The ratio is then corrected for the following:

   a. Background incorporation of the incorrect [^{32}P]dNTP. This is calculated from the ratio of the signals seen when the control templates for each allele are incubated with the incorrect versus correct [^{32}P]dNTP. The ratio should be $\leq$1%.

   b. Different efficiency of incorporation with each [^{32}P]dNTP. This is determined from the ratio seen when equal amounts of each control template are incubated with the corresponding [^{32}P]dNTP.

TROUBLESHOOTING

- *Irreproducible results.* Generally the SNuPE assay is quite reproducible. Occasional problems are encountered when [^{32}P]dNTPs are not used within several days of arrival.

- *Increasing background.* In some cases, background has been found to increase upon repeated freezing and thawing of reagents. This problem can be avoided by freezing the SNuPE primer and template DNA in aliquots, or by storing them at 4°C.

- *Bands of incorrect size.* Occasionally minor bands are seen, of incorrect size; although the origin of these bands is not clear, only the signal from the major band(s), representing the SNuPE product of the expected size(s), is used in calculation.

DISCUSSION

Variation in Background

The background varies with different mismatches. We have found measurement of C in an excess of A (C/A) to be sensitive down to the 0.001% level; other mismatches giving a low background level include C/T, T/C, and A/G. G/C, C/G, and G/A mismatches give a background of approximately 1%.

Absolute Quantitation

When the assay is used for absolute quantitation, an internal standard, differing by one base from the sequence of interest, is added at the time of amplification. This then becomes the equivalent of an allelic variant in the SNuPE assay. Because the internal standard can be added at known concentration, the ratios determined by the SNuPE assay can be used to estimate the concentration of any given DNA or RNA sequence (Buzin et al. 1994).

ACKNOWLEDGMENTS I am grateful to Drs. Jeanne M. LeBon, Carolyn Buzin, Arthur D. Riggs, and Piroska Szabó for their valuable contributions.

REFERENCES

Buzin, C.H., J.R. Mann, and J. Singer-Sam. 1994. Quantitative RT-PCR assays show *Xist* RNA levels are low in mouse female adult tissue, embryos, and embryoid bodies. *Development* **120:** 3529–3536.

Kuppuswamy, M.N., J.W. Hoffmann, C.K. Kasper, S.G. Spitzer, S.L. Groce, and S.P. Bajaj. 1991. Single nucleotide primer extension to detect genetic diseases: Experimental application to hemophilia B (factor IX) and cystic fibrosis genes. *Proc. Natl. Acad. Sci.* **88:** 1143–1147.

Maniatis, T., E.F. Fritsch, and J. Sambrook. 1982. *Molecular cloning: A laboratory manual.* Cold Spring Harbor Laboratory, Cold Spring Harbor, New York.

Singer-Sam, J. and A.D. Riggs. 1993. Quantitative analysis of messenger RNA levels: Reverse transcription-polymerase chain reaction single nucleotide primer extension assay. *Methods Enzymol.* **225:** 344–351.

Singer-Sam, J., J.M. LeBon, A. Dai, and A.D. Riggs. 1992. A sensitive, quantitative assay for measurement of allele-specific transcripts differing by a single nucleotide. *PCR Methods Appl.* **1:** 160–163.

Trapping Internal and 3′-Terminal Exons

Paul E. Nisson,[1] Abdul Ally,[1] and Paul C. Watkins[2]

[1]Life Technologies, Inc., Gaithersburg, Maryland 20877
[2]Sequana Therapeutics, La Jolla, California 92037

INTRODUCTION

The cloning of genes based on their genetic map position (positional cloning) has proven to be a powerful tool in the growing field of gene discovery (Collins 1992). The list of heritable human disease genes successfully isolated by using this strategy is rapidly growing. Positional cloning can be described as a four-step process: (1) Finding genetic linkage between a disease phenotype and genetic markers of known map position; (2) obtaining this region as a set of physically overlapping pieces of cloned DNA using cosmid, P1, P1 artificial chromosome (PAC), bacterial artificial chromosome (BAC), or yeast artificial chromosome (YAC) clones (Green et al. 1991; Yokobata et al. 1991; Pierce et al. 1992; Shizuya et al. 1992; Ioannou et al. 1994); (3) screening the set of overlapping clones (known as a contig) for expressed sequences; and (4) using these captured bits of expressed sequence to identify a full-length cDNA clone and the mutation that distinguishes the normal from the mutant allele associated with the genetic disease.

Construction of genetic and physical maps of the human and mouse genomes is progressing rapidly; thus, perhaps the most difficult part of the positional cloning strategy is the actual isolation of candidate genes from the mapped region. There are several methods for screening genomic clones for expressed sequences, including multispecies blotting, cDNA library screening, and CpG island identification; however, direct cDNA selection and exon trapping are becoming the preferred methods (Bird 1986; Rommens et al. 1989; Auch and Reth 1990; Duyk et al. 1990; Elvin et al. 1990; Buckler et al. 1991; Lovett et al. 1991; Parimoo et al. 1991; Hamaguchi et al. 1992; Krizman and Berget 1993; Ozawa et al. 1993; Church et al. 1994). Direct cDNA selection usually captures larger bits of expressed sequence than does exon trapping, but it requires a high-quality cDNA source (primary cDNA or cDNA library) in which the gene of interest is expressed (Lovett et al. 1991; Parimoo et al. 1991). Exon trapping, however, extracts exons directly from cloned genomic DNA without the need a priori for cDNA. For this reason, exon trapping can be considered a very useful tool for gene discovery and transcript mapping because the selection for expressed sequences is based solely on the presence of splicing recognition sequences. Methods are presented here for the isolation of internal and 3′-terminal exons from YACs and cosmids containing cloned human DNA.

Trapping Internal Exons Using pSPL3

The exon-trapping vector pSPL3 is an improved version of the plasmid pSPL1 developed by Buckler and colleagues (Buckler et al. 1991; Church et al. 1994). pSPL3 contains a multiple cloning site and other modifications that result in a reduction of cryptic splicing as well as a higher transfection efficiency due to the incorporation of restriction enzyme (*Bst*XI) half-sites flanking the splice donor and acceptor sites of pSPL3 (Church et al. 1994). Although most published reports on exon trapping have used cosmids as targets (Church et al. 1994; Nisson and Watkins 1994), at least one paper has described a method for trapping exons from YACs (Gibson et al. 1994). We have focused on using pSPL3 for exon trapping YACs, highlighting their special requirements.

A flow diagram describing the internal exon trapping procedure for YACs is shown in Figure 1: A YAC clone is propagated, and DNA is purified after pulsed-field gel electrophoresis, then subcloned into the plasmid pSPL3. The subclones are then propagated in *Escherichia coli* and arrayed in pools. After mini-prep DNA is made from subclone pools, the DNA is transfected into COS-7 cells. After transient expression, the RNA is harvested, and first-strand cDNA is reverse-transcribed using a vector-specific oligonucleotide. Following digestion of the RNA template by RNase H, an initial round of PCR is performed. *Bst*XI is then added, resulting in the removal of PCR products that do not contain exons. A second round of PCR is performed followed by rapid cloning into a phagemid vector using uracil DNA glycosylase (Nisson et al. 1991). The following sections describe the details of each step.

Purification and Preparation of YAC DNA for Subcloning into the Exon-trapping Vector pSPL3

By encapsulation of yeast cells in agarose beads prior to treatment with yeast lytic enzyme and lysis buffers, YAC DNA is protected from the shear forces associated with the manipulations required by this procedure. Following pulsed-field gel electrophoresis and isolation

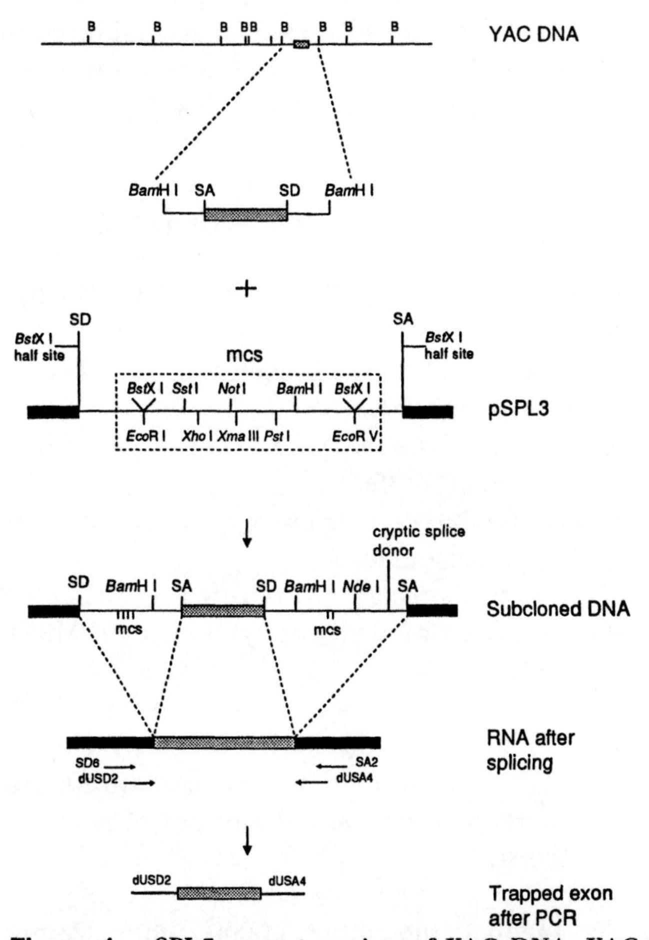

Figure 1 pSPL3 exon trapping of YAC DNA. YAC DNA is digested with a restriction enzyme, i.e., *Bam*HI (B). In this example, a single exon (*gray box*) is contained in a *Bam*HI fragment. The exon is flanked by splicing acceptor (SA) and splicing donor (SD) sites. The pSPL3 vector is prepared for subcloning at the same restriction site. The multiple cloning site (mcs) of pSPL3 contains several restriction enzyme sites. The *Eco*RI and *Eco*RV sites are contained within *Bst*XI sites. The black boxes represent HIV-*tat* and rabbit β-globin exon sequences in the pSPL3 vector. The splicing donor and acceptor sites of pSPL3 contain *Bst*XI half-sites. These sites form a complete *Bst*XI site after splicing events occur that fail to capture a "trapped" exon and can therefore be used to eliminate vector background in the subsequent PCR. After subcloning genomic DNA into pSPL3, DNA is isolated and transfected into COS-7 cells. Cytoplasmic RNA is isolated for RNA-based PCR analysis. After generation of cDNA, the first round of PCR is performed using the outside primer pair SD6 (TCT GAG TCA CCT GGA CAA CC) and SA2 (ATC TCA GTG GTA TTT GTG AGC). A second round of PCR is performed using the nested primers dUSD2 (CUA CUA CUA CUA GTG AAC TGC ACT GTG ACA AGC TGC) and dUSA4 (CUA CUA CUA CUA CAC CTG AGG AGT GAA TTG GTC G) to provide specificity and to install DNA sequences that allow UDG cloning of the PCR products. Trapped exons are recognized after gel analysis of the PCR products. Occasional but infrequent use of the cryptic splice donor site can be detected by digestion with *Nde* I.

using β-agarase, the resulting intact YAC DNA is an ideal high-molecular-weight substrate for partial or complete restriction enzyme digestion. Because the yeast chromosomes are enclosed in very small agarose beads, the material is easily pipetted.

REAGENTS

AHC broth (a rich -Ura, -Trp [double drop-out] medium for the routine growing of YACs)
SE buffer (75 mM NaCl, 25 mM EDTA, pH 8.0)
Light mineral oil
LMP (low melting point) agarose
β-Mercaptoethanol
YLE (yeast lytic enzyme, Sigma, cat. no. L-5263) solution, 5 mg YLE/ml SE buffer
Proteinase buffer (1% sarcosyl [w/v], 25 mM Na_2EDTA, pH 8.0, 500 mg/ml proteinase K)
ET buffer (50 mM EDTA, 10 mM Tris-HCl, pH 8.0)
0.1 mM phenylmethylsulfonyl fluoride (PMSF)

PROTOCOL

1. Inoculate 200 ml of AHC broth with 10 ml of an overnight culture started from a single YAC-containing yeast colony from an AHC plate or from a small amount of material from a frozen glycerol stock.

2. Incubate the culture at 200 rpm for 24 hours at 30°C on a shaking incubator.

3. Decant the culture into two 50-ml sterile polypropylene centrifuge tubes and pellet the cells by centrifuging at 1600*g* for 10 minutes at room temperature.

   Note: While the cells are being pelleted, prepare the following:

   a. Prewarm a 50-ml sterile polypropylene centrifuge tube in a 45°C water bath.

   b. Prepare 10 ml of a 1.5% solution of LMP agarose in SE buffer and equilibrate in a 45°C water bath.

   c. Equilibrate 25 ml of light mineral oil in a 45°C water bath. Equilibrate 100 ml of SE buffer in a 250-ml glass beaker containing a stir bar to 4°C by placing the beaker in an ice bucket containing an ice-water slurry. Place this on a stir plate set to medium speed. Allow at least 10 minutes for equilibration of the SE buffer to 4°C.

4. Wash the cells three times with 10 ml of SE buffer. Centrifuge the cells at 1600*g* for 5 minutes at room temperature.

5. Resuspend the combined yeast pellets in a total of 4 ml of SE buffer.

6. Add the yeast cell suspension to a prewarmed 50-ml sterile polypropylene centrifuge tube in the 45°C water bath. Incubate for 3–5 minutes.

7. Add 2.5 ml of 1.5% LMP agarose in SE buffer (45°C) to the yeast cell suspension. Gently swirl to mix.

8. Add 12 ml of light mineral oil (45°C) to the yeast cell suspension. Shake the tube back and forth vigorously for 5 seconds to form an emulsion.

9. Steadily pour the emulsion into the cold SE buffer (100 ml at 4°C) while stirring. (Do not pour into the vortex; pour just inside the beaker. The bead size is influenced by temperature and degree of swirling. Faster swirling and increasing the temperature of the SE buffer from 4°C to 9°C will result in smaller beads.)

10. Stir for 5 minutes. Transfer to two 50-ml centrifuge tubes and centrifuge at 1600*g* for 10 minutes at room temperature.

11. Remove the mineral oil from the top of each tube with a pipette. Disperse the bead pellet (both at the bottom of the tube and immediately below the mineral oil layer at the top) by pipetting through a 1-ml pipette tip after removing a few millimeters from the tip with a razor blade to increase the bore size.

12. Centrifuge again. Remove the supernatant with a pipette. Combine the agarose bead pellets using a large-bore pipette tip prepared as above.

13. Centrifuge once more. Remove any remaining supernatant.

14. Add 0.5 ml of β-mercaptoethanol and 0.5 ml of YLE solution to the agarose bead pellet, and adjust the volume to 10 ml with SE buffer. Disperse the beads and incubate for 2 hours at 37°C.

15. Centrifuge the tubes at 1600*g* for 10 minutes at room temperature.

16. Remove the supernatant with a pipette. Resuspend in 10 ml of proteinase buffer and incubate for 18 hours at 50°C.

17. Centrifuge the tubes at 1600*g* for 10 minutes at room temperature. Remove the supernatant with a pipette. Resuspend the bead pellet in ET buffer.

18. Centrifuge again and resuspend the beads in 10 ml of ET buffer containing 0.1 mM PMSF.

19. Centrifuge again. Wash twice in 10 ml of ET buffer and store at 4°C.

Pulsed-field Gel Electrophoresis of YAC DNA

In many cases, a YAC can be distinguished from the yeast chromosomes after electrophoresis and staining with ethidium bromide. In those cases where the YAC comigrates with a yeast chromosome, a human or mouse YAC can be detected and sized by transferring the pulsed-field gel to a nylon membrane and hybridizing it with labeled high-molecular-weight human or mouse Cot-1 DNA. YAC DNA can be purified after pulsed-field gel electrophoresis in the following manner: A sample of beads prepared as described above can be electrophoresed using 1/2X TBE buffer, 200 V, with a switch time of 20 seconds in a 1% LMP agarose gel for 24 hours to resolve a YAC in the 250-kb size range from neighboring yeast chromosomes. The switch time can be altered to suit the separation requirements—longer times for resolving larger YACs and shorter times for smaller YACs. After electrophoresis, the YAC band is cut out of the gel and purified as follows.

REAGENTS

Cot-1 DNAs (Life Technologies, human Cot-1 DNA, cat. no. 18376-012, mouse Cot-1 DNA, cat. no. 18440-016)
20X TBE pulsed-field gel buffer (Life Technologies, cat. no. 15559-016)
β-Agarase (Life Technologies, cat. no. 10195-014)
10X β-Agarase buffer (100 mM bis-Tris-HCl, pH 6.5, 10 mM EDTA)
5 M NaCl
7.5 M Ammonium acetate
Isopropanol
70% Ethanol

PROTOCOL

1. Cut out the YAC from the pulsed-field gel with a clean scalpel.

2. Add one volume of 1X β-agarase buffer to the gel slice. Incubate for 30 minutes at room temperature.

3. Remove the liquid from the gel slice and add β-agarase buffer to a final concentration of 1x and NaCl to a final concentration of 100 mM; melt for 10 minutes at 65°C. Measure the volume, and then place for 5 minutes at 45°C.

4. Add 1 μl of β-agarase per 200 μl of liquified gel solution and mix thoroughly. Place for 1–2 hours at 40°C.

5. Centrifuge at 9000*g* for 15 minutes at 4°C to pellet any undigested agarose.

6. Remove the supernatant to a new tube, and add 0.5 volume of ammonium acetate and 0.8 volume of isopropanol. Place on ice for 1 hour, then centrifuge at 9000*g* for 20 minutes at 4°C to pellet the DNA. Alternatively, overnight incubation at 4°C may be preferred if the amount of DNA being precipitated is small; that is, <0.1 μg.

7. Wash the pellet with 500 μl of 70% ethanol at –20°C, and centrifuge the tube for 5 minutes in a microcentrifuge.

8. Decant the ethanol. Add TE to the pellet and let it sit at room temperature for 1 hour to allow the DNA to dissolve (especially if fragments are larger than 100 kb) or overnight at 4°C.

Subcloning YAC DNA into pSPL3

REAGENTS

Plasmid pSPL3 (Exon Trapping System, Life Technologies, cat. no. 18449-017)
Phenol/chloroform/isoamyl alcohol (25:24:1)
T4 DNA ligase and buffer (Life Technologies, cat. no. 15224-017)
Human Cot-1 DNA (Life Technologies, cat. no. 15279-011)
*Pvu*II restriction enzyme
LB plus ampicillin (100 μg/ml) plates

PROTOCOL

1. Digest the YAC and pSPL3 DNA with the same or compatible restriction endonucleases. There are a large number of unique restriction enzyme sites within the multiple cloning site (mcs) of pSPL3 (Fig. 2), enabling a great deal of flexibility in the choice of cloning site.

2. Phenol-extract YAC and pSPL3 DNA after restriction digestion: Add 1 volume of phenol/chloroform/isoamyl alcohol, vortex, and centrifuge at 14,000*g* for 2 minutes at room temperature to separate the phases. Carefully remove the upper aqueous phase and transfer to a fresh microcentrifuge tube.

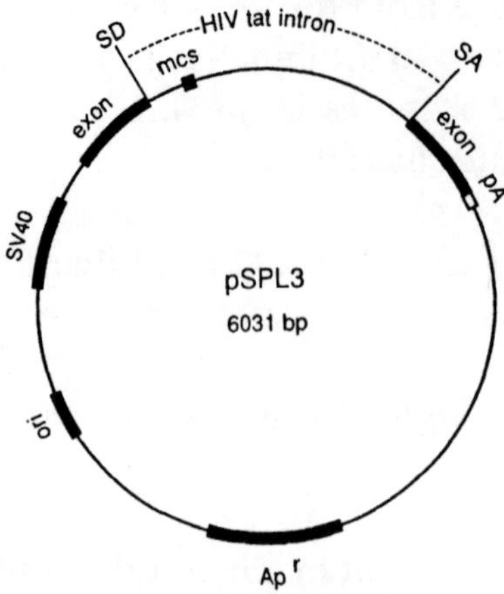

Figure 2 Restriction map of pSPL3. The vector contains sequences that enable replication in *E. coli* and COS-7 cell hosts (bacterial and SV40 origins of replication are present). An ampicillin resistance marker (Ap^r) allows selection of subclones. A multiple cloning site (mcs) that interrupts the HIV-*tat* intron provides several restriction sites for the subcloning of genomic DNA. After transfection of COS-7 cells, transcription occurs at high levels facilitated by the SV40 promoter (SV40). Processing of the transcript results in removal of the HIV-*tat* intron via splicing in which the vector exons (which contain sequences from HIV and rabbit β-globin) are combined at the splice donor (SD) and splice acceptor (SA) sites. Cytoplasmic RNA is polyadenylated [pA=SV40 poly(A) addition recognition sequence]. Exons are "trapped" from genomic DNA cloned into pSPL3 as a result of interaction of the vector splice sites (derived from HIV-*tat*) with splice sites flanking exons contained in genomic DNA.

3. Precipitate the DNA with 0.5 volume of 7.5 M ammonium acetate and 2 volumes of absolute ethanol. Mix, hold on wet ice for 20 minutes, microcentrifuge for 30 minutes, wash with 70% ethanol, air-dry, and resuspend each in TE. Resuspend a gel's worth of YAC DNA in no more than 10 µl. (The yield per gel depends on the system being used; with the Life Technologies Hex-A-Field apparatus, we obtained between 100 and 500 ng/gel.)

4. Dephosphorylate the linearized pSPL3 DNA using calf intestinal alkaline phosphatase according to the supplier's recommendations.

5. Add distilled water to the following components for a final volume of 5 µl: 5-10 ng of pSPL3 restriction enzyme-digested dephosphorylated DNA, 10 ng of restriction enzyme-digested YAC DNA, 1 µl of 5x T4 DNA ligase buffer, and 0.5 µl (at least 0.5 unit) of T4 DNA ligase. Include a control (all of the above components without YAC DNA) to assess the level of vector self-ligation.

6. Mix gently and incubate for 1 hour at room temperature.

7. Transform *E. coli* with the ligated DNA using either chemically competent or electrocompetent cells.

8. Subdivide the transformation into at least five equivalent aliquots and inoculate each into a 10X volume of LB plus ampicillin medium for an overnight growth. Isolate DNA from each culture by the alkaline lysis procedure (Moore 1987). The efficiency of the shotgun subcloning can be assessed by cutting the subcloned DNA with the same restriction enzyme that was used to subclone it and comparing it with the original cloned DNA by performing a "Cot-blot" (P.E. Nisson and P.C. Watkins, unpubl.) In this method, the digested YAC and the pSPL3 library DNA are electrophoresed, transferred to a filter, and hybridized with labeled high-molecular-weight human Cot-1 DNA. If the hybridization patterns of the digested pSPL3 library DNA and the digested YAC DNA are similar, this indicates that most of the YAC is represented in the pSPL3 library. In addition, the frequency of recombinants can be qualitatively determined by digesting the pSPL3 library DNA with *Pvu*II, which cuts pSPL3 twice, yielding a 1.8- and a 4.2-kb band. The mcs is embedded within the 1.8-kb band (Fig. 2). The presence of an intense 4.2-kb vector band, together with a variety of other bands along with the absence of a strong 1.8-kb band, indicates that the nonrecombinant frequency is quite low.

9. Plate 0.1 and 0.01 ml of each transformation on LB plus ampicillin plates to assess the degree of vector religation. Repreparation of dephosphorylated vector is recommended if greater than 10% of the transformants are nonrecombinants.

Transfection of COS-7 Cells, Preparation of Total Cellular RNA

COS-7 cells, a commonly used tissue culture cell line, have a doubling time of ~20 hours. When ordered from the ATCC, the cells should be passaged at least once to establish the appropriate doubling time. A T-70 flask of COS-7 cells is maintained and passaged by trypsinizing the cells and diluting them 1:50 in fresh medium whenever they reach 100% confluence (twice each week). Although electroporation can be used to introduce the genomic DNA subcloned in pSPL3 into COS-7 cells, transfection with the cationic lipid LipofectACE has been found to be comparable in efficiency and less costly in time and materials (Rose et al. 1991). Positive and negative (no DNA) transfection controls are recommended. A positive control plasmid containing an internal exon is included in the Life Technologies Exon Trapping System mentioned above.

REAGENTS

COS-7 cells (ATCC, cat. no. CRL-1651)
LipofectACE (Life Technologies, cat. no. 18301-010)
Supplemented DMEM (Dulbecco's modified Eagle medium, 10% fetal bovine serum, 1X glutamine, 1X nonessential amino acids, 1X penicillin and streptomycin; all available from Life Technologies)
Serum-free medium (Opti-MEM, Life Technologies, cat. no. 31985-021)
3.5-cm tissue culture dishes
TRIzol (Life Technologies, cat. no. 15596-026)
Chloroform
Isopropanol
75% Ethanol

PROTOCOL

1. Add 3×10^5 to 5×10^5 COS-7 cells in supplemented DMEM to 3.5-cm tissue culture dishes 16 hours before starting the transfection.

2. For each transfection, add an optimized amount of LipofectACE (generally about 3–10 μl) to 100 μl of Opti-MEM, mix gently, and hold for 5 minutes at room temperature. The amount of LipofectACE and DNA, as well as the media and the confluence of COS-7 cells used, should be optimized according to the supplier's recommendations.

3. For each transfection, place 1–2 μg of DNA into 100 μl of Opti-MEM.

4. To form DNA-LipofectACE complexes, combine the two solutions prepared in steps 2 and 3, mix gently, and hold at room temperature for 15 minutes.

5. Prepare the COS-7 cells for transfection by aspirating the supplemented DMEM from the tissue culture dishes, adding 1 ml of Opti-MEM to each dish, and incubating for 5 minutes at 37°C in 5% CO_2.

6. After the DNA-LipofectACE complexes are formed, add 0.8 ml of Opti-MEM to each DNA-cationic lipid mixture from step 4.

7. Remove the Opti-MEM from each dish.

8. Add one DNA-LipofectACE complex per dish.

9. Incubate for 5–6 hours at 37°C in a 5% CO_2 incubator.

10. Remove the DNA-LipofectACE complexes by aspiration and add 2 ml of supplemented DMEM to each dish.

11. Incubate the cell cultures for 16 hours at 37°C in a 5% CO_2 incubator before isolating total RNA. The dishes will become nearly 100% confluent after 16 hours' growth.

The following procedure uses TRIzol reagent, a monophasic solution of phenol and guanidine isothiocyanate; it maintains the integrity of RNA while disrupting cells and dissolving cell components (Chomczynski and Sacchi 1987; Simms et al. 1993).

Note: Do not wash the cells prior to adding the TRIzol reagent.

1. Aspirate medium from well.

2. Add 1 ml of TRIzol reagent per 3.5-cm dish.

3. Pass the cell lysate several times through a pipette and transfer to an autoclaved microcentrifuge tube.

4. Hold the sample for 5 minutes at room temperature.

5. Add 0.2 ml of chloroform per ml of TRIzol reagent and mix vigorously for 15 seconds.

6. Hold the sample for 2–3 minutes at room temperature.

7. Centrifuge at 12,000*g* for 15 minutes at 4°C.

8. Transfer the upper aqueous phase containing the RNA to a fresh tube.

9. Add 0.5 ml of isopropanol per ml of TRIzol reagent originally added and mix well.

10. Hold for 5–10 minutes at room temperature.

11. Centrifuge at 12,000*g* for 10 minutes at 4°C. The RNA forms a gel-like pellet on the side and bottom of the tube.

12. Remove the supernatant.

13. Add at least 1 ml of 75% ethanol per ml of TRIzol reagent originally added and mix well.

14. Centrifuge at 12,000*g* for 2 minutes at 4°C.

15. Air-dry the pellet.

    Note: Avoid drying completely to enhance subsequent solubilization.

16. Dissolve the RNA in 50 µl of DEPC-treated distilled water. A 5-minute incubation at 65°C followed by a brief vortexing is recommended. RNA yields of 40–50 µg per 3.5-cm dish are obtained.

RNA-PCR: Internal Exon Trapping

REAGENTS

All of the following except the *Taq* DNA polymerase, buffer, and $MgCl_2$ are in the Exon Trapping System, Life Technologies, catalog number 18449-017.

Oligonucleotides SA2, SD6, dUSA4, dUSD2
SuperScript II reverse transcriptase
RNase H
5x First-strand buffer (250 mM Tris-HCl, pH 8.3, 375 mM KCl, 15 mM $MgCl_2$)
0.1 M DTT
10 mM dNTP
*Bst*XI
$MgCl_2$
Taq DNA polymerase
10x *Taq* DNA polymerase buffer
0.5-ml microcentrifuge tubes
pAMP10 cloning vector
Uracil DNA glycosylase

First-strand cDNA Synthesis

The cDNA synthesis reaction is catalyzed by SuperScript II RNase H-Reverse Transcriptase (SuperScript II RT). This enzyme has been modified to eliminate the RNase H activity (normally present in reverse transcriptases) that degrades mRNA during first-strand cDNA synthesis. The use of this enzyme results in higher yields of cDNA (Gerard et al. 1992). Because SuperScript II RT is not inhibited significantly by ribosomal and transfer RNA, it may be used effectively to synthesize cDNA from a total RNA preparation. The remaining RNA should be stored at –20°C or –70°C.

In preparation for first-strand cDNA synthesis, set up a program on a thermal cycler that will run for 10 minutes at 70°C, 30 minutes at 42°C, and 15 minutes at 55°C. Preheat the cycler to 70°C and hold. Alternatively, water baths may be used for the following procedures.

PROTOCOL

1. Add the following components to a microcentrifuge tube: 1 μl of oligonucleotide SA2 (20 μM), RNA (1–3 μg), and DEPC-treated water (to a final volume of 12.0 μl).

2. Incubate the mixture for 5 minutes at 70°C, and then place on ice for 1 minute. Collect the contents of the tube by a brief centrifugation and add the following components at room temperature: 4 μl of 5X first-strand buffer, 2 μl of 0.1 M DTT, and 1 μl of 10 mM dNTP.

3. Mix gently, centrifuge briefly, and incubate for 2 minutes at 42°C.

4. Add 1 μl (200 units) of SuperScript II RT per reaction, mix gently, and incubate for 30 minutes at 42°C.

5. Incubate for 5 minutes at 55°C.

6. Add 1 μl (2 units) of RNase H, mix gently, and incubate for 10 minutes at 55°C.

   Note: RNase H digestion is conducted at a temperature at which reverse transcriptase is inactive. This eliminates the possible synthesis of snap-back structures and second-strand products.

7. Collect the reverse transcription reaction mixture by a brief centrifugation and place on ice. Remove 5 μl for the primary PCR amplification described below.

Primary PCR

In this portion of the protocol, the first-strand cDNA is amplified by PCR using vector-directed oligonucleotides. Following six rounds of amplification, the double-strand PCR products are digested with *Bst*XI, resulting in the elimination of two classes of background. PCR product background can result from pSPL3-derived molecules that contain only vector sequences, or from products resulting from use of the cryptic splice donor (Fig. 2). Vector-only molecules result from pSPL3 subclones that contain partial or alternatively spliced exons, exons in the antisense orientation, or introns; or from subclones that do not contain an insert. Vector-only mRNA containing the *Bst*XI site is generated via the pairing of the vector splice donor and acceptor sequences. These molecules are converted into cDNA, made double-strand, and digested with *Bst*XI to greatly reduce the entry of vector-only molecules into the secondary PCR. This permits a higher complexity of DNA to be used per transfection than with the original vector, pSPL1 (Buckler et al. 1991). pSPL3 also contains *Bst*XI sites at either end of the multiple cloning site. Digestion with *Bst*XI at these sites results in the reduction of products that result from the utilization of the cryptic splice donor site. Normally when an exon is trapped, the mcs is spliced out, unless the cryptic splice donor site located 3′ to the mcs is used (Fig. 2).

PROTOCOL

1. Add the following components to a 0.5-ml tube (this reaction and the secondary reaction can be scaled down to 10 μl using 0.2-ml tubes): 5 μl of first-strand cDNA reaction, 5 μl of 10x *Taq* DNA polymerase buffer, 1.5 μl of 50 mM $MgCl_2$, 1 μl of 10 mM dNTP, 2.5 μl of oligonucleotide SA2 (20 μM), 2.5 μl of oligonucleotide SD6 (20 μM), and sterile water to a final volume of 49.5 μl.

2. Mix the contents of the tube and overlay with a drop of mineral oil.

3. Place the tubes in a thermal cycler that has been preheated to 94°C and incubate for 5 minutes.

4. Lower the temperature to 80°C, hold, and add 2.5 units of *Taq* DNA polymerase.

5. Cycle as follows: 6 cycles (1 minute at 94°C, 1 minute at 60°C, and 5 minutes at 72°C); 1 cycle (10 minutes at 72°C).

6. Hold the reactions at 55°C, add 25 units of *Bst*XI to each, and incubate overnight at 55°C. *Bst*XI restricts the DNA in 1x PCR buffer.

7. Add 5 units of *Bst*XI to each reaction and incubate for 2 hours at 55°C.

Secondary PCR

PROTOCOL

1. Add the following components to a 0.5-ml microcentrifuge tube: 5 μl of *Bst*XI-treated primary PCR product, 4.5 μl of 10x *Taq* DNA polymerase buffer, 1.5 μl of $MgCl_2$, 1 μl of 10 mM dNTP, 1 μl of oligonucleotide dUSA4 (20 μM), 1 μl of oligonucleotide dUSD2 (20 μM), and sterile water to a final volume of 49.5 μl.

2. Mix the contents of the tube and overlay with a drop of mineral oil.

3. Place the tubes in a thermal cycler that has been preheated to 94°C and incubate for 5 minutes.

4. Lower the temperature to 80°C, hold, and add 2.5 units of *Taq* DNA polymerase.

5. Cycle as follows: 30 cycles (1 minute at 94°C, 1 minute at 60°C, and 3 minutes at 72°C); 1 cycle (10 minutes at 72°C).

6. Electrophorese 3–5 μl of each reaction on a 2% agarose gel to identify which reactions contain potential internal exons.

Cloning the Secondary PCR Products

There are several methods for cloning PCR products, including the incorporation of restriction enzyme sites in the oligonucleotides, TA cloning, or using uracil DNA glycosylase (UDG) and modified primers (Nisson et al. 1991). The UDG cloning procedure takes about 30 minutes and avoids the risk of inefficient restriction digestion and the cutting of restriction sites within an exon.

PROTOCOL

1. Add the following components to a microcentrifuge tube: 1–2 μl of secondary PCR product (100 ng), 2 μl of pAMP10 cloning vector (50 ng), 1 μl of 10x *Taq* DNA polymerase buffer, 1 μl of UDG (1 unit), and distilled water to a final reaction volume of 10 μl.

2. Mix, and incubate for 30 minutes at 37°C.

3. Transform 100 μl of competent *E. coli* cells with 5 μl of the annealing reaction and plate on LB plus ampicillin.

Evaluation of Transformants by Colony-PCR

This is a rapid method to determine the size of the PCR product that has been cloned.

REAGENTS

(see above, RNA-PCR: Internal Exon Trapping)

PCR reagents
Oligonucleotides dUSD2, dUSA4

PROTOCOL

1. Pick each colony into 50 μl of the following solution in a 0.5-ml tube (this procedure can also be scaled down to 10 μl using 0.2-ml tubes): 5 μl of 10x *Taq* DNA polymerase buffer, 1.5 μl of $MgCl_2$, 1 μl of 10 mM dNTP, 1 μl of oligonucleotide dUSA4 (20 μM), 1 μl of oligonucleotide dUSD2 (20 μM), and 0.5 μl of *Taq* DNA polymerase (2.5 units). Add sterile water to a final volume of 50 μl.

2. Place a drop of mineral oil on each sample.

3. Place tubes in a thermal cycler preheated to 94°C and hold for 5 minutes.

4. Cycle as follows: 30 cycles (45 seconds at 94°C, 30 seconds at 55°C, and 1 minute at 72°C); 1 cycle (10 minutes at 72°C).

5. Analyze the PCR products on a 2% agarose gel.

   Note: The presence of a faint band at 177 bp results from the amplification of pSPL3 sequences; however, bands larger than 177 bp should be analyzed because they may be internal exons.

Internal Exon Confirmation and Further Uses

After cloning candidate internal exons, a number of steps should be taken to eliminate some common forms of background, including *Alu* and HIV-*tat* sequences. For convenience, individual colonies should be picked and stored in microtiter dishes to increase the speed with which a large number of subclones can be screened for the presence of repetitive sequences such as *Alu* in human and B1 in the mouse. Colonies are propagated and stored in arrays that can be replicated onto filters and subsequently tested by hybridization to label high-molecular-weight human or mouse Cot-1 DNA. A filter set should be hybridized against a labeled HIV-*tat* intronic sequence (a template for labeling can be generated by PCR amplification of the intronic region of pSPL3). Following identification of repetitive and HIV-*tat* sequences, pairwise hybridization of filters containing candidate internal exons against labeled colony-PCR-amplified candidate clones will rapidly identify a minimal set of clones that can be hybridized to a filter containing either yeast chromosomes or total yeast DNA that has been digested, electrophoresed, and transferred to a filter. Following the elimination of yeast sequences, the minimal set can be sequenced and the database can be checked using BLAST (Altschul et al. 1990). The BLAST search will not only eliminate those clones that may be derived from the YAC cloning vector, but will also identify sequences in the DNA or protein sequence database that the internal exon-trapping protocol may have identified.

A critical step after the elimination of the various forms of background is determining whether a putative exon is expressed as part of an mRNA. This can be accomplished in one of several ways, including zoo blotting, Northern blotting, and RNA-PCR. Expressed sequences tend to be conserved among related species; therefore, a human exon may hybridize to primate, bovine, or even rodent DNA. A zoo blot is therefore a useful indicator of whether a sequence is expressed or not. A more rigorous test for expression, however, is provided by Northern blot analysis. Alternatively, if oligonucleotides are made to a putative exon, positive RNA-PCR results can provide a definitive answer. Once a source of RNA is known in which a novel exon is expressed, the cDNA sequence can be lengthened using

RACE, or used as a hybridization probe to screen a cDNA library directly to obtain a full-length cDNA using a new method called GeneTrapper available from Life Technologies.

Trapping 3′-Terminal Exons Using pTAG4

Most genes contain only one, or at most a few, 3′-terminal exons. This is in contrast to internal exons, which vary greatly in number. Very large genes can have a surprisingly large number of internal exons. For example, there are 79 exons in the dystrophin gene (Roberts et al. 1992). 3′-Terminal exons also tend to be larger than internal exons (Hawkins 1988), and therefore make better hybridization probes than internal exons. The protocol described here is based on the analysis of a cosmid clone; however, the method can be adapted for the analysis of other (P1, YAC, etc.) cloned DNAs. We discuss recent improvements in the procedure including DNase I treatment of COS-7 RNA and restriction digestion after primary PCR which resulted in a significant increase in the efficiency of 3′ exon capture.

3′-Terminal exon trapping, like internal exon trapping, consists of four operations: subcloning cloned genomic DNA, transfecting the subcloned DNA into COS-7 cells, purifying the mRNA from the COS-7 cells, RNA-PCR with vector-directed oligonucleotide primers, and UDG cloning. The procedure is depicted in Figure 3. Note that 3′-terminal exons are trapped because they provide flanking splice acceptor and poly(A)$^+$ signals. This is in contrast to internal exon trapping, which traps sequences flanked by a splice donor and splice acceptor.

Subcloning into pTAG4

pTAG4 contains a multiple cloning site that can be uniquely cut with a variety of restriction endonucleases (see Fig. 4). For subcloning, the vector should be linearized, dephosphorylated, and ligated to the target cloned genomic DNA. To test for the presence of functional 3′-terminal exons, either individual subclones or pools of subclones can be used. Use the same protocols to prepare vector and target DNA as described above for internal exon trapping. When subcloning human or mouse genomic DNA that is cloned in a vector that uses ampicillin resistance for selection, digested target DNA should be dephosphorylated and ligated into linearized (nondephosphorylated) pTAG4. Alternatively, one can purify the vector away from the insert or choose an enzyme that will inactivate the ampicillin-resistance gene (e.g., *Pvu*I). Otherwise, a major fraction of the subclones will be religated genomic vector or pTAG4 containing additional copies of the β-lactamase gene from the starting vector (Connors et al. 1994).

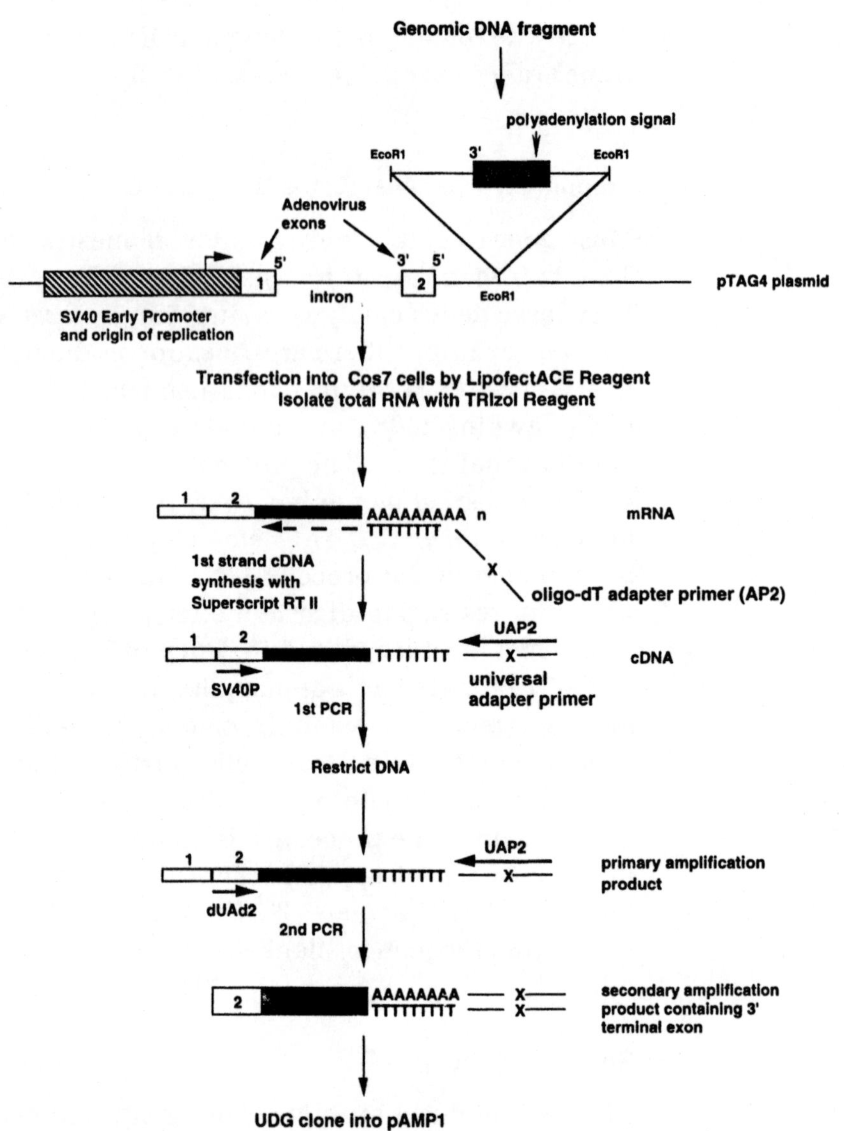

Figure 3 Flow diagram of 3′-terminal exon trapping of a cosmid using pTAG4. YAC DNA is digested with a restriction enzyme, i.e., *Eco*RI. In this example, a single exon (*gray box*) is contained in an *Eco*RI fragment. The exon is flanked by splicing acceptor (3′) and polyadenylation sites. The pTAG4 vector is prepared for subcloning at the same restriction site. The multiple cloning site (mcs) of pTAG4 contains many restriction enzyme sites. After subcloning genomic DNA into pTAG4, DNA is isolated and transfected into COS-7 cells. Cytoplasmic RNA is isolated for RNA-based PCR analysis. After generation of cDNA with adapter primer AP2 (AAG GAT CCG TCG ACA TC $(T)_{17}$, the first round of PCR is performed using the outside primer pair UAP2 (CUA CUA CUA CUA AAG GAT CCG TCG ACA TC) and SV40P (AGC TAT TCC AGA AGT AGT GA). A second round of PCR is performed using primer UAP2 and the nested primer dUAd2 (CAU CAU CAU CAU CAG TAC TCT TGG ATC GGA) to provide specificity and to install DNA sequences that allow UDG cloning of the PCR products. Trapped exons are recognized after gel analysis of the PCR products.

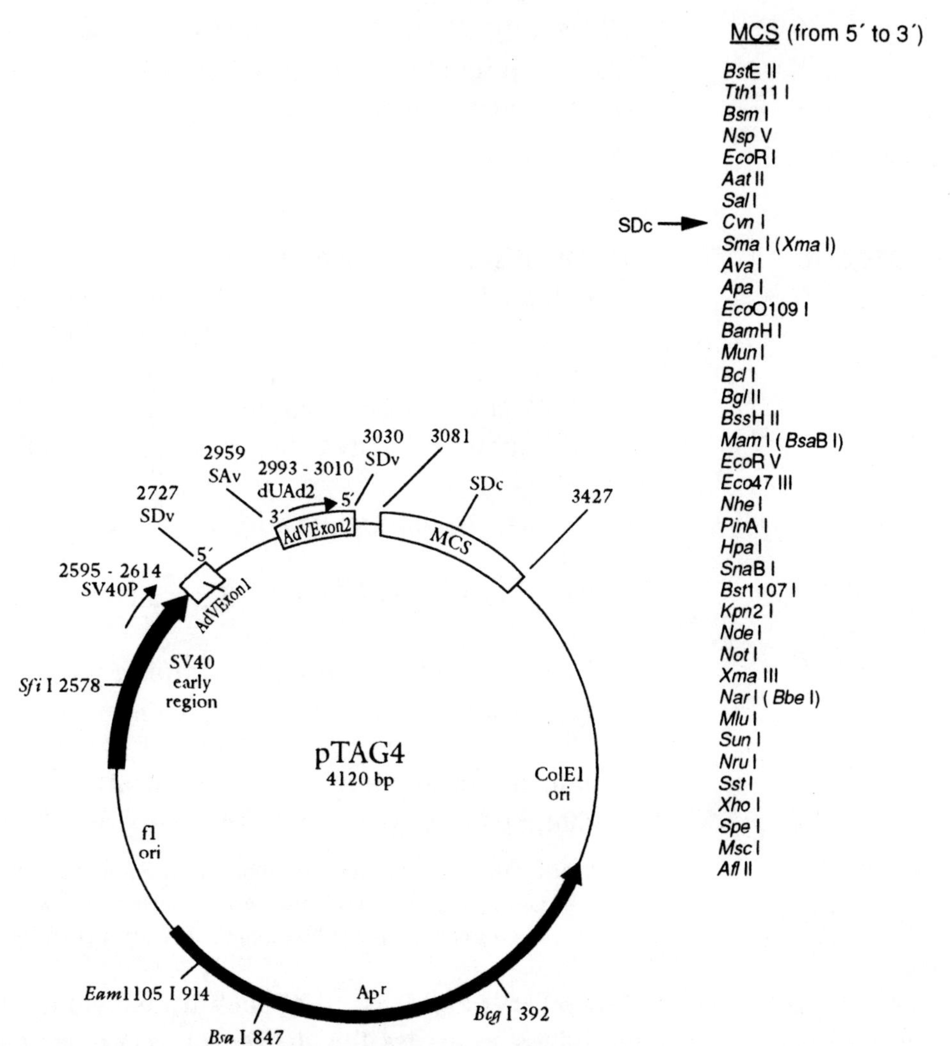

Figure 4 Restriction map of pTAG4. The vector contains sequences that enable replication in *E. coli* and COS-7 cell hosts (bacterial and SV40 origins of replication are present). An ampicillin resistance marker (Ap^r) allows for the selection of subclones. A multiple cloning site (mcs), located in the intron to the right of the adenovirus-2 exon 2, provides many restriction sites for the subcloning of genomic DNA (see sidebar). After transfection of COS-7 cells, transcription occurs at high levels facilitated by the SV40 promoter (SV40 early region). Processing of the transcript results in removal of adenovirus introns via splicing in which the vector exons are combined at the splice donor (5′) and splice acceptor (3′) sites and the appropriate vector splice donor (SDv or SDc) pairs with a splice acceptor in the subcloned genomic DNA (SDv is used if DNA is subcloned into sites 5′ of *Cvn*I in the mcs, SDc is used if DNA is subcloned into sites 3′ of *Cvn*I). If the subcloned genomic DNA contains a polyadenylation site, RNA is polyadenylated. 3′-Terminal exons are "trapped" from genomic DNA cloned into pTAG4 as a result of the interaction between a vector splice donor site (derived from adenovirus-2) and a splice acceptor site flanking an exon contained in genomic DNA.

REAGENTS

pTAG4 DNA, part of the 3′-Exon Trapping System (Life Technologies, cat. no. 18439-018)
T4 DNA ligase and buffer (Life Technologies, cat. no. 15224-17)
LB plus ampicillin (100 μg/ml) plates
Restriction enzymes

PROTOCOL

1. Add distilled water to the following components for a final volume of 5 μl: 50–100 ng of restriction enzyme-digested dephosphorylated pTAG4 DNA, 100–200 ng of restriction enzyme-digested cosmid DNA, 1 μl of 5x T4 DNA ligase buffer, and 0.5 μl (at least 0.5 unit) of T4 DNA ligase. Include a control (all of the above components without cosmid DNA) to assess the level of vector self-ligation.

2. Mix gently and incubate for 1 hour at room temperature.

3. Transform *E. coli* with the ligated DNA.

4. Recover subclone pools by scraping the transformation plates with media and a glass spreader, using at least 100 colonies for every cosmid to be tested. Alternatively, inoculate the transformation into liquid medium as described above. Recover DNA from subclone pools or liquid cultures using the alkaline lysis method.

   Note: The efficiency of the shotgun subcloning can be assessed by cutting the pTAG4 library DNA with the same restriction enzyme that was used to subclone it, and comparing it with the digested cosmid DNA by agarose gel electrophoresis.

5. Plate 0.1 and 0.01 ml of each transformation on LB plus ampicillin plates to assess the degree of vector religation. Repreparation of the dephosphorylated vector is recommended if the nonrecombinant frequency is >10%.

The methods for COS-7 transfection and preparation of total RNA are the same as was used for the internal exon trapping method described above. Include mock and pTAG4 DNA transfections as controls.

DNase I Treatment

DNase I treatment of total RNA in combination with restriction digestion following the first PCR (see below) eliminates the major form of background that has been observed using the trapping vector pTAG4 (P.E. Nisson, unpubl.).

REAGENTS

Amplification-grade DNase I (Life Technologies, cat. no. 18068-015)
10X DNase I buffer (200 mM Tris-HCl, pH 8.3, 500 mM KCl, 25 mM $MgCl_2$, 1 mg/ml BSA)

PROTOCOL

1. Resuspend RNA from a 3.5-cm dish (~10^6) cells in 50 µl of DEPC-treated water.

2. Combine 17 µl of the RNA with 2 µl of 10X DNase I buffer and 1 µl of DNase I.

3. Incubate for 30 minutes at room temperature and 5 minutes at 65°C.

4. Phenol/chloroform-extract, ethanol-precipitate, wash with 70% ethanol, dry, and resuspend in 17 µl of DEPC-treated water.

RNA-PCR: 3′-Exon Trapping

REAGENTS

All components needed for RNA-PCR are listed in the Protocol for RNA-PCR: Internal Exon Trapping, except the oligonucleotides that are specific for 3′-exon trapping.

Oligonucleotides AP2, UAP2, SV40P, dUAd2

cDNA Synthesis

RNA samples that provide positive signals can be useful as controls for cDNA synthesis and primary PCR.

PROTOCOL

1. Add the following components to a microcentrifuge tube.

Adapter primer (AP2; 10 µM)	2 µl
RNA	1–3 µg
DEPC-treated water to a final volume of 12 µl	

2. Incubate the mixture for 5 minutes at 70°C to denature the RNA, then chill 1 minute on ice. Collect the contents of the tube by brief centrifugation and add the following components:

5X First-strand buffer	4 µl
0.1 M DTT	2 µl
10 mM dNTP mix	1 µl

3. Mix gently and collect the solution by brief centrifugation. Incubate for 2 minutes at 42°C.

4. Add 1 μl (200 units) of SuperScript II RT, mix gently, and incubate for 30 minutes at 42°C.

5. Incubate for 5 minutes at 55°C.

6. Add 1 μl of RNase H, mix gently, and incubate for 10 minutes at 55°C.

   Note: RNase H digestion is conducted at a temperature at which SuperScript II RT is inactive. This eliminates the possible synthesis of snap-back structures and second-strand products.

7. Collect the reverse transcription reaction mixture by brief centrifugation and place on ice. Remove 1 μl for the primary PCR amplification.

Primary PCR

PROTOCOL

1. Add the following components to a microcentrifuge tube:

Reverse transcriptase reaction mixture	1 μl
10X *Taq* buffer	5 μl
50 mM $MgCl_2$	1.5 μl
10 mM dNTP mix	1 μl
Oligonucleotide SV40P (10 μM)	2.5 μl
Universal amplification primer (UAP2; 10 μM)	2.5 μl
Distilled water	36 μl

2. Mix the contents of the tube and overlay with a drop of mineral oil.

3. Place the tube in a thermal cycler that has been preheated to 94°C and incubate for 5 minutes.

4. Reduce the temperature to 80°C; add 2.5 units (0.5 μl) of *Taq* DNA polymerase per tube.

5. PCR for 10 cycles as follows: denature for 45 seconds at 94°C, anneal for 45 seconds at 55°C, extend for 1 minute at 72°C.

6. Incubate for an additional 10 minutes at 72°C and then hold the reaction at 4°C.

Restriction Digestion of Primary PCR Products

A second modification to the original protocol for 3′-terminal exon trapping is restriction enzyme digestion following the first PCR (Krizman and Berget 1993). A major form of background, presumably derived from the contamination of the COS-7 RNA with pTAG4 subclone DNA, can be controlled by two treatments: (1) DNase I treatment of RNA (see Protocol DNase I Treatment) and (2) restriction enzyme digestion following the first PCR. If the intervening sequence from the pTAG4 vector is present, then the subcloning restriction site will be present and susceptible to restriction cleavage. This cleavage will prevent these molecules from being amplified in the secondary PCR.

Secondary PCR

There are two reasons for performing this reaction: (1) The secondary PCR amplification uses a primer (dUAd2) that is nested 3′ to SV40P, which helps ensure the specificity of the amplified exons; and (2) the secondary primers contain dUMP residues for efficient directional UDG cloning into the pAMP1 vector (Nisson et al. 1991).

PROTOCOL

1. Add the following components to a 0.5-ml microcentrifuge tube:

Primary PCR mixture	1 μl
10X *Taq* buffer	5 μl
50 mM $MgCl_2$	1.5 μl
10 mM dNTP mix	1 μl
UAP2 (10 μM)	2 μl
Oligonucleotide dUAd2 (10 μM)	2 μl
Distilled water	37.5 μl

2. Mix the contents of the tube and overlay with a drop of mineral oil.

3. Place the tube in a thermal cycler that has been preheated to 94°C and incubate for 5 minutes.

4. Lower the temperature to 80°C, and add 2.5 units (0.5 μl) of *Taq* DNA polymerase per tube.

5. Perform 30 cycles of PCR amplification as follows: denature for 45 seconds at 94°C, anneal for 45 seconds at 55°C, extend for 1 minute at 72°C.

6. Incubate for an additional 10 minutes at 72°C and then hold the reaction at 4°C.

Agarose Gel Electrophoresis and Analysis of Secondary PCR Products

Electrophorese at least 3 μl of each reaction in a 2% agarose gel. Compare the PCR products in the experimental lanes with the pTAG4 and mock transfection lanes. Occasionally, a faint pTAG4 PCR product may appear as a ~0.9-kb band with a smear. In some cases, products derived from COS-7 RNA will be seen. All products other than the pTAG4 and COS-7 products are potential 3′ exons and can be subcloned into pAMP1; alternatively, individual PCR fragments can be isolated. PCR products can be reamplified, sized, and labeled for use as hybridization probes using colony-PCR.

Cloning and Analyzing the Secondary PCR Products

The same options described above for internal exon trapping are available for the analysis of secondary PCR products from 3′-exon trapping. After subcloning of putative 3′ exons, sequence analysis will indicate whether the trapped sequences are real. The first criterion is that one of the pTAG4 splice donors has paired with a splice acceptor in a novel sequence. The second criterion is observing one of the major polyadenylation signal sequences (AATAAA or ATTAAA) 10–30 nucleotides 5′ to the polyadenylation site. Finally, if a novel sequence is expressed as mRNA, as shown by Northern blotting or RNA-PCR, the case will be strengthened that a trapped sequence is a real 3′ exon.

ACKNOWLEDGMENTS We acknowledge the supportive information provided by conversations with Alan Buckler, Deanna Church, Eric Green, David Krizman, and Michael North.

REFERENCES

Altschul, S.F., W. Gish, W. Miller, E.W. Myers, and D.J. Lipman. 1990. Basic local alignment search tool. *J. Mol. Biol.* **215:** 403–410.

Auch, D. and M. Reth. 1990. Exon trap cloning: Using PCR to rapidly detect and clone exons from genomic DNA fragments. *Nucleic Acids Res.* **18:** 6743–6744.

Bird, A. 1986. CpG-rich islands and the function of DNA methylation. *Nature* **321:** 209–213.

Buckler, A.J., D.D. Chang, S.L. Graw, J.D. Brook, D.A. Haber, P.A. Sharp, and D.E. Housman. 1991. Exon amplification: A strategy to isolate mammalian genes based on RNA splicing. *Proc. Natl. Acad. Sci.* **88:** 4005–4009.

Chomczynski, P. and N. Sacchi. 1987. Single-step method of RNA isolation by acid guanidinium thiocyanate-phenol-chloroform extraction. *Anal. Biochem.* **161:** 156–159.

Church D.M., C.J. Stotler, J.L. Rutter, J.R. Murrell, J.A. Trofatter, and A.J. Buckler. 1994. Isolation of genes from complex sources of mammalian genomic DNA using exon amplification. *Nature Genet.* **6:** 98–105.

Collins, F.S. 1992. Positional cloning: Let's not call it reverse anymore. *Nature Genet.* **1:** 3–6.

Connors, T., T. Burn, and G. Landes. 1994. Exon trapping vector pSPL3-CAM: Improved shotgun subcloning of cosmid-derived fragments. *Focus* **16:** 111–112.

Duyk, G.M., S. Kim, R.M. Myers, and D.R. Cox. 1990. Exon trapping: A genetic screen to identify candidate transcribed sequences in cloned mammalian genomic DNA. *Proc. Natl. Acad. Sci.* **87:** 8995–8999.

Elvin, P., G. Slynn, D. Black, A. Graham, R. Butler, J. Riley, R. Anand, and A.F. Markham. 1990. Isolation of cDNA clones using yeast artificial chromosome probes. *Nucleic Acid Res.* **18:** 3913–3917.

Gerard, G.F., B.J. Schmidt, M.L. Kotewicz, and J.H. Campbell. 1992. cDNA synthesis by Moloney murine leukemia virus RNase H-minus reverse transcriptase possessing full DNA polymerase activity. *Focus* **14:** 91–93.

Gibson, F., H. Lehrach, A.J. Buckler, S.D.M. Brown, and M.A. North. 1994. Isolation conserved sequences from yeast artificial chromosomes by exon amplification. *Bio/Techniques* **16:** 453–459.

Green, E.D., H.C. Riethman, J.E. Dutchik, and M.V. Olson. 1991. Detection and characterization of chimeric yeast artificial-chromosome clones. *Genomics* **11:** 658–669.

Hawkins, J.D. 1988. A survey on intron and exon lengths. *Nucleic Acids Res.* **16:** 9893–9908.

Hamaguchi, M., H. Sakamoto, H. Tsuruta, H. Sasaki, T. Muto, T. Sugimura, and M. Terada. 1992. Establishment of a highly sensitive and specific exon-trapping system. *Proc. Natl. Acad. Sci.* **89:** 9779–9783.

Ioannou, P.A., C.T. Amemiya, J. Garnes, P.M. Kroisel, H. Shizuya, C. Chen, M.A. Batzer, and P.J. de Jong. 1994. A new bacteriophage P1-derived vector for the propagation of large human DNA fragments. *Nature Genet.* **6:** 84–89.

Krizman, D.B. and S.M. Berget. 1993. Efficient selection of 3′-terminal exons from vertebrate DNA. *Nucleic Acids Res.* **21:** 5198–5202.

Lovett, M., J. Kere, and L.M. Hinton. 1991. Direct selection: A method for the isolation of cDNAs encoded by large genomic regions. *Proc. Natl. Acad. Sci.* **88:** 9628–9632.

Moore, D. 1987. Minipreps of plasmid DNA. In *Current protocols in molecular biology* (ed. F.M. Ausubel et al.), vol. 1, pp. 1.6.2–1.6.4. Wiley, New York.

Nisson, P.E. and P.C. Watkins. 1994. Isolation of exons from cloned DNA by exon trapping. In *Current protocols in human genetics* (ed. N.C. Dracopoli), vol. 1, pp. 6.1.1–6.1.14. Wiley, New York.

Nisson, P.E., A. Raschtian, and P.C. Watkins. 1991. Rapid and efficient cloning of Alu-PCR products using uracil DNA glycosylase. *PCR Methods Appl.* **1:** 120–123.

Ozawa, N., T. Kano, C. Taga, M. Hattori, Y. Sakaki, and J. Suzuki. 1993. An exon-trapping system with a newly constructed trapping vector pEXT2; its application to the proximal region of the human chromosome 21 long arm. *FEBS Lett.* **325:** 303–308.

Parimoo, S., S.R. Patanjali, H. Shulka, D.D. Chaplin, and S.M. Weissman. 1991. cDNA selection: Efficient PCR approach for the selection of cDNAs in large genomic DNA fragments. *Proc. Natl. Acad. Sci.* **88:** 9623–9627.

Pierce, J.C., B. Sauer, and N. Sternberg. 1992. A positive selection vector for cloning high molecular weight DNA by the bacteriophage P1 system: Improved cloning efficacy. *Proc. Natl. Acad. Sci.* **89:** 2056–2060.

Roberts, R.G., A.J. Coffey, M. Bobrow, and D.R. Bentley. 1992. Determination of the exon structure of the distal portion of the dystrophin gene by vectorette PCR. *Genomics* **13:** 942–950.

Rommens, J.M., M.C. Iannuzzi, B.-S. Kerem, M.L. Drumm, G. Melmer, M. Dean, R. Rozmahel, J.L. Cole, D. Kennedy, N. Hidaka, M. Zsiga, M. Buchwald, J.R. Riordan, L.-C. Tsui, and F.S. Collins. 1989. Identification of the cystic fibrosis gene: Chromosome walking and jumping. *Science* **245:** 1059–1065.

Rose, J.K., L. Buonocore, and M.A. Whitt. 1991. A new cationic liposome reagent mediating nearly quantitative transfection of animal cells. *Bio/Techniques* **10:** 520–525.

Shizuya, H., B. Birren, U. Kim, V. Mancino, T. Slepak, Y. Tachiiri, and M. Simon. 1992. Cloning and stable maintenance of 300-kilobase-pair fragments of human DNA in *Escherichia coli* using an F-factor-based vector. *Proc. Natl. Acad. Sci.* **89:** 8794–8797.

Simms, S., P.E. Cizdiel, and P. Chomczynski. 1993. TRIzol: A new reagent for optimal single-step isolation of RNA. *Focus* **15:** 99–102.

Yokobata, K., B. Trenchak, and P.J. de Jong. 1991. Rescue of unstable cosmids by in vitro packaging. *Nucleic Acids Res.* **19:** 403–404.

Expression-PCR

David E. Lanar[1] and Kevin C. Kain[2]

[1]Department of Immunology, Walter Reed Army Institute of Research, Washington, DC 20307-5100

[2]Tropical Disease Unit, Division of Infectious Diseases, The Toronto Hospital and The University of Toronto, Toronto M5G 2C4, Ontario

INTRODUCTION

In vitro transcription and translation are powerful tools to examine the structure-function relationships of proteins. Plasmid vectors containing the bacteriophage promoters T7, T3, and SP6 are available; genes can be cloned into them and then transcribed in the presence of an appropriate RNA polymerase. However, efficient in vitro translation of these RNA transcripts often requires the insertion of an appropriate untranslated leader sequence downstream from the promoter to provide a suitable context for ribosomal binding and initiation of protein synthesis. Standard methods for in vitro transcription and translation are further limited by their requirements for cloning, bacterial amplification, DNA extraction, and restriction enzyme digestion before the desired DNA template can be transcribed and translated.

We thought it would be of great advantage to express functional proteins from DNA without these constraints. Because of the ease of obtaining adequate quantities of any gene segment by the use of PCR, we designed a method called Expression-PCR (E-PCR) (Kain et al. 1991) to modify this DNA quickly and to allow its expression without having to go through the rigors of cloning. E-PCR is a rapid simple method for the in vitro production of proteins without cloning. The resulting radiochemically pure proteins are useful for a variety of purposes, including studies on the subunit structure of proteins, epitope mapping, and protein mutagenesis.

Design of the Universal Promoter

The key to E-PCR is the design of a small DNA cassette that has all of the functional regions needed by RNA polymerase to initiate transcription of a downstream DNA segment into a RNA molecule that could be translated into protein. Four functional regions are needed for this upstream segment (see Fig. 1): (1) an RNA polymerase-binding region; (2) an untranslated leader sequence; (3) a Kozak sequence; and (4) an initiation of translation codon. We incorporated these four regions into a single unit, which is called a universal promoter (UP) because it permits the transcription of any DNA segment spliced to it.

The binding domain of T7 RNA polymerase was chosen because it is well studied and is commercially available. Nine bases 5′ to the T7 promoter were added because, although the recognition sequence of 17 bp is required, the T7 RNA polymerase also needs a number of bases 5′ of these to attach stably. Footprint analysis indicates that the sequence of the upstream fragment is not critical; however, at least 5 nucleotides are needed to stabilize the protein-DNA interaction of the polymerase with the promoter site (Ikeda and Richardson 1986). We took advantage of these base requirements to add the nucleotides of a restriction enzyme site for *Hin*dIII in case cloning of a final construct was desired.

In vitro transcription requires little more than the presence of a bacteriophage promoter upstream of the cloned DNA of interest; however, translating the desired gene can be more problematic. In vitro translation, unlike transcription, is often dependent on the presence of, and characteristics of, an untranslated leader (UTL) sequence 5′ to the initiation codon. The efficiency of translation may be poor if the AUG initiation codon lies too close to, or too far from, the 5′ end of the RNA (Struhl 1987), resides in poor sequence context (Kozak 1986), is inaccessible due to the secondary structure of the mRNA (Pelletier and Sonenbert 1985), or if there are increased requirements for translation initiation factors (Browing et al. 1988). To avoid these problems, it is often necessary to replace the normal 5′ UTL with a UTL from an efficiently translated protein. Jobling and Gehrke (1987) have previously shown that replacement of a gene's native UTL with the UTL sequence derived from the coat protein of the alfalfa mosaic virus (AMV) can increase translation efficiency as much as 35-fold. For these reasons, we have followed the T7 promoter site by the 33 nucleotides of the UTL of this AMV coat protein.

5′ CCAAGCTTCTAATACGACTCACTATAGGGTTTTTATTTTTAATTTTCTTTCAAATACTTCCACC ATG GCA CTG 3′
T7 promoter (UTL from AMV) KS Met Ala Leu

SINGLE STRANDED UNIVERSAL PROMOTER

5′ CCAAGCTTCTAATACGACTCACTATAGGG 3′
Hind III Site T7 promoter
H3T7 PRIMER

Figure 1 The sequence of the single-strand universal promoter used in E-PCR and its 5′-specific H3T7 primer.

The eukaryotic ribosome initiates translation at the first AUG codon in a favorable context on the mRNA strand. Upstream and downstream sequence requirements have been identified by Kozak (1986) and are termed Kozak sequences (KSs). Therefore, following the UTL sequence and upstream of the ATG, the sequence CCACC was added as a consensus KS. Immediately following the ATG, the G^{+4} was added to strengthen further the consideration of the ATG as an initiator codon. This G^{+4} requirement is not absolute.

Although the above domains up to and including the ATG are needed for efficient translation, an additional extension of the 3′ end of the UP was designed to allow the universal promoter to be installed upstream of a desired gene fragment using a process called splicing by overlap extension (SOE) (Horton et al. 1989). The G^{+4} was incorporated into the codon (GCA) for the amino acid alanine, a small hydrophobic amino acid, and the codon (CTG) for leucine was added so that the experimenter would have the choice of adding either ^{35}S-methionine or ^{3}H-leucine as a radioactive tag in the final translation product. The bases chosen for the codons for alanine and leucine maximized the GC content to allow a higher annealing temperature during the SOE reaction needed to link the UP to the DNA sequence of interest.

Design, Synthesis, and Purification of the UP and Specific Primers

Any PCR primers can be used to amplify the DNA of interest, and many computer programs exist that help select optimum primer pairs. The only requirement is that the forward primer have added to its 5′ end the last 9 bases (ATGGCACTG) of the UP with the first codon (ATG) lined up in the desired open reading frame (ORF). The 73-base UP and the 29-nucleotide H3T7 (Fig. 1) primer can be made as single-stranded oligonucleotides using a standard DNA synthesizer. Purification of full-length products either by acrylamide gel or by using the "trityl on" step in the last synthesis cycle followed by an OPC column (Perkin-Elmer, Applied Biosystems Div.) is acceptable.

REAGENTS

DNA containing gene of interest
Gene-specific forward primer, ATGGCACTG on 5′ end
Gene-specific reverse primer
Universal promoter, 30 fmoles/μl
H3T7 primer, 50 pmoles/μp
dNTP mix, 1.25 mM each
10x *Taq* DNA polymerase buffer with Mg^{++}

Taq DNA polymerase
PCR optimizer kit (Invitrogen, cat. no. K1220-01)
MEGAscript T7 transcription kit (Ambion, cat. no. 1314)
Retic Lysate IVT kit (Ambion, cat. no. 1200)
TNT coupled reticulocyte lysate system (Promega, cat. no. L4610)
^{35}S-Methionine, translation grade (NEN, cat. no. 0009T)
Complete amino acid mixture (if making protein for animal injection) (Promega, cat. no. L446A)

PROTOCOL

The technique is outlined in Figure 2.

1. Amplify the gene or DNA segment to be expressed by a standard PCR. Design the 5′ primer to contain at its 5′ end a 9-nucleotide sequence identical to the 9 bases on the 3′ of the UP. Keep amplifications to a maximum of 10–15 cycles.

   It is important to obtain a clean PCR band on the initial PCR. This may require the testing of several different thermophilic polymerase enzymes and/or buffer conditions. The use of a PCR optimizer kit can help if initial attempts do not yield acceptable results. If only a few extra bands are observed on gel analysis, then low-melt agarose can be used to separate and excise the DNA band of interest. The DNA can be used directly from the low-melt agarose.

2. Splice the UP to the gene of interest by a two-step PCR, which includes an initial overlap extension program followed by a secondary PCR. In the overlap extension program, add 1–10 ng of the primary PCR product to 30 fmoles of the single-strand UP (higher concentrations of both can be used without adverse effects) to a 100-µl reaction containing 1x PCR buffer, 50 µM dNTPs, 1 mM $MgCl_2$, 2.5 units of *Taq* DNA polymerase, but no primers. Denature at 94°C for 5 minutes followed by five cycles at 94°C for 30 seconds, 25°C for 30 seconds, and 72°C for 1–6 minutes. Link to a soak file at 80°C.

3. When the temperature has equilibrated to 80°C, add 50 pmoles of the H3T7 primer, complementary to the 5′ end of the UP, and 50 pmoles of antisense gene-specific primer.

4. Denature the reaction at 94°C for 5 minutes followed by 20–30 cycles at 94°C for 30 seconds, approximately 50°C for 30 seconds, and 72°C for 1–6 minutes, depending on the T_m of the gene-specific primer and the DNA template length.

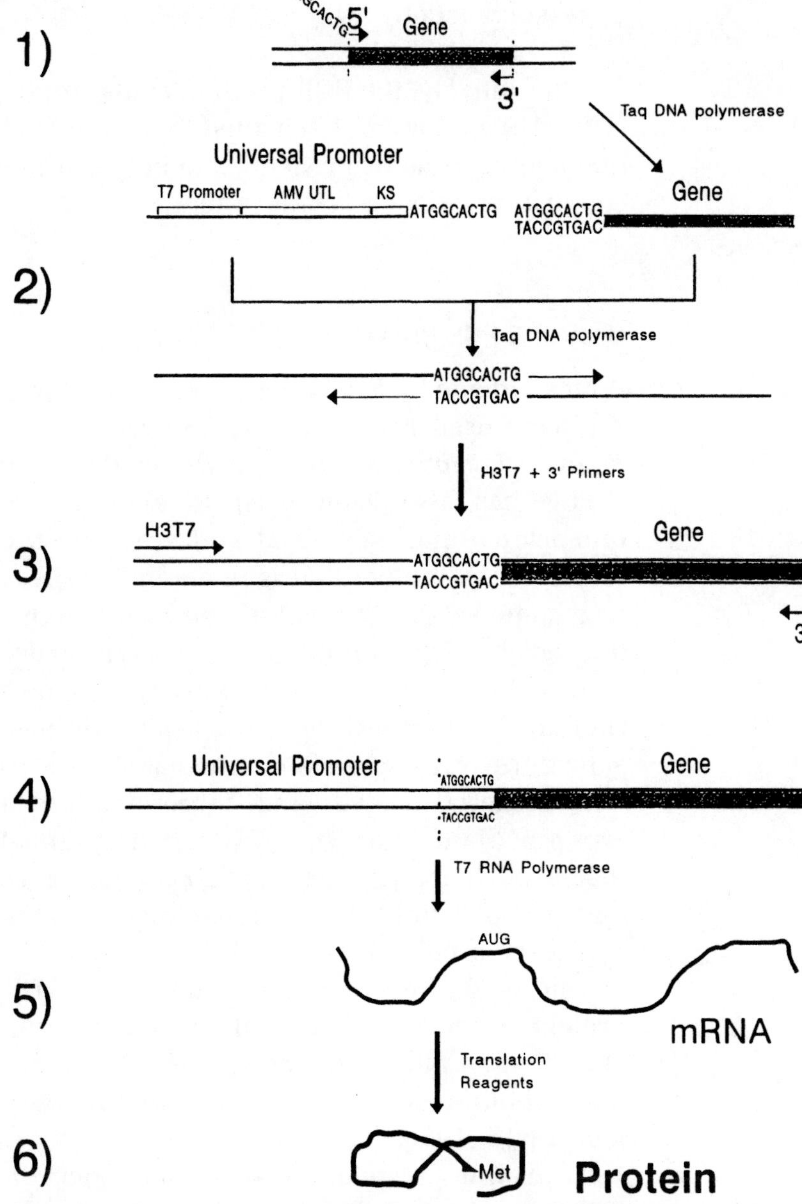

Figure 2 Principles of E-PCR. A gene or gene segment can be translated into usable protein without cloning.

5. Extract the DNA products with chloroform, precipitate, and resuspend in 10 µl of RNase-free water. This product is enough for at least ten in vitro transcription reactions.

6. Transcribe the gene linked to the UP into RNA by the use of T7 RNA polymerase. Unlike cloned genes in plasmid-based systems, the DNA does not have to be CsCl-banded or cut with restriction enzymes to give high yields of RNA.

7. Translate the mRNA into protein in a separate rabbit reticulocyte lysate reaction.

Alternatively, the PCR product containing the gene segment linked to the UP can be directly added to a transcription/translation cocktail that allows these two reactions to be performed in a single tube (TNT coupled reticulocyte lysate system).

Uses of E-PCR Protein

Proteins made by E-PCR have been used for a variety of experiments. We have used E-PCR to map the red blood cell receptor of a *Plasmodium falciparum* protein down to 40 amino acids (Kain et al. 1993). E-PCR has also been used to epitope-map the binding sites of monoclonal and polyclonal antisera (Burch et al. 1993; Tropak and Roder 1994). Protein that is made by E-PCR can also have functional enzymatic activity. Thornton and Rashtchian (1991) have synthesized the enzyme chloramphenicol acetyltransferase (CAT) and have shown that it has the same activity as enzyme made from purified CAT mRNA synthesized using a plasmid-based system. Because the genes are transcribed and translated in vitro, problems associated with bacterial growth and expression are avoided. Carole Long's group at Hahneman University (Farley and Long 1995) has shown that a polypeptide made by E-PCR has a conformationally correct epitope as defined by a monoclonal antibody, whereas bacterially expressed protein does not fold correctly.

One of the most intriguing uses of E-PCR protein is for making proteins to inject into animals for the production of antibodies (Kain et al. 1993). In this case, the synthesis of protein from mRNA is carried out without the use of radioactive methionine or leucine. Mice are injected with 100 μl of a translation mix, and rabbits are injected with 1000 μl. One potential advantage of generating antibodies against E-PCR proteins is the use of a protein synthesized in a rabbit reticulocyte lysate to immunize rabbits. In this instance, only the newly synthesized protein is "foreign" to the rabbit, and thus a low complement of background antibodies is seen. An advantage of in vitro transcription and translation is the ability to produce mutant protein by altering the DNA template. By incorporating E-PCR with the site-directed mutagenesis procedure of Higuchi et al. (1988), it is possible to generate mutant polypeptides in 1 day that can then be screened for biologic activity or used as immunogens.

In summary, E-PCR makes it possible to move from PCR product to functional translated protein in less than 8 hours. This approach offers significant advantages for researchers performing domain-mapping, epitope-mapping, and site-directed mutagenesis, because it offers the potential to rapidly identify biologically important domains and constructs for further analysis.

ACKNOWLEDGMENTS K.C.K. was supported in part by the UNDP/World Bank/WHO Special Programme for Research and Training in Tropical Diseases and the Medical Research Council of Canada (grant MT-12665).

REFERENCES

Browning, K.S., S.R. Lax, J. Humphreys, J.M. Ravel, S.A. Jobling, and L. Gehrke. 1988. Evidence that the 5′-untranslated leader of mRNA affects requirement for wheat germ initiation factors 4A, 4F, and 4G. *J. Biol. Chem.* **263:** 9630–9634.

Burch, H.B., E.V. Nagy, K.C. Kain, D.E. Lanar, F.E. Carr, L. Wartofsky, and K.D. Burman. 1993. Expression polymerase chain reaction for the in vitro synthesis and epitope mapping of autoantigen: Application to the human thyrotropin receptor. *J. Immunol. Methods* **158:** 123–130.

Farley, P.J. and C.A. Long. 1995. *Plasmodium yoelii yoelii* 17XL MSP-1: Fine-specificity mapping of a discontinuous, disulfide-dependent epitope recognized by a protective monoclonal antibody using expression-PCR (E-PCR). *Exp. Parasitol.* **80:** 328–332.

Higuchi R., B. Drummel, and R.K. Saiki. 1988. A general method of in vitro preparation and specific mutagenesis of DNA fragments: Study of protein and DNA interactions. *Nucleic Acids Res.* **16:** 7351–7367.

Horton, R.M., H.D. Hunt, S.N. Ho, J.K. Pullen, and L.R. Pease. 1989. Engineering hybrid genes without the use of restriction enzymes: Gene splicing by overlap extension. *Gene* **77:** 61–68.

Ikeda, R.A. and C.C. Richardson. 1986. Interactions of the RNA polymerase of bacteriophage T7 with its promoter during binding and initiation of transcription. *Proc. Natl. Acad. Sci.* **83:** 3614–3618.

Jobling S.A. and L. Gehke. 1987. Enhanced translation of chimaeric messenger RNAs containing a plant viral untranslated leader sequence. *Nature* **325:** 622–625.

Kain, K.C., P.A. Orlandi, and D.E. Lanar. 1991. Universal promoter for gene expression without cloning: Expression-PCR. *BioTechniques* **10:** 366–374.

Kain, K.C., P.A. Orlandi, J.D. Haynes, B.K.L. Sim, and D.E. Lanar. 1993. Evidence for two stage binding by the 175-kD erythrocyte binding antigen of *Plasmodium falciparum. J. Exp. Med.* **178:** 1497–1505.

Kozak, M. 1986. Point mutations define a sequence flanking the AUG initiator codon that modulates translation by eukaryotic ribosomes. *Cell* **44:** 283–292.

Pelletier, J. and N. Sonenbert. 1985. Insertion mutagenesis to increase secondary structure within the 5′ noncoding region of a eukaryotic mRNA reduces translational efficiency. *Cell* **40:** 515–526.

Struhl, K. 1987. Synthesizing proteins in vitro by transcription and translation of cloned genes. In *Current protocols in molecular biology* (ed. R.M. Ausubel et al.), suppl. 8 (1989), pp. 10.17.1–10.17.5. Greene/Wiley, New York.

Thornton C.G. and A. Rashtchian. 1991. Expression-PCR: A rapid method for in vitro analysis of gene products. *Focus* **14:** 86–90.

Tropak, M.B. and J.C. Roder. 1994. High-resolution mapping of GenS3 and B11F7 epitopes on myelin-associated glycoproteins by expression PCR. *J. Neurochem.* **62:** 854–862.

PCR-mediated Cloning

Under standard PCR conditions, sufficient sequence information from a template is required to design two primers that hybridize to each strand of the DNA. This, in turn, will result in the exponential amplification of the template. When attempting to clone a previously uncharacterized cDNA or gene fragment, a limited quantity of nucleic acid sequence may be available, and thus only one primer can be designed. Under these circumstances, a site for the annealing of a second primer must be created. The protocols in this section describe different alternatives for the creation of this second site for primer annealing. They include the ligation of DNA or RNA linkers and the addition of homopolymer tails with terminal deoxynucleotidyl transferase (TdT). When employing these protocols, a PCR product that includes previously uncharacterized 5′- or 3′-end sequences of cDNA or genomic DNA is obtained.

"Panhandle PCR" in this section and related techniques, such as vectorette PCR and targeted inverted repeat amplification, are used to extend and clone fragments of genomic DNA. These techniques, as well as inverse PCR, are useful for determining virus, transposon, and transgene integration sites and for cloning promoter regions.

"Rapid Amplification of cDNA Ends" (RACE) describes a one-sided PCR technique that is designed to obtain full-length cDNAs when starting from limited nucleotide or amino acid sequence. This chapter includes two alternative one-sided PCR protocols, classic and new RACE. The classic RACE protocol, and a third commonly used strategy known as AmpliFinder RACE (which is not described in this section), are available as kits from Life Technologies and CLONTECH, respectively.

The screening of phage, cosmid, or yeast artificial chromosome (YAC) libraries can be accomplished by PCR. When screening by PCR, the time and effort necessary to isolate the desired clone are significantly reduced in comparison to screening with labeled nucleic acid probes. The preparation of the template as well as the generation of pools of clones for screening phage and YAC libraries are detailed in "A PCR-based Method for Screening DNA Libraries" and "Screening of YAC Libraries with Robotic Support."

PCR is also useful when attempting to clone uncharacterized cDNAs for which no nucleic acid or amino acid sequence is available. Cloning of the desired cDNA is therefore dependent on its association with a particular phenotype. This section includes two of such PCR-based protocols: "Detection and Identification of Expressed Genes by Differential Display" and "Construction of a Subtractive cDNA Library Using Magnetic Beads and PCR," respectively. Differential display consists of a modified form of RNA-PCR that employs a set of primers of arbitrary sequence together with primers containing oligo(dT). cDNA(s) of interest are cloned after comparing PCR products obtained following the reverse transcription of RNA isolated from two different cell types. The latter protocol allows the construction of subtractive cDNA libraries using oligo(dT)-magnetic beads followed by PCR. For subtractive cDNA library construction, mRNAs and cDNAs obtained from two different cell lines are hybridized to each other. The mRNAs that do not anneal to complementary cDNAs are reverse-transcribed and amplified after ligation to linkers. As in differential display, the cDNAs of interest are obtained after two populations of mRNAs are compared.

"Phagemid Display Libraries Derived from PCR-immortalized Rearranged Immunoglobulin Genes" describes the PCR-mediated cloning of immunoglobulin sequences that results in a library of Fab fragments. The generated library is displayed as active antibody fragments in *E. coli*, and clones that bind to the antigen of interest can be selected after screening. Therefore, through the design of primer sets for the heavy- and light-chain genes, PCR allows the generation of Fab fragments as well as the amplification of cDNAs encoding monoclonal antibodies produced by hybridoma cell lines.

Rapid Amplification of cDNA Ends

Michael A. Frohman

Department of Pharmacology, University Medical Center at Stony Brook, New York 11794-8651

INTRODUCTION

Most attempts to identify and isolate a novel cDNA result in the acquisition of clones that represent only a part of the mRNA's complete sequence. The missing sequence (cDNA ends) can be cloned by PCR, using a technique variously called *r*apid *a*mplification of *c*DNA *e*nds (RACE) (Frohman et al. 1988), anchored PCR (Loh et al. 1989), or one-sided PCR (Ohara et al. 1989). Since the initial reports of this technique, many labs have developed significant improvements on the basic approach (Frohman 1989, 1993, 1994; Frohman and Martin 1989; Dumas et al. 1991; Fritz et al. 1991; Borson et al. 1992; Jain et al. 1992; Rashtchian et al. 1992; Schuster et al. 1992; Bertling et al. 1993; Monstein et al. 1993; Templeton et al. 1993). I describe here the most recent hybrid version of the relatively simple, classic RACE and a more powerful, but technically more challenging, "new RACE" protocol that is adapted from the work of a number of laboratories (see Fig. 1) (Tessier et al. 1986; Mandl et al. 1991; Volloch et al. 1991; Brock et al. 1992; Bertrand et al. 1993; Fromont-Racine et al. 1993; Liu and Gorovsky 1993; Sallie 1993). Commercial RACE kits are available from BRL (Schuster et al. 1992) and CLONTECH. They are convenient but not as powerful as the most recent versions of classic and new RACE.

Why use PCR (RACE) at all instead of screening (additional) cDNA libraries? RACE cloning is advantageous for several reasons: First, it takes weeks to screen cDNA libraries, obtain individual cDNA clones, and analyze the clones to determine if the missing sequence is present; using PCR, such information can be generated within a few days. As a result, it becomes practical to modify RNA preparation and/or reverse transcription conditions until full-length cDNAs are generated

Figure 1 Schematic representation of the setting in which RACE is useful in cDNA cloning strategies. Depicted is an mRNA for which a cDNA representing only an internal portion of the transcript has been obtained. Such circumstances often arise; one such example is when closely related genes are cloned using PCR amplification with degenerate primers encoding sequences homologous to amino acids found in all known members of the gene family.

and observed. In addition, essentially unlimited numbers of independent clones can be generated using RACE, unlike library screens in which generally a single to a few cDNA clones are recovered. The availability of large numbers of clones provides confirmation of nucleotide sequence and allows the isolation of unusual transcripts that are alternately spliced or that begin at infrequently used promoters.

Classic RACE

PCR is used to amplify partial cDNAs representing the region between a single point in a mRNA transcript and its 3′ or 5′ end (Fig. 2). A short internal stretch of sequence must already be known from the mRNA of interest. From this sequence, gene-specific primers are chosen that are oriented in the direction of the missing sequence. Extension of the partial cDNAs from the unknown end of the message back to the known region is achieved using primers that anneal to the preexisting poly(A) tail (3′ end) or to an appended homopolymer tail (5′ end). Using RACE, enrichments on the order of 10^6- to 10^7-fold can be obtained. As a result, relatively pure cDNA "ends" are generated that can be easily cloned or rapidly characterized using conventional techniques (Frohman et al. 1988).

To generate "3′-end" partial cDNA clones, mRNA is reverse-transcribed using a "hybrid" primer (Q_T) that consists of 17 nucleotides of oligo(dT) followed by a unique 35-base oligonucleotide sequence (Q_I-Q_O; Fig. 2a, c), which in many reports is denoted as an "anchor" primer. Amplification is then performed using a primer containing part of this sequence (Q_O) that now binds to each cDNA at its 3′ end, and using a primer derived from the gene of interest (GSP1). A second set of amplification cycles is then carried out using "nested" primers (Q_I and GSP2) to quench the amplification of nonspecific products. To generate "5′-end" partial cDNA clones, reverse transcription (primer extension) is carried out using a gene-specific primer (GSP-RT; Fig. 2b) to generate first-strand products. Then, a poly(A) tail is appended

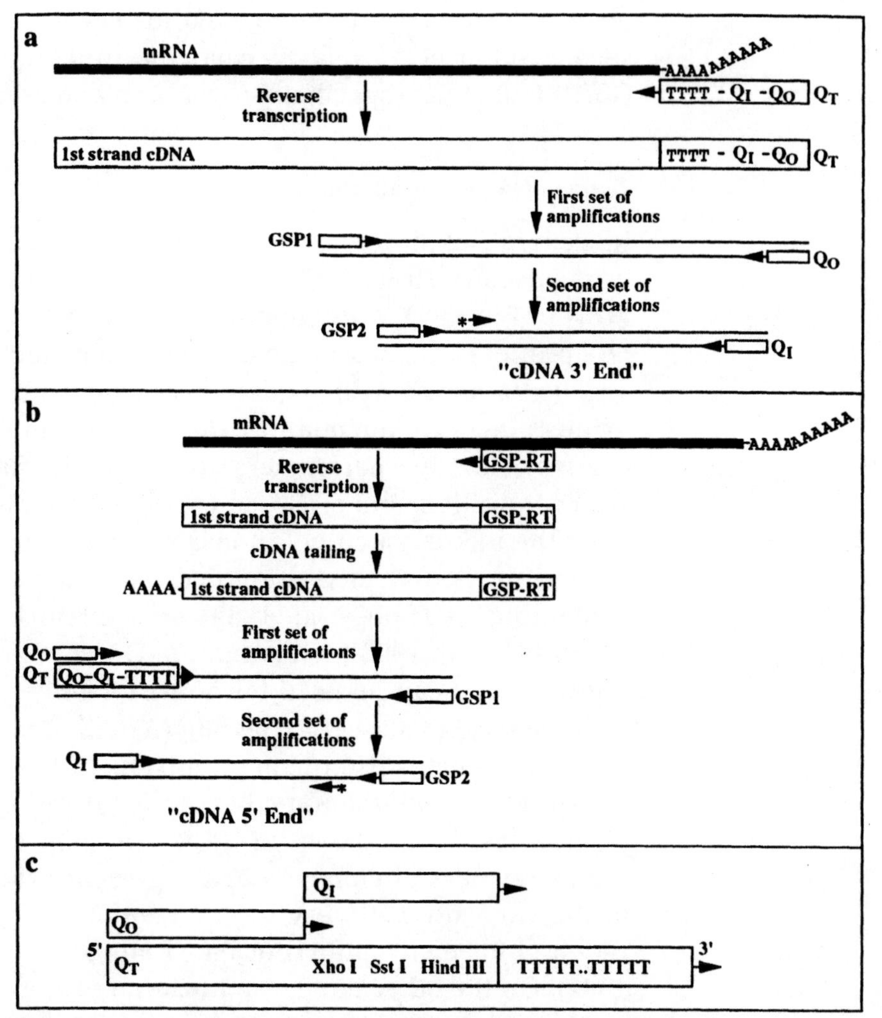

Figure 2 Schematic representation of classic RACE. Explanations are given in the text. At each step, the diagram is simplified to illustrate only how the new product formed during the previous step is utilized. (GSP1) Gene-specific primer 1; (GSP2) gene-specific primer 2; (GSP-RT) gene-specific primer used for reverse transcription; (*→) GSP-Hyb/Seq or gene-specific primer for use in hybridization and sequencing reactions. (*a*) Amplification of 3′-partial cDNA ends. (*b*) Amplification of 5′-partial cDNA ends. (*c*) Schematic representation of the primers used in classic RACE. The 52-nucleotide Q_T primer (5′-Q_O-Q_I-TTTT-3′) contains a 17-nucleotide oligo-(dT) sequence at the 3′end followed by a 35-nucleotide sequence encoding *Hin*dIII, *Sst*I, and *Xho*I recognition sites. The Q_I and Q_O primers overlap by 1 nucleotide; the Q_I primer contains all three of the recognition sites.
Primers: Q_T: 5′-CCAGTGAGCAGAGTGACGAGGACTCGAGCTCAAGCTTT-TTTTTTTTTTTTT-3′; Q_O: 5′- CCAGTGAGCAGAGTGACG-3′; Q_I: 5′-GAGGA-CTCGAGCTCAAGC-3′.

using terminal deoxynucleotidyltransferase (TdT) and dATP. Amplification is then achieved using (1) the hybrid primer Q_T to form the second strand of cDNA, (2) the Q_O primer, and (3) a gene-specific primer upstream of the one used for reverse transcription. Finally, a second set of PCR cycles is carried out using nested primers (Q_I and GSP2) to increase specificity (Frohman and Martin 1989).

Classic RACE Variations

In general, as described above, the gene-specific primer is derived from a short stretch of sequence that is already known from the mRNA of interest. A frequent question is whether degenerate primers, i.e., primers directed against a predicted nucleotide sequence based on known amino acid sequence, can be used instead. Although such primers increase the quantity of spurious amplification, the approach can work, if other parameters are favorable (i.e., message abundance, GC composition, and cDNA end size; see Monstein et al. 1993).

At the unknown end of the cDNA, the 5′ end can be tailed with Cs instead of As and then amplified using a hybrid primer with a tail containing Gs (Loh et al. 1989) or a mixture of Gs and inosines (I) (Schuster et al. 1992). Although the G:I approach entails synthesizing a primer that can be used for 5′ RACE only (since a T-tailed primer must be used to anneal to the poly(A) tail of the 3′ end), there may be sufficient benefits from using a mixed G:I tail to justify the cost, since the G:I region should anneal at temperatures similar to those of other primers normally used in PCR. In contrast, it is believed that homopolymers of either Ts or Gs present problems during PCR, due to the very low and very high annealing temperatures, respectively, required for their optimal usage (Frohman et al. 1988; Schuster et al. 1992). On the other hand, the inosine residues function as degenerate nucleotides and lead to higher spurious amplification, so the magnitude of the benefit of using a mixed G:I primer is unknown.

To minimize the length of homopolymer tail actually amplified, a lock-docking primer was developed by Borson et al. (1992). In this approach, the final 2 nucleotides on the 3′end of the primer are degenerate. For example, to amplify cDNAs linked to an A-tail, the lock-docking primer would look like:

$$5' \; \text{XXXXXXXXXX-AAAAAAAAA-} \begin{pmatrix} G \\ T \\ C \end{pmatrix} \begin{pmatrix} G \\ A \\ T \\ C \end{pmatrix} \; 3'$$

where X represents (e.g.) one or more restriction sites at the 5′ end of the primer. The advantage of this approach is that it forces the primer to anneal to the junction of the natural or appended homopolymer tail and the cDNA sequence. The disadvantage is that it is necessary to synthesize four primers, because most synthesizers can only synthesize primers starting from an unambiguous 3′ end.

In another variation, the location of the anchor primer is changed from the end of the unknown region of sequence to random points within the unknown region (Fritz et al. 1991). This is accomplished using a primer containing an anchor region followed by six random nucleotides (5′-XXXXXXX-NNNNNN-3′) either for reverse transcription (3′ RACE) or for creation of the second strand of cDNA (5′ RACE). This approach is valuable when the 3′ or 5′ ends lie so far away from the region of known sequence that the entire unknown region cannot be amplified effectively. Using this approach, cDNA ends of defined sizes are not generated; instead, one obtains a library of randomly sized fragments, all of which initiate at the gene-specific primer. The largest fragments can be cloned and characterized, extending the length of the known sequence, and the process (or standard RACE) can be repeated until the real unknown end is identified. The development of "long" PCR may make this approach unnecessary.

New RACE

The most technically challenging step in classic 5′ RACE is to cajole reverse transcriptase to copy the mRNA of interest in its entirety into first-strand cDNA. Because prematurely terminated first-strand cDNAs are tailed by terminal transferase just as effectively as full-length cDNAs, cDNA populations composed largely of prematurely terminated first strands result primarily in the amplification and recovery of cDNA ends that are not full length either (Fig. 3a). This problem is encountered routinely for vertebrate genes, which are often quite GC-rich at their 5′ ends and thus frequently contain sequences that hinder reverse transcription. A number of laboratories have developed steps or protocols designed to approach the problem (Tessier et al. 1986; Mandl et al. 1991; Volloch et al. 1991; Brock et al. 1992; Bertrand et al. 1993; Fromont-Racine et al. 1993; Liu and Gorovsky 1993; Sallie 1993); the "new RACE" protocol described here is for the most part a composite adapted from the cited reports.

New RACE departs from classic RACE in that the "anchor" primer is attached to the 5′ end of the mRNA *before* the reverse transcription step; hence, the anchor sequence becomes incorporated into the first-strand cDNA if and only if the reverse transcription proceeds through the entire length of the mRNA of interest (and through the relatively short anchor sequence), as shown in Figure 3b.

Before beginning new RACE (Fig. 4a), the mRNA is subjected to a dephosphorylation step using calf intestinal phosphatase (CIP). This step actually does nothing to full-length mRNAs, which have methyl-G caps at their termini, but it does dephosphorylate degraded mRNAs, which are uncapped at their termini (Volloch et al. 1991). This makes

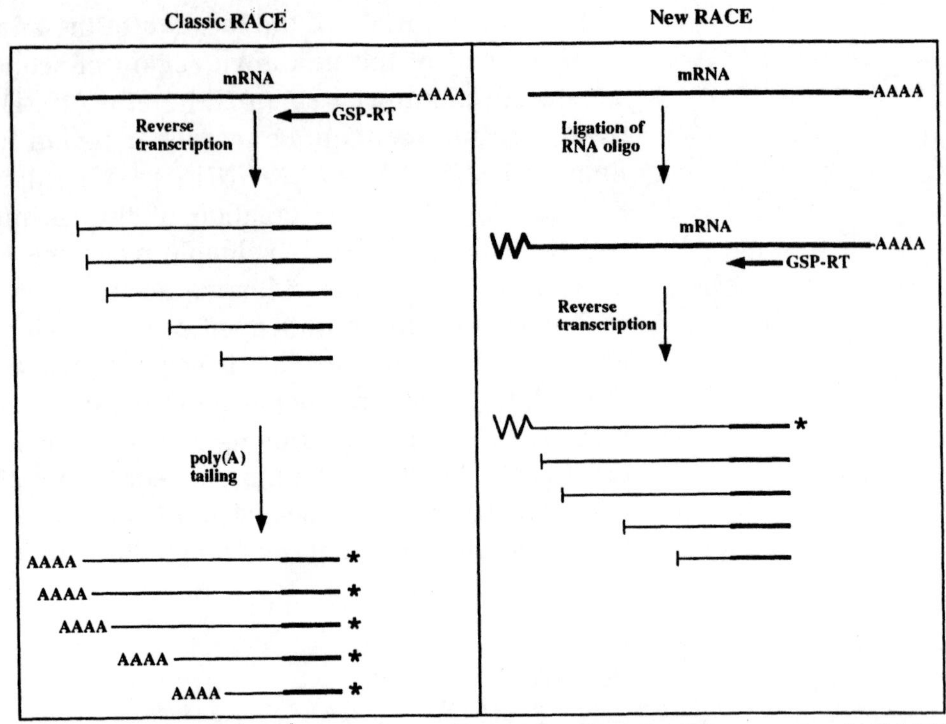

Figure 3 Depiction of the advantage of using new RACE over classic RACE. In classic RACE, premature termination in the reverse transcription step results in polyadenylation of less-than-full-length first-strand cDNAs, all of which can be amplified using PCR to generate less-than-full-length cDNA 5′ ends. * indicates cDNA ends created that will be amplified in the subsequent PCR. In new RACE, less-than-full-length cDNAs are also created, but are not terminated by the anchor sequence, and hence cannot be amplified in the subsequent PCR.

the degraded RNA biologically inert during the ensuing ligation step, because the phosphate group is required to drive the reaction. The full-length mRNAs are then decapped using tobacco acid pyrophosphatase (TAP), which leaves them with an active and phosphorylated 5′ terminus (Mandl et al. 1991; Fromont-Racine et al. 1993). Using T4 RNA ligase, this mRNA is then ligated to a short synthetic RNA oligonucleotide that has been generated by in vitro transcription of a linearized plasmid (Fig. 4b) (Tessier et al. 1986). The RNA oligonucleotide-mRNA hybrids are then reverse-transcribed using a gene-specific primer or random primers to create first-strand cDNA. Finally, the 5′ cDNA ends are amplified in two nested PCR procedures using additional gene-specific primers and primers derived from the sequence of the RNA oligonucleotides.

The new RACE approach can also be used to generate 3′ cDNA ends (Volloch et al. 1991; see also related protocols in Mandl et al. 1991 and Brock et al. 1992) and is useful in particular for non-polyadenylated RNAs. In brief, cytoplasmic RNA is dephosphorylated

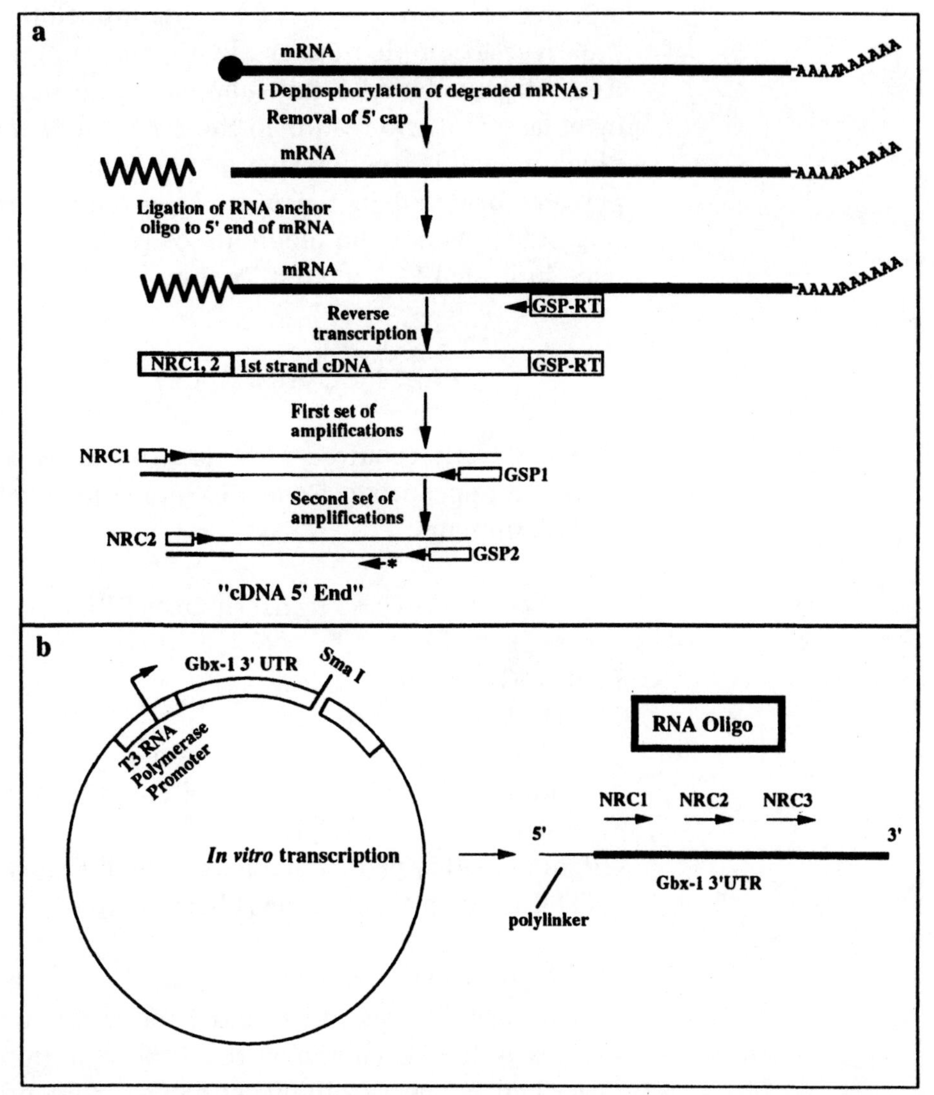

Figure 4 Schematic representation of new RACE. Explanations are given in the text. At each step, the diagram is simplified to illustrate only how the new product formed during the previous step is utilized. See legend to Fig. 2 for description of some primers. (*a*) Amplification of 5′-partial cDNA ends. (*b*) In vitro synthesis of the RNA oligonucleotide used for ligation in new RACE and schematic representation of the corresponding required primers. A 132-nucleotide RNA oligonucleotide is produced by in vitro transcription of the plasmid depicted using T3 RNA polymerase. Primers NRC-1, -2, and -3 are derived from the sequence of the oligonucleotide but do not encode restriction sites. To assist in the cloning of cDNA ends, the sequence ATCG is added to the 5′ end of NRC-2, as described in the cloning section of the text.

and ligated to a short synthetic RNA oligonucleotide as described above. Although ligation of the oligonucleotide to the 5′ end of the RNA was emphasized above, RNA oligonucleotides actually ligate to both ends of the cytoplasmic RNAs. For the reverse transcription step, a primer derived from the RNA oligonucelotide sequence is used (e.g., the reverse complement of NRC-3, Fig. 4). Reverse transcription of the RNA oligonucleotides that happen to be ligated to the 3′ end of the cytoplasmic RNAs results in the creation of cDNAs that have the RNA oligonucleotide sequence appended to their 3′ end. Gene-specific primers oriented in the 5′→ 3′ direction and new RACE primers (e.g., the reverse complements of NRC-2 and NRC-1, Fig. 4) can be used in nested PCR to amplify the 3′ ends.

PART I: CLASSIC RACE

REAGENTS

The materials required for this procedure can be purchased, along with the appropriate 5X or 10X enzyme reaction buffers, from most major suppliers.

SuperScript II reverse transcriptase (BRL)
RNase H (BRL)
RNasin (Promega)
Taq DNA polymerase
TdT (BRL or Boehringer Mannheim)
10X buffer: 670 mM Tris-HCl, pH 9.0, 67 mM $MgCl_2$, 1700 μg/ml BSA, and 166 mM $(NH_4)_2SO_4$
Oligonucleotide primer sequences as in Figure 2 legend
dNTPs as 100 mM solutions (Pharmacia or Boehringer Mannheim)

Enzymes are used as directed by the suppliers, except for *Taq* DNA polymerase. Instead of using the recommended reaction mixture, use the 10X buffer listed above; reaction conditions are altered as further described below (Frohman et al. 1988; Frohman 1994). Primers can be used "crude" except for Q_T, which should be purified to ensure that it is uniformly full length.

PROTOCOLS

3′-End cDNA Amplification

STEP 1. REVERSE TRANSCRIPTION TO GENERATE cDNA TEMPLATES

1. Assemble the following reverse transcription components on ice: 4 μl of 5X reverse transcription buffer (5X buffer contains 250 mM Tris-HCl, pH 8.3, 375 mM KCl, 15 mm $MgCl_2$), 1 μl of dNTPs (stock concentration is 10 mM of each dNTP), 2 μl of 0.1 M DTT, 0.5 μl of Q_T primer (100 ng/μl), and 0.25 μl (10 units) of RNasin.

2. Heat 1 μg of poly(A)+ RNA or 5 μg of total RNA in 13 μl of water for 3 minutes at 80°C, cool rapidly on ice, and spin for 5 seconds in a microfuge.

3. Add poly(A)+ RNA or total RNA to the reverse transcription components. Add 1 μl (200 units) of SuperScript II reverse transcriptase, and incubate for 5 minutes at room temperature, 1 hour at 42°C, and 10 minutes at 50°C.

4. Incubate for 15 minutes at 70°C to inactivate reverse transcriptase. Spin for 5 seconds in a microfuge.

5. Add 0.75 μl (1.5 units) of RNase H to the tube and incubate for 20 minutes at 37°C to destroy the RNA template.

6. Dilute the reaction mixture to 1 ml with TE (10 mM Tris-HCl, pH 7.5, 1 mM EDTA) and store at 4°C (3′-end cDNA pool).

Poly(A)+ RNA is preferentially used for reverse transcription to decrease the background, but it is unnecessary to prepare it if only total RNA is available. An important factor in the generation of full-length 3′-end partial cDNAs concerns the stringency of the reverse transcription reaction. Reverse transcription reactions were historically carried out at relatively low temperatures (37–42°C) using a vast excess of primer (~1/2 the mass of the mRNA, which represents an ~30:1 molar ratio). Under these low-stringency conditions, a stretch of A residues as short as 6–8 nucleotides suffices as a binding site for an oligo(dT)-tailed primer. This may result in cDNA synthesis being initiated at sites upstream of the poly(A) tail, leading to truncation of the desired amplification product. One should suspect that this has occurred if a canonical polyadenylation signal sequence (Wickens and Stephenson 1984) is not found near the 3′ end of the cDNAs generated. This can be minimized by controlling two parameters: primer concentration and reaction temperature. The primer concentration can be reduced dramatically without significantly decreasing the amount of cDNA synthesized (Coleclough 1987) and will begin to bind preferentially to the longest A-rich stretches present (i.e., the poly(A) tail). The quantity recommended above represents a good starting point; it can be reduced fivefold further if significant truncation is observed.

In the protocol described above, the incubation temperature is raised slowly to encourage reverse transcription to proceed through regions of difficult secondary structure. Since the half-life of reverse transcriptase rapidly decreases as the incubation temperature increases, the entire reaction cannot be carried out at elevated temperatures. Alternatively, the problem of difficult secondary structure (and nonspecific reverse transcription) can be approached using heat-stable reverse transcriptases, which are now available from several suppliers (Perkin-Elmer, Amersham, Epicentre, and others). As in

PCR, the stringency of reverse transcription can be controlled by adjusting the temperature at which the primer is annealed to the mRNA. The optimal temperature depends on the specific reaction buffer and reverse transcriptase used and should be determined empirically, but it usually is in the range of 48–56°C for a primer terminated by a 17-nucleotide oligo(dT)-tail.

STEP 2. AMPLIFICATION

First round:

1. Add an aliquot of the cDNA pool (1 μl) and primers (25 pmoles each of GSP1 and Q_O) to 50 μl of the PCR cocktail (1x *Taq* DNA polymerase buffer [described above], each dNTP at 1.5 mM, and 10% DMSO) in a 0.5-ml microfuge tube.

2. Heat in a DNA thermal cycler for 5 minutes at 98°C to denature the first-strand products. Cool to 75°C. Add 2.5 units of *Taq* DNA polymerase, overlay the mixture with 30 μl of mineral oil (Sigma 400-5; preheat it in the thermal cycler to 75°C), and incubate for 2 minutes at the appropriate annealing temperature (52–60°C). Extend the cDNAs for 40 minutes at 72°C.

3. Carry out 30 cycles of amplification using a step program (1 minute at 94°C, 1 minute at 52–60°C, 3 minutes at 72°C), followed by a 15-minute final extension at 72°C. Cool to room temperature.

Second round:

4. Dilute 1 μl of the amplification products from the first round into 20 μl of TE.

5. Amplify 1 μl of the diluted material with primers GSP2 and Q_I using the procedure described above, but eliminate the initial 2-minute annealing step and the 40-minute 72°C extension step.

It is important to add the *Taq* DNA polymerase *after* heating the mixture to a temperature above the T_m of the primers (hot start PCR). Addition of the enzyme prior to this point allows one "cycle" to take place at room temperature, promoting the synthesis of nonspecific background products dependent on low-stringency interactions.

An annealing temperature close to the effective T_m of the primers should be used. The Q_I and Q_O primers work well at 60°C under the PCR conditions recommended here, although the actual optimal temperature may depend on the PCR machine used. Gene-specific primers of similar length and GC content should be chosen. Computer programs to assist in the selection of primers are widely available and should be used. An extension time of 1 minute/kb of expected product should be allowed during the amplification cycles. If the expected length of product is unknown, try 3–4 minutes initially.

Very little substrate is required for the PCR. A 1-μg amount of poly(A)$^+$ RNA typically contains ~5 × 10^7 copies of *each* low-abundance transcript. The PCR described here works optimally when 10^3–10^5 templates (of the desired cDNA) are present in the starting mixture; thus, as little as 0.002% of the reverse transcription mixture suffices for the PCR! The addition of too much starting material to the amplification reaction leads to the production of large amounts of nonspecific product and should be avoided. The RACE technique is particularly sensitive to this problem because every cDNA in the mixture, both desired and undesired, contains a binding site for the Q_O and Q_I primers.

It was found empirically that allowing extra extension time (40 minutes) during the first amplification round (when the second strand of cDNA is created) sometimes resulted in increased yields of the specific product relative to background amplification, and in particular, increased the yields of long cDNAs versus short cDNAs when specific cDNA ends of multiple lengths were present (Frohman et al. 1988). Prior treatment of cDNA templates with RNA hydrolysis or a combination of RNase H and RNase A infrequently improves the efficiency of amplification of specific cDNAs.

For some applications intended for cloned PCR products, such as expressing cDNAs to generate proteins, it is critically important to minimize the rate at which mutations occur during amplification. In other applications, such as using the cloned DNA as a probe in hybridization experiments, the presence of a few mutations is relatively unimportant and thus it is most convenient to use PCR conditions that maximize the likelihood of generating the desired product the first time a set of primers is used. Unfortunately, PCR conditions that result in a minimum of mutations are finicky, and often the desired product cannot be generated until the PCR conditions have been optimized, whereas PCR conditions that reliably produce desired products result in a relatively high mutation rate (~1% after 30 rounds). Thus, appropriate conditions must be chosen to generate the PCR products required prior to undertaking cloning steps. PCR conditions that result in a minimum of mutations require the use of nucleotides (dNTPs) at low concentrations (0.2 mM). Using the conditions recommended for *Taq* DNA polymerase by Perkin-Elmer results in an error rate of about 0.05% after 30 rounds of amplification. However, the conditions recommended often have to be optimized, meaning that the pH of the buffer and the concentration of magnesium have to be adjusted until the desired product is observed. In addition, the inclusion of DMSO or formamide may be required. For those who do not wish to prepare their own reagents to carry out optimization experiments, such kits are commercially available (e.g., from Invitrogen and Stratagene). PCR conditions that work much more frequently in the absence of optimization steps require the use of DMSO, ammonium sulfate, and relatively high concentrations (1.5 mM) of dNTPs, as described above. It should be noted that the inclusion of DMSO to 10% decreases primer melting temperatures (and thus optimal annealing temperatures) by about 5–6°C.

5′-End cDNA Amplification

STEP 1. REVERSE TRANSCRIPTION TO GENERATE cDNA TEMPLATES

1. Assemble the following reverse transcription components on ice: 4 μl of 5x reverse transcription buffer (5x buffer contains 250 mM Tris-HCl, pH 8.3, 375 mM KCl, 15 mM $MgCl_2$), 1 μl of dNTPs (stock concentration is 10 mM of each dNTP), 2 μl of 0.1 M DTT, and 0.25 μl (10 units) of RNasin.

2. Heat 0.5 ml of GSP-RT primer (100 ng/μl) and 1 μg of poly(A)$^+$ RNA or 5 μg of total RNA in 13 μl of water for 3 minutes at 80°C, cool rapidly on ice, and spin for 5 seconds in a microfuge. Add to the reverse transcription components.

3. Add 1 μl (200 units) of SuperScript II reverse transcriptase, and incubate for 1 hour at 42°C, and 10 minutes at 50°C.

4. Incubate for 15 minutes at 70°C to inactivate reverse transcriptase. Spin for 5 seconds in a microfuge.

5. Add 0.75 μl (1.5 units) of RNase H to the tube and incubate for 20 minutes at 37°C to destroy the RNA template.

6. Dilute the reaction mixture to 400 μl with TE (10 mM Tris-HCl, pH 7.5, 1 mM EDTA) and store at 4°C (5′-end nontailed cDNA pool).

Many of the remarks made above in the section on reverse transcribing 3′-end partial cDNAs are also relevant here and should be noted. There is, however, one major difference. The efficiency of cDNA extension is now critically important, because each specific cDNA, no matter how short, is subsequently tailed and becomes a suitable template for amplification (Fig. 3a). Thus, the PCR products eventually generated directly reflect the quality of the reverse transcription reaction. Extension can be maximized by using clean, intact RNA; by selecting the primer for reverse transcription to be near the

5′ end of region of known sequence; and (in theory) by using heat-stable reverse transcriptase at elevated temperatures or a combination of SuperScript II and heat-stable reverse transcriptase at multiple temperatures. Synthesis of cDNAs at elevated temperatures should diminish the amount of secondary structure encountered in GC-rich regions of the mRNA. Random hexamers (50 ng) can be substituted for GSP-RT to create a "universal" 5′-end cDNA pool. A universal pool can be used for amplification of the 5′ end of any cDNA created in the reverse transcription. Correspondingly, however, each cDNA is present at a much lower level than if created using the gene-specific reverse transcription described above. If using random hexamers, insert a room temperature 10-minute incubation period after mixing everything together.

STEP 2. APPENDING A POLY(A) TAIL TO FIRST-STRAND cDNA PRODUCTS

1. Remove excess primer using MICROCON-100 spin filters (Amicon) or an equivalent product, following the manufacturer's instructions. Wash the material by spin filtration twice more using TE. The final volume recovered should not exceed 10 μl. Adjust volume to 10 μl using water.

2. Add 4 μl of 5x tailing buffer (125 mM Tris-HCl, pH 6.6, 1 M potassium cacodylate, and 1250 μg/ml BSA), 1.2 μl of 25 mM $CoCl_2$, 4 μl of 1 mM dATP, and 10 units of TdT.

3. Incubate for 5 minutes at 37°C and then 5 minutes at 65°C.

4. Dilute to 500 μl with TE (5′-end tailed cDNA pool).

To attach a known sequence to the 5′ end of the first-strand cDNA, a homopolymeric tail is appended using TdT. We prefer appending poly(A) tails rather than poly(C) tails for several reasons. First, the 3′-end strategy is based on the naturally occurring poly(A) tail; thus, the same adapter primers can be used for both ends, decreasing both variability in the protocol and cost. Second, since A:T binding is weaker than G:C binding, longer stretches of A residues (~2x) are required before the oligo(dT)-tailed Q_T primer will bind to an internal site and truncate the amplification product. Third, vertebrate coding sequences and 5′-untranslated regions tend to be biased toward G/C residues; thus, use of a poly(A) tail further decreases the likelihood of inappropriate truncation.

Unlike many other situations in which homopolymeric tails are appended, the actual length of the tail added here is unimportant, as long as it exceeds 17 nucleotides. This is because although the oligo(dT)-tailed primers subsequently bind all along the length of the appended poly(A) tail, only the innermost primer becomes incorporated into the amplification product, and consequently, the remainder of the poly(A) tail is lost (Frohman et al. 1988). The truncation appears to occur because *Taq* DNA polymerase is unable to resolve branched structures (efficiently). The conditions described in the procedure above result in the addition of 30–400 nucleotides.

STEP 3. AMPLIFICATION

First round:

1. Add an aliquot of the 5′-end tailed cDNA pool (1 μl) and primers (25 pmoles each of GSP1 and Q_O [shown in Fig. 3b], and 2 pmoles of Q_T) to 50 μl of the PCR cocktail (1x *Taq* DNA polymerase buffer [described above], each dNTP at 1.5 mM, and 10% DMSO) in a 0.5-ml microfuge tube.

2. Heat in a DNA thermal cycler for 5 minutes at 98°C to denature the first-strand products. Cool to 75°C. Add 2.5 units of *Taq* DNA polymerase, overlay the mixture with 30 μl of mineral oil (Sigma 400-5; preheat it in the thermal cycler to 75°C), and incubate for 2 minutes at the appropriate annealing temperature (48–52°C). Extend the cDNAs for 40 minutes at 72°C.

3. Carry out 30 cycles of amplification using a step program (1 minute at 94°C, 1 minute at 52–60°C, 3 minutes at 72°C), followed by a 15-minute final extension at 72°C. Cool to room temperature.

Second round:

4. Dilute 1 μl of the amplification products from the first round into 20 μl of TE.

5. Amplify 1 μl of the diluted material with primers GSP2 and Q_I using the procedure described above, but eliminate the initial 2-minute annealing step and the 40-minute 72°C extension step.

Many of the remarks made above in the section on amplifying 3′-end partial cDNAs are also relevant here and should be noted. There is, however, one major difference. The annealing temperature in the first step (48–52°C) is lower than that used in successive cycles (52–60°C). This is because cDNA synthesis during the first round depends on the interaction of the appended poly(A) tail and the oligo(dT)-tailed Q_T primer, whereas in all subsequent rounds, amplification can proceed using the Q_O primer, which is composed of about 60% GC and which can anneal at a much higher temperature to its complementary target.

PART II: NEW RACE

REAGENTS

See the classic RACE section above for sources of some of the required materials.

Calf intestinal phosphatase (CIP) (Boehringer Mannheim)
Proteinase K (Boehringer Mannheim)
Tobacco acid pyrophosphatase (TAP) (Epicentre)
RNA transcription kit (Epicentre)
10X Enzyme reaction buffers (Epicentre)
T4 RNA ligase (New England Biolabs or Boehringer Mannheim)

Note that 10X T4 RNA ligase buffers supplied by some manufacturers contain too much ATP (Tessier et al. 1986). Check the composition of any commercially supplied 10X buffer and make your own if it contains more than 1 mM ATP (final 1X concentration should be 0.1 mM), as described below.

PROTOCOLS

5′-End cDNA Amplification

The procedure detailed below uses relatively large amounts of RNA and can be scaled down if RNA quantities are limiting. The advantage of starting with large amounts of RNA is that aliquots can be electrophoresed quickly after each step of the procedure to confirm that detectable degradation of the RNA has not occurred, and the dephosphorylated, decapped, ligated RNA can be stored indefinitely for many future experiments.

STEP 1. DEPHOSPHORYLATION OF DEGRADED RNAs

In general, follow the manufacturer's recommendations for use of the phosphatase.

1. Prepare a reaction mixture containing 50 μg of RNA in 41 μl of water, 5 μl of 10X buffer, 0.5 μl of 100 mM DTT, 1.25 μl of RNasin (40 units/μl), and 3.5 μl of CIP (1 unit/μl).

2. Incubate the reaction for 1 hour at 50°C.

3. Add proteinase K to 50 μg/ml and incubate for 30 minutes at 37°C.

4. Extract the reaction with a mixture of phenol/chloroform, extract again with chloroform, and precipitate the RNA using 1/10th volume of 3 M NaOAc and 2.5 volumes of ethanol. Resuspend the RNA in 43.6 μl of water.

5. Electrophorese 2 μg (1.6 μl) on a 1% TAE agarose gel adjacent to a lane containing 2 μg of the original RNA preparation, stain the gel

with ethidium bromide, and confirm visually that the RNA remained intact during the dephosphorylation step.

STEP 2. DECAPPING OF INTACT RNAs

1. Prepare a reaction mixture containing 38 μg of RNA in 42 μl of water (this is the RNA recovered from step 1), 5 μl of 10X TAP buffer, 1.25 μl of RNasin (40 units/μl), 1 μl of 100 mM ATP, and 1 μl of TAP (5 units/μl).

2. Incubate the reaction for 1 hour at 37°C, and then add 200 μl of TE.

3. Extract the reaction with a mixture of phenol/chloroform, extract again with chloroform, and precipitate the RNA using 1/10th volume of 3 M NaOAc and 2.5 volumes of ethanol. Resuspend the RNA in 40 μl of water.

4. Electrophorese 2 μg of the RNA on a 1% TAE agarose gel adjacent to a lane containing 2 μg of the original RNA preparation, stain the gel with ethidium bromide, and confirm visually that the RNA remained intact during the decapping step.

   Note: Most protocols call for much more TAP. The enzyme is very expensive and it is not necessary to use more!

STEP 3. PREPARATION OF RNA OLIGONUCLEOTIDE

Choose a plasmid that can be linearized at a site approximately 100 bp downstream from a T7 or T3 RNA polymerase site (see Fig. 4b). Ideally, a plasmid containing some insert cloned into the first polylinker site is optimal, because primers made from palindromic polylinker DNA do not perform well in PCR. For my experiments, I use the 3′ UTR of the mouse gene Gbx-1 (Frohman et al. 1993), which is cloned into the *Sst* I site of pBS-SK (Stratagene); I linearize with *Sma* I and transcribe with T3 RNA polymerase to produce a 132-nucleotide RNA oligonucleotide, of which all but 17 nucleotides are from Gbx-1. Note that adenosines are the best "acceptors" for the 3′ end of the RNA oligonucleotide to ligate to the 5′ end of its target, if an appropriate restriction site can be found. The primers subsequently used for amplification are all derived from the Gbx-1 3′ UTR sequence. Interested investigators are welcome to the Gbx-1 NRC primer sequences and plasmid upon request.

Carry out a test transcription to make sure that everything is working; then scale up. The oligonucleotide can be stored at –80°C indefinitely for many future experiments, and it is important to synthesize enough oligonucleotide so that losses due to purification and spot checks along the way leave plenty of material at the end of the procedure.

1. Linearize 25 μg of the plasmid that is to be transcribed (the plasmid should be reasonably free of RNases).

2. Treat the digestion reaction with 50 μg/ml of proteinase K for 30 minutes at 37°C, followed by 2x phenol/$CHCl_3$ extractions, one $CHCl_3$ extraction, and ethanol precipitation.

3. Resuspend the template DNA in 25 μl of TE, pH 8.0, for a final concentration of about 1 μg/μl.

4. For transcription, mix the reagents at room temperature in the following order:

	test scale or	prep scale
DEPC-water	4 μl	80 μl
5x Buffer	2	40
0.1 M DTT	1	20
10 mM UTP	0.5	10
10 mM ATP	0.5	10
10 mM CTP	0.5	10
10 mM GTP	0.5	10
Restricted DNA (1 μg/μl)	0.5	10
RNasin (40 units/μl)	0.25	5
RNA polymerase (20 units/μl)	0.25	5

Incubate for 1 hour at 37°C.

5. Prepare the DNase template by adding 0.5 μl of DNase (RNase-free) for every 20 μl of reaction volume and incubate for 10 minutes at 37°C.

6. Run 5 μl of test or prep reaction on a 1% TAE agarose gel to check. Expect to see a diffuse band at about the right size (or a bit smaller) in addition to some smearing all up and down the gel.

7. Purify the oligonucleotide by extracting with phenol/$CHCl_3$ and $CHCl_3$, and then rinse three times using water and a MICROCON spin filter (prerinsed with water). MICROCON 30 spin filters have a cutoff size of 60 nucleotides, and MICROCON 100 spin filters have a cutoff size of 300 nucleotides. MICROCON 10 spin filters are probably most appropriate if the oligonucleotide is smaller than 100 nucleotides, and MICROCON 30 spin filters should be used for anything larger.

8. Run another appropriately sized aliquot on a 1% TAE agarose gel to check the integrity and concentration of the oligonucleotide.

STEP 4. RNA OLIGONUCLEOTIDE - CELLULAR RNA LIGATION

1. Set up two tubes, one with TAPped cellular RNA, and the other with unTAPped cellular RNA.

11.25 μl	water
3 μl	10X buffer = 500 mM Tris-HCl, pH 7.9, 100 mM $MgCl_2$, 20 mM DTT, 1 mg/ml BSA
0.75 μl	RNasin (40 units/μl)
2 μl	4 μg of RNA oligonucleotide (3–6 molar excess over target cellular RNA)
10 μl	10 μg TAPped (or unTAPped) RNA
1.5 μl	2 mM ATP
1.5 μl	T4 RNA ligase (20 units/μl)
30 μl	

2. Incubate for 16 hours at 17°C.

3. Purify the ligated oligonucleotides-RNA using MICROCON 100 spin filtration (X3 in water; prerinse filter with RNase-free water). The volume recovered should not exceed 20 μl.

4. Run one-third of the ligation on a 1% TAE agarose gel to check the integrity of the ligated RNA. It should look about as it did before ligation.

STEP 5. REVERSE TRANSCRIPTION

1. Assemble the following reverse transcription components on ice: 4 μl of 5X reverse transcription buffer (5X buffer contains 250 mM Tris-HCl, pH 8.3, 375 mM KCl, 15 mM $MgCl_2$), 1 μl of dNTPs (stock concentration is 10 mM of each dNTP), 2 μl of 0.1 M DTT, and 0.25 μl (10 units) of RNasin.

2. Heat 1 μl of antisense-specific primer (20 ng/μl) or random hexamers (50 ng/μl) and the remaining RNA (~6.7 μg) in 13 μl of water for 3 minutes at 80°C, cool rapidly on ice, and spin for 5 seconds in a microfuge. Add to the reverse transcription components.

3. Add 1 μl (200 units) of SuperScript II reverse transcriptase, and incubate for 1 hour at 42°C, and 10 minutes at 50°C. If using random hexamers, insert a room temperature 10-minute incubation period after mixing everything together.

4. Incubate for 15 minutes at 70°C to inactivate reverse transcriptase. Spin for 5 seconds in a microfuge.

5. Add 0.75 μl (1.5 units) of RNase H to the tube and incubate for 20 minutes at 37°C to destroy the RNA template.

6. Dilute the reaction mixture to 100 μl with TE (10 mM Tris-HCl, pH 7.5, 1 mM EDTA) and store at 4°C (5′-end oligonucleotide-cDNA pool).

STEP 6. AMPLIFICATION

First round:

1. Add an aliquot of the 5′-end oligonucleotide-cDNA pool (1 μl) and primers (25 pmoles each of GSP1 and NRC-1) to 50 μl of the PCR cocktail (1× *Taq* DNA polymerase buffer [described above], each dNTP at 1.5 mM, and 10% DMSO) in a 0.5-ml microfuge tube.

2. Heat in a DNA thermal cycler for 5 minutes at 98°C to denature the first-strand products. Cool to 75°C. Add 2.5 units of *Taq* DNA polymerase, overlay the mixture with 30 μl of mineral oil (Sigma 400-5; preheat it in the thermal cycler to 75°C), and incubate for 2 minutes at the appropriate annealing temperature (52–60°C). Extend the cDNAs for 40 minutes at 72°C.

3. Carry out 35 cycles of amplification using a step program (1 minute at 94°C, 1 minute at 52–60°C, 3 minutes at 72°C), followed by a 15-minute final extension at 72°C. Cool to room temperature.

Second round:

4. Dilute 1 μl of the amplification products from the first round into 20 μl of TE.

5. Amplify 1 μl of the diluted material with primers GSP2 and NRC-2 using the procedure described above, but eliminate the initial 2-minute annealing step and the 40-minute 72°C extension step.

   Note: Many of the remarks made above in the Classic RACE section on amplifying 5′-end partial cDNAs are also relevant here and should be noted.

PART III. ANALYSIS OF AMPLIFICATION PRODUCTS

The production of specific partial cDNAs by the RACE protocol is assessed using Southern blot hybridization analysis. After the second set of amplification cycles, the first and second set of reaction products are electrophoresed in a 1% agarose gel, stained with ethidium bromide, denatured, and transferred to a nylon membrane. After hybridization with a labeled oligomer or gene fragment derived from a region contained within the amplified fragment (e.g., GSP-Hyb/Seq in Fig. 2a,b), gene-specific partial cDNA ends should be detected easily. Yields of the desired product relative to nonspecific amplified cDNA in the first-round products should vary from <1% of the amplified material to nearly 100%, depending largely on the stringency of the amplification reaction, the amplification efficiency of the specific cDNA end, and the relative abundance of the specific transcript within the mRNA source. In the second set of amplification cycles, about 100% of the cDNA detected by ethidium bromide staining should represent specific product. If specific hybridization is not observed, then troubleshooting steps should be initiated.

Information gained from this analysis should be used to optimize the RACE procedure. If low yields of specific product are observed because nonspecific products are being amplified efficiently, then annealing temperatures can be raised gradually (~2° at a time) and sequentially in each stage of the procedure until nonspecific products are no longer observed. Alternatively, some investigators have reported success using the touchdown PCR procedure to optimize the annealing temperature without trial and error (Don et al. 1991). Optimizing the annealing temperature is also indicated if multiple species of specific products are observed, which could indicate that truncation of specific products is occurring. If multiple species of specific products are observed after the reverse transcription and amplification reactions have been fully optimized, then alternate splicing or promoter use may be occurring.

- *Classic RACE only:* If a nearly continuous smear of specific products is observed up to a specific size limit after 5′-end amplification, polymerase pausing may have occurred during the reverse transcription step. To obtain nearly full-length cDNA ends, the amplification mixture should be electrophoresed and the longest products recovered by gel isolation. An aliquot of this material can then be reamplified for a limited number of cycles.

- *New RACE only:* Expect to see one or two extra nucleotides insert between the RNA oligonucleotide 3′ end and the 5′ end of the gene of interest. These nucleotides come from the transcription step using T7, T3, or SP6, which can add an extra nucleotide or two to oligonucleotides past the end of the template (template-independent transcription).

Compare the results from unTAPped RNA versus TAPped RNA. Junction sites (where the oligonucleotide is connected to the 5′ end of the gene) in common arise from ligation of the oligonucleotide to degraded RNA; unique junctions in the TAPped RNA population represent candidate transcription start sites. If you have RNA degradation sites (e.g., TTT↓AAA) in your 5′ RNA end, you may have substantial numbers of clones that begin at exactly the same nucleotide but arise from ligation of the oligonucleotide to degraded RNA molecules, not from ligation of the oligonucleotide to the true 5′ end of the RNA.

Look for TATA, CCAAT, and initiator element (Inr) sites at or around the candidate transcription site in the genomic DNA sequence if it is available; usually either a TATA or an Inr can be found.

PROTOCOL

Cloning

RACE products can be cloned like any other PCR products.

OPTION 1

To clone the cDNA ends directly from the amplification reaction (or after gel purification, which is recommended), ligate an aliquot of the products to plasmid vector encoding a 1-nucleotide 3′ overhang consisting of a T on both strands. Such vector DNA is available commercially (TA Kit; Invitrogen) or can be easily and cheaply prepared (Holton and Graham 1991; Kovalic et al. 1991; Marchuk et al. 1991; Mead et al. 1991; Frohman 1994).

OPTION 2

The classic RACE Q_I primer encodes *Hin*dIII, *Sst* I, and *Xho*I restriction enzyme sites. Products can be efficiently cloned into vectors that have been double-cut with one of these enzymes and with a blunt-cutting enzyme such as *Sma*I (*Note:* Remember to "polish" the amplification products with Klenow enzyme or T4 DNA polymerase and separate them from residual *Taq* DNA polymerase and dNTPs before carrying out the restriction enzyme digest). If clones are not obtained, determine whether the restriction enzyme chosen is cutting the amplified gene fragment a second time, at some internal location in the new and unknown sequence. A somewhat easier strategy is to append a restriction site (not *Hin*dIII, *Sst* I, or *Xho*I) onto the 5′ end of the GSP2 primer to allow the creation of overhanging strands at both ends of the amplified product.

OPTION 3

A safer and very effective approach is to modify the ends of the primers to allow the creation of overhanging ends using (1) T4 DNA polymerase to chew back a few nucleotides from the amplified product in a controlled manner and (2) Klenow enzyme (or Sequenase) to

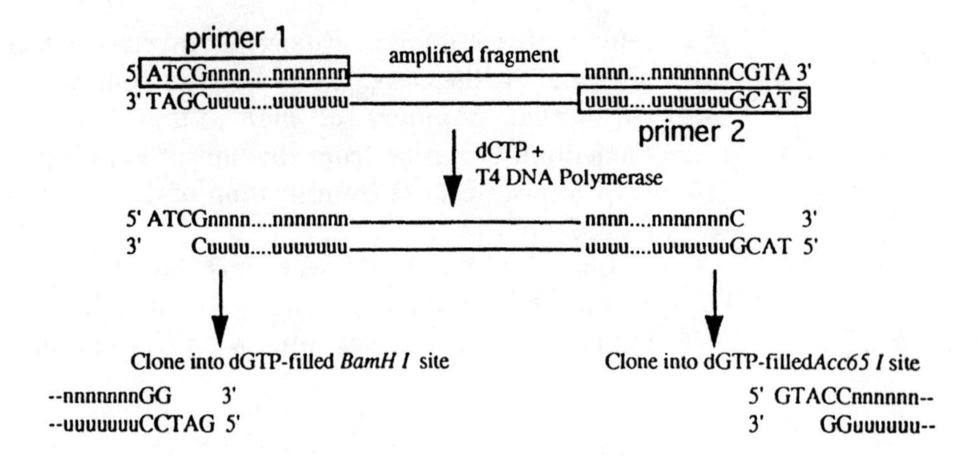

(Note - Acc65 I recognizes the same sequence that Kpn I does, but creates a 5' overhang instead of a 3' one.)

Figure 5 An approach to modify the primers to create overhanging ends using (1) T4 DNA polymerase to chew back a few nucleotides from the amplified product in a controlled manner and (2) Klenow enzyme (or Sequenase) to partially fill in restriction-enzyme-digested overhanging ends on the vector. (Adapted from Stoker 1990 and Iwahana et al. 1994.)

partially fill in restriction-enzyme-digested overhanging ends on the vector, as shown in Figure 5. For another conceptual variation, see Rashtchian et al. (1992).

The advantages of this approach are that (1) it eliminates the possibility that the restriction enzymes chosen for the cloning step will cleave the cDNA end in the unknown region and (2) vector dephosphorylation is not required because vector self-ligation is no longer possible, which means that insert kinasing (and polishing) is not necessary, which means that insert multimerization and fusion clones are not observed either. In addition, the procedure is more reliable than "TA" cloning.

1. For insert preparation, select a pair of restriction enzymes for which you can synthesize half-sites appended to PCR primers that can be chewed back to form the appropriate overhangs, as shown for *Bam*HI and *Acc* 65I, above. *Not* I and *Eag* I are also compatible with *Bam*HI and *Acc* 65I. For RACE cloning, add ATCG to the 5′ end of Q_1 or NRC-2, and add TACG to the 5′ end of GSP2. Carry out PCR as usual.

2. After PCR, add proteinase K (10 mg/ml stock) to the PCR product to a concentration of 50 μg/ml and incubate for 30 minutes at 37°C to remove sticky *Taq* DNA polymerase from the amplified DNA (Crowe et al. 1991).

3. Extract the PCR products with phenol/$CHCl_3$, then $CHCl_3$ (but don't precipitate!), to remove the proteins.

4. Filter the PCR products through a MICROCON 100 spin column (or MICROCON 30 if the product is <150 bp) three times using TE (not water) as the wash buffer, to remove unwanted organics, primers, and dNTPs.

5. On ice, add the selected dNTP(s) (e.g., dCTP) to a final concentration of 0.2 mM, 1/10th volume of 10X T4 DNA polymerase buffer, and 1–2 units of T4 DNA polymerase.

6. Incubate for 15 minutes at 12°C, and then for 10 minutes at 75°C to heat-inactivate the T4 DNA polymerase. (Optional: Gel-isolate the DNA fragment of interest, depending on the degree of success of PCR amplification.)

7. For vector preparation, digest the vector (e.g., pGem-7ZF [Promega]) using the selected enzymes (e.g., *Acc* 65I and *Bam*HI) under optimal conditions, in a volume of 10 µl.

8. Add a 10-µl mixture containing the selected dNTP(s) (e.g., dGTP) at a final concentration of 0.4 mM, 1 µl of the restriction buffer used for digestion, 0.5 µl of Klenow, and 0.25 µl of Sequenase.

9. Incubate for 15 minutes at 37°C, and then for 10 minutes at 75°C to heat-inactivate the polymerases.

10. Gel-isolate the linearized vector fragment.

11. For ligation, use equal molar amounts of vector and insert.

Sequencing

RACE products can be sequenced directly on a population level using a variety of protocols, including cycle sequencing, from the end at which the gene-specific primers are located. Note that the classic RACE products cannot be sequenced on a population level using the Q_I primer at the unknown end, because individual cDNAs contain different numbers of A residues in their poly(A) tails and, as a consequence, the sequencing ladder falls out of register after reading through the tail. 3′-End products can be sequenced from their unknown end using the following set of primers (TTTTTTTTTTTTTTTTTA, TTTTTTTTTTTTTTTTTG, and TTTTTTTTTTTTTTTTTC). The non-T nucleotide at the 3′ end of the primer forces the appropriate primer to bind to the inner end of the poly(A) tail (Thweatt et al. 1990). The other two primers do not participate in the sequencing reaction. Individual cDNA ends, once cloned into a plasmid vector, can be sequenced from either end using gene-specific or vector primers.

HYBRIDIZATION PROBES

RACE products are generally pure enough that they can be used as probes for RNA and DNA blot analyses. It should be kept in mind that small amounts of contaminating nonspecific cDNAs are always present. It is also possible to include a T7 RNA polymerase promoter in one or both primer sequences and to use the RACE products in in vitro transcription reactions to produce RNA probes (Frohman and Martin 1989). Primers encoding the T7 RNA polymerase promoter sequence do not appear to function as amplification primers as efficiently as the ones listed in Figure 2 (personal observation). Thus, as a general rule, the T7 RNA polymerase promoter sequence should not be incorporated into RACE primers.

CONSTRUCTION OF FULL-LENGTH cDNAs

It is possible to use the RACE protocol to create overlapping 5′ and 3′ cDNA ends that can later, through judicious choice of restriction enzyme sites, be joined together through subcloning to form a full-length cDNA. It is also possible to use the sequence information gained from acquisition of the 5′ and 3′ cDNA ends to make new primers representing the extreme 5′ and 3′ ends of the cDNA, and to employ them to amplify a de novo copy of a full-length cDNA directly from the 3′-end cDNA pool. Despite the added expense of making two more primers, there are several reasons that the second approach is preferred.

First, a relatively high error rate is associated with the PCR conditions for which efficient RACE amplification takes place, and numerous clones may have to be sequenced to identify one without mutations. In contrast, two specific primers from the extreme ends of the cDNA can be used under inefficient but low-error-rate conditions (Eckert and Kunkel 1990) for a minimum number of cycles to amplify a new cDNA that is likely to be free of mutations. Second, convenient restriction sites are often not available, thus making the subcloning project difficult. Third, by using the second approach, the synthetic poly(A) tail (if present) can be removed from the 5′ end of the cDNA. Homopolymer tails appended to the 5′ ends of cDNAs have in some cases been reported to inhibit translation. Finally, if alternate promoters, splicing, and polyadenylation signal sequences are being used and result in multiple 5′ and 3′ ends, it is possible that one might join two cDNA halves that are never actually found together in vivo. Employing primers from the extreme ends of the cDNA as described confirms that the resulting amplified cDNA represents an mRNA actually present in the starting population.

TROUBLESHOOTING

Problems with Reverse Transcription and Prior Steps

- *Damaged RNA.* Electrophorese RNA in a 1% formaldehyde minigel and examine the integrity of the 18S and 28S ribosomal bands. Discard the RNA preparation if the ribosomal bands are not sharp.

- *Contaminants.* Ensure that the RNA preparation is free of agents that inhibit reverse transcription, e.g., LiCl and SDS (see Sambrook et al. 1989, regarding the optimization of reverse transcription reactions).

- *Bad reagents.* To monitor reverse transcription of the RNA, add 20 μCi of [^{32}P]dCTP to the reaction, separate newly created cDNAs using gel electrophoresis, wrap the gel in plastic wrap, and expose it to X-ray film. Accurate estimates of cDNA size can best be determined using alkaline agarose gels, but a simple 1% agarose minigel will suffice to confirm that reverse transcription took place and that cDNAs of reasonable length were generated. Note that adding [^{32}P]dCTP to the reverse transcription reaction results in the detection of cDNAs synthesized both through the specific priming of mRNA and through RNA self-priming. When a gene-specific primer is used to prime transcription (5′-end RACE) or when total RNA is used as a template, the majority of the labeled cDNA will actually have been generated from RNA self-priming. To monitor extension of the primer used for reverse transcription, label the primer using T4 DNA kinase and [γ-^{32}P]ATP prior to reverse transcription. Much longer exposure times will be required to detect the labeled primer-extension products than when [^{32}P]dCTP is added to the reaction.

  To monitor reverse transcription of the gene of interest, one may attempt to amplify an internal fragment of the gene containing a region derived from two or more exons, if sufficient sequence information is available.

Problems with Tailing

- *Bad reagents.* Tail 100 ng of a DNA fragment of approximately 100–300 bp in length for 30 minutes. In addition, mock-tail the same fragment (add everything but the TdT). Run both samples in a 1% agarose minigel. The mock-tailed fragment should run as a tight band. The tailed fragment should have increased in size by 20–200 bp and should appear to run as a diffuse band that trails off into higher-molecular-weight products. If this is not observed, replace reagents.

- *Experimental control.* Mock-tail 25% of the cDNA pool (add everything but the TdT). Dilute to the same final concentration as the tailed cDNA pool. This serves two purposes. First, although amplification products will be observed using both tailed and untailed cDNA templates, the actual pattern of bands observed should be different. In general, discrete bands are observed using untailed templates after the first set of cycles, and a broad smear of amplified cDNA accompanied by some individual bands is typically observed using tailed templates. If the two samples appear different, this confirms that tailing took place and that the oligo(dT)-tailed Q_T primer is annealing effectively to the tailed cDNA during PCR. Second, observing specific products in the tailed amplification mixture that are not present in the untailed amplification mixture indicates that these products are being synthesized off the end of an A-tailed cDNA template, rather than by annealing of the dT-tailed primer to an A-rich sequence in or near the gene of interest.

Problems with Amplification

- *No product.* If no products are observed for the first set of amplifications after 30 cycles, add fresh *Taq* DNA polymerase and carry out an additional 15 rounds of amplification (extra enzyme is not necessary if the entire set of 45 cycles is carried out without interruption at cycle 30). Product is always observed after a total of 45 cycles if efficient amplification is taking place. If no product is observed, carry out a PCR using control templates and primers to ensure the integrity of the reagents.

- *Smeared product from the bottom of the gel to the loading well.* This is caused by too many cycles or too much starting material.

- *Nonspecific amplification, but no specific amplification.* Check the sequences of cDNA and primers. If all are correct, examine primers (using a computer program) for secondary structure and self-annealing problems. Consider ordering new primers. Determine whether too much template is being added, or if the choice of annealing temperatures could be improved.

  Alternatively, secondary structure in the template may be blocking amplification. Consider adding formamide (Sarker et al. 1990) or ^{7}aza-GTP (in a 1:3 ratio with dGTP) to the reaction to assist polymerization. ^{7}aza-GTP can also be added to the reverse transcription reaction.

- *The last few base pairs of the 5′-end sequence do not match the corresponding genomic sequence.* Be aware that reverse transcriptase and T7 and T3 RNA polymerase can add on a few extra template-independent nucleotides.

- *Inappropriate templates.* To determine whether the amplification products observed are being generated from cDNA or whether they derive from residual genomic DNA or contaminating plasmids, pretreat an aliquot of the RNA with RNase A.

CONCLUSIONS

The RACE protocol offers several advantages over conventional library screening to obtain additional sequence for cDNAs that are already partially cloned. RACE is cheaper, much faster, requires very small amounts of primary material, and provides rapid feedback on the generation of the desired product. Information regarding alternate promoters, splicing, and polyadenylation signal sequences can be obtained and a judicious choice of primers (e.g., within an alternately spliced exon) can be used to amplify a subpopulation of cDNAs from a gene for which the transcription pattern is complex. Furthermore, differentially spliced or initiated transcripts can be separated by electrophoresis and cloned separately, and essentially unlimited numbers of independent clones can be generated to examine rare events. Finally, for 5′-end amplification, the ability of reverse transcriptase to extend cDNAs all the way to the ends of the mRNAs is greatly increased, because a primer extension library is created instead of a general purpose library.

REFERENCES

Bertling, W.M., F. Beier, and E. Reichenberger. 1993. Determination of 5′ ends of specific mRNAs by DNA ligase-dependent amplification. *PCR Methods Appl.* **3:** 95–99.

Bertrand, E., M. Fromont-Racine, R. Pictet, and T. Grange. 1993. Visualization of the interaction of a regulatory protein with RNA in vivo. *Proc. Natl. Acad. Sci.* **90:** 3496–3500.

Borson, N.D., W.L. Salo, and L.R. Drewes. 1992. A lock-docking oligo(dT) primer for 5′ and 3′ RACE PCR. *PCR Methods Appl.* **2:** 144–148.

Brock, K.V., R. Deng, and S.M. Riblet. 1992. Nucleotide sequencing of 5′ and 3′ termini of bovine viral diarrhea virus by RNA ligation and PCR. *J. Virol. Methods* **38:** 39–46.

Coleclough, C. 1987. Use of primer-restriction end adapters in cDNA Cloning. *Methods Enzymol.* **154:** 64–83.

Crowe, J.S., H.J. Cooper, M.A. Smith, M.J. Sims, D. Parker, and D. Gewert. 1991. Improved cloning efficiency of polymerase chain reaction (PCR) products after proteinase K digestion. *Nucleic Acids Res.* **19:** 184.

Don, R.H., P.T. Cox, B.J. Wainwright, K. Baker, and J.S. Mattick. 1991. "Touchdown" PCR to circumvent spurious priming during gene amplification. *Nucleic Acids Res.* **19:** 4008.

Dumas, J.B., M. Edwards, J. Delort, and J. Mallet. 1991. Oligodeoxyribonucleotide ligation to single-stranded cDNAs: A new tool for cloning 5′ ends of mRNAs and for constructing cDNA libraries by in vitro amplification. *Nucleic Acids Res.* **19:** 5227–5233.

Eckert, K.A. and T.A. Kunkel. 1990. High fidelity DNA synthesis by the *Thermus aquaticus* DNA polymerase. *Nucleic Acids Res.* **18:** 3739–3745.

Fritz, J.D., M.L. Greaser, and J.A. Wolff. 1991. A novel 3′ extension technique using random primers in RNA-PCR. *Nucleic Acids Res.* **119:** 3747.

Frohman, M.A. 1989. Creating full-length cDNAs from small fragments of genes: Amplification of rare transcripts using a single gene-specific oligonucleotide primer. In *PCR protocols and applications: A laboratory manual* (ed. M. Innis et al.), pp. 28–38. Academic Press, New York.

——. 1993. Rapid amplification of cDNA for generation of full-length cDNA ends: Thermal RACE. *Methods Enzymol.* **218:** 340–356.

——. 1994. Cloning PCR products: Strategies and tactics. In *PCR: The polymerase chain reaction* (ed. K.B. Mullis et al.), pp. 14–37. Birkhauser, Boston.

Frohman, M.A. and G.R. Martin. 1989. Rapid amplification of cDNA ends using nested primers. *Technique* **1:** 165–173.

Frohman, M.A., M.K. Dush, and G.R. Martin. 1988. Rapid production of full-length cDNAs from rare transcripts by amplification using a single gene-specific oligonucleotide primer. *Proc. Natl. Acad. Sci.* **85:** 8998–9002.

Frohman, M.A., M.E. Dickinson, B.L.M. Hogan, and G.R. Martin. 1993. Localization of two new and related homeobox-containing genes to chromosomes 1 and 5, near the phenotypically similar mutant loci dominant *hemimelia (Dh)* and *hemimelic extra-toes (Hx)*. *Mouse Genome* **91:** 323–325.

Fromont-Racine, M., E. Bertrand, R. Pictet, and T. Grange. 1993. A highly sensitive method for mapping the 5′ termini of mRNAs. *Nucleic Acids Res.* **21:** 1683-1684.

Holton, T.A. and M.W. Graham. 1991. A simple and efficient method for direct cloning of PCR products using ddT-tailed vectors. *Nucleic Acids Res.* **19:** 1156.

Iwahana, H., N. Mizusawa, S. Ii, K. Yoshimoto, and M. Itakura. 1994. An end-trimming method to amplify adjacent cDNA fragments by PCR. *BioTechniques* **16:** 94–98.

Jain, R., R.H. Gomer, and J.J.J. Murtagh. 1992. Increasing specificity from the PCR-RACE technique. *BioTechniques* **12:** 58–59.

Kovalic, D., J.H. Kwak, and B. Weisblum. 1991. General method for direct cloning of DNA fragments generated by the polymerase chain reaction. *Nucleic Acids Res.* **19:** 4650.

Liu, X. and M.A. Gorovsky. 1993. Mapping the 5′ and 3′ ends of *Tetrahymena-thermophila* mRNAs using RNA ligase mediated amplification of cDNA ends (RLM-RACE). *Nucleic Acids Res.* **21:** 4954–4960.

Loh, E.L., J.F. Elliott, S. Cwirla, L.L. Lanier, and M.M. Davis. 1989. Polymerase chain reaction with single sided specificity: Analysis of T cell receptor delta chain. *Science* **243:** 217–220.

Mandl, C.W., F.X. Heinz, E. Puchhammer-Stockl, and C. Kunz. 1991. Sequencing the termini of capped viral RNA by 5′-3′ ligation and PCR. *BioTechniques* **10:** 484–486.

Marchuk, D., M. Drumm, A. Saulino, and F.S. Collins. 1991. Construction of T-vector, a rapid and general system for direct cloning of unmodified PCR products. *Nucleic Acids Res.* **19:** 1154.

Mead, D.A., N.K. Pey, C. Herrnstadt, R.A. Marcil, and L.A. Smith. 1991. A universal method for direct cloning of PCR amplified nucleic acid. *BioTechnology* **9:** 657–663.

Monstein, H.J., J.U. Thorup, R. Folkesson, A.H. Johnsen, and J.F. Rehfeld. 1993. cDNA deduced procionin-structure and expression in protochordates resemble that of procholecystokinin in mammals. *FEBS Lett.* **331:** 60–64.

Ohara, O., R.I. Dorit, and W. Gilbert. 1989. One-sided PCR: The amplification of cDNA. *Proc. Natl. Acad. Sci.* **86:** 5673–5677.

Rashtchian, A., G.W. Buchman, D.M. Schuster, and M.S. Berninger. 1992. Uracil DNA glycosylase-mediated cloning of PCR-amplified DNA: Application to genomic and cDNA cloning. *Anal. Biochem.* **206:** 91–97.

Sallie, R. 1993. Characterization of the extreme 5′ ends of RNA molecules by RNA ligation-PCR. *PCR Methods Appl.* **3:** 54–56.

Sambrook, J., E.F. Fritsch, and T, Maniatis. 1989. *Molecular cloning: A laboratory manual*, 2nd edition. Cold Spring Harbor Laboratory Press, Cold Spring Harbor, New York.

Sarker, G., S. Kapelner, and S.S. Sommer. 1990. Formamide can dramatically improve the specificity of PCR. *Nucleic Acids Res.* **18:** 7465.

Schuster, D.M., G.W. Buchman, and A. Rastchian. 1992. A simple and efficient method for amplification of cDNA ends using 5′ RACE. *Focus* **14:** 46–52.

Stoker, A.W. 1990. Cloning of PCR products after defined cohesive termini are created with T4 DNA polymerase. *Nucleic Acids Res.* **18:** 4290.

Templeton, N.S., E. Urcelay, and B. Safer. 1993. Reducing artifact and increasing the yield of specific DNA target fragments during PCR-RACE or anchor PCR. *BioTechniques* **15:** 48–50.

Tessier, D.C., R. Brousseau, and T. Vernet. 1986. Ligation of single-stranded oligodeoxyribonucleotides by T4 RNA ligase. *Anal. Biochem.* **158:** 171–178.

Thweatt, R., S. Goldstein, and R.J.S. Reis. 1990. A universal primer mixture for sequence determination at the 3′ ends of cDNAs. *Anal. Biochem.* **190:** 314.

Volloch, V., B. Schweizer, X. Zhang, and S. Rits. 1991. Identification of negative-strand complements to cytochrome oxidase subunit III RNA in *Trypanosoma brucei. Biochemistry* **88:** 10671–10675.

Wickens, M. and P. Stephenson. 1984. Role of the conserved AAUAAA sequence: Four AAUAAA point mutants prevent mRNA 3′end formation. *Science* **226:** 1045–1050.

Panhandle PCR

Douglas H. Jones

Department of Pediatrics, University of Iowa College of Medicine,
Iowa City, Iowa 52242

INTRODUCTION

PCR permits highly specific DNA amplification in vitro. PCR occurs by primer extension from each end of a sequence and therefore requires knowledge of those ends (Mullis et al. 1986). Several methods have been developed for the PCR amplification of unknown DNA that flanks one end of a known sequence (Frohman et al. 1988; Ochman et al. 1988; Triglia et al. 1988; Loh et al. 1989; Ohara et al. 1989; Pfeifer et al. 1989; Shyamala and Ames 1989; Silver and Keerikatte 1989; Kalman et al. 1990; Riley et al. 1990; Rosenthal and Jones 1990; Roux and Dhanarajan 1990; Arnold and Hodgson 1991; Edwards et al. 1991; Lagerstrom et al. 1991; MacGregor and Overbeek 1991; Parker et al. 1991; Parks et al. 1991; Jones and Winistorfer 1992, 1993a,b; Tormanen et al. 1992; Troutt et al. 1992; Sarkar et al. 1993). These methods extend the application of PCR to the retrieval of DNA where only one end of the DNA sequence is known, so that one can use PCR to "walk" along an uncharacterized stretch of DNA without screening a library for overlapping clones. None of the existing methods, apart from panhandle PCR, has permitted the highly specific PCR amplification of more than 3.0 kb of human genomic DNA that flanks a known site. This paper describes a protocol for panhandle PCR that permits the highly specific PCR amplification directly from bulk human genomic DNA of more than 3.0 kb, and up to 4.4 kb, of DNA that flanks a known site.

REAGENTS

Restriction endonucleases (New England Biolabs)
Calf intestinal alkaline phosphatase (Boehringer Mannheim)
GENECLEAN II (BIO 101)
TE (10 mM Tris-HCl, pH 8.0, 1 mM EDTA)
T4 DNA ligase (Boehringer Mannheim)

AmpliTaq polymerase 5 units/μl (Perkin-Elmer)
10x *Taq* DNA polymerase buffer (500 mM KCl, 100 mM Tris-HCl, pH 8.3, 15 mM $MgCl_2$, 0.1% w/v gelatin)
100 mM dATP, 100 mM dCTP, 100 mM dGTP, 100 mM dTTP (Boehringer Mannheim)
Oligonucleotides (Midland Certified Reagent)
Agarose
Ethidium bromide
TAE buffer (Sambrook et al. 1989)
Long thin micropipette tips (gel loader tips T-010; Phenix Research Products)

PROTOCOL

Panhandle PCR is illustrated in Figure 1 and its steps are outlined as follows:

1. Digest genomic DNA with a restriction endonuclease to generate a 5′ overhang; dephosphorylate with calf intestinal alkaline phosphatase.

2. Use glass bead extraction to remove the alkaline phosphatase. With T4 DNA ligase join the cleaved DNA to a single-stranded oligonucleotide to create 3′-end extensions.

3. These 3′-end extensions are complementary to the known sequence located upstream of the unknown DNA of interest. Remove unligated oligonucleotides with glass bead extraction. Denature the mixture and reanneal under dilute conditions and high stringency in *Taq* DNA polymerase buffer with *Taq* DNA polymerase and the four dNTPs. Those single strands that contain the complement to the 3′-end extension can form a stem-loop with a recessed 3′ end that primes template-directed DNA polymerization. This polymerization attaches known DNA to the uncharacterized end of the flanking unknown DNA. This polymerization results in known DNA being positioned on both sides of the unknown flanking DNA.

4. Amplify the unknown DNA with PCR.

5. Perform nested PCR. Two primers are used during each PCR amplification. Amplification using a single primer that anneals to a sequence unique to the strand of interest works poorly because relatively large inverted repeats in PCR product ends impede amplification. The steps outlined above are detailed below. Figure 2 shows the positioning and sequences of a set of oligonucleotides

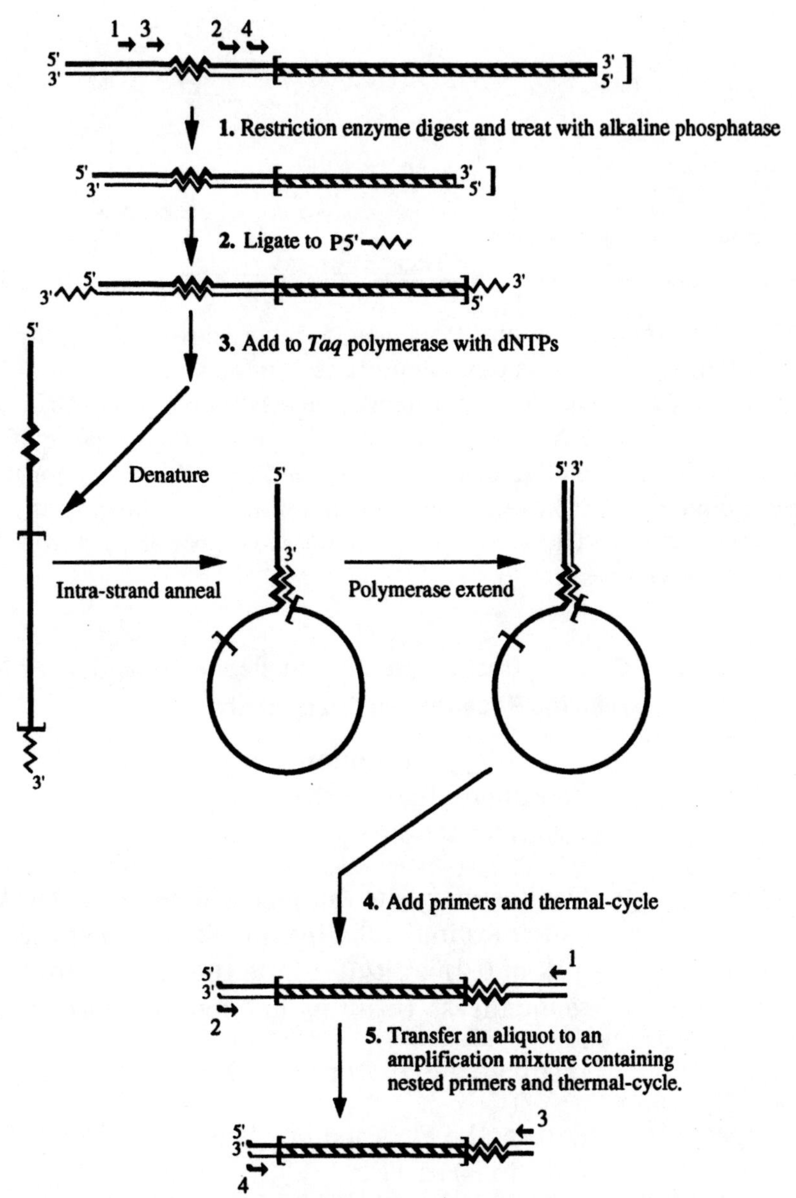

Figure 1 Panhandle PCR. The numbered steps correspond to the numbered steps in the Protocol. The two complementary strands of genomic DNA are represented by thin and thick lines. The unknown region of genomic DNA is enclosed by brackets and is striped when double-stranded. The jagged portion of the thick line represents the annealing region for the ligated oligonucleotide. The PCR primers are numbered arrows. The locations of the primers in relation to the relevant strands of genomic DNA are shown on top of the diagram for Step 1, and the primers are not used until Step 4. One or two nucleotides are added to the 5′ ends of primers 2 and 4 that are not complementary to their template and are represented by upended 5′ ends.

that were used to amplify 4.4 kb of DNA flanking the primer annealing sites following the digestion of human genomic DNA with the restriction endonuclease *Bcl*I (Jones and Winistorfer 1993a).

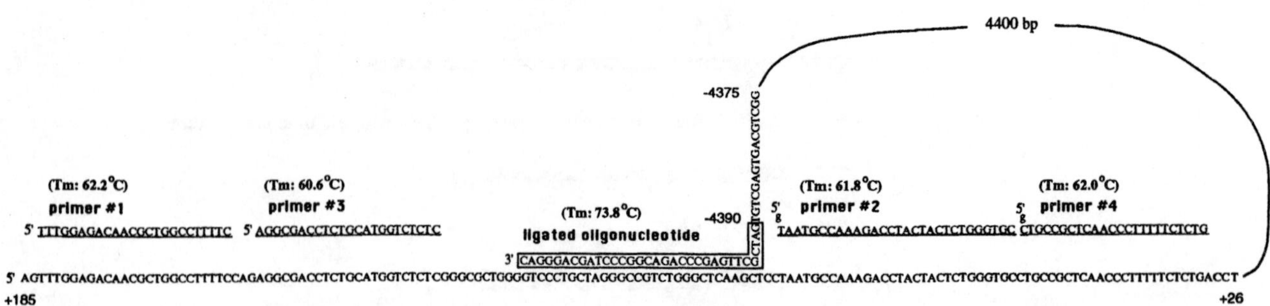

Figure 2 Ligated oligonucleotide and PCR primer sequences used to amplify human genomic DNA lying upstream of the cystic fibrosis transmembrane conductance regulator cDNA sequence (Jones and Winistorfer 1993a). The template is numbered according to the cDNA sequence, with the first nucleotide of the published cDNA sequence in the +1 position (Riordan et al. 1989). The PCR primers are underlined and the ligated oligonucleotide (referred to as the phosphorylated oligonucleotide prior to ligation to genomic DNA) is boxed. The corresponding T_m values for annealing to the template are above each oligonucleotide. The single 5′ nucleotides of primers 2 and 4 that do not anneal to the template are lowercase letters.

Step 1. Restriction Endonuclease Digestion Followed by Calf Intestinal Alkaline Phosphatase Treatment

1. Digest 5 μg of human genomic DNA with 20–40 units of a restriction endonuclease that yields a 5′ overhang, in a final volume of 100 μl for 2 hours.

2. Dephosphorylate the digested genomic DNA by adding 0.05 unit of calf intestinal alkaline phosphatase (Boehringer Mannheim; add 5 μl of 0.01 unit/μl; dilute the stock 1 unit/μl 1:100 in TE) directly to the digest, resulting in a total volume of 105 μl.

3. Incubate the mixture for 30 minutes at 37°C.

4. Extract the DNA using GENECLEAN II (BIO 101), and suspend the DNA in 50 μl TE.

5. Freeze 25 μl of the GENECLEAN II product for later use as the template control.

   Notes: For convenience we purchase our genomic DNA from CLONTECH; this DNA is made according to the method of Blin and Stafford (1986). The restriction endonucleases typically have a 6-nucleotide recognition domain and generate a 4-nucleotide-long 5′-end extension. Enzymes used to generate panhandle PCR products have included *Hin*dIII, *Bam*HI, *Bcl* I, *Bgl* II, *Avr* II, and *Xba*I (Jones and Winistorfer 1992, 1993a). Glass bead extraction using GENECLEAN II removes small DNA fragments and the calf intestinal alkaline phosphatase.

Step 2. Ligation of the Phosphorylated Oligonucleotide

1. Ligate the remaining 25 μl of genomic DNA (containing ~2.5 μg of DNA) to 50-fold molar excess of a 5′ phosphorylated oligonucleotide in T4 DNA ligase buffer (50 mM Tris-HCl, pH 7.6,

10 mM $MgCl_2$, 0.5 mM ATP, 10 mM dithiothreitol) using one Weiss U T4 DNA ligase (Boehringer Mannheim) for 4 hours at 23°C or overnight at 16°C.

2. Purify the ligation mixture using GENECLEAN II and suspend it in 25 µl of TE. This is the template for Step 3.

Notes: A restriction endonuclease with a 6-nucleotide recognition domain cuts genomic DNA to an average size of roughly 4.0 kb. Fifty molar excess of a single-stranded 33-nucleotide-long phosphorylated oligonucleotide (as shown in Fig. 2) requires about 516 ng of the oligonucleotide. These calculations are rough estimates, and it is not necessary to make a more precise calculation that compensates for poor GENECLEAN II recovery of DNA shorter than 500 bp and for a restriction endonuclease cutting frequency that varies with the nucleotide composition of the restriction endonuclease recognition domain (due, in part, to the A/T predominance [59%] of mammalian genomes [Normore et al. 1976]).

The oligonucleotide with a phosphorylated 5′ end is generated during oligonucleotide synthesis by the addition of a phosphorylated phosphoramidite during oligonucleotide synthesis. Alternatively, it can be 5′-phosphorylated using T4 polynucleotide kinase as follows: Incubate 2 µg of the oligonucleotide with 10 units of T4 polynucleotide kinase (New England Biolabs) in kinase buffer (50 mM Tris-HCl, pH 7.6, 10 mM $MgCl_2$, 1 mM ATP, 5 mM dithiothreitol) for 30 minutes at 37°C. Then, inactivate the T4 polynucleotide kinase by heating for 10 minutes at 68°C and store it at –20°C until it is used.

The 5′-phosphorylated oligonucleotide has a 5′ end that is complementary to the single-stranded ends of restriction endonuclease-digested genomic DNA generated in Step 1. A single phosphorylated oligonucleotide can be used with genomic DNA that has undergone digestion with one of several restriction enzymes. For instance, an oligonucleotide whose 5′ end consists of 5′P-GATC can be ligated to genomic DNA that has been digested with either *Bam*HI, *Bcl*I, or *Bgl*II. An oligonucleotide whose 5′ end consists of 5′P-CTAG can be ligated to genomic DNA that has been digested with either *Xba*I, *Avr* II, *Nhe* I, or *Spe* I.

The phosphorylated oligonucleotide is designed so that the T_m of the domain that undergoes annealing during panhandle formation (Step 3) is about 74°C. T_m values for the phosphorylated oligonucleotide and primers are calculated using the computer program OLIGO (National BioSciences).

One application of this method is the amplification of a gene promoter from human genomic DNA using cDNA sequence data. In this application, the cDNA sequence is used to design primers that will amplify the genomic DNA located 5′ to the cDNA sequence. It is important to consider that the cDNA is usually not intact in the genome, because it is interrupted by one or more introns. The presence of an intron within a primer annealing domain will prevent this method from working. Therefore, it is advisable to amplify the primer annealing domains (the DNA sequence to which primers 1–4 and the phosphorylated oligonucleotide will anneal) directly from genomic DNA using conventional PCR prior to carrying out panhandle PCR, to ensure that the predicted piece (typically 150–200 bp) is obtained. This precaution will ensure that this stretch of DNA is intact in the genome, and not interrupted by an intron.

Step 3. Panhandle Formation

1. Add 18 µl of H_2O to a 25-µl aliquot of 2x PCR mix (1.25 units of *Taq* DNA polymerase (AmpliTaq, Perkin-Elmer), 100 mM KCl, 20 mM

Tris-HCl, pH 8.3, 3 mM $MgCl_2$, 0.02% w/v gelatin, 400 mM each dNTP). Layer with approximately 50 µl of mineral oil.

2. Preheat the tube to 80°C prior to the addition of 2 µl of template.

3. Place the mixture in a thermal cycler and proceed with the following temperature transitions (Perkin-Elmer): for 1 minute at 95°C, followed by a 2-minute-long cooling to 72°C, followed by a 30-second incubation and then a rapid transition to an 80°C soak.

Notes: The 2x PCR mix aliquots can be pre-aliquoted and stored at –20°C for at least 6 months prior to use. Maintaining the tube at 80°C before adding the template prevents nonspecific annealing and polymerization (Mullis 1991).

Since the genomic DNA concentration is less than 4 ng/µl, these denaturation and reannealing steps can result in intrastrand annealing of the ligated synthetic oligonucleotide to its complementary sequence in the genomic DNA (Triglia et al. 1988). This is followed by template-directed polymerase extension of the recessed 3′ end.

Two additional tubes, one containing the template control (genomic DNA that has been restriction-endonuclease-digested but not ligated to the phosphorylated oligonucleotide) and the other containing a reagent control (no DNA), should be processed concurrently as controls.

Step 4. Initial Amplification

1. Add 12.5 pmoles of each primer (primers 1 and 2 in Fig. 1) in a total volume of 5 µl of H_2O under the mineral oil while each tube from the previous step remains in the heat block at 80°C.

2. Cycle (30 seconds at 95°C, 30 seconds at 60°C, and 4 minutes at 72°C) for 30 cycles followed by a final extension for 7 minutes at 72°C. Then go to an 80°C soak.

Notes: Final concentrations are 0.25 µM for each primer with 200 µM for each dNTP.

Following thermal cycling, bringing the reagents to 4°C (as opposed to keeping the tubes at 80°C) at the end of this step also works well.

A hot start is carried out by not allowing the temperature of the fully constituted polymerase reactants to drop below 80°C prior to a primer annealing step. This eliminates priming at low stringency, preventing the generation of nonspecific products (Mullis 1991).

Primers in each amplification are designed so that their T_m values are near 62°C (see Fig. 2). The annealing temperature during the thermal cycling is approximately 2°C below the primer T_m values. Primer 2 in this amplification and primer 4 in Step 5 have 1–2 nucleotides added to each 5′ end that do not anneal to the original template, and the T_m values used have not included these added nucleotides. These nucleotides are added as a precaution to prevent the possibility of a short-circuiting of the amplification. Short-circuiting could occur by the annealing of the 3′ end of one strand of a short nonspecific PCR product to the template, this strand being complementary to the strand into which one of these primers was incorporated. It is an inexpensive precaution whose necessity has not been tested.

Step 5. Nested Amplification

1. Remove 1 μl of the unpurified initial PCR product by inserting a long thin pipette tip through the mineral oil layer, and place it in a corresponding PCR tube containing nested primers (primers 3 and 4 in Fig. 1) preheated to 80°C with the same enzyme, reagents, and primer concentrations as in the first PCR amplification.

2. Tubes with the nested primers undergo 35 PCR thermal cycles with cycling parameters identical to those used in the initial amplification.

   Note: PCR products are detected by agarose gel electrophoresis followed by ethidium bromide staining. If multiple bands are seen, raising the annealing temperature in this nested amplification by 1°C or 2°C will frequently eliminate the shorter products. The PCR product can either be directly sequenced or cloned prior to sequencing (Costa and Weiner 1994; Jones 1994). Once the unknown end is sequenced, a new primer can be made so the sequence can be amplified from genomic DNA using conventional PCR.

DISCUSSION

Panhandle PCR amplifies more than 3.0 kb of human genomic DNA that flanks a known site. High specificity occurs because full nested amplifications can be carried out. This has permitted panhandle PCR to amplify near the limit of what can be accomplished using *Taq* DNA polymerase and routine buffer conditions.

The other method that amplifies more than 2.0 kb of human genomic DNA that flanks a known site is the related targeted inverted repeat amplification method (Jones and Winistorfer 1993b). Targeted inverted repeat amplification shares with panhandle PCR the restriction endonuclease digestion of DNA followed by the ligation of an oligonucleotide that is complementary to the known sequence. In contrast to panhandle PCR, the 5′ ends of genomic DNA are extended by ligation to an oligonucleotide instead of the 3′ ends, so that denaturation and intrastrand annealing result in the formation of a stem-loop structure with a recessed and phosphorylated 5′ end that can undergo template-directed ligation to a second oligonucleotide using a heat-stable ligase. These two oligonucleotides can then be used sequentially in single primer amplifications of the unknown flanking DNA.

Targeted inverted repeat amplification requires fewer primers than panhandle PCR and can be carried out using a shorter stretch of known sequence. Panhandle PCR has fewer steps and has amplified longer fragments using *Taq* DNA polymerase and routine buffer conditions. Panhandle PCR is, at this time, the preferred method unless a limited stretch of known sequence (<150–200 nucleotides) precludes its use. Applications of panhandle PCR include chromosome walking, retrieval of promoters and gene regulatory domains using cDNA data, determination of viral and transposon integration sites, and the retrieval and sequencing of unclonable DNA.

ACKNOWLEDGMENTS This work was supported by National Institutes of Health grant R01 HG-00569 and the Roy J. Carver Charitable Trust.

REFERENCES

Arnold, C. and I.J. Hodgson. 1991. Vectorette PCR: A novel approach to genomic walking. *PCR Methods Appl.* **1:** 39–42.

Blin, N. and D.W. Stafford. 1986. A general method for isolation of high molecular weight DNA from eukaryotes. *Nucleic Acids Res.* **3:** 2303–2308.

Costa, G.L. and M.P. Weiner. 1994. Protocols for cloning and analysis of blunt-ended PCR-generated DNA fragments. *PCR Methods Appl.* **3:** S95–S106.

Edwards, J.B., J. Delort, and J. Mallet. 1991. Oligodeoxyribonucleotide ligation to single-stranded cDNAs: A new tool for cloning 5′ ends of mRNAs and for constructing cDNA libraries by in vitro amplification. *Nucleic Acids Res.* **19:** 5227–5232.

Frohman, M.A., M.K. Dush, and G.R. Martin. 1988. Rapid production of full-length cDNAs from rare transcripts: Amplification using a single gene-specific oligonucleotide primer. *Proc. Natl. Acad. Sci.* **85:** 8998–9002.

Jones D.H. 1994. DNA mutagenesis and recombination in vivo. *PCR Methods Appl.* **3:** S141–S148.

Jones, D.H. and S.C. Winistorfer. 1992. Sequence specific generation of a DNA panhandle permits PCR amplification of unknown flanking DNA. *Nucleic Acids Res.* **20:** 595–600.

——. 1993a. Genome walking with 2- to 4-kb steps using panhandle PCR. *PCR Methods Appl.* **2:** 197–203.

——. 1993b. A method for the amplification of unknown flanking DNA: Targeted inverted repeat amplification. *BioTechniques* **15:** 894–904.

Kalman, M., E.T. Kalman, and M. Cashel. 1990. Polymerase chain reaction (PCR) amplification with a single specific primer. *Biochem. Biophys. Res. Commun.* **167:** 504–506.

Lagerstrom, M., J. Parik, H. Malmgren, J. Stewart, U. Pettersson, and U. Landegren 1991. Capture PCR: Efficient amplification of DNA fragments adjacent to a known sequence in human and YAC DNA. *PCR Methods Appl.* **1:** 111–119.

Loh, E.Y., J.F. Elliott, S. Cwirla, L.L. Lanier, and M.M. Davis. 1989. Polymerase chain reaction with single sided specificity: Analysis of T cell receptor δ chain. *Science* **243:** 217–220.

MacGregor, G.R. and P.A. Overbeek. 1991. Use of a simplified single-site PCR to facilitate cloning of genomic DNA sequences flanking a transgene integration site. *PCR Methods Appl.* **1:** 129–135.

Mullis, K.B. 1991. The polymerase chain reaction in an anemic mode: How to avoid cold oligodeoxyribonuclear fusion. *PCR Methods Appl.* **1:** 1–4.

Mullis, K., F. Faloona, S. Scharf, R. Saiki, G. Horn, and H. Erlich. 1986. Specific enzymatic amplification of DNA in vitro: The polymerase chain reaction. *Cold Spring Harbor Symp. Quant. Biol.* **51:** 263–273.

Normore, W.M., H.S. Shapiro and P. Setlow. 1976. *CRC handbook of biochemistry and molecular biology*, (ed. G.D. Fastman). CRC Press, Boca Raton, Florida.

Ochman, H., A.S. Gerber, and D.L. Hartl. 1988. Genetic applications of an inverse polymerase chain reaction. *Genetics* **120:** 621–623.

Ohara, O., R.L. Dorit, and W. Gilbert. 1989. One-sided polymerase chain reaction: The amplification of cDNA. *Proc. Natl. Acad. Sci.* **86:** 5673–5677.

Parker J.D., P.S. Rabinovitch, and G.C. Burmer. 1991. Targeted gene walking polymerase chain reaction. *Nucleic Acids Res.* **19:** 3055–3060.

Parks C.L., L.-S. Chang, and T. Shenk. 1991. A polymerase chain reaction mediated by a single primer: Cloning of genomic sequences adjacent to a serotonin receptor protein coding region. *Nucleic Acids Res.* **19:** 7155–7160.

Pfeifer, G.P., S.D. Steigerwald, P.R. Mueller, B. Wold, and A.D. Riggs. 1989. Genomic sequencing and methylation analysis by ligation mediated PCR. *Science* **246:** 810–813.

Riley, J., R. Butler, D. Ogilvie, R. Finniear, D. Jenner, S. Powell, R. Anand, J.C. Smith, and A.F. Markham. 1990. A novel, rapid method for the isolation of terminal sequences from yeast artificial chromosome (YAC) clones. *Nucleic Acids Res.* **18:** 2887–2890.

Riordan, J.R., J.M. Rommens, B.-S. Kerem, N. Alon, R. Rozmahel, Z. Grzelczak, J. Zielenski, S. Lok, N.

Plavsic, J.-L. Chou, M.L. Drumm, M.C. Iannuzzi, F.S. Collins, and L.-C. Tsui. 1989. Identification of the cystic fibrosis gene: Cloning and characterization of complementary DNA. *Science* **245:** 1066–1073.

Rosenthal, A. and D.S.C. Jones. 1990. Genomic walking and sequencing by oligo-cassette mediated polymerase chain reaction. *Nucleic Acids Res.* **18:** 3095–3096.

Roux, K.H. and P. Dhanarajan. 1990. A strategy for single site PCR amplification of dsDNA: Priming digested cloned or genomic DNA from an anchor-modified restriction site and a short internal sequence. *BioTechniques* **8:** 48–57.

Sambrook, J., E.F. Fritsch, and T. Maniatis. 1989. *Molecular cloning: A laboratory manual*, 2nd edition, p. B.23. Cold Spring Harbor Laboratory Press, Cold Spring Harbor, New York.

Sarkar, G., R.T. Turner, and M.E. Bolander. 1993. Restriction-site PCR: A direct method of unknown sequence retrieval adjacent to a known locus by using universal primers. *PCR Methods Appl.* **2:** 318–322.

Shyamala, V. and G.F.L. Ames. 1989. Genome walking by single-specific-primer polymerase chain reaction: SSP-PCR. *Gene* **84:** 1–8.

Silver, J. and V. Keerikatte. 1989. Novel use of polymerase chain reaction to amplify cellular DNA adjacent to an integrated provirus. *J. Virol.* **63:** 1924–1928.

Tormanen, V.T., P.M. Swiderski, B.E. Kaplan, G.P. Pfeifer, and A.D. Riggs. 1992. Extension product capture improves genomic sequencing and DNase I footprinting by ligation-mediated PCR. *Nucleic Acids Res.* **20:** 5487–5488.

Triglia, T., M.G. Peterson, and D.J. Kemp. 1988. A procedure for *in vitro* amplification of DNA segments that lie outside the boundaries of known sequences. *Nucleic Acids Res.* **16:** 8186.

Troutt, A.B., M.G. McHeyzer-Williams, B. Pulendran, and G.J.V. Nossal. 1992. Ligation-anchored PCR: A simple amplification technique with single-sided specificity. *Proc. Natl. Acad. Sci.* **89:** 9823–9825.

Detection and Identification of Expressed Genes by Differential Display

Peter Warthoe,[1] David Bauer,[2] Mikkel Rohde,[1] and Michael Strauss[1,2]

[1]Danish Cancer Society, Division for Cancer Biology, Department of Cell Cycle and Cancer, DK-2100 Copenhagen Ø, Denmark
[2]Max-Planck-Gesellschaft, Research Group on Cell Division and Gene Substitution, Humboldt Universität, D-13122 Berlin-Buch, Germany

INTRODUCTION

The need to analyze changes in gene expression patterns occurring in response to developmental or physiological factors, mutations, or simply transfection of particular genes has stimulated the search for proper methods to identify the actual differences between two well-defined biological situations. For a long time, differential hybridization was the only method that could isolate genes active in either of the two situations. There was no method that could indicate the total number of genes subject to up- or down-regulation in a particular setting.

In 1992, two different but related strategies were introduced for fingerprinting of expressed RNAs as cDNA tags. One was called differential display (DD) (Liang and Pardee 1992) or, according to the PCR nomenclature, DDRT-PCR (Bauer et al. 1993; Liang et al. 1995a) and the other was named RNA arbitrary primed PCR (RAP-PCR) (Welsh et al. 1992; McClelland and Welsh 1994). Both methods use primers of arbitrary sequence, previously used for DNA fingerprinting (Welsh and McClelland 1990; Williams et al. 1990), to generate cDNA tags. DDRT-PCR uses combinations of 10-mer arbitrary primers with anchored cDNA primers and generates fragments that originate mostly from the poly(A) tail and extend about 50–600 nucleotides upstream (Liang and Pardee 1992; Bauer et al. 1993; Liang et al. 1993, 1994a, 1995a). Both variants of RNA fingerprinting are able to detect differences in gene expression of a certain percentage of expressed genes. However, only the original differential display (DDRT-PCR) technique (Liang and Pardee 1992; Bauer et al. 1993) is able to generate a largely complete pattern of all mRNAs expressed in a particular cell using a reasonable number of primer pairs (Bauer et al. 1993). Under the appropriate conditions, the pattern of fragments derived from one type of cells is reproducible and can be compared with that of another cell type.

DDRT-PCR has already been applied successfully by several groups, mainly by using only a limited set of primers, to isolate genes differentially expressed in various biological situations, including cancers (Liang et al. 1992, 1994b; Sager et al. 1993; Zhang and Medina 1993; Mok et al. 1994; Sun et al. 1994; Watson and Fleming 1994), heart disease (Russel et al. 1994; Utans et al. 1994), diabetes (Aiello et al. 1994; Nishio et al. 1994), embryogenesis (Zimmermann and Schultz 1994), brain development (Watson and Margulies 1993; Joseph et al. 1994), and growth factor regulation (Hsu et al. 1993; Donohue et al. 1994). The method will be applied to a large variety of biological and medical processes in the near future and will lead to the identification of hundreds of new genes that are crucially involved in regulating or executing essential cellular functions. Here we present a detailed protocol that is applicable to any set of two or more comparable eukaryotic cell types and will result in reproducible patterns of PCR products that correspond to expressed genes.

METHOD STRATEGY

The strategy of the method consists of three basic and two additional steps, as shown in Figure 1: (1) reverse transcription in fractions using a set of anchored primers; (2) amplification of cDNA species from each fraction using a set of arbitrary primers and anchored primers; (3) electrophoretic separation of the resulting fragments; (4) reamplification of fragments that are different between two situations, cloning, and sequencing; and (5) confirmation of differential expression by an independent RNA analysis technique (Northern blotting, RNase protection, and/or nuclear run-on).

The method uses either cytoplasmic or mRNA as the starting material; it is first reverse-transcribed to yield single-stranded cDNA. The step of reverse transcription is already an inherent part of the method. A clever subdivision of the total number of mRNAs is required to be able to display as many different expressed genes as possible. Subdivision of the mRNAs should, at the same time, provide an anchor sequence for the subsequent PCR amplification. Liang and Pardee (1992) originally suggested using 12 different primers of the type $T_{11}VN$, where V can be A, G, or C, and N can be any of the four nucleotides. By using these primers, one could generate 12 subfractions of cDNA, which should represent almost equally 1/12 of the ex-

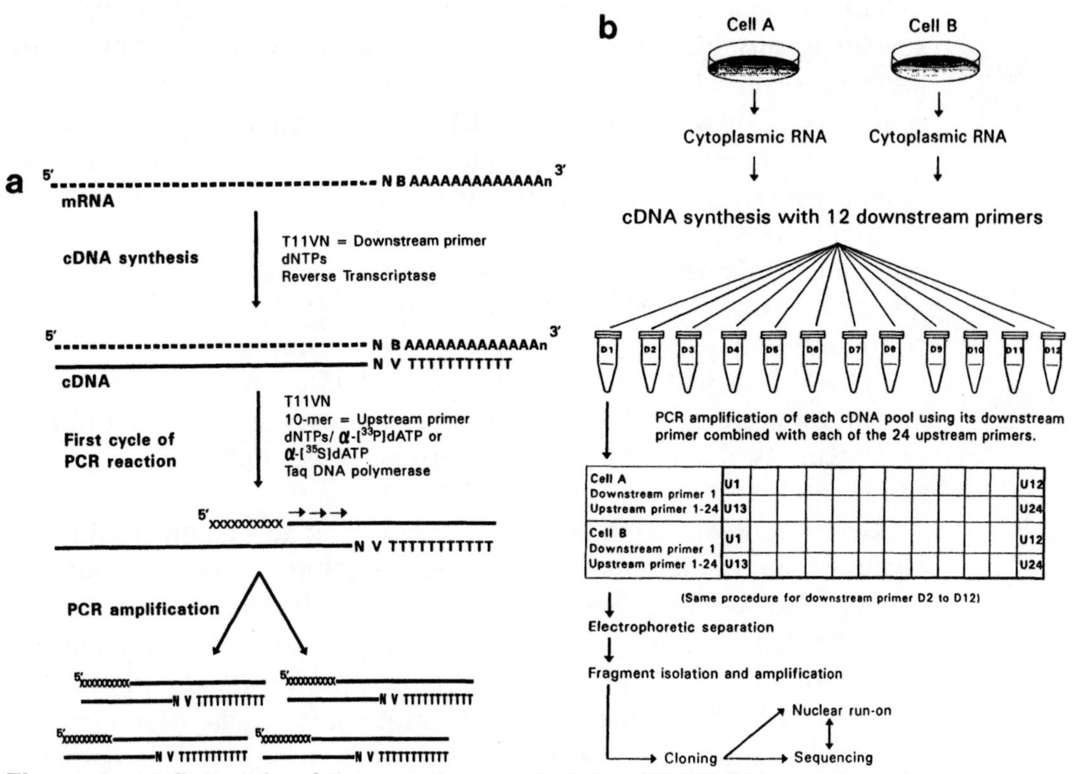

Figure 1 (*a*) Principle of the reactions underlying DDRT-PCR analysis. N represents any of the four nucleotides G, A, T, or C. V can be any nucleotide except T. B can be any nucleotide except A. (*b*) Flow chart of DDRT-PCR analysis with subsequent confirmation of differential regulation and characterization of differentially expressed genes.

pressed genes of a particular cell. Assuming 15,000 genes to be expressed in a cell at one time, one subfraction of cDNA would contain cDNA species representing 1,250 different genes. Because the terminal 3′ base of the primer provides most of the specificity, Liang and Pardee (1992) suggested reducing the number of cDNA subfractions to only four by using a $T_{12}MN$ primer, where M is a degenerate mixture of A, G, or C. Recently, the same group showed that the use of three one-base anchored oligo-dT primers solves most of the problems related to the use of the two-nucleotide redundant primers (Liang et al. 1995b). This reduces the number of cDNA fractions to only 3 instead of 4 or 12, but does not necessarily reduce the number of PCR procedures required to display most cDNA species present in the pool. On the contrary, it reduces the theoretical chance of identifying any cDNA species that are actually present. According to our experience, to obtain an almost complete picture of the expressed genes, use 12 fractions. However, to have a "first glance" at the situation or to identify new genes, use the reduced number of fractions.

The next step is the actual amplification of as many of the cDNA species as possible in a PCR incubation. The anchored primers used for reverse transcription serve as the downstream primers for this step. The upstream primers are 10-mer oligonucleotides of arbitrary sequence; not every 10-mer is suitable and must be tested experimentally; the lack of self-complementarity and the same GC content (50%) of all primers are important factors (Bauer et al. 1993). From our theoretical calculations, the chance for every mRNA species to be displayed requires that 24–26 different primers (Table 1) must be used in combination with every downstream primer, amounting to 288 or 312 individual PCR procedures (Bauer et al. 1993). Optimal conditions for annealing of the primers are required to obtain the maximal number of fragments displayed in the following electrophoretic separation (Bauer et al. 1993; Liang et al. 1995a).

Electrophoresis can be done on sequencing gels (Liang and Pardee 1992; Liang et al. 1993), on nondenaturing polyacrylamide gels (Bauer et al. 1993), or even on agarose gels (Hsu et al. 1993; Sokolov and Prockop 1994). We recommend the use of nondenaturing polyacrylamide gels because they reduce the artificial complexity of the patterns often observed with sequencing gels due to strand separation and incomplete addition of a terminal nucleotide by certain thermostable DNA polymerases; they also make processing of the bands much easier. Because radioactive nucleotides are the preferred labeled substrates, gels are first exposed to autoradiography with X-ray film. The bands of interest are cut out from the gel and reamplified, and the fragments are cloned into a suitable vector. Because of the possibility of more than one cDNA species within one band, isolation of several colonies that should be further characterized is advisable. We also recommend repeating the DDRT-PCR analysis with the pairs of primers that led to the detection of differences, ideally once with the same RNA and once or twice with a new RNA preparation. This increases the certainty of detecting real differences in gene expression.

Characterization of the cloned fragments should be done in two ways. First of all, a fragment(s) of interest should be analyzed to determine whether it corresponds to a known or an unknown gene. This requires sequencing of the clone(s). However, one also has to make sure that the fragment is indeed differentially regulated and not just an artifact. This requires using another method for detecting differences in gene expression. Northern blot analysis or RNase protection assays are the most obvious choice. Disadvantages of these methods include not only their labor-intensive and time-consuming performance, but also the requirement for labeling multiple probes corresponding to the differentially displayed cloned candidates. We prefer to verify the differential expression of the PCR-obtained frag-

Table 1 List of 10-mer Upstream Primers Used for DDRT-PCR

No.	Sequence (5′ to 3′)
1.	T A C A A C G A G G
2.	T G G A T T G G T C
3.	C T T T C T A C C C
4.	T T T T G G C T C C
5.	G G A A C C A A T C
6.	A A A C T C C G T C
7.	T C G A T A C A G G
8.	T G G T A A A G G G
9.	T C G G T C A T A G
10.	G G T A C T A A G G
11.	T A C C T A A G C G
12.	C T G C T T G A T G
13.	G T T T T C G C A G
14.	G A T C A A G T C C
15.	G A T C C A G T A C
16.	G A T C A C G T A C
17.	G A T C T G A C A C
18.	G A T C T C A G A C
19.	G A T C A T A G C C
20.	G A T C A A T C G C
21.	G A T C T A A C C G
22.	G A T C G C A T T G
23.	G A T C T G A C T G
24.	G A T C A T G G T C
25.	G A T C A T A G C G
26.	G A T C T A A G G C

In the standard setup only the first 24 primers are used.

ments by nuclear run-on assays, which require only one labeling reaction for every cell type studied and can incorporate an almost unlimited number of samples. We recommend that six clones from every fragment of interest should be included in the nuclear run-on analysis. This analysis can determine which clone derived from a particular fragment actually corresponds to the regulated gene. Sequencing is carried out afterward only for the clones that are derived from differentially expressed genes.

The whole DDRT-PCR analysis of two cell types, including 576 PCR incubations and 12 gels with 48 lanes each, can be carried out by one person within 8 working days. Repetition of DDRT-PCR with selected primer pairs takes another 2–6 days. The time required for analysis of the bands by nuclear run-on and subsequent sequencing is hard to estimate. To introduce the beginner to the method and to provide the advanced user with a handy guide, we here provide a day-by-day protocol that is easy to follow.

REAGENTS

Cytoplasmic RNA Preparation

Phosphate-buffered saline (PBS)
Guanidinium thiocyanate
N-Lauroyl-sarcosine, sodium salt
Nonidet, NP-40
β-Mercaptoethanol
Diethyl pyrocarbonate (DEPC)
Phenol
Chloroform
Isoamyl alcohol
Sodium acetate
Sodium chloride
DNase, RNase-free (Boehringer Mannheim)
RNasin

SOLUTIONS

Extraction buffer:
- 0.14 M NaCl
- 10 mM Tris-HCl, pH 7.5
- 1.5 mM $MgCl_2$

Solution D (Chomczynski and Sacchi 1987): Guanidinium thiocyanate lysis solution. Dissolve 100 g of guanidinium thiocyanate (Fluka) in 117 ml of water, add 7.0 ml of 0.75 M sodium citrate (pH 7.0), add 10.5 ml of 10% Sarkosyl, dissolve at 65°C, and store at room temperature for several months. Before use, add 72 μl of concentrated β-mercaptoethanol to 10 ml of lysis solution.

PROTOCOL

Basically, any kind of RNA preparation can be used. However, we have obtained the best results using DNA-free cytoplasmic RNA preparations that are carried out as follows:

1. Wash approximately 4×10^6 cells with PBS (lacking Ca^{++}).

2. Gently resuspend cells in 450 μl of extraction buffer.

3. Dropwise, add 50 μl of 5% Nonidet NP-40 and mix. Leave on ice for 2 minutes.

4. Spin at 550*g* for 5 minutes at 4°C.

5. Transfer the supernatant into a fresh tube containing 4.5 ml of solution D and mix; leave 10 minutes on ice.

6. Add 450 μl of 2 M sodium acetate, pH 5, and 4.5 ml of acid phenol; shake for 5 minutes, and add 1.5 ml chloroform/isoamyl alcohol (49:1).

7. Spin at 4000*g* for 10 minutes.

8. Take the upper phase, add 5.7 ml of isopropanol, and leave on ice for 20 minutes.

9. Spin at 4500 rpm for 30 minutes; remove supernatant carefully and discard.

10. Resuspend the pellet in 0.3 ml of solution D, transfer to an Eppendorf tube, and incubate on ice for 15 minutes.

11. Precipitate the RNA with 1 ml of 96% ethanol at –70°C.

12. Spin at full speed for 10 minutes, wash the pellet with 1 ml 70% ethanol, and spin again.

13. Remove the supernatant and resuspend the pellet in 200 μl of DEPC-treated water, add 10 μl of 1 M Tris-HCl, pH 7.5, 2 μl of 1 M $MgCl_2$, 2 μl of RNase-free DNase, and 0.5 μl RNasin; incubate for 15 minutes at 37°C.

14. Add 200 μl of phenol/chloroform (1:1), vortex, and spin to separate the two phases.

15. Take the supernatant and add 20 μl of 2 M sodium acetate, pH 4, and 1 ml 96% ethanol; cool down to –70°C for 10 minutes, spin for 10 minutes, and wash the pellet with 70% ethanol.

16. Resuspend the pellet in 10–20 μl of DEPC-treated distilled water and measure the concentration.

17. RNA should be stored in ethanol/acetate at –70°C.

RNA preparation and cDNA synthesis can normally be done within one day.

Single-strand cDNA Synthesis

The cDNA synthesis reaction is designed for running 6 x 96 PCR incubations, which enables the complete DDRT-PCR analysis for two cell types using 12 downstream and 24 upstream primers. Each cell type will yield 12 cDNA pools, which will be incubated with 24 different upstream primers. For two cell types, prepare 2 x 12 cDNA reactions. The volumes in the master mix are adjusted accordingly.

REAGENTS

Single-strand cDNA synthesis kit (Life Technologies, cat. no. 8053SA) including 5× buffer, DTT, and SuperScript reverse transcriptase, RNase H-minus

dNTP stock solution (Pharmacia)

12 Anchored primers ($T_{11}VN$)

cDNA master mix:

150 μl of 5× first-strand cDNA synthesis buffer
75 μl of 0.1 M DTT
150 μl of 100 μM dNTP mix
19 μl of 40 units/μl RNasin
19 μl of sterile DEPC-treated water

PROTOCOL

For cDNA synthesis, 24 reaction tubes are prepared as follows:

1. Add 3.0 μl of 25 μM downstream primer to each tube. Because there are 12 different downstream primers, 12 tubes are set up per cell line: 1–12 for cell line A and 13–24 for cell line B.

2. Add 3.0 μl (100–300 ng) of cytoplasmic RNA from cell A to each of tubes 1–12 and of RNA from cell B to each of tubes 13–24.

3. Add 7.5 μl of sterile water to all 24 tubes.

4. Heat for 10 minutes at 70°C and place on ice immediately.

5. Add 16.5 μl of the cDNA master mix to all 24 tubes.

6. Incubate for 2 minutes at room temperature.

7. Add 1.5 μl (300 units) of SuperScript RNase H-minus reverse transcriptase.

8. Incubate for 8 minutes at room temperature followed by 1 hour at 37°C; heat for 5 minutes to 95°C, place on ice immediately, and store at –70°C for further use.

DDRT-PCR and Gel Electrophoresis of DNA Fragments

This PCR protocol is designed for 96 PCR procedures for analyzing two cDNA subfractions from two different cell types with two dT11VN downstream primers and 24 upstream primers within one day (see Fig. 1b). (This has to be repeated five times within the next days using the other 10 downstream primers.)

REAGENTS

[^{33}P]dATP, 10.0 mCi/ml (NEN)
Taq DNA polymerase
dNTP stock solution (100 μM)
24 Arbitrary 10-mer primers (see Table 1)

DDRT-PCR master mix:
- 234.0 μl of 10X PCR buffer (500 mM KCl, 100 mM Tris-HCl [pH 9 at 25°C], 0.1% gelatin, 1% Triton X-100)
- 140.4 μl of 25 mM $MgCl_2$ (to be optimized depending on the *Taq* DNA polymerase)
- 11.7 μl of [^{33}P]dATP
- 46.8 μl of 100 μM dNTP mix
- 22.0 μl of *Taq* DNA polymerase (5 units/μl)
- 949.1 μl of sterile water (DEPC-treated)

To each of four tubes, transfer 324 μl of this master mix and supplement with downstream primer and cDNA as follows:

DDRT-PCR setup:

Tube no.	1	2	3	4
Total no. of reactions	24	24	24	24
Master mix (μl)	324	324	324	324
Downstream primer no.	1	1	2	2
Downstream pr. (μl, 25 μM)	54	54	54	54
cDNA from cell type	A	B	A	B
Amount of cDNA (μl)	27	27	27	27
Total μl	405	405	405	405

96 PCR mixes are prepared in a microtiter plate or in tubes arranged in a microtiter format (Perkin-Elmer) as follows:

1. Distribute 15 μl of the premixed components from each of the four tubes into 24 wells or tubes.

2. Add 5 μl (2 μm) of the 24 different upstream primers using a 12-channel micropipette. Each of the 24 wells or tubes will only receive one of the 24 different upstream primers.

3. Spin briefly in a centrifuge that holds microtiter plates to mix the upstream primers with the premixed components.

4. Run PCR in a suitable thermal cycler (e.g., GeneAmp PCR system 9600, Perkin-Elmer) equipped with a heated lid.

PROTOCOL

Amplification

30 seconds at 94°C
60 seconds at 40°C*

60 seconds at 72°C
40 cycles

*Background smear can be avoided if this lower stringency is used only for the first 1–5 cycles and temperature is then raised to 45°C for the remaining 35-39 cycles.

The gels can be prepared during the PCR incubations. For 96 reactions, two gels can be run with 48 samples each covering exactly all reactions from one cDNA subfraction of two cell types to be compared. If more cell types are to be compared, it is advisable to run the related reactions (generated using the same pair of primers) side by side.

SOLUTIONS

Electrophoresis

20x TTE: 215 g of Tris base, 71.3 g of Taurine, 20 ml of 0.5 M EDTA; make up to 1 liter with water

6% Gel solution: 14.5 ml of 40% stock 19:1 acrylamide:bisacrylamide, 5.0 ml of 20x TTE, 80.5 ml of water; pass through a 0.2-μm filter

Use 50 ml of the 6% gel solution for a 35 x 40 x 0.2 cm gel; add 40 μl of *N,N,N′,N′*-tetramethylenediamine (TEMED) and 200 μl of 10% ammonium persulfate.

Concentrate half of the DDRT-PCR incubation mixture (10 μl) by vacuum and heat (10 minutes) to 2 μl. Adjust with glycerol to 5%, xylene cyanol FF, and bromphenol blue. Load 2 μl onto a prerun 6% polyacrylamide gel without urea and run in TTE or TBE buffer at 50 watts. The run is stopped when the bromphenol blue runs out of the gel. Dry the gel on filter paper and expose to X-ray film overnight.

Reamplification, Cloning, and Sequencing of DNA Fragments

Cut the bands of interest, e.g., those differing between the patterns from control cells and the test cells, from the filter paper and transfer to Eppendorf tubes. Add 100 μl of DEPC-treated water and boil for 10 minutes. Take out and discard the filter paper.

Reamplification PCR

26.5 μl of eluted DNA
5 μl of 10x PCR buffer (as above)
3 μl of 25 mM $MgCl_2$
5 μl of 500 μM dNTP mix
5 μl of 2 μM downstream primer*
5 μl of 2 μM upstream primer*
0.5 μl of *Taq* DNA polymerase (5 units/μl)

*Both primers should be the same as those used to generate the original PCR product.

PROTOCOL

1. Amplify using the cycle protocol given for DDRT-PCR. Run 10 μl of the reaction mixture on a 2% agarose gel to check the size.

2. Clone fragments of interest using the TA cloning kit (Invitrogen) and 1–3 μl of the reamplification solution.

3. Pick six white colonies from each cloned fragment and make plasmid minipreparations.

4. Cut plasmids with *Eco*RI and check for inserts by agarose gel electrophoresis.

   Note: At this point, the inserts could be sequenced. However, because it is still unclear which of the potentially different cDNA species contained within one band actually corresponds to a regulated gene, if there is a regulated one at all, it is advisable to confirm the actual difference in expression by another RNA detection method.

5. Use nuclear run-on analysis to process multiple fragments (the six clones isolated of each fragment) at the same time using only one labeling reaction per cell type.

   Note: If the primary interest is to find out as fast as possible whether the fragment corresponds to an unknown gene or a gene of special interest, then sequence the DNA immediately. DNA sequence analysis can be carried out using the fmole sequencing kit (Promega) or the cycle sequencing kit (USB).

Nuclear Run-on Analysis

To confirm differential regulation of individual candidate bands, we suggest using nuclear run-on assays instead of Northern blot hybridizations. Because vector sequences can create considerable nonspecific signals, we prefer to probe the cloned fragments. The DNA fragments to be tested are amplified from 1 ng of fragment using two primers specific for flanking sequences in the TA vector (5′-TAGTAACGGCCGCCAGTGT and 5′-GCCGCCAGTGTGATGGATA) in a standard PCR using the Perkin-Elmer 9600 thermal cycler (30 seconds at 94°C, 30 seconds at 65°C, 60 seconds at 72°C, 40 cycles; 5 minutes at 72°C). Here we give our standard method (Strauss et al. 1990) which is relatively fast and can be carried out within 3 days.

SOLUTIONS

Lysis buffer:
10 mM Tris-HCl, pH 7.4
5 mM $MgCl_2$
10 mM KCl
0.5% NP-40

Nuclear freezing buffer:
- 50 mM Tris-HCl, pH 7.4
- 5 mM $MgCl_2$
- 40% glycerol
- 0.5 mM DTT

Transcription buffer:
- 25 mM Tris-HCl, pH 8.0
- 12.5 mM $MgCl_2$
- 750 mM KCl
- 1.25 mM NTP (without CTP)

10X STE:
- 10% SDS
- 0.1 M Tris-HCl, pH 7.5
- 50 mM EDTA

20X SSPE:
- 3.6 M NaCl
- 200 mM NaH_2PO_4, pH 7.4
- 20 mM EDTA, pH 7.4

PROTOCOLS

Preparation of Nuclei

1. Approximately 5 X 10^7 cells are required for one assay. Harvest the cells, wash with PBS, and resuspend in 10 ml of lysis buffer.

2. A Dounce homogenizer (type B) is moved up and down eight times in the cell suspension.

3. Spin the extract at 2000 rpm for 10 minutes and remove the supernatant completely.

4. Resuspend the nuclei in 210 μl of nuclear freezing buffer and mix by pipetting up and down several times with a yellow tip.

5. Freeze immediately at –70°C even if you continue with the next steps. Upon thawing, the quality of the nuclei becomes obvious. The nuclei should be dispersed easily by pipetting with a yellow tip. Big clumps, which are difficult to disperse, indicate bad quality.

Nuclear Transcription

1. Add 60 μl of transcription buffer and 30 μl of [^{32}P]CTP (300 μCi) to 210 μl of nuclei.

2. Vortex briefly and gently and incubate for 30 minutes at 30°C; shake every 10 minutes.

Preparation of Labeled RNA

1. Add 10 µl of DNase (230 units, RNase-free) to the nuclear transcription reaction and incubate for 30 minutes.

2. Add 36 µl of 10X SET buffer and 10 µl of proteinase K (10 mg/ml). Heat to 65°C for 3 minutes and incubate at 37°C for 45 minutes.

3. Extract with 360 µl of phenol/chloroform, reextract the interphase with 100 µl of SET buffer, combine aqueous phases (460 µl), and add 203 µl of 7.5 M ammonium acetate, pH 5. Add the same volume of isopropanol, put in dry ice for 20 minutes, spin for 10 minutes, and reprecipitate the RNA with ethanol.

4. Resuspend the pellet in 180 µl of 10 mM Tris-HCl, pH 7.5, and 1 mM EDTA; add 20 µl of 2N NaOH, and place the tube on ice for 5 minutes. After adding 200 µl of 0.48 M HEPES (free acid), precipitate the RNA with 900 µl of ethanol at –20°C overnight.

Filter Hybridization

1. Amplify the inserts out of the cloning vector using primers specific for flanking vector sequences. Denature 2 µg of each fragment (per filter) with 0.25 N NaOH for 10 minutes at room temperature and apply via individual slots to a HYBOND N filter (Amersham) in a slot blot device (Schleicher and Schuell). Immediately wash the slots with 1 M ammonium acetate (5–10 volumes) to neutralize the DNA solution. Prepare identical strips with multiple samples for each hot RNA probe.

2. Hybridize the filters with the total lot of hot RNA (1–5 x 10^7 cpm) at 55°C for 2 days in the SSCP buffer system (10 ml). Use the same amount of label for each nuclear RNA preparation.

3. After washing, expose the filters to X-ray film for about 1 week or analyze on a phosphor imager.

ANALYSIS OF RESULTS AND TROUBLESHOOTING

A typical result of DDRT-PCR is shown in Figure 2. In this particular experiment, tumor cells lacking a functional retinoblastoma (Rb) tumor suppressor gene (BT549) were compared with BT549 cells transfected with the Rb gene (P. Warthoe et al., in prep.). In addition, both types of cells were either treated with transforming growth factor β (TGF-β) or untreated, resulting in four sources of RNA. Four candidate cDNA fragments were identified in this part of a gel. The three candidates in the upper half of the figure seem to be induced by the transfected Rb gene but repressed by additional TGF-β treatment. The

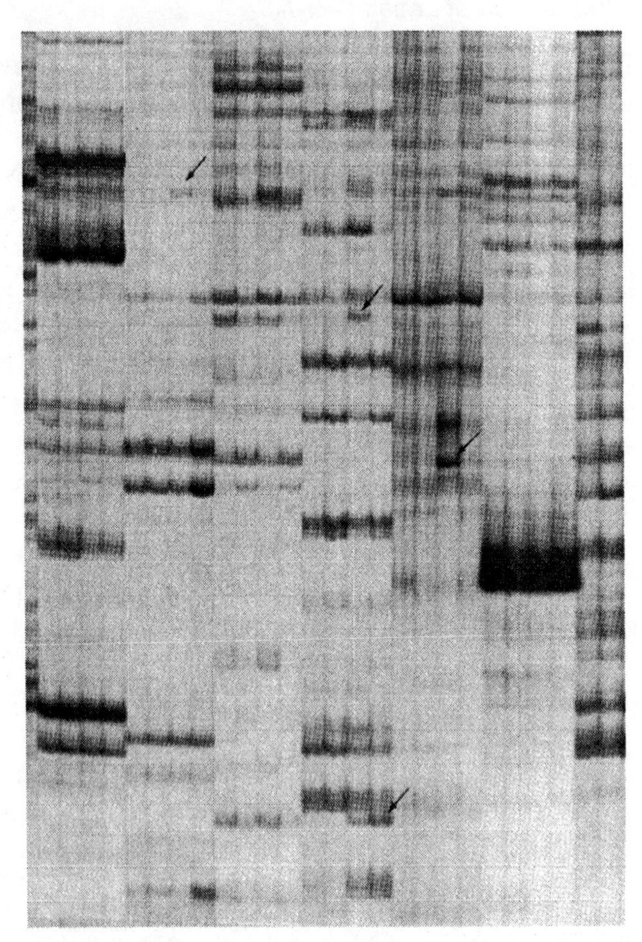

Figure 2 Electrophoretic DDRT-PCR pattern of part of one nondenaturating gel. This part of the gel compares four different cells with one downstream primer and 6 upstream primers. In every set of four cell types, the two left lanes represent the Rb-deficient mammary tumor cell line BT549 after mock transfection and the two right lanes represent the same cell line expressing the Rb gene. Cells in the second and in the fourth lanes of each set were treated with TGF-β for 8 hours. Arrows indicate cDNA bands corresponding to potential differentially regulated genes.

candidate gene in the lower part of the figure is obviously induced by the Rb gene and might be additionally stimulated by TGF-β. However, the quantitative difference between the two corresponding bands of the two samples could be due to experimental variation. This kind of quantitative variation is normal and would usually not be considered as a real difference between a control RNA and the test RNA. In this particular case, it could be real because one would expect TGF-β to enhance the activity of the Rb gene product. Therefore, confirmation by repetition of the DDRT-PCR analysis and additional confirmation of differential regulation by an independent method are required. Because Northern blotting frequently fails to give a signal at all, or sometimes shows twofold differences that are difficult to verify, we prefer to use hybridization of the candidate fragments to nuclear run-on transcripts.

A typical result from slot blot hybridization of candidate fragments (after cloning) with nuclear run-on transcripts is given in Figure 3. It clearly shows that 10 (including the four bands from Fig. 2) out of 11 candidate genes are indeed expressed at different levels in the two cell lines compared. Only 8 of these candidates were detectable by Northern blotting and only 5 of them appeared to be differentially regulated by Northern blot analysis. This result is probably typical for situations in which genes are activated or repressed for a short period of time but the stability of the transcript is high. In these cases, when transcription is stimulated, e.g., tenfold for a short time, the total steady-state level of the particular RNA may increase only twofold, a level that is hard to detect by Northern blot analysis.

The evaluation of quantitative differences between bands in two samples is one of the major difficulties in DDRT-PCR. In principle, the difference between both strong versus weak bands or present versus absent bands needs to be confirmed. According to our experience, a considerable number of differentially displayed bands of one analysis

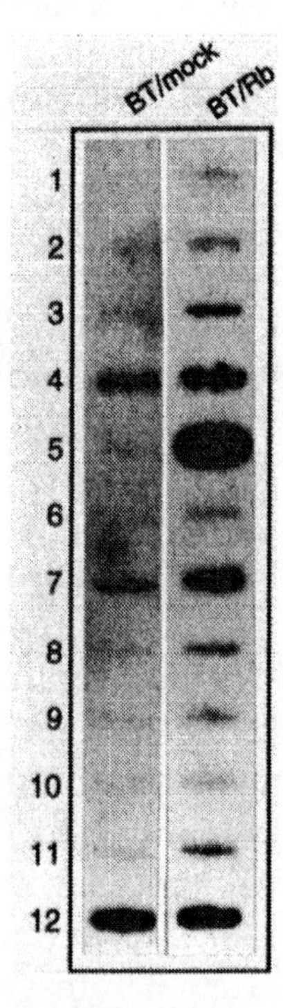

Figure 3 Nuclear run-on analysis of fragments differentially displayed on nondenaturing gels in the DDRT-PCR analysis of Rb-regulated genes. Hybridized filters were analyzed using a phosphor imager. BT/mock is the Rb-deficient tumor cell line BT549 after mock transfection; BT/Rb is BT549 transfected with the Rb gene (P. Warthoe et al., in prep.). Lane *12* shows GADPH as a nonregulated control gene.

are not reproducible (20–40%). However, if a particular difference is reproduced once, it is 90% certain that it can be reproduced in every other repetition. In our experiments with completely different cell types and biological situations, we could always confirm differential regulation if the DDRT-PCR analysis detected the difference three times. Other problems and pitfalls with DDRT-PCR are those that can occur with standard PCR or sequencing techniques. Even the source and quality of the reaction vessels can have an influence on the results (Chen et al. 1994).

DISCUSSION

The DDRT-PCR method allows the generation of highly reproducible patterns of bands from a particular cell line. We have used DDRT-PCR in both our laboratories for more than 2 years in studies of liver regeneration and in the analysis of genes regulated by tumor suppressors. We have identified a large number of differentially regulated genes in both projects (D. Bauer et al.; P. Warthoe et al.; both in prep.). We stress that although confirmation of differentially displayed bands as such by repetition of the DDRT-PCR analysis is advisable, confirmation of the actual differential regulation by another method like nuclear run-on analysis is required. A reverse-Northern blot technique has recently been shown to be superior over standard Northerns (Mou et al. 1994). For detection of short-term effects on gene activities, nuclear run-on is the only reliable method.

It must be stressed that DDRT-PCR is not a totally quantitative method so far. Even drastic differences between two situations (e.g., a strong band from one cell type and the absence of this band in the control cell) can occur when the actual difference in expression levels is only fivefold and, sometimes, slightly different intensities in a particular band position are related to dramatic differences in gene expression. The latter can occur when a constitutively expressed RNA fragment runs in the same gel position as one that corresponds to a regulated gene. The major problem is the wide spectrum of abundance of individual mRNAs. Under the standard conditions, probably only the less abundant class of RNA is amplified in the linear range. If the number of cycles is reduced to 30, only a very few bands are detectable. Probably one would have to run at least three sets of analyses (30, 35, and 40 cycles) or three different concentrations of cDNA to compensate for this problem. However, this would make the method slow and expensive. Other improvements are imaginable. Because most of the 10-mer arbitrary primers anneal with only 6 or 7 nucleotides perfectly matched at their 3′ end (Bauer et al. 1993; Liang et al. 1994a, 1995a) in the first round but would function as perfect 10-mer primers in the next rounds, we have changed the standard amplification protocol by raising the annealing temperature after the first or after five cycles to 45°C (see protocol). Very few bands would be detectable if annealing were carried out at 45°C from the beginning (probably those where the primers match the target sequence in 9 or 10 nucleotides), whereas only a very few bands are lost from the standard pattern by raising the annealing temperature after one to five cycles. This modification leads to clear patterns without background smear, which is due to continuous low-stringency priming of primers that do not match perfectly. The modification will probably further reduce the number of false positives, which are actually no major problem with the protocols described here.

In summary, DDRT-PCR is the most flexible and comprehensive method available for the detection of almost all genes expressed in a particular cell and for the identification of differences in gene expression between different cell types. This technique has at least four advantages over differential hybridization or differential libraries: (1) it allows simultaneous display of all differences, (2) it detects up-regulation and down-regulation of genes at the same time, (3) it allows comparison of more than two situations, and (4) it is faster. Future improvements of the technique could be easier detection of weak differentially expressed bands and automatic analysis using fluorescently labeled primers (Bauer et al. 1993; Ito et al. 1994). The method is already established in numerous basic research laboratories, and we anticipate its use in applied research and diagnostic laboratories in the near future.

REFERENCES

Aiello, L.P., G.S. Robinson, Y.W. Lin, Y. Nishio, and G.L. King. 1994. Identification of multiple genes in bovine retinal pericytes altered by exposure to elevated levels of glucose by using mRNA differential display. *Proc. Natl. Acad. Sci.* **91:** 6231–6235.

Bauer, D., H. Müller, J. Reich, H. Riedel, V. Ahrenkiel, P. Warthoe, and M. Strauss. 1993. Identification of differentially expressed mRNA species by an improved display technique (DDRT-PCR). *Nucleic Acids Res.* **21:** 4272–4280.

Chen, Z., K. Swisshelm, and R. Sager. 1994. A cautionary note on the reaction tubes for differential display and cDNA amplification in thermal cycling. *BioTechniques* **16:** 1003–1006.

Chomczynski, P. and N. Sacchi. 1987. Single-step method of RNA isolation by acid guanidinium thiocyanate-phenol-chloroform extraction. *Anal. Biochem.* **167:** 157–159.

Donohue, P.J., G.F. Alberts, B.S. Hampton, and J.A. Winkles. 1994. A delayed-early gene activated by fibroblast growth factor-1 encodes a protein related to aldose reductase. *J. Biol. Chem.* **269:** 8606–8609.

Hsu, D.K., P.J. Donohue, G.F. Alberts, and J.A. Winkles. 1993. Fibroblast growth factor-1 induces phosphofructokinase, fatty acid synthase and Ca(2+)-ATPase mRNA expression in NIH 3T3 cells. *Biochem. Biophys. Res. Commun.* **197:** 1483–1491.

Ito, T., K. Kito, N. Adati, Y. Misui, H. Hagiwara, and Y. Sakaki. 1994. Fluorescent differential display: Arbitrarily primed RT-PCR fingerprinting on an automated DNA sequencer. *FEBS Lett.* **351:** 231–236.

Joseph, R., D. Dou, and W. Tsang. 1994. Molecular cloning of a novel mRNA (neuronatin) that is highly expressed in neonatal mammalian brain. *Biochem. Biophys. Res. Commun.* **201:** 1227–1234.

Liang, P. and A.B. Pardee. 1992. Differential display of eukaryotic messenger RNA by means of the polymerase chain reaction. *Science* **257:** 967–971.

Liang, P., L. Averboukh, and A.B. Pardee. 1993. Distribution and cloning of eukaryotic mRNAs by means of differential display: Refinements and optimization. *Nucleic Acids Res.* **21:** 3269–3275.

——. 1994a. Method of differential display. *Methods*

Mol. Genet. **5:** 3–16.

Liang, P., L. Averboukh, W. Zhu, and A. B. Pardee. 1994b. Ras activation of genes: Mob-1 as a model. *Proc. Natl. Acad. Sci.* **91:** 12515–12519.

Liang, P., L. Averboukh, K. Keyomarsi, R. Sager, and A.B. Pardee. 1992. Differential display and cloning of messenger RNAs from human breast cancer versus mammary epithelial cells. *Cancer Res.* **52:** 6966–6968.

Liang, P., D. Bauer, L. Averboukh, P. Warthoe, M. Rohrwild, H. Müller, M. Strauss, and A.B. Pardee. 1995a. Analysis of altered gene expression by differential display. *Methods Enzymol.* **254:** (in press).

Liang, P., W. Zhu, X. Zhang, Z. Guo, R.P. O'Connell, L. Averboukh, F. Wang, and A.B. Pardee. 1995b. Differential display using one-base anchored oligo-dT primers. *Nucleic Acids Res.* **22:** 5763–5764.

McClelland, M. and J. Welsh. 1994. RNA fingerprinting by arbitrarily primed PCR. *PCR Methods Appl.* **4:** S66–S81.

Mok, S.C., K.K. Wong, R.K. Chan, C.C. Lau, S.W. Tsao, R.C. Knapp, and R.S. Berkowitz. 1994. Molecular cloning of differentially expressed genes in human epithelial ovarian cancer. *Gynecol. Oncol.* **52:** 247–252.

Mou, L., H. Miller, J. Li, E. Wang, and L. Chalifour. 1994. Improvements to the differential display method for gene analysis. *Biochem. Biophys. Res. Commun.* **199:** 564–569.

Nishio, Y., L.P. Aiello, and G.L. King. 1994. Glucose induced genes in bovine aortic smooth muscle cells identified by mRNA differential display. *FASEB J.* **8:** 103–106.

Russel, M.E., U. Utans, A.F. Wallace, P. Liang, M.J. Arceci, M.J. Karnovsky, L.R. Wyner, Y. Yamashita, and C. Tarn. 1994. Identification and upregulation of galactose/N-acetylgalactosamine macrophage lectin in rat cardiac allografts with arteriosclerosis. *J. Clin. Invest.* **94:** 722–730.

Sager, R., A. Anisowicz, M. Neveu, P. Liang, and G. Sotiropoulou. 1993. Identification by differential display of alpha 6 integrin as a candidate tumor suppressor gene. *FASEB J.* **7:** 964–970.

Sokolov, B.P. and D.J. Prockop. 1994. A rapid and simple PCR-based method for isolation of cDNAs from differentially expressed genes. *Nucleic Acids Res.* **22:** 4009–4015.

Strauss, M., S. Hering, L. Lübbe, and B.E. Griffin. 1990. Immortalization and transformation of human fibroblasts by regulated expression of polyoma T antigens. *Oncogene* **5:** 1223–1229.

Sun, Y., G. Hegamyer, and N.H. Coleburn. 1994. Molecular cloning of five messenger RNAs differentially expressed in preneoplastic or neoplastic JB6 mouse epidermal cells: One is homologous to human tissue inhibitor of metalloproteinase. *Cancer Res.* **54:** 1139–1144.

Utans, U., P. Liang, L.R. Wyner, M.J. Karnovsky, and M. Russel. 1994. Chronic cardiac rejection: Identification of five upregulated genes in transplanted hearts by differential mRNA display. *Proc. Natl. Acad. Sci.* **91:** 6463–6467.

Watson, M.A. and T.P. Fleming. 1994. Isolation of differentially expressed sequence tags from human breast cancer. *Cancer Res.* **54:** 4598–4602.

Watson, J.B. and J.E. Margulies. 1993. Differential cDNA screening strategies to identify novel state-specific proteins in the developing mammalian brain. *Dev. Neurosci.* **15:** 77–86.

Welsh, J. and M. McClelland. 1990. Fingerprinting genomes using PCR with arbitrary primers. *Nucleic Acids Res.* **18:** 7213–7218.

Welsh, J., K. Chada, S.S. Dalal, R. Cheng, D. Ralph, and M. McClelland. 1992. Arbitrarily primed PCR fingerprinting of RNA. *Nucleic Acids Res.* **20:** 4965–4970.

Williams, J.K.G., A.R. Kubelik, K.J. Livak, J.A. Rafalski, and S.V. Tingey. 1990. DNA polymorphisms amplified by arbitrary primers are useful as genetic markers. *Nucleic Acids Res.* **18:** 6531–6535.

Zhang, L. and D. Medina. 1993. Gene expression screening for specific genes associated with mouse mammary tumor development. *Mol. Carcinog.* **8:** 123–126.

Zimmermann, J.W. and R. Schultz. 1994. Analysis of gene expression in preimplantation mouse embryo: Use of mRNA differential display. *Proc. Natl. Acad. Sci.* **91:** 5456–5460.

Construction of a Subtractive cDNA Library Using Magnetic Beads and PCR

Anders Lönneborg,[1] Praveen Sharma,[1] and Peter Stougaard [2]

[1]Norwegian Forest Research Institute, Høgskoleveien 12, N-1432 Ås, Norway
[2]Biotechnological Institute, Anker Engelonds Vej 1, DK-2800 Lyngby, Denmark

INTRODUCTION

Differentially expressed genes have been studied in many biological systems, such as the analysis of differences between closely related species or varieties (Darasse et al. 1994), genes expressed differentially in different tissues (Maréchal et al. 1993), and for parallel structure gene expression in relation to external stimuli (e.g., addition of hormones, drugs, or pathogens; elevated temperature; or changes in pH) (Hara et al. 1991; Sharma et al. 1993; Barilá et al. 1994; Wu et al. 1994). The cloning and characterization of genes that are differentially expressed have long been troublesome and laborious, because it is necessary to isolate large amounts of mRNA for the construction of a cDNA library, and traditionally, this library is screened with two probes, for example, ones specific for the induced and noninduced genes (Sambrook et al. 1989). Alternatively, cDNA libraries can be screened with subtracted probes prepared from labeled cDNA corresponding to mRNA from treated cells subtracted with mRNA from untreated cells (Sambrook et al. 1989). Such an experiment requires large amounts of mRNA for cDNA library construction and for probe construction and is rather time-consuming and laborious, because each protocol includes several steps.

Recent advances in molecular biology have provided powerful tools for the construction of cDNA libraries and subtractive hybridization. PCR has proved to be a very efficient method for the amplification of DNA to be used in cDNA library construction (Hara et al. 1991; Houge 1993; Jepson et al. 1991; Raineri et al. 1991), and the use of oligo(dT) coupled to magnetic beads has in several cases been shown to facilitate handling of small amounts of mRNA and cDNA (Raineri et al. 1991; Jacobsen et al. 1990; Schraml et al. 1993). The primary advantage of using subtractive libraries is that the size of the library can be reduced, as constitutively expressed housekeeping genes and repetitive sequences have a very low representation in such libraries. However, induced genes or pseudogenes having homology with constitutively expressed genes are not represented in the subtracted libraries. Induced genes with a very low constitutive expression level may not be represented in the library either. Other problems may arise from the use of PCR in library construction, because PCR preferentially amplifies short cDNA sequences or sequences without the potential of forming secondary structures. Thus, a cDNA library constructed by PCR may not represent the complete pool of mRNA molecules. However, in many cases, the advantages of using PCR overshadow the disadvantages, because PCR allows the manipulation of very small amounts of cDNA.

We have combined these two powerful methods and have developed a strategy for the construction of cDNA libraries (Fig. 1) (Sharma et al. 1993). This method, which is described in detail below, was first used to construct a subtracted cDNA library enriched for cDNA clones unique to roots of Norway spruce infected with the fungal root pathogen *Pythium dimorphum* (Sharma et al. 1993). The protocol has been used successfully to construct a subtracted cDNA library from giant cells in tomato roots infected by a root-knot nematode (R. Potter, pers. comm.). There is no reason to believe that this method should not work for organisms other than plants as well.

REAGENTS

Equipment

Agarose gel electrophoresis equipment
Glass homogenizer
Magnetrack
Microcentrifuge
Mortar and pestle
Pipettes
Rotary hybridization oven
Thermal cycler
Water bath

Enzymes

AMV RT or MoMLV RT
E. coli ribonuclease H
E. coli DNA polymerase I
Human placental ribonuclease inhibitor
Taq DNA polymerase
T4 DNA ligase
T4 DNA polymerase

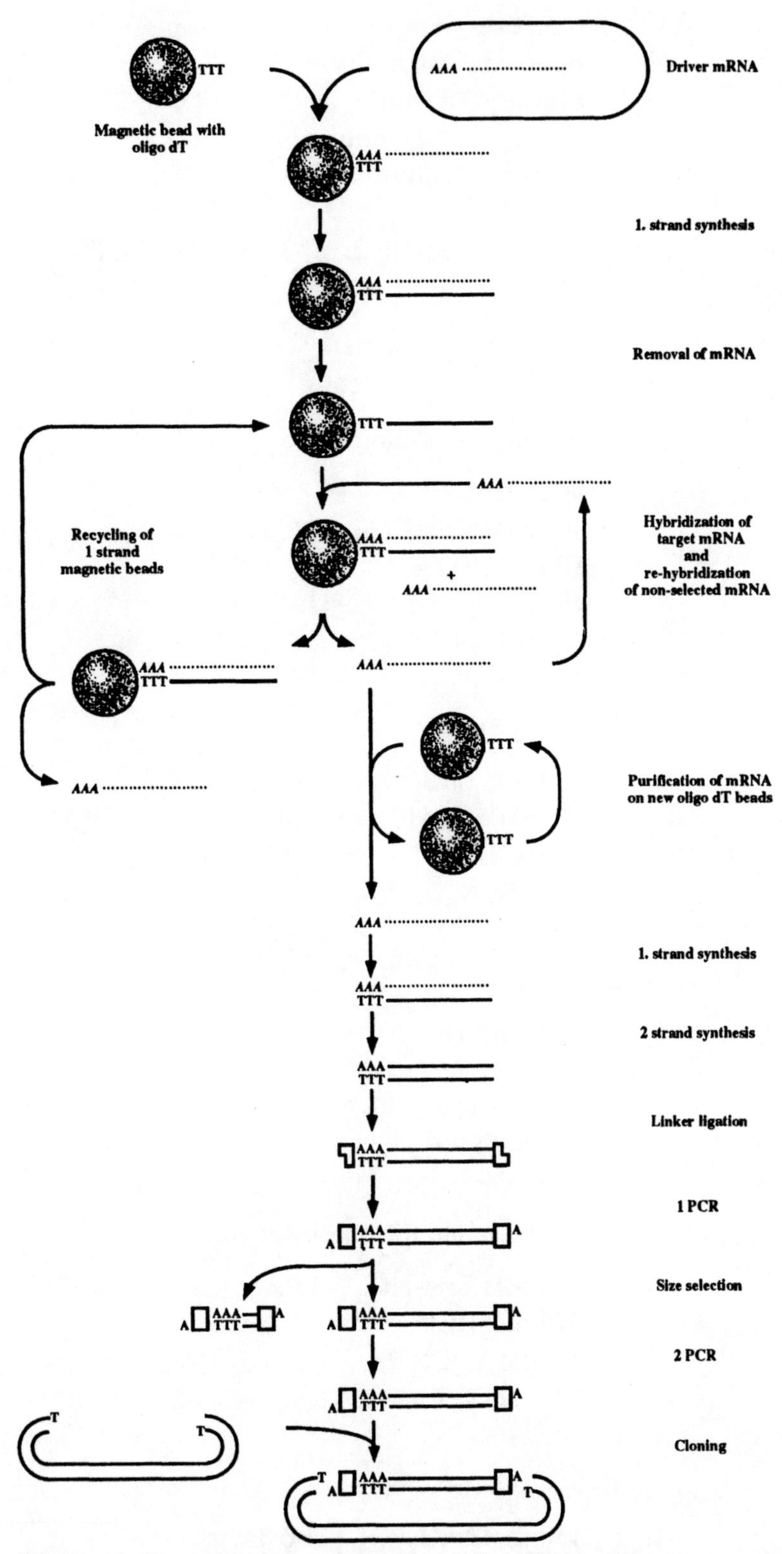

Figure 1 Protocol for subtraction of mRNA and construction of a PCR-amplified cDNA library. (*Broken lines*) mRNA; (*solid lines*) DNA; (*large shaded circles*) oligo(dT) magnetic beads. For further details, see text.

Supplies

Agarose
Agarose, low melting
Liquid nitrogen
Oligo$(dT)_{25}$ Dynabeads (DYNAL)
Oligonucleotide 1: 5′-ATG CTT AG**G AAT TC**C GAT TTA GCC TCA TA-3′
Oligonucleotide 2: 5′-TAT GAG GCT AAA-3′
Mineral oil
PCR II vector
Spin column

BUFFERS

Lysis/Poly(A)-binding Buffer

500 mM LiCl
100 mM Tris-HCl, pH 8.0
10 mM EDTA
1% Lithium dodecylsulfate
5 mM DTT

Washing Buffer

150 mM LiCl
10 mM Tris-HCl, pH 8.0
1 mM EDTA
0.1% SDS

dNTP Mix

7 mM dATP
7 mM dCTP
7 mM dGTP
7 mM dTTP

Reverse Transcriptase Buffer, 5X

250 mM Tris-HCl, pH 8.3
375 mM KCl
15 mM $MgCl_2$
50 mM DTT

Hybridization Buffer

120 mM NaH_2PO_4, pH 6.8
820 mM NaCl
1 mM EDTA
0.1% SDS

Second-strand Buffer, 5X

100 mM Tris-HCl, pH 7.5
500 mM KCl
25 mM $MgCl_2$
0.25 mg/ml BSA
50 mM $(NH_4)_2SO_4$

STE

100 mM NaCl
10 mM Tris-HCl, pH 7.5
0.1 mM EDTA

PCR Buffer, 10X

500 mM KCl
100 mM Tris-HCl, pH 8.3
15 mM $MgCl_2$
0.1% (w/v) gelatin

Ligase Buffer, 10X

50 mM Tris-HCl, pH 7.8
10 mM $MgCl_2$
10 mM DTT
25 μg/ml BSA

PROTOCOLS

Isolation of RNA

This protocol, described by Jacobsen et al. (1990), is an effective method for isolating mRNA directly from plant tissue without first having to isolate total RNA. If preferred, this method can easily be adjusted to isolate mRNA from total RNA.

1. In a mortar, grind 1.75 g of tissue, or 5×10^6 cells when starting from cultured cells, to be used for subtraction (driver) to a fine powder in liquid nitrogen.

2. Grind 0.25 g of tissue, or 7×10^6 cells when starting from cultured cells, to be subtracted (target) in the same way.

3. Homogenize the samples in a glass homogenizer with 1.2 ml of lysis/poly(A)-binding buffer. As an alternative, the samples can be transferred to another mortar and ground with lysis/poly(A)-binding buffer.

4. Spin the samples at 10,000g for 30 seconds in a microcentrifuge.

5. Transfer the supernatant of both samples to new microcentrifuge tubes containing 2 mg of oligo$(dT)_{25}$ Dynabeads (DYNAL) suspended in 100 μl of lysis/poly(A)-binding buffer.

6. Incubate 4–5 minutes on ice to let the RNA anneal to the magnetic beads.

7. Place the tube in a rack with a magnet (i.e., DYNAL MPC-E) at room temperature so that the magnetic beads are collected on the microcentrifuge tube wall. Depending on the viscosity, allow 2–5 minutes for the collection of magnetic beads.

8. Remove the supernatant while the beads are still attached to the tube wall.

9. Wash the magnetic beads three times at room temperature in 1 ml of washing buffer. Depending on the tissue used, 2–4 μg of mRNA should be attached to the magnetic beads.

10. Add 30 μl of distilled RNase-free water and incubate the tube with the attached target mRNA for 2 minutes at 65°C. Immediately place the tube in a magnetrack, allow 30 seconds for the collection of magnetic beads, and transfer the supernatant containing eluted target mRNA to a new tube. Driver RNA is used for first-strand cDNA synthesis directly on the Dynabeads.

    Note: The amount of RNA available varies depending on the material used. The average yield of RNA per gram of tissue is reported to be 8 mg in liver, 1–1.5 mg in muscle tissue, 0.5 mg in pea seedlings, and 0.2 mg in *Arabidopsis* plants. Because mRNA usually constitutes 1–5% of total RNA, the average yield of extractable mRNA per gram of tissue is about 2–400 μg. Using the information from the supplier that 2 mg of Dynabeads binds 4 μg of mRNA, this protocol can be adjusted accordingly. We prefer Dynabeads to other magnetic particles because of their higher capacity to bind mRNA and their small and uniform size, which makes them easier to keep in suspension.

First-strand cDNA Synthesis

ALTERNATIVE 1

This method is described in the cDNA Synthesis System Plus kit (Amersham). The buffers and enzymes provided in this kit can be used in exchange for those listed here.

1. Wash the magnetic beads with bound driver mRNA twice in 1 ml of 5x reverse transcriptase (RT) buffer at room temperature.

2. Resuspend the beads in 5x RT buffer to a calculated concentration of ~0.25 μg/μl of mRNA.

3. Add in listed order to a tube on ice:

driver mRNA (1 μg) in 5x RT buffer	4 μl
sodium pyrophosphate solution (80 mM)	1 μl
human placental ribonuclease inhibitor (200 units/μl)	1 μl
dNTP mix	1 μl
H_2O	13 μl
	20 μl

4. Mix and spin in a microcentrifuge for a few seconds.

5. Add 20 units of avian myeloblastosis virus (AMV) RT.

6. Incubate for at least 40 minutes at 42°C. Maintain the beads in suspension by incubating in a rotary hybridization oven.

7. Place the microcentrifuge tubes in a magnetrack, allow 30 seconds for the collection of magnetic beads containing RNA and first-strand cDNA, and discard the supernatant.

8. Wash the beads with 20 μl of 2 mM EDTA for 3 minutes at 95°C in a water bath to remove the RNA. Immediately place the microcentrifuge tubes in a magnetrack and allow 30 seconds for the collection of beads. Discard the supernatant. Repeat the washing step once.

ALTERNATIVE 2

This alternative protocol is essentially as described previously by Raineri et al. (1991). The main difference from Alternative 1 is that Moloney murine leukemia virus (MoMLV) RT is used instead of AMV RT.

1. Wash the magnetic beads with bound driver mRNA twice in 1 ml of 5x RT buffer at room temperature.

2. Resuspend the beads in 5x RT buffer to a calculated concentration of ~0.25 μg/μl of mRNA.

3. Add in listed order to a tube on ice:

driver mRNA (1 μg) in RT buffer	4 μl
DTT (0.1 M)	2 μl
human placental ribonuclease inhibitor (200 units/μl)	1 μl

dNTP mix	2 µl
H_2O	11 µl
	20 µl

4. Mix and spin in a microcentrifuge for a few seconds.

5. Add 25–50 units of MoMLV RT.

6. Incubate for 1 hour at 42°C. Maintain the beads in suspension by incubating in a rotary hybridization oven.

7. Wash the bands with 20 µl of 2 mM EDTA for 3 minutes at 95°C to remove the mRNA. Repeat the washing step once.

Subtractive Hybridization

1. Capture the magentic beads carrying the first-strand driver cDNA with a magnet, discard the EDTA solution, and resuspend the beads in 90 µl of hybridization buffer.

2. Add 15 µl of the previously isolated target mRNA.

3. Overlay the hybridization mix with mineral oil.

4. Denature the sample for 3 minutes at 95°C.

5. Hybridize for 24 hours at 65°C in a rotary hybridization oven to keep the beads suspended.

6. Capture the cDNA-containing beads with hybridized mRNA using a magnet.

7. Remove the mineral oil with a pipette.

8. Collect and save the supernatant containing unhybridized mRNA, and transfer the beads to a fresh microcentrifuge tube.

9. Add 1 ml of water and incubate at 95°C. After 1 minute of incubation, capture the beads with a magnet and discard the supernatant containing mRNA that hybridized to the cDNA coupled to magnetic beads. Repeat the elution once.

10. Wash the recycled beads twice in 1 ml of washing buffer at room temperature.

11. Add the collected supernatant from step 8 to the recycled cDNA-containing magnetic beads and repeat the hybridization three times.

12. Unhybridized mRNA after the third hybridization is captured with 1 mg of fresh oligo$(dT)_{25}$ Dynabeads in 50 µl of lysis/poly(A)-binding buffer.

13. Elute the bound mRNA in 10 µl of water for 5 minutes at 55°C.

14. Repeat step 13 and pool the supernatants.

At a temperature of 65°C, hybridization between a poly(A) tail and the oligo(dT) on the beads does not occur. Thus, the risk of capturing driver mRNA for use in the next subtraction cycles is avoided.

The number of hybridization cycles necessary for subtractive hybridization is difficult to predict but may be determined empirically. Increasing the number of cycles will enrich for unique target transcripts. cDNA corresponding to transcripts present in high amounts in the target mRNA population and, at the same time, also present in smaller amounts as driver mRNA will be underrepresented or even absent in the subtracted cDNA library. The efficiency of the subtraction after each cycle may be monitored using a hybridization probe that recognizes transcripts present in the driver and target mRNA pools. If subtraction is effective, this probe should not give detectable signal after three to four cycles.

The recycled first-strand driver cDNA can also be used for purposes other than subtractive hybridization. For example, it can be labeled and used later as a hybridization probe in the characterization of the library and of isolated clones.

Second-strand Synthesis

This method is carried out according to the protocol for the cDNA Synthesis System Plus (Amersham). The buffers and enzymes provided in the kit can be used in place of those listed here. The first strand is synthesized as described previously, but only to step 6. *Do not add any EDTA.* Then proceed as follows:

1. Mix in a microcentrifuge tube on ice:

first-strand cDNA reaction mix	20 µl
second-strand buffer, 5X	37.5 µl
Escherichia coli ribonuclease H	0.8 µl
E. coli DNA polymerase I	23 µl
H_2O	x µl
	100 µl

2. Mix gently and incubate for 60 minutes at 12°C.

3. Incubate for 60 minutes at 22°C.

4. Incubate for 10 minutes at 70°C.

5. Spin in a microcentrifuge for a few seconds.

6. Place on ice. Add 2 units of T4 DNA polymerase per microgram of original mRNA template.

7. Mix and incubate for 10 minutes at 37°C.

8. Add 4 μl of 0.25 M EDTA, pH 8.0, per 100 μl of final reaction mix to stop the reaction.

Adapter Ligation

For adapter ligation (Jepson et al. 1991), the following oligonucleotides work well, but other oligonucleotides with other restriction sites may be preferred when cloning in other vectors. For example, introducing a site for a rare cutter like *Not*I would reduce the risk of internal digests when digestion is needed or preferred for cloning.

Oligonucleotide 1 (29-mer); 5′-ATG CTT AG**G AAT TC**C GAT TTA GCC TCA TA-3′ (*Eco*RI)

Oligonucleotide 2 (12-mer); 5′-TAT GAG GCT AAA-3′

1. Dissolve 40 μg of oligonucleotide 1 and 100 μg of oligonucleotide 2 in 50 μl of STE.

2. Heat the oligonucleotide mixture for 2 minutes at 70°C. Cool slowly, for 2 hours to 30°C.

3. Add the following to a microcentrifuge tube and mix:

cDNA	10 μl
annealed oligonucleotide	2.5 μl
ATP, 10 mM	2.5 μl
ligase buffer, 10X	2 μl
H_2O	2.5 μl
T4 DNA ligase (2.5 units/ml)	1 μl
	20 μl

4. Incubate overnight at 16°C.

5. Extract with phenol/chloroform.

6. Remove excess adapters on a spin column, e.g., Miniprep Spin Column (Pharmacia). Prior to loading the column, adjust the sample volume to 100 μl with STE.

7. Apply the sample to an STE-equilibrated column and spin.

8. Precipitate the sample with ethanol.

9. Resuspend the pellet in 50 μl of H_2O.

PCR Amplification of cDNA

For PCR amplification of cDNA, see Jepson et al. (1991).

1. Add to a microcentrifuge tube and mix:

PCR buffer, 10x	5.0 μl
oligonucleotide 1, 600 ng	6.0 μl
dNTP stock, 10 mM each dNTP	1.0 μl
cDNA	5.0 μl
H_2O	33.0 μl
	50.0 μl

2. Overlay the PCR mix with a few drops of mineral oil.

3. Boil sample for 10 minutes.

4. Place the tube in a thermal cycler at 80°C.

5. After a minute, add 1 μl of *Taq* DNA polymerase (AmpliTaq) and return the sample immediately to the thermal cycler.

6. Run 35 cycles for 1.1 minutes at 68°C, 3.0 minutes at 73°C, and 0.8 minute at 94°C.

7. Run 10 μl of the PCR mixture on a 1% low-melting agarose (NuSieve) gel.

8. Cut away fragments smaller than ~500 bp.

9. Run the gel in reverse orientation with the same voltage as in step 7 until the remaining fraction is well concentrated. This is achieved when the sample dye is close to the well.

10. Cut out the concentrated fraction from the gel and recover the cDNA in 20 µl of H_2O. Alternatively, the gel piece containing the cDNA may be melted and an aliquot used directly in the next step.

11. Repeat the PCR amplification with 5 µl of the recovered cDNA fraction.

12. Run 5 µl of the sample to check the size distribution of the amplified DNA by agarose gel electrophoresis (1% agarose).

13. Extract with phenol/chloroform to inactivate the *Taq* DNA polymerase.

14. Ethanol-precipitate and resuspend in 20 µl of H_2O.

The cDNA after the second PCR is not purified on an agarose gel, to save the A overhang produced by the *Taq* DNA polymerase. If this overhang is lost, it severely affects the efficiency of ligation in the next step when the cDNAs are cloned into the PCR II vector (TA Cloning System, Invitrogen).

Cloning the cDNA into the Vector

1. Add in a microcentrifuge tube and mix:

cDNA	x µl
PCR II vector	y µl
ligase buffer	1.0 µl
ATP, 10 mM	1.0 µl
H_2O	z µl
T4 DNA ligase, 2.5 units/ml	0.5 µl
	10.0 µl

2. Incubate overnight at 16°C.

3. Heat for 10 minutes at 75°C.

4. Dilute 2 µl of ligation mixture with 10 µl of water.

5. Transform competent *E. coli* with the diluted ligation mixture and plate the transformed *E. coli* at different dilutions to estimate the number of clones in the library.

The ratio between vector and cDNA ends should be ~1. However, because the concentration of cDNA ends may be difficult to calculate, it may be necessary to use different amounts of cDNA in the ligation.

The cDNA can also be digested with *Eco*RI utilizing the restriction site in the adapter. In this way, other vectors (e.g., λ vectors) can be used. However, we try to keep the enzymatic manipulations to a minimum, as they all reduce the yield of the final product, the cDNA clones. In addition, the risk of fragmenting of cDNAs because of internal *Eco*RI sites is reduced.

TROUBLESHOOTING

It is essential for all later work to extract high-quality mRNA from the tissue of interest. The method described is very quick, which minimizes the risk of degradation. However, it is also of great importance to handle and store the tissue in an optimal way to avoid RNase activity. If the mRNA cannot be extracted at once after tissue sampling, it should be frozen immediately in liquid nitrogen and stored at −80°C until extraction. For the same reason, all buffers coming in contact with the RNA should be free of RNase activity.

To avoid the unintended binding of target mRNA to free oligo(dT) sites on the magnetic beads during the subtractive hybridization, it is important to perform the hybridization at a stringent temperature. The use of a stringent temperature also avoids nonhomologous hybridization. It has been reported that high temperatures (65–68°C) may cause breaks and/or degradation of mRNA (Wu et al. 1994). The use of formamide in the hybridization solution avoids this problem and makes it possible to perform a high-stringency hybridization at 42–45°C (Wu et al. 1994). The same workers also recommend the use of autoclaved mineral oil to avoid nuclease contamination (Wu et al. 1994).

We have utilized the adenosine overhang produced by *Taq* DNA polymerase in our cloning strategy. Using this overhang, it is possible to clone the cDNAs directly into the vector without further modification of the ends with restriction enzymes. This has two advantages: (1) All extra manipulations reduce the yield of cloned cDNA and (2) the risk of degradation of the cDNA because of existing internal sites for the enzyme is avoided. However, the A overhang is rather unstable and is easily lost when separating the cDNA in an agarose gel. Therefore, it is important to avoid such a separation after the last PCR amplification.

CONCLUSION

PCR and oligo(dT) magnetic beads have proved to be very powerful tools in the handling of small amounts of mRNA and DNA. Using magnetic beads and PCR, we have constructed subtractive libraries enriched for differentially expressed genes from 250 mg of tissue. After cDNA synthesis, the recycled first-strand cDNA coupled to magnetic beads can be used for another round of subtraction and for the synthesis of the probes used in the screening of the libraries.

REFERENCES

Barilá, D., C. Murgia, F. Nobili, S. Gaetani, and G. Perozzi. 1994. Subtractive hybridization cloning of novel genes differentially expressed during intestinal development. *Eur. J. Biochem.* **223:** 701–709.

Darasse, A., A. Koutoujansky, and Y. Bertheau. 1994. Isolation by genomic subtraction of DNA probes specific for *Erwinia carotovora* subsp. *atroseptica. Appl. Environ. Microbiol.* **60:** 298–306.

Hara, E., T. Kato, S. Nakada, S. Sekiya, and K. Oda. 1991. Subtractive cDNA cloning using oligo(dT)$_{30}$-latex and PCR: Isolation of cDNA clones specific to undifferentiated human embryonal carcinoma cells. *Nucleic Acids Res.* **19:** 7097–7104.

Houge, G. 1993. Simplified construction of a subtracted cDNA library using asymmetric PCR. *PCR Methods Appl.* **2:** 204–209.

Jacobsen, K.S., E. Breivold, and E. Hornes. 1990. Purification of mRNA directly from crude plant tissues in 15 minutes using magnetic oligo dT microspheres. *Nucleic Acids Res.* **18:** 3669.

Jepson, I., J. Bray, G. Jenkins, W. Schuch, and K. Edwards. 1991. A rapid procedure for the construction of PCR cDNA libraries from small amounts of plant tissue. *Plant Mol. Biol. Reporter* **9:** 131–138.

Marechal, D., C. Forceille, D. Breyer, D. Delapierre, and A. Dresse. 1993. A subtractive hybridization method to isolate tissue-specific transcripts: Application to the selection of new brain-specific products. *Anal. Biochem.* **208:** 330–333.

Raineri, I., C. Moroni, and H.P. Senn. 1991. Improved efficiency for single-sided PCR by creating a reusable pool of first-strand cDNA coupled to a solid phase. *Nucleic Acids Res.* **19:** 4010.

Sambrook, J., E.F. Fritsch, and T. Maniatis. 1989. *Molecular cloning: A laboratory manual*, 2nd edition. Cold Spring Harbor Laboratory Press, Cold Spring Harbor, New York.

Schraml, P., R. Shipman, P. Stultz, and C.U. Ludwig. 1993. cDNA subtraction library construction using a magnet-assisted subtraction technique (MAST). *Trends Genet.* **9:** 70–71.

Sharma, P., A. Lönneborg, and P. Stougaard. 1993. PCR-based construction of subtractive cDNA library using magnetic beads. *BioTechniques* **15:** 610–611.

Wu, G., S. Su, and R.C. Bird. 1994. Optimization of subtractive hybridization in construction of subtractive cDNA libraries. *Genet. Anal. Tech. Appl.* **11:** 29–33.

A PCR-based Method for Screening DNA Libraries

David I. Israel

Pharmaceutical Peptides, Inc., Cambridge, Massachusetts 02139

INTRODUCTION

Screening DNA libraries of high complexity for rare sequences is one of the fundamental techniques of molecular biology. In a typical genomic library with an average insert size of 20,000 bp from an organism with a haploid genome size of 2×10^9 bp, the occurrence of a single-copy gene will be approximately $1/10^5$. Likewise, for cDNA clones within a highly complex library derived from a tissue or cell line that expresses many different genes, a particular clone may occur with a similarly low frequency. Screening of libraries of high complexity by techniques such as filter hybridization (Benton and Davis 1977) or expression cloning (Wong et al. 1985) is therefore a labor-intensive and time-consuming process due to the large number of clones that must be screened to obtain the clone of interest.

PCR results in the amplification of a given nucleic acid sequence by many orders of magnitude (Saiki et al. 1985). When applied to the screening of highly complex DNA libraries contained within either bacteriophage or plasmid vectors, PCR offers the opportunity to identify rare DNA sequences in complex mixtures of molecular clones by increasing the abundance of a particular sequence, thereby allowing the easy identification of a particular clone in a portion of the library. This is accomplished by subdividing the original library into pools of decreased complexity and screening each pool or groups of pools for a given DNA sequence (Fig. 1). A pool that contains the desired clone is subsequently subdivided into smaller pools, each of which is screened using the same PCR protocol that was used for the primary screen. After several cycles of subdividing and screening, the initially rare clone is greatly enriched and can be easily obtained as a pure clone (Israel 1993).

A method for screening highly complex DNA libraries using PCR is described in this chapter. This method allows a library of high complexity to be screened in a short time and provides an alternative to more traditional and time-consuming screening methods that entail plaque or colony hybridization, or methods that require the expression of a functional gene product. The main advantages of this screening technique are its speed, sensitivity, and ease. In addition, the method as described below requires more than one oligonucleotide to anneal correctly to the template DNA and/or PCR product, thereby providing a high degree of stringency for a true positive signal.

The main limitation of this technique is the need for precise sequence information from the target gene for the design of specific and efficient PCR primers. In some cases, such as using sequences from one species to clone the corresponding gene from a different species, or using sequence information from one gene to clone other genes in a multigene family, the precise sequence within the target clone may not be known. Hybridization conditions of reduced stringency or using mixtures of degenerate oligonucleotide probes are commonly employed in these circumstances when screening by plaque or colony hybridization (Goeddel et al. 1980; Toole et al. 1984). Using annealing conditions of reduced stringency or mixtures of degenerate oligonucleotide primers may present technical difficulties using PCR, because the occurrence of false-positive signals increases as the annealing specificity is decreased. In addition, due to the exquisite sensitivity of PCR, careful laboratory technique must be practiced diligently (see "Setting Up A PCR Laboratory" in Section 1) to avoid cross-contamination of samples that can result in false positives. However, with reliable sequence information for primer synthesis and careful experimental design and technique, this method provides an efficient means to screen libraries with a high probability of success.

REAGENTS

DNA library in phage vector
Bacterial strain and media to propagate library
Oligonucleotides
- 2 for PCR primers
- 1 for hybridization

96-Well U-bottomed plates (Corning Costar)
Acetate plate sealers (Dynatech Laboratories)
Taq DNA polymerase, or equivalent
PCR cocktail: For 1-ml reaction cocktail
- 200 nmoles each dNTP
- 1x *Taq* buffer
- 2.5 μmoles $MgCl_2$
- 2 nmoles each primer
- (PCR cocktail can be frozen in aliquots)

PCR master mix
1 μl of *Taq* DNA polymerase
99 μl of PCR cocktail
(PCR master mix should be prepared fresh for each reaction)

PROTOCOL

Prior to screening the DNA library, several parameters should be investigated to establish efficient experimental conditions. These include:

- *PCR conditions.* Use the primers that will be used to screen the library and vary annealing temperature and cycle number to optimize the specificity and yield of the product. The source of template can be either the library to be screened (10^6 phage particles) or 10 ng of total genomic DNA. During the PCR, phage DNA is released and serves as template. Therefore, phage DNA does not need to be purified prior to the reaction. Typical PCR primers (16–24 nucleotides long, 50% G+C content) yielding a product 0.1–1.0 kb in length can be used. If necessary (for example, using primers from different exons that span a large intron[s]), target sequences may be greater than 1 kb apart, but the yield of product may be lower.

- *Determination of the frequency of the gene in the library.* Using the PCR conditions established above, titer the library by varying the amount of input phage. The minimum number of phage that yields a PCR product is the experimentally determined frequency of the gene in the library. Genomic libraries with an average insert size of 20 kb should have a complexity of greater than 10^5 to assure that the target gene is present at least once. Use of 10^4–10^6 phage as template from a typical library of high complexity should indicate the frequency of the gene in the library. The number of input phage that contains one or two copies of the target gene should be used in the screen of the library.

Once the PCR conditions and gene frequency have been determined, the library can be screened as follows:

1. Mix 0.5 ml of a fresh overnight bacterial culture grown in L broth with 0.5 ml of SM and add phage containing the library. Incubate for 20 minutes at room temperature.

2. Add 20 ml of L broth containing 10 mM $MgSO_4$ and dispense 100 μl/well in a 96-well plate in an 8 X 8 matrix. Seal the plate with acetate sealing tape and incubate for 5–6 hours at 37°C, shaking at 225 rpm. This allows amplification of the phage within the sub-

pool of the library. (Note that approximately 1/3 of the culture is aliquoted into the 96-well plate. The number of input phage used in step 1 should take this into account). Phage titers should increase to approximately 10^9/ml following amplification.

3. Combine phage from 8 wells across a row or 8 wells down a column (25 μl/well) using a multiwell pipette (see Analysis of Results and Discussion for alternative formats). Special care must be taken at this step when removing the plate sealer and when pipetting to avoid cross-contamination of samples. Brief centrifugation of plates should help clear liquid from the plate sealer if cross-contamination is a problem. Reseal the plate with fresh acetate sealing tape and store at 4°C.

4. Dilute pooled phage 1:1 with glass-distilled water. The phage are now ready to use as PCR templates.

5. Perform PCR using conditions established above by adding 0.5 μl of PCR template (pooled phage) to 24.5 μl of PCR master mix. Each experiment should contain a negative control (no template) and positive control (i.e., 10 ng of total genomic DNA or an aliquot of the starting library known to yield a positive signal).

6. Analyze PCR products by agarose gel electrophoresis. Stain gel with ethidium bromide and photograph.

7. Dry the gel in vacuo at 70°C, denature the DNA, and hybridize the oligonucleotide probe (end-labeled with $[^{32}P]$phosphate) directly to the dried gel using standard DNA hybridization conditions. Wash, and perform autoradiography. (Technical details for this step can be found in Israel 1993.) This step is optional if the specific PCR product can be readily visualized by ethidium bromide staining, as discussed below. Hybridization can also be performed after the transfer of DNA to a nitrocellulose or nylon filter using standard techniques.

The data from step 6 and/or step 7 should allow the identification of a subpool of the library containing the gene of interest (see the next section). The primary screen is now complete, and the gene within the positive subpool is now enriched relative to the starting library. Subsequent screening cycles are iterations of steps 1–7 and can be performed after titration of the phage in the amplified subpool.

8. Determine the phage titer from the positive well by plaque formation.

9. Initiate the next round of screening by infecting bacteria with approximately 30-fold fewer phage than were used in the previous round of screening.

10. Repeat steps 2–9.

ANALYSIS OF RESULTS AND DISCUSSION

For a screen that yields a single positive clone, the positive well within a row is located at the column that also yields a positive signal (Fig. 1). For screens that yield two or more positive clones, a second PCR using phage from individual wells within the positive column or row will definitively locate the positive well(s). Alternative formats for screening libraries are discussed below and in the legend to Figure 1.

Depending on the purity and yield of the specific PCR product, the data from the agarose gel may be sufficient to identify positive pools. For greater sensitivity and specificity, a radioactive oligonucleotide probe specific for sequences between the PCR primers can be hybridized to the PCR products in the gel, as detailed in step 7. This step is optional, particularly at the secondary and tertiary stages of screening. Figure 2A shows the analysis of a primary screen by agarose gel electrophoresis visualized by ethidium bromide staining. Due to both the complexity of the PCR products and the low amount of specific product, hybridization is required to identify positive pools (Fig. 2B). As the target gene becomes more enriched at the secondary and tertiary levels of screening, the abundance and purity of the PCR product increases (compare Fig. 2A, lane 2 to Fig. 3A, lane 9). When this occurs, the hybridization step becomes dispensable. Once pure phage containing the target gene have been isolated, the correct PCR product predominates over side products (Fig. 3).

The frequency of the gene after any round of screening can be estimated by PCR titration of the positive subpool as described in step 2 of the protocol or by plaque hybridization. This frequency should increase at each round of screening and can be used to check the effectiveness of the enrichment procedure. Once the gene is sufficiently enriched, the clone can be obtained as a pure plaque by performing PCR on individually picked plaques or by plaque hybridization.

Several modifications of this technique can be made in certain situations to make cloning more efficient or for other applications. These include:

- *Changing the format of sublibrary plating or pooling.* Pooling in an 8 x 8 matrix is one of many formats that can be used with this technique. At one extreme, the pooling of phage can be dispensed with, and PCR can be performed using phage from individual wells. This approach is now more feasible with the recent advent of 96-well

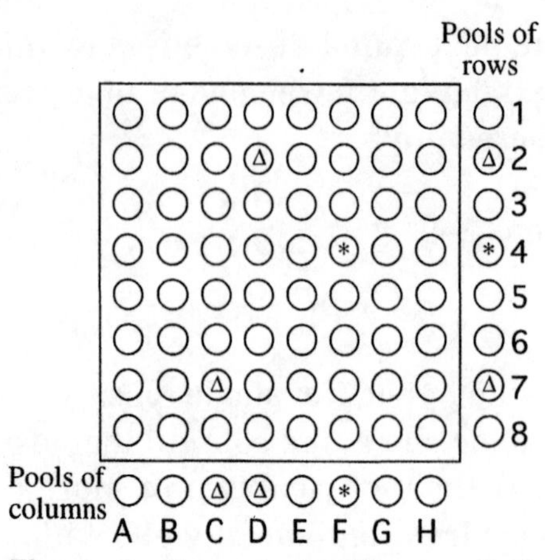

Figure 1 Formats for library plating and PCR screening. The figure shows the schematic division of a library into 64 subpools where the target gene occurs once (*) or three times (* and Δ). In a typical primary screen, each well is seeded with 1000 phage in an 8 x 8 format. A portion of the library with complexity of 64,000 is therefore subdivided into 64 subpools, each containing 1000 independent phage. After amplification of phage, PCR can be performed in a number of formats. (1) Pooling strategy: Eight wells are pooled across rows and down columns. PCR is then performed on the 16 pools of phage. For a single positive (*), pool F and pool 4 will yield the correct PCR product, identifying well 4F as the source of the positive clone. However, for three occurrences of the target gene (* and Δ), pools C, D, F, 2, 4, and 7 will all be positive. The precise identification of the positive well(s) therefore requires a second PCR using individual wells within a positive pool. (2) Two-step strategy: Phage are pooled in only one direction, yielding 8 pools which are used as template in the first PCR. The 8 individual wells within a positive pool are then analyzed in a second PCR. For a single positive, the pooling strategy is more efficient, because the 16 samples are analyzed in a single PCR. For multiple positives, the two-step format is more efficient because it requires less pooling and the analysis of fewer samples. (3) Analysis of individual wells: Phage from single wells are used as template. No pooling is required, and although the number of individual samples is higher (64 versus 16 for the pooling and two-step strategies), a single experiment should unambiguously identify positive wells. This strategy becomes more efficient as the number of positive clones in the screen increases, and if PCR is performed in a 96-well format.

PCR plates and thermal cyclers, thereby allowing both phage amplification and PCR to be performed using multiwell pipettes and the same plate formats. However, the PCR product still needs to be analyzed by gel electrophoresis, and the number of individual samples will be much higher if the pooling strategy is not employed. An intermediate approach is to pool amplified phage in one dimension (i.e., only columns or only rows), perform PCR, and then use phage from individual wells within a positive pool for a second PCR. The advantage of this approach is that it requires only

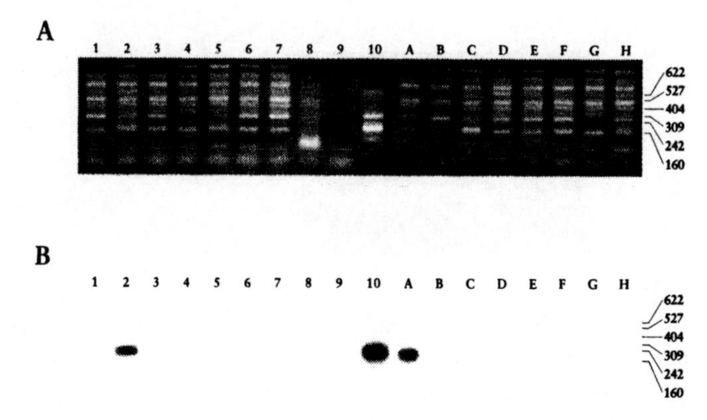

Figure 2 PCR screening of pooled phage. Sixty-four wells were inoculated with 1000 phage/well and screened for the murine M-CSF gene (Israel 1993) using the pooling strategy. The agarose gel in *A* was vacuum-dried and used for direct hybridization to an internal M-CSF oligonucleotide. (*A*) Ethidium bromide staining of PCR products; (*B*) hybridization to M-CSF-specific oligonucleotide. The templates for each reaction were: (lanes *1–8*) pools of rows; (lane *9*) no template; (lane *10*) 10 ng of mouse genomic DNA; (lanes *A–H*) pools of columns. The migration of *Msp*I-digested pBR322 is indicated on the right (bp).

eight PCR analyses for the pools and eight more PCR analyses for individual wells, and yields the exact location of the positive sample. However, this strategy requires two PCR protocols to be done serially, adding time to the overall procedure.

- *Screening for genes within plasmid vectors.* DNA libraries, particularly cDNA libraries, are often contained in plasmid vectors. This PCR screening technique has been used to clone a cDNA gene from a library within a plasmid vector (Israel 1993). To screen plasmid libraries, the plasmid-bearing bacteria should be allowed to grow for 16 hours in the 96-well plate. This assures that the bacterial number will be sufficiently high to allow for representation of all the clones in the pool during the PCR. The other steps of analysis are as described for phage libraries.

- *Priming from the vector.* When limited sequence information is available, or when purposefully screening for a clone that contains a particular sequence toward one end of the insert, one of the primers can be complementary to sequences within the vector (Jansen et al. 1989). This modification may yield a higher amount of incorrect product because one PCR primer will anneal to all phage within the library. In addition, the amount of PCR product will be greatly decreased if the amplified sequence is greater than 2 kb. As a consequence, the apparent frequency of a particular clone in a library will be lower. The recent description of methods for

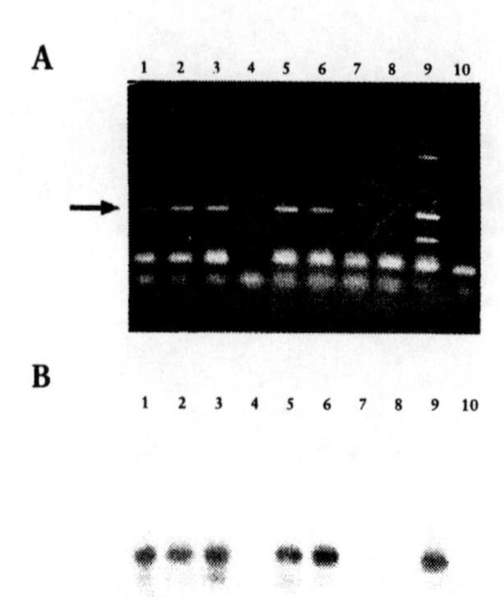

Figure 3 PCR screening of individual plaques. (*A*)Phage from a positive well following the tertiary screen were picked from random plaques (lanes *1–8*) and used as template. The arrow indicates the correct M-CSF PCR product. (Lane *9*) Phage from the tertiary positive well; (lane *10*) no template. (*B*) Hybridization to the M-CSF-specific oligonucleotide.

obtaining long PCR products (Cheng et al. 1994) should make this approach more feasible.

The basic PCR protocol for screening highly complex DNA libraries, and several modifications, have been described in this chapter. This strategy for screening libraries should consistently yield positive results with the use of good reagents and careful technique.

ACKNOWLEDGMENT I thank Dr. Ezra Abrams for his helpful comments on this manuscript.

REFERENCES

Benton, W.D. and R.W. Davis. 1977. Screening λgt recombinant clones by hybridization to single plaques in situ. *Science* **196:** 180–182.

Cheng, S., C. Fockler, W.M. Barnes, and R. Higuchi. 1994. Effective amplification of long targets from cloned inserts and human genomic DNA. *Proc. Natl. Acad. Sci.* **91:** 5695–5699.

Goeddel, D.V., E. Yelverton, A. Ullrich, H.L. Heyneker, G. Miozzari, W. Holmes, P.H. Seeburg, T. Dull, L. May, N. Stebbing, R. Crea, S. Maeda, M. McCandliss, A. Sloma, J.M. Tabor, M. Gross, P.C. Familletti, and S. Pestka. 1980. Human leukocyte interferon produced by *E. coli* is biologically active. *Nature* **287:** 411–416.

Israel, D.I. 1993. A PCR-based method for high stringency screening of DNA libraries. *Nucleic Acids Res.* **21:** 2627–2631.

Jansen, R., F. Kalousek, W.A. Fenton, L.E. Rosenberg, and F.D. Ledley. 1989. Cloning of full-length methylmalonyl-CoA mutase from a cDNA library using the polymerase chain reaction. *Genomics* **4:** 198.

Saiki, R.K., S. Scharf, F. Faloona, K.B. Mullis, G.T. Horn, H.A. Erlich, and N. Arnheim. 1985. Enzymatic amplification of β-globin genomic sequences and restriction site analysis for diagnosis of sickle cell anemia. *Science* **230:** 1350–1354.

Toole, J.J., J.L. Knopf, J.M. Wozney, L.A. Sultzman, J.L. Buecker, D.D. Pittman, R.J. Kaufman, E. Brown, C. Shoemaker, E.C. Orr, G.W. Amphlett,

W.B. Foster, M.L. Coe, G.J. Knutson, D.N. Fass, and R.M. Hewick. 1984. Molecular cloning of a cDNA encoding human antihaemophilic factor. *Nature* **312:** 342–347.

Wong, G.G., J.S. Witek, P.A. Temple, K.M. Wilkens, A.C. Leary, D.P. Luxenberg, S.S. Jones, E.L. Brown, R.M. Kay, E.C. Orr, C. Shoemaker, D.W. Golde, R.J. Kaufman, R.M. Hewick, E.A. Wang, and S.C. Clark. 1985. Human GM-CSF: Molecular cloning of the complementary DNA and purification of the natural and recombinant proteins. *Science* **228:** 810–815.

Screening of YAC Libraries with Robotic Support

Mary M. Blanchard and Volker Nowotny

Center for Genetics in Medicine, School of Medicine, Washington University, St. Louis, Missouri 63110

INTRODUCTION

The major landmarks for the physical mapping of the human genome are sequence tagged sites (STSs) (Olson et al. 1989) markers defined by primer pairs and PCR products (Saiki et al. 1985, 1988; Li et al. 1988). The placement of an STS at every 100 kb along the genome (or 30,000 in all) is one of the stated goals for the Human Genome Project. Several groups are generating and ordering STSs along chromosomes or the genome, thus using STSs to construct as well as to format maps. To generate an STS-based map (STS content mapping; Green and Green 1991; Kere et al. 1992), a library or libraries of large-insert clones (e.g., *y*east *a*rtificial *c*hromosomes [YACs]; [Burke et al. 1987]) are screened recursively with STS markers. The overlap of two clones is then determined by their common STS content.

Because libraries of YACs (Schlessinger 1990) can number many thousands of clones, screening for a single STS involves a large number of PCR assays. The considerable variation of optimal PCR assay conditions for different primer pairs has made high-throughput PCR a challenging task. We have standardized screening of STSs. The reactions are run under uniform temperature conditions by adjusting the stringency of the reactions through optimization of the ion conditions for the reaction of a particular primer pair (Blanchard et al. 1993). This uniform temperature approach has made robot-assisted screening feasible.

To meet the demand for STS-content mapping, an approach has been implemented at the Center for Genetics in Medicine (CGM) using a robotic workstation, a high-throughput PCR machine, and complementary steps of combinatorial and tree-based screening of DNA pools of decreasing complexity to recover clones from YAC libraries. The demand for ever-increasing numbers of PCR procedures has brought PCR machines compatible with 96-well microtiter trays and pipetting stations (e.g., BIOMEK; Beckman) into laboratories. In addition, multichannel pipettor technology has been improved. This will allow a broader community to use the described methods for the screening of YAC libraries by PCR.

Here, we provide a complete set of methods for YAC library screening, starting from the growth of the library clones to a detailed description of the screening process. The process is applicable for converting libraries with several thousand or several tens of thousands of yeast clones into a form that allows a rapid, easy, and robust screening routine. In the cases where this set of methods has been implemented, a single library setup has provided material for the placement of several thousand markers. The process for incorporation of a library for screening (which includes cell growth and pooling, as well as testing and DNA sample allocation) can be completed by a single trained person in less than 2 months.

REAGENTS

Template DNA preparation:

AHC medium: 0.17% (w/v) yeast nitrogen base w/o amino acids or $(NH_4)_2SO_4$ (Difco, cat. no. 0335-15); 38 mM $(NH_4)_2SO_4$; 0.10% (w/v) casein acid hydrolysate (USB/Amersham, cat. no. 12852); 0.11 mM adenine hemi-sulfate (Sigma, cat. no. A-9126), pH 5.8; 2% (w/v) glucose

Bubble lids (Beckman, cat. no. 267005)

Deep-well tray (polypropylene, Beckman, cat. no. 267006)

Hard lids (microtiter tray lids, FALCON, cat. no. 3917)

Lysing solution: 10 mM Tris-HCl, pH 8.3; 5 mM NH_4Cl; 100 mM KCl; 1.5 mM $MgCl_2$; 1% NP-40; and proteinase K (f.c.=720 μg/ml; BRL, cat. no. 5530U A, 20 mAnson/100 mg)

RNAse: Ribonuclease A (Worthington Biochemical, cat. no. LS02131)

Spheroplasting solution: 1 M sorbitol; 0.1 M sodium citrate, pH 7.0; 0.01 M EDTA; 0.6 M β-mercaptoethanol; YLE (yeast lytic enzyme; final concentration 2.8 mg/ml, ICN, cat. no. 152270, 70 units/mg). Solutions containing enzymes are made immediately before use and should be used on the same day.

PCR assay:

dNTP-solution: 2.5 mM each nucleotide (100 μM final assay concentration)

primer-pair solution: 5 μM each oligonucleotide in water (0.4 μM final assay concentration)

Taq DNA polymerase: AmpliTaq DNA (Perkin-Elmer N801-0060) (5 units/μl)

10x TNK-50 buffer: 100 mM Tris-HCl (pH 8.6 at room temperature), 15 mM $MgCl_2$, 50 mM NH_4Cl, 0.5 M KCl

10x TNK-100 buffer: 100 mM Tris-HCl (pH 8.6 at room temperature), 15 mM $MgCl_2$, 50 mM NH_4Cl, 1.0 M KCl

PROTOCOLS

In all methods described in this chapter, a common theme is the use of the 96-well format for reaction and storage vessels. Many steps are performed, or material is stored, in the deep-well version of this format. The wells in these deep-well trays (DWTs) hold a volume of slightly more than 1 ml. Two flowcharts give an overview of the series of steps required to generate DNA pools from yeast cells (Fig. 1) and of the associated screening process (Fig. 2). The major steps involved in preparing the DNA reagents for screening are the extraction of DNA from yeast cells and the pooling of the DNA samples. The DNA pools are then diluted and used as templates in PCR experiments for the screening process.

Even though the procedures described can be handled in a straightforward manner, time should be spent, before beginning the process, to assemble a detailed work schedule. In the description of the methods, time estimates are included that will, together with the flowchart in Figure 1, be helpful for this planning phase.

As an example, to help provide an estimate of the labor involved, assume we have a library containing 30,000 clones. This library

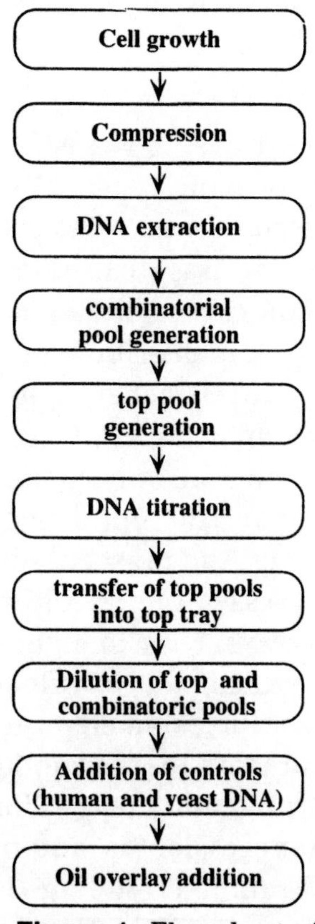

Figure 1 Flowchart of library preparation for screening by PCR. In addition to the isolation of DNA from YAC libraries, steps are required to generate DNA that is used in the screening process as template DNA for the PCR assays. The different steps are explained in detail in the text.

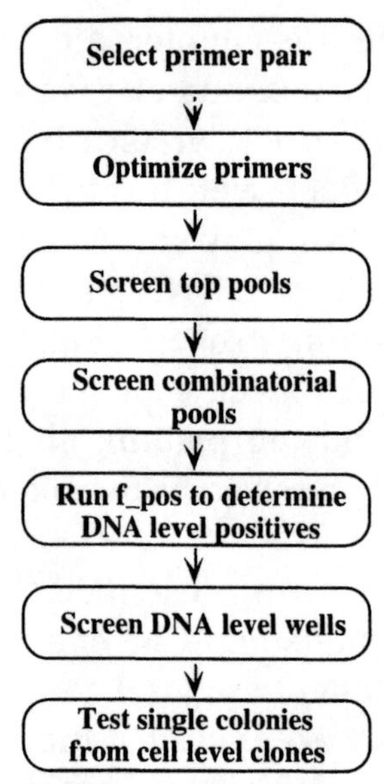

Figure 2 Flowchart of the screening process. Screening of libraries with several thousand clones is a complex process that is most efficiently organized if broken up into several steps. Several of these steps are performed with the help of a robotic workstation. The general steps are described in the text. Special aspects of the automation can be found in the Appendix.

would reside in 313 96-well trays. A realistic number of trays to inoculate on a daily basis for growing yeast cells is 40 trays of 1-ml cultures. This task would stretch across 8 days. The cell growth should not last more than 4 days. The trays can then be stored at 4°C until extracted. A BIOMEK method can easily be implemented to fill these DWTs using the multibulk tool to distribute media from bottles.

Once the workload is reduced because the inoculation phase is over, focus should be placed on cell compression. The DWTs containing the cells are centrifuged and the supernatant is decanted. The tray content is compressed to reduce the number of DNA extractions. For the assumed number of clones, the most reasonable compression factor of 12 (1 deep well compressed into 1 column of a new deep well) leads to 26 DNA extraction trays. Four to eight of these DNA trays can be generated per day with reasonable workload on a BIOMEK workstation. The output of this compression step (pooled cell trays) is handled by a DNA extraction process that is semiautomatic and requires pipetting and several centrifugation steps. Thus, with a 1-day offset, and assuming that the compression is automated with a pipetting workstation requiring only minimal tending (loading in the morning and unloading in the evening), the DNA should be ready after about 1 week for the combinatorial pooling process. This combinatorial pooling step is quite slow. We usually set aside 9 hours of BIOMEK processing time for a set of 4 DNA source trays. This pipetting step has been automated for the BIOMEK environment and requires, even if in a more manual mode (the user loads and unloads the tablet), only limited user time. The top-level DNA pools, however, are generated by the user with a 12-channel hand-held pipettor. This step can be completed within 1 hour because the user combines all the rows from only 4 trays following a very simple pattern into either specialized labware or into a single row of a DWT. The user then combines these destination wells in a second step with a single-tip pipettor. The process is finished with another week of work comprising steps such as DNA concentration measurements and adjustments, preparation of working DNA trays for screening, and storage of the master copies of the three DNA level trays.

Cell Growth and Compression for DNA Extraction

With this rapid and simple method, yeast DNA is prepared and functions in PCR as robustly as highly purified preparations (Blanchard and Nowotny 1994). For this process, yeast cells are inoculated from the library trays with a 96-prong tool into DWTs containing 800 μl of AHC medium per well. The cells are grown for 4 days at 30°C. After growth, the cells are spun at 1400*g* for 20 minutes and the supernatant is decanted. The trays are kept refrigerated (they can be left for several days if precautions are taken not to dry out the cells, e.g., capping with bubble lids).

The later steps of DNA extraction and DNA pool preparation are significantly simplified and the labware costs can be drastically reduced for larger libraries if the total number of trays is reduced to a number smaller than 30 2/3 trays. This number represents the maximal number that is feasible by the proposed pooling scheme. This reduction is done by combining the cell contents of wells after completion of the growth. We have utilized compression factors of 2, 3, and 12, respectively. A compression is the transfer of multiple columns (e.g., 2, 3, or 12 columns of 8 wells each) with the multibulk tool into a single column of a new tray (schemes for compression factors 2 and 3 can be found in Fig. 3). This new tray becomes the source for the DNA extraction. For a list of steps performed with the BIOMEK workstation, see the Appendix to this chapter.

DNA Isolation

The described method of DNA extraction is a combination of cell lysis and proteinase K digestion. The DNA provides active PCR templates over a wide concentration range and this activity remains stable over more than a year of storage at 4°C. The resulting DNA is sheared and considered unsuitable for restriction enzyme analysis or other procedures that require high-molecular-weight material.

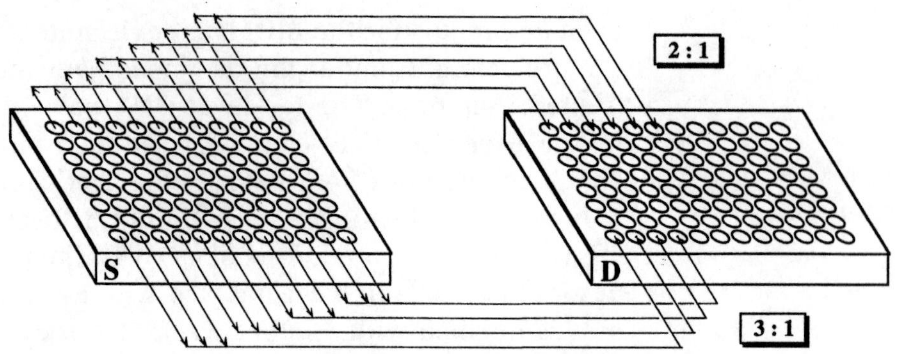

Figure 3 Compression of clones before DNA extraction. Two compression factors are depicted in this scheme. With an 8-channel pipettor, columns from the source tray (S) are transferred into the destination tray (D), thus combining two (upper part) or three (lower part) wells into a single well.

PROTOCOLS

The DNA extractions are performed in DWTs starting from pelleted yeast cells. The DNA remains in these trays throughout the procedure. It is important to use polypropylene rather than polystyrene trays, because polystyrene will not withstand the temperatures required for the extraction procedure. A small hole should be drilled into the DWT (3 mm diameter) approximately an inch from the bottom right corner to prevent the trapping of air bubbles underneath the tray when it is placed in water baths for spheroplasting and lysing.

Conversion of Cells into Spheroplasts

1. Add 162 μl of spheroplasting solution to the pelleted cells in each well.
2. Cover the DWTs with bubble lids and gently vortex until the cells are in solution.
3. Incubate for 1 hour at 37°C in a water bath with the water level adjusted to keep the height approximately one-half that of the DWT to ensure adequate heat transfer to the wells.
4. Centrifuge at 700*g* for 20 minutes. Discard the supernatants carefully.

Cell Lysis

1. Add 125 μl of lysing solution to each well.
2. Gently vortex the trays to resuspend the cells in lysis solution.

3. Incubate for 90 minutes at 60°C.

4. Incubate for 30 minutes in a 95°C water bath to inactivate proteinase K.

   Note: Bubble lids will not withstand this temperature and should be replaced with hard lids for this step.

5. Cool the trays to room temperature.

DNA Precipitation

1. Add 255 μl of ethanol to each well. Place bubble lids on the trays.

2. Vortex gently to facilitate precipitation.

3. Keep extractions for at least 1 hour at –75°C.

4. Collect the precipitate by centrifugation at 850*g* for 30 minutes at 4°C.

5. Very carefully decant the supernatants by careful inversion of the DWTs. Remove the residual ethanol by placing the trays, without lids, under a mild vacuum.

6. Add to the dried samples 300 μl of sterile distilled water containing 5×10^{-3} units/μl of RNase A.

7. Dissolve the pellets by gentle vortexing and shaking for 30 minutes at room temperature (or until the solutions appear homogeneous).

8. Quantitate the DNA fluorimetrically by the Hoechst-dye method (Labarca and Paigen 1980).

For the subsequent robotic pooling process, transfer portions of the partially purified DNA samples to standard round-bottomed 96-well microtiter plates and dilute to a concentration of about 50 ng of DNA/μl. Each well should contain at least 100 μl of solution. In practice, more than threefold higher DNA concentrations result in unreliable PCR yields. However, a very precise concentration adjustment in this step is not important because of the robustness of the PCR assay. Before use in the screening process, a DNA titration with appropriate primer pairs provides information about the optimal range of DNA concentrations for PCR. This concentration is then used for the dilution of DNA in the creation of working trays used for screening (in general by a factor of 2–5). The remaining stock DNA solutions are stored at –80°C.

All pipetting steps for the DNA extraction are performed by a Perkin-Elmer PRO/PETTE; trays are moved manually for the incubation and mixing steps. Alternatively, a BIOMEK could be set up to handle the liquid transfers. Even a hand-held multichannel pipetting scheme is feasible. Up to twelve 96-well trays of YACs can comfortably be extracted by one person in 1 day.

Currently, DNA from five different YAC libraries, totaling over 85,000 clones, has been extracted by this method and successfully used for PCR-based screening.

Pooling Scheme and Pooling Process

The pooling scheme was designed to meet two criteria: to minimize the overall number of PCR procedures that need to be performed for a screen and to maximize the speed and ease of the PCR assembly. Providing inherent redundancy (i.e., overdetermining the address of the positives) within the pooling scheme offers a convenient way to be able either to determine addresses if multiple positives are present in a combinatorial set (here 128 DNA wells) or to help resolve difficulties in the encounter of false-positive or -negative signals. This redundancy and the inclusion of control reactions at every screening level are indispensable to render the screening process sufficiently robust if high throughput is to be sustained. The BIOMEK's 8-channel tool is used for the PCR assay preparation with the DNA pools conveniently set up in three columns of 96-well DWTs. The underlying pooling scheme has been described in more detail previously (Nowotny et al. 1993; Sloan et al. 1994). Once the combinatorial pooling is completed, the DNA is titrated and assayed by PCR to determine the optimal concentration for creation of the working trays that are used for screening.

In general, a YAC screen goes through three major stages. The top level comprises the most complex DNA pools:

1. A set of top-level DNA pools contains up to 23 DNA pools that are tested with every primer pair dedicated for a full library screen (see Appendix to this chapter for details). A top-level pool should contain less than 1/3 of a positive, otherwise the rate of multiple positives within a given top-level pool leads to serious problems in deciphering and following a substantial part of the positive signals through the combinatorial level. The first three columns of a DWT contain the top-level DNA pools with well 24 either filled with DNA for control reactions (e.g., well 24 contains total human DNA used as a positive control), or left empty. The numbering scheme follows the one for the combinatorial level DNA-template trays shown in Figure 4. It is generally true that most of the DNA pools act as negative controls.

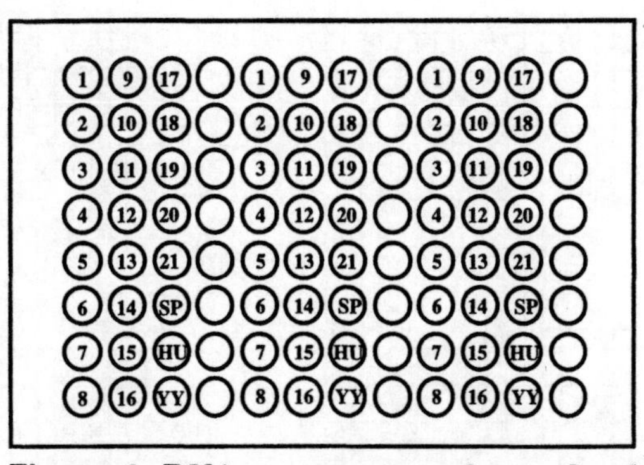

Figure 4 DNA source tray scheme for the combinatorial part of the screening. DNA samples from 128 wells are combined according to the pooling scheme (see Fig. 5). Thus, one DNA source tray contains DNA from three fractions referring to three top pools. The numbered wells are those from the pools in Fig. 6. PCR assays determine the address of positives from within a family of 128 members. A part of the scheme yields redundant information. The controls in the wells YY are 33 ng/μl yeast DNA, HU 50 ng/μl total human DNA, and SP 33 ng/μl yeast DNA mixes with 6 ng/μl total human DNA.

2. The result from the top level leads into the "combinatorial" level. Each top pool is underlaid by a set of 21 combinatorial pools. At this combinatorial level, the screening identifies groups of positives that determine addresses from within 128 members. Figure 5 shows the decoding table for the 21 combinatorial second-level pools. These pools are located in three columns in a 96-well DWT, with the assignment of well numbers in the DNA-template trays shown in Figure 4. The remaining 3 wells contain positive and negative controls, respectively. The 128 members are traced to the DNA-extraction trays with the scheme shown in Figure 6.

3. A positive from the combinatorial level requires the testing of yeast colonies, the number of which is determined by the compression factor of libraries for DNA extraction. This final level is the single-colony screening level and yields at its end the location of the YACs positive for the screened STS. For the cell-screening level, yeast colonies stored in 96-well format working trays at –80°C are accessed. The advantage is that this colony-PCR scheme accesses those DNAs that are necessary, eliminating the need to have this material available as DNA at all times. Simple prong devices are used to pick the complete compressed set at once and transfer the yeast cells into provided 96-well trays for analysis by PCR. Prongs with 2, 3, 4, 6, or 12 pins are nicely compatible with microtiter trays.

The flowchart in Figure 1 describes the steps that lead to DNA pools from the DNA extracted with the method described above. For

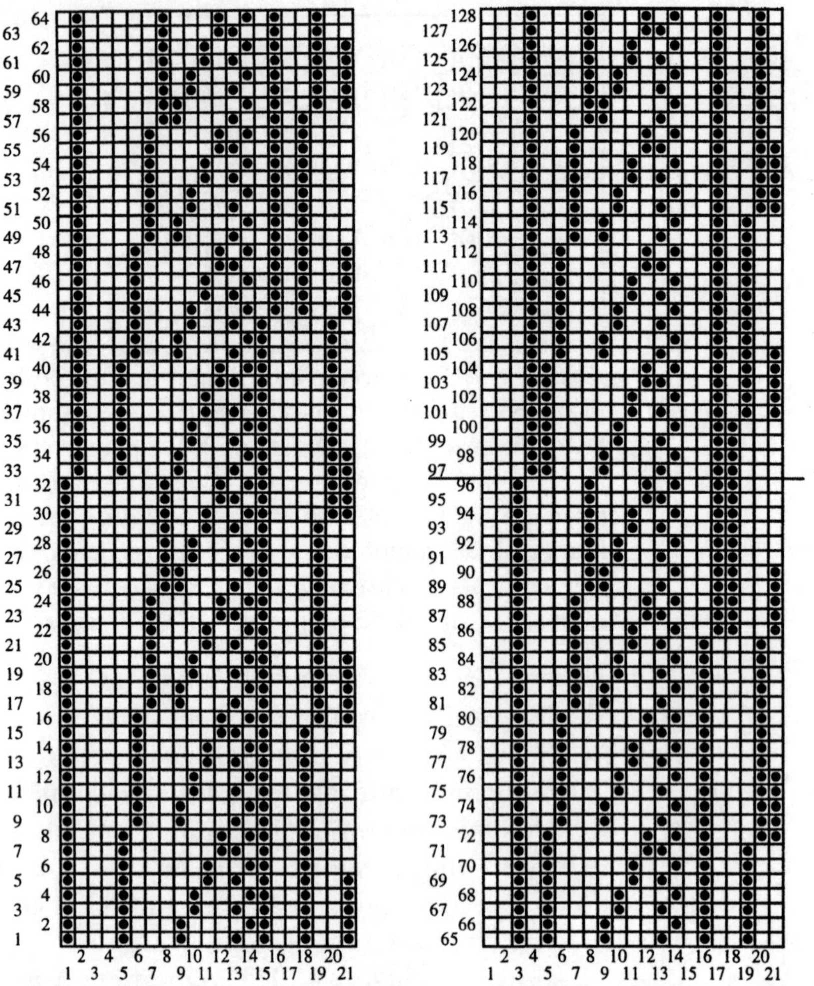

Figure 5 Pooling scheme. DNA samples from 128 wells are combined into 21 pools according to the depicted scheme. Pools 1–14 are necessary to find the address of a positive signal (pools are based on factor 2 or 4). Pools 15–21 yield redundant information useful in case of false-positive or false-negative signals (pools are based on factor 3).

this part, a BIOMEK is indispensable. Schemes relying on manual pool combination do not provide the required robustness. A plate handler like the BIOMEK SL option provides additional convenience; however, most demanding is the number of tipbox changes.

For the creation of a pool set from four round-bottomed 96-well microtiter trays of DNA (each well containing about 100 μl of solution with about 50 ng of DNA/μl), 27 tipboxes are required. The layout of the four source trays is shown in Figure 6. The DNA is collected into a single destination DWT. With the BIOMEK, a sterile DWT is used as destination tray. The pools are collected with a single-tip tool, cycling through the four DNA-source trays. After completion of the pooling, the three sets with 21 members each reside in three sets of three columns, leaving an empty column between sets (see Fig. 4). A few

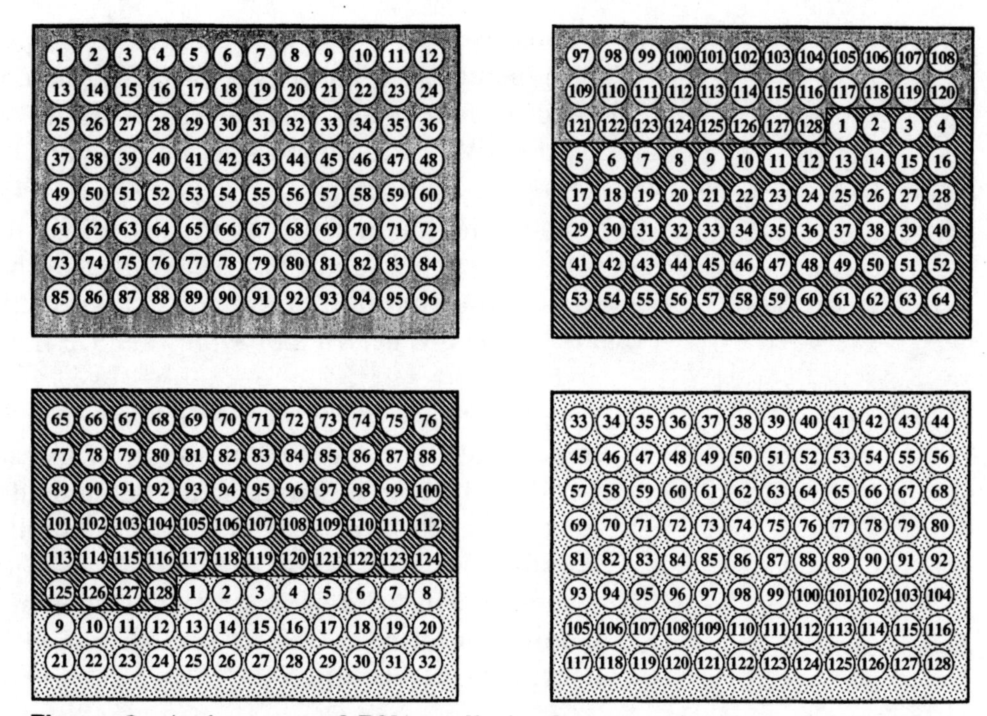

Figure 6 Assignment of DNA wells in the extraction trays for the combinatorial part of the screening. The numbers refer to the numbers in the pooling scheme (Fig. 5). DNA from these four trays is pooled into a single tray, the scheme of which is shown in Fig. 4. This scheme also serves as a guide for the combination of well contents for the top-level pools with a multichannel pipettor.

steps are required to finish the setup routine and to adjust the pooled DNA to the "correct" concentration. These steps are described in the Appendix.

To create working trays that are used for the daily screening process, concentrations are adjusted to about 7–35 ng/μl of DNA for both the top pools and the combinatorial pools. The stock solutions are capped and stored at −80°C until the working pools are depleted. DNA prepared and stored in this manner has functioned in PCR without significant loss of performance for more than a year.

PCR Conditions

A complete PCR assay that is used for the screening process requires several reagents. Blanchard et al. (1993) have described a buffer system (TNK) that allows the control of stringency of PCR assays through changes in the potassium concentration of the buffer. The formulated PCR conditions meet simple standards: (1) fixed temperature and cycling parameters for PCR assays in library screenings and (2) easy identification of product bands, with minimal background from yeast or extraneous human DNA species.

A complete PCR assay is combined from 4 µl of PCR mix and 1 µl of DNA solution. Because we run PCR in open vessels in the integrated system, each reaction is overlaid with at least 20 µl of mineral oil. We found that the reactions perform more robustly if the combined assay mixtures are kept close to 0°C and are not prepared at room temperature. Thus, on the BIOMEK tablet the position for the PCR tray is cooled by an integrated heat-exchange plate. Completed assays are transferred into a small station within the robot's envelope that is kept at 6°C before they enter the PCR machine.

1. To prepare the PCR trays, fill each well with at least 20 µl of mineral oil to prevent evaporation of the sample from the wells. For the following pipetting steps, the tip ends should reach into the oil phase. The precision of pipetting small volumes is increased because the oil has the tendency to facilitate the release of the small droplets from tip ends. Tips have to be discarded to avoid accumulation of oil in the tip, which would lead to subsequent inaccuracies in pipetting.

2. In a standard PCR temperature regimen, preheat the samples for 4 minutes above 94°C to melt the template DNA.

3. Then cycle as follows: 35 cycles proceed for 1 minute at 94°C, 2 minutes at 55°C, and 2 minutes at 72°C. After the last cycle, reduce the temperature to 0°C. The temperature cycling process is finished after a run time of about 140 minutes. However, the runs with yeast cells as template donors (cell-PCR) work reliably only with extended incubation times.

PCR Assay Optimization

PCR assay conditions vary with primer pairs. Most of this variation could be overcome by determining optimal primer sequences with computer searches by applying thermodynamic parameters or a rule-based system. The remaining variation requires assay optimization. This is done here by a potassium concentration variation in the assay buffer rather than by adapting the annealing temperatures. For PCR tray preparation, see above.

1. Prepare the PCR-mix source tray.

2. For each primer pair to be optimized, prepare two mixes with either TNK-50 or TNK-100 buffer.

Mix 6.0 μl of TNK buffer
2.4 μl of dNTP solution
4.8 μl of primer-pair solution
34.2 μl of H_2O
0.6 μl of *Taq* DNA polymerase

3. If distributing cocktails manually, place 4 μl of the TNK-50 mixes into each well of the first column of the assay tray, 4 μl of the TNK-100 mixes in the second, and so on. The DNA can be distributed into the PCR trays with the use of a hand-held 8-channel pipettor.

In our setup, the pipetting of cocktail and DNA is performed by the BIOMEK. Each cocktail is placed in a well of a 96-well round bottomed microtiter tray (FALCON, cat. no. 3918) (i.e., TNK-50 cocktail of primer 1 in well 1A1, TNK-100 primer 1 in 1A2, TNK-50 primer 2 in 1B1, etc.). A maximum of 32 primer pairs are tested in a single run. Thus, not more than eight columns will be filled with PCR mix. The last column of this tray is filled with control DNA. The procedure we have programmed uses a single-tip tool (P20) for the distribution of the PCR mix. The DNA is brought into the PCR tray with the 8-channel MP-20 tool. This process is rather slow. However, PCR optimization is only a relatively small part of the overall process and speedup of this process will not result in a significant increase of the overall speed and throughput of screening.

PCR Assay Setup for Screening

At this point, all the prerequisites for screening are in place and the central process can be initiated. Top-level and combinatorial-level screening share the same format. Each primer requires three columns of a 96-well tray, prepared as described above.

1. For both levels, for the PCR-mix source trays, mix

   21.0 μl of TNK-x buffer
   8.4 μl of dNTP solution
   16.8 μl of primer-pair solution
   119.7 μl of H_2O
   2.1 μl of *Taq* DNA polymerase

2. Distribute, for each individual primer, each cocktail equally into one column of 8 wells of a round-bottomed 96-well tray. This is the PCR source tray from which the cocktails will be pipetted.

3. For the second-level screen, fill a multiple of this value (i.e., multiply each component of the above recipe) according to the required number of positives that are to be screened.

Because in our system a maximum of six 96-well trays can be processed in parallel, no more than two PCR-mix trays need to be prepared for a single run. The BIOMEK transfers the PCR reagent mix from the source tray to the PCR assay tray with the 8-channel modified MP-20 pipetting tool. It then adds the appropriate pools of DNA to the PCR trays.

For six 96-well PCR trays, this entire process takes approximately 1–2 hours. This time includes the pipetting of the PCR reagent mix and DNA and the transfer of the required labware, tipboxes, DNA trays, reagent-mix trays, and PCR trays.

ANALYSIS OF THE PCR RESULTS WITH HIGH-DENSITY GELS

Gel analysis of the PCR products is a reliable technology. To speed and simplify this process, thus allowing for throughput increase, multichannel loading is used. Agarose gels are cast with wells formed in predetermined positions by inserting combs that register in the casting tray (see Fig. 7). A single (15 cm x 25 cm) gel with four tiers of wells and 24 sample-plus-marker wells in each tier provides the frame for this technology. The 4.5-mm well-to-well spacing accepts two rows of PCR products from the PCR trays.

AUTOMATION HARDWARE AND SOFTWARE

Additional information about this topic can be found in the Appendix. More detailed information has been published for some of the involved modules by Nowotny et al. (1991, 1993) and Sloan et al. (1994) or is available upon request.

A BIOMEK with a side-loader arm provides pipetting and labware transfer capabilities. The DNA pools are kept in a three-door refrigerator that is furnished with an x-y-z robot retrieving and transferring trays in and out of the storage positions. A custom-built thermal cycler (SP-PCR machine) holds a stack of up to six 96-well PCR plates. A mini-refrigerator with eight shelves provides a cold environment (at 6°C) for PCR trays that are waiting to be subjected to the thermal-cycling process.

Our software environment, mRCS (Nowotny et al. 1995), provides support for the user interaction with the machines, machine control, protocol generation, and real-time error control. The described methods and mentioned programs are installed under mRCS control. mRCS is a set of machine-specific processes (BIOMEK, SP-PCR, DNA storage, etc.) that interact with each other via interprocess communication.

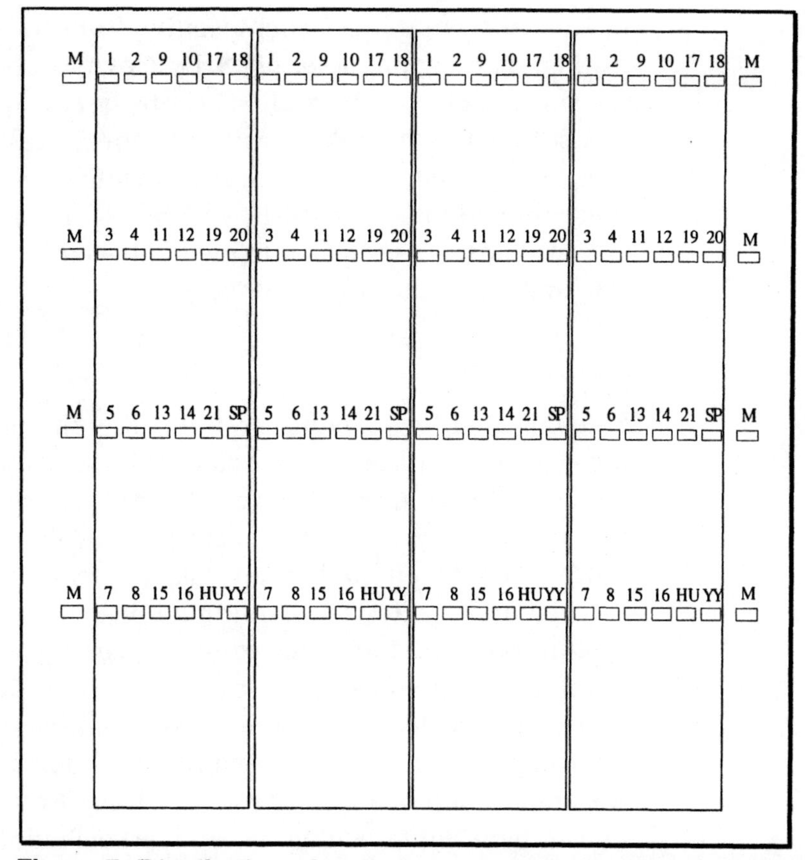

Figure 7 Distribution of wells on an analytical agarose gel. The products of the PCR procedure are loaded with 12-channel pipettors onto agarose gels. The four tiers allow the contents of one 96-well tray to be processed on a single gel. Added marker wells to the left and right of the gel parts filled with samples can hold size standards. The numbering refers to the one shown in Fig. 1, and this change in orientation results from the DNA source being pipetted with an 8-channel tool into the PCR trays, but the samples are then applied with a 12-channel tool to the gel.

During the execution of a script-generating program, the user provides data in answer to program queries. This, in turn, produces a control code for mRCS adapted to the specific requirements of the user. A set of programs has been created to cover the tasks for library preparation and screening. The three major programs include a pooling program for creating the combinatorial pools used for screening, a program for the testing of primer pairs to optimize screening conditions, and the most commonly used, a screening program that automates the screening process (for details, see Appendix).

For the top-level screen, each primer pair requires 24 PCR assays (the wells in three columns from a DNA tray, 23 top-level pools, and 1 positive control) to be run. The user is asked to specify the library to be screened and the number of primer pairs (max. 24). For the combinatorial second level, each positive from the previous level is traced further with, again, a set of 24 PCR experiments (21 for determining the address of a positive, plus 3 controls; see Figs. 4 and 5).

Users place their PCR trays and their PCR reagent-mix trays into their designated shelf stack located in the robotic environment. After the completion of the PCR, the trays are brought back into these stacks. Thus, a user has a defined entry and exit point for the material into and out of the workstation.

DISCUSSION

Stations used for pipetting and the assembly of assays in laboratories have become more widespread. They are an attempt to keep pace with the increasing demand to analyze large numbers of samples rapidly (usually a highly repetitive task that can be accommodated more easily by equipping laboratories with robotics than by increasing personnel). However, automation and robotics are only useful if their implementation is based on or accompanied by the development of robust biochemical methods.

The described DNA extraction procedure is sufficiently simple, and throughput of around 1000 extractions a day per person is feasible for the preparation of a library for screening. At the same time, this method can be accommodated by the work schedule of a standard laboratory. Only minimal automation is required and all the required reagents are easily available. The involved cost is manageable considering the enormous quantity of work that can be done. The use of polypropylene DWTs allows higher centrifugal forces than polystyrene trays. However, compared to standard protocols, these forces do not lead to very solid pellets. Thus, human interaction during the decanting steps is necessary to maintain the fragile sediments so that they are not accidentally discarded during the process. Aiding the PCR-based mapping is the formulation of the TNK buffer system, which provides an environment sufficiently stable for carrying out human chromosome mapping projects with a wide variety of STSs (Kere et al. 1992; Vollrath et al. 1992).

The generation of pools is an essential tool for a rapid screening process (Bruno et al. 1995). Pooling schemes can be devised that make use of multichannel pipetting and generate DNA pools much more rapidly than the pooling scheme presented above. In general, however, DNA pooling is a rare process and is required only at the very start of the screening, when a library is first set up for the screening process. A single isolation provides template DNA sufficient for more than 6000 complete screens. The advantage of the scheme presented here is the time saved by allowing the PCR assays to be assembled with 8-channel tools. The small reaction volumes do not generate excess PCR product because the complete assay is transferred to a well and analyzed on an agarose gel. Throughout the two levels covered by automation, the format of PCR assays for a primer pair is kept constant, and three columns of a 96-well tray contain a set of experiments. This subdivides a PCR tray into four quadrants and simplifies programming by allowing the user to mix top- and second-level screens during one run. The scheme is maintained for all libraries that we have prepared for screening.

If the goal is to go to large throughput, some consideration should be given to the cost of a single reaction. Not much can be done about the cost of oligonucleotides that enter STS content mapping in the form of primer pairs. Next to oligonucleotide costs that enter the equation as a constant, the cost of other ingredients (*Taq* DNA polymerase, template DNA, labware) rises linearly with assay size and number. The assay size used throughout the described methods is 5 μl and is made possible only by modifications of the involved pipetting tools (Nowotny et al. 1993). The screening process has been split into limited tasks taken on by a set of machines. This approach combines a flexible use of vendor and custom hardware and software that provide modular solutions to adapt to the many changes demanded by new technology.

The combination of pipetting station, transfer station, and storage, with a high-throughput PCR machine, allows considerable automation of PCR. The workstation can generate at least a throughput sufficient to provide completed PCR assays to two users, each with 576 samples a day. With the station as described, in a straightforward approach screenings are performed in batches of 24 primer pairs, thus reducing inherent bookkeeping problems.

Automation of the process—selecting a DNA sample from a given DNA collection, preparation of the reaction, and thermal cycling—saves effort and eliminates important sources of error. It can be completed by including PCR product determination. Eventually, YAC library screening for a specific STS sequence or PCR-based screening of DNA samples from populations for the occurrence of a disease gene will become a matter of machine time.

ACKNOWLEDGMENTS We thank Drs. David Schlessinger and Philip Green for their constant support and discussions. Dan Stanglein and Bob Varwig from Automation Technology (St. Louis, Missouri) and John Kreitler, Richard Mac Donald, and David S. Sloan from Washington University contributed to the construction of the hardware; and Frank W. Burough from Washington University contributed to the design and production of the software for this project. This work was supported by National Institutes of Health grant HG-00201.

APPENDIX

Automation Aspects of Library Screening

The automation of PCR assay setup and processing is a direct result of the increased demand for throughput. In addition to providing PCR assay conditions and template DNA for the screening process, significant resources were spent to enhance machines to sustain the tedious, repetitive, and high-precision requirement resulting from the demand for robust handling of the screening process. Most of the steps described in the previous parts can also be performed manually by a determined person. The compression of clones (see Fig. 3) to provide material for DNA extraction is, however, a good example where automation can rapidly have a significant impact on the routine work. The second, even more impressive, example is the pooling of DNA, which is literally impossible to perform manually.

The following steps are implemented on the CGM workstation in a fully automated version and can be taken as suggestions for a program setup or a manual process. The BIOMEK tablet has four positions to hold pipetting tools that can be exchanged automatically and four labware positions. One of the four labware positions is taken for tipboxes, and its footprint is not compatible with microtiter trays. The tipbox position is position 0. The labware in all four positions can be brought in, exchanged, or discarded by an optional robotic arm.

Library Compression, the Step Prior to DNA Extraction

1. Load the BIOMEK tablet.
   a. Place a tipbox into the tablet (remove the lid).
   b. Place a single-well tray in position 1; fill it with 170 ml of water.
   c. Place a deep-well tray in position 2.
   d. Place the source tray in position 3.
2. For the liquid transfer:
   a. With a MP200 tool pick up a set of eight tips and transfer 80 μl of water from the single-well tray into each source well (neither touch the yeast cells nor apply tip touch to avoid a change of tips).
   b. With the same tips pick up 100 μl from the bottom of the source well.
   c. Mix three times (if you can control it, offset the tip slightly; yeast cells can stick quite tightly).

d. Transfer into the column of a destination tray (tip touch, keep tips).

e. Repeat until the compression factor is reached. Change tips with each successive set for compression.

Not included in this description are tray changes to bring in the new source trays nor the tipbox handling.

3. Finish the process as described. Proceed by spinning the destination trays at 3000*g* for 10 minutes. Decant the supernatants and repeat the process a second time in case the cell transfer seems incomplete.

Steps to Generate the Top-level and Combinatorial-level DNA Trays

1. Top-pool generation:

   a. Combine 10 μl from each of the appropriate wells (see Fig. 3) of the DNA source trays into an appropriate container with a 12-channel multipipettor (volume of each of the three top pools after this step is 1.28 ml).

   b. Transfer 80 μl from a top pool into the appropriate well of a DWT.

   c. Store the remaining portion as top-level master mix at –80°C.

   d. Add 320 μl of water and 40 μl of mineral oil (or dilute the DNA to the appropriate concentration as determined by a PCR titration experiment).

2. Adjust the DNA concentration for the combinatorial-level (destination) trays:

   a. Add water to the appropriate level according to Table 1. This brings all pools in the DWT to the same volume.

   b. Transfer 80 μl of the combinatorial pool DNA to a fresh DWT (with BIOMEK), retaining the exact pattern.

   c. Store the remaining portion as combinatorial-level master mix at –80°C.

   d. Dilute the DNA by adding 320 μl of water to each DNA well (or dilute second-level DNA to the appropriate concentration as determined by a PCR titration experiment).

Table 1 The Pools When Prepared according to the Pooling Scheme in Figure 5 Have Different End Volumes

To pools	1–12	15, 17	16, 19, 20	18	21
Add water (μl)	320	210	220	200	190

Water is added to have the same volume in all 21 pools to the level of pools 13 and 14 (640 μl). This step keeps the concentration of DNA species constant. For production screening, the DNA concentration in the pools is reduced further by a factor of 3 to 5. However, only aliquots are used for this dilution (for details see text). For the distribution of pools in DWT, see Fig. 4.

e. Fill in the controls.

f. Add 40 μl of mineral oil to each DNA well.

g. Add a standard polystyrene lid (Becton Dickinson Labware 3071) to protect the DWTs during storage at 4°C.

Automation Hardware and Software

The present workstation has been assembled using commercially available components when possible. The following descriptions and schematics are in summary form (for specific modules, more detailed information has been published [Sloan et al. 1994; Nowotny et al. 1991, 1993] or is available upon request).

Pipetting and transfer station: For pipetting functions, the BIOMEK eight-channel tools are used extensively. Modified P20 and MP20 (Beckman) tools have enhanced precision at the low end (down to 0.3 μl; with an error below 10%) (Nowotny et al. 1993). To preserve the activity in PCR assay mixtures before thermal cycling, position 3 of the BIOMEK 1000 tablet is cooled by an installed cooling plate.

Storage: Storing the DNA solutions at 4°C rather than frozen makes DNA pools readily available to the system. An x-y-z robot inside a refrigerator retrieves and stores trays on demand of the control system. A shuttle slide brings trays in and out through the back wall of the refrigerator into the envelope of the side loader.

Thermal cycler and incubation station: The thermal cycler design is described in detail elsewhere (Nowotny et al. 1991). The PCR trays are loaded into a stack of seven heat-transfer plates located in the arms' envelope. A Programmable Logic Controller executes the control of the thermal cycling process, including cycling valves for the temperature changes and opening and closing of the stack.

PCR trays: The PCR trays are gold-coated aluminum plates that were molded to contain 96 wells in a standard microplate array. A polypropylene lining is vacuum-formed into the 96-well aluminum trays. Alternatively, the trays can be filled with the more expensive MicroAmp tubes (Perkin-Elmer) or compatible 96-well PCR trays.

mRCS, the software environment for robot control: The software environment, mRCS (Nowotny et al. 1995), is a set of machine-specific processes (BIOMEK, SP-PCR, DNA storage, etc.) that interact with each other via interprocess communication. The modular design of mRCS allows straightforward adaptation to the constantly changing biochemical requirements. Tasks are fed into mRCS as ASCII files containing arbitrary command lists. A list is fed into a process called a router. The router maintains the command queues back and forth to the different processes and maintains the protocol. The ASCII file (script) that determines the action is provided off-line. However, we allow for a mode where the user successively types commands into a command window. This mode is used to start script execution and more extensively in error conditions. The main development was done on a UNIX system that provides a stable multiuser, multitasking environment. Programming and debugging environments in this operating system are complete, well documented, and well supported.

Script Generation

Software for script generation reduces for the user the complexity of interaction and control of the system. A limited set of simple-to-enter data is requested from the user (e.g., which library is to be screened) to produce control code for mRCS adapted to the specific requirements of the user.

POOL: During the preparation of libraries for screening, the isolated DNA is combined into DNA pools according to the 128-sample pooling scheme (see Fig. 5). Liquid transfers are done with a single-tipped tool. After each pipetting step, the tip is replaced. A complete library set contains five to eight sets of four 96-well trays containing the DNA (see Fig. 4). The setup for the DNA pools with respect to this pipetting alone requires a week and several thousand pipette tips. POOL includes a mode in which the user can start the pooling from an arbitrary point somewhere inside the scheme. The counterpart for this script generator is the program F_POS (find positives) that is used to determine the positives from the PCR results and the pooling-scheme table and to compute the trays and wells the user has to go to for the cell PCR sources.

P-TEST: P-TEST is a script generator developed to test and optimize the performance of up to 36 primer pairs in a single run with mRCS prior to initiating the automatic screening process.

SCREEN: Two of the three major parts of screening are handled by mRCS:

1. A set of top-level DNA pools is tested with every primer pair dedicated for a full library screen. Each primer pair requires 24 PCR procedures (the wells in three columns from a DNA tray, 23 top-level pools, and one positive control) to be run. The user specifies the library to be screened and the number of primer pairs (max. 24).

2. The result from the top level leads into a second, combinatorial level that traces the positives from a set of 128 pools. For the combinatorial second level, each positive from the previous level is traced further with, again, a set of 24 PCR experiments (21 for determining the address of a positive, plus 3 controls; see also Fig. 4). During execution of this part of SCREEN, the user enters information about the library to be screened and specifies the identified numbers of the positives found within the top level for the respective primer pair. This program also offers a path for the user to mix top-level and second-level screens using any of the libraries kept in the refrigerator and allows access to several parameters for pipetting as well as parameters for the PCR run.

3. This final level is the single-colony screening level that yields at its end the location of the YACs positive for the screened STS. For the cell-level part of screening, yeast colonies stored in 96-well format working trays at –80°C are accessed.

Users enter their PCR trays and their PCR reagent-mix trays into the system via a designated shelf stack located in the robotic environment. After the completion of the PCR, the trays are brought back into these stacks. This defines the entry and exit point of the user's material for the workstation.

The combination of machine-control software and script-generation software into the mRCS system package has been shown to be quite reliable and is now relieving staff of hours previously spent in the manual preparation of PCR. The system allows users to load tipboxes and PCR trays, start a method, and then go on with other tasks. mRCS takes over preparing and running sets of PCR, and after the completion of the process, the trays are brought back into the user stacks ready for gel analysis. The use of ASCII files for all command and control files offers easy access and adaptation of the system to new needs (with hardware moved in or out of the system). Following the concept of task distribution, the implementation of mRCS is highly modular. This helps in further adaptation to support users with more automation to acquire higher throughput for highly repetitive tasks. The implementation of mRCS as a collection of independent tasks linked by interprocess communication permits parts of the system to be present on different computers. This is useful for a future implementation where processes can be monitored and accessed from remote locations to provide information to users about the status of the robotic workstation.

An enormous number of commands have to be issued during a complete process. The user cannot be expected to keep in mind the sequence of required commands to complete a process, nor should the user carry the burden of translating between the different numbering schemes involved in a multilevel screen and accessing a set of different libraries (each on average 30,000 clones). Script generation has integrated the required scheduling and syntax checking and provides easy variation of the assay setup to the user.

The user interaction with the script generator is further simplified by defaults that have been chosen for each response. By accepting default values, the user can generate a script rapidly with a few keystrokes.

Automation is consistent in its demand for extensive investment. The gain in speed and precision alone is not always sufficient justification for these investments. Thus, the focus has to be kept on versatility and flexibility in automation setups. If efforts in automation for applications similar to the one described are to succeed, the methods to be automated have to be thoroughly tested and must be performing robustly. Only then will automation free skilled labor for more creative and challenging tasks and leave the repetitive and tedious part of the work to the co-worker machine.

REFERENCES

Blanchard, M.M. and V. Nowotny. 1994. High throughput rapid yeast DNA extraction. Application to yeast artificial chromosomes as polymerase chain reaction templates. *Genet. Anal. Tech. Appl.* **11:** 7–11.

Blanchard, M.M., P. Nowotny, P. Taillon-Miller, and V. Nowotny. 1993. PCR buffer optimization with uniform temperature regimen to facilitate automation. *PCR Methods Appl.* **2:** 234–240.

Bruno, W.J., E. Knill, D.J. Balding, D.C. Bruce, N.A. Doggett, W.W. Sawhill, R.L. Stallings, C.C. Whittaker, and D.C. Torney. 1995. Efficient pooling designs for library screening. *Genomics* (in press).

Burke, D.T., G.F. Carle, and M.V. Olson. 1987. Cloning of large segments of exogenous DNA into yeast by means of artificial chromosome vectors. *Science* **236:** 806–812.

Green, E. and P. Green. 1991. Sequence tagged site (STS) content mapping of human chromosomes: Theoretical considerations and early experiences. *PCR Methods Appl.* **1:** 77–90.

Kere, J., R. Nagaraja, S. Mumm, A. Ciccodicola, M. D'Urso, and D. Schlessinger. 1992. Mapping human chromosomes by walking with sequence-tagged sites from end fragments of yeast artificial chromosome inserts. *Genomics* **14:** 241–248.

Labarca, C. and K. Paigen. 1980. A simple, rapid, and sensitive DNA assay procedure. *Anal. Biochem.* **102:** 344–352.

Li, H., U.B. Gyllensten, X. Cui, R.K. Saiki, H.A. Ehrlich, and N. Arnheim. 1988. Amplification and analysis of DNA sequences in single human sperm and diploid cells. *Nature* **335:** 414–417.

Nowotny, V., M.M. Blanchard, F.W. Burough, and D.D. Sloan. 1991. Automation for the mapping of the human genome. In *Proceedings of the International Symposium on Laboratory Automation and Robotics* (ed. J.N. Little et al.), pp. 333–348. Zymark Corporation, Hopkinton, Massachusetts.

Nowotny, V., M.M. Blanchard, D.D. Sloan, and F.W. Burough. 1995. PCR-based genome mapping with a multitasking robotic control system. In *Laboratory robotics and automation.* VCH Publishers, New York, New York.

Nowotny, V., F.W. Burough, D.D. Sloan, and M.M. Blanchard. 1993. Human genome mapping: Progress with robot-aided automation. In *Proceedings of the International Symposium on Laboratory Automation and Robotics* (ed. J.N. Little et al.), pp. 277–298. Zymark Corporation, Hopkinton, Massachusetts.

Olson, M.V., L. Hood, C. Cantor, and D. Botstein. 1989. A common language for physical mapping of the human genome. *Science* **245:** 1434–1435.

Saiki, R.K., S. Scharf, F. Faloona, K.B. Mullis, G.T. Horn, H.A. Erlich, and N. Arnheim. 1985. Enzymatic amplification of β-globin genomic sequences and restriction site analysis for diagnosis of sickle cell anemia. *Science* **230:** 1350–1354.

Saiki, R.K., D.H. Gelfand, S. Stoffel, S.J. Scharf, R. Higuchi, G.T. Horn, K.B. Mullis, and H.A. Erlich. 1988. Primer-directed enzymatic amplification of DNA with a thermostable DNA polymerase. *Science* **239:** 487–491.

Schlessinger D. 1990. Yeast artificial chromosomes: Tools for mapping and analysis of complex genomes. *Trends Genet.* **6:** 248–258.

Sloan, D.D., M.M. Blanchard, F.W. Burough, and V. Nowotny. 1994. Screening yeast artificial chromosome libraries with robot-aided automation. *Genet. Anal. Tech. Appl.* **10:** 128–143.

Vollrath, D., S. Foote, A. Hilton, L.G. Brown, P. Beer-Romero, J.S. Bogan, and D. Page. 1992. The human Y chromosome: A 43-interval map based on naturally occurring deletions. *Science* **258:** 52–59.

Expression Libraries Derived from PCR-immortalized Rearranged Immunoglobulin Genes

Holly H. Hogrefe[1] and Bob Shopes[2]

[1]Stratagene Cloning Systems, La Jolla, California 92037
[2]Tera Biotechnology, La Jolla, California 92037

INTRODUCTION

The expression of active antibody fragments in *Escherichia coli* has greatly advanced the identification of antigen-binding clones using molecular biology techniques (Better et al. 1988; Huse et al. 1989; Larrick et al. 1989; Orlandi et al. 1989; Skerra and Plunkthun 1989; Ward et al. 1989; Mullinex et al. 1990). When combined with phage display technology (Parmley and Smith 1988; Scott and Smith 1990), Fab fragment libraries containing up to 10^8 unique specificities may be readily constructed and screened. In the prokaryotic-derived phagemid display libraries, the phenotype (e.g., binding activity) is physically linked to the genotype (the antibody genes) in analogy with the surface antibody of B cells during the maturation process. Phagemid display clones that bind to a particular antigen may be isolated by biopanning libraries of recombinant phagemid that display Fab on the surface as a fusion with the minor coat protein cpIII of M13 (Barbas et al. 1991; Burton et al. 1991; Garrard et al. 1991; Hoogenboom et al. 1991; Zebedee et al. 1992; Hogrefe et al. 1993a,b; Orum et al. 1993). The phagemid contains the light-chain (LC) and heavy-chain (HC) genes of the displayed Fab, thereby allowing further characterization and manipulation of antigen-binding activity at the molecular level (Barbas et al. 1992; Gram et al. 1992; Riechmann and Weill 1993).

To construct Fab display libraries, the diversity of the immunoglobulin repertoire is effectively immortalized by PCR. Light-chain and heavy-chain genes are PCR-amplified from reverse-transcribed transcripts (cDNA) of the mRNA derived from rearranged immunoglobulin genes. To construct comprehensive Fab libraries, exhaustive sets of PCR primers have been designed by us (Hogrefe et al. 1993a,b; Hogrefe and Shopes 1994) and by many other workers

(LaBoeuf et al. 1989; Larrick et al. 1989b; Orlandi et al. 1989; Sastry et al. 1989; Marks et al. 1991a; Orum et al. 1993; Sassano et al. 1994; Zhou et al. 1994). These primers correspond to the amino termini of the variable regions of known immunoglobulins and the carboxyl termini of the C_H1 domains of the heavy-chain isotypes or the κ and λ light chains. For the expression of Fab, a bicistronic operon is assembled by randomly ligating the light-chain and heavy-chain PCR products at a rare, nonpalindromic restriction site (Hogrefe et al. 1993a,b; Hogrefe and Shopes 1994).

The methods described here allow the construction of highly diverse phage display Fab expression libraries. Specific Fab clones may be isolated from phage display libraries by biopanning, an iterative process involving the capture and elution of binding clones from antigen-coated surfaces (petri dishes, affinity columns). We have previously published the isolation of antigen-binding Fab clones from a number of phage display libraries using biopanning techniques (Hogrefe et al. 1993a,b).

Cloning Strategy for Phage Display Fab Libraries

To construct comprehensive phage display libraries of human Fab fragments, we use the SurfZAP λ vector (Stratagene) (Hogrefe et al. 1993a) shown in Figure 1. A Fab gene cassette, which encodes random combinations of diverse sets of light-chain and heavy-chain sequences, is assembled as described in Figure 2, and then cloned into the *Not*I and *Spe*I cloning sites of SurfZAP. In our Fab fragment library constructs, the heavy-chain Fd portion (V_H–C_H1 domains) is expressed as a fusion with a truncated version (amino acids 198–406) of the M13 minor coat protein cpIII.

The use of λ packaging extracts to introduce the SurfZAP λ DNA into bacterial host cells allows the efficient construction of phage display libraries containing up to 10^8 unique Fab clones. Subsequent in vivo excision (Short et al. 1988) of the SurfZAP λ phage library with helper phage gives rise to pSurfScript SK(–) phagemids that encode an ampicillin-resistance gene, the colE1 origin, and the Fab gene cassettes. In cells harboring the pSurfScript plasmid, the *pel*B leader sequences direct the light-chain and the Fd-cpIII fusion protein across the inner bacterial membrane into the periplasmic space. Functional Fab molecules assemble in the periplasm with the cpIII portion of the Fd-cpIII fusion embedded in the inner membrane. When these cells are superinfected with M13 helper phage, phagemid particles are generated that bear a Fab molecule on the surface and contain the pSurfScript DNA encoding the displayed antigenic specificity. Libraries of phagemid displaying Fab fragments may then be subjected to biopanning to enrich for antibody clones with desired binding activities.

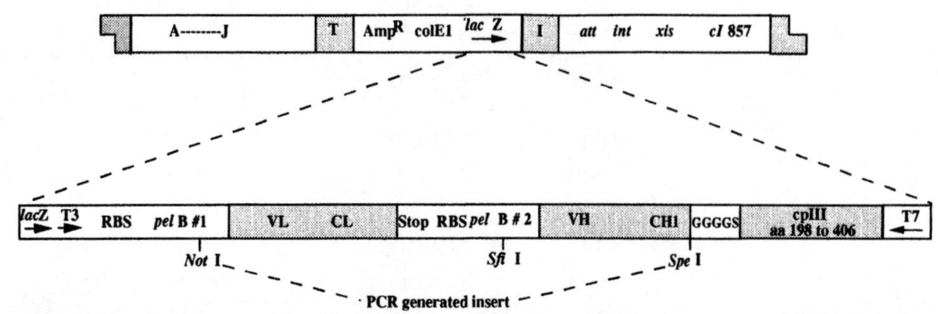

Figure 1 The SurfZAP λ vector containing a Fab gene cassette insert. The SurfZAP λ vector contains the I (initiator) and T (terminator) sequences of the F1 origin for excision of the pSurfScript plasmid (Short et al. 1988). The left arm of the SurfZAP vector encodes the *lacZ* promoter, a ribosome-binding site (RBS) for translation of the light chain, and a partial *pel*B leader sequence (#1) to direct the light chain to the periplasm. The Fab gene cassette, assembled as shown in Fig. 3, contains the remaining carboxy-terminal portion of the *pel*B leader (#1), followed by the light-chain gene sequence, a stop codon, a second RBS, a downstream *pel*B leader sequence (#2), and finally, the Fd gene sequence. The right arm of the SurfZAP vector encodes a 5-amino-acid spacer sequence ([Gly]$_4$-Ser), amino acids 198–406 of cpIII, and a stop codon. After ligation, the vector expresses a soluble light chain and an Fd/cpIII fusion protein as a dicistronic message from the *lac*Z promoter.

Construction of the Fab Gene Cassette

To prepare the Fab gene cassette, light-chain and heavy-chain Fd sequences are PCR-amplified separately from human B-cell cDNA using the PCR primers listed in Table 1. To increase diversity of the com-

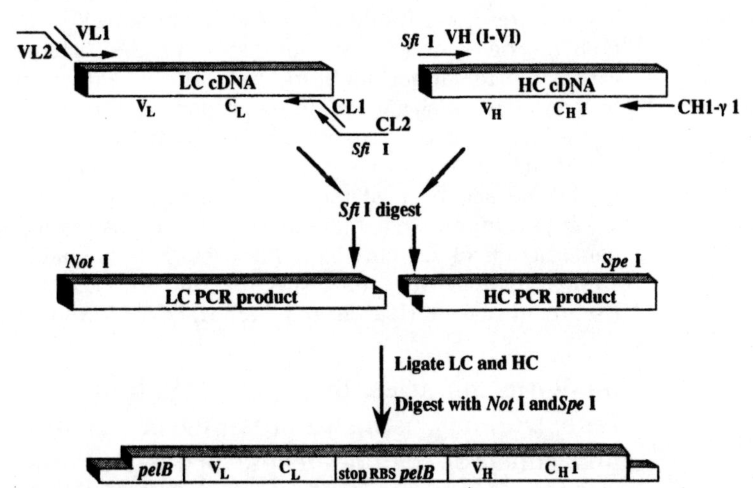

Figure 2 Preparation of the light-chain-Fd insert for the SurfZAP vector. Light-chain and Fd sequences are PCR-amplified from B-cell cDNA using the primer sets listed in Table 1. The PCR primers introduce *Sfi*I sites used to randomly combine light-chain and Fd sequences, as well as *Not*I and *Spe*I cloning sites, the light-chain stop codon, and a RBS and *pel*B leader sequence for translation and translocation of the Fd-cpIII fusion protein.

Table 1 Primers Used for PCR Amplification of Human Heavy- and Light-chain Genes

A. Human heavy-chain primers	
VH-I	5′- TCGCGGCCCAACCGGCCATGGCCCAGGTGC A GCTGGTGCAG-3′
VH-II	5′- TCGCGGCCCAACCGGCCATGGCCCAGGTCA A CTAAGGGAG-3′
VH-III	5′- TCGCGGCCCAACCGGCCATGGCCCAGGTGC A GCTGGTGGAG-3′
VH-IV	5′- TCGCGGCCCAACCGGCCATGGCCCAGGTGC A GCTGCAGGAGTCG-3′
VH-V	5′- TCGCGGCCCAACCGGCCATGGCCCAGGTGC A GCTGGTGCAG-3′
VH-VI	5′- TCGCGGCCCAACCGGCCATGGCCCAGGTAC A GCTGCAGCACTCA-3′
CH1-γ	5′- CGGACTAGTACAAGATTTGGGCTCTGCTTT-3′
CH1-μ	5′- AGCATCACTAGTGGCAATCACTGGAAGAGG-3′
B. Human light-chain primers	
VL1-κI	5′- GCCCAACCAGCCATGGCCGACATCCAGATG A CCCAGTC-3′
VL1-κII	5′- GCCCAACCAGCCATGGCCGATATTGTGATG A CTCAG-3′
VL1-κDeg.	5′- GCCCAACCAGCCATGGCCGATATTGTGATG A CCCAGTCT-3′
VL2	5′- GAAATCACTCCCAATTAGCGGCCGCTGGAT T GTTATTACTCGCTGCCC AACCAGCCATGGCC-3′
CL1-κ	5′- ATGACTGTCTCCTTGAAGCTTTCATTAACA C TCTCCCCTGTTGAAGCT CTTTGTGACGGGCGAACTC-3′
CL2	5′- CATGGCCGGTTGGGCCGCGAGTAATAACAA T CCAGCGGCTGCCGTAGG CAATAGGTATTTCATTATGACTGTCTCCTTG-3′

(*A*) The Fd portion (V_H-D-J-C_H1) of heavy-chain genes is PCR-amplified with primer sets designated V_H-(I-VI) and C_H1-γ1 or C_H1-μ (Fig. 3A). The 3′ ends of the V_H-(I-VI) primers are homologous to the sense strand of the first few codons of the six V_H families designated in Kabat et al. (1987). The 5′ ends of the V_H primers introduce the *Sfi*I restriction site for assembly of the Fab gene cassette (underlined) and encode the 3′ half of the *pel*B leader sequence #2. The 3′ ends of the C_H1-γ_1 and the C_H1-μ primers are homologous to the antisense strand of the last few codons of the C_H1 domain gene of human IgG_1 and human IgM, respectively. The C_H1 primers also introduce the *Spe*I cloning site (underlined). (*B*) κ light chains are amplified from cDNA with the V_L1-κ and C_L1-κ primer pairs, and these genes are then reamplified with the V_L-2 and C_L-2 primers to add restriction sites and leader sequences (Fig. 3B). The 3′ end of the V_L1-κ primers are homologous to the sense strand of the –4 to +8 codons of V_Lκ families I and II and the consensus human V_L sequence compiled in Kabat et al. (1987). The V_L-2 primer introduces the *Not*I cloning site (underlined) and the carboxy-terminal portion of the *pel*B. The 5′ end of the C_L1-κ primer is homologous to the antisense strand of the last codons of human κ from Kabat et al. (1987). The C_L2 primer adds the 5′ portion of the second *pel*B leader sequence #2 and introduces the *Sfi*I common restriction site (underlined). Conservative changes were made in the nucleotide sequence of the downstream *pel*B leader (partially encoded by primer C_L-2) to minimize recombination with the upstream *pel*B leader sequences. For the construction of λ light-chain libraries, λ light-chain genes can be amplified in a similar fashion with family-specific V_L1-λ and C_L1-λ primer pairs (data not shown). (Reproduced from Hogrefe and Shopes, *PCR Methods Appl. 4:* S112 [1994].)

binatorial libraries, the κ and λ light-chain genes are PCR-amplified independently using all possible combinations of each family-specific and constant region-specific primer. Similarly, separate PCR assays are performed for each combination of family-specific V_H primer and constant region primer to generate a comprehensive collection of Fd sequences. The light-chain PCR products are pooled and digested with *Sfi*I, and then ligated with the similarly pooled and digested Fd PCR products to generate the Fab gene cassette.

Preparation of Phage Display Fab Libraries

The Fab gene cassette is cloned into the *Not*I and *Spe*I cloning sites of the SurfZAP λ vector. Primary SurfZAP libraries are packaged into λ phage and introduced into bacterial cells by infection. The primary libraries are then amplified by lytic growth to increase the representation of each Fab clone. Amplified libraries are then converted to pSurfScript phagemid libraries by in vivo mass excision (Short et al. 1988), the excised phagemid libraries are amplified, and recombinant phagemid particles for biopanning are prepared by superinfection with M13 helper phage.

The following protocol is used to prepare a Fab gene cassette for the SurfZAP λ vector. These specific methods allow the construction of a human Fab fragment phage display library. These protocols can be readily adapted to the construction of mouse Fab libraries with the use of PCR primers that are specific for mouse immunoglobulin sequences and available from Tera Biotechnology. Methods for biopanning the SurfZAP phagemid display libraries and analyzing the binding activities of enriched clones have been published elsewhere (Hogrefe et al. 1993a,b; Hogrefe and Shopes 1994).

All vectors, enzymes, and reagents are from Stratagene Cloning Systems unless otherwise noted. Restriction enzymes for cloning were usually obtained from New England Biolabs. All primers used were from Tera Biotechnology. Buffer recipes can be obtained from the manufacturers.

First-strand cDNA Synthesis

To increase the likelihood of enriching antigen-binding human Fab clones, RNA should be recovered from the B cells of actively immunized humans, mice, or rabbits. The example below details the cloning of human Fab fragments, but the process is essentially the same for other species, although the primer design usually differs. Lymph node, spleen, bone marrow, and peripheral blood of various species have been used as sources of activated B cells. For the construction of human Fab libraries from peripheral blood lymphocyte RNA, white blood cells should be isolated from at least 10 ml of heparinized blood (HistoPaque, Pharmacia) 5–6 days post-boost, when the concentration of antigen-specific B cells in the circulation is maximal (Stevens et al. 1979; Falkoff et al. 1983). Total RNA is isolated (RNA Isolation kit, Stratagene, or the researcher's method of choice) and stored in ethanol at –20°C until use. It is not necessary to isolate the mRNA; in fact, the ribosomal RNA in the total RNA preparation may serve to protect the valuable mRNA from degradation. It is important to avoid RNase in the preparation and handling of RNA. The use of DEPC-treated water is recommended as a minimum precaution. For the preparation of first-strand cDNA from total RNA preparations, we have found the best results using the Superscript kit from GIBCO/BRL followed by RNase H treatment (GIBCO/BRL).

PROTOCOL

1. Add 10–20 mg of total RNA to each of three autoclaved 0.5-ml microfuge tubes.

2. For κ light-chain cDNA synthesis, add 400 ng of oligo$(dT)_{18}$ to one tube. To the second and third tubes, add 400 ng of the appropriate C_H1 primer to prime heavy-chain cDNA synthesis. Add DEPC-treated water to bring the final volume to 70 μl, and mix the reactants gently.

3. Heat the mixtures for 5 minutes at 70°C, and then cool slowly to 37°C.

4. Collect the contents of the tubes by centrifuging briefly. Add the following reagents to each tube:

   20 μl of 5x Moloney Reverse Transcriptase buffer (GIBCO/BRL)
   5 μl of 1 M DTT
   4 μl of dNTP mix (25 mM each)
   1 μl (80 units) of RNase H^- Moloney LV Reverse Transcriptase (GIBCO/BRL)

5. Mix the reactants gently. Incubate the tubes for 60 minutes at 37°C, and for 30 minutes at 42°C.

6. Add 2 units of RNase H (GIBCO/BRL) and incubate the tubes for an additional 10 minutes at 55°C.

7. Treat the reaction mixtures with 5 μl of StrataClean resin to remove protein. The cDNA may be stored for up to 2 weeks at 4°C.

PCR Amplification of Light-chain and Fd Sequences

Human light-chain and Fd sequences are PCR-amplified in a series of individual reactions that employ combinations of the forward human family-specific primers and reverse human constant region primers (e.g., each V_H with C_H1-γ or C_H1-μ; each V_L1-κ with C_L1-κ). Typically, the amount of PCR product obtained varies for each family-specific variable region and constant region primer pair depending on the particular cDNA sample employed (Fig. 3). PCR is performed to generate a total of ≥1 μg of each light-chain product and ≥1 μg of each Fd product. This specific protocol describes the amplification of human κ

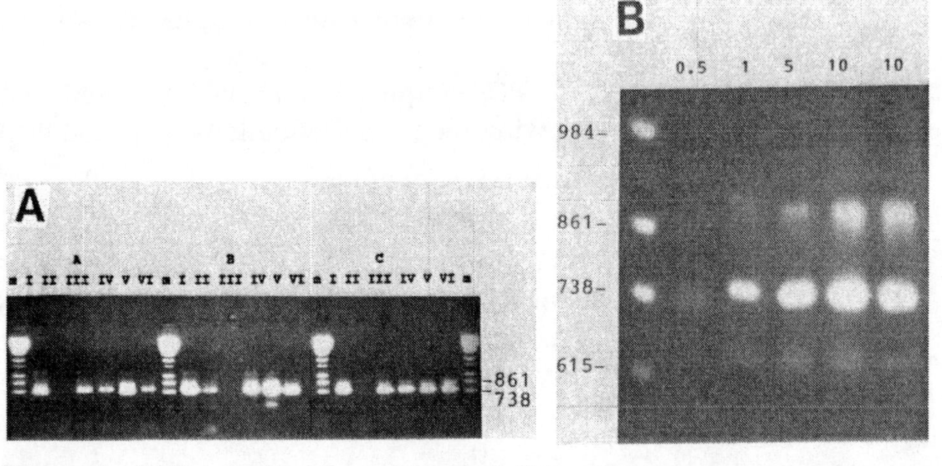

Figure 3 PCR amplification of Fd and light-chain sequences. cDNA was prepared as described in the text from human peripheral blood and lymph node tissue. In *A*, Fd sequences were PCR-amplified with the V_H(I-VI) and C_H1-γ1 primers using cDNA prepared from three different donors (A,B,C). DNA standards were run in lanes denoted *m*, and the molecular weights are given in base pairs. In *B*, κ light-chain sequences, previously amplified with the V_L1-κ and C_L1-κ primers are re-amplified with the V_L2 and C_L2 primers. The amount of light-chain product obtained after 10 rounds of PCR is shown using 0.5-, 1-, 5-, and 10-μl aliquots of the original light-chain PCR product as template. Molecular-weight standards were run in the first lane on the left and the marker bp are indicated in the left margin.

light-chain genes and human γ and μ Fd sequences for the construction of κ/γ and κ/μ libraries. The PCR assays are set up in replicates to allow separate ligation reactions of the κ light-chain sequences with the μ Fd sequences and the γ Fd sequences. This PCR protocol can be readily adapted to the amplification of λ light chains and/or Fd sequences of other immunoglobulin classes.

PROTOCOL

1. To PCR-amplify human κ light-chain sequences, add the following reagents sequentially to three 0.5-ml PCR tubes:

Component	volume (μl)		
Water	82.2	82.2	82.2
10X *Taq* buffer	10	10	10
25 mM dNTPs	0.8	0.8	0.8
PCR primers (100 ng/μl)			
V_L1-κI	1		
V_L1-κ2		1	
V_L1-κDeg.			1
C_L1-κ	1	1	1
LC cDNA	5	5	5
Taq DNA polymerase (5 units/μl)	0.5	0.5	0.5

(The polymerase may be added at 94°C if a hot start is desired.)

2. To PCR-amplify human γ heavy-chain Fd sequences, add the following reagents sequentially to 0.5-ml PCR tubes:

Human V_H family	I	II	III	IV	V	VI
component			volume (μl)			
Water	82	82	82	82	82	82
10X *Taq* buffer	10	10	10	10	10	10
25 mM dNTPs	0.8	0.8	0.8	0.8	0.8	0.8
PCR primers (100 ng/μl)						
V_H-I	1					
V_H-II		1				
V_H-III			1			
V_H-IV				1		
V_H-V					1	
V_H-VI						1
C_H1-γ	1	1	1	1	1	1
HC cDNA	5	5	5	5	5	5
Taq DNA polymerase (5 units/μl)	0.5	0.5	0.5	0.5	0.5	0.5

(The polymerase may be added at 94°C if a hot start is desired.)

3. To PCR-amplify human μ heavy-chain Fd sequences, perform the same amplifications as in step 2, but substitute the C_H1-μ primer for the C_H1-γ primer.

4. As a negative control to test for the contamination of reagents, we strongly recommend performing a mock amplification with all of the components while omitting the cDNA.

5. Mix the reactants gently and layer two drops (~100 μl) of mineral oil over the reactions.

6. To anneal the PCR primers to the cDNA template, heat the PCR mixtures for 5 minutes at 94°C, and then for 5 minutes at 54°C. Perform 40 cycles of PCR amplification: Extend for 3 minutes at 72°C, denature for 1.5 minutes at 92°C, anneal for 2.5 minutes at 54°C. Last, extend for 10 minutes at 72°C to fill in partial products completely.

7. Analyze 10 μl of each PCR product using agarose gel electrophoresis and the appropriate molecular-size standards. The light-chain and Fd heavy-chain PCR products should migrate with apparent molecular weights of ~700 bp.

8. To introduce additional sequences at the 5′ and 3′ ends of the light-chain PCR products, amplify 10 ng (1–10 μl). *Each* successful light-chain PCR product is reamplified for 10 cycles with primers V_L2 and C_L2 using the conditions given in step 6. The example assumes 10 ng/μl for the starting product. The PCR mixtures are set up as given below and the resulting PCR products are analyzed as in step 7 (see Fig. 3b).

Component	volume (μl)
Water	86
10X *Taq* buffer	10
25 mM dNTPs	0.8
PCR primers (100 ng/μl)	
V_L2	1
C_L2	1
LC PCR product	1
Taq DNA polymerase (5 units/μl)	0.5

Sfi I Digestion and Purification of the Light-chain and Fd PCR Products

At this step, the successful heavy-chain γ products may be pooled together in one 1.5-ml polypropylene tube or set of tubes (up to a final volume of 500 μl per tube), the heavy-chain μ PCR products may be combined in a second set of tubes, and the light-chain PCR products may be pooled together in a third set of tubes.

Note: Do not combine the light-chain and Fd PCR products at this step. However, one may wish to keep separate the PCR products obtained with different combinations of the V_L1 and C_L1 primers and/or the V_H and C_H1 primers. The latter approach would allow one to monitor and regulate the composition of a combinatorial library, which may be of particular interest when the immune response to a particular antigen is known and dominated by one family of antibody genes or class of immunoglobulins. To ensure that a sufficient amount of the Fab gene cassette insert is produced prior to cloning into the SurfZAP λ arms, one should begin these procedures with a total of at least 1 μg of each pooled sample of light-chain PCR product that is to be ligated to each pooled sample of heavy chain product.

PROTOCOL

1. Extract the pooled samples with an equal volume of phenol/chloroform. Transfer the upper aqueous layer to a fresh 1.5-ml polypropylene tube, leaving behind the white interface. Extract the aqueous layer with chloroform to remove any residual phenol, and again transfer the upper layer to a new tube.

2. Precipitate the DNA with ethanol by adding 1/10th the volume of 3 M sodium acetate and 2.5X the volume of 100% ethanol. Spin the tubes at 14,000 rpm for 20 minutes at 4°C. Remove the supernatant, being careful not to dislodge the pellet. Wash the DNA pel-

let with 0.5 ml of 70% ethanol and respin. Dry down and resuspend the pellet in 25 μl of TE (10 mM Tris-HCl, pH 7.5, 1 mM EDTA).

3. Determine the approximate concentration of the purified PCR products by spotting a dilution of each sample on ethidium bromide agar plates (Sambrook et al. 1989) and compare to known standards (e.g., DNA markers diluted to 1, 2, 5, and 10 ng/μl).

4. Digest the light-chain and Fd PCR products separately with *Sfi*I by combining the following reagents (we assume 100 ng/μl concentration for the purified PCR products):

Component	volume (μl)
Fd (or LC) PCR product (1 μg)	10
Water	425
10X Buffer #3 (Stratagene)	50
*Sfi*I (4 units/μl; New England Biolabs)	15

5. Digest the PCR products for 2 hours at 50°C. Efficiency of the *Sfi*I digestion may be assessed by adding 10 μl of each PCR product digestion, just after enzyme addition, to 0.2 μg of a test plasmid that contains an *Sfi*I restriction site. If the test plasmid can be cut, one can assume that the PCR product is free of potential contaminants that inhibit digestion, and that the enzyme is present in sufficient excess to cut the PCR product completely.

6. Prior to loading on the preparative gel, it is necessary to precipitate the DNA with ethanol. Purify the *Sfi*I-digested light-chain and Fd PCR products by preparative gel electrophoresis using 1% SeaKem GTG (FMC) agarose gels. Excise the ~700-bp bands for the light-chain and the heavy-chain Fd. Remove the agarose with GENECLEAN (BIO 101).

7. Precipitate the gel-purified DNA samples with ethanol, wash the pellet with 70% ethanol, and dry. Resuspend the pellet with 5 μl of water and determine the DNA concentrations on ethidium bromide plates.

Preparation of the *Not* I and *Spe* I-digested Fab Gene Cassette

The light-chain and Fd PCR products are now ligated to each other at the common *Sfi*I restriction site (Fig. 4). Depending on the strategy employed for library construction, one may ligate the pooled light-chain and the pooled heavy-chain Fd (μ plus γ) sequences together, or one may have prepared separate ligation reactions of selected combi-

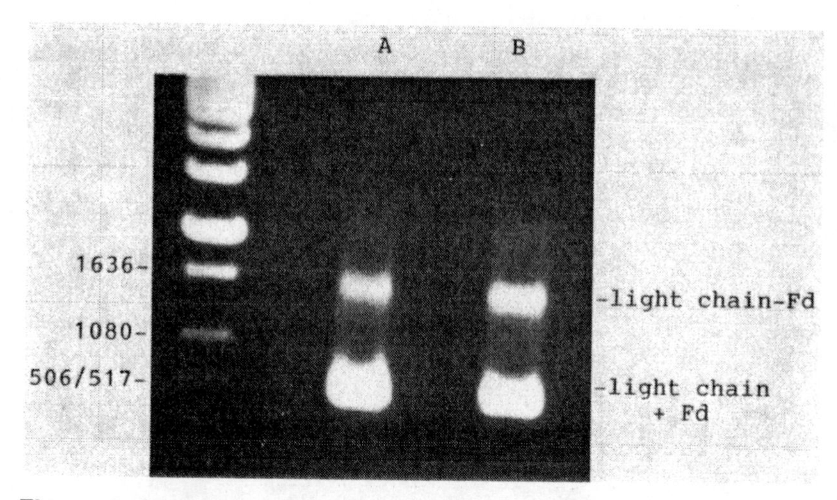

Figure 4 Ligation of light-chain and heavy-chain PCR products at the common *Sfi*I restriction site. Light-chain and Fd PCR products amplified from two different cDNA sources (A,B) were digested with *Sfi*I, gel-isolated, and ligated together as described in the text. The intended light-chain-Fd cassette migrates at ~1400 bp, whereas the unligated light-chain and Fd PCR products comigrate at ~700 bp. The molecular weights of DNA standards, run in the first lane on the left, and the marker bp are indicated in the left margin.

nations of family-specific and constant-region-specific light-chain and Fd PCR products.

For optimal ligation efficiencies, equimolar amounts of the Fd and light-chain products are added to the ligation reaction in a volume of ≤15 μl. Efficiencies of this particular ligation are typically 30–50%. At least 34 ng of digested insert is needed per microgram of λ vector. To obtain the greatest yield of light-chain + Fd insert possible, the ligation products are analyzed only after the Fab gene cassette has been digested and subjected to preparative gel electrophoresis (step 9 below).

PROTOCOL

1. Combine the following reagents in the order listed (we assume 100 ng/μl concentration for the *Sfi*I cut and purified PCR products).

Component	volume (μl)
*Sfi*I-digested Fd PCR product	5
*Sfi*I-digested LC PCR product	5
10× Ligase buffer	1.5
10 mM rATP	1.5
Water	1.5
T4 DNA ligase (4 units/μl)	0.5

2. Incubate the ligation reaction overnight at 4°C.

3. Inactivate the ligase by heating for 10 minutes at 65°C.

4. Extract the ligation reaction and purify the DNA as in steps 1 and 2 of the *Sfi*I digestion protocol, except resuspend the pellet in 20 μl of water.

5. Digest the ligation products with *Spe*I and *Not*I by combining the following reagents in a 1.5-ml polypropylene tube:

Component	volume (μl)
LC-Fd ligation product	20
10x NEB buffer #2	10
100x NEB aceylated BSA	1
Water	64
*Spe*I (3 units/μl; New England Biolabs)	5

6. Incubate for 2 hours at 37°C.

7. Add 1.1 μl of 5 M NaCl and 4.5 μl of 1 M Tris-HCl, pH 7.5, to adjust to a final molarity 100 mM NaCl, 50 mM Tris-HCl. Add 7 μl of *Not*I (20 units/μl; New England Biolabs). Incubate overnight (~16 hours) at 37°C.

8. If desired, the efficiencies of the *Spe*I and *Not*I digestions may be determined by removing a small aliquot of the ligation mixture digest (1/20 volume), just after enzyme additions, and incubating this small aliquot to 0.1 μg of supercoiled control plasmids that contain *Spe*I or *Not*I restriction sites.

9. Purify the 1.4-kb *Not*I/*Spe*I-digested light-chain + Fd insert (see Fig. 4) by preparative gel electrophoresis, followed by GENE-CLEAN, as described in step 6 of the *Sfi*I digestion procedure.

10. Precipitate, wash, and dry the DNA pellet with ethanol. Resuspend the pellet in 5 μl of water. Quantify the recovery of the double-digested Fab gene cassette with ethidium bromide plates as described above.

Ligation of the Fab Gene Cassette into the SurfZAP λ Vector

For optimal ligation efficiencies, an equimolar amount of digested insert and λ arms should be ligated in a small volume of ≤10 μl (5 μl is optimal). For the 1.4-kb Fab gene cassette described here, 34 ng of insert should be ligated per microgram of λ vector. We also strongly recommend positive (vector with test insert supplied by Stratagene) and negative (vector only; no insert) controls for the vector ligation reactions.

PROTOCOL

1. Ligate the digested Fab gene cassette into the SurfZAP λ arms by adding the following reactants to sterile 0.5-ml polypropylene tubes in the order listed:

Component volume (μl)	Sample	Test insert	Vector only
SurfZAP arms (1 μg/ml) (*NotI/SpeI*-digested and dephosphorylated)	1	1	1
Digested light-chain–Fd insert (34 ng/μl)	1		
Test insert (Stratagene)		1	
10X Ligase buffer	0.5	0.5	0.5
10 mM rATP	0.5	0.5	0.5
Water	1.5	1.5	2.5
T4 DNA ligase (4 units/μl)	0.5	0.5	0.5

2. Incubate the ligation reaction overnight at 4°C. (The ligation reaction may be stored for several days at 4°C.)

3. Package 1 μl of each ligation reaction with λ packaging extract. We recommend Gigapack II Gold (Stratagene) following the manufacturer's procedure.

4. To estimate the size of the primary λ library, titer the packaged λ phage. Serially dilute 1 μl of the λ library in SM (Sambrook et al. 1989) (10^{-3}, 10^{-4}, 10^{-5}, 10^{-6}) and mix 10 μl of each dilution with 200 μl of freshly prepared XL1-Blue cells ($OD_{600nm} = 0.5$). Incubate the phage and bacteria 15 minutes at 37°C, add 3 ml of melted top agar (cooled to 50°C), and then immediately pour onto NZY (Sambrook et al. 1989) agar plates. Incubate the plates overnight at 37°C and count the plaques. The size of the primary library is calculated from the number of plaque-forming units (pfu) by extrapolation. Typically, the test insert control yields ~10^7 pfu, and the vector-only control yields ~10^4 pfu. The yield of the LC-Fd insert depends on the quality of the preparation. If the library size is deemed sufficient (e.g., >10^6), then proceed with the following protocols.

Preparation of the SurfZAP λ Library

To prevent contamination of the λ library with filamentous helper phage, and hence excision of the pSurfScript phagemid, λ phage stocks should be stored in the presence of 1% chloroform. It is fairly easy to contaminate λ or phagemid libraries. To minimize the chance of cross-contamination, scrupulous sterile laboratory techniques must be used at all times; the use of aerosol-resistant tips, changing pipette tips with every use, and frequent sterilization of bench tops and equipment are highly recommended practices.

To construct the SurfZAP λ phage library, the ligation reactions are packaged directly with λ packaging extracts. If the prepared Fab gene cassette is free of contaminants and contains a high percentage of ligatable ends, between 2×10^6 and 5×10^7 recombinant plaques (clones) should be recovered per microgram of λ vector when using high-efficiency packaging extract and well-prepared vector.

The resulting primary libraries should be amplified by lytic growth to produce a large, stable quantity of high-titer stock. To minimize biasing clonal diversity, the amplification in liquid culture should not exceed 8 hours, nor should the primary library be amplified more than once. As an alternative, we recommend amplification of the λ library on plates rather than in liquid culture. Several 1-ml aliquots of the λ library should be frozen in 7% DMSO and stored at –80°C.

PROTOCOL

1. Package the remainder of the LC-Fd insert ligation reaction with λ packaging extract.

   Note: Do not package the remainder of the ligation until you are ready to amplify the resulting λ library; it is preferable to leave the ligation reaction at 4°C for a few days rather than to store the packaged but unamplified λ library.

2. Pool the packaging reactions and titer. Combine the primary library phage with 1 ml of fresh, early-log LE392 cells (resuspended in 10 mM $MgSO_4$ to an OD_{600nm} of 5.00) for every 5×10^6 pfu of phage and incubate for 15 minutes at 37°C. Fresh, early-log cells are obtained by diluting an overnight culture 50-fold in LB (Sambrook et al. 1989) and incubating at 37°C, shaking at 250 rpm, for 2–3 hours until OD_{600nm} ≈0.3. Cells are centrifuged and resuspended to the desired OD in 10 mM $MgSO_4$. Old or late-log cultures tend to have nonviable cells that act as unproductive sinks for phage or phagemid.

3. Add 50 ml of NZY media and incubate the culture at 37°C, shaking at 250 rpm, for 4–6 hours until lysis is complete. The culture will become clear and particulate cell debris will be evident.

Alternative Step 3: Plate amplification (also see Sambrook et al. 1989)

a. Plate the primary library phage in fresh cells at a density of 5×10^5 pfu/plate on 150-mm NZY agar plates with 10 ml of top agar.

b. Incubate for 3–5 hours at 37°C. Check the plates every 15 minutes after a hazy lawn appears. Remove the plates after pinhead-sized plaques appear.

c. Add 10 ml of SM to each plate and gently rock overnight at 4°C (~16 hours). Collect the liquid in polypropylene or glass tubes.

4. Add 0.5 ml of chloroform to the amplified phage and incubate for 15 minutes at 37°C with agitation.

5. Centrifuge at 5000 rpm for 10 minutes at 4°C. Transfer the supernatant to a fresh tube and store at 4°C. Discard the pellet.

6. Titer dilutions of the amplified λ library as described above. Titers are typically in the range of 10^9–10^{11} pfu/ml. λ phage libraries are fairly stable stored at 4°C. Typically, the titer drops 10-fold in the first week and 10-fold every 6 months thereafter.

7. If desired, the percentage of clones with Fab inserts may be determined in a plaque lift assay as described previously (Mullinax et al. 1990). The expression of κ light chain is detected by probing nitrocellulose plaque lifts with rabbit anti-human κ light chain (Cappell), followed by alkaline phosphatase-coupled goat anti-rabbit IgG and a color-developing substrate.

In Vivo Mass Excision of the pSurfScript Phagemid Library

The amplified λ phage libraries are converted to pSurfScript phagemid libraries by coinfecting XL1-Blue cells with λ phage and ExAssist helper phage (in vivo mass excision) (Hogrefe et al. 1993b; Short and Sorge 1993). The representation of the library should not be significantly altered if excision times are limited to ≤3 hours.

PROTOCOL

1. Grow overnight cultures of XL1-Blue cells and SOLR (supO-λ resistant strain) cells in LB.

2. Spin and resuspend the SOLR cells in 10 mM $MgSO_4$ to an OD_{600nm} of 1.0 and store at 4°C until needed in step 6. Dilute an aliquot of the XL1-Blue cells 1/50 in fresh LB media and grow for several hours to early-log (OD_{600nm} ≈0.3). Spin and resuspend the fresh, early-log XL1-Blue cells to an OD_{600nm} of 5.0.

3. In a 50-ml polypropylene conical tube, combine 10^9 pfu of amplified λ phage library with 2.5 ml of fresh, early-log XL1-Blue cells. Add a minimum of 2×10^{10} pfu of ExAssist helper phage, and incubate for 15 minutes at 37°C. The ratio of λ:cells:M13 should be 1:10:20.

4. Add 20 ml of LB media and incubate for 15 minutes at 37°C with no agitation and then for 3 hours at 37°C, with gentle shaking, to limit breakage of the *E. coli* pili.

5. Incubate the culture for 20 minutes at 70°C. Spin at 5000 rpm for 10 minutes at 4°C and transfer the supernatant to a fresh tube. The excised phagemid particles may be stored at 4°C for up to several months.

6. The titer of the excised pSurfScript phagemid library is determined as the total number of ampicillin-resistant colonies (cfu) in the excision supernatant. Dilute a small aliquot of the excision supernatant in TE and mix 10 μl with 200 μl of SOLR cells. Incubate for 15 minutes at 37°C and spread onto LB plates containing 100 μg/ml ampicillin. Incubate overnight at 37°C and count the resulting colonies.

Preparation of M13 Particles Displaying Fab for Biopanning

A high-titer stock of recombinant phagemid particles for biopanning is prepared by superinfecting cells harboring the pSurfScript phagemid libraries with helper phage. The M13 phagemid particles produced by this procedure should display Fab and contain the DNA encoding the light and heavy chains. Specific antigen-binding clones may then be isolated by biopanning the phagemid display libraries. Biopanning may be performed by a number of techniques described elsewhere (Hogrefe et al. 1993a,b; Hogrefe and Shopes 1994).

PROTOCOL

1. Grow an overnight culture of SOLR cells.

2. Prepare fresh, early-log SOLR cells, spin and resuspend in 10 mM $MgSO_4$ to an OD_{600nm} of 1.0.

3. In a 250-ml flask, combine a portion of the excision supernatant representing 10^8 cfu (use at least tenfold more excised phagemid than members of the primary λ library) with 1 ml of SOLR cells (use at least twofold more cells than excised phagemid). Incubate for 15 minutes at 37°C. Add 50 ml of LB media containing 100 μg/ml carbanicillin (similar to ampicillin; Sigma) and 70 μg/ml kanamycin.

4. Grow the culture at 37°C, 250 rpm, to an OD_{600nm} of 0.1.

5. Add approximately 10^{10} pfu VCSM13 helper phage (Stratagene). Incubate at 37°C for 15 minutes, then overnight with shaking at 30°C.

6. Spin the single-strand rescue culture at 5000 rpm for 10 minutes at 4°C. Transfer equal volumes of 25 ml each to two polypropylene centrifuge tubes (Nalgene #3100-0500 or the equivalent).

7. Precipitate the phagemid particles by adding 5 ml 30% PEG-8000 (in 1.5 M NaCl) dropwise to each tube. Invert the tubes several times to thoroughly mix. Incubate at room temperature for 15 minutes. Centrifuge at 10,000 rpm for 20 minutes at 4°C (SS-34 rotor, Sorvall). Carefully decant the supernatant and resuspend the pellet in 1 ml of 10 mM Tris-HCl, pH 7.5/1 mM EDTA (TE). Spin at 10,000 rpm in a 1.5-ml polypropylene tube for 10 minutes to pellet residual cell debris. Discard the pellet and transfer the supernatant to a fresh 1.5-ml polypropylene tube.

8. If desired, reprecipitate the phage by adding 200 µl of 30% PEG-8000 (in 1.5 M NaCl) to the 1 ml of supernatant. Spin down the phage pellet in a microfuge at 10,000 rpm and decant the supernatant. Resuspend the pellet in 100 µl of TE.

9. Aliquot the phagemid display library and store at 4°C. Titer the phagemid particles as ampicillin-resistant cfu/ml in XL1-Blue cells.

Enrichment of Binding Clones in the Phage Display Library

Although it is straightforward to provide protocols for the creation of a library, it is somewhat more difficult to give guidance on enrichment procedures because the constraints vary from experiment to experiment. Every phagemid display library enrichment protocol differs. The design of the experiment depends on the antigen and the source of the antibody library. The process of enrichment can be divided into three parts: binding of phagemids to a target, desorption of nonspecifically bound phagemid, and recovery of specifically bound phagemid.

Our group and many others have found that biopanning, using ligand- or antigen-coated polystyrene (plates or dishes) or even whole cells, can be useful for the enrichment of phagemid display libraries. However, we have found column affinity chromatography of phagemid libraries, or biochromatography, to be both efficient and versatile for enriching binding clones. In general, biochromatography is most successful with purified protein antigens or with small ligands. A column format satisfies most of the criteria for a successful enrichment. The elution of phagemid can be accomplished by changing to a condition that disfavors the binding interaction. Low pH, high pH, free ligand, chaotropic agents, and detergents may serve to elute the bound phagemids.

PROTOCOL

The following is a suggested protocol for biochromatography but likely will need to be adjusted for your particular antigen:

1. Couple the antigen to Sepharose CNBr-CL-4B (Pharmacia) at 1 mg/ml according to the manufacturer's protocol.

2. Pour a 0.2-ml column of protein-Sepharose. A disposable column is convenient; for example, a 10-ml Econo-Column (Bio-Rad Laboratories). Wash with 25 ml of high-salt buffer, 20 mM Tris-HCl, 500 mM NaCl, pH 7.5 (TBHS).

3. Load 1 ml of concentrated display phagemid from step 7 of the previous protocol (~10^{12}/ml). Collect flowthrough and reload twice more. Wash with 200 ml of TBHS.

4. Elute the bound phagemid with 2 ml of 0.1 N HCl (to pH 2.2 w/glycine), 0.1% BSA. Collect fractions and neutralize each eluate to pH 7.5 with small additions of 2 N Tris base.

5. Titer for cfu/ml as above. If another round of enrichment is desired, add the eluted phage sample to 1 ml of fresh early-log XL1-Blue cells (OD_{600nm} = 1.0). Incubate for 15 minutes at 37°C and add 9 ml of LB broth containing 100 mg/ml of carbanicillin. Incubate for 1 hour at 37°C. Add VCSM13 helper phage to the culture at a minimum moi of 1:1 to 10:1 helper phage to host cells. Incubate for 1 hour at 37°C. Add kanamycin to a final concentration of 50 μg/ml. Continue the incubation at 30°C until late-log phase is reached (OD_{600nm} ≈ 1.0 in ~16–24 hours).

6. PEG-precipitate the phagemid particles as described above to a volume of ~1 ml.

7. Repeat the biopanning or screen the amplified phagemid samples for evidence of enrichment.

SUMMARY

The methodologies described here and elsewhere (Huse et al. 1989; McCafferty et al. 1990; Mullinax et al. 1990; Barbas et al. 1991, 1992; Burton et al. 1991; Garrard et al. 1991; Hoogenboom et al. 1991; Marks et al. 1991b; Gram et al. 1992; Zebedee et al. 1992; Hogrefe et al. 1993a,b; Orum et al. 1993; B. Shopes, in prep.) for constructing phage display libraries have allowed the identification of a number of specific, high-affinity Fab clones when the immunoglobulin sequences are derived by PCR from B-cell RNA.

ACKNOWLEDGMENTS

The authors thank Amy Lovejoy, Bev Hay, Jeff Amberg, and Martin Gore, all formerly of Stratacyte Corporation, for their contributions to this and previous work. Antibody primers are frequently revised, and the latest versions can be obtained from Tera Biotechnology Corporation. B.S. acknowledges assistance from the National Institutes of Health (AI-32822, AI-33250, CA-59077, and CA-57040).

REFERENCES

Barbas, C.F., J.D. Bain, D.M. Hoekstra, and R.A. Lerner. 1992. Semisynthetic combinatorial antibody libraries: A chemical solution to the diversity problem. *Proc. Natl. Acad. Sci.* **89:** 4457–4461.

Barbas, C.F., A.S. Kang, R.A. Lerner, and S.J. Benkovic. 1991. Assembly of combinatorial antibody expression libraries on phage surfaces: The gene III site. *Proc. Natl. Acad. Sci.* **88:** 7978–7982.

Better, M., C.P. Chang, R.R. Robinson, and A.H. Horwitz. 1988. *Escherichia coli* secretion of an active chimeric antibody fragment. *Science* **240:** 1041–1043.

Burton, D.R., C.F. Barbas, M.A.A. Persson, S. Koenig, R.M. Chanock, and R.A. Lerner. 1991. A large array of human monoclonal antibodies to type I human immunodeficiency virus from combinatorial libraries of asymptomatic seropositive individuals. *Proc. Natl. Acad. Sci.* **88:** 10134–10137.

Falkoff, R.J.M., L.-P. Zhu, and A.S. Fauci. 1983. The relationship between immunization and circulating antigen-specific plaque-forming cells. *Cell. Immunol.* **78:** 392–399.

Garrard, L.J., M. Yang, M.P. O'Connell, R.F. Kelley, and D.J. Henner. 1991. Fab assembly and enrichment in a monovalent phage display system. *BioTechnology* **9:** 1373–1377.

Gram, H., L.-A. Marconi, C.F. Barbas, T.A. Collet, R.A. Lerner, and A.S. Kang. 1992. *In vitro* selection and affinity maturation of antibodies from a naive combinatorial immunoglobulin library. *Proc. Natl. Acad. Sci.* **89:** 3576–3580.

Hogrefe, H.H. and B. Shopes. 1994. Construction of phagemid display libraries with PCR-amplified immunoglobulin sequences. *PCR Methods Appl.* **4:** S109–S122.

Hogrefe, H.H., J.R. Amberg, B.N. Hay, J.A. Sorge, and B. Shopes. 1993a. Cloning in a bacteriophage lambda vector for the display of binding proteins on filamentous phage. *Gene* **137:** 85–91.

Hogrefe, H.H., R.L. Mullinax, A.E. Lovejoy, B.N. Hay, and J.A. Sorge. 1993b. A bacteriophage lambda vector for the cloning and expression of immunoglobulin Fab fragments on the surface of filamentous phage. *Gene* **128:** 119–126.

Hoogenboom, H.R., A.D. Griffiths, K.S. Johnson, D.J. Chiswell, P. Hudson, and G. Winter. 1991. Multi-subunit proteins on the surface of filamentous phage: Methodologies for displaying antibody (Fab) heavy and light chains. *Nucleic Acids Res.* **19:** 4133–4137.

Huse, W.D., L. Sastry, S.A. Iverson, A.S. Kang, M. Alting-Mees, D.R. Burton, S.J. Benkovic, and R.A. Lerner. 1989. Generation of a large combinatorial library of the immunoglobulin repertoire on phage lambda. *Science* **246:** 1275–1281.

Kabat, E.A., T.T. Wu, M. Reid-Miller, H.M. Perry, and K.S. Gottesmann. 1987. *Sequences of proteins of immunological interest*, 4th edition. U.S. Department of Health and Human Services, U.S. Government Printing Services.

Larrick, J.W., L. Danielsson, C.A. Brenner, M. Abrahamson, K.E. Fry, and C.A.K. Borrebaeck. 1989a. Rapid cloning of rearranged immunoglobulin genes from human hybridoma cells using mixed primers and the polymerase chain reaction. *Biochem. Biophys. Res. Commun.* **160:** 1250–1256.

Larrick, J.W., L. Daniellson, C.A.B. Brenner, E.F. Wallace, M. Abrahamson, K.E. Fry, and C.A.K. Borrebaeck. 1989b. Polymerase chain reaction using mixed primers: Cloning of human monoclonal antibody variable region genes from single hybridoma cells. *BioTechnology* **7:** 934–938.

LeBoeuf, R.D., F.S. Galin, S.K. Hollinger, S.C. Peiper, and J.E. Blalock. 1989. Cloning and sequencing of immunoglobulin variable-region genes using degenerate oligodeoxyribonucleotides and polymerase chain reaction. *Gene* **82:** 371–377.

Marks, J.D., M. Tristem, A. Karpus, and G. Winter. 1991a. Oligonucleotide primers for polymerase chain reaction amplification of human immunoglobulin variable genes and design of family-specific oligonucleotide probes. *Eur. J. Immunol.* **21:** 985–991.

Marks, J.D., H.R. Hoogenboom, Y.P. Bonnert, J. McCafferty, A.D. Griffiths, and G. Winter. 1991b. By-passing immunization: Human antibodies from V-gene libraries displayed on phage. *J. Mol. Biol.* **222:** 581–597.

McCafferty, J., A.D. Griffiths, G. Winter, and D.J. Chiswell. 1990. Phage antibodies: Filamentous phage displaying antibody variable domains. *Nature* **348:** 552–554.

Mullinax, R.L., E.A. Gross, J.R. Amberg, B.N. Hay, H.H. Hogrefe, M.M. Kubitz, A. Greener, M. Alting-Mees, D. Ardourel, J.M. Short, J.A. Sorge, and B. Shopes. 1990. Identification of human antibody fragment clones specific for tetanus toxoid in a bacteriophage lambda immunoexpression system. *Proc. Natl. Acad. Sci.* **87:** 8095–8099.

Orlandi, R., D.H. Gussow, P.T. Jones, and G. Winter. 1989. Cloning immunoglobulin variable domains for expression by the polymerase chain reaction. *Proc. Natl. Acad. Sci.* **86:** 3833–3837.

Orum, H., P.S. Andersen, A. Oster, L.K. Johansen, E. Riise, M. Bjornvad, I. Svendsen, and J. Engberg. 1993. Efficient method for constructing comprehensive murine Fab antibody libraries displayed on phage. *Nucleic Acids Res.* **21:** 4491–4498.

Parmley, S.F. and G.P. Smith. 1988. Antibody-selectable filamentous fd phage vectors: Affinity purification of target genes. *Gene* **73:** 305–318.

Riechmann, L. and M. Weill. 1993. Phage display and selection of a site-directed randomized single-chain antibody Fv fragment for its affinity improvement. *Biochemistry* **32:** 8848–8855.

Sambrook, J., E.F. Fritsch, and T. Maniatis. 1989. *Molecular cloning: A laboratory manual*, 2nd edition. Cold Spring Harbor Laboratory Press, Cold Spring Harbor, New York.

Sassano, M., M. Repetto, G. Cassani, and A. Corti. 1994. PCR amplification of antibody variable regions using primers that anneal to constant regions. *Nucleic Acids Res.* **22:** 1768–1769.

Sastry, L., M. Alting-Mess, W.D. Huse, J.M. Short, J.A. Sorge, B.N. Hay, K.D. Janda, S.J. Benkovic, and R.A. Lerner. 1989. Cloning of the immunological repertoire in *Escherichia coli* for generation of monoclonal catalytic antibodies: Construction of a heavy chain variable region-specific cDNA library. *Proc. Natl. Acad. Sci.* **86:** 5728–5732.

Scott, J.K. and G.P. Smith. 1990. Searching for peptide ligands with an epitope library. *Science* **249:** 386–390.

Short, J.M. and J.A. Sorge. 1993. *In vivo* excision properties of bacteriophage lambda ZAP[R] expression vectors. *Methods Enzymol.* **216:** 495–508.

Short, J.M., J.M. Fernandez, J.A. Sorge, and W.D. Huse. 1988. Lambda ZAP: A bacteriophage lambda expression vector with *in vivo* excision properties. *Nucleic Acids Res.* **16:** 7583–7600.

Skerra, A. and A. Pluckthun. 1989. Assembly of a functional immunoglobulin Fv fragment in *Escherichia coli. Nature* **341:** 1038–1041.

Stevens, R.H., E. Macy, C. Morrow, and A. Saxon. 1979. Characterization of a circulating subpopulation of spontaneous anti-tetanus toxoid antibody producing B cells following *in vivo* booster immunization. *J. Immunol.* **122:** 2498–2504.

Ward, E.S., D. Gussow, A.D. Griffiths, P.T. Jones, and G. Winter. 1989. Binding activities of a repertoire of single immunoglobulin variable domains secreted from *Escherichia coli. Nature* **341:** 544–546.

Zebedee, S.L., C.F. Barbas, Y.-L. Hom, R.H. Caothien, R. Graff, J. Degraw, J. Pyati, R. Lapolla, D.R. Burton, R.A. Lerner, and G.B. Thornton. 1992. Human combinatorial antibody libraries to hepatitis B surface antigen. *Proc. Natl. Acad. Sci.* **89:** 3175–3179.

Zhou, H., R.J. Fisher, and T.S. Papas. 1994. Optimization of primer sequences for mouse scFv repertoire display library construction. *Nucleic Acids Res.* **22:** 888–889.

PCR Sequencing

The standard DNA sequencing protocol based on chain termination using dideoxynucleotide triphosphates, first described by Sanger, has been modified by the availability of thermostable DNA polymerases. The two major advantages of using a thermostable enzyme in what is known as "cycle sequencing" over other DNA polymerases, such as Sequenase, used in standard DNA sequencing are the decreased amounts of template required to obtain a readable sequence ladder and the high temperature at which the reactions are performed. Carrying out sequencing reactions at higher temperatures facilitates the sequencing of double-strand templates, including PCR products, plasmids, lambda DNA, and cosmids, and is preferred for templates with a high G-C base content.

Two differences between cycle sequencing and a standard PCR should be noted. In cycle sequencing, only one primer is required and both dideoxynucleotide triphosphates and deoxynucleotide triphosphates are employed. Through the use of a single primer, the amplification of the template DNA is linear, as opposed to exponential. This linear amplification generates a signal corresponding to the template positions at which molecules terminate.

The DNA present in crude samples can be sequenced using a two-step process in which the DNA of interest is first PCR-amplified, followed by a cycle sequencing reaction. One of the two primers used to amplify this DNA, or a third primer that hybridizes to the PCR product, can be used for sequencing. This approach, described in detail in "Direct Sequencing of PCR-amplified DNA," bypasses the need to clone the amplified DNA prior to sequencing and is particularly useful when amplification reactions are performed with thermostable polymerases that do not possess proofreading activity. Direct sequencing of the amplified DNA guarantees that the obtained sequence corresponds to that of the template, because any particular mutation is "diluted" within the overall population of molecules. Alternatively, when a PCR product is cloned prior to sequencing, more than one clone should be sequenced, because mutations generated by nucleotide misincorporations during the PCR may be carried by the clone selected. Also, when cloning a PCR product prior to sequencing, the use of thermostable polymerases that have proofreading activity is recommended to minimize nucleotide misincorporations.

"Cycle Sequencing" describes the preparation of different DNA templates for cycle sequencing. It also discusses modifications of basic cycle sequencing techniques aimed at maximizing the length of the readable sequence data.

Direct Sequencing of PCR-amplified DNA

Venigalla B. Rao

Department of Biology, Institute for Biomolecular Studies, The Catholic University of America, Washington, DC 20064

INTRODUCTION

The amplification of target DNA by PCR (Mullis et al. 1986) followed by the direct sequencing of amplified DNA (Gyllensten 1989) has emerged as a powerful strategy for rapid molecular genetic analysis. Using this strategy, time-consuming cloning steps can be completely bypassed and the sequence of the target DNA can be determined directly from a crude biological sample. The crude sample can be cultured cells, bacteria, or a viral preparation. Furthermore, the copy number of target DNA in the sample can be as low as one to a few molecules of genomic DNA among a vast excess of contaminating nontarget DNA (Mullis 1991; Saunders et al. 1993).

This review describes two sequencing strategies that allow the generation of the DNA sequence from almost any PCR-amplified DNA template. In addition, factors that influence the sequencing reactions are discussed, which allow the manipulation of these strategies for specific sequencing needs. For a more comprehensive analysis of the direct sequencing strategy, the reader is referred to reviews by Gyllensten (1989) and Rao (1994).

SEQUENASE STRATEGY

This strategy consists of three steps (Fig. 1) (Tabor and Richardson 1987). In the first step, the double-stranded, PCR-amplified DNA is denatured to single strands and the sequencing primer is annealed to the complementary sequence on one of the template strands. In the second step, the annealed primer is extended by 20–80 nucleotides by DNA polymerase, incorporating multiple radioactive labels into the newly synthesized DNA. This step is performed under nonoptimal reaction conditions so that the enzyme acts in a low-processive fashion, synthesizing only short stretches of DNA. In the third step, the labeled DNA chains are extended and terminated by the incorporation of the ddNMP (Fig. 1).

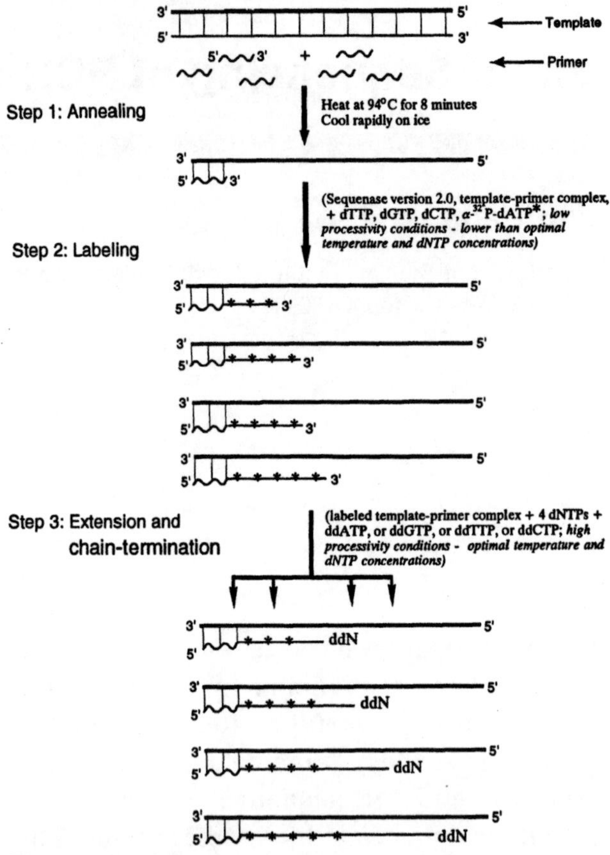

Figure 1 Schematic diagram of the Sequenase protocol. The wavy line represents the primer. The template strand for sequencing of the PCR-amplified DNA is represented as a thick line. The second strand is represented as a thin line. The second strand is shown only in the template DNA at the top of the figure. It is not included in the rest of the figure although it is present in the reaction mixture. The cross bars between the strands of the template DNA, and between the template and the primer, represent the classical Watson-Crick hydrogen bonds. The asterisks represent the radioactive labels incorporated as [^{32}P]dAMP into the newly synthesized DNA chains.

REAGENTS

DNA template: The PCR-amplified DNA should be first separated from the unused dNTPs and primers by QIAquick ion-exchange column chromatography (QIAGEN). The concentration of recovered DNA template can be estimated by agarose gel electrophoresis and ethidium bromide staining of a small aliquot of the purified DNA and comparing the intensity of the stained band with that of a standard DNA of known concentration. Alternately, the DNA concentration can be estimated by using a DNA dipstick (Invitrogen). A DNA concentration in the range of about 1 pmole is desirable to generate high-intensity sequence ladders. However, concentrations as low as 0.1 pmole can be used.

Sequencing primer: A 20-nucleotide DNA primer synthesized by an oligonucleotide synthesizer can be used directly as a sequencing primer without any purification steps involving high-performance liquid chromatography or polyacrylamide gel electrophoresis. The concentration of the primer is estimated by spectroscopy at 260 nm (one absorbance unit is equivalent to a primer concentration of 33 μg/ml).

DNA polymerase: Sequenase, version 2.0 (USB/Amersham), is the most suitable enzyme for performing this protocol. Sequenase, version 2.0, is a genetically modified phage T7 DNA polymerase that has no 3′→ 5′ exonuclease activity. Sequenase is active at low temperatures and incorporates multiple radioactive labels efficiently to generate high-intensity sequence ladders.

Sequencing buffer: The composition of the sequencing buffer varies with the polymerase used. For Sequenase, the buffer is 40 mM Tris-HCl, pH 7.5, 20 mM $MgCl_2$, and 50 mM NaCl.

Radiolabeled dNTP: Approximately 5 μCi of either [α-^{32}P]dATP, [α-^{33}P]dATP, or [α-^{35}S]dATP is used for each set of sequencing reactions. However, [α-^{32}P]dATP is the preferred radionucleotide for this protocol.

PROTOCOL

The following is a Sequenase protocol (USB/Amersham) modified for sequencing PCR-amplified DNA.

1. Set up the extension-termination reaction mixtures. Transfer 2.5 μl of each of the four dNTP/ddNTP mixtures to four independent tubes. Each tube receives 80 μM of each of the four dNTPs and 8 μM of either ddATP, ddTTP, ddGTP, or ddCTP. Preincubate these tubes for 5 minutes at 37°C before initiating the extension and chain-termination reactions (see step 4 below).

   For sequencing GC-rich templates, dGTP should be replaced with 7-deaza-dGTP (USB/Amersham) to overcome compression artifacts that are known to occur during sequencing gel electrophoresis (Barr et al. 1986). dGTP can also be replaced with dITP, but dITP may not be a good substrate for all DNA polymerases.

2. Mix about 1 pmole of PCR-amplified DNA, 10 pmoles of sequencing primer, 2 μl of 5X sequencing buffer, and H_2O in a total reaction volume of 10 μl. The addition of nonionic detergents such as Tween 20 and NP-40 to a final concentration of 0.5% has been reported to improve the specificity of annealing of the sequencing primer to the template (Bachmann et al. 1990). Incubate the samples in a heat block for 8 minutes at 94–96°C. Chill the tubes on ice for 1 minute (Kusukawa et al. 1990). Centrifuge in a micro-

fuge at 10,000 rpm for 10 seconds. Transfer the tubes to ice, and immediately proceed to the next step.

Note that the above annealing conditions are applicable only to double-stranded, PCR-amplified DNA templates. For single-stranded templates, such as the products of asymmetric PCR (Gyllensten and Erlich 1988), the gradual cooling technique is preferred. In this procedure, the above mixture is incubated for 6 minutes in a 65°C heat block and is gradually cooled to 30°C by turning off the heat block.

3. To the above mixture on ice, add 2 µl of cold dNTP mix containing 0.75 µM each of dTTP, dGTP, and dCTP, and 5 µCi of [α-^{32}P]dATP (sp. act. 3000 Ci/mmole, Amersham), 1 µl of 0.1 M dithiothreitol, H_2O, and 2 units of freshly diluted Sequenase, version 2.0, to a final reaction volume of 15.5 µl. Incubate the reaction mixture on ice for 2 minutes to label the DNA.

4. Transfer 3.5 µl of the above mixture to each of the four dNTP/ddNTP tubes that have been preincubated for 5 minutes at 37°C (from step 1). Allow the extension and chain-termination reactions to proceed for 5 minutes at 37°C.

5. Stop the extension and chain-termination reactions by adding 4 µl of Stop buffer containing 95% formamide, 20 mM EDTA, 0.05% bromophenol blue, and 0.05% xylene cyanol FF. If the samples are not used immediately, they can be stored frozen at –70°C for about a week. However, it is preferable to use the samples within 2 days. Heat the samples for 3 minutes in an 80°C heat block, and load a 2- to 3-µl aliquot of the sample in each lane of a sequencing gel.

Important Characteristics of This Protocol

1. The labeling step is the most critical step in this protocol and should be performed under well-controlled conditions. In this step, the annealed primer is extended by only 20–80 nucleotides to incorporate multiple radioactive labels into the newly synthesized DNA. Because the Sequenase enzyme is highly processive and synthesizes several thousand nucleotides (~4000 nucleotides with version 2.0) at a stretch before dissociating from the complex, this step should be performed under low-processivity conditions such as low temperature and low concentrations of dNTPs (Tabor and Richardson 1987).

2. A common pitfall of this procedure is the appearance of either very faint sequence ladders or no sequence ladders on the final autoradiogram. However, an intense band is seen in the high-

molecular-weight position. This characteristic pattern is due to a lack of control at the labeling step. If the labeling reactions are not controlled well, the Sequenase enzyme, instead of synthesizing a short stretch of DNA, predominantly synthesizes the full-length product. Consequently, the sequence ladders representing the shorter chain-terminated products constitute a minor fraction and therefore appear as faint bands. This can be controlled by modifying the labeling conditions in such a way that the enzyme acts in a low-processive manner. These modifications include decreasing the enzyme concentration, decreasing the unlabeled dNTP concentration, reducing the reaction time, and increasing the template concentration.

3. This protocol requires the incorporation of at least one radioactive label into the newly synthesized DNA to visualize a band on the autoradiogram. Therefore, the sequence of the first nucleotide that can be determined by this protocol depends on the distance between the 3′ end of the primer and the first labeled [^{32}P]dAMP incorporated. If the template is known to be, or suspected to be, a GC-rich template, the use of [α-^{32}P]dCTP rather than [α-^{32}P]dATP is recommended.

CYCLE SEQUENCING STRATEGY

Based on the report by Murray (1989), a number of PCR-directed sequencing strategies (Lee 1991; Ruano and Kidd 1991; Rao and Saunders 1992), referred to as cycle sequencing strategies, have been developed for the direct sequencing of PCR-amplified DNA. These strategies take advantage of the powerful automated cycling capability of thermal cyclers to amplify chain-terminated sequencing products and generate high-intensity sequence ladders. Each sequencing cycle consists of three steps (Fig. 2). First, the PCR-amplified DNA is denatured to single strands. This is followed by the annealing of a ^{32}P-labeled sequencing primer (or a biotinylated primer) to the complementary sequence on one of the strands. In the final step, the annealed primer is extended and chain-terminated by a thermostable DNA polymerase. The resulting partially double-stranded chain-terminated product is then denatured in the next sequencing cycle, releasing the template strand for another round of priming reactions, while accumulating chain-terminated products in each cycle. These steps are repeated for 20–40 cycles to amplify the chain-terminated products in a linear fashion (Fig. 2).

REAGENTS

DNA template: As in the Sequenase protocol, the PCR-amplified DNA should be first purified by QIAquick column chromatography (QIAGEN Inc.) to remove unused dNTPs and primers. It is possible to use 1–2 μl of the unpurified PCR product directly for cycle sequenc-

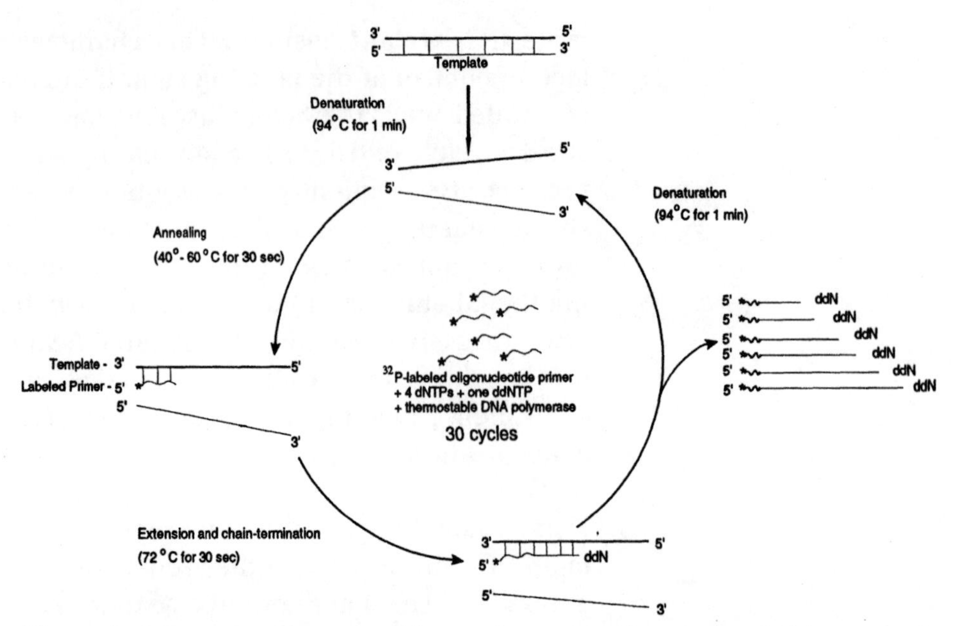

Figure 2 Schematic diagram of the cycle sequencing protocol. The wavy line represents the primer. The template strand for sequencing of the PCR-amplified DNA is represented as a thick line. The second strand is represented as a thin line. The cross bars between the strands of the template DNA, and between the template and the primer, represent the classical Watson-Crick hydrogen bonds. The asterisks represent the ^{32}P-label incorporated into the 5′ end of the primer as a result of phosphorylation by T4 polynucleotide kinase.

ing. However, the overall quality of the sequence ladders generated is not as good as with the purified DNA. Therefore, the purification of amplified DNA is recommended. When sequencing a large number of templates routinely, the purification step can be bypassed by using the recently described modified cycle sequencing strategies (Ruano and Kidd 1991; Rao and Saunders 1992). In these strategies, lower concentrations of dNTPs (10–20 μM of each dNTP) and primers (10 pmoles of each primer) are used for the amplification of the target DNA. As a result, the carryover of unused dNTPs and primers into the sequencing reactions is minimized. It is desirable to have a template DNA concentration of at least 0.1 pmole for each set of sequencing reactions to generate high-intensity sequencing ladders. Lower template concentrations can be used, but the sequence ladders generated will be of low intensity.

Sequencing primer: The sequencing primer is labeled with ^{32}P at the 5′ end using [γ-^{32}P]ATP and T4 polynucleotide kinase. The ratio of primer, kinase, and [γ-^{32}P]ATP should be maintained at an optimal level to achieve high-specific-activity labeling of the sequencing primer (see below). High specific activity of the primer, but not the total activity, is critical for generating high-intensity sequence ladders. Alternatively, it is possible to use [α-^{32}P]dATP or [α-^{33}P]dATP to label newly synthesized DNA, rather than using a 5′-labeled primer (Promega). However, this generates higher background (see below). In addition, there is a greater degree of nonuniformity in the intensity of the bands generated. For example, chain-terminated products of high molecular weight incorporate radioactive labels at severalfold higher frequency when compared to those of low molecular weight.

DNA polymerase: Any thermostable DNA polymerase that lacks 3′→ 5′ exonuclease activity can be used. *Taq* DNA polymerase (Perkin-Elmer), which has no 3′→ 5′ exonuclease activity, is the most widely used enzyme (Innis et al. 1988). Other enzymes that are used for cycle sequencing include *Pfu* polymerase (exonuclease-minus) (Stratagene), Vent polymerase (exonuclease-minus) (New England Biolabs), and *Tub* polymerase (Amersham).

PROTOCOL

The following is a basic protocol for *Taq* DNA polymerase. The same can be used as well for other thermostable polymerases, but the buffer conditions and dNTP/ddNTP ratios should be modified accordingly for each polymerase used. Many factors affect the optimal dNTP/ddNTP ratios required for generating high-intensity sequence ladders with low background (see below). Therefore, it is highly recommended that the dNTP/ddNTP ratios be optimized for a given sequencing application.

1. For labeling the primer, mix 10–15 pmoles of sequencing primer, 50 μCi of [γ-^{32}P]ATP (sp. act. 6000 Ci/mmole or >6000 Ci/mmole [NEG-035C, NEN]), 5 μl of 10X kinase buffer (70 mM Tris-HCl, pH 7.5, 10 mM $MgCl_2$, and 5 mM dithiothreitol, or buffer supplied by the vendor), and H_2O, to a final reaction volume of 50 μl. Preincubate the reaction mixture for 5 minutes at 37°C. Add 1 μl of freshly diluted T4 polynucleotide kinase (10 units) and incubate for 30 minutes at 37°C. Add an additional 1-μl aliquot of freshly diluted kinase (10 units) and continue incubation for 30 minutes at 37°C. The removal of unincorporated [γ-^{32}P]ATP, although not critical, is recommended. This can be accomplished by gel filtration through a Biospin-10 column (Bio-Rad Laboratories). The labeled primer can be stored at –70°C for at least 2 weeks.

2. Transfer 2 μl of each of the four dNTP/ddNTP extension-termination mixtures to four tubes. Each tube receives 30 μM of each of four dNTPs and either 1.2 mM of ddATP, 1.2 mM of ddTTP, 90 μM of ddGTP, or 600 μM of ddCTP. As in the Sequenase protocol, dGTP should be replaced with 7-deaza-dGTP (USB/Amersham) for

GC-rich templates to overcome the compression artifacts that are known to occur during sequencing gel electrophoresis (Barr et al. 1986).

3. Mix 0.1–0.2 pmole of QIAGEN-purified PCR-amplified DNA, 1–2 pmoles of 5′ ^{32}P-labeled sequencing primer, 3 µl of 10X sequencing buffer (500 mM Tris-HCl, pH 8.8, and 20 mM $MgCl_2$, or buffer supplied by the vendor), and 5 units of *Taq* DNA polymerase in a total reaction volume of 20 µl. Reagents such as dimethyl sulfoxide, Triton X-100, Tween 20, or NP-40 may be added to this mixture to enhance the quality of sequence ladders (Bachmann et al. 1990).

4. Transfer 4 µl of the above mixture to each of the four tubes containing dNTP/ddNTP mixtures (from step 2). Mix the contents and layer with 20 µl of mineral oil.

5. Place the tubes in a thermal cycler that has been preheated to 94°C to initiate thermal cycling. Each cycle consists of denaturation for 1 minute at 94°C, annealing for 30 seconds at 40–60°C, and extension and chain-termination for 30 seconds at 72°C. These steps are repeated for 20–40 cycles.

6. After the completion of thermal cycling, add 4 µl of Stop buffer containing 95% formamide, 20 mM EDTA, 0.05% bromophenol blue, and 0.5% xylene cyanol FF to the reaction mixture. Mix and centrifuge for a few seconds in a microfuge to separate the layers. If the samples are not used immediately, they can be stored frozen at −70°C for about a week. However, it is preferable to use the samples within 2 days. Heat the samples for 3 minutes in an 80°C heat block, and load a 2–3 µl aliquot of the aqueous phase in a single well of the sequencing gel.

Manipulation of the Size Range of Sequence Ladders

Most of the compositions of dNTP/ddNTP mixtures reported in the literature or available commercially are optimized to generate sequence ladders of high intensity and uniformity in the range of 50–200 nucleotides from the 3′ end of the primer. Generally, the sequence ladders closer to the primer (1–50 nucleotides from the 3′ end of the primer) are of low intensity under these conditions. If high-intensity sequence ladders are desired very close to the primer (in the 1- to 100-nucleotide range), Mn^{++} can be added to the sequencing reactions (Tabor and Richardson 1989). DNA polymerases incorporate ddNTPs about five- to tenfold more frequently in the presence of Mn^{++}. Because Mn^{++} effects are seen in the presence of Mg^{++}, no changes to the basic protocols are required other than the addition of Mn^{++}. Different polymerases require different Mn^{++} concentrations for optimal results (Tabor and Richardson 1989; Rao and Saunders 1992). A Mn^{++} concentration of 3.5 mM is optimal for Sequenase for increasing the frequency of terminations by about fivefold. (Mn^{++}, prepared as a stock solution of 100 mM $MnCl_2$ in 150 mM sodium isocitrate, is available from USB/Amersham.) If it is desired to generate sequence ladders far from the primer (in the 200- to 400-nucleotide range), the dNTP/ddNTP ratios should be increased by simply adding an appropriate aliquot (this varies depending on the polymerase used) of a dNTP stock solution to the extension-termination mixture. This decreases the frequency of terminations, and therefore, increases the average length of the chain-terminated products.

Carryover Nucleotides and Primers

In a standard PCR, about 200 μM each of four dNTPs and 50 pmoles each of two primers are used to amplify several micrograms of target DNA. Of these, >97% of dNTPs and >90% of primers remain unused at the end of PCR (Rao and Saunders 1992; Rao 1994). Unless these are removed, direct use of PCR-amplified DNA as a sequencing template results in the carryover of unused dNTPs and primers to the sequencing reactions. These interfere with the sequencing reactions in the following ways: (1) The carryover dNTPs alter the dNTP/ddNTP ratios that are required for optimal chain-termination reactions; (2) since the carryover primers can also prime DNA synthesis, they compete with the sequencing primer and titrate out polymerase as well as dNTPs and ddNTPs; and (3) as a result of priming by carryover primers, a mixture of sequence ladders is generated by the Sequenase protocol because the DNA synthesized from the PCR primers is also radioactively labeled.

As a consequence of the above interferences, the sequence ladders generated are of low intensity with high background. It is difficult to decipher the DNA sequence from such ladders; therefore, it is essential to remove the unused dNTPs and primers from PCR mixtures. A number of strategies have been reported for the separation of low-molecular-weight dNTPs and primers from the high-molecular-weight, PCR-amplified DNA (Rao 1994). These include differential precipitation, ion-exchange chromatography, gel filtration, and streptavidin chromatography. Of these, ion-exchange chromatography may be the best way to remove quantitatively the low-molecular-weight dNTPs and primers from the high-molecular-weight PCR-amplified DNA. In our experience, QIAquick spin column separation gives clean DNA templates and generates high-intensity sequence ladders. These columns are designed to resolve either the single-stranded or the double-stranded PCR-amplified DNA from dNTPs and PCR primers that are less than 50 nucleotides long.

Heterogeneity of DNA Template

PCR-amplified DNA is a product of virtually billions of in vitro priming and extension reactions. Therefore, inherent in the PCR process is the amplification of heterogeneous DNA in addition to the unique target DNA. These heterogeneous DNA molecules differ in sequence and size, and arise as a result of secondary reactions (Rao 1994). These include (1) partial products generated by the premature termination of DNA synthesis, (2) mosaic products generated by random intermolecular recombination between target DNA strands, and (3) multiple products generated as a result of priming at sequences that have either accidental homology or functional relatedness to the target DNA of interest.

Most of the standard PCRs amplify predominantly the unique target DNA, whereas the heterogeneous DNA constitutes a minor fraction. This fraction usually appears as a smear upon agarose gel electrophoresis and ethidium bromide staining. However, it is not uncommon to see that a major portion of the amplified DNA is constituted by heterogeneous DNA. This happens particularly when the copy number of starting sample is low or when PCRs are performed under low-stringency conditions (Arnheim and Erlich 1992; Saunders et al. 1993). The heterogeneous DNA, depending on the amount present in the PCR-amplified DNA, accordingly contributes to the background in the final sequence ladders.

Two major factors that lead to the formation of heterogeneous DNA are strand annealing and random priming during PCR amplification (see below). Therefore, careful consideration should be given in designing the parameters for PCR amplification. In particular, primer design, annealing temperature, and conditions or treatments prior to the initiation of the first PCR cycle should be stringently controlled to amplify only the target DNA. In general, a high-quality PCR-amplified product is generated by choosing highly specific PCR primers with a GC-content of >50%, using a high annealing temperature that is very close to the estimated T_m value of the primers (in the range of 50–60°C), initiating PCR by hot start, and if necessary, performing a second nested PCR with an internal set of primers (Arnheim and Erlich 1992). In addition, PCR should be performed for as few cycles as possible using a high copy number of target DNA. Despite these modifications, if considerable background still exists, or when multiple products are amplified, it is essential to purify the desired amplified product by agarose gel electrophoresis. The DNA fragment is extracted from the agarose gel by the procedure given below (Rao 1994), or by any standard procedure, and is used as a template for sequencing reactions.

PROTOCOL

Purification of PCR-amplified DNA by Agarose Gel Electrophoresis

1. Extract the aqueous phase (100 μl) of the PCR mix twice with an equal volume of chloroform, and concentrate it to ~25 μl by extracting twice with an equal volume of *n*-butanol. More than one PCR mixture can be pooled and concentrated in this way. Extract once with an equal volume of water-saturated ether to remove traces of *n*-butanol. Evaporate the ether for 3 minutes at 67°C.

2. Load the entire sample into a single well of an 0.8% low-melting-temperature agarose gel containing 1 μg/ml of ethidium bromide, and electrophorese at 150 volts for 1 hour. Excise the DNA band(s) of interest and suspend each agarose slice in about 300 μl of a buffer containing 10 mM Tris-HCl, pH 8.0, 1 mM EDTA, and 100 mM NaCl. Melt the gel for 20–30 minutes at 67°C. It is critical to melt the gel completely. When the tube is tapped, the solution should appear clear, and no fine gel particles should remain. Otherwise, poor recovery of DNA will result.

3. Add an equal volume of phenol to the melted gel while it is still at 67°C. Mix immediately by vortexing for a few seconds. Centrifuge in a microfuge at the maximum speed for 15 minutes. The melted gel material forms a thick precipitate at the interface. Remove the top aqueous phase and extract it once again with phenol, twice with an equal volume of chloroform, and once with an equal volume of water-saturated ether. Each extraction is done for a few seconds either by tapping the tube vigorously or by vortexing. Between extractions, the aqueous phase is recovered by centrifugation in a microfuge for a few seconds at the maximum speed. Residual ether after the last extraction is removed by heating the sample for 3 minutes at 67°C.

4. Precipitate the DNA with three volumes of ethanol for 30 minutes to overnight at −70°C. Sediment the DNA by centrifugation at the maximum speed for 15 minutes. Wash the DNA pellet twice with 1 ml each of 80% ethanol. Remove residual ethanol by drying the pellet for 2 minutes at 37°C. Dissolve the DNA in 20 μl of sterile H_2O. Some insoluble material may remain in the DNA pellet, but this does not interfere with the sequencing reactions. A 1- to 3-μl aliquot of the DNA is used for the sequencing reactions.

   Occasionally, the final DNA preparation may contain some inhibitors (most likely residual phenol), which inactivate the Sequenase enzyme during DNA synthesis and generate considerable ddNTP-independent terminations. These appear as background bands in all four lanes. This background can be eliminated by repeating the extractions starting from the chloroform extraction in step 3.

Random Priming

Random priming, as opposed to specific priming by the sequencing primer, refers to the priming of DNA synthesis at random points on any DNA strand in the reaction mixture. Random priming is facilitated by the interaction of a few nucleotides at the 3′ end of the sequencing primer with a short stretch of complementary sequence at a random position on the DNA (Rao 1994). Such complexes, although inherently unstable, are rapidly stabilized by the addition of a few nucleotides by the highly active DNA polymerase (the in vitro synthetic rate of *Taq* DNA polymerase is on the order of 5000 nucleotides per minute). Random priming events are also facilitated by the ends of PCR-amplified DNA. PCR-amplified DNA, because it is a linear molecule, consists of numerous 3′ ends, each having a free hydroxyl group. These 3′ ends, especially the ends of short partial products formed during PCR, are capable of priming DNA synthesis in a random fashion. As a result, a mixture of random chain-terminated products are generated in addition to the specific chain-terminated products. The random products appear as background bands in all lanes when [α-^{32}P]dATP is used for labeling DNA, either by the Sequenase strategy or by the cycle sequencing strategy.

The random priming events can be minimized by the following considerations. First, use of a 5′-labeled primer rather than [α-^{32}P]dATP in the cycle sequencing protocol eliminates this background. This is because the random chain-terminated products are not radioactively labeled and therefore do not appear on the final autoradiogram. However, even under these conditions, extensive random priming should be avoided, since the random events, by competing with the specific events for polymerase and dNTPs, reduce the intensity of the specific bands.

Random priming can also be minimized by choosing a stringent annealing temperature. In general, a very low temperature of annealing (snap cooling [Kusukawa et al. 1990] as in the Sequenase protocol), or a high temperature of annealing (close to the estimated T_m value of the primers as in the cycle sequencing protocol), improves the specificity of priming. However, a high annealing temperature (50–60°C) as in the cycle sequencing protocol is preferred over the snap cooling procedure because a temperature shift from 94°C (denaturation temperature) to 50–60°C (annealing temperature) can be accomplished by a thermal cycler in less than 30 seconds. This minimizes random priming events as well as the strand-annealing events (see below).

Strand Annealing

PCR-amplified DNA is a linear, double-stranded molecule several hundred nucleotides in length. However, the DNA polymerases used for in vitro DNA synthesis cannot replicate a double-stranded template because they lack the accessory proteins such as single-stranded DNA-binding protein, helicase, etc. Therefore, the double-stranded PCR product must first be converted to a single-stranded form. This is accomplished by heating PCR-amplified DNA for several minutes at 94°C. The denatured single strands tend to reassociate, reconstituting the double-stranded form, which has two consequences.

First, if this happens during the annealing step, binding of the primer to the template and priming of DNA synthesis are inefficient, leading to the appearance of very low intensity sequence ladders. Therefore, the two protocols described above are designed to minimize the time required for the annealing step. This favors the kinetics of annealing of a short primer to the template rather than that of the opposite strand that is several hundred nucleotides long. The presence of detergents such as NP-40 and Tween 20, and possibly also dimethyl sulfoxide, may further inhibit strand annealing and improve the quality of the sequence ladders generated (Bachmann et al. 1990).

A second problem associated with strand annealing involves the extension step. Even when the annealing of primer to template occurs, extension by polymerase can be hindered due to strand annealing in the regions downstream from the replication fork. Therefore, DNA synthesis is terminated randomly, resulting in the dissociation of polymerase from the replication complex. These random stops, resulting from ddNTP-independent terminations, appear as background bands in all four lanes. Indeed, this is one of the primary reasons for the appearance of high background in the sequence ladders generated from PCR-amplified DNA. In addition to the random stops, stops can also occur at a high frequency at specific points on the template. For instance, GC-rich regions of the template tend to form double-stranded complexes readily and generate a specific terminated product at a much higher frequency than a random product. These "strong-stop" products appear as intense bands in all four lanes at specific positions on the autoradiogram.

These strand-annealing problems can be overcome by the following modifications: (1) Sequencing reactions should be performed under high-stringency conditions, for example, the use of high annealing and extension temperatures for short time periods (clearly, the cycle sequencing protocol is best suited for maintaining high-stringency conditions during sequencing reactions); (2) the concentrations of ddNTPs required for chain termination should be optimized to generate predominantly ddNTP-specific terminations that would far outweigh the nonspecific terminations; and (3) because the products of ddNTP-independent terminations have a free 3′-hydroxyl group, these can be further extended by terminal dideoxynucleotidyl transferase in a template-independent manner; this enzyme converts the background bands to high-molecular-weight products that are retained at the top of the sequencing gel, thereby reducing the background (Fawcett and Bartlett 1990). Finally, strand-annealing problems can be eliminated by generating a single-stranded DNA template either by asymmetric PCR (Gyllensten and Erlich 1988) or by converting the double-stranded PCR-amplified DNA into single-stranded half-templates using phage T7 gene *6* exonuclease (USB/Amersham) (Fuller 1989).

Errors in the DNA Sequence Generated from PCR-amplified DNA

It is known that the *Taq* DNA polymerase, which is the most widely used enzyme for PCR, lacks proofreading 3′→5′ exonuclease activity and incorporates errors at a high frequency. Estimates of error frequencies under PCR conditions range from 1 error in 4000 nucleotides synthesized to 1 error in 400 nucleotides synthesized (Saiki et al. 1988; Ho et al. 1989; Eckert and Kunkel 1990). Therefore, it is possible that almost every molecule in a 1-kb size PCR-amplified DNA could have an error. The question then is whether direct sequencing of the mutant PCR-amplified DNA results in the generation of an inaccurate DNA sequence. The answer to this question is clear when one considers a worst-case scenario in which a hypothetical target DNA is amplified starting from a single DNA molecule and assuming that an error is incorporated in the very first cycle. Consequently, after the first cycle, one of the four DNA strands has a mutation. Upon further amplification of these four strands for about 25 cycles, the mutant sequence constitutes about 25% of the final amplified product. When this DNA is used as a template for sequencing, the final autoradiogram shows the mutant band only at one-third the intensity of the correct nucleotide band. Because the DNA sequence is read by subtracting the background, the nucleotide sequence deduced from such a sequence ladder is that of the correct nucleotide.

The above scenario is one that is encountered rarely because, in most PCR experiments, the copy number of starting DNA is on the order of 10^3–10^5 molecules. Any errors incorporated in the initial cycles are randomized, and therefore, any specific mutant sequence constitutes only a minuscule fraction of the total product. The DNA sequence generated is a consensus sequence from millions of template

DNA molecules, thus these errors are not accounted for in the final sequence. Therefore, in practice, the direct sequencing of PCR-amplified DNA does not result in the incorporation of errors. The sequence generated is accurate, despite the low fidelity of *Taq* DNA polymerase.

Indeed, it is desirable to generate the DNA sequence by directly sequencing the PCR-amplified DNA rather than from a clone of the amplified DNA, because during cloning, a single DNA molecule out of billions of amplified molecules is selected. It is highly probable that the cloned molecule will contain a mutation as a result of PCR. Therefore, the sequence generated from a cloned DNA should be confirmed by comparing it with the sequence generated directly from the PCR-amplified product to ascertain the accuracy of the cloned sequence.

Other thermostable DNA polymerases such as the Vent polymerase (New England Biolabs) and the *Pfu* polymerase (Stratagene) exhibit proofreading exonuclease activity and reportedly incorporate errors 15- to 30-fold less frequently than the *Taq* DNA polymerase (Lundberg et al. 1991; Mattila et al. 1991). Therefore, these polymerases should generate a better-quality PCR product, particularly if the cloned DNA is also used for the expression and functional characterization of gene products.

There are, however, situations in which the direct sequencing of PCR-amplified DNA is not the answer. For example, if the starting sample contains one normal allele and one deleted allele, the deleted allele is masked and cannot be deduced from the sequence generated by direct sequencing. Similarly, if the sample contains multiple alleles, direct sequencing generates a composite sequence ladder. It would be hard, if not impossible, to decipher the sequence of individual alleles from such a composite ladder. In such situations, the PCR-amplified DNA should be cloned first and the DNA sequence of a number of clones should be determined to generate the DNA sequence of individual alleles.

CONCLUSIONS

The direct sequencing of PCR-amplified DNA bypasses the time-consuming cloning steps and rapidly generates accurate DNA sequence information from small quantities of precious biological samples. Although this approach generates considerable background, in most cases it does not preclude the researcher from generating complete and accurate DNA sequencing information. As discussed above, this background can be minimized by performing both the PCR and DNA sequencing reactions under stringent conditions. The two strategies described above allow the determination of the DNA sequence from almost any type of PCR-amplified DNA template. Although both protocols are expected to generate high-intensity sequence ladders, the cycle sequencing strategy is preferred over the Sequenase strategy because it is very convenient to set up, especially when it is necessary to generate DNA sequences routinely from a large number of templates. In addition, reaction parameters, such as random priming and strand annealing, can be controlled stringently using the cycle sequencing strategy. This is particularly useful for generating sequences from templates having a high GC content, because the extension and chain-termination reactions can be performed at an elevated temperature, which destabilizes secondary structures in the DNA template.

In summary, high-intensity, low-background sequence ladders can be consistently generated directly from PCR-amplified DNA by (1) performing PCR under stringent conditions, (2) purifying the PCR-amplified DNA from unused dNTPs and primers by QIAGEN chromatography, (3) performing cycle sequencing under stringent conditions, and (4) using a high-specific-activity, 5′-labeled sequencing primer for amplifying the chain-terminated products.

ACKNOWLEDGMENT The author thanks Mark Miller from the Center for Advanced Training in Cell and Molecular Biology, Department of Biology, The Catholic University of America, for his expert assistance in the preparation of schematics.

REFERENCES

Arnheim, N. and H. Erlich. 1992. Polymerase chain reaction strategy. *Annu. Rev. Biochem.* **61:** 131–156.

Bachmann, B., W. Luke, and G. Hunsmann. 1990. Improvement of PCR amplified DNA sequencing with the aid of detergents. *Nucleic Acids Res.* **18:** 1309.

Barr, P.J., R.M. Thayer, P. Laybourn, R.C. Najarian, F. Sela, and D.R. Tolan. 1986. 7-Deaza-2′- deoxyguanosine-5′-triphosphate: Enhanced resolution in M13 dideoxy sequencing. *BioTechniques* **4:** 428–432.

Eckert, K.A. and T.A. Kunkel. 1990. High fidelity DNA synthesis by the *Thermus aquaticus* DNA polymerase. *Nucleic Acids Res.* **18:** 3739–3744.

Fawcett, T.W. and S.G. Bartlett. 1990. An effective method for eliminating "artifact banding" when sequencing double-stranded DNA templates. *BioTechniques* **8:** 46–48.

Fuller, C.W. 1989. Using T7 gene 6 exonuclease to prepare single stranded templates for sequencing. *U.S. Biochemical Editorial Comments* **16:** 1–8.

Gyllensten, U.B. 1989. PCR and DNA sequencing. *BioTechniques* **7:** 700–708.

Gyllensten, U.B. and H.A. Erlich. 1988. Generation of single stranded DNA by the polymerase chain reaction and its application to direct sequencing of the HLA-DQα locus. *Proc. Natl. Acad. Sci.* **85:** 7652–7656.

Ho, S.N., H.D. Hunt, R. Horton, J.K. Pullen, and L.R. Pease. 1989. Site-directed mutagenesis by overlap extension using the polymerase chain reaction. *Gene* **77:** 51–59.

Innis, M.A., K.B. Myambo, D.H. Gelfand, and M.A.-D. Brow. 1988. DNA sequencing with *Thermus aquaticus* polymerase and direct sequencing of polymerase chain reaction-amplified DNA. *Proc. Natl. Acad. Sci.* **85:** 9436–9440.

Kusukawa, N., T. Uemori, K. Asada, and I. Kato. 1990. Rapid and reliable protocol for direct sequencing of material amplified by the polymerase chain reaction. *BioTechniques* **9:** 66–71.

Lee, J.-S. 1991. Alternative dideoxy sequencing of double-stranded DNA by cyclic reactions using *Taq* polymerase. *DNA Cell Biol.* **10:** 67–73.

Lundberg, K.S., D.D. Shoemaker, M.W.W. Aetams, J.M. Short, J.A. Sorge, and E.J. Mathur. 1991. High-fidelity amplification using a thermostable DNA polymerase isolated from *Pyrococcus furiosus*. *Gene* **108:** 1–6.

Mattila, P., J. Korpela, T. Tenkanen, and P. Pitkanen. 1991. Fidelity of DNA synthesis by the *Thermococcus litoralis* DNA polymerase—An extremely heat stable enzyme with proofreading activity. *Nucleic Acids Res.* **19:** 4967–4973.

Mullis, K. 1991. The polymerase chain reaction in an anemic mode: How to avoid cold oligodeoxyribonuclear fusion. *PCR Methods Appl.* **1:** 1–5.

Mullis, K., F. Faloona, S. Scharf, R. Saiki, G. Horn, and H. Erlich. 1986. Specific enzymatic amplification of DNA in vitro: The polymerase chain reaction. *Cold Spring Harbor Symp. Quant. Biol.* **51:** 263–273.

Murray, V. 1989. Improved double-stranded DNA sequencing using the linear polymerase chain reaction. *Nucleic Acids Res.* **17:** 8889.

Rao, V.B. 1994. Direct sequencing of polymerase chain reaction amplified DNA. *Anal. Biochem.* **216:** 1–14.

Rao, V.B. and N.B. Saunders. 1992. A rapid polymerase chain reaction-directed sequencing strategy using a thermostable DNA polymerase from *Thermus flavus*. *Gene* **113:** 17–23.

Ruano, G. and K.K. Kidd. 1991. Coupled amplification and sequencing of genomic DNA. *Proc. Natl. Acad. Sci.* **88:** 2815–2819.

Saiki, R.K., D.H. Gelfand, S. Stoffel, S.F. Scharf, R. Higuchi, R.T. Horn, K.B. Mullis, and H.A. Erlich. 1988. Primer-directed enzymatic amplification of DNA with a thermostable DNA polymerase. *Science* **239:** 487–491.

Saunders, N., W. Zollinger, and V.B. Rao. 1993. A rapid and sensitive strategy employed for amplification and sequencing of *por A* from a single colony forming unit of *Neisseria meningitidis*. *Gene* **137:** 153–162.

Tabor, S. and C.C. Richardson. 1987. DNA sequence analysis with a modified bacteriophage T7 DNA polymerase. *Proc. Natl. Acad. Sci.* **84:** 4767–4771.

———. 1989. Effect of manganese ions on the incorporation of dideoxy-nucleotides by bacteriophage T7 DNA polymerase and *Escherichia coli* DNA polymerase I. *Proc. Natl. Acad. Sci.* **86:** 4076–4080.

Cycle Sequencing

Keith Kretz, Walter Callen, and Valerie Hedden

Stratagene Cloning Systems, La Jolla, California 92037

INTRODUCTION

PCR is being used in an ever-increasing number of new techniques. One related technique that has emerged is cycle sequencing, or linear amplification sequencing, as it is sometimes known (Carothers et al. 1989; Murray 1989). Cycle sequencing employs a thermostable DNA polymerase in a temperature cycling format to perform multiple rounds of dideoxynucleotide sequencing on the template (Fig. 1). The result of the temperature cycling is linear amplification of the sequencing product, leading to an increase in the signal generated during the sequencing reaction when compared to standard sequencing protocols.

Cycling the sequencing reactions results in several advantages: (1) The amount of template necessary for the sequencing reaction is greatly reduced when compared to Sequenase sequencing (200 ng vs. 2 μg), (2) screening reactions can be performed on minimally prepared templates, (3) the high temperature at which the sequencing reactions are run allows the DNA polymerase to synthesize through areas of secondary structure, and (4) the multiple heat-denaturation steps allow double-stranded templates like plasmids, cosmids, lambda DNA, and PCR products to be sequenced without a separate denaturation step.

This chapter describes procedures for cycle sequencing using manual detection methods, including radioactive labeling and hapten-mediated nonradioactive detection.

REAGENTS

Cyclist Exo$^-$*Pfu* DNA sequencing kit (Stratagene)
Taq DNA sequencing kit (Stratagene or others [Roberts 1992])
Deoxynucleotide triphosphates (Pharmacia)
Dideoxynucleotide triphosphates (Pharmacia)
6% Polyacrylamide gel mix

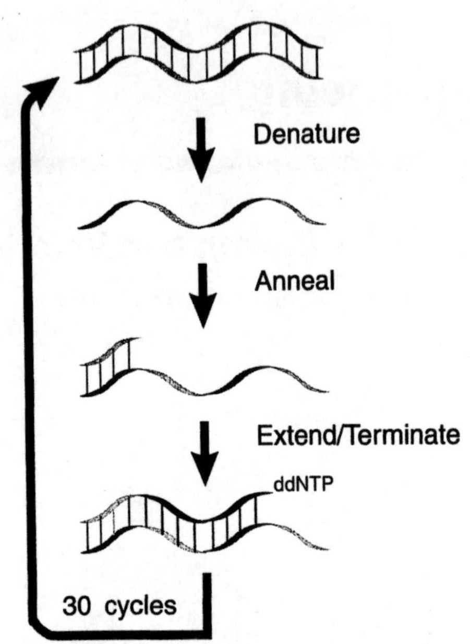

Figure 1 Outline of the cycle sequencing reaction. Double- or single-stranded templates are cycled through a three-temperature cycling reaction consisting of high-temperature denaturation, primer annealing, and primer extension/termination.

Acrylamide
Bis-acrylamide
Urea
TEMED (*N*,*N*,*N*′,*N*′-tetramethylenediamine)
Ammonium persulfate

SOLUTIONS

10× Exo^{-}*Pfu* sequencing buffer
- 200 mM Tris-HCl, pH 7.5
- 100 mM KCl
- 20 mM $MgSO_4$
- 100 μM $(NH_4)_2SO_4$
- 1% Triton X-100
- 1 mg/ml bovine serum albumin
- 20 μM dATP
- 50 μM dCTP
- 50 μM dGTP
- 50 μM dTTP

10× *Taq* sequencing buffer
- 100 mM Tris-HCl, pH 8.8
- 500 mM KCl
- 40 mM $MgCl_2$
- 0.01% gelatin
- 20 μM dATP

50 μM dCTP
50 μM dGTP
50 μM dTTP

Exo⁻*Pfu* termination mixes
ddATP (1.5 mM)
ddCTP (1.5 mM)
ddGTP (1.5 mM)
ddTTP (1.5 mM)

Taq termination mixes
ddATP (600 μM)
ddCTP (600 μM)
ddGTP (100 μM)
ddTTP (1000 μM)

Stop dye
95% formamide
20 mM EDTA
0.05% bromophenol blue
0.05% xylene cyanol

PROTOCOLS

Sequencing Template Preparation

Cycle sequencing enables researchers to sequence much smaller amounts of DNA with little or no purification. However, it should be noted that the purity of the template will affect the quality of the sequence data, particularly when automated sequencing protocols are employed. Previously inconsistent DNA templates such as PCR products, cosmids, and lambda DNA can now be sequenced routinely using standard laboratory techniques. The sequencing of PCR products presents unique problems and opportunities, as described in "Direct Sequencing of PCR-amplified DNA" in Section 7.

PLASMIDS

Plasmids purified by any standard small-scale (boiling mini-prep or alkaline lysis) or large-scale protocol (cesium chloride banding) are suitable templates (Sambrook et al. 1989). In addition, several solid-phase plasmid purification kits are available (Stratagene, QIAGEN, or Promega) that provide high-quality sequencing templates. Typically 50–200 fmoles of plasmid (100–400 ng of a 3-kb plasmid) are used in the sequencing reactions; however, 1–500 fmoles of plasmid template have been used successfully.

Alternatively, high-copy-number plasmids may be sequenced directly from colonies using the following protocol (Krishnan et al. 1991).

1. Pick a colony from the plate with an inoculating loop and boil in 25 μl of TE (5 mM Tris-HCl, pH 8.0, 0.1 mM EDTA) for 5 minutes.

2. Vortex well and cool on ice. Centrifuge for 1 minute in a microfuge to remove cell debris.

3. Use 10 μl of the supernatant as the template in subsequent sequencing reactions.

4. Use ^{32}P for an overnight exposure or ^{33}P for a 2–3-day exposure.

If the plasmid is a low-copy-number plasmid, it may still be sequenced directly using the following protocol.

1. Prepare a 1-ml overnight liquid culture.

2. Collect 100–200 μl of the cells by centrifugation for 1 minute in a microfuge.

3. Resuspend the cell pellet in 25 μl of TE.

4. Boil the cells for 5 minutes.

5. Vortex well and cool on ice. Then centrifuge for 1 minute in a microfuge to remove cell debris.

6. Use 10 μl of the supernatant as the template in subsequent sequencing reactions.

7. Use ^{32}P for an overnight exposure or ^{33}P for a 2–3-day exposure.

Although the use of plasmid DNA obtained by boiling cells in this manner will not yield high-quality sequence data (only ~100–200 bases of readable sequence), the technique is particularly useful when screening large numbers of site-specific mutagenesis products or screening cloning products.

M13, COSMIDS, AND LAMBDA DNA

M13, cosmids, and lambda DNA prepared by standard protocols (Sambrook et al. 1989) are also suitable templates. For M13 and lambda DNA, 10–100 fmoles should be used as template, whereas 50–200 fmoles of cosmid should be used.

The DNA present in M13 or lambda plaques may be sequenced directly in a manner analogous to that for plasmids from colonies (Krishnan et al. 1991).

1. Cut closely around the plaque of interest with a scalpel and peel off *just the top agarose layer.*

2. Boil the top agarose containing the plaque in 25 μl of TE for 5 minutes.

3. Vortex well and cool on ice. Remove any remaining pieces of agarose by centrifuging for 1 minute.

4. Use 10 μl of the supernatant as the template in subsequent sequencing reactions.

5. Use ^{32}P for an overnight exposure or ^{33}P for a 2–3-day exposure.

As with boiled cells above, the use of phage DNA obtained by boiling plaques will not yield high-quality sequence data (only ~100–200 bases of readable sequence), but it is particularly useful when screening large numbers of clones.

BACTERIAL GENOMIC DNA

Unfractionated, high-molecular-weight *Escherichia coli* genomic DNA has been used as the template in a cycle sequencing reaction (Kretz et al. 1994). Useful sequence information for 100–150 bases may be obtained from this reaction. Approximately 1 μg of high-quality genomic DNA should be used as the template, and slightly longer sequencing primers (20–25 nucleotides) should be used at the highest possible annealing temperature. Radioactive labeling with ^{32}P is required. No other deviations from the standard protocol are required. This template DNA may be prepared by the method of Ausubel et al. (1987).

Detection

Cycle sequencing products can be labeled in a number of different ways. If radioactive labeling is used, either incorporation of an α-labeled dNTP or end-labeled primers can be used. ^{32}P and ^{35}S are the most common radiolabels used in sequencing. ^{32}P has the advantage of stronger signal strength but ^{35}S has the advantage of sharper bands and a longer half-life. One disadvantage to the use of ^{35}S in cycle sequencing is the low labeling efficiency obtained using *Taq* DNA polymerase. For those who wish to use ^{35}S, a non-*Thermus sp.* DNA polymerase-based kit is better. Recently, an exonuclease-deficient DNA polymerase from *Pyrococcus furiosus* (Exo^{-}*Pfu*) was shown to provide much stronger signals than *Taq* DNA polymerase when ^{35}S-labeled dATP was used as the radiolabel (Hedden et al. 1992). ^{33}P was recently introduced as an alternative radioactive label for use in molecular biology (Zagursky et al. 1991). The β-emission of this new label has physical characteristics intermediate between ^{35}S and ^{32}P, making it ideal for DNA sequencing applications. It gives strong, sharp bands and allows long read lengths with shorter exposure times, often 2 hours. This isotope is used efficiently by Exo$^-$*Pfu* polymerase and *Taq* DNA polymerase; it can be used in both incorporation and end-labeling formats with no deviations from the standard protocols.

RADIOLABELING OF SEQUENCING PRIMERS

Radiolabeling is carried out according to the method of Sambrook et al. (1989). Label 2–5 pmoles of primer (10–25 ng of a 17-mer) for each sequencing reaction. A master mix of labeling components may be made and aliquoted to tubes containing the primers.

1. Prepare a master labeling mix containing the following components (multiply reagent amounts by number of labeling reactions).

   2.5 μl of H_2O
   0.5 μl of 10X T4 polynucleotide kinase buffer (50 mM Tris-HCl, pH 7.5, 7 mM $MgCl_2$, and 1.0 mM DTT)
   0.5 μl of [γ-^{33}P]ATP or [γ-^{32}P]ATP (NEN or Amersham), 10 μCi/μl; 1000–5000 μCi/mmole
   0.5 μl of T4 polynucleotide kinase (5–10 units/μl)

2. Aliquot 4 μl of master mix to each reaction tube containing 1 μl of primer (2–5 pmoles; 10–25 ng of a 17-mer).

3. Incubate for 15 minutes at 37°C.

4. Heat-kill the kinase for 10 minutes at 80°C.

5. Use the labeled primer, without purification, in the sequencing reaction.

NONRADIOACTIVE LABELING

Procedures for performing fluorescence-based dideoxy sequencing reactions are available from Applied Biosystems, Inc. and rely on four dye-labeled dideoxynucleotides, which are combined in a single tube. This cycle sequencing protocol requires specific equipment for running and analysis of the sequencing gel. For manual nonradioactive detection, hapten-labeled primers can be used and detection is accomplished enzymatically. Most oligonucleotide synthesis suppliers can prepare these modified nucleotides. For those who synthesize their own oligonucleotides, the appropriate modified phosphoramidites can be purchased from most major suppliers. If a biotinylated oligonucleotide is used as the sequencing primer, the sequencing products are transferred to a nylon or nitrocellulose membrane by capillary action or electroblotting. After washing and blocking the membrane, a streptavidin-alkaline phosphatase (or horseradish peroxidase) conjugate is applied. This enzyme then generates either chemiluminescent or colored products from specialized substrates. Alternatively, other haptens such as fluorescein may be added to the end of the primer during synthesis. The fluorescein molecules are detected with an anti-fluorescein antibody conjugated to alkaline phosphatase (or horseradish peroxidase) and detected with appropriate substrates.

TRANSFER OF SEQUENCING REACTION PRODUCTS TO MEMBRANE

1. Perform the cycle sequencing reactions using a haptenated sequencing primer.

2. Separate the reaction products by polyacrylamide gel electrophoresis.

3A. Transfer the sequencing reaction products to a membrane by capillary transfer.

   a. Separate the glass plates of the electrophoresis gel.
   b. Overlay the gel with a nylon membrane (Stratagene).
   c. Overlay the membrane with two sheets of Whatman 3MM paper and cover with the glass plate that was removed in step a. Place approximately 5 pounds of weight on the glass plate.
   d. Allow transfer to proceed for 1 hour at room temperature.

3B. Alternatively, the sequencing reaction products can be electroblotted onto a membrane using the Gene Sweep (Hoefer).

4. Disassemble the transfer setup and fix the DNA to the membrane by drying and/or UV cross-linking.

5. Develop the gel with appropriate detection reagents. Kits for the detection of biotinylated primers are available from several suppliers (USB/Amersham, New England Biolabs, Tropix).

Cycle Sequencing Reaction (Cyclist Exo$^-$*Pfu* DNA Sequencing Kit)

The reaction components are described below.

1. Add 3 μl of the appropriate termination mix to each of four tubes. Cap the tubes and keep on ice.

2. For each DNA template, combine the following reaction components on ice and mix by pipetting:

   water to a final volume of 30 μl
   template (see above for amount depending on the type of template used)
   primer (2–5 pmoles; 10–25 ng of a 17-mer)
   4 μl of 10X sequencing buffer
   10 μl of [α-^{32}P]dATP, [α-^{33}P]dATP, or [α-^{35}S]-labeled dATP (not included if labeled primers are used) (NEN or Amersham; 10 μCi/μl; 1000–5000 μCi/mmole)
   2 units of Exo$^-$*Pfu* DNA polymerase
   4 μl of DMSO (add last and mix thoroughly)

3. Aliquot 7 μl of the reaction mixture from step 2 into each of the four tubes containing termination mix from step 1. Mix reagents well.

4. Overlay the reactions with 25 μl of silicone oil or mineral oil (Sigma) if necessary.

5. Optimum cycling parameters will vary depending on the template and primer. A useful starting profile is given below:

   30 cycles of:
   30 seconds at 95°C
   30 seconds at 50°C
   60 seconds at 72°C

6. Add 5 μl of stop dye below the oil overlay and mix by pipetting.

7. Heat-denature the samples for 2–5 minutes at ≥80°C immediately prior to loading 1–3 μl onto a sequencing gel (6% polyacrylamide, 7 M urea, 1X TBE).

If *Taq* DNA polymerase is the enzyme of choice for the cycle sequencing reactions, the following changes should be made in the protocol.

1. Substitute *Taq* termination mixes for the Exo$^-$*Pfu* termination mixes (step 1).

2. Substitute 10X *Taq* sequencing buffer for the 10X Exo$^-$*Pfu* sequencing buffer (step 2).

3. Do not add DMSO to the reaction mix (step 2).

4. Substitute 2 units of *Taq* DNA polymerase for the Exo⁻*Pfu* DNA polymerase (step 2).

Manipulating the Sequencing Reaction

Note that this particular protocol is designed to allow the user to vary the dNTP/ddNTP ratio for special sequencing needs. If it is necessary to read a sequence close to the primer, it is recommended that end-labeled primers be used. With this change, the sequence can generally be determined to within 5–10 bases of the primer. Decreasing the ratio of dNTP/ddNTP is also helpful. To increase the intensity of bands further from the primer, it is suggested that the ratio of dNTP/ddNTP be increased by a factor of 2–5. Changes in the dNTP/ddNTP ratio can be accomplished by using more or less of the termination mixes in the reaction (1–5 µl). No other changes to the protocol are required.

The protocol described above is designed to use labeled dATP for radioactive detection by incorporation. If a different labeled nucleotide is to be used, the amount of the corresponding unlabeled nucleotide should be decreased to 20 µM in the 10x sequencing buffer and the amount of dATP should be increased to 50 µM.

If end-labeled primers are used for either radioactive or nonradioactive detection, the only change to the above protocol is the omission of the α-labeled dATP.

TROUBLESHOOTING

- *Light or blank film.* (1) The primer is not annealing efficiently. Reduce the annealing temperature or redesign the primer. (2) Not enough template DNA was used. (3) No radioactivity was added, or the radiolabel is old. (4) One of the reaction components is missing, or the reaction components were not thoroughly mixed. (5) The template DNA is contaminated. Make sure excess salt and EDTA have not contaminated the preparation.

- *High background on the sequencing gel.* There is too much template DNA. This can be a serious problem with short PCR products. Titrate down the amount of DNA added to the reaction.

- *Bands in multiple lanes.* (1) The primer annealed at multiple sites. A higher annealing temperature may help. (2) Multiple templates are in the sequencing reaction. PCR products may be a particular problem if multiple products or primer-dimer artifacts are present and one of the PCR primers is used as a sequencing primer. Gel purification should alleviate the problem. (3) Gel compression artifacts are present. Increase the gel temperaure (up to 60°C) or add formamide to the gel. (4) Template DNA is contaminated.

- *Blurry or smeared bands.* (1) Old or improperly prepared acrylamide solutions were used. (2) The samples were not fully denatured prior to gel electrophoresis.

- *Little or no sample volume remaining after cycling.* There was sample evaporation, possibly due to insufficient oil overlay. Add at least 25 μl of silicone or mineral oil and briefly centrifuge before cycling.

DISCUSSION

The traditional sequencing templates (purified M13 and plasmids) are excellent templates for cycle sequencing, but the use of the temperature cycling format now makes the sequencing of PCR products, cosmids, and lambda DNA routine as well. In addition, direct sequencing of plasmid DNA from colonies and phage DNA from plaques is a valuable timesaving technique in situations where larger numbers of clones need to be screened. Future improvements in this technique should allow sequences to be obtained from ever-longer templates.

REFERENCES

Ausubel, F.M., R. Brent, R.E. Kingston, D.D. Moore, J.G. Seidman, J.A. Smith, and K. Struhl, eds. 1987. *Current protocols in molecular biology.* Wiley, New York.

Carothers, A.M., G. Urlaub, J. Mucha, D. Grunberger, and L.A. Chasin. 1989. Point mutation analysis in a mammalian gene: Rapid preparation of total RNA, PCR amplification of cDNA, and *Taq* sequencing by a novel method. *BioTechniques* **7:** 494–499.

Hedden, V., M. Simcox, W. Callen, B. Scott, J. Cline, K. Nielson, E. Mathur, and K. Kretz. 1992. Superior sequencing: Cyclist Exo^{-}*Pfu* DNA sequencing kit. *Strat. Mol. Biol.* **5:** 79.

Kretz, K., W. Callen, and V. Hedden. 1994. Cycle sequencing. *PCR Methods Appl.* **3:** S107–S112.

Krishnan, B.R., R.W. Blakesley, and D.E. Berg. 1991. Linear amplification DNA sequencing directly from single phage plaques and bacterial colonies. *Nucleic Acids Res.* **19:** 1153.

Murray, V. 1989. Improved double-stranded DNA sequencing using the linear polymerase chain reaction. *Nucleic Acids Res.* **17:** 8889.

Roberts, S.S. 1992. Thermostable DNA polymerases heat up DNA sequencing. *J. NIH Res.* **4:** 89–94.

Sambrook, J., E.F. Fritsch, and T. Maniatis. 1989. *Molecular cloning: A laboratory manual,* 2nd edition. Cold Spring Harbor Laboratory Press, Cold Spring Harbor, New York.

Zagursky, R.J., P.S. Conway, and M.A. Kashdan. 1991. Use of ^{33}P for Sanger DNA sequencing. *BioTechniques* **11:** 36–38.

Cloning of PCR Products

Following PCR, it is often necessary to clone the amplified fragment of DNA into a plasmid. The value of cloning an amplicon into a vector depends on the amplicon's future use. Cloning is generally indicated when it constitutes the end-point of a number of complex enzymatic steps (such as those performed in one-sided PCR) or of labor-intensive protocols (such as differential display). Cloning an amplicon avoids having to repeat the reaction every time the product of an amplification is needed. This is particularly important when the template is of limited availability or when the PCR product is difficult to obtain, for example, because of its length. Cloning a PCR product into a vector is also convenient when the amplified DNA fragment will be used as a probe or as a positive control in future PCR procedures. In certain instances, a PCR product may be cloned into a vector suitable for in vivo expression studies. Although expression-PCR (see Section 5) bypasses the need to clone an amplicon for further expression, the desired protein can only be produced by in vitro transcription-translation, not in vivo. A number of strategies are available for cloning PCR products and are described in "Cloning and Analysis of PCR-generated DNA Fragments" and "Strategies for Cloning PCR Products."

A particular cloning strategy should be chosen before the actual PCR is performed, because in a number of approaches, specific sequences need to be added to the 5′ ends of the primers. These added nucleotide sequences are compatible with a particular vector after enzymatic treatment of various types (i.e., T4 DNA polymerase, uracil DNA glycosylase, and restriction enzyme digestion). The cloning strategy will also determine the choice of thermostable DNA polymerase. Two types of thermostable DNA polymerases are available: polymerases with 3′→ 5′ exonuclease (proofreading) activity, and polymerases without this activity. The activities of the thermostable DNA polymerases are shown in Table 1 of the introduction to Section 1. Only enzymes without proofreading activity will add an extra adenosine residue at the 3′ end of a PCR product and therefore should be chosen when cloning the amplicon into vectors that contain overhangs of thymidine residues.

Some cloning strategies provide a means of directionally placing the amplicon within a vector, whereas others are random with respect to amplicon orientation. Directional cloning of a PCR product is beneficial in applications where in vivo and in vitro expression are required.

An important consideration applicable to all cloning methods is the need to optimize the amplification protocol so that the accumulation of spurious DNAs is avoided. These extraneous DNA fragments may compete with the desired PCR product in the ligation reaction, resulting in a reduced number of the desired clones containing the correct insert. To be successful, it may be necessary to add an extra purification step to eliminate these DNA fragments prior to initiating the ligation step.

Strategies for Cloning PCR Products

Robin Levis

Department of Pathology, Uniformed Services University of the Health Sciences, Bethesda, Maryland 20814

INTRODUCTION

Generating PCR products is often only the first step in a series of experiments such as producing specific cDNA clones or making mutations in a particular regulatory sequence or in an open reading frame. Whether the starting material is RNA or DNA, it is important to have an efficient method for cloning PCR products that will facilitate subsequent studies. A variety of methods have been developed to clone the DNA products of PCR amplification (Costa and Weiner 1994a). The method selected to clone PCR fragments is determined by several factors. This might depend on whether the sequence of the PCR product is known, what the unique restriction sites in the vector are, and what will be done with the PCR product; for example, sequencing, mutagenesis, or expression in bacterial or eukaryotic cells.

For some applications, PCR products can be used directly without subcloning. These include sequencing, in vitro transcription, and some mutagenesis studies. For most applications, however, it is essential to subclone a PCR product before further manipulation or analysis. This chapter summarizes several systems available for the preparation and subcloning of PCR products.

After generating a PCR product, it is important to run a small amount of the sample on an agarose gel to verify the size of the product and to ensure that contamination does not exist. For most PCR procedures, it is possible to analyze 5–10% of the product by electrophoresis through agarose and to visualize the DNA by ethidium bromide staining. Depending on the homogeneity of the PCR product, it may be possible to continue directly with cloning.

Several subcloning protocols require that the PCR mix be purified on a column to remove excess primers and nucleotides. This can also be accomplished by doing a selective ammonium acetate precipitation of the sample. If further enzymatic manipulation is required before cloning, it is recommended that the sample be extracted once or twice with an equal volume of phenol:chloroform/isoamyl alcohol and once or twice with chloroform. After extraction, the sample should be ethanol-precipitated either with sodium acetate (pH 5.5) or with ammonium acetate. Crowe et al. (1991) also describe an increase in cloning efficiency of PCR products if the completed PCR product is first treated with proteinase K and then extracted and precipitated.

If the PCR products are heterogeneous when examined by gel electrophoresis, it is important to gel-purify the desired fragment or fragments before cloning. This can be accomplished by using low-melting-temperature agarose, by electroelution, or by dissolution of the gel and purification of the DNA on glass beads (GENECLEAN Kit, BIO-101 or Elu Quik Kit, Schleicher and Schuell).

Once the sample has been analyzed and, if necessary, purified, the DNA can be subcloned into a bacterial plasmid and used to transform competent bacteria. This chapter describes five methodologies for the cloning of PCR products, several of which are facilitated by the availability of commercially prepared cloning vectors. These methods are restriction endonuclease site incorporation (Kaufman and Evans 1990; Scharf 1990; Lorens 1991), uracil DNA glycosylase (UDG) cloning (Duncan 1981; Friedberg et al. 1981; Nisson et al. 1991; Buchman 1992, 1993), ligation-independent cloning (LIC-PCR) (Aslanidis and de Jong 1990; Shuldiner et al. 1990; Haun et al. 1992), T/A cloning (Clark 1988; Holton and Graham 1991; Marchuk et al. 1991; Mead et al. 1991; Hu 1993), and blunt-end cloning (Liu and Schwartz 1992; Weiner 1993; Costa and Weiner 1994a,b; Costa et al. 1994). With the exception of restriction endonuclease site incorporation, all of these techniques rely on the use of a specific vector in which to ligate the PCR product. The construction and availability of these vectors are described.

RESTRICTION ENDONUCLEASE SITE INCORPORATION

One of the primary methods for cloning a PCR product is to digest the product with restriction endonucleases whose recognition sequences are present in the amplified DNA (Scharf 1990). Cloning of the digested product is straightforward and no special reagents or vectors are required. For some applications, where the sequence of the PCR product is known, it may be possible to utilize restriction endonuclease sites within the DNA to clone the PCR product. However, it is also possible to incorporate restriction endonuclease sites into the product by designing primers that contain specific or unique restriction sites at their 5′ termini. The resulting PCR product will be flanked by the same restriction endonuclease site at both termini or by different sites at either end. Digestion of the DNA with the appropriate enzyme or enzymes will yield an insert with compatible ends to the desired vector. The vector and insert DNAs are then ligated and transformed into competent bacteria.

Several factors must be considered in designing PCR primers that contain restriction endonuclease sites. First, if the sequence of the PCR product is not known, the product DNA may contain one of the sites utilized in the primers. This would lead to internal cleavage of the product, giving rise to deleted clones. Several restriction endonucleases are now available which have 8-bp recognition sequences (*Asc* I, *Not* I, *Pac* I, *Pme* I, *Sfi* I, *SgrA* I, *Srf* I, *Sse8387* I, and *Swa* I). It is suggested that these enzyme sites be used for cloning when the sequence of the template RNA or DNA is not known. The frequency with which these enzymes cleave DNA is significantly lower than those with recognition sequences of six or less base pairs.

A second consideration when using restriction endonuclease site incorporation as a method for cloning PCR products is the efficiency with which the desired enzyme will cleave the DNA product, if the enzyme recognition site is near the terminus of the DNA. The table on page 542 shows the efficiency of enzyme digestion for various restriction endonucleases when the substrate DNA is a short region. As shown, enzymes vary considerably in their ability to cleave small fragments of DNA. To increase the efficiency of restriction endonuclease digestion of PCR products, it is important to incorporate several extra nucleotides at the 5′ end of the primer. This table can serve as a guide for the number of nucleotides required in addition to any particular restriction endonuclease recognition sequence to ensure efficient digestion. Kaufman and Evans (1990) have also described the efficiency of restriction endonuclease cleavage of restriction sites in the polylinker of the pBluescript II (Stratagene) cloning vector.

Lorens (1991) has described a technique that circumvents the difficulties associated with inefficient cleavage due to a restriction endonuclease site too close to the end of the DNA product. Briefly, the procedure utilizes Klenow, T4 polynucleotide kinase, and T4 DNA ligase. By kinasing the PCR product, filling in any overhangs, and ligating, a concatamer of PCR products is formed. This multimer can then be efficiently cleaved by the desired restriction endonuclease. The linear product is purified and ligated into a vector with compatible ends.

A third consideration when using this technique for cloning PCR products is that each restriction endonuclease has a specific requirement for salt. It is therefore necessary to purify the PCR product on a column, or by extracting and precipitating the DNA before digesting. This ensures that the buffer from the PCR does not interfere with restriction endonuclease digestion. This also prevents the polymerase present in the PCR from filling in the overhangs created by restriction endonuclease digestion. It is important to gel-purify, or to re-extract the sample after digestion to remove the restriction endonuclease. This will increase the efficiency of insert ligation with the vector DNA.

The use of restriction endonuclease site incorporation to clone PCR products requires several additional steps after PCR. Each additional

Cleavage close to the End of DNA Fragments

ENZYME	OLIGO SEQUENCE	CHAIN LENGTH	% CLEAVAGE 2 HR	20 HR
Acc I	GGTCGACC	8	0	0
	CGGTCGACCG	10	0	0
	CCGGTCGACCGG	12	0	0
Afl III	CACATGTG	8	0	0
	CCACATGTGG	10	>90	>90
	CCCACATGTGGG	12	>90	>90
Asc I	GGCGCGCC	8	>90	>90
	AGGCGCGCCT	10	>90	>90
	TTGGCGCGCCAA	12	>90	>90
Ava I	CCCCGGGG	8	50	>90
	CCCCCGGGGG	10	>90	>90
	TCCCCCGGGGGA	12	>90	>90
BamH I	CGGATCCG	8	10	25
	CGGGATCCCG	10	>90	>90
	CGCGGATCCGCG	12	>90	>90
Bgl II	CAGATCTG	8	0	0
	GAAGATCTTC	10	75	>90
	GGAAGATCTTCC	12	25	>90
BssH II	GGCGCGCC	8	0	0
	AGGCGCGCCT	10	0	0
	TTGGCGCGCCAA	12	50	>90
BstE II	GGGT(A/T)ACCC	9	0	10
BstX I	AACTGCAGAACCAATGCATTGG	22	0	0
	AAAACTGCAGCCAATGCATTGGAA	24	25	50
	CTGCAGAACCAATGCATTGGATGCAT	27	25	>90
Cla I	CATCGATG	8	0	0
	GATCGATC	8	0	0
	CCATCGATGG	10	>90	>90
	CCATCGATGGG	12	50	50
EcoR I	GGAATTCC	8	>90	>90
	CGGAATTCCG	10	>90	>90
	CCGGAATTCCGG	12	>90	>90
Hae III	GGGGCCCC	8	>90	>90
	AGCGGCCGCT	10	>90	>90
	TTGCGGCCGCAA	12	>90	>90
Hind III	CAAGCTTG	8	0	0
	CCAAGCTTGG	10	0	0
	CCCAAGCTTGGG	12	10	75
Kpn I	GGGTACCC	8	0	0
	GGGGTACCCC	10	>90	>90
	CGGGGTACCCCG	12	>90	>90
Mlu I	GACGCGTC	8	0	0
	CGACGCGTCG	10	25	50
Nco I	CCCATGGG	8	0	0
	CATGCCATGGCATG	14	50	75
Nde I	CCATATGG	8	0	0
	CCCATATGGG	10	0	0
	CGCCATATGGCG	12	0	0
	GGGTTTCATATGAAACCC	18	0	0
	GGAATTCCATATGGAATTCC	20	75	>90
	GGGAATTCCATATGGAATTCCC	22	75	>90
Nhe I	GGCTAGCC	8	0	0
	CGGCTAGCCG	10	10	25
	CTAGCTAGCTAG	12	10	50
Not I	TTGCGGCCGCAA	12	0	0
	ATTTGCGGCCGCTTTA	16	10	10
	AAATATGCGGCCGCTATAAA	20	10	10
	ATAAGAATGCGGCCGCTAAACTAT	24	25	90
	AAGGAAAAAAGCGGCCGCAAAAGGAAAA	28	25	>90
Nsi I	TGCATGCATGCA	12	10	>90
	CCAATGCATTGGTTCTGCAGTT	22	>90	>90
Pac I	TTAATTAA	8	0	0
	GTTAATTAAC	10	0	25
	CCTTAATTAAGG	12	0	>90
Pme I	GTTTAAAC	8	0	0
	GGTTTAAACC	10	0	25
	GGGTTTAAACCC	12	0	50
	AGCTTTGTTTAAACGGCGCGCCGG	24	75	>90
Pst I	GCTGCAGC	8	0	0
	TGCACTGCAGTGCA	14	10	10
	AACTGCAGAACCAATGCATTGG	22	>90	>90
	AAAACTGCAGCCAATGCATTGGAA	24	>90	>90
	CTGCAGAACCAATGCATTGGATGCAT	26	0	0
Pvu I	CCGATCGG	8	0	0
	ATCGATCGAT	10	10	25
	TCGCGATCGCGA	12	0	10
Sac I	CGAGCTCG	8	10	10
Sac II	GCCGCGGC	8	0	0
	TCCCCGCGGGGA	12	50	>90
Sal I	GTCGACGTCAAAAGGCCATAGCGGCCGC	28	0	0
	GCGTCGACGTCTTGGCCATAGCGGCCGCGG	30	10	50
	ACGCGTCGACGTCGGCCATAGCGGCCGCGGAA	32	10	75
Sca I	GAGTACTC	8	10	25
	AAAAGTACTTTT	12	75	75
Sma I	CCCGGG	6	0	10
	CCCCGGGG	8	0	10
	CCCCCGGGGG	10	10	50
	TCCCCCGGGGGA	12	>90	>90
Spe I	GACTAGTC	8	10	>90
	GGACTAGTCC	10	10	>90
	CGGACTAGTCCG	12	0	50
	CTAGACTAGTCTAG	14	0	50
Sph I	GGCATGCC	8	0	0
	CATGCATGCATG	12	0	25
	ACATGCATGCATGT	14	10	50
Stu I	AAGGCCTT	8	>90	>90
	GAAGGCCTTC	10	>90	>90
	AAAAGGCCTTTT	12	>90	>90
Xba I	CTCTAGAG	8	0	0
	GCTCTAGAGC	10	>90	>90
	TGCTCTAGAGCA	12	75	>90
	CTAGTCTAGACTAG	14	75	>90
Xho I	CCTCGAGG	8	0	0
	CCCTCGAGGG	10	10	25
	CCGCTCGAGCGG	12	10	75
Xma I	CCCCGGGG	8	0	0
	CCCCCGGGGG	10	25	75
	CCCCCCGGGGGG	12	50	>90
	TCCCCCCGGGGGGA	14	>90	>90

To test the ability of a restriction endonuclease to cleave a site that lies within a few bases of the end of a DNA fragment, a series of short, double-stranded oligonucleotides that contain the restriction endonuclease recognition site were digested. The table above illustrates the varying requirements restriction endonucleases have for the number of bases flanking their recognition sequences. Enzyme recognition sites appear in bold print. (Reprinted, with permission, from the 1993/94 New England Biolabs Catalog.)

step takes time and can lead to a loss in DNA recovery, potentially reducing cloning efficiency. Several of the cloning methods available for inserting PCR products into plasmids are optimized for speed and efficiency. They require as little manipulation as possible after the completion of the PCR. These methods, ligation-independent cloning, blunt-end cloning, and T/A cloning, do not require extraction and precipitation of the PCR products or lengthy enzyme digestion of the vector or the insert.

CLONING WITHOUT LIGATION

Several methods are available for cloning PCR products without having to perform a ligation reaction. These methods increase the efficiency of insertion of the amplified DNA into the vector and do not require a lengthy ligation reaction. The PCR products to be cloned contain long single-stranded tails, usually ≥12 bases, that are annealed to a linearized vector that contains complementary single-stranded tails. Because of the length of the corresponding complementary overhangs, no ligase reaction is required. After annealing, the vector and insert are transformed directly into competent bacteria. It is possible to generate either symmetrical or asymmetrical overhangs for either bidirectional or unidirectional cloning, respectively, depending on the intended use of the PCR product. Several methods are available for generating single-stranded tails that are long enough to form a stable interaction between the vector and the insert such that ligation is not required for transformation. These are uracil DNA glycosylase (UDG) cloning (Duncan 1981; Friedberg et al. 1981; Nisson et al. 1991; Buchman et al. 1992, 1993), ligation-independent cloning (Aslanidis and de Jong 1990; Haun et al. 1992), and PCR-induced subcloning (Shuldiner et al. 1990).

UDG Cloning

UDG cloning takes advantage of the biological properties of UDG, an enzyme involved in the TTP biosynthetic pathway that cleaves the *N*-glycosylic bond between the deoxyribose moiety and uracil. This cleavage yields abasic dU residues, which lead to a disruption of DNA base pairing (Duncan 1981; Friedberg et al. 1981). To utilize this enzyme for the cloning of PCR products, primers must be designed that contain dUMP in the 5′-terminal nucleotides. These nucleotides correspond to the complementary overhang in the vector sequences (Nisson et al. 1991). UDG cleaves the incorporated UMP residues in the final PCR products and generates a single-stranded 3′ overhang. It is important to generate a 3′ tail that contains a sufficient number of nucleotides so that efficient and stable annealing occurs between the vector and PCR insert DNAs. It is also important to design primers such that a dUMP is inserted at the junction between the sequences complementary to the vector and the PCR template sequences, as well as to incorporate several dUMP residues into the 5′-terminal nucleotides of the oligonucleotide primer so that after cleavage with UDG, the other base-paired oligonucleotides lose contact with the opposite strand, thereby generating a single-stranded overhang.

To generate unique vector- and insert-specific overhangs, it is necessary to design primers for the cloning vector that are complementary to the overhangs generated for the insert fragment. These primers are then used to PCR-amplify and generate a linearized vector with compatible overhangs. In this method, both the vector and the insert are treated with UDG, annealed, and transformed into bacteria (Nisson et al. 1991). The advantage of synthesizing unique vector- and insert-specific overhangs is the ability to customize the nucleotides in the primers to contain sequences specific to a template DNA, or the incorporation of specific restriction sites that allow further manipulation after the PCR product is cloned. The disadvantages of generating both vector- and insert-specific primers are the requirement for two additional oligonucleotide primers for the vector, and the fact that mistakes in the PCR amplification of the vector may lead to alterations of important vector sequences, such as the *lacZ* gene, the bacterial antibiotic resistance gene, the origin of replication, or other regulatory sequences in the vector, such as eukaryotic promoters. These mutations can lead to a decrease in the cloning efficiency and cause problems with subsequent uses of the plasmid.

Several vectors are commercially available (Life Technologies) which facilitate the cloning of products that have dUMP incorporated into the 5′-terminal nucleotides on both strands. The available vectors are CLONEAMP pAMP 1, CLONEAMP pAMP 10, CLONEAMP pAMP 18, and CLONEAMP pAMP 19 (Fig. 1A). These vectors are provided as linear DNA molecules with defined 3′ overhangs, and to use these vectors, all oligonucleotide primers must have the appropriate complementary nucleotides at the 5′ end (Fig. 1B). As described above, after PCR, the products are treated with UDG in the presence of vector, annealed for 30 minutes at 37°C, and then transformed into competent bacteria (Fig. 1C).

Figure 1 shows the important features for each of these vectors, sequences at the polylinker cloning site, and a schematic diagram of the UDG cloning protocol. CLONEAMP pAMP 1 has asymmetrical 12-base overhangs, 5′-CTACTACTACTA-3′ and 5′-CATCATCATCAT-3′, which are nested within a polylinker region flanked by either the SP6 or T7 phage promoters. CLONEAMP pAMP 10 is identical to CLONEAMP pAMP 1 except that the cloning region comprises two symmetrical 12-nucleotide overhangs, 5′-CTACTACTACTA-3′. Inserts anneal in both orientations in the CLONEAMP pAMP 10 vector. CLONEAMP pAMP 18 and CLONEAMP pAMP 19 are identical to pUC 18 and pUC 19, respectively, except for the addition of several new restriction endonuclease sites within the polylinker region that make up the single-stranded cloning region (Fig. 1A). These vectors are provided as linear DNA with 12-base overhangs generated within the polylinker region. Figure 1B shows the 5′-terminal primer sequences needed for each of the vectors.

Ligation-independent Cloning

The second system for cloning that does not require ligation is based on generating PCR products that have one of the dNTPs missing in the 5′-terminal nucleotides (Aslanidis and de Jong 1990; Haun et al. 1992). This is accomplished by designing primers that have only three of the nucleotides present in the first 12 or more bases, depending on the desired length of the overhang. After generating a PCR product, the sample is treated with T4 DNA polymerase in the presence of the dNTP, which is not present in the 5′-terminal sequences. T4 DNA polymerase has exonuclease activity as well as polymerase activity. In the absence of the necessary dNTPs, the exonuclease activity removes nucleotides in a 5′→ 3′ direction from the ends of the PCR product until the enzyme reaches the first nucleotide in the amplified DNA that corresponds to the dNTP present in the reaction mix. The presence of this dNTP inhibits further exonuclease activity, and the single-stranded 3′ overhang is stable. The enzymatic activity is killed by heat-inactivation and the insert is annealed to a linearized vector treated in a similar way and transformed into competent bacteria.

A commercially available vector, pDIRECT (CLONTECH), has been modified for the ligation-independent cloning (LIC) of products that have been treated with T4 DNA polymerase. Figure 2 shows a schematic map of the vector, its polylinker, and the single-stranded region for annealing with insert DNA (Fig. 2A), as well as a diagram of the LIC protocol (Fig. 2B). The single-stranded ends of the DNA are asymmetrical, 5′-GGGCCGAACCAG-3′ and 5′-GGGCGAGCGAG-3′, and therefore the insert is cloned unidirectionally. This vector has the T3 and T7 phage promoters flanking the cloning region to allow direct sequencing of the insert or in vitro transcription of the insert from either strand.

When using this vector, it is essential to design PCR primers that contain the specific nucleotide sequence that corresponds to the complementary sequences of the pDIRECT vector. In this vector, there are no T residues in the single-stranded region; therefore, dTTP must be included in the T4 DNA polymerase reaction to block exonuclease activity at the first T in the vector or in the insert. This vector is supplied as a linear DNA with the appropriate buffers and T4 DNA polymerase. As described above, after treatment of the PCR products with T4 DNA polymerase, the vector and insert are annealed and then transformed into competent bacteria. The cloning site is within the *lacZ* gene, and a color selection can be used to identify clones that are positive for the insert DNA.

PCR-induced Subcloning

The third ligation-independent method for cloning PCR products is described by Shuldiner et al. (1990). This strategy is called PCR-induced (ligase-free) subcloning and utilizes primers that have 24 additional nucleotides at the 5′ end that correspond to the 3′ ends of the linearized vector. After initial amplification, the PCR product is divided into two tubes and mixed with linearized vector. A second set of PCR amplifications is performed using the 24 vector-specific bases in the original PCR product as the primer at one end and a second set of primers, one in each reaction, that correspond to the complementary strands at the other termini of the vector. The second PCR results in the generation of two full-length vector- and insert-containing clones, one with the insert upstream of the vector and one with the insert downstream from the vector. The two samples are mixed, chemically denatured, and then reannealed. The annealed product contains a long region of overlap and can therefore be transformed directly into competent bacteria. The advantages of this sys-

Figure 1 CLONEAMP pAMP vectors and the UDG cloning strategy. (*A*) Schematic maps of pAMP vectors. pAMP 1 and pAMP 10 are identical vectors except for the sequence of the single-stranded region (shown here, and described in the text) required for annealing with insert DNA. pAMP 1 is for unidirectional cloning of insert DNA and pAMP 10 is for bidirectional cloning. pAMP 1 and pAMP 10 contain a bacterial origin of replication, the ampicillin resistance gene for the selection of transformed bacteria, and the f1 intergenic region from M13 phage for the production of single-stranded DNAs. The polylinker is within the *lacZ* gene, permitting color selection of positive colonies. Vectors pAMP 18 and pAMP 19 are modified pUC18 and pUC19 vectors. They have a single-stranded region added in the polylinker of the original vectors for annealing to complementary PCR product inserts. The single-stranded cloning region contains the restriction endonuclease sites *Nco*I and *Bgl* II on one strand, and *Mlu* I and *Spe* I on the other strand. The two plasmids are identical except for the orientation of the polylinker region (shown here), which is inverted. The single-stranded tails are the same for both clones and can be used for directional cloning of an insert. The vectors contain a bacterial origin of replication and the ampicillin resistance gene for the selection of transformed bacteria. The polylinker, within the *lacZ* gene, permits color selection of positive colonies. (*B*) dUMP PCR primer tails for pAMP 1, pAMP 10, pAMP 18, and pAMP 19. The sequences shown for each clone are required at the 5′ end of oligonucleotide primers to be used for PCR amplification. (*C*) pAMP cloning protocol. PCR primers with pAMP-specific sequences at the 5′ end are used to amplify DNA. PCR products are treated with UDG in the presence of a compatible pAMP vector (supplied as a linear molecule). The vector and the insert are then annealed, and the recombinant plasmid is transformed into competent bacteria. Primer-specific sequences for pAMP 1 are shown in this diagram. The protocol is identical for all of the pAMP clones. (Adapted, with permission, from Life Technologies Inc.)

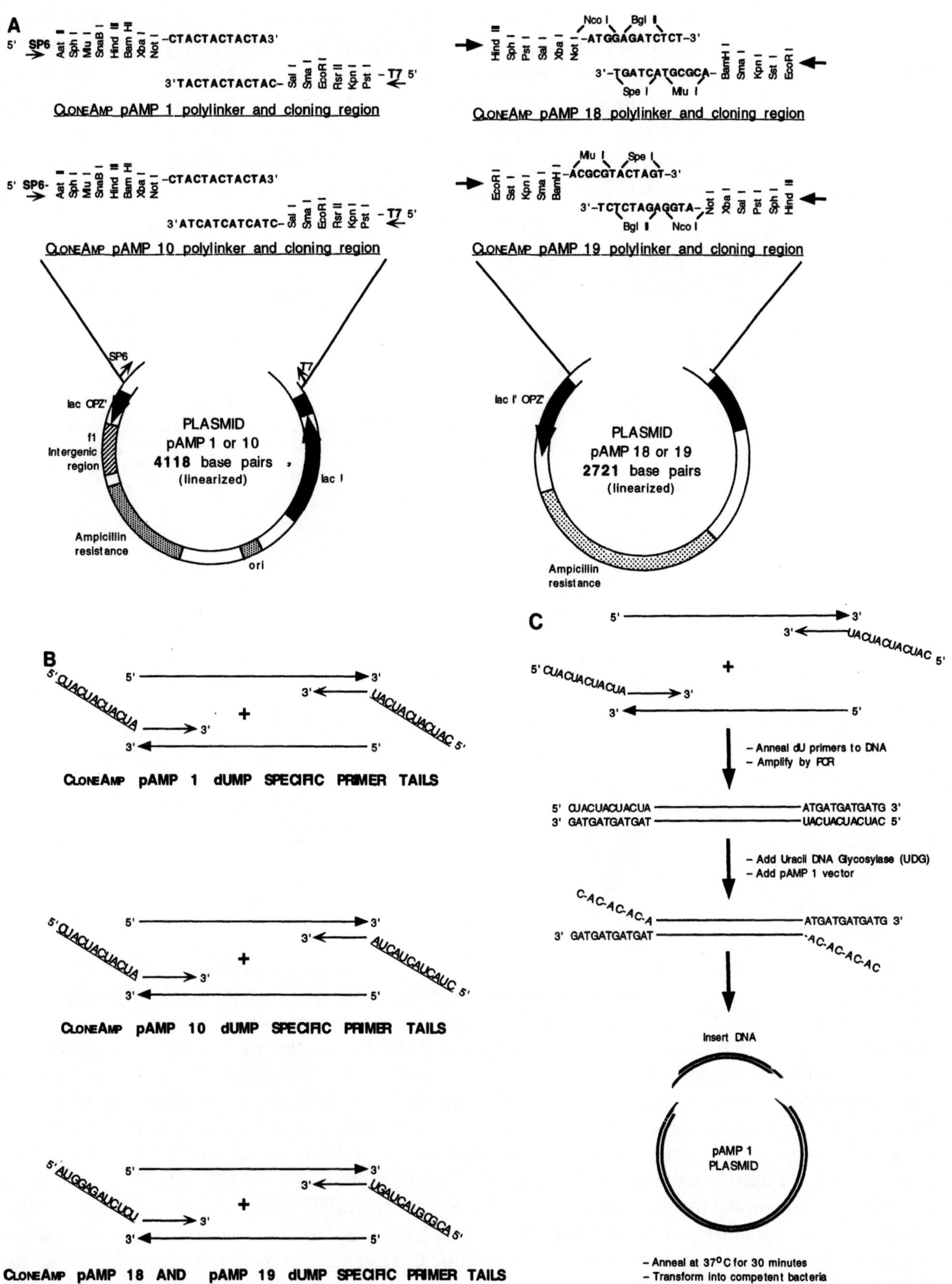

Figure 1 (*See facing page for legend.*)

tem are that any vector can be used and the cloning is directional. However, although this procedure does not require any enzymatic alteration of the PCR products, it does require the production of very long oligonucleotide primers for the first PCR amplification and an additional set of primers for the secondary PCR. Also, because the vector is being reamplified, the potential problems associated with misincorporation, as described above, apply.

T/A CLONING

One method for the direct cloning of PCR products utilizes linearized vectors that contain a single 3′ thymidine (T) overhang and inserts

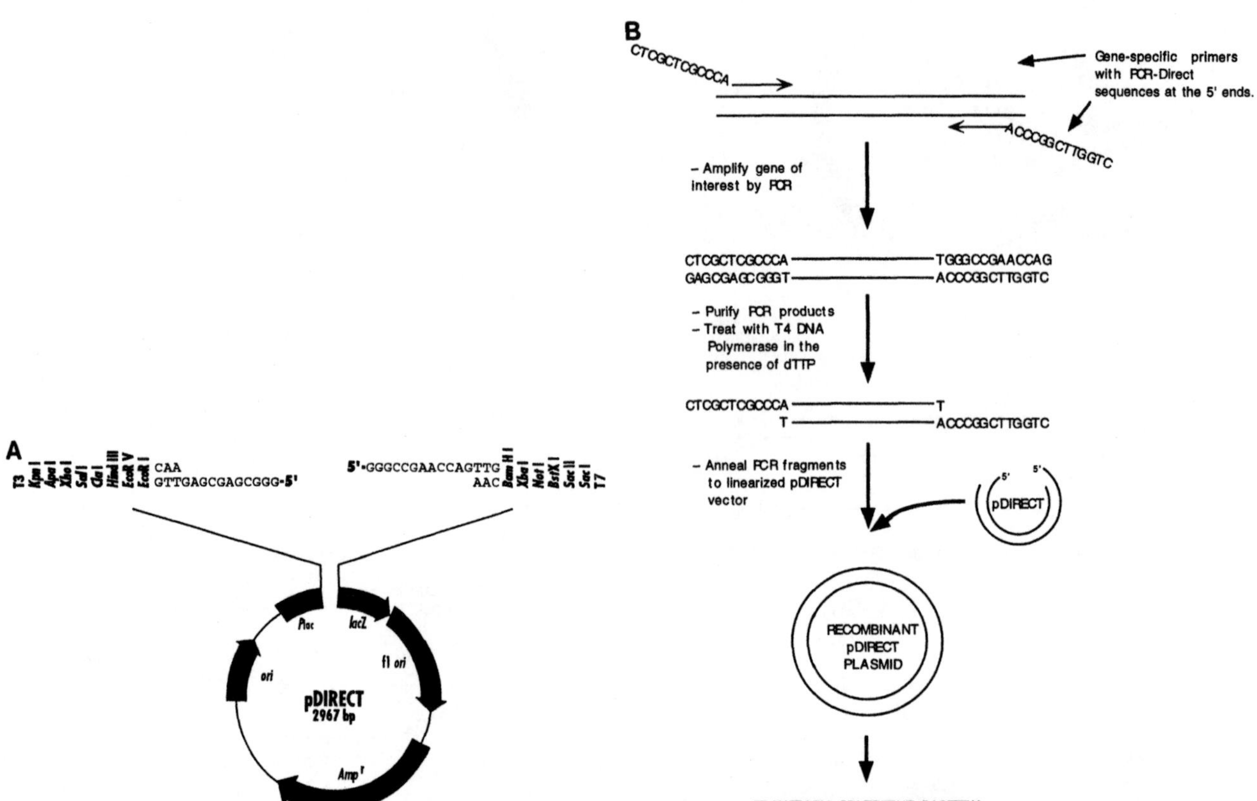

Figure 2 pDIRECT vector and ligase-independent cloning strategy. (*A*) Schematic map of pDIRECT. The linearized pDIRECT vector contains noncomplementary 5′ single-stranded ends for cloning PCR products using the PCR-Direct system. In addition, pDIRECT contains the f1 origin of replication for the synthesis of single-stranded DNA upon coinfection with M13 helper phage. The cloning site and the multiple restriction enzyme sites flanking it are located in the *lacZ* gene, allowing blue/white color screening for recombinant plasmids. T3 and T7 promoters flank the cloning site in opposite orientations. (*B*) Schematic diagram of pDIRECT cloning procedure. PCR primers with PCR-Direct sequences at the 5′ end are used to amplify the desired DNA. The PCR product is purified and then treated with T4 DNA polymerase in the presence of dTTP. The PCR insert fragment is annealed with the pDIRECT vector, which is supplied as a linear DNA molecule, and the recombinant plasmid is used to transform competent bacteria. (Adapted, with permission, from CLONTECH Laboratories, Inc. [PCR Research Tools Catalog].)

with a single 3′ adenosine (A) overhang. This system is called T/A cloning and takes advantage of the extendase activity that several of the DNA polymerases have (Clark 1988; Hu 1993). Extendase activity is defined as the non-template-dependent addition of a single nucleotide at the 3′ end of an extended PCR product. For *Thermus aquaticus*, *Thermus flavus*, and *Thermococcus litoralis*, this nucleotide is generally an A residue; however, the added nucleotide differs according to the terminal nucleotide in the template-dependent product (Hu 1993).

Several techniques are available for generating linear vectors that have a single 3′ T overhang so that the PCR products containing a 3′ A residue can be directly ligated into the vectors. One method uses vectors that have been linearized with enzymes that leave a single T nucleotide overhang (Mead et al. 1991). Three such restriction endonucleases are available: *Mbo*II, *Xcm*I, and *Hph*I. *Hph*I and *Mbo*II each have a 5-bp recognition sequence and a cleavage site 8 nucleotides downstream that generates a single 3′ nucleotide overhang. *Xcm*I has a 15-bp recognition sequence that is cleaved internally to generate a single-base overhang. It is possible to design each of these restriction sites so that the single-base overhang is a T residue. It is essential to insert two inverted copies of the restriction site in the vector, such that a T residue is present on both strands after digestion.

Two additional methods are available for generating vectors with a single 3′ T residue tail. These rely on the addition of a single T residue to a vector that has been linearized by a restriction enzyme that generates a blunt end (Holton and Graham 1991; Marchuk et al. 1991). The first method adds a dideoxy-thymidine triphosphate (ddTTP) using terminal transferase. Because a dideoxy-nucleotide is used, only a single T residue is added to the vector template (Holton and Graham 1991). The presence of a ddTMP nucleotide at the 3′ ends of the vector does not inhibit ligation of the vector with the PCR product. The second method takes advantage of the extendase activity described for *Taq* DNA polymerase (Marchuk et al. 1991). *Taq* DNA polymerase preferentially adds an A residue to the 3′ end of DNA templates; however, the polymerase does add other nucleotides at a lower efficiency. To add a single 3′ T residue, the blunt-end vector is incubated with *Taq* polymerase in the presence of high levels of dTTP. In the absence of any other nucleotide, a single 3′ T residue is added to the vector, generating compatible ends for ligation with PCR products containing an extra 3′ A residue.

Several commercially available vectors (Invitrogen) are designed to facilitate the cloning of PCR products containing an additional 3′ A overhang. The vectors are called T/A cloning vectors (Fig. 3) and are supplied as linear DNAs with a single 3′ T nucleotide overhang on each strand. PCR products can be ligated directly into the vectors

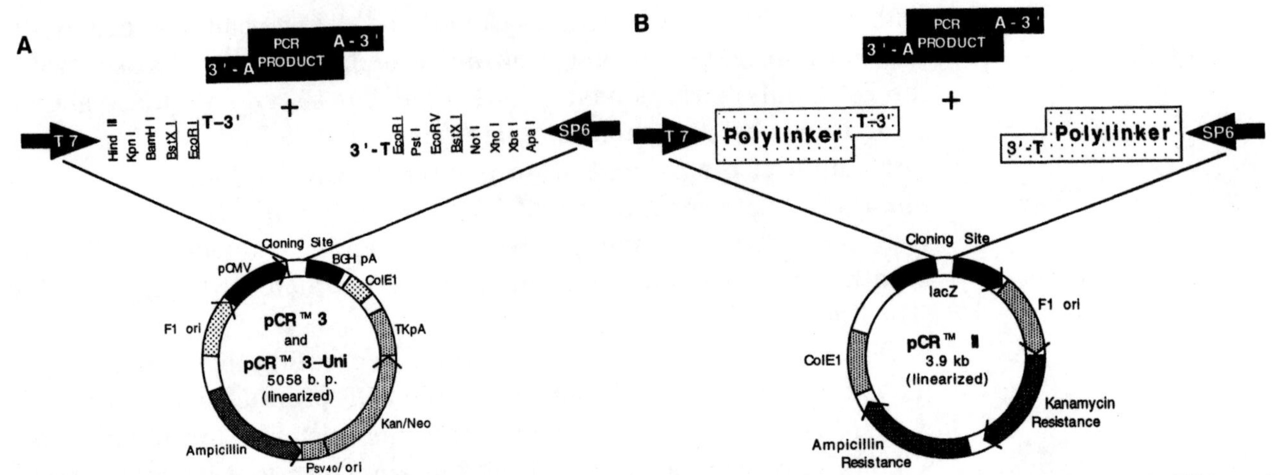

Figure 3 pCR™3, pCR™3-Uni, and pCR™II vectors for T/A cloning. (*A*) Vector map of pCR™3 and pCR™3-Uni. These two vectors are identical, except that pCR™3-Uni has been modified so that the linear vector is hemiphosphorylated. This allows the directional cloning of PCR product DNAs that are also hemiphosphorylated. The polylinker and T/A cloning sites in these vectors are downstream from the immediate-early CMV promoter to allow the direct expression of the DNA insert in eukaryotic cells. Downstream from the cloning site is the bovine growth hormone (BGH) polyadenylation signal. These vectors also contain the neomycin/kanamycin resistance gene, under the control of the SV40 early promoter, for the selection of stable, G418-resistant, eukaryotic cell lines. Other features of these vectors are the bacterial origin of replication, the ampicillin and kanamycin resistance genes for the selection of transformed bacteria, the f1 origin from M13 phage for the production of single-stranded DNA, and the T7 and SP6 phage promoters flanking the cloning site for the in vitro transcription of sense and antisense strands. These vectors are supplied as linear molecules and the PCR product is ligated directly into these vectors without modifications. (*B*) Vector map of pCR™II. Vector pCR™II contains a bacterial origin of replication, the ampicillin and kanamycin resistance genes for the selection of transformed bacteria, the f1 origin from M13 phage for the production of single-stranded DNA, and the T7 and SP6 phage promoters flanking the cloning site for the in vitro transcription of sense and antisense strands. The polylinker region is in the *lacZ* gene, allowing blue/white colony selection to identify positive clones. pCR™II is supplied as a linear molecule, and the PCR product is ligated directly into this vector without modifications. (Adapted, with permission, from Invitrogen Corporation.)

without further enzymatic modification, or if necessary, the insert DNA can be gel-purified before ligation.

Two vectors, pCR™3 and pCR™3-Uni (Fig. 3A), are designed as eukaryotic expression vectors. In addition to the T/A cloning site and polylinker region, these vectors contain the immediate-early cytomegalovirus (CMV) promoter upstream of the cloning site. This promoter directs efficient transcription in most eukaryotic cell lines. Downstream from the cloning site is the bovine growth hormone (BGH) poly(A) signal and the transcription termination signal. This sequence allows the efficient polyadenylation of the expressed open reading frame (ORF) in eukaryotic cells and aids message stability. These two vectors also contain the SV40 promoter/origin upstream of the neomycin/kanamycin resistance gene for the selection of stable cell lines expressing the inserted ORF. The SV40 origin directs episomal replication in cell lines expressing the SV40 large T antigen.

The vector pCR^{TM}3-Uni has been modified so that the 5′ phosphate has been removed from the upstream, or left, arm of the vector. To ensure that the insert is ligated in only one orientation, the upstream or forward primer must be phosphorylated at the 5′ end. Using this primer and an unphosphorylated downstream, or reverse, primer, a hemiphosphorylated PCR product is generated. Because the vector and the insert each have only one strand with a 5′ phosphate, the insert only ligates in one orientation. Although the cloning efficiency is reduced, more than 90% of the plasmids with inserts have the PCR product in the correct orientation.

The second T/A cloning vector, pCR^{TM}II (Fig. 3B), has a polylinker surrounding the T/A cloning site that is flanked by the SP6 and T7 phage promoters for in vitro transcription and for sequencing of the inserts. The polylinker region for this vector is in the *lacZ* gene, so that a color selection can be used to identify clones with inserts. The vector contains both the ampicillin and the kanamycin resistance genes for the selection of vector-containing bacteria.

BLUNT-END CLONING

A method for cloning PCR fragments that does not rely on the addition of specific "primer-tails" or the incorporation of restriction endonuclease sites at the 5′ ends of the oligonucleotide primers is the direct ligation of PCR products into vectors linearized with a restriction endonuclease that generates a blunt end. This method is more inefficient than other cloning procedures, because the ligation of blunt ends favors the direct recircularization of the vector; however, there are several methods that increase the efficiency of insert ligation. One is to have a large molar excess of insert in the ligation reaction. A second method to increase the efficiency of blunt-end ligation is to treat the vector with alkaline phosphatase to remove the 5′ phosphates from both ends of the vector. This generates termini that do not serve as substrates for T4 DNA ligase and therefore prevent recircularization of the vector. When ligating a PCR product with a phosphatased vector, it is essential to generate PCR products that are phosphorylated at both 5′ ends. This can be achieved by directly kinasing the amplified DNA or by using primers that have been kinased or have a phosphorylated nucleotide incorporated at the 5′ end.

Another important factor that leads to the inefficient cloning of unmodified PCR products is the incorporation of a single nucleotide (usually an A residue) at the 3′ end of the amplified PCR products. Several of the thermostable DNA polymerases have template-independent extendase activity that incorporates an additional nucleotide at the 3′ end of the DNA product (Clark 1988; Hu 1993). Products that contain an extra nucleotide ligate with a very low efficiency into blunt-end vectors. Some cloning strategies, described above, take advantage of the extendase activity of DNA polymerases, but for blunt-end cloning, it is important to use a thermostable DNA

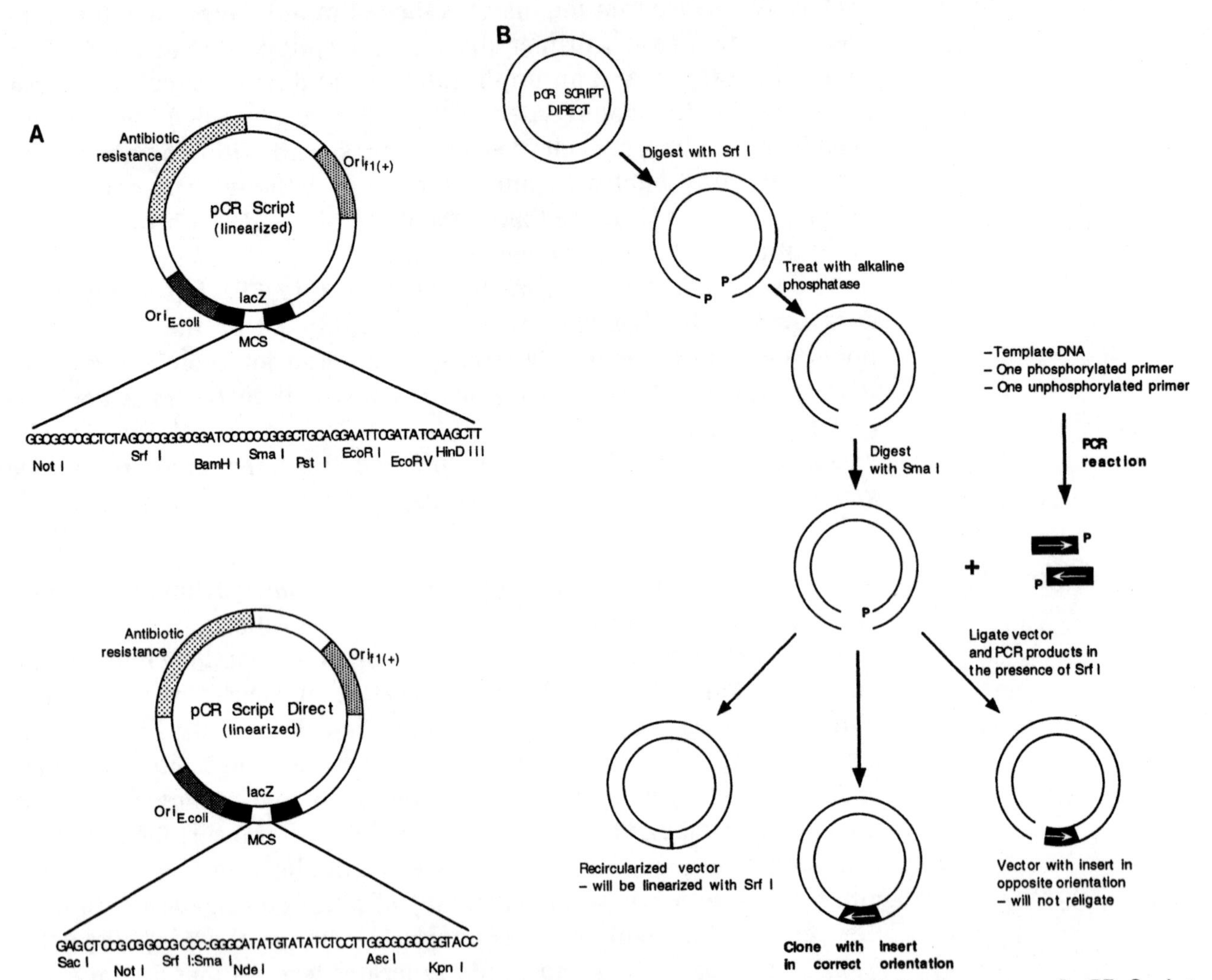

Figure 4 pCR-Script vectors and blunt-end directional cloning strategy. (*A*) Vector maps of pCR-Script and pCR-Script Direct. The pCR-Script vectors contain a bacterial origin of replication, an antibiotic resistance gene for selection of transformed bacteria, and the f1 origin from M13 phage for the synthesis of single-stranded DNA. These vectors differ in the restriction endonuclease sites in the polylinker region. The polylinker region is within the *lacZ* gene, allowing blue/white color selection of positive colonies. pCR-Script Direct has been modified so that the linear DNA is hemiphosphorylated. DNA inserts that are also hemiphosphorylated ligate in one orientation into this vector. (*B*) pCR-Script Direct cloning protocol. This diagram shows the steps involved in the preparation of pCR Script Direct and the ligation with insert DNA. After digestion with restriction endonuclease *Srf* I, the vector is treated with alkaline phosphatase to remove 5′ phosphates from the ends of the DNA. The vector is subsequently digested with restriction endonuclease *Sma*I, generating a blunt-end product that contains a single 5′ phosphate. This vector, supplied as a linear molecule, is ligated directly with a hemiphosphorylated PCR product. The recombinant plasmid is then transformed into competent bacteria. (Adapted, with permission, from Costa and Weiner 1994b [Stratagene Cloning Systems].)

polymerase that has proofreading activity, and therefore reduced extendase activity, or to remove any extra nucleotides from the PCR product. This can be accomplished by incubating the amplified DNA with a DNA polymerase that has exonuclease activity. If the reaction is done in the presence of the four deoxynucleotide triphosphates (dNTPs), only the extra 3′ nucleotide is removed, generating a blunt-end product (Liu and Schwartz 1992). Two DNA polymerases, T4 DNA polymerase and *Pfu* polymerase (from *Pyrococcus furiosus*), are recommended for removing terminal 3′ nucleotides from PCR products (Costa and Weiner 1994b).

Two commercially available vectors, pCR Script and pCR Script Direct (Stratagene), have been optimized for cloning blunt-end PCR products (Costa and Weiner 1994a; Costa et al. 1994b). Inserts are ligated bidirectionally into pCR Script plasmids and unidirectionally into pCR Direct. Figure 4 shows a schematic diagram of the two vectors (Fig. 4A) and a diagram of the protocol for using the pCR Script Direct clone (Fig. 4B). The next chapter in this section ("Cloning and Analysis of PCR-generated DNA Fragments) describes in detail the basis for this blunt-end cloning protocol.

Each of the techniques described here has different advantages and disadvantages. The method chosen for cloning PCR products depends on the type and amount of PCR product being generated, the ultimate use of the PCR product, and the resources available to the researcher. Under normal reaction conditions, a large amount of PCR product is produced and the efficiency of obtaining clones, regardless of the cloning procedure used, is very high. It becomes necessary to optimize the cloning of PCR products when very small amounts of insert DNA are generated. For each of the methodologies described here, it is important to optimize conditions to ensure the generation of clones with the correct insert, which can then be used for further studies.

ACKNOWLEDGMENTS The author thanks Drs. Gabriela Dveksler and Alison McBride for their critical reviews of this article. I also thank New England Biolabs, Life Technologies, CLONTECH, Invitrogen Corporation, and Stratagene Cloning Systems (Drs. G. Costa and M. Weiner) for allowing the reproduction of the table and figures. The statements and assertions herein are those of the authors and do not represent the opinions of the Uniformed Services University of the Health Sciences or the Department of Defense.

REFERENCES

Aslanidis, C. and P.J. de Jong. 1990. Ligation-independent cloning of PCR products (LIC-PCR). *Nucleic Acids Res.* **18:** 6069–6074.

Buchman, G.W., P. Booth, and A. Rashtchian. 1993. The CLONEAMP pUC18 system: Cloning of the brain-derived neurotrophic factor gene. *Focus* **15:** 36–41.

Buchman, G.W., D.M. Schuster, and A. Rashtchian. 1992. Rapid and efficient cloning of PCR products using the CLONEAMP system. *Focus* **14:** 41–45.

Clark, J.M. 1988. Novel non-templated nucleotide addition reactions catalyzed by prokaryotic and eucaryotic DNA polymerases. *Nucleic Acids Res.* **16:** 9677–9686.

Costa, G.L. and M.P. Weiner. 1994a. Protocols for cloning and analysis of blunt-ended PCR-generated DNA fragments. *PCR Methods Appl.* (suppl.) **3:** S95–S106.

——. 1994b. Polishing with T4 or *Pfu* polymerase increases the efficiency of cloning of PCR fragments. *Nucleic Acids Res.* **22:** 2423.

Costa, G.L., A. Grafsky, and M.P. Weiner. 1994. Cloning and analysis of PCR-generated DNA fragments. *PCR Methods Appl.* **3:** 338–345.

Crowe, J.S., H.J. Cooper, M.A. Smith, M.J. Sims, D. Parker, and D. Gewert. 1991. Improved cloning efficiency of polymerase chain reaction (PCR) products after proteinase K digestion. *Nucleic Acids Res.* **19:** 184.

Duncan, B.K. 1981. DNA glycosylases. In *The enzymes*, 3rd edition, part A (ed. P. Boyer), pp. 565–586. Academic Press, New York.

Friedberg, E.C., T. Bonura, E.H. Radany, and J.D. Love. 1981. Enzymes that incise damaged DNA. In *The enzymes*, 3rd edition, part A (ed. P. Boyer), pp. 251–279. Academic Press, New York.

Haun, R.S., I.M. Serventi, and J. Moss. 1992. Rapid, reliable ligation-independent cloning of PCR products using modified plasmid vectors. *BioTechniques* **13:** 515–518.

Holton, T.A. and M.W. Graham. 1991. A simple and efficient method for direct cloning of PCR products using ddT-tailed vectors. *Nucleic Acids Res.* **19:** 1156.

Hu, G. 1993. DNA polymerase-catalyzed addition of non-templated extra nucleotides to the 3′ end of a DNA fragment. *DNA Cell Biol.* **12:** 763–770.

Kaufman, D.L. and G.A. Evans. 1990. Restriction endonuclease cleavage at the termini of PCR products. *BioTechniques* **9:** 304–305.

Liu, Z. and L.M. Schwartz. 1992. An efficient method for blunt-end ligation of PCR products. *BioTechniques* **12:** 28–30.

Lorens, J.B. 1991. Rapid and reliable cloning of PCR products. *PCR Methods Appl.* **1:** 140–141.

Marchuk, D., M. Drumm, A. Saulino, and F.S. Collins. 1991. Construction of T-vectors, a rapid and general system for direct cloning of unmodified PCR products. *Nucleic Acids Res.* **19:** 1154.

Mead, D.A., N.K. Pey, C. Herrnstadt, R.A. Marcil, and L.M. Smith. 1991. A universal method for the direct cloning of PCR amplified nucleic acid. *BioTechnology* **9:** 656–663.

Nisson, P.E., A. Rashtchian, and P.C. Watkins. 1991. Rapid and efficient cloning of Alu-PCR products using uracil DNA glycosylase. *PCR Methods Appl.* **1:** 120–123.

Scharf, S.J. 1990. Cloning with PCR. In *PCR protocols: A guide to methods and applications* (ed. M.A. Innis et al.), pp. 84–91. Academic Press, San Diego, California.

Shuldiner, A.R., L.A. Scott, and J. Roth. 1990. PCR-induced (ligase-free) subcloning: A rapid reliable method to subclone polymerase chain reaction (PCR) products. *Nucleic Acids Res.* **18:** 1920.

Weiner, M.P. 1993. Directional cloning of blunt-ended PCR products. *BioTechniques* **15:** 502–505.

Cloning and Analysis of PCR-generated DNA Fragments

Gina L. Costa and Michael P. Weiner

Stratagene Cloning Systems, La Jolla, California 92037

INTRODUCTION

This chapter presents methods for the improved blunt-end cloning of PCR-generated DNA fragments (Costa et al. 1994a). We show that *Pfu* DNA polymerase polishing of *Taq* DNA polymerase-generated PCR fragments increases the yield and efficiency of cloning. Using a triple primer set consisting of two outside, asymmetrically distanced primers and one fragment-specific primer, one can determine both the presence and orientation of cloned inserts. Application of these methods allows one to generate and clone a fragment in 1 day and to analyze putative clones the next, thereby saving a substantial amount of both time and effort.

REAGENTS

Preparation of Cloning Vector

Cloning kits:

- pCR-Script SK(+) Amp cloning system (Stratagene, cat. no. 211190)
- pCR-Script SK(+) Cam cloning system (Stratagene, cat. no. 211192)
- pCR-Script Direct SK(+) cloning system (Stratagene, cat. no. 211194)

Reagents required:

- Cloning vector (10 ng/μl)
- Blunt-end restriction endonuclease (10–20 units)
  - *Srf*I restriction endonuclease (Stratagene; cat. no. 501064)
  - *Sma*I restriction endonuclease
- 10X Universal buffer (1 M KOAc, 250 mM Tris-acetate, pH 7.6, 100 mM MgOAc, 5 mM β-mercaptoethanol, 100 μg/ml BSA)
- Alkaline phosphatase (0.1–0.2 units)
- Phenol (Tris-buffered)
- Chloroform-isoamyl alcohol (24:1)

Lithium chloride (LiCl; 10 M)
TE buffer (5 mM Tris-HCl, pH 8.0, 0.1 mM EDTA)

Several vectors and vector derivatives have been created for PCR cloning. These include the standard pBluescript-type multiple cloning sites (e.g., pBluescript II, pBK, pBC, pCR-Script Amp, and pCR-Script Cam; Stratagene) and the abbreviated multiple cloning sites as contained in the pCR-Script Direct plasmids (Stratagene) (see Fig. 1) (Bauer et al. 1992; Costa and Weiner 1994a,b; Costa et al. 1994b,c). The abbreviated multiple cloning sites allow the end user to incorporate common restriction enzyme sites into the PCR primer sets without the problem of having the same target sequence occurring in the plasmid vector. It is recommended that the chloramphenicol (cam) derivatives be used when subcloning DNA fragments generated from ampicillin (amp)-resistance-encoding plasmids. This ensures that after *Escherichia coli* transformation recombinant colonies are not the result of parental plasmid transformation.

PROTOCOLS

Bidirectional Cloning Vectors

Blunt-end cloning procedures utilize cloning vectors with blunt ends to capture DNA fragments for bidirectional insertion. Therefore, blunt-end cloning vectors do not require nucleotide overhangs for clonal insertion. Subsequently, blunt DNA fragments, such as PCR products, may be cloned directly. Because blunt-end cloning does not require the addition of extra bases to the primer sets, preexisting primers may be used to generate and clone a DNA fragment. By itself, blunt-end cloning is an inefficient method, with recombinant insertion generally accounting for less than 10% of all transformants (see Fig. 2A). Increased efficiency can be achieved by the inclusion of a restriction enzyme in the ligation reaction as in the pCR-Script method (see Fig. 2B) (Liu and Schwartz 1992).

1. Digest the vector DNA with restriction endonuclease in a 20-μl reaction mixture containing ddH_2O, the appropriate reaction buffer, plasmid DNA (1 μg), and restriction enzyme (10–20 units). Allow the digestion to incubate at the recommended temperature for 1 hour. *Optional:* A 1-μl aliquot of the digestion can be run on an agarose gel to check for linearization of the vector DNA.

2. Phenol/chloroform-extract the restriction digestion. Add an equal volume of Tris-buffered phenol, vortex, and transfer the aqueous top phase to a new tube. Add an equal volume of chloroform to the tube and vortex. Transfer the aqueous top phase to a new tube. Heat-treat the extracted DNA for 20 minutes at 65°C to remove any remaining chloroform (the boiling point of chloroform is 55°C).

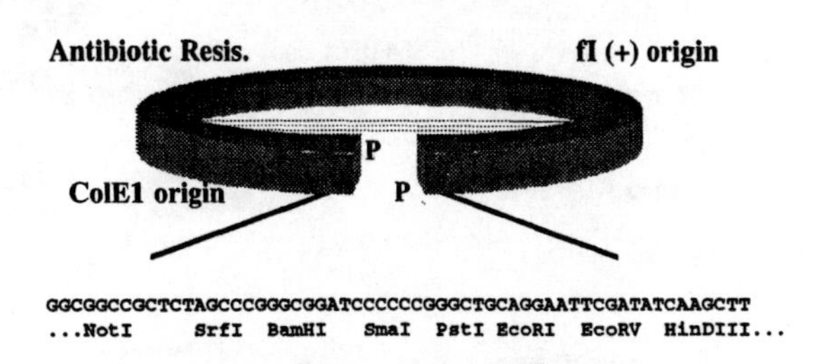

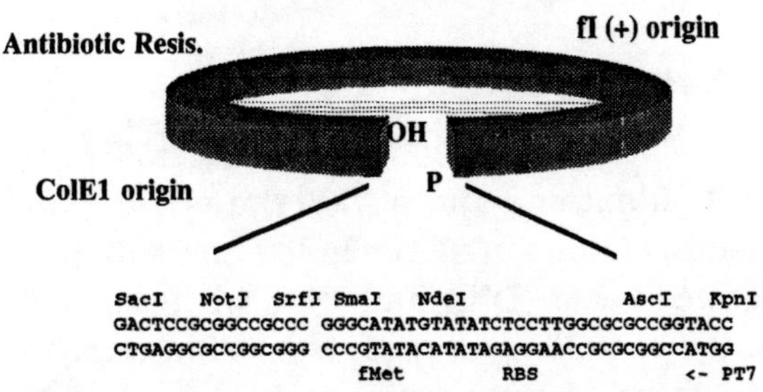

Figure 1 Several vectors have been developed for directional and bidirectional cloning. These include derivatives that encode either chloramphenicol or ampicillin resistance with the modified multiple cloning sites optimized for specific cloning operations (e.g., general subcloning or protein expression).

3. Precipitate the DNA extracted in step 2 with 0.1 volume of 10 M LiCl and 2.5 volumes of ice-cold 100% ethanol. Mix gently and centrifuge at room temperature at 12,000*g* for 10 minutes.

4. Following centrifugation, decant the supernatant. Dry the DNA pellet in vacuo for 10 minutes.

5. Resuspend the DNA in 50 μl of TE buffer. When resuspended into 50 μl of TE, the final concentration of the 1 μg of digested DNA should be approximately 10 ng/μl. Store the predigested vector at 4°C.

Directional Cloning Vectors

We relied on previous characterization of T4 DNA ligase and *E. coli* transformation to create a directional cloning method that does not require the addition of extra bases to the primers. First, T4 DNA ligase requires both a 5′ phosphate and a 3′ hydroxyl group to ligate two strands of DNA together efficiently. Second, linear DNA transforms *recBC*-proficient hosts of *E. coli* at a greatly reduced efficiency (it is decreased by approximately four orders of magnitude). It was therefore reasoned that directional cloning could be achieved by creating a

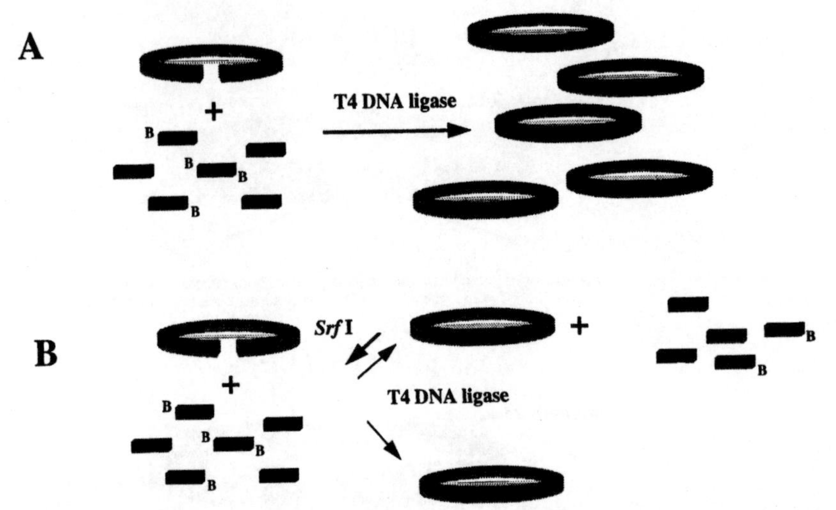

Figure 2 Blunt-end and pCR-Script cloning methods. Methods for standard blunt-end cloning (*A*) include incubation of the PCR product with predigested vector DNA and T4 DNA ligase. More efficient methods (*B*) include the addition of the restriction enzyme (in this example, *Srf* I endonuclease) to regenerate the linearized vector from the self-ligated vector during the ligation reaction.

monophosphorylated vector and a monophosphorylated insert. In the desired orientation, the ligation would result in a single-nicked, circular molecule. In the undesired, opposite orientation, the ligation would result in a linear molecule that would transform *E. coli* with a drastically reduced efficiency. A monophosphorylated vector is created by enzymatically treating the vector with a restriction endonuclease, removing the exposed 5′ phosphates with an alkaline phosphatase, and subsequently digesting the vector with a second restriction endonuclease. Degenerate restriction endonucleases may also be used.

Proper DNA sequence manipulation will enable the enzymatically processed vector to be used in a pCR-Script-type reaction, whereby self-ligated vector is susceptible to restriction by the endonuclease present in the ligation reaction, and the reading frame of the reporter gene is conserved. Owing to the importance of recreating a restriction enzyme site following vector self-ligation, the necessity of using highly purified enzymes for performing the directional and bidirectional cloning protocols as described cannot be overstated. Nuclease contamination must be determined and eliminated prior to performing the described experiments.

In a specific example, using the pCR-Script Direct directional cloning method, we enzymatically processed an SK(+) multiple cloning site that was engineered to contain both an *Srf* I (5′-GCCC|GGGC-3′) site and an *Sma*I (5′-TCCC|GGGC-3′; where the *Sma*I target sequence is underlined) site (Weiner 1993). The vector was first digested with *Srf* I, followed by removal of the 5′ phosphates with alkaline phosphatase and a second digestion with *SmaI*. Removal of the short DNA fragment after the *Srf* I-*SmaI* digestions results in the retention of an *Srf* I site (see Fig. 3). Phenotypic selection can still be used, since the reading frame is conserved. The monophosphorylated vector is produced by the general guidelines below.

1. Digest the appropriate vector DNA with the first blunt-end restriction endonuclease in a 50-μl reaction mixture containing ddH_2O, 1× Universal buffer, plasmid DNA (1 μg), and enzyme (10–20 units). Allow the digestion to incubate at the recommended temperature for 1 hour. *Optional:* A 1-μl aliquot of the reaction can be run on an agarose gel to check for linearization of the vector DNA.

   Note: A buffer that is compatible with the first restriction endonuclease digestion as well as the alkaline phosphatase dephosphorylation should be used to optimize the enzymatic processing of the vector DNA.

2. Inactivate the restriction enzyme by incubating the reaction for 20 minutes at 65°C. Remove to ice.

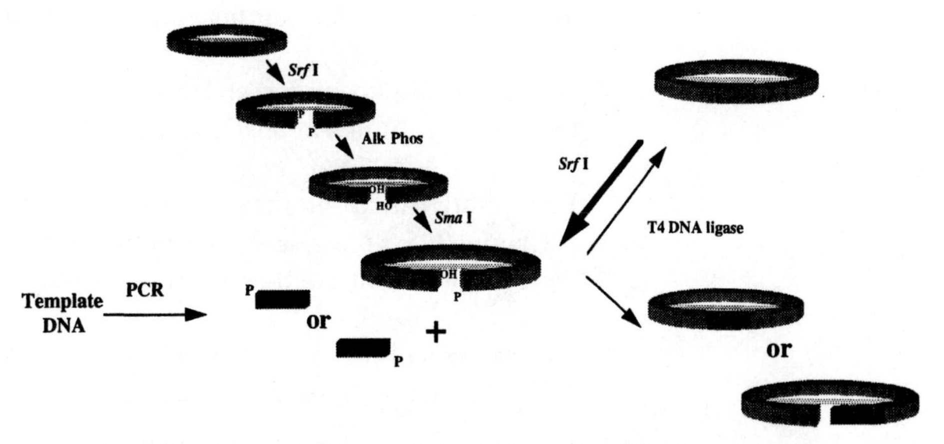

pCR-Script Direct SK(+) Multiple Cloning Site

```
PβGal  ->
SacI     NotI SrfI PstI BamHI  SmaI    NdeI             AscI   KpnI
GACTCCGCGGCCGCCCGGGCTGCAGGATCCCGGGCATATGTATATCTCCTTGGCGCGCCGGTACC
CTGAGGCGCCGGCGGGCCCGACGTCCTAGGGCCCGTATACATATAGAGGAACCGCGCGGCCATGG
                                  fMet          RBS          <- PT7
```

Figure 3 Monophosphorylation and pCR-Script Direct cloning. The plasmid pCR-Script Direct is digested with the restriction enzyme *Srf* I, treated with alkaline phosphatase to remove the 5′ phosphate groups, and then digested with a second restriction enzyme *SmaI*. Ethanol precipitation was used to remove the small (15-bp) linker. The insert fragment was created using either a machine-synthesized 5′-phosphorylated or kinase-treated primer. The monophosphorylated primer and vector are incubated in the presence of both *Srf* I and T4 DNA ligase. After room temperature incubation, the DNA was used to transform *E. coli.*

3. Add the alkaline phosphatase enzyme (0.1–0.2 units) directly to the heat-treated reaction mixture and incubate according to the manufacturer's guidelines.

   Note: Commercially available molecular biology-grade alkaline phosphatase often contains nuclease contamination. We recommend the use of bacterial alkaline phosphatase that has been purified devoid of contaminating nucleases and specifically quality-controlled for use in the pCR-Script assay.

4. Phenol/chloroform-extract the restriction-digested, alkaline-phosphatase-treated plasmid DNA. Add an equal volume of Tris-buffered phenol, vortex, and transfer the aqueous top phase to a new tube. Add an equal volume of chloroform to the tube and vortex. Transfer the aqueous top phase to a new tube. Heat-treat the extracted DNA for 20 minutes at 65°C to remove any remaining chloroform.

5. Set up a second 30-μl restriction enzyme digestion containing the processed alkaline-phosphatase-treated vector DNA with the downstream blunt-end restriction endonuclease by adding a 15-μl aliquot of phenol/chloroform-extracted DNA, ddH_2O, 1x Universal buffer, and enzyme (10–20 units). Allow this digestion to incubate at the recommended temperature for 1 hour.

6. Inactivate the second restriction enzyme by incubating the reaction for 20 minutes at 65°C. Remove to ice.

7. Precipitate the monophosphorylated DNA with 0.1 volume of 10 M LiCl and 2.5 volumes of ice-cold 100% ethanol. Mix gently and centrifuge at room temperature at 12,000*g* for 10 minutes.

8. Following centrifugation, decant the supernatant. Dry the DNA pellet in vacuo for 10 minutes.

9. Resuspend the DNA in 25 μl of TE buffer. When resuspended into 25 μl of TE, the final concentration of the monophosphorylated DNA should be approximately 10 ng/μl. The monophosphorylated vector can be stored at 4°C.

REAGENTS

Preparation of Insert

Synthetic oligonucleotide primers

Deoxynucleotide triphosphate mix (dNTP; 10 mM and 100 mM; Pharmacia, cat. no. 27-2094-01)

PCR optimization buffers

- Opti-Prime PCR optimization kit (Stratagene, cat. no. 200422)
- The PCR Optimizer (Invitrogen, cat. no. K1220-01)

Thermostable DNA polymerases (5–10 units)
Taq DNA polymerase
native *Pfu* DNA polymerase (Stratagene, cat. no. 600135)
cloned *Pfu* DNA polymerase (Stratagene, cat. no. 600153)
Taq Extender PCR additive (Stratagene, cat. no. 600148)
T4 polynucleotide kinase (10 units)
Kinase buffer (10 mM $MgCl_2$, 100 mM Tris-HCl, pH 7.5, 5 mM dithiothreitol)
ATP (10 mM)
TE buffer
STE buffer (1 M NaCl, 200 mM Tris-HCl, pH 7.5, 100 mM EDTA)
Mineral oil (Sigma, cat. no. M-3516)
Ammonium acetate (NH_4OAc; 4 M)

PCR Primer Design Considerations

Recent studies have shown that many species of DNA polymerases (e.g., T7, modified T7, *Taq*, Vent, *Tth*, and Klenow) exhibit terminal deoxynucleotidyl transferase (TdT) activity (Clark 1988; Hu 1993). The 3′-end nucleotide extension of PCR products by DNA polymerases has been found to be both nucleotide- and polymerase-specific. For example, *Taq* DNA polymerase-generated PCR products would be preferentially modified as follows (+ for extension; – for nonaddition).

3′ Nucleotide extensions associated with *Taq*-generated PCR products (Hu 1993)

3′-End nucleotide	3′-End extension
A	+A (at very low efficiency)
C	+A > +C
G	+G > +A > +C
T	–T > +A

There appears to be no consistent pattern by which bases are added by the polymerase. Therefore, it cannot be assumed that all DNA polymerases can be used to create blunt-end DNA fragments. However, for certain DNA polymerases, the expected 3′-end nucleotide of a PCR product can be controlled by the 5′-end nucleotide of the PCR primer (Hu 1993; Costa and Weiner 1994d).

For directional cloning using a monophosphorylated vector, insert monophosphorylation can be achieved by kinase-treating one primer prior to the PCR. Preferably, this could be achieved by synthesizing a PCR primer with a 5′ phosphate group chemically attached. Synthesis of a PCR primer with a 5′-terminal phosphate group ensures that all single-stranded DNA has been monophosphorylated. An advantage to kinase treatment is that all preexisting primer sets can be modified for use in directional cloning using a monophosphorylated vector. T4 polynucleotide kinase treatment is a simple and rapid technique.

PROTOCOLS

Primer-kinasing Treatment of a DNA Primer

1. Add the following to a microcentrifuge tube:

   3 μl of 10X kinase buffer
   0.5 μl of 10 mM ATP
   1 μl of T4 DNA kinase (10 units)
   5 μg of primer
   Deionized, distilled H_2O (ddH_2O) to a final volume of 30 μl

2. Incubate for 1 hour at 37°C.

3. Boil the reaction at 95°C to inactivate the T4 DNA kinase.

PCR Parameters

Use of the following standard conditions will amplify most target sequences, although it is recommended that conditions be optimized for each PCR application. Because no specific guidelines exist for choosing which buffer conditions to use for the various types of DNA primer-template systems, it is often advantageous to test a range of PCR buffers. Recently, a number of PCR optimization kits have been created that enable one to test several different buffer compositions (e.g., Opti-Prime PCR Optimization Kit [Stratagene] and The PCR Optimizer [Invitrogen]). By modifying specific buffer components of a PCR, it is possible to improve the yield and specificity of the desired PCR products. In addition, *Taq* Extender PCR additive (Stratagene) improves the PCR amplification of difficult templates and increases the reliability and yield of many PCR targets up to 10 kb in length (Nielson et al. 1994). The *Taq* Extender PCR additive increases the efficiency at which *Taq* DNA polymerase performs extension reactions on specific DNA segments in each cycle of PCR, thus resulting in a greater percentage of the extension reactions reaching completion. The *Taq* Extender should be added to amplification reactions in a unit-equivalent amount equal to that of *Taq* DNA polymerase; the standard *Taq* 10X reaction buffer may be replaced with an optimized *Taq* Extender 10X reaction buffer. Cycling is performed using standard PCR conditions.

PCR Amplification

1. Set up a 100-μl reaction in a 0.5-μl sterile, autoclaved microcentrifuge tube by adding in order:

ddH_2O (for a final volume of 100 ml)	75–84 μl
10X DNA polymerase buffer (for a final 1X volume)	10 μl
Template DNA (10–500 ng of plasmid DNA or 10^5–10^6 target molecules*)	1–10 μl
dNTP mix (250 nM of each dNTP)	2 μl
1 μg of upstream primer (T_m >55°C preferred)	1 μl
1 μg of downstream primer (T_m >55°C preferred)	1 μl
Thermostable DNA polymerase (5 units)	1 μl
suggested:	
Taq DNA polymerase (5 units)	1 μl
Taq Extender PCR additive (5 units)	1 μl

*For 3 X 10^5 targets: 1 μg of human single-copy genomic DNA
10 ng of yeast DNA
1% of an M13 plaque

2. Mix well and overlay with approximately 75 μl of mineral oil. PCR amplification should be conducted immediately.

3. Perform PCR using the following suggested temperature profile:

For $n = 1$ cycle		
denaturation	4 minutes	94°C
primer annealing	2 minutes	50°C
primer extension	2 minutes	72°C
Follow with $n = 25$–30 cycles		
denaturation	1 minute	94°C
primer annealing	2 minutes	54°C
primer extension	1 minute	72°C
Extend $n = 1$ cycle		
primer extension	10 minutes	72°C

The reactions are stopped by chilling to 4°C.

4. Following thermal cycling, check the PCR products for fidelity and yield by agarose gel analysis. A 10-μl aliquot of PCR product can be monitored by ethidium bromide staining of the DNA fragments following agarose gel electrophoresis. Known amounts of control DNAs should be run as markers for PCR product size and concentration.

Optional PCR Product Purification

The removal of excess PCR primers with selective ammonium acetate precipitation before proceeding with cloning protocols has been shown to increase the percentage of recombinants. An aliquot of the PCR product can be salted out of solution by the following protocol.

1. Add 0.1 volume of 10X STE buffer.

2. Add an equal volume of 4 M NH_4OAc to the sample.

3. Add 2.5 volumes of room-temperature 100% ethanol.

4. Immediately spin in a centrifuge at 12,000*g* for 10 minutes at room temperature to pellet the DNA. *Carefully* decant the supernatant.

5. Add 200 μl of 70% (v/v) ethanol.

6. Spin in a centrifuge at 12,000*g* for 10 minutes at room temperature. *Carefully* decant the supernatant. Dry the pellet in vacuo.

7. Resuspend the DNA in the original volume using TE buffer. Store at 4°C until further use.

End-polishing PCR Products with *Pfu* DNA Polymerase

Optimizing primer design in accordance with the specific DNA polymerase used can only minimally increase the number of blunt-end fragments produced following PCR. The traditional Klenow polymerase should be ***absolutely avoided*** for end-polishing because it retains a substantial amount of extendase activity. Fortunately, T4 and *Pfu* DNA polymerases were found not to exhibit any DNA extendase or TdT activity and can be used to create blunt-end fragments following PCR (see Fig. 4) (Hu 1993; Costa and Weiner 1994c,d). PCR polishing is used to remove the 3′-end nucleotide extensions placed on completed PCR products by DNA polymerases. The resulting *Pfu*-polished molecules will ligate into blunt-end cloning vectors at high efficiency in the presence of T4 DNA ligase. End-polishing of PCR products prior to ligation has been shown to increase overall recombinant cloning efficiencies (Costa and Weiner 1994c,d; Weiner et al. 1994).

Pfu DNA polymerase is essentially inactive at temperatures below 50°C. This allows ligation reactions to be done at 4–25°C and to be set up directly from the 72°C *Pfu* polishing step. This eliminates the need to extract the enzyme prior to ligation, as would be required if T4 DNA polymerase were used for polishing. *Pfu* polishing of PCR products generates high-fidelity, blunt-end DNA fragments in 30 minutes using only a small aliquot of the PCR product. *Pfu* polishing is outlined below and can be performed directly following the PCR or following the purification of the desired PCR product.

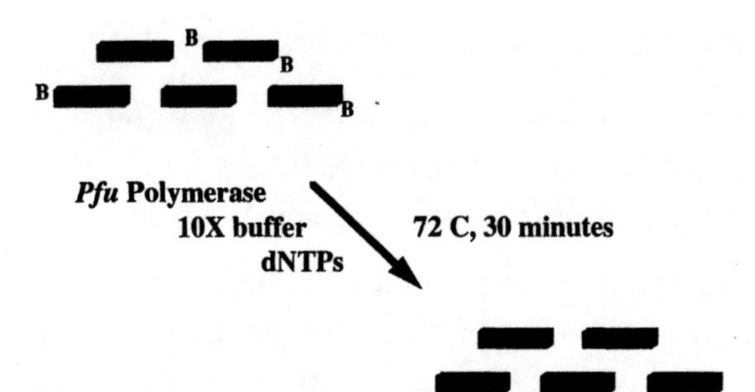

Figure 4 End-polishing of PCR-generated DNA fragments with *Pfu* DNA polymerase is used for increasing the amount of blunt-end DNA available for cloning. See text for protocol.

PROTOCOLS

PCR Polishing: Use of *Pfu* DNA Polymerase

Note: Before PCR polishing, it may be advantageous to verify the PCR products by agarose gel analysis to estimate the approximate concentration and to ensure that the correct PCR products have been created following thermal amplification. PCR polishing is conducted using an aliquot of the PCR amplification reaction and will therefore polish the ends of all DNA fragments present. In a typical 100-μl PCR amplification reaction, 5 μl of product can be used for PCR polishing.

Because routine PCR cloning procedures require the use of a small amount of DNA insert (1–4 μl), end-polishing reactions can be set up directly from the amplification using only 10 μl of PCR product. When end-polishing directly from the PCR, the remaining dNTPs and reaction buffer following thermal cycling are adequate for the polishing reaction (see below). Precipitation of the PCR products may be used for even more efficient end-polishing.

Pfu Polishing of Unpurified PCR Product

1. For polishing PCR-generated DNA fragments, transfer an aliquot of the PCR product directly from the reaction tube into a sterile 0.5-ml microcentrifuge tube and add the following reagents, in order:

   5–10 μl of PCR product
   1 μl of cloned *Pfu* DNA polymerase (2.5 units)
   ddH_2O to a final volume of 10 μl

   Gently mix the components and add a mineral oil overlay.

2. Incubate the polishing reaction for 30 minutes at 72°C.

3. Following the 30-minute incubation, remove the reaction to ice.

4. End-polished DNA fragments may be added directly to a ligation reaction.

Pfu Polishing of Purified PCR Product

1. For polishing purified PCR product, transfer an aliquot of the precipitated PCR product into a sterile 0.5-ml microcentrifuge tube and add the following reagents, in order:

   5–10 μl of precipitated PCR product
   1 μl of 10X cloned *Pfu* DNA polymerase buffer
   1 μl of dNTP mix (10 mM total; 2.5 mM each nucleotide triphosphate)
   1 μl of cloned *Pfu* DNA polymerase (2.5 units)
   ddH_2O to a final volume of 10 μl

   Gently mix the components and add a mineral oil overlay.

2. Incubate the polishing reaction for 30 minutes at 72°C.

3. Following the 30-minute incubation, remove the reaction to ice.

4. End-polished DNA fragments may be added directly to a ligation reaction.

REAGENTS

Efficient Clonal Ligation and Transformation of Blunt-end PCR Products

Cloning kits:
- pCR-Script SK(+) Amp cloning system (Stratagene, cat. no. 211190)
- pCR-Script SK(+) Cam cloning system (Stratagene, cat. no. 211192)
- pCR-Script Direct SK(+) cloning system (Stratagene, cat. no. 211194)

Reagents required:
- Cloning vector (10 ng/μl)
- Blunt-end insert (100 ng/μl)
- Blunt-end restriction endonuclease (5 units)
  - *Srf*I restriction endonuclease (Stratagene, cat. no. 501064)
- 10X Ligation buffer (250 mM Tris-HCl, pH 7.5, 100 mM $MgCl_2$, 100 mM DTT, 200 μg/ml BSA)
- ATP (10 mM)
- T4 DNA ligase (4 units)
- ddH_2O
- Competent *E. coli* cells
  - XL1-Blue (Stratagene, cat. no. 200236)
  - XL1-Blue MRF′ Kan (Stratagene, cat. no. 200248)
- SOC medium (see Media Preparation, below)
- Luria broth (LB) agar plates (see Media Preparation, below)
- Ampicillin-methicillin LB plates (see Media Preparation, below)
- Chloramphenicol LB plates (see Media Preparation, below)

5-bromo-4-chloro-3-indoyl-β-D-galactopyranoside (XGal; 100 mg/ml)
Isopropyl-β-D-thio-galactopyranoside (IPTG; 100 mM)
FALCON 2059 polypropylene tubes

MEDIA PREPARATION

SOC medium (per liter)
20 g tryptone
5 g yeast extract
0.5 g NaCl
water to 900 ml
Autoclave
Mix the following separately:
2.03 g $MgCl_2$
1.2 g $MgSO_4$
3.6 g glucose
Add water to a final volume of 100 ml
Filter-sterilize and add to cooled, autoclaved medium

LB agar (per liter)
10 g NaCl
10 g bacto-tryptone
5 g bacto-yeast extract
20 g bacto-agar
Adjust pH to 7.0 with 5 N NaOH
Add deionized H_2O to a final volume of 1 liter
Autoclave
Pour into petri dishes (~25 ml/100-mm plate)

LB-ampicillin-methicillin agar (per liter)
(Use for reduced satellite colony formation)
1 liter of LB agar
Autoclave
Cool to 55°C
Add 20 mg of filter-sterilized ampicillin
Add 80 mg of filter-sterilized methicillin
Pour into petri dishes (~25 ml/100-mm plate)

LB-chloramphenicol agar (per liter)
1 liter of LB agar
Autoclave
Cool to 55°C
Add 30 mg of filter-sterilized chloramphenicol
Pour into petri dishes (~25 ml/100-mm plate)

Ligation Reaction

To increase the efficiency of blunt-end cloning of PCR-generated fragments, it was found that a restriction enzyme added in a functional-unit excess relative to the units of T4 DNA ligase increases the efficiency of the ligation reaction (Liu and Schwartz 1992). This simultaneous restriction digestion and ligation reaction results in an increased efficiency of the blunt-end cloning by two mechanisms. First, as long as the PCR fragment does not, when ligated with the vector, create a restriction enzyme target site, the available circular vector is removed from the overall reaction by recombinant insertion. An increased amount of linear vector is made available during the ligation reaction by the restriction enzyme on self-ligated vector molecules. Second, because linear DNA molecules transform *E. coli* at a greatly reduced efficiency, they do not significantly contribute to the number of colonies observed after transformation. Both of these mechanisms result in a reduced overall transformation efficiency, but because only the linearized, nonrecombinant plasmids are reduced, the overall recombinant efficiency actually increases.

The pCR-Script method uses the restriction enzyme *Srf* I (Simcox et al. 1991). *Srf* I has an octanucleotide recognition sequence (5′-GCCC|GGGC-3′) that is rare and would occur on an average of 1 in 65,000 bp (because of the bias against CpG sequences in some DNA its actual occurrence in mammalian DNA is closer to 1 in 100,000 bp). The target site is blunt-ended and contains an internal 6-base recognition site (5′-CCC|GGG-3′) that can be recognized by another blunt-end restriction enzyme (*Sma*I). This was important to the development of the pCR-Script Direct method because the actual PCR cloning with directionality occurs in a reaction identical to that described for pCR-Script.

The addition of a restriction endonuclease to the ligation reaction allows an overall fourfold increase in clonal efficiency, along with a greatly reduced background. For the bidirectional cloning of PCR-generated DNA fragments, it is recommended to use a pCR-Script-type reaction containing a predigested vector DNA and *Pfu* DNA polymerase-generated or *Pfu*-polished inserts (Costa and Weiner 1994c,d). For the directional cloning of PCR-generated DNA fragments, it is recommended to use a pCR-Script Direct-type cloning reaction with a *Pfu* DNA polymerase-generated or *Pfu*-polished monophosphorylated insert. The procedure is the same for both the pCR-Script bidirectional and the pCR-Script Direct directional cloning.

PROTOCOL

Ligation Procedure

1. In an autoclaved, sterile 1.5-ml tube, set up the pCR-Script reaction by adding the following reagents in order:

1 μl of cloning vector
1 μl of 10x ligation buffer
0.5 μl of ATP (10 mM)
1–4 μl of *Pfu*-polished PCR product insert*
1 μl of *Srf* I restriction endonuclease (5 units)
1 μl of T4 DNA ligase (4 units)
ddH_2O to a final volume of 10 μl

**Note:* For ligation, the ideal ratio of insert-to-vector DNA is variable. For sample DNA, a range from 5:1 (when using polished inserts) to 100:1 (when using unpolished inserts) may be necessary. A greater insert-to-vector ratio is necessary for unpolished inserts because there will be a decreased occurrence of PCR fragments with both ends blunted. It may be advantageous to optimize conditions for a particular insert using the following equation:

$$\text{pmole ends}/\mu\text{g of DNA} = 2 \times 10^6/\text{number of bp}$$

2. Mix gently and incubate for 1–2 hours at room temperature.

3. Heat-treat the sample for 10 minutes at 65°C.

4. Store the sample on ice until transformation into competent *E. coli.*

E. coli Transformation

Competent cells are very sensitive to even small variations in temperature and must be stored at –80°C. Repetitive freeze-thawing will result in a loss of efficiency and should be avoided. It is important to use FALCON 2059 tubes for the transformation procedure, as the critical incubation period during the heat-pulse step described below is calculated for the thickness and shape of the FALCON 2059 tube. Also, β-mercaptoethanol has been shown to increase transformation efficiencies two- to threefold. Upon transformation, there seems to be a defined "window" of highest efficiency resulting from the heat pulse. Optimal efficiencies are observed when cells are heat-pulsed for 45–60 seconds. Supercompetent cells can be purchased commercially that yield extremely high efficiencies upon transformation.

PROTOCOL

Transformation Guidelines

1. Thaw competent cells on ice.

2. Gently mix the cells by swirling. Aliquot 40 μl of cells into a prechilled 15-ml FALCON 2059 tube.

3. Add β-mercaptoethanol (for a final 25 mM concentration) to the 40 μl of bacteria.

4. Swirl gently. Place on ice for 10 minutes; swirl gently every 2 minutes.

5. Add 2 µl of DNA from the heat-treated ligation reaction (step 4, Cloning Procedure).

6. Place on ice for 30 minutes.

7. Heat-pulse for 45 seconds in a 42°C water bath. The length of the heat pulse is critical for the highest efficiencies.

8. Place the transformation mixture on ice for 2 minutes.

9. Add 450 µl of preheated (42°C) SOC medium and incubate with shaking at 225–250 rpm for 1 hour at 37°C.

10. Plate 50–200 µl of the transformation mixture (100 µl is standard) using a sterile spreader to place the mixture onto the appropriate antibiotic-containing agar plates (a chromogenic substrate may be added to the LB plates to detect recombinants; see also β-Galactosidase Color Selection, below).

    Note: If plating ≥100 ml, the cells can be spread directly onto the plates. If plating <100 µl of the transformation mixture, increase the volume of the transformation mixture to be plated to a total volume of 200 µl using SOC medium.

11. Incubate the plates overnight (≥16 hours) at 37°C.

12. Choose white colonies for examination, avoiding colonies with a light-blue appearance or colonies with a blue center.

    Note: Colonies containing inserts that were initially white may turn very light blue after 2–5 days on the plate.

β-Galactosidase Color Selection

Phenotypic selection by disruption of the β-galactosidase (βGal) gene is often used to detect recombinants (Maniatis et al. 1982). Such phenotypic selection is monitored by the appearance of recombinants as white colonies and nonrecombinants as blue transformant colonies on XGal-containing agar plates. IPTG is often used as an inducer in conjuction with XGal.

PROTOCOL

Phenotypic Color Selection

1. Prepare a 100 mg/ml solution of XGal in dimethylformamide (DMF).

2. Prepare a 100 mM solution of IPTG in sterile, ddH_2O.

   Note: The XGal and IPTG solutions can be spread directly onto antibiotic-containing plates. Avoid mixing XGal and IPTG, as these chemicals will precipitate.

3. Add a 20-μl aliquot each of XGal and IPTG solutions onto agar plates. Spread immediately in an evenly distributed manner (a slight precipitate may be apparent).

4. Allow the plates to dry for 15–30 minutes before spreading transformants.

REAGENTS

Analysis of Cloned Recombinants

Recombinant screening kit
- ScreenTest recombinant screening kit (Stratagene, cat. no. 301800)

DNA minipreparation kits:
- ClearCut miniprep kit (Stratagene, cat. no. 400732)
- QIAGEN Plasmid Mini kit (QIAGEN, cat. no. 12123)
- Wizard Minipreps (Promega, cat. no. A7100)

DNA sequencing kits:
- Cyclist exo$^-$ *Pfu* DNA Sequencing kit (Stratagene, cat. no. 200326)
- Sequenase Version 2.0 DNA Sequencing Kit (USB/Amersham, cat. no. 70770)

Reagents required:
- Synthetic oligonucleotide primers
- Deoxynucleotide triphosphate mix (dNTP; 100 mM)
- PCR optimization buffers
  - Opti-Prime PCR optimization kit (Stratagene, cat. no. 200422)
  - The PCR Optimizer (Invitrogen, cat. no. K1220-01)
- Thermostable DNA polymerases (5–10 units)
  - *Taq* DNA polymerase
  - native *Pfu* DNA polymerase (Stratagene, cat. no. 600135)
  - cloned *Pfu* DNA polymerase (Stratagene, cat. no. 600153)
- *Taq* Extender PCR additive (Stratagene, cat. no. 600148)
- Ethidium bromide
- DNA sequencing primers
- 10x Cycle sequencing buffer (200 mM Tris-HCl, pH 8.8, 100 mM KCl, 200 mM $MgSO_4$, 100 μM $(NH_4)_2SO_4$, 1% Triton, 1 mg/ml BSA, 20 μM dATP, 50 μM dCTP, 50 μM dGTP, 50 μM dTTP)
- Stop dye mix (80% formamide, 50 mM Tris-HCl, pH 8.3, 1 mM EDTA, 0.1% bromophenol blue, 0.1% xylene cyanol)
- Mineral oil (Sigma, cat. no. M-3516)
- Sterile toothpicks

Rapid Recombinant Screening Analysis by Colony-PCR

Recombinant insert analysis of colonies resulting from transformed cells can be performed in 1 day using colony-PCR (Costa and Weiner 1994e,f). Recombinant PCR screening allows the rapid and efficient detection of cloned inserts from most ColE1-based plasmids. By using primers asymmetrically distanced from the clonal insertion site, it is possible to discern both insert presence and orientation from the resulting PCR product (see Fig. 5). One can also conduct PCR using a triple primer set containing the two asymmetric primers and an additional, fragment-specific primer from the set used to generate the original fragment. Agarose gel analysis of the PCR using such a three-primer set confirms both the presence and the orientation of the cloned insert without the need for further restriction enzyme digestion.

Further characterization of the cloned inserts can be done using restriction enzyme analysis of the colony-PCR product. Restriction enzyme digestion of the recombinant-screen PCR products can be performed directly from the amplification reaction (Costa and Weiner 1994f). In addition to restriction analysis, the recombinant-screen PCR products can be further characterized by cycle sequencing (Hedden et al. 1992; Costa and Weiner 1994f; Kretz et al. 1994). Because the

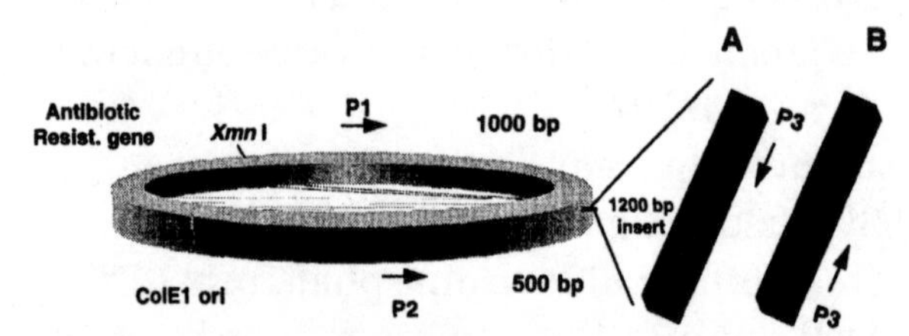

Figure 5 Recombinant screening method. Shown on the plasmid map are the PCR primers (P1 and P2) used in a colony-PCR screening procedure. PCR products produced using the colony-PCR method are analyzed by agarose gel electrophoresis. In a representative experiment in which a 1.2-kb insert is cloned into a plasmid, colonies selected directly from the transformation plates are inoculated into colony-PCR reaction mixtures, and PCR is conducted followed by agarose gel analysis. Nonrecombinants, which do not contain the insert, exhibit a 1.5-kb PCR product. Recombinants, which do contain the 1.2-kb insert, exhibit a 2.7-kb PCR product. Directionality of the cloned insert can be determined in a separate reaction by adding a third, insert-specific primer (P3) to the colony-PCR reaction mixture. Bidirectional cloning produces recombinants that contain inserts cloned in both directions, and the use of a third primer in the reaction mixture confirms the orientation of the cloned fragment (orientation A or B). An example of the recombinant screening of clones with a 1.2-kb insert by colony-PCR in the presence of a third primer indicates the orientation of the cloned insert after agarose gel analysis with a constitutive 2.7-kb PCR product and either a 1.7-kb PCR product (orientation A) or a 2.2-kb PCR product (orientation B).

primer set is designed to flank the polylinker by a distance of ≥500 bases on either side of the multiple cloning site, there is a retention of common priming sites used for DNA sequencing. Colony-PCR procedures that result in a single product (no spurious bands or primer-dimers) may be diluted and used in a cycle-sequencing reaction. High-resolution sequences have been consistently obtained using an aliquot of a 1:50 dilution of ScreenTest PCR products.

PCR-mediated clonal analysis allows one to screen numerous clones in a simple, rapid, and highly efficient manner. The procedure for recombinant screening by colony-PCR is outlined below. The recombinant screening protocol is outlined in four sections: preparation of the colony-PCR mixture; PCR parameters; analysis of the PCR results; and troubleshooting. It may be beneficial to evaluate the considerations in the troubleshooting section as a precautionary measure before proceeding with the PCR-based screening protocol. Then further experimentation and characterization of the PCR-generated products by restriction endonuclease analysis and DNA cycle sequencing are addressed.

PROTOCOL

Preparation of the Colony-PCR Mixture

As a positive control for this method, use nonrecombinant DNA (vector that does not contain insert). This will provide negative internal control colony-PCR. However, one may also transform the nonrecombinant, vector DNA and inoculate colonies from the transformation plate into the standard reaction mixture to serve as a positive control for the colony-PCR.

According to the number of reactions or multiples of reactions needed, prepare the PCR cocktail master mix in a single microcentrifuge tube *on ice* by adding the components *in the order indicated* below. After preparing the PCR cocktail master mix, follow the suggested guidelines outlined below.

Colony-PCR master mix

For the *PCR cocktail master mix:*

Sterile ddH_2O	40.6 µl
10X ScreenTest buffer	5 µl
dNTP mix (25 mM of each dNTP)	0.4 µl
Recombinant screening primer set (100 ng/µl)	2 µl
Taq Extender PCR additive (5 units/µl)	0.5 µl
Taq DNA polymerase (5 units/µl)	0.5 µl
Total reaction volume	49.0 µl

For the *control reaction:*

Nonrecombinant DNA (no insert; 1–5 ng/ml)	1 µl

For *insert orientation:*
Third, insert-specific primer (100 ng/μl) 1 μl

For *recombinant screening:*
Single colony toothpick inoculation

1. Aliquot 49 μl of the PCR cocktail master mix into each microcentrifuge tube on ice.

2. For control reaction(s), add 1 μl of the nonrecombinant DNA.

3. *Optional:* For insert orientation reaction(s), add a third, insert-specific primer to the appropriate reaction tubes.

4. For recombinant screening, stab the transformed colonies with a sterile toothpick and swirl the colony material into the appropriate reaction tubes. Immediately following inoculation into each reaction mixture, remove the toothpick and score onto antibiotic-containing patch plates for future reference.

   Note:
   - PCR inhibition may result in the event of excess colony material in the colony-PCR. It is important to note that only a small amount of colony material is necessary to perform the recombinant screening method. Stab the colonies from the transformation plate using the smaller ends of sterilized toothpicks. When performing the recombinant screening method from patch plates, only "touch" onto the patch and inoculate directly into the reaction tubes.
   - Toothpicks have the ability to "wick" liquid, so it is important to remove the toothpick immediately after inoculation.
   - Archive the screened colony material by using the inoculation toothpick to score an antibiotic-containing LB-agar plate.
   - See also Troubleshooting for PCR-mediated recombinant screening, below.

5. Mix each reaction gently.

6. Overlay each reaction with 30 μl of mineral oil.

7. Perform PCR using the recommended cycling parameters.

PCR Parameters

Depending on the needs of the investigator and the performance characteristics of the thermal cycler, sensitivity can be altered by changing both the number of cycles and the annealing temperature of segment 2. Also, depending on the oligonucleotide primer that is used, it may be advantageous to calculate its optimal annealing temperature. Several of the oligonucleotide primer selection programs listed in the Appendices (see pp. 681–686) accurately calculate the melting temperature. Once a melting temperature is determined, a revised segment 2 can be constructed. Alternatively, this equation, which is valid for oligonucleotides shorter than 20 bases, and overestimates the melting temperature, can be used.

$$T_m = 2^oC\ (A+T) + 4^oC\ (G+C)$$

The following PCR program has been successfully used with 30-base oligonucleotide primers. The sensitivity of the program is determined by segment 2 and may need to be reoptimized when using a third, insert-specific oligonucleotide primer that is <30 bases.

Segment	Cycles	Time	Temperature
1	1	4 minutes	94°C
		2 minutes	50°C
		2 minutes	72°C
2	30	1 minute	94°C
		2 minutes	56°C
		1 minute	72°C
3	1	5 minutes	72°C

Analysis of PCR Results

The PCR products are analyzed using standard agarose gel electrophoresis. It is recommended to use a 1.0–1.5% (w/v) agarose gel for optimal resolution of the expected 500- to 6000-bp PCR products. Typically, 15 µl of each PCR is analyzed utilizing ethidium bromide staining. Images may be archived using conventional instant photography or computer-based imaging software.

TROUBLESHOOTING

Under optimal conditions, colony-PCR provides an adequate amount of DNA template that will yield maximum signal in the PCR. Undoubtedly, there will be variations in thermal cyclers and reagents that may contribute to signal differences in the experiments. The following are guidelines for troubleshooting these variations in PCR-mediated recombinant screening.

- *Low signal with control DNA.* Suboptimal reagents (e.g., *Taq* DNA polymerase) and/or the thermal cycler used in conducting the assay may account for the results. The positive control is a good indicator of amplification efficiency and, when using 1–5 ng of DNA according to the specified guidelines, has been calculated to yield amounts of PCR product approaching plateau levels.

- *Low signal in the screening samples.* PCR inhibition may result in the event of excess colony material in the colony-PCR. It is important to note that only a small amount of colony material is necessary to perform the recombinant screening method. Stab the colonies from the transformation plate using the smaller ends of sterilized toothpicks. When perfoming the recombinant screening method from patch plates, only "touch" onto the patch and inoculate directly into the reaction tubes.

- *Loss of sample volume.* Reduced sample volume results when toothpicks are left in the reaction mixtures. Toothpicks "wick" the solution out of the reaction tubes, and for this reason, removal of the toothpicks shortly after the PCR cocktail master mix inoculation is strongly recommended.

- *Excessive signals in the samples.* This PCR-mediated screening method has been optimized on thermal cyclers whose temperature profiles are very exact and reproducible. Thermal cyclers whose transition times are very long inadvertantly add time to the PCR program and may result in excessive signals in both test samples and controls. In an attempt to reduce the signal, it may be advantageous to reoptimize segment 2 in the PCR program.

- *Multiple banding patterns.* This screening method has been designed with parameters optimized for use in colony-PCR where limited amounts of colony material are present. In the schematic representation of the recombinant screening method (see Fig. 5), two potential PCR products can be produced in the presence of a third, insert-specific oligonucleotide primer (P3). One PCR product is generated by P3+P1 or by P3+P2, and a second constitutive PCR product is generated by P1+P2. The method relies on the fact that, when limited amounts of template DNA are available, the smallest PCR product will be preferentially amplified. In cases in which pure, miniprep, or cesium-banded DNA is used, such purified DNA provides an optimally accessible template in cyclic amplification procedures, thereby producing both "expected" PCR products. Therefore, it is very important to calculate the expected PCR products when using a third, insert-specific oligonucleotide primer in directionality studies for the determination of insert orientation.

PROTOCOLS

Restriction Analysis of the PCR-generated Products

Upon deciphering insert presence, further characterization—such as the orientation of cloned inserts—can be achieved by restriction enzyme analysis. Restriction endonuclease digestion of the PCR-generated products can be performed directly from the amplification reaction.

1. Aliquot the following into a restriction enzyme digestion reaction as outlined below:

   5 μl of colony-PCR product
   2 μl of 10x enzyme-compatible buffer
   10–15 units of restriction endonuclease
   ddH_2O to a final volume of 20 μl

2. Incubate digestion at the recommended enzyme-specific temperature for 30–60 minutes.

3. Following incubation, load 10 μl of the digestion onto a 1.0–1.5% (w/v) agarose gel and analyze the PCR restriction digestion products by ethidium bromide staining.

DNA Cycle Sequencing of the PCR-generated Products

The PCR products from the recombinant screening procedure may be ultimately characterized by DNA cycle sequencing. Use of the Cyclist Exo^- *Pfu* DNA sequencing system (Stratagene) in combination with a radioactive label (e.g., α-^{33}P) provides high-resolution sequencing of the recombinant screening PCR products. Using the standard cycle-sequencing guidelines, PCR products may be diluted for use as follows:

1. Add 3 μl of the appropriate ddNTP to each of four termination tubes. Cap the tubes and keep on ice.

2. Prepare a 1:50 dilution of the recombinant screening PCR products into TE buffer.

3. For each PCR-generated template, combine the following reaction components *on ice.*

   per cycle-sequencing reaction:
      10 μl of diluted recombinant screen PCR product (~200 fmoles)
      1 μl of sequencing primer (~1 pmole*)
      4 μl of 10x cycle-sequencing buffer
      1 μl of [α-^{33}P]dATP (10 μCi)
      1 μl of Exo^- *Pfu* DNA polymerase (2.5 units)

   *Weight of DNA equal to 1 pmole:
      $0.33 \times N =$ ng of ssDNA, where $N =$ number of bases in primer

4. In a separate microcentrifuge tube mix together:

   13 μl of ddH_2O
   4 μl of dimethylsulfoxide (DMSO)

5. Add the 17 µl of ddH_2O-DMSO mixture to the above cycle-sequencing reaction solution to yield a final volume of 30 µl.

6. Aliquot 7 µl of the cycle-sequencing reaction mixture from step 5 into each of the four termination tubes already containing 3 µl of ddNTP. Mix thoroughly, making sure the reaction mix and the dideoxynucleotide mix are at the bottom of the tube.

7. Overlay the reactions with 20 µl of mineral oil.

8. Cycle the sequence reaction through an appropriate temperature profile. Optimum cycling parameters will vary depending on the template and primer, as well as the type of machine used. A useful PCR cycling profile is given below.

Segment	Cycles	Time	Temperature
1	30	30 seconds	95°C
		30 seconds	60°C
		60 seconds	72°C

9. Add 5 µl of stop mix below the mineral overlay and mix by pipetting.

10. Heat-denature the samples for 2–5 minutes at 80°C, then immediately load 2–4 µl of the samples onto a sequencing gel.

11. Using standard procedures, dry the gel and expose to autoradiograph film.

TROUBLESHOOTING

- *Little or no sample after cycling.* This may be due to sample evaporation, possibly resulting from insufficient mineral oil overlay. Add at least 20 µl of mineral oil and briefly centrifuge before cycling.

- *Faint bands or blank film.* (1) No radioactivity was added, or the radiolabel is old. (2) One of the reaction components is missing, or the reaction components were not thoroughly mixed. (3) The primer did not anneal efficiently. Reduce annealing temperature or redesign the primer. (4) Not enough template was used. (5) The template DNA is contaminated. Make sure excess salt and EDTA have not contaminated the preparation. The PCR product may need to be precipitated prior to template dilution (see Optional PCR Product Purification, above).

- *High background on the sequencing gel.* Too much template DNA. This can be a serious problem with short PCR products. Titrate down the amount of DNA added to the reaction.

- *Bands in multiple lanes.* (1) The primer annealed at multiple sites. A higher annealing temperature may help. (2) There are multiple templates in the sequencing reaction. PCR products containing multiple sequences or primer-dimer artifacts are present and one of the PCR primers was used as a sequencing primer. Gel purification should alleviate the problem. (3) There are gel compression artifacts. Increase the gel temperature (up to 60°C) or add formamide to the gel. (4) The template DNA is contaminated.

- *Blurry or smeared bands.* (1) Samples were not fully denatured prior to gel electrophoresis. (2) Old or improperly prepared acrylamide solutions were used. (3) Bad DMSO was used. Do not freeze-thaw DMSO more than once.

Traditional DNA Minipreparation and Restriction Enzyme Analysis

Alternatively, one could conduct routine plasmid DNA isolation after overnight incubation and determine both the insert size and orientation following restriction enzyme digestion and agarose gel analysis. A number of commercially available kits can be used that produce high-quality miniprep plasmid DNA (see Reagents: DNA minipreparation kits, above). Non-cycle procedures for DNA sequencing can be used from these minipreparations (Sequenase, USB).

CONCLUSION

PCR has both simplified and accelerated the process for cloning DNA fragments. It is now possible to synthesize primers and perform the PCR, cloning, and transformation reactions in a single day. The analysis of putative clones by colony-PCR and cycle sequencing can be completed the following day. The methods presented allow PCR cloning operations to exhibit more than 50% recombinant efficiency and facilitate PCR screening methods that can be completed in a rapid and highly efficient manner.

ACKNOWLEDGMENTS

The authors thank John Bauer, Steve Wells, Tim Sanchez, Mark Kaderli, and Bruce Jerpseth for their substantial contributions in experimental design. We also thank Loretta Callan for assistance with the manuscript outline. G.L.C. especially thanks Dr. Keith Kretz for his invaluable cycle-sequencing procedures.

REFERENCES

Bauer, J., D. Deely, J. Braman, J. Viola, and M.P. Weiner. 1992. pCR-Script SK(+) cloning system: A simple and fast method for PCR cloning. *Strat. Mol. Biol.* **5:** 62–65.

Clark, J.M. 1988. Novel non-templated nucleotide addition reactions catalyzed by procaryotic and eucaryotic DNA polymerases. *Nucleic Acids Res.* **16:** 9677–9686.

Costa, G.L. and M.P. Weiner. 1994a. Improved PCR cloning. *Strat. Mol. Biol.* **8:** 8.

——. 1994b. pCR-Script SK(+) cloning system: Questions and answers. *Strat. Mol. Biol.* **7:** 53–54.

——. 1994c. Increased cloning efficiency with the PCR polishing kit. *Strat. Mol. Biol.* **7:** 47–48.

——. 1994d. Polishing with T4 or *Pfu* polymerase increases the efficiency of cloning PCR fragments. *Nucleic Acids Res.* **22:** 2423.

——. 1994e. ScreenTest recombinant screening in one day. *Strat. Mol. Biol.* **7:** 35–37.

——. 1994f. ScreenTest colony-PCR screening: Questions and answers. *Strat. Mol. Biol.* **7:** 80–82.

Costa, G.L., A. Grafsky, and M.P. Weiner. 1994a. Cloning and analysis of PCR-generated DNA fragments. *PCR Methods Appl.* **3:** 338–345.

Costa, G.L., T.R. Sanchez, and M.P. Weiner. 1994b. pCR-Script Direct SK(+) vector for directional cloning of blunt-ended PCR products. *Strat. Mol. Biol.* **7:** 5–7.

——. 1994c. New chloramphenicol-resistant version of pCR-Script vector. *Strat. Mol. Biol.* **7:** 52.

Hedden, V., M. Simcox, B. Scott, J. Cline, K. Nielson, E. Mathur, and K. Kretz. 1992. Superior sequencing: Cyclist Exo$^-$ *Pfu* DNA sequencing kit. *Strat. Mol. Biol.* **5:** 79.

Hu, G. 1993. DNA polymerase-catalyzed addition of nontemplated extra nucleotides to the 3′ end of a DNA fragment. *DNA Cell Biol.* **12:** 763–770.

Kretz, K., W. Callen, and V. Hedden. 1994. Cycle sequencing. *PCR Methods Appl.* **3:** S107–112.

Liu, Z. and L. Schwartz. 1992. An efficient method for blunt-end ligation of PCR product. *BioTechniques* **12:** 28–30.

Maniatis, T., E. Fritsch, and J. Sambrook. 1982. *Molecular cloning: A laboratory manual.* Cold Spring Harbor Laboratory, Cold Spring Harbor, New York.

Nielson, K.B., W. Schoettlin, J.C. Bauer, and E. Mathur. 1994. *Taq*Extender PCR additive for improved length, yield and reliability of PCR products. *Strat. Mol. Biol.* **7:** 27.

Simcox, T., S. Marsh, E. Gross, W. Lernhardt, S. Davis, and M. Simcox. 1991. *Srf* I, a new type-II restriction endonuclease that recognizes the octanucleotide sequence, 5′-GCCCGGGC-3′. *Gene* **109:** 121–123.

Weiner, M.P. 1993. Directional cloning of blunt-ended PCR products. *BioTechniques* **15:** 502–505.

Weiner, M.P., G.L. Costa, W. Schoettlin, J. Cline, E. Mathur, and J.C. Bauer. 1994. Site-directed mutagenesis of double-stranded DNA by the polymerase chain reaction. *Gene* **151:** 119–123.

Mutagenesis by PCR

A mutagenized PCR product is an amplicon that contains base changes from the nucleotide sequence of its template. PCR-mediated nucleotide changes, deletions, or insertions can be accomplished by the different protocols grouped in this section.

Two basic types of mutagenized PCR products can be obtained. In the first type, mutations are introduced randomly through inaccurate copying of the DNA template by a thermostable DNA polymerase. Certain thermostable DNA polymerases, like *Taq* DNA polymerase, have an intrinsic error rate due to the lack of $3' \rightarrow 5'$ exonuclease activity. These enzymes are biased toward incorporating GC pairs in a reaction product when reading AT pairs in a template. In the protocol "Mutagenic PCR," a series of conditions (including modifications in the amount of nucleotides, enzyme, and magnesium, as well as the addition of manganese chloride) are recommended to generate a library of mutants that does not exhibit substantial sequence bias. Use of this approach is based on the availability of an efficient and rapid screening method to select the mutation(s) of interest in the pool or library of mutagenized products.

A second type of mutagenized PCR product is obtained by introducing single or multiple base changes, insertions, and deletions at predefined positions within the DNA template. There are many strategies for this type of site-directed mutagenesis. Although they differ considerably, these protocols share the principle of "primer modification" to achieve a mutagenized product (for further details, see the mutagenesis primer section of "Design and Use of Mismatched and Degenerate Primers" in Section 3).

In some cases, primers are modified to introduce one or more nucleotide changes in the product of the amplification reaction. The mismatches in the primer are introduced at the 5′ end because perfect base-pairing between template and primer is required at the 3′ end. This same principle applies when deleting or inserting nucleotides in a PCR product. Nucleotides are either added or skipped at the 5′ ends of the primers to obtain PCR products with insertions or deletions, respectively.

A common feature of PCR-mediated mutagenesis protocols that differs from traditional site-directed mutagenesis methods is the absence of a single-stranded DNA intermediate. This eliminates the use of M13-based bacteriophage vectors or the need to rescue single-stranded DNA with helper virus, reducing the time required for completion of an experiment.

The different protocols for PCR-mediated mutagenesis presented in this section are based on in vivo recombination of PCR-generated homologous DNA ends ("PCR Mutagenesis and Recombination In Vivo"), differential susceptibility of the original template versus the mutagenized product to enzymatic digestion ("Rapid PCR Site-directed Mutagenesis"), and the extension by the polymerase of fragments that overlap in sequence ("Mutagenesis and Synthesis of Novel Recombinant Genes Using PCR"). Because these protocols can employ the same set of primers, we recommend that all the possible strategies presented in this section be reviewed before designing and synthesizing a set of oligonucleotide primers for a specific use.

Mutagenic PCR

R. Craig Cadwell[1] and Gerald F. Joyce

Departments of Chemistry and Molecular Biology, The Scripps Research Institute, La Jolla, California 92037

INTRODUCTION

Most practitioners of PCR prefer to carry out DNA amplification in an accurate manner, introducing as few base substitutions as possible. This is especially critical when one is studying clonal isolates and must distinguish natural variation from artifactual variation that is introduced by polymerase error. Fortunately, thermostable DNA polymerases are available that operate with high fidelity due to an intrinsic 3′→ 5′ exonuclease activity (for review, see Cha and Thilly 1993). Manipulation of the PCR conditions can lead to further improvement of copying accuracy.

Here we consider the other side of the fidelity issue—those instances where promiscuity is a virtue. Oftentimes in probing the structure or function of a protein or nucleic acid, one wishes to generate a library of mutants and apply a screening method to isolate individuals that exhibit a particular property. For mutations over a short stretch of nucleotides within a cloned gene, it is appropriate to replace a portion of the gene with a synthetic DNA fragment that contains random or partially randomized nucleotides (Matteucci and Heyneker 1983; Wells et al. 1985; Oliphant et al. 1986). For mutations over a longer segment, up to the size of an entire gene, it may be preferable to scatter random mutations over the entire sequence, typically at a frequency of one or a few mutations per molecule. In such cases, it is most convenient to introduce random mutations through inaccurate copying by a DNA polymerase, especially if the polymerase is a thermostable enzyme that can operate in the context of PCR. Each pass of the polymerase during PCR allows the possibility of mutation, so that the cumulative error rate can become substantial.

[1]Present address: Department of Biological Sciences, University of California, Santa Barbara, California 93106.

The error rate of *Taq* DNA polymerase is the highest of the known thermostable DNA polymerases, in the range of 0.1×10^{-4} to 2×10^{-4} per nucleotide per pass of the polymerase, depending on the reaction conditions (Eckert and Kunkel 1990; Ling et al. 1991). Over the course of PCR, in which the polymerase makes an average of 20–25 passes, the cumulative error rate is roughly 10^{-3} per nucleotide. In most cases, this is insufficient to generate a diverse library of variant sequences, especially over a region shorter than 1000 nucleotides. A further drawback is that the errors made by *Taq* DNA polymerase under standard PCR conditions are heavily biased toward AT→GC changes (Keohavong and Thilly 1989). We have devised a mutagenic PCR that has an overall error rate of about 7×10^{-3} per nucleotide and does not exhibit substantial sequence bias (Cadwell and Joyce 1992).

REAGENTS

Native *Taq* or AmpliTaq DNA polymerase (Perkin-Elmer or licensed supplier); do not substitute any other thermostable DNA polymerase

High-purity deoxynucleoside 5′-triphosphates (Pharmacia, USB/Amersham)

PROTOCOL

The top priority of mutagenic PCR is to introduce the various types of mutations in an unbiased fashion rather than to achieve a high overall level of amplification. The DNA input in a 100-μl reaction mixture consists of 10^{10} molecules (20 fmoles), which are amplified about 1000-fold to yield 10^{13} molecules (20 pmoles). This modest amplification requires an average of 10 passes of the polymerase. However, 30 cycles of PCR are carried out to ensure that mismatched termini have ample opportunity to become extended to produce complete copies. The large input prevents the PCR products from being influenced by the effects of clonal expansion. Even if a mutation occurs in the first pass of the polymerase and is passed along to all of the descendant molecules, there is very little chance that any two molecules isolated from the final population will carry the same mutation as a consequence of their being derived from a common ancestor.

The protocol for mutagenic PCR is derived from "standard" PCR conditions (Coen 1991): 1.5 mM $MgCl_2$; 50 mM KCl; 10 mM Tris-HCl, pH 8.3 at 25°C; 0.2 mM each dNTP; 0.3 μM each primer; and 2.5 units of *Taq* DNA polymerase in a 100-μl volume. Incubate for 30 cycles for 1 minute at 94°C, 1 minute at 45°C, and 1 minute at 72°C in a conventional thermal cycler. The following changes are made to enhance the mutation rate:

- The $MgCl_2$ concentration is increased to 7 mM to stabilize noncomplementary pairs (Eckert and Kunkel 1991; Ling et al. 1991).

- 0.5 mM $MnCl_2$ is added to diminish the template specificity of the polymerase (Beckman et al. 1985; Leung et al. 1989).
- The concentration of dCTP and dTTP is increased to 1 mM to promote misincorporation (Leung et al. 1989; Cadwell and Joyce 1992).
- The amount of *Taq* DNA polymerase is increased to 5 units to promote chain extension beyond positions of base mismatch (Gelfand and White 1990).

The experimental protocol is as follows:

1. Prepare a 10X mutagenic PCR buffer containing 70 mM $MgCl_2$; 500 mM KCl; 100 mM Tris-HCl, pH 8.3 at 25°C; and 0.1% (w/v) gelatin.
2. Prepare a 10X dNTP mix containing 2 mM dGTP, 2 mM dATP, 10 mM dCTP, and 10 mM dTTP.
3. Prepare a solution of 5 mM $MnCl_2$. DO NOT combine with the 10X PCR buffer, which would result in the formation of a precipitate that disrupts PCR amplification.
4. Combine 10 µl of 10X mutagenic PCR buffer, 10 µl of 10X dNTP mix, 30 pmoles of each primer, 20 fmoles of input DNA, and an amount of H_2O that brings the total volume to 88 µl. Mix well.
5. Add 10 µl of 5 mM $MnCl_2$. Mix well and confirm that a precipitate has not formed.
6. Add 5 units of *Taq* DNA polymerase, and bring the final volume to 100 µl. Mix gently. Cover with mineral oil or a wax bead, if desired.
7. Incubate for 30 cycles for 1 minute at 94°C, 1 minute at 45°C, and 1 minute at 72°C. Do not employ a hot start procedure or a prolonged extension time at the end of the last cycle.
8. Purify the reaction products by extraction with chloroform/isoamyl alcohol (24/1, v/v) and subsequent ethanol precipitation.
9. Run a small portion of the purified products on an agarose gel stained with ethidium bromide to confirm a satisfactory yield of full-length material. Mutagenic PCR should be carried out in parallel with standard PCR (omitting the four changes listed above); the yields of full-length DNA should be comparable.

TROUBLESHOOTING

By employing a DNA of ordinary nucleotide composition, mutagenic PCR introduces errors at a frequency of 0.66% ± 0.13% per position over the course of the PCR (95% confidence interval) (Cadwell and Joyce 1992). Nearly all of these changes are base substitutions. The combined frequency of insertions and deletions is less than 0.05% (one-tailed test, 95% confidence interval). The number of mutations per DNA copy follows a Poisson distribution. The probability of mutating each of the four bases is approximately equal except for a 1.5-fold enhanced probability of mutating T residues, which is significant at the 99% confidence level. The most common specific mutations are A→T and T→A changes, which we attribute to TT mismatches that manifest as either A→T changes in the same strand or T→A changes in the opposing strand. The least common mutations are G→C and C→G changes, which presumably reflects the difficulty in forming and extending GG and CC mismatches. Summing over all types of mutations and correcting for the base composition of the mutated gene, the ratio of AT→GC to GC→AT changes is 1.0 (0.6–1.7, 95% confidence interval).

The most common difficulty with mutagenic PCR stems from the fact that 30 temperature cycles are employed, even though *Taq* DNA polymerase makes an average of only 10 passes along the DNA. As noted above, this provides ample opportunity for extension of mismatched termini, which is necessary to lock in mutations. However, it also favors the occurrence of amplification artifacts (Mullis 1991). Compounding the problem is the markedly elevated $MgCl_2$ concentration, which lowers the stringency of primer hybridization, thereby promoting the formation of nonspecific amplification products. As a general rule, one should begin mutagenic PCR with either cDNA or a double-stranded DNA fragment that encompasses only the region of interest. It is risky to employ plasmid DNA and hopeless to begin with a genomic library. We limit the use of mutagenic PCR to DNAs no longer than about 1000 nucleotides. For longer target sequences, the DNA can be divided into two or more fragments that are mutagenized separately.

Because mutagenic PCR enhances primer mishybridization, certain combinations of primers and target sequences will inevitably give rise to short amplification products that outcompete the full-length DNA. These artifacts are best seen by carrying out the reaction with a radiolabeled primer and separating the products on a nondenaturing polyacrylamide gel. On the basis of their size and (if necessary) sequence, it should be possible to discern the site of primer mishybridization and redesign the primers accordingly. Alternatively, it may be preferable to use nonmutagenic PCR to attach "well-behaved" primer-binding sites to the ends of the DNA, then carry out mutagenic PCR using primers that hybridize to the attached sites.

Another source of difficulty is the urge to make slight modifications of the protocol without evaluating their consequences. If, for example, C→G changes are less frequent than T→A changes, then why not double the concentration of dGTP to 0.4 mM? Doing so, it turns out, results in a fourfold increase in the ratio of AT→GC to GC→AT changes (Cadwell and Joyce 1992). We encourage others to explore alternative reaction conditions that may lead to an improved PCR mutagenesis procedure. However, in view of the extreme sensitivity of *Taq* DNA polymerase to dNTP concentrations and other aspects of the reaction conditions, general users are encouraged to follow the protocol to the letter.

DISCUSSION

The distribution of variants that results from mutagenic PCR depends on the error rate and the length of the sequence that is being randomized. The probability P of having k mutations in a sequence of length n is given by:

$P(k,n,\varepsilon) = \{n! / [(n-k)!\ k!]\}\ \varepsilon^k (1-\varepsilon)^{n-k}$, where ε is the error rate per position.

In the present case, the error rate is 0.66% per position over the course of the PCR ($\varepsilon = 0.0066$). Thus, for a target sequence of 500 nucleotides, the resulting population of variants would consist of about 4% wild type, 12% one-error mutants, 20% two-error mutants, 22% three-error mutants, 18% four-error mutants, 12% five-error mutants, and 12% mutants with six or more errors. The number of distinct sequences with k errors, N_k, increases exponentially with increasing k:

$$N_k = \{n! / [(n-k)!\ k!]\}\ 3^k$$

Thus, 20 pmoles of material resulting from the mutagenesis of a 500-nucleotide target sequence would contain all possible one-, two-, three-, and four-error mutants, but only about 2% of the possible five-error mutants and a progressively sparser sampling of higher-error mutants. These calculations refer to the composition of the DNA; for the corresponding protein they must be modified to take into account the degeneracy of the genetic code.

For some purposes an error rate of 0.66% per position will be insufficient. It is possible to carry out successive rounds of mutagenic PCR to double or even triple the overall error rate. However, two potential pitfalls must be avoided. First, if a small aliquot of one reaction mixture is used to seed the next, there is an increased chance that molecules isolated from the final pool will be related by descent. Taking one-thousandth of the products from a first mutagenic PCR to seed a second should not be a problem, but taking one-thousandth of the second to seed a third would reduce diversity to an unacceptably low level. This problem could be remedied by scaling up the third reaction mixture to a 10-ml volume, preferably in multiple reaction vessels containing 100 μl each. A second potential pitfall is the risk of generating nonspecific amplification products, made more likely by the increased number of temperature cycles. It may be necessary to gel-purify full-length DNA after the first mutagenic PCR before proceeding with the second.

Until a more error-prone thermostable DNA polymerase is found in nature or developed through enzyme engineering, *Taq* DNA polymerase provides the most effective way to generate a library of DNAs that contain random mutations over a stretch of 100–1000 nucleotides. If one is interested in a library of RNAs, then a promoter sequence for T7 RNA polymerase can be included near the 5′ end of the appropriate PCR primer, allowing the DNA products to serve as templates in an in vitro transcription reaction (Chamberlin and Ryan 1982). If one is interested in a library of proteins, then the PCR primers can be designed to include either restriction sites for cloning into a suitable expression vector or a ribosome binding site and start codon for in vitro translation.

An important advantage of mutagenic PCR is that it allows repeated randomization of a population of nucleic acids without isolating clones and obtaining sequence information. After one has generated a library of mutants and applied a screening method to obtain individuals that exhibit a particular property, the selected individuals can then be used directly as input for a second mutagenic PCR. Repeating the cycle of selection and mutagenic amplification allows one to carry out in vitro evolution of nucleic acids, including those that have catalytic function (Beaudry and Joyce 1992). Similarly, a population of protein-encoding DNAs, harvested from a selected subset of cells or viral particles, can be treated as an ensemble and subjected to mutagenic PCR to produce variants of the selected variants.

REFERENCES

Beaudry, A.A. and G.F. Joyce. 1992. Directed evolution of an RNA enzyme. *Science* **257:** 635–641.

Beckman, R.A., A.S. Mildvan, and L.A. Loeb. 1985. On the fidelity of DNA replication: Manganese mutagenesis *in vitro*. *Biochemistry* **24:** 5810–5817.

Cadwell, C. and G.F. Joyce. 1992. Randomization of genes by PCR mutagenesis. *PCR Methods Appl.* **2:** 28–33.

Cha, R.S. and W.G. Thilly. 1993. Specificity, efficiency, and fidelity of PCR. *PCR Methods Appl.* **3:** S18–S29.

Chamberlin, M. and T. Ryan. 1982. Bacteriophage DNA-dependent RNA polymerases. *Enzymes* **15:** 85–108.

Coen, D.M. 1991. The polymerase chain reaction. In *Current protocols in molecular biology* (ed. F.M. Ausubel et al.), pp. 151.1–151.7. Wiley Interscience, New York.

Eckert, K.A. and T.A. Kunkel. 1990. High fidelity DNA synthesis by the *Thermus aquaticus* DNA polymerase. *Nucleic Acids Res.* **18:** 3739–3744.

——. 1991. DNA polymerase fidelity and the polymerase chain reaction. *PCR Methods Appl.* **1:** 17–24.

Gelfand, D.H. and T.J. White. 1990. Thermostable DNA polymerases. In *PCR protocols: A guide to methods and applications* (ed. M.A. Innis et al.), pp. 129–141. Academic Press, San Diego, California.

Keohavong, P. and W.G. Thilly. 1989. Fidelity of DNA polymerases in DNA amplification. *Proc. Natl. Acad. Sci.* **86:** 9253–9257.

Leung, D.W., E. Chen, and D.V. Goeddel. 1989. A

method for random mutagenesis of a defined DNA segment using a modified polymerase chain reaction. *Technique* **1:** 11–15.

Ling, L.L., P. Keohavong, C. Dias, and W.G. Thilly. 1991. Optimization of the polymerase chain reaction with regard to fidelity: Modified T7, *Taq*, and Vent polymerases. *PCR Methods Appl.* **1:** 63–69.

Matteucci, M.D. and H.L. Heyneker. 1983. Targeted random mutagenesis: The use of ambiguously synthesized oligonucleotides to mutagenize sequences immediately 5′ of an ATG initiation codon. *Nucleic Acids Res.* **11:** 3113–3121.

Mullis, K.B. 1991. The polymerase chain reaction in an anemic mode: How to avoid cold oligodeoxyribonuclear fusion. *PCR Methods Appl.* **1:** 1–4.

Oliphant, A.R., A.L. Nussbaum, and K. Struhl. 1986. Cloning of random-sequence oligodeoxynucleotides. *Gene* **44:** 177–183.

Wells, J.A., M. Vasser, and D.B. Powers. 1985. Cassette mutagenesis: An efficient method for generation of multiple mutations at defined sites. *Gene* **34:** 315–323.

PCR Mutagenesis and Recombination In Vivo

Douglas H. Jones

Department of Pediatrics, University of Iowa College of Medicine, Iowa City, Iowa 52242

INTRODUCTION

Site-directed mutagenesis is an underpinning of the recombinant DNA revolution. For example, site-directed mutagenesis is used to modify protein-coding domains and to characterize regulatory DNA elements. Even the routine subcloning of an insert into a plasmid is a site-directed insertional mutagenesis, and the creation of a recombinant construct such as a gene chimera is a site-directed mutagenesis where one DNA segment replaces another. Technology for the site-directed mutagenesis of DNA is vital for genetic engineering.

PCR is best known as a method for the retrieval and detection of a specific DNA sequence. More recently, PCR has also become a popular method for the in vitro modification of a DNA sequence. Modification of DNA can occur because the primers are incorporated into the ends of the amplification product, permitting primer-directed modification of such ends. Alteration of a DNA sequence can be accomplished by incorporating a nucleotide mismatch within a primer-annealing domain or by using a primer whose 5′ end is not determined by the original template (Mullis et al. 1986). The variety of PCR-based methods that have been developed for the site-directed mutagenesis of DNA, including the generation of recombinant constructs, attests to the continuing search for better or simpler methods (White 1993).

Previous investigators have shown that DNA ends containing short regions of homology undergo intramolecular recombination in vivo in *Escherichia coli*, including the RecA-minus *E. coli* strains used routinely for cloning (Conley et al. 1986a,b; Sung and Zahab 1987), and that *E. coli* can mediate intermolecular recombination between a short, single-stranded oligonucleotide and a restriction endonuclease-digested plasmid (Mandecki 1986). Recombination PCR is a method for making DNA joints in vivo by the recombination of PCR-generated homologous DNA ends in *E. coli* (Jones and Howard 1991; Jones and Winistorfer 1992). This section details two recombination-PCR protocols in which intermolecular recombination in vivo of PCR-generated DNA ends mutates a plasmid.

In brief, each protocol involves the generation of two products in two separate PCR amplifications. Each end of one PCR product is designed to be homologous to a different end of the other PCR product. These two unpurified PCR products are combined, and this single sample is used to transform *E. coli.* Transformation is accomplished by recombination in vivo between the PCR-generated homologous ends, resulting in recombinant circles. If these recombinant circles contain plasmid sequences that permit replication and a selectable marker such as an antibiotic resistance gene, the *E. coli* can be transformed by the recombinant. Two protocols are detailed, one for the point mutagenesis of a plasmid and one for the generation of a recombinant construct in which a DNA segment from one plasmid is seamlessly and directionally inserted into a specific locus of another plasmid. In addition, a modification of the second protocol is described for subcloning any PCR product.

REAGENTS

AmpliTaq DNA polymerase (5 units/μl) (Perkin-Elmer)

Routine 10x *Taq* DNA polymerase buffer (500 mM KCl, 100 mM Tris-HCl, pH 8.3, 15 mM $MgCl_2$, 0.1% gelatin)

100 mM dATP, 100 mM dCTP, 100 mM dGTP, 100 mM dTTP (Boehringer Mannheim)

Primers: (a) Two pairs of primers (four altogether), each pair used in a separate amplification. Each primer in one pair has 24 nucleotides of complementarity to a different primer in the other pair. (b) Sequencing primers that flank the insert or mutated region, used to characterize the resulting construct.

Agarose

Ethidium bromide

TAE buffer (Sambrook et al. 1989a)

Long, thin micropipette tips (Gel loader tips T-010; Phenix Research Products)

MAX Efficiency DH5α-competent *E. coli* (BRL, Life Technologies)

SOC media (Sambrook et al. 1989b)

Top agar (Sambrook et al. 1989c)

LB plates with 100 μg/ml ampicillin (Sambrook et al. 1989c)

Luria-Bertani medium (LB broth) (Sambrook et al. 1989d)

QIAGEN Midi-columns (QIAGEN)

Sequenase 2.0 Sequencing Kit (USB/Amersham)

PROTOCOLS

Point Mutagenesis

Point mutagenesis using recombination PCR is illustrated in Figure 1. The plasmid carrying the insert that is to be mutated is linearized by restriction endonuclease digestion prior to each PCR amplification. This plasmid is amplified and mutated in two separate PCR amplifications. In each of the two amplifications, the identical base pair is mutated, so that the mutated ends of each product are homologous to each other. The nonmutating primers are also designed to produce ends that are homologous to each other. Both unpurified PCR products are combined to transform *E. coli*, resulting in clones containing the mutation of interest.

1. PCR mutagenesis

   a. In each PCR amplification, start with a template plasmid that has been restriction-endonuclease-digested outside the region to be amplified. Use 2 ng of linearized plasmid, 25 pmoles of each primer, 200 μM of each dNTP, 1X PCR buffer, and 1.25 units of *Taq* DNA polymerase in a total volume of 50 μl.

   b. Pipette 50 μl of mineral oil on top of each reaction mix prior to amplification.

   c. Amplification parameters are as follows: Initial denaturation for 1 minute at 94°C, 14–20 amplification cycles (for 30 seconds at 94°C, 30 seconds at 50°C, 1 minute at 72°C per kb of PCR product), and a final extension step for 7 minutes at 72°C.

   Note: Each amplification uses a plasmid template that has undergone restriction-endonuclease digestion in a different "side" of the plasmid, relative to the site targeted for mutagenesis. Usually, unique restriction-endonuclease-recognition sites in the original vector are used, so that the two linearized templates can be used to mutate any site in the insert. Each primer is designed to generate 15–45 bp of homology between each end of one PCR product relative to the other PCR product. Using 24 nucleotides of homology works very well. Decreasing the length homology from 25 to 12 bp in an early protocol that entails a single recombination event decreases the transformation efficiency four- to fivefold. Single point mismatches lie no closer than 6 nucleotides from the 3′ end of a primer, and are frequently placed toward the middle. Placing point mutations near the 5′ end of each mutating primer will generate two PCR products whose mutated ends have less than 24 bp of homology. Primers that generate point mismatches are typically 25–35 nucleotides long. Multiple point mismatches should be placed in the middle or toward the 5′ end of a primer, with primer lengths long enough to create 24 bp of homology between the mutated ends of the two PCR products. Primers that are nonmutating (primers 2 and 4 in Fig. 1) are generally 20–30 nucleotides long. The nonmutating primers can be designed to anneal to the β-lactamase gene, so that they can be used with many different plasmids.

   The mutating primer and nonmutating primers are frequently designed to be perfect complements to each other. In Figure 1, primer 3 is the complement to primer 1 and primer 4 is the complement to primer 2. If the mis-

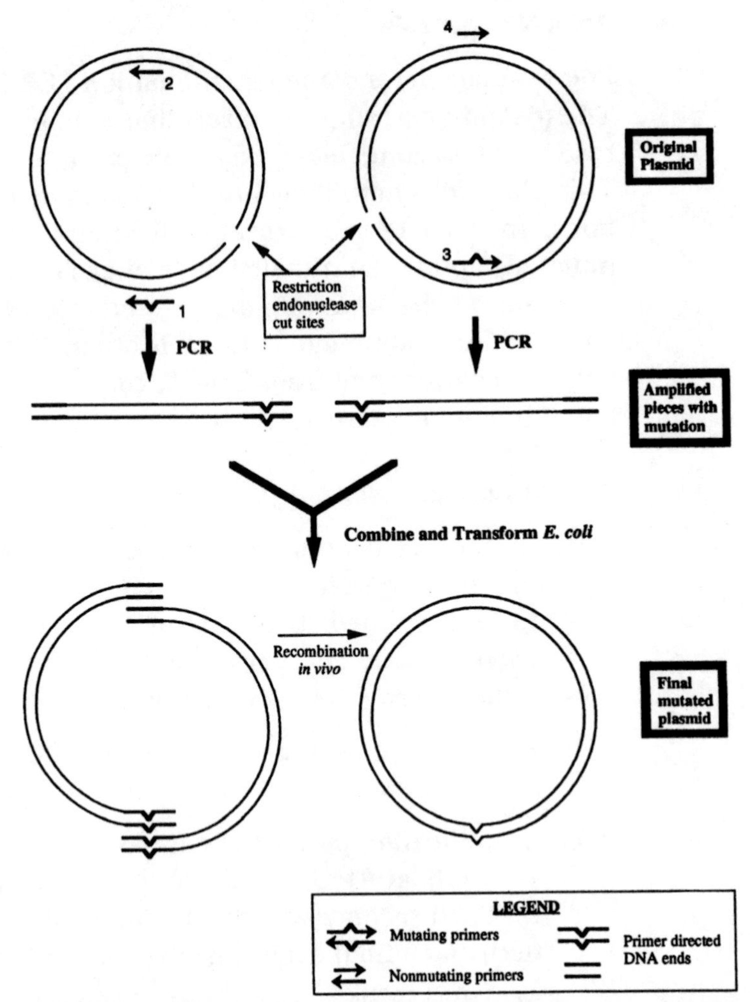

Figure 1 Point mutagenesis by recombination in vivo of two PCR products. The primers are numbered hemiarrows. The notches designate point mismatches in the primers and resulting mutations in the PCR products. Primer 1 is complementary to primer 3 and primer 2 is complementary to primer 4. Two unique restriction enzyme recognition sites used to linearize the plasmid bracket the insert. There is no purification of the PCR products. For each additional single site-specific mutagenesis reaction, only new primers 1 and 3 need to be synthesized, and the same cut templates can be used.

matched (mutagenesis) site of each mutating primer (primers 1 and 3) is positioned toward the 3′ end of each primer, the primers will generate homologous ends that are somewhat longer than the length of each primer, and this also works well. The nonmutating primers (primers 2 and 4) can anneal at positions that generate PCR product ends with long homology by positioning these primers so that their 3′ ends are directed toward each other (in the separate amplifications). Positioning the nonmutating primers to generate long homology between the nonmutated PCR product ends is not necessary and, surprisingly, has decreased the transformation efficiency using recombination PCR (Jones and Winistorfer 1992).

2. PCR product detection

   a. Visualize each product by electrophoresis on an agarose minigel.

   Note: If 5 μl of the PCR product can be clearly viewed following ethidium bromide staining (≥15 ng/5 μl), there is enough product.

3. Transformation of *E. coli*

   a. Insert a long, thin micropipette tip through the mineral oil, withdraw 2.5 μl from each PCR tube (typically 10–60 ng per 2.5 μl; maintaining an even ratio of one product to another is not necessary), combine the two samples, and transform MAX Efficiency DH5α-competent *E. coli* (transfection efficiency >1 x 10^9 transformants/μg of monomer pUC19) with the 5 μl.

   Note: Transformation is done following the manufacturer's (BRL) protocol with the following modifications: (1) Use 50 μl of *E. coli* for each transformation, as this is effective and less expensive than the 100 μl recommended. After incubation at 37°C in a shaker for 1 hour, do not dilute the sample prior to plating; place the entire sample onto an LB plate containing 100 μg/ml ampicillin. To keep the sample on the plate, add 2 ml of top agar, prewarmed to 42°C, to each sample immediately prior to pouring it onto the plate. Once an aliquot of bacteria is thawed, it is not used subsequently.

   Typically, we set up the following five plates and transform using the following PCR samples and controls:

   Plate A: 2.5 μl of PCR #1 + 2.5 μl of PCR #2
   Plate B: 2.5 μl of PCR #1 + 2.5 μl of TE
   Plate C: 2.5 μl of PCR #2 + 2.5 μl of TE
   Plate D: 1.0 ng of a supercoiled template in 5 μl of TE
   Plate E: 5 μl of TE

   Only 25 μl of *E. coli* is used for the control plates D and E, so that only one BRL tube, which contains 200 μl of bacteria, needs to be used per mutagenesis.

   The yield of colonies from plate A will be greater than 2x that from plates B + C, confirming a high percentage of recombinants in plate A. Plate D is a transformation control, and should yield a thick lawn of colonies. Plate E is an antibiotic control, and should yield no colonies, since the bacteria that have not been transformed are sensitive to ampicillin.

4. Plasmid screening

   a. Place individual colonies in 2 ml of LB broth containing 100 μg/ml of ampicillin and grow for 6–24 hours at 37°C. The plasmids are screened using PCR by a modification of a previously described method (Liang and Johnson 1988) as follows: Remove 2 μl of the LB broth, place directly in the PCR tube, and amplify for 25 cycles using primers that flank the mutated site or insert (e.g., M13 primers). Usually, the mutated site can be designed to either insert or remove a restriction endonuclease site in the amplified product. For example, the degenerate amino acid code can be used to make the desired amino acid change and at the same time create a restriction endonuclease recognition site to facilitate screening.

b. Screen for the mutation by adding 3 units of the restriction endonuclease and 1 µl of the appropriate 10X restriction buffer directly to 5 µl of the unpurified PCR product in a total volume of 10 µl. Creation or elimination of the restriction endonuclease recognition site can then be directly assessed by minigel analysis.

c. Purify the plasmid and sequence the mutated region. QIAGEN columns can be used for plasmid purification and Sequenase 2.0 can be used for sequencing following the manufacturers' instructions.

Notes:

- Plasmids can be screened immediately by placing a colony in 5 µl of LB broth, vortexing, and then amplifying 2 µl as described. The remaining LB sample can later be grown up for isolation of the plasmid and storage of the colony.
- If the mutagenesis cannot be designed to eliminate or create a restriction enzyme recognition site, the PCR screening primers can be designed to preferentially amplify the mutant by creating a perfect match between the 3′ end of one of the primers and the mutated site, such that plasmids containing the original, nonmutated template sequence are not amplified in sufficient quantity to be detected by ethidium bromide staining (Sommer and Tautz 1989). Alternatively, the nonmutating primers that anneal to the original vector (primers 2 and 4 in Fig. 1) can be designed to mutate a nucleotide that creates or eliminates a restriction endonuclease recognition site. Screening for mutagenesis of this site will effectively screen for the site of interest, since both sites will be mutated concurrently.

Recombining DNA Segments

A protocol for amplifying a portion of a donor plasmid and placing it in a recipient plasmid at a defined location and orientation, with the simultaneous removal of a DNA segment in the recipient plasmid, is illustrated in Figure 2. The conditions for PCR amplification and transformation are identical to those detailed above, and are not restated. In Figure 2, the donor plasmid DNA is shown on the left side and the recipient plasmid is shown on the right side. The DNA segment that is to be inserted into the recipient construct is amplified from the donor plasmid using primers 1 and 2. In a separate PCR amplification, the recipient plasmid is amplified with primers 3 and 4. The 5′ regions of primers 1 and 2 are complementary to primers 3 and 4. Each plasmid is linearized by restriction-enzyme digestion prior to PCR amplification. In this figure, the 5′ regions of primers 1 and 2 contain regions that are homologous to the recipient-plasmid sequences to which primers 3 and 4 anneal. The only requirement for

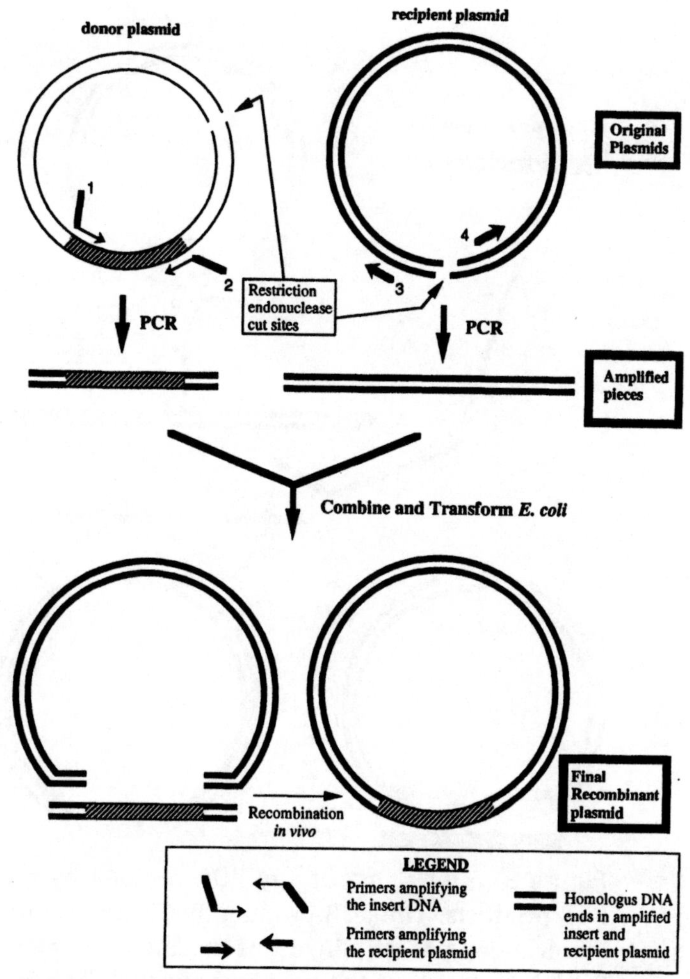

Figure 2 Generation of a recombinant construct by recombination in vivo of two PCR products. The insert is cross-hatched. Thin circles represent the DNA strands of the donor plasmid and thick circles represent the DNA strands of the recipient plasmid. The 5′ regions of primers 1 and 2 are complementary to primers 3 and 4.

this method is that primers 1 and 2 must have regions of complementarity to primers 3 and 4. As in the point mutagenesis protocol, the homologous ends between the PCR products are approximately 24 bp long. A similar strategy, in which the recipient vector is modified by PCR to contain ends that are homologous to a given PCR product, can be used for the rapid subcloning of any PCR product. This is illustrated in Figure 3.

TROUBLESHOOTING

In Protocols 1 and 2, specificity for the construct of interest is high, but the transformation efficiency is low, averaging 10 colonies with the mutation per nanogram of total DNA transfected. Therefore, only highly competent *E. coli* (transfection efficiency $>1 \times 10^9$ trans-

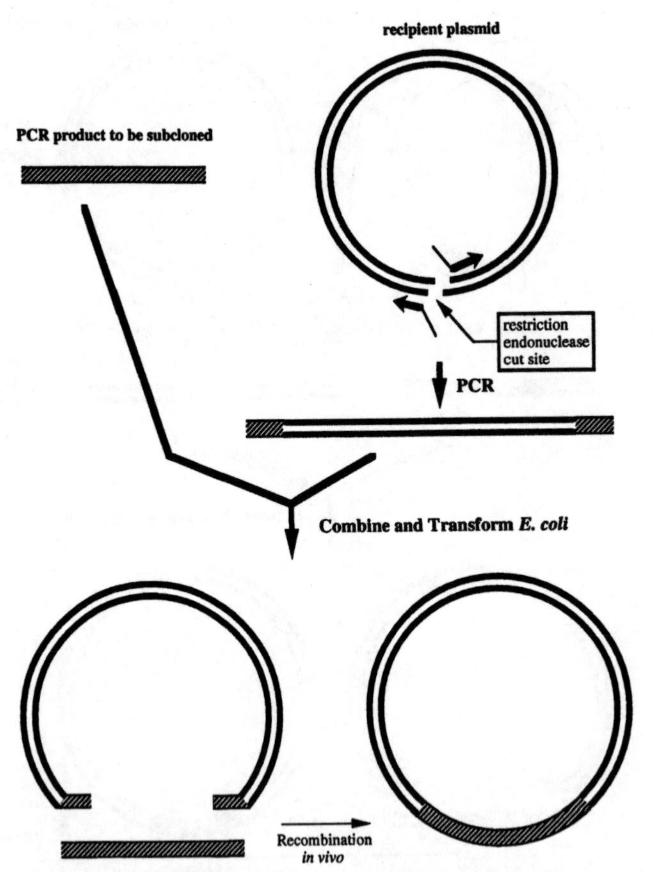

Figure 3 Subcloning of any PCR product by recombination in vivo of two PCR products. The PCR product that is to be subcloned is cross-hatched, and the primers that amplified this product are not shown. The circles represent the DNA strands of the recipient plasmid. The 5′ regions of the primers that amplify the recipient plasmid are complementary to the primers used to amplify the PCR product to be subcloned.

formants/μg of monomer pUC19) are used. The proportion of clones containing the recombinant of interest is ≥50%. Difficulty may be encountered in Protocol 2 when attempting to insert a DNA segment to generate a large direct repeat in a plasmid, because recombination between the large direct repeats could compete with recombination between PCR product ends.

If a plasmid cannot be linearized outside the region to be amplified by PCR (for instance, in the right-side portion of Fig. 2), the PCR product must be purified from the plasmid because supercoiled plasmids have a very high transformation efficiency. This can readily be accomplished by agarose gel electrophoresis. Extraction of the PCR product from the agarose gel may be done using GENECLEAN (BIO 101). When a supercoiled plasmid is used, more amplification cycles may be necessary, because PCR product yields are lower when using a supercoiled template than when using a linearized template.

DISCUSSION

In Protocol 1, primers 2 and 4 can be re-used for any mutagenesis of the insert, so that only two new primers need to be generated for each new site targeted for mutagenesis (primers 1 and 3). Furthermore, only approximately one-half of the length of the entire template needs to be amplified in each of the two PCR amplifications, facilitating the mutagenesis of large constructs. Recombination PCR has been used to mutate constructs up to 7.1 kb (Yao et al. 1992).

Because the region of complementarity of one primer to another need not anneal to the original template sequence, a variety of DNA sequence modifications can be carried out during a single transformation. For instance, in Protocol 2, as long as primers 1 and 2 contain regions that are complementary to regions of primers 3 and 4, the PCR products will contain ends that are homologous to each other, and these primer-determined DNA ends do not need to be determined by the original donor or recipient templates. Therefore, it is clear that point mutations can be placed in the recipient plasmid simultaneously with insertion of an amplified fragment (Jones et al. 1994).

In the original description of the recombination-PCR method, 2 clones of 19 sequenced contained a base deletion in a primer sequence, and we speculated that this may have resulted from the recombination in vivo that generated the construct of interest (Jones and Howard 1991). Since that paper was published, we utilized recombination PCR and recombinant-circle PCR multiple times to generate a new plasmid (Jones et al. 1994). There were six sequence errors in the resulting plasmid, and all of these errors were base substitutions. Thirty-four primer sequences were incorporated into this new plasmid using recombination-PCR Protocol 1 or Protocol 2 described above. These primer sequences constitute 36% of the resulting plasmid, and the cumulative regions of homology between the PCR product ends generated using these primers that recombined to generate this plasmid constitute 34% of this plasmid sequence. Two of the six sequence errors (33%) in the plasmid reside within these primer sequences as well as within the regions of homology generated by these primers. Therefore, the distribution of sequence errors did not cluster within the primer sequences or the short regions of homology generated by these primers, suggesting that sequence errors due to recombination in vivo between short regions of homology generated by primers are rare. The most likely cause for the majority of the errors that did occur in the final plasmid is nucleotide misincorporation during the 196 sequential polymerase extensions (11 recombination-PCR and 3 recombinant-circle-PCR protocols with 14 cycles per PCR) used to generate this plasmid. Since there is always the possibility of a sequence error in a single clone following PCR amplification, a mutated region should be sequenced, and one may choose to clone a restriction fragment containing the mutation into a construct that has not undergone PCR amplification.

A strategy similar to the original recombination-PCR protocol for subcloning has been described that substitutes a restriction-enzyme-digested plasmid for a PCR-amplified plasmid (Jones and Howard 1991; Bubeck et al. 1993; Oliner et al. 1993). A modification of Protocol 2, illustrated in Figure 3, is now used here to subclone PCR products (Jones and Winistorfer 1993). The PCR product to be subcloned is not purified, and the molar ratio of insert to the amplified recipient plasmid can vary widely. Furthermore, the PCR product to be subcloned is not amplified by primers containing 5′-end extensions that do not anneal to the original template. This is desirable when optimizing primer lengths and T_ms to avoid spurious products in protocols for amplifying a rare sequence from a complex mixture.

ACKNOWLEDGMENTS This work was supported by National Institutes of Health grant R01 HG-00569, the University of Iowa Hospitals and Clinics through funds generated by the Children's Miracle Network Telethon, and the Roy J. Carver Charitable Trust.

REFERENCES

Bubeck, P., M. Winkler, and W. Bautsch. 1993. Rapid cloning by homologous recombination in vivo. *Nucleic Acids Res.* **21:** 3601–3602.

Conley, E.C., V.A. Saunders, and J.R. Saunders. 1986a. Deletion and rearrangement of plasmid DNA during transformation of *Escherichia coli* with linear plasmid molecules. *Nucleic Acids Res.* **14:** 8905–8917.

Conley, E.C., V.A. Saunders, V. Jackson, and J.R. Saunders. 1986b. Mechanism of intramolecular recyclization and deletion formation following transformation of *Escherichia coli* with linearized plasmid DNA. *Nucleic Acids Res.* **14:** 8919–8932.

Jones, D.H. and B.H. Howard. 1991. A rapid method for recombination and site-specific mutagenesis by placing homologous ends on DNA using polymerase chain reaction. *BioTechniques* **10:** 62–66.

Jones, D.H. and S.C. Winistorfer. 1992. Recombinant circle PCR and recombination PCR for site-specific mutagenesis without PCR product purification. *BioTechniques* **12:** 528–535.

——. 1993. Genome walking with 2- to 4-kb steps using panhandle PCR. *PCR Methods Appl.* **2:** 197–203.

Jones, D.H., A.N. Riley, and S.C. Winistorfer. 1994. Production of a vector to facilitate DNA mutagenesis and recombination. *BioTechniques* **16:** 694–701.

Liang, W. and J.P. Johnson. 1988. Rapid plasmid insert amplification with polymerase chain reaction. *Nucleic Acids Res.* **16:** 3579.

Mandecki, W. 1986. Oligonucleotide-directed double-strand break repair in plasmids of *Escherichia coli:* A method for site-specific mutagenesis. *Proc. Natl. Acad. Sci.* **83:** 7177–7181.

Mullis, K., F. Faloona, S. Scharf, R. Saiki, G. Horn, and H. Erlich. 1986. Specific enzymatic amplification of DNA in vitro: The polymerase chain reaction. *Cold Spring Harbor Symp. Quant. Biol.* **51:** 263–273.

Oliner, J.D., K.W. Kinzler, and B. Vogelstein. 1993. In vivo cloning of PCR products in *E. coli. Nucleic Acids Res.* **21:** 5192–5197.

Sambrook, J., E.F. Fritsch, and T. Maniatis. 1989a. *Molecular cloning: A laboratory manual,* 2nd edition, p. B.23. Cold Spring Harbor Laboratory Press, Cold Spring Harbor, New York.

——. 1989b. *Molecular cloning: A laboratory manual,* 2nd edition, p. A.2. Cold Spring Harbor Laboratory Press, Cold Spring Harbor, New York.

——. 1989c. *Molecular cloning: A laboratory manual,* 2nd edition, p. A.4. Cold Spring Harbor Laboratory Press, Cold Spring Harbor, New York.

——. 1989d. *Molecular cloning: A laboratory manual,* 2nd edition, p. A.1. Cold Spring Harbor Laboratory Press, Cold Spring Harbor, New York.

Sommer, R. and D. Tautz. 1989. Minimal homology

requirements for PCR primers. *Nucleic Acids Res.* **17:** 6749.

Sung, W.L. and D.M. Zahab. 1987. Site-specific recombination directed by single-stranded crossover linkers: Specific deletion of the amino-terminal region of the β-galactosidase gene in pUC plasmids. *DNA* **6:** 373–379.

White, B., ed. 1993. *PCR protocols: Current methods and applications.* Humana Press, Clifton, New Jersey.

Yao, Z., D.H. Jones, and C. Grose. 1992. Site-directed mutagenesis of herpesvirus glycoprotein phosphorylation sites by recombination polymerase chain reaction. *PCR Methods Appl.* **1:** 205–207.

Mutagenesis and Synthesis of Novel Recombinant Genes Using PCR

Abbe N. Vallejo, Robert J. Pogulis, and Larry R. Pease

Department of Immunology, Mayo Clinic/Foundation, Rochester, Minnesota 55905

INTRODUCTION

Since its description, PCR (Mullis et al. 1986) has become a fundamental analytical tool in cellular and molecular biology. The literature is replete with the use of this technique in the identification of new genes or members of gene families, even those from distantly related species. PCR permitted the introduction of mutations into DNA sequences (Higuchi et al. 1988) to allow assessment of the biological function of genes. PCR can also be used to simplify site-directed mutagenesis and to generate recombinant or chimeric gene constructs (Ho et al. 1989; Horton et al. 1990). Mutagenesis by PCR is accomplished by incorporating the desired genetic changes in custom-made primers used in amplification reactions. Because these mutagenizing primers have terminal complementarity, two separate DNA fragments amplified from a target gene may be fused into a single product (see below) by primer extension without relying on restriction endonuclease sites or ligation reactions.

The relative ease of rapidly combining two DNA fragments by overlap extension led to its application in the production of chimeric genes. As long as pairs of primers used in amplification reactions have complementary terminal regions, overlapping strands of PCR-amplified DNA fragments from various sources may be spliced together by the extension of overlapping strands (hence, the term *s*plicing by *o*verlap *e*xtension or gene SOEing) in a subsequent reaction. The ability to generate genetic recombinants by SOEing is limited only by the knowledge of the mechanisms of folding and/or assembly of the encoded novel protein when transfected into cells. SOE has been used routinely to generate mutant molecules, complex hybrid genes, engineer proteins, and create mutant libraries (Hunt et al. 1990a; Daugherty et al. 1991; Davis et al. 1991; Pullen et al. 1991; Cai and Pease 1992; Gobius et al. 1992; Bobek et al. 1993; Kirchoff and Desrosiers 1993; Pease et al. 1993; Yun et al. 1994).

Concept of Overlap Extension

Overlap extension uses PCR both for introducing site-specific mutations and for generating recombinant gene constructs. Details of the strategy have been described previously (Ho et al. 1989; Horton et al. 1990). Briefly, mutagenesis is achieved by PCR with the use of specially designed oligonucleotide primers that include in their sequence the desired changes (i.e., substitutions, insertions, or deletions) to be incorporated in the gene construct (Fig. 1). Because the mutagenizing primers also have terminal complementarity, two overlapping fragments can be fused together in a subsequent extension reaction. The inclusion of outside primers in the extension reaction amplifies the fused product by PCR. In principle, the primers may be moved anywhere along the targeted gene to introduce mutations. Thus, the region of the gene containing the introduced mutations may be lengthened in a single reaction.

The ability to fuse two DNA fragments by overlap extension can be exploited further to splice (or SOE together) two or more DNA fragments from different genes to generate a chimeric product. Like the

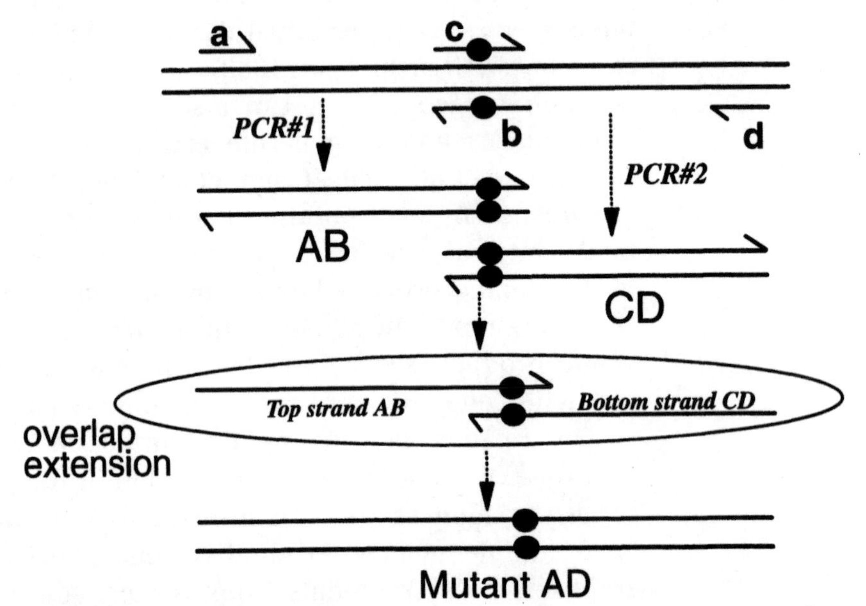

Figure 1 Mutagenesis by overlap extension. Two segments of a gene are PCR-amplified independently and then fused together in a subsequent reaction. Mutations are introduced into a targeted region with the use of specially designed mutagenizing primers (b and c), which contain nucleotide mismatches (represented by the solid circles) in the center of the primers. Because these primers are complementary, strands of PCR products generated independently with these primers will have an overlap that can be extended in a subsequent reaction to form the mutant product. When outside primers (a and d) are added to this latter reaction, the fused mutant product is amplified as soon as it is formed.

mutagenizing primers, SOEing primers have terminal complementarity. Thus, DNA fragments generated by PCR can be spliced together by primer extension and amplified to yield a recombinant product (Fig. 2).

A limitation of SOE, however, is the difficulty of manipulating large DNA segments, i.e., those more than 1–2 kb. To circumvent this difficulty, a cassette system may be developed wherein shorter DNA segments, typically around 500–1000 bp, can be easily targeted, modified by SOE, and reinserted using restriction endonuclease sites designed into the cassette structure (Horton et al. 1990; Hunt et al. 1990b). Thus, specific segments of genes can be manipulated at will. This cassette approach also allows easy shuffling or replacement of gene segments. Such a cassette system has been a very valuable tool in dissecting the structure-function relationship of class I major histocompatibility complex molecules (Horton et al. 1990; Hunt et al. 1990a; Cai and Pease 1992).

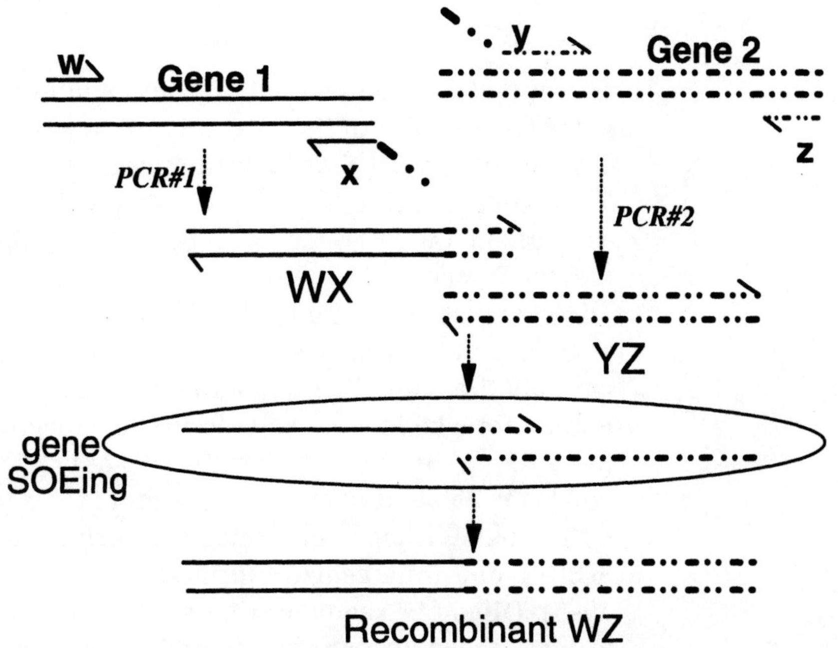

Figure 2 Gene SOEing. The reactions involved in SOEing are similar to that depicted in Fig. 1, with two exceptions. The first is that the PCR products to be fused are derived from two different genes. Second, the fusion is mediated by an overlap of the two strands of PCR products that were created with the use of SOEing primers (x and y). In this case, the 5′ region of primer x, used in the amplification of gene 1, is complementary to a segment of gene 2 (i.e., the 5′ region of primer y). A recombinant product is therefore formed when this overlap is extended in a subsequent reaction, and this recombined product may be amplified with the inclusion of outside primers (w and z) in this latter extension reaction.

Design of Oligonucleotide Primers

SOEing oligomers have two sequence regions; namely, a "priming" region and an "overlap" region. The priming region is the 3′ end of the oligomer, which serves as the PCR primer, and the overlap region is the 5′ end of the oligomer, which is complementary to a sequence of a DNA fragment that will be fused with it in the overlap extension reaction. As depicted in Figure 2, primer x contains a priming sequence for the amplification of gene 1 and also has an overlap sequence at its 5′ end that is complementary to a segment of gene 2.

Mutagenizing oligomers can have their priming and overlap regions completely coinciding with each other. The centers of the oligomers contain the mismatches or deletions that will be incorporated by overlap extension (Fig. 1).

In designing the oligomers, the actual lengths of the priming and overlap regions may depend on the particular situation. One approach is to determine the length by a simple estimate of the melting temperature (Horton and Pease 1991; Suggs et al. 1981), which is calculated as follows:

$$T_m = ([G+C] \times 4) + ([A+T] \times 2)$$

This quantity is an estimate of the denaturation temperature (in degrees Centigrade) of the oligomer. As a rule of thumb, our oligomers are designed such that both their priming and overlap regions have T_m values around 50°C. This 50°C rule generally gives a length estimate of 13–20 bp for both the priming and the overlap regions of the oligomers. The mismatches incorporated in mutagenizing oligomers are not included in this estimation.

It must be emphasized, however, that T_m does not estimate accurately the annealing temperature of any specific oligomer, because it does not take into consideration the concentration of salts present in the cycling reaction. Nevertheless, our 50°C T_m rule has usually yielded reliable oligomers that anneal at this temperature.

The usual rules in designing PCR primers apply when synthesizing SOEing and mutagenizing primers (Horton and Pease 1991; Rychlik 1993). Oligomer sequences must always be checked for the potential to form secondary structures and for complementarity to primers in the same reaction. Proper priming is achieved when at least the five bases of the 3′-terminal end of the primer are complementary to the template. Introducing mismatches too close to the 3′ end of the primer could result in the lack of amplification.

REAGENTS

Thermal cycling (PCR) machine
Taq DNA polymerase (Perkin-Elmer)
10X PCR buffer containing 100 mM Tris-HCl, pH 8.3, 500 mM KCl, and

10 mM $MgCl_2$ (see Note 4, below, for determining the appropriate concentration of Mg^{++})

10X dNTPs: Prepare a 2 mM stock (10X) solution of each of the dNTPs. The pH of the solution must be adjusted to 7.5, otherwise polymerase activity will be inhibited. This pH adjustment is not necessary when using commercially available dNTP solutions (Pharmacia) that already come as aqueous stocks with the correct pH.

Mineral oil (Sigma)

Agarose: The concentration of agarose needed to resolve DNA fragments depends on the fragment size to be isolated. In our experience, 0.8–1% agarose is generally satisfactory in resolving DNA fragments from 100 bp to 2 kb. Greater resolution also may be achieved by mixing ultragrade agaroses like NuSieve with the less-expensive "working" agaroses like LE or SeaKem agarose (FMC).

Agarose gel electrophoresis units and supplies

GENECLEAN or MERMAID kits (BIO 101) or electroelution apparatus (see Note 5, below).

PROTOCOLS

PCR Mutagenesis by Overlap Extension

1. Set up the PCR assays to produce products AB and CB (refer to Fig. 1) in individual microcentrifuge tubes. These reaction mixtures can be assembled at room temperature.

Reaction no.	1	2
PCR product	AB	CD
Template	parental template	parental template
5′ primer	a	c*
3′ primer	b*	d
10X PCR buffer	10 μl	10 μl
10X dNTPs	10 μl	10 μl
Taq DNA polymerase	2.5 units	2.5 units
Sterile water	to 100 μl	to 100 μl

(*) denotes mutagenizing primers.

2. Overlay the PCR mixture with 2–3 drops of mineral oil.

3. Cycle for 15–25 rounds (30 seconds at 94°C, 2 minutes at 50°C, and 1 minute at 72°C).

4. Run the entire reactions on agarose gels.

5. Cut out the bands of interest (i.e., products AB and CD), and recover the DNA fragments by GENECLEAN, MERMAID, electroelution, or freeze-squeeze (see Note 5, below).

6. Set up the overlap extension reaction in a new microcentrifuge tube.

Product	AD
Template 1	AB
Template 2	CD
5′ primer	a
3′ primer	d
10X PCR buffer	10 μl
10X dNTPs	10 μl
Taq DNA polymerase	2.5 units
Sterile water	to 100 μl

Note: This reaction is not very sensitive to the amounts of templates added. Typically 10–100 ng of each template may be used. Primers *a* and *d* may also be added from the beginning of this PCR. Alternatively, 3–5 cycles of extension can first be carried out to allow for the formation of the mutant product AD followed by amplification upon the addition of primers.

7. Overlay the reaction mixture with mineral oil and amplify by PCR as in step 3, above.

8. Run the reactions on agarose gels, cut out the band of interest (i.e., product AD), and recover the DNA fragment as in step 5, above.

9. The purified mutant product AD may then be digested with the appropriate restriction enzymes and ligated to an appropriate cloning vector for subsequent transformation and sequencing (see Note 6, below).

Gene SOEing

1. Set up PCR in individual microcentrifuge tubes to generate products WX and YZ from genes 1 and 2, respectively (Fig. 2).

Reaction no.	1	2
PCR product	WX	YZ
Template	gene 1	gene 2
5′ primer	w	y*
3′ primer	x*	z
10X PCR buffer	10 μl	10 μl
10X dNTPs	10 μl	10 μl
Taq DNA polymerase	2.5 units	2.5 units
Sterile water	to 100 μl	to 100 μl

(*) denotes SOEing primers

2. Overlay the reaction with 2–3 drops of mineral oil.

3. Cycle for 15–25 rounds (30 seconds at 94°C, 2 minutes at 50°C, 1 minute at 72°C).

4. Run the entire reactions on agarose gels.

5. Cut out the bands of interest (i.e., products WX and YZ), and recover the DNA fragments by GENECLEAN, MERMAID, electroelution, or freeze-squeeze (see Note 5, below).

6. Set up the SOEing reaction in a new microcentrifuge tube.

Product	WZ
Template 1	WX
Template 2	YZ
5′ primer	w
3′ primer	z
10X PCR buffer	10 μl
10X dNTPs	10 μl
Taq DNA polymerase	2.5 units
Sterile water	to 100 μl

7. Overlay the reaction mixture with mineral oil and amplify by PCR as in step 3, above.

8. Run the reaction on agarose gels, cut out the band of interest (i.e., product WZ), and recover the DNA fragment as in step 5, above.

9. The purified recombinant product WZ may then be digested with the appropriate restriction enzymes and ligated to an appropriate cloning vector for subsequent transformation and sequencing (see Note 6, below).

NOTES/COMMENTS

1. *Oligonucleotide purification.* Oligomers used in this laboratory are synthesized using a DNA Synthesizer (Perkin-Elmer, Applied Biosystems Division) in our institutional core facility, and the final material comes as an aqueous ammonia solution. The solution is dried under vacuum (SpeedVac SC100, Savant). The residue is resuspended in 1 ml of sterile water and desalted in Sephadex G-25 (NAP-10 columns, Pharmacia). Fractions of 500 μl are collected, and the absorbance of each fraction is determined at 260 nm. The fractions containing the first peak are pooled (typically fractions 3–5 from NAP-10 columns), and the absorbance of the pooled material is measured at 260 and 300 nm. The concentration of the

purified oligomer is determined by the formula: μg/ml = (A_{260nm}−A_{300nm}) × dilution factor × 33, where the number 33 is the approximate concentration of oligonucleotides (μg/ml) per unit of absorbance at 260 nm (Sambrook et al. 1989). The concentration of the stock oligomer may also be expressed in molar terms as follows: μM = ([A_{260nm}−A_{300nm}) × dilution factor] ÷ ε260) × 10^6; where ε260, the extinction coefficient, equals the number of bases times 10,000. The purified oligomer is aliquoted into smaller quantities and stored at −20°C.

2. *Primer concentration.* In a standard reaction of 100 μl, we find that 100 pmoles of each primer is optimal. This gives a final concentration of 1 pmole/μl (or 1 μM). It is recommended, however, that the amounts of primers be determined empirically. The important point is that large amounts of primers increase amplification errors caused by mispriming.

3. *Template concentration.* In theory, larger amounts of template reduce the number of cycles required to generate enough product and, therefore, lessen the chance of *Taq* polymerase-induced base misincorporations. However, high concentrations of template tend to impede successful amplification and fewer rounds of amplification produce more open-ended strands (i.e., single-stranded products that extend beyond the length of the primer), which contribute to the background. Thus, it is necessary to titrate the amount of template to determine the optimal concentration. We generally find that 500 ng of cloned template or as much as 1–2 μg of genomic DNA yields the PCR products of interest without significant background amplification.

4. *Magnesium concentration.* Perhaps the single most important parameter for PCR is the amount of Mg^{++} present in the amplification reaction. High concentrations of Mg^{++} generally produce background amplification and increase *Taq*-induced errors that can be reduced by lowering the amount of Mg^{++} in the reaction. Therefore, the optimal concentration of Mg^{++} must first be determined. Experimental protocols involving PCR begin with ascertaining optimal amplification conditions. Stocks of 10× PCR buffers are prepared containing various concentrations of $MgCl_2$, which, when used at 1× concentration, give a final concentration of 0.5–2.5 mM of Mg^{++}. In most of our PCR experiments, we find that 1 mM Mg^{++} is optimal.

5. *Purification of PCR and SOEing products.* Agarose gel purification of the intermediate PCR products leads to cleaner SOE reactions

(Ho et al. 1989). This purification step removes templates as well as open-ended strands that could generate unwanted products. Similarly, gel purification of SOE products ensures cleaner fragments for subsequent cloning.

Depending on the size of the DNA fragments, amplification products may be purified by either GENECLEAN or MERMAID kits (BIO 101). The latter are specifically designed for the purification of products less than or equal to 200 bp. Alternatively, purification may also be achieved by electroelution (Sambrook et al. 1989) or by a modification of the freeze-squeeze technique (Tautz and Renz 1983).

To purify DNA by freeze-squeeze, the band containing the DNA fragment of interest is excised from an agarose gel, macerated, transferred to a microcentrifuge tube, and frozen at −70°C for 1 hour (or overnight at −20°C). The frozen gel is thawed at 37°C for 1 hour, transferred to an Ultrafree-MC filter unit (Millipore) or a Spin-X filter unit (Corning Costar), and centrifuged at 12,000*g* for 20 minutes at room temperature. The filtrate is collected and subjected to two rounds of ethanol precipitation (Sambrook et al. 1989).

6. *Cloning and sequencing of SOEing products.* Mutant and/or recombinant genes generated by SOE must always be cloned and sequenced to determine the accuracy of the introduced genetic changes and to ascertain that there have been no amplification errors. We routinely sequence DNA using the Sequenase kit (USB/Amersham) or an automated sequencer (Model 373A, Perkin-Elmer, Applied Biosystems Division).

   Cloning of the PCR-generated genes could be facilitated by designing amplification primers (i.e., a-b and w-z primer pairs in Figs. 1 and 2, respectively) with unique flanking restriction enzyme sites (details of the strategy are discussed elsewhere) (Kwok et al. 1994). The product of the third PCR (mutant AD or recombinant WZ) may then be digested with these enzymes and cloned into a suitable vector or exchanged with the appropriate fragment of a plasmid cassette (Horton et al. 1990; Hunt et al. 1990b).

REFERENCES

Bobek, L.A., H. Tsai, and M.J. Levine. 1993. Expression of human salivary histatin and cystatin/histatin chimeric cDNAs in *Escherichia coli*. *Crit. Rev. Oral Biol. Med.* **4:** 581–590.

Cai, Z. and L.R. Pease. 1992. Structural and functional analysis of three D/L-like class I molecules from H-2^{v}: Indications of an ancestral family of D/L genes. *J. Exp. Med.* **175:** 583–596.

Daugherty, B.L., J.A. deMartino, M.F. Law, D.W. Kawka, I.I. Singer, and G.E. Mark. 1991. Polymerase chain reaction facilitates the cloning, CDR-grafting and rapid expression of a murine monoclonal antibody directed against the CD18 component of leukocyte integrins. *Nucleic Acids Res.*

19: 2471–2476.

Davis, G.T., W.D. Bedzyk, E.W. Voss, and T.W. Jacobs. 1991. Single chain antibody (SCA) encoding genes: One step construction and expression in eukaryotic cells. *BioTechnology* **9:** 165–169.

Gobius, K.J., S.W. Rowlinson, R. Barnard, J.S. Mattick, and M.J. Waters. 1992. The first disulfide loop of the rabbit growth hormone receptor is required for binding to the hormone. *J. Mol. Endocrinol.* **9:** 213–220.

Higuchi, R., B. Krummel, and R.K. Sakai. 1988. A general method of in vitro preparation and specific mutagenesis of DNA fragments: Study of protein and DNA interactions. *Nucleic Acids Res.* **15:** 7351–7367.

Ho, S.N., H.D. Hunt, R.M. Horton, J.K. Pullen, and L.R. Pease. 1989. Site-directed mutagenesis by overlap extension using polymerase chain reaction. *Gene* **77:** 51–59.

Horton, R.M. and L.R. Pease. 1991. Recombination and mutagenesis of DNA sequences using PCR. In *Directed mutagenesis: A practical approach* (ed. M.J. McPherson), pp. 217–247. Oxford University Press, England.

Horton, R.M., Z. Cai, S.N. Ho, and L.R. Pease. 1990. Gene splicing by overlap extension: Tailor-made genes using the polymerase chain reaction. *BioTechniques* **8:** 528–535.

Hunt, H.D., J.K. Pullen, R.F. Dick, J.A. Bluestone, and L.R. Pease. 1990a. Structural basis of K^{bm8} alloreactivity: Amino acid substitutions on the β-pleated floor of the antigen recognition site. *J. Immunol.* **145:** 1456–1462.

Hunt, H.D., J.K. Pullen, Z. Cai, R.M. Horton, S.N. Ho, and L.R. Pease. 1990b. Novel MHC variants spliced by overlap extension. In *Transgenic mice and mutants in MHC research* (ed. I.K. Egorov and C.S. David), pp. 47–55. Springer-Verlag, Berlin.

Kirchoff, F. and R.C. Desrosiers. 1993. A PCR-derived library of random mutations within the V3 region of simian immunodeficiency virus. *PCR Methods Appl.* **2:** 301–304.

Kwok, S., S.Y. Chang, J.J. Sninsky, and A. Wang. 1994. A guide to the design and use of mismatched and degenerate primers. *PCR Methods Appl.* **3** S39–S47.

Mullis, K., F. Faloona, S. Scharf, R. Saiki, G. Horn, and H. Erlich. 1986. Specific enzymatic amplification of DNA in vitro: The polymerase chain reaction. *Cold Spring Harbor Symp. Quant. Biol.* **51:** 263–273.

Pease, L.R., R.M. Horton, J.K. Pullen, H.D. Hunt, T.J. Yun, E.M. Rohren, J.L. Prescott, S.M. Jobe, and K.S. Allen. 1993. Amino acid changes in the peptide binding site have structural consequences at the surface of class I glycoproteins. *J. Immunol.* **150:** 3375–3381.

Pullen, J.K., H.D. Hunt, and L.R. Pease. 1991. Peptide interactions with the K^b antigen recognition site. *J. Immunol.* **146:** 2145–2151.

Rychlik, W. 1993. Selection of primers for polymerase chain reaction. In *Methods in molecular biology, PCR protocols: Current methods and applications* (ed. B.A. White), vol. 15, pp. 31–40. Humana Press, Totowa, New Jersey.

Sambrook, J., E.F. Fritsch, and T. Maniatis. 1989. *Molecular cloning: A laboratory manual*, 2nd edition, pp. 6.28–6.29; E.10–E.15. Cold Spring Harbor Laboratory Press, Cold Spring Harbor, New York.

Suggs, S.V., T. Hirose, T. Miyaki, E. Kawashima, M.J. Johnson, K. Itakura, and R.B. Wallace. 1981. Use of synthetic oligodeoxyribonucleotides for the isolation of cloned DNA sequences. In *Developmental biology using purified genes* (ed. D.D. Brown and C.F. Fow), pp. 683–693. Academic Press, New York.

Tautz, D. and M. Renz. 1983. An optimized freeze-squeeze method for the recovery of DNA fragments from agarose gels. *Anal. Biochem.* **132:** 14–19.

Yun, T.J., M.D. Tallquist, E.M. Rohren, J.M. Sheil, and L.R. Pease. 1994. Minor pocket B influences peptide binding, peptide presentation and alloantigenicity of H-$2K^b$. *Intl. Immunol.* **6:** 1037–1047.

Rapid PCR Site-directed Mutagenesis

Michael P. Weiner and Gina L. Costa

Stratagene Cloning Systems, La Jolla, California 92037

INTRODUCTION

In vitro site-directed mutagenesis is an invaluable technique for studying protein structure-function relationships, gene expression, and vector modification. Several methods for performing site-directed mutagenesis have appeared in the literature, but these methods generally require single-stranded DNA (ssDNA) as the template (Kunkel 1985; Taylor et al. 1985; Vandeyar et al. 1988; Sugimoto et al. 1989). PCR-mediated methods have been developed that allow the use of double-stranded DNA (dsDNA). PCR-mediated methods use denaturation by heat to separate complementary strands of DNA and thus allow the use of double-stranded molecules (Jones and Winistorfer 1992; Watkins et al. 1993; Picard et al. 1994). Often, these procedures require the use of multiple pairs of primers. To circumvent the disadvantages of preexisting PCR-based methods of site-directed mutagenesis (SDM), we have developed a PCR-SDM method that allows site-specific mutation in virtually any double-stranded plasmid, thus eliminating the need for subcloning into M13-based bacteriophage vectors and for ssDNA rescue (Weiner et al. 1994). The PCR-SDM system uses a single mutagenesis primer set and generates mutants with >50% efficiency in approximately 3 hours. The protocol is simple to perform and uses either miniprep or cesium chloride-purified DNA.

The PCR-SDM system (see Fig. 1) uses increased template concentration and reduced cycling number to decrease potential second-site mutations during the PCR. The polymerase adjunct, *Taq* Extender PCR additive, is added to the PCR to increase reliability (Nielson et al. 1994). The *Dpn*I endonuclease (target sequence: 5′-G^{m6}ATC-3′) is specific for methylated and hemimethylated DNA and is used to digest parental DNA and select for mutation-containing amplified DNA. DNA isolated from almost all *Escherichia coli* strains is Dam-methylated and therefore susceptible to *Dpn*I digestion. DNA isolated from *dam*-deficient *E. coli* or other host organisms can be methylated in vitro using Dam methylase. Cloned *Pfu* DNA polymerase is used prior to end-to-end ligation of the linear template to remove any bases extended onto the 3′ ends of the product by *Taq* DNA polymerase (Costa and Weiner 1994a,b). The recircularized vector DNA incorporating the desired mutations is then ligated and transformed into *E. coli.* A single buffer (SDM buffer) has been developed that can be used for all of the steps involved in the procedure.

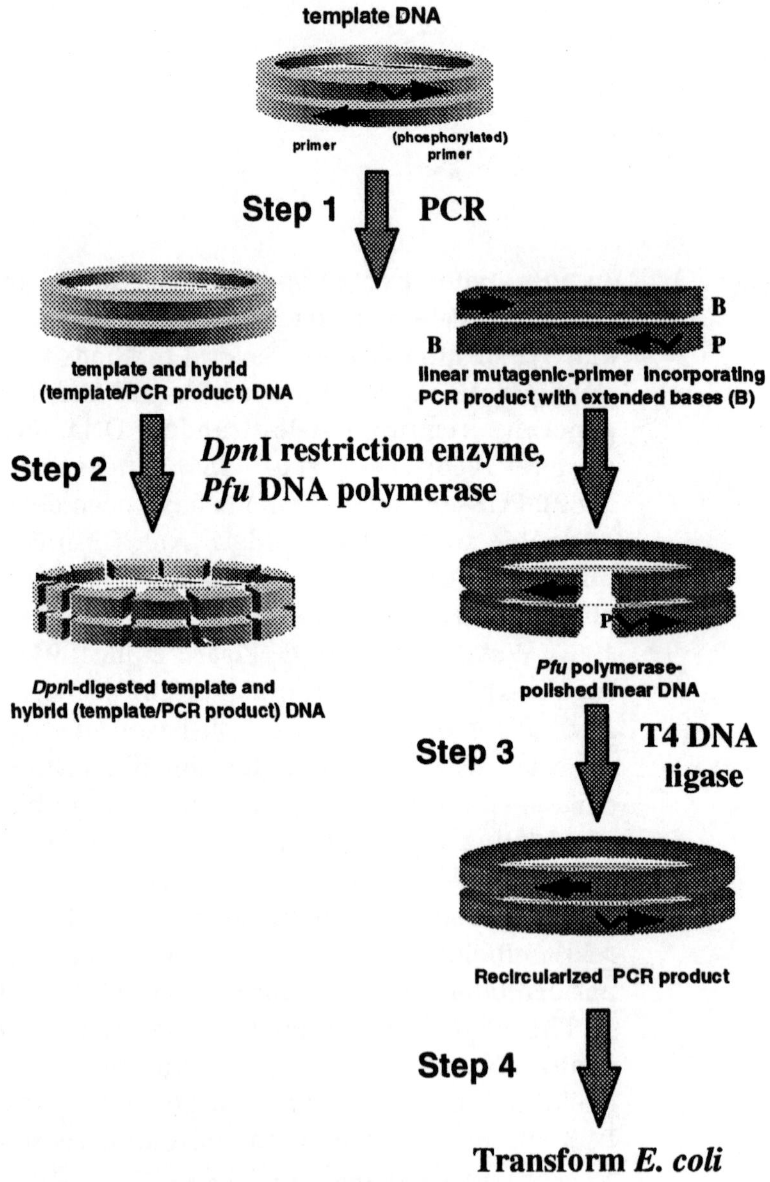

Figure 1 *(See facing page for legend.)*

REAGENTS

Taq DNA polymerase
Cloned *Pfu* DNA polymerase (Stratagene, cat. no. 600153)
Taq Extender PCR additive (Stratagene, cat. no. 600148)
*Dpn*I restriction endonuclease
T4 DNA ligase
ATP (10 mM)
dNTP mixture (25 mM total; 6.25 mM of each nucleotide triphosphate)
Competent *E. coli* cells
Mutagenesis primers
Template DNA
SDM buffer (20 mM Tris-HCl, pH 7.5, 8 mM $MgCl_2$, 40 μg/ml BSA)
FALCON 2059 tubes

Primer Considerations

Mutagenesis primers introduce chosen mutations, so mutagenesis primer oligonucleotides for use in this protocol must be designed individually according to the desired mutation. The following considerations should be taken into account in designing mutagenesis primers:

1. Both the mutagenesis and the second primers must anneal to different strands of the plasmid.

Figure 1 Schematic of the method for SDM using PCR. Template DNA is treated by the PCR-SDM protocol for a limited number of PCR cycles. The resulting mixture of template, newly synthesized, and hybrid parental/newly synthesized DNA is treated with *Dpn*I (target site 5′G^{m6}ATC) and *Pfu* DNA polymerase. (P) 5′ phosphate; (B) 3′-terminal extended base(s). The end-polished PCR product is then intramolecularly ligated together and transformed into *E. coli.* Specific methods: Template DNA (~0.5 pmole) is added to a 25-μl PCR cocktail containing 1x SDM buffer, 15 pmoles of each primer, 250 μM of each dNTP, 2.5 units of *Taq* DNA polymerase, and 2.5 units of *Taq* Extender (Stratagene). The PCR cycling parameters were: 1 cycle of 4 min at 94°C, 2 min at 50°C, and 2 min at 72°C, followed by 8 cycles of 1 min at 94°C, 2 min at 56°C, and 1 min at 72°C (*step 1*). The parental template DNA and the linear, mutagenesis primer incorporating the newly synthesized DNA are treated with *Dpn*I (10 units) and *Pfu* DNA polymerase (2.5 units, Stratagene). This results in the *Dpn*I digestion of the in vivo-methylated parental template and hybrid DNA (Nelson and McClelland 1992) and the removal, by *Pfu* DNA polymerase, of the *Taq* DNA polymerase-extended base(s) on the linear PCR product. The reaction is incubated at 37°C for 30 min and then transferred to 72°C for an additional 30 min (*step 2*). SDM buffer (115 μl, containing 0.5 mM ATP) is added to the *Dpn*I-digested, *Pfu* DNA polymerase-polished PCR product. The solution is mixed, and 10 μl is removed to a sterile microcentrifuge tube and T4 DNA ligase (4 units) is added. The ligation is incubated for >60 min at 37°C (*step 3*). The ligase-treated DNA is then transformed into competent *E. coli* (*step 4*).

2. The distance between the primers is not crucial, but the primers should not overlap.

3. Primers should be >20 bases in length (shorter primers may be used; see 4, below).

4. The mismatched portions should be at the 5′ end of one or both of the primers with ≥15 bases of correct sequence on the 3′ side.

5. The primers should be capable of synthesizing a PCR product (see PCR Considerations, below).

6. One or both of the primers must be 5′ phosphorylated (Sambrook et al. 1989).

PCR Considerations

It may be desirable to optimize the PCR before beginning the PCR-SDM procedure. Researchers should establish the PCR conditions needed to synthesize full-length products. *Taq* Extender (Stratagene) is a PCR adjunct that increases the efficiency and reliability of *Taq* DNA polymerase-generated PCR products. Pilot PCR reactions should be performed with reduced (50 ng) template concentrations and increased cycle numbers (30–40 cycles). Other reagent concentrations (including the primer set) should be kept constant as described below in the protocol section. These concentrations should be used to optimize denaturing, annealing, and extension times. The parameters that generate full-length, linear template molecules as the major amplification product should be used in the PCR-SDM method. After these conditions have been established, the template concentration should be increased, the cycle number reduced, and the PCR-SDM protocol followed.

Introduction of Translationally Silent Mutations

For easy analysis and later manipulations, it is often desirable to incorporate translationally silent restriction sites into the mutagenesis primer during site-specific mutagenesis (Weiner and Scheraga 1989; Shankarappa et al. 1992). Tables 1 and 2 can be used for reverse translation of protein-coding regions to determine where, in a particular sequence, a translationally silent restriction site can be inserted.

PROTOCOL

For an overview of the PCR-SDM protocol, see Figure 1. Prepare the PCR-SDM reaction mixture by adding the following reagents to a reaction tube:

Table 1 Introducing Restriction Endonuclease Sites by Silent Mutations

R.E.[a]	Target Sequence	Reading frame 1 (amino acids) 1	2	Reading frame 2 (amino acids) 1	2	3	Reading frame 3 (amino acids) 1	2	3
*Alw*441	GTGCAC	V	H	CRSG	A	LPHQR	LSXWPQRMTKVAEG	C	T
*Apa*I	GGGCCC	G	P	WRG	A	LPHQR	LSXWPQRMTKVAEG	G	P
*Bam*HI	GGATCC	G	S	WRG	I	LPHQR	LSXWPQRMTKVAEG	D	P
*Bcl*I	TGATCA	X	S	LMV	I	IMTNKSR	FSYCLPHRITNVADG	D	HQ
*Bgl*II	AGATCT	R	S	XQKE	I	FLSYXCW	LSXPQRITKVAEG	D	L
*Bsp*MII	TCCGGA	S	G	FLIV	R	IMTNKSR	FSYCLPHRITNVADG	P	DE
*Cla*I	ATCGAT	I	D	YHND	R	FLSYXCW	LSXPQRITKVAEG	S	IM
*Eco*RI	GAATTC	E	F	XRG	I	LPHQR	LSXWPQRMTKVAEG	N	S
*Eco*RV	GATATC	D	I	XRG	Y	LPHQR	LSXWPQRMTKVAEG	I	S
*Hind*III	AAGCTT	K	L	XQKE	A	FLSYXCW	LSXPQRITKVAEG	S	FL
*Hpa*I	GTTAAC	V	N	CRSG	X	LPHQR	LSXWPQRMTKVAEG	L	T
*Kpn*I	GGTACC	G	T	WRG	Y	LPHQR	LSXWPQRMTKVAEG	V	P
*Mlu*I	ACGCGT	T	R	YHND	A	FLSYXCW	LSXPQRITKVAEG	R	V
*Msc*I	TGGCCA	W	P	LMV	A	IMTNKSR	FSYCLPHRITNVADG	G	HQ
*Nae*I	GCCGGC	A	G	CRSG	R	LPHQR	LSXWPQRMTKVAEG	P	A
*Nar*I	GGCGCC	G	A	WRG	R	LPHQR	LSXWPQRMTKVAEG	A	P
*Nco*I	CCATGG	P	W	SPTA	M	VADEG	FSYCLPHRITNVADG	H	G
*Nde*I	CATATG	H	M	SPTA	Y	VADEG	FSYCLPHRITNVADG	I	CXW
*Nhe*I	GCTAGC	A	S	CRSG	X	LPHQR	LSXWPQRMTKVAEG	L	A
*Nru*I	TCGCGA	S	R	FLIV	A	IMTNKSR	FSYCLPHRITNVADG	R	DE
*Pst*I	CTGCAG	L	Q	SPTA	A	VADEG	FSYCLPHRITNVADG	C	SR
*Pvu*I	CGATCG	R	S	SPTA	I	VADEG	FSYCLPHRITNVADG	D	R
*Pvu*II	CAGCTG	Q	L	SPTA	A	VADEG	FSYCLPHRITNVADG	S	CXW
*Sal*I	GTCGAC	V	D	CRSG	R	LPHQR	LSXWPQRMTKVAEG	S	T
*Sma*I	CCCGGG	P	G	SPTA	R	VADEG	FSYCLPHRITNVADG	P	G
*Spe*I	ACTAGT	T	S	YHND	X	FLSYXCW	LSXPQRITKVAEG	L	V
*Sph*I	GCATGC	A	C	CRSG	M	LPHQR	LSXWPQRMTKVAEG	H	A
*Sst*I	GAGCTC	E	L	XRG	A	LPHQR	LSXWPQRMTKVAEG	S	S
*Sst*II	CCGCGG	P	R	SPTA	A	VADEG	FSYCLPHRITNVADG	R	G
*Stu*I	AGGCCT	R	P	XQKE	A	FLSYXCW	LSXPQRITKVAEG	G	L
*Xba*I	TCTAGA	S	R	FLIV	X	IMTNKSR	FSYCLPHRITNVADG	L	DE
*Xho*I	CTCGAG	L	E	SPTA	R	VADEG	FSYCLPHRITNVADG	S	SR
*Xma*III	CGGCCG	R	P	SPTA	A	VADEG	FSYCLPHRITNVADG	G	R

Adapted from Weiner and Scheraga (1989) (for Macintosh-based computers); and Shankarappa et al. (1992) (for IBM PC-based computers).

Supplementary material containing these computer programs written for an Apple MacIntosh computer is available from the Quantum Chemistry Program Exchange (QCPE), Department of Chemistry, Indiana University, Bloomington, Indiana 47405 (Program No. QMAC006).

[a](R.E.) Restriction endonuclease.

Table 2 The Genetic Code

	Abbreviation letters		
Amino acid	1	3	Codon[a]
Alanine	A	Ala	(GCN)
Arginine	R	Arg	(CGN) or (AGR)
Asparagine	N	Asn	(AAY)
Aspartic acid	D	Asp	(GAY)
Cysteine	C	Cys	(TGY)
Glutamine	Q	Gln	(CAR)
Glutamic acid	E	Glu	(GAR)
Glycine	G	Gly	(GGN)
Histidine	H	His	(CAY)
Isoleucine	I	Ile	(ATH)
Leucine	L	Leu	(CTN) or (TTR)
Lysine	K	Lys	(AAR)
Methionine	M	Met	(ATG)
Phenylalanine	F	Phe	(TTY)
Proline	P	Pro	(CCN)
Serine	S	Ser	(TCN) or (AGY)
Threonine	T	Thr	(ACN)
Tryptophan	W	Trp	(TGG)
Tyrosine	Y	Tyr	(TAY)
Valine	V	Val	(GTN)

[a](N) Any base; (R) purine; (Y) pyrimidine; (H) A, C, or T.

0.5 pmole of template DNA (0.5 pmole template = 0.33 μg/kb X size of template [kb])
2.5 μl of 10X SDM buffer
1 μl of dNTP mix
15 pmoles of each primer (15 pmoles of primer = [5 ng/base] X size of primer [base])
ddH_2O to a final volume of 24 μl

Note: (1) It is critical that there be a 5′ phosphate in one or both primers. The primer(s) can either be phosphorylated with T4 polynucleotide kinase or synthesized with a 5′-terminal phosphate. (2) Template DNA must contain methylated $G^{m6}ATC$ sequences; if not, in vitro methylation with Dam methylase must be done prior to the initiation of PCR-SDM.

PCR

1. Add 2.5 units each of *Taq* DNA polymerase and 2.5 units of *Taq* Extender PCR additive.

   Note: These enzymes can be mixed together and stored as a 1:1 (v/v) mixture at –20°C for at least 3 months.

2. Overlay with 20 μl of mineral oil, and thermal cycle the DNA using at least 7, but no more than 12, cycles. (For an initial general reac-

Table 3 PCR-SDM Cycling Parameters

Segment	Cycles	Temperature	Time
1	1	94°C	4 minutes
		50°C	2 minutes
		72°C	2 minutes
2	8	94°C	1 minute
		56°C	2 minutes
		72°C	1 minute
3	1	72°C	5 minutes

tion, the parameters in Table 3 are suggested; see also PCR Considerations, above).

Digesting and Polishing the PCR-SDM Product

1. Following the PCR, place the reaction on ice for 2 minutes to cool the reaction to ≤37°C.

2. Add the following components directly to the 25-μl amplification reaction below the mineral oil overlay.

   1 μl of the *Dpn*I restriction enzyme (10 units)
   1 μl of cloned *Pfu* DNA polymerase (2.5 units)

   Note: It is important to insert the pipette tip below the mineral oil overlay when adding additional components to the reaction tube in the digestion, polishing, and ligation steps.

3. Gently mix and spin the reaction in a microcentrifuge for 1 minute. Immediately incubate the reaction for 30 minutes at 37°C.

4. Incubate the reaction for an additional 30 minutes at 72°C.

Ligating the PCR-SDM Product

1. Add the following components to the *Dpn*I, cloned *Pfu* DNA polymerase-treated product:

   100 μl of ddH_2O
   10 μl of 10X SDM buffer
   5 μl of 10 mM ATP

2. Gently mix and spin the reaction in a microcentrifuge for 1 minute. *Optional:* An 8-μl aliquot may be stored and analyzed by standard agarose gel electrophoresis. To verify the integrity of the PCR-SDM product, a single band should be apparent.

3. Remove 10 μl of the above reaction to a sterile microcentrifuge tube and add 1 μl of T4 DNA ligase (4 units).

   Note: There seem to be considerable differences in the efficiency of various lots of T4 DNA ligase to ligate blunt-end DNA molecules. Different lots from the same manufacturer may yield anywhere from 30% to 70% mutagenesis efficiency in this assay. Once a useful lot has been identified, it is recommended that it be held in reserve for use in PCR-SDM.

4. Incubate the reaction for 1 hour at 37°C.

Transforming into Competent Cells

RAPID TRANSFORMATION PROTOCOL (used for XL1-Blue-competent cells from Stratagene)

1. Gently thaw the competent cells on ice, and aliquot 80 μl of the cells to a prechilled FALCON 2059 polypropylene tube.

2. Add 2 μl of the ligase-treated DNA to the cells, swirl gently, and incubate for 30 minutes on ice.

3. Heat-pulse for 45 seconds at 42°C and place on ice for 2 minutes.

   Note: This heat pulse has been optimized for the FALCON 2059 tubes.

4. Immediately plate the entire volume of transformed cells.

5. Incubate the plates overnight at 37°C.

CONCLUSIONS

The advantages of the PCR-SDM method include increased template concentration to allow reduced cycling; the use of *Taq* Extender, which provides increased reliability in generating longer PCR products; the use of *Dpn*I restriction endonuclease to reduce the number of parental molecules; the use of *Pfu* DNA polymerase to remove undesired base extensions; and the efficient method for blunt-end ligation. The complete protocol is extremely fast and does not require cleanup or precipitation procedures between steps. PCR-SDM can be used for both large deletions and insertions.

ACKNOWLEDGMENTS

The authors thank Dr. Joseph Sorge, Jr. and John Bauer for their support with this research. We thank Daniel McMullan, Barbara McGowan, and Mila Angert for their substantial contributions in experimental design. The authors have also benefited from discussions with the Genetics Systems group at Stratagene.

REFERENCES

Costa, G.L. and M.P. Weiner. 1994a. Protocols for cloning and analysis of blunt-ended PCR generated DNA fragments. *PCR Methods Appl.* **3:** S95–S106.

——. 1994b. Polishing with T4 or Pfu polymerase increases the efficiency of cloning PCR fragments. *Nucleic Acids Res.* **22:** 2423.

Jones, D.H. and S.C. Winistorfer. 1992. Recombinant circle PCR and recombination PCR for site-directed mutagenesis without PCR product purification. *BioTechniques* **12:** 528–533.

Kunkel, T. 1985. Rapid and efficient site-specific mutagenesis without phenotypic selection. *Proc. Natl. Acad. Sci.* **82:** 488–492.

Nelson, M. and M. McClelland. 1992. The use of DNA methyltransferase/endonuclease enzyme combinations for megabase mapping of chromosomes. *Methods Enzymol.* **216:** 279–303.

Nielson, K.B., W. Schoettlin, J.C. Bauer, and E. Mathur. 1994. *Taq* Extender PCR additive for improved length, yield and reliability of PCR products. *Strategies Mol. Biol.* **7:** 27.

Picard, V., E. Ersdal-Badju, A. Lu, and S.C. Bock. 1994. A rapid and efficient one-tube PCR-based mutagenesis technique using Pfu DNA polymerase. *Nucleic Acids Res.* **22:** 2587–2591.

Sambrook, J., E.F. Fritsch, and T. Maniatis. 1989. *Molecular cloning: A laboratory manual*, 2nd edition. Cold Spring Harbor Laboratory Press, Cold Spring Harbor, New York.

Shankarappa, B., K. Vijayananda, and G. Ehrlich. 1992. Silmut: A computer program for the identification of regions suitable for silent mutagenesis to introduce restriction enzyme recognition sequences. *BioTechniques* **12:** 882–884.

Sugimoto, M., N. Esaki, H. Tanaka, and K. Soda. 1989. A simple and efficient method for oligonucleotide-directed mutagenesis using plasmid DNA template and phosphorothioate-modified nucleotide. *Anal. Biochem.* **179:** 309–311.

Taylor, J.W., J. Ott, and F. Eckstein. 1985. The rapid generation of oligonucleotide-directed mutations at high frequency using phosphorothioate-modified DNA. *Nucleic Acids Res.* **13:** 8765–8785.

Vandeyar, M., M.P. Weiner, C. Hutton, and C. Batt. 1988. A simple and rapid method for the selection of oligodeoxynucleotide-directed mutants. *Gene* **65:** 129–133.

Watkins, B., A.E. Davis, F. Cocchi, and M.S. Reitz, Jr. 1993. A rapid method for site-specific mutagenesis using larger plasmids as templates. *BioTechniques* **15:** 700–704.

Weiner, M.P. and H.A. Scheraga. 1989. A set of Macintosh programs for the design of synthetic genes. *Comput. Appl. Biosci.* **5:** 191–198.

Weiner, M.P., G.L. Costa, W. Schoettlin, J. Cline, E. Mathur, and J.C. Bauer. 1994. Site-directed mutagenesis of double-stranded DNA by the polymerase chain reaction. *Gene* **15:** 119–123.

Alternative Amplification Technology

Nucleic acid amplification methods are employed to detect a defined sequence, often present at very low levels, within a complex population of molecules. Ideally, this amplification should be specific, with minimal to nonexistent background from the nonhomologous molecules, and, when possible, quantitative. Although the majority of this manual is devoted to PCR, there are three other methods of DNA amplification available. Two of these—ligase chain reaction (LCR; for a complete review, see Barany 1991b) and nucleic acid sequence-based amplification (NASBA), also known as self-sustained sequence replication (3SR; for complete review, see Fahy et al. 1991)—are enzymatic cycles of nucleic acid amplification. The third method, branched DNA (bDNA; Horn and Urdea 1989; Urdea 1993), is a hybridization capture and detection-based amplification method. Each of these methods, including PCR, has distinct advantages and disadvantages, which are outlined in Table 1 and discussed briefly below.

Ligase Chain Reaction

The cornerstone of LCR is the ligation, using a thermostable DNA ligase, of two adjacent oligonucleotides that hybridize to a target in a sequence-dependent manner. LCR systems are designed to detect specific nucleic acid sequence variations and do not provide quantitative information about the sequence in question. The sequence variation detected, such as the single-base difference between β^A-globin and β^S-globin, is localized at the 3′ end of one of the oligonucleotides. Following ligation, this base is localized in the middle of the joined

Table 1 Amplification Methods

	Polymerase chain reaction, RNA-PCR	Ligase chain reaction	Nucleic acid sequence-based amplification	Branched DNA assay
Starting nucleic acid	DNA/RNA	DNA/RNA	RNA	DNA/RNA
Relative cost	medium	low	medium	high
Special equipment	PCR machine, oligonucleotide primers, detection system, thermostable DNA polymerase, RT also required for detection of RNA	oligonucleotides, detection system, 2 enzymes–Stoffel fragment and thermostable DNA ligase	3 enzymes–RT, RNase H, and T7 RNA polymerase	large number of oligonucleotides, DNA-coated plates, luminometer
Quantitation	yes	no	yes; detection system	yes
Detection of sequence variation	yes	yes	yes; detection system specific	no
Commercialization	yes	yes	yes	yes
Special problems	contamination control	high background limit of detection	contamination control	cost; this can easily drop tenfold. Limitations on detection sensitivity have been resolved.

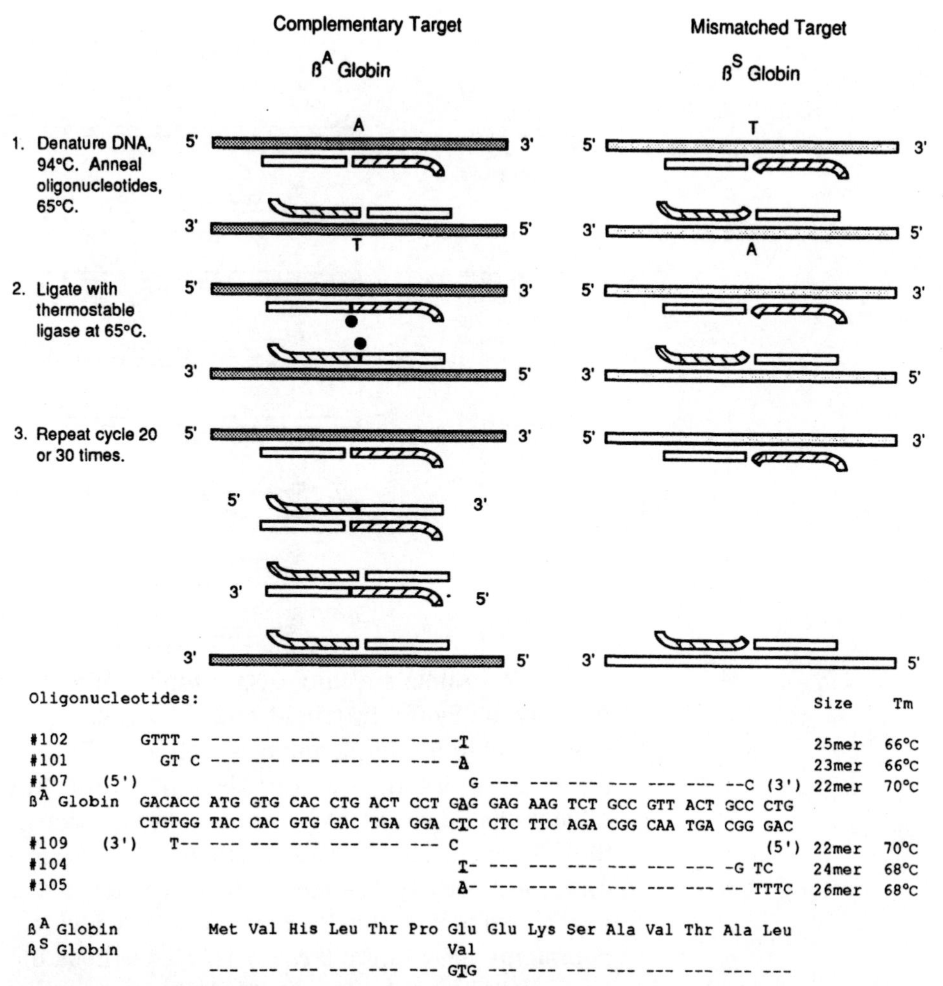

Figure 1 (*Top*) Allele-specific DNA amplification and detection using LCR. DNA is heat-denatured (94°C), and four complementary oligonucleotides anneal to the target at a temperature (65°C) near their melting temperature (T_m). Thermostable ligase will covalently attach only adjacent oligonucleotides that are perfectly complementary to the target (*left*). Products from one round of ligation become targets for the next round, and thus the products increase exponentially. Oligonucleotides containing a single-base mismatch at the junction do not ligate efficiently, and therefore do not amplify the product (*right*). The diagnostic oligonucleotides (*striped*) have the discriminating nucleotide on the 3′ end for both the top and bottom strands. Thus, target-independent, four-way ligation would require sealing an (unfavorable) single-base 3′ overhang. (*Bottom*) Nucleotide sequence and corresponding translated sequence of the oligonucleotides used in detecting β^A and β^S globin genes. Oligonucleotides 101 and 104 detect the β^A target, and oligonucleotides 102 and 105 detect the β^S target when ligated to labeled oligonucleotides 107 and 109, respectively. The diagnostic oligonucleotides (101, 102, 104, and 105) contain slightly different-length tails to minimize aberrant ligation products and to facilitate the discrimination of various products when separated on a polyacrylamide denaturing gel. Oligonucleotides have calculated T_m values of 66–70°C (calculated as described in Miyada and Wallace 1987, or for a more precise determination, see Wetmur 1991), at or slightly above the ligation temperature. (Adapted from Barany 1991a.)

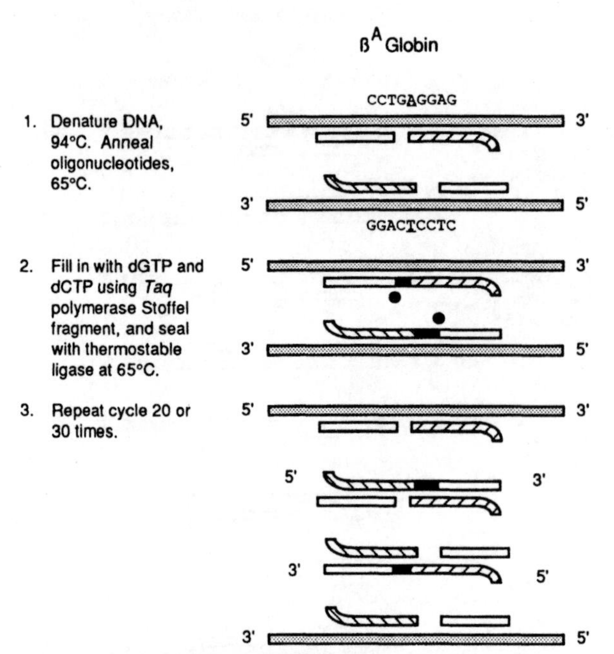

Figure 2 Allele-specific DNA amplification and detection using the *Taq* polymerase Stoffel fragment and *Taq* ligase (pLCR). DNA is heat-denatured (94°C) and four complementary (but not adjacent) oligonucleotides anneal to the target at a temperature (65°C) near their melting point. The diagnostic oligonucleotides (*striped strands*) contain the discriminating nucleotide on their 3′ ends. The *Taq* polymerase Stoffel fragment, which lacks 5′→3′ exonuclease activity, is added in the presence of 1 μM dGTP and dCTP to fill the gap between oligonucleotides (*black*), provided that the last nucleotide is complementary to the target DNA. Thermostable ligase would subsequently covalently link only the extended diagnostic oligonucleotide to the newly adjacent (*white strand*) oligonucleotide. The specificity of this type of amplification depends on the fidelity of polymerase extension at a mismatch. (Reprinted from Barany 1991b.)

Figure 3 Strategy of the 3SR amplification scheme. The 3SR reaction consists of continuous cycles of reverse transcription and RNA transcription designed to replicate a nucleic acid (RNA target in the figure) using a double-stranded cDNA intermediate. Oligonucleotides A and B prime DNA synthesis, producing a double-stranded cDNA containing a functional T7 promoter (steps 1–6). Complete cDNA synthesis is dependent on the digestion of RNA in the intermediate RNA-DNA hybrid by RNase H (step 3). Transcription-competent cDNAs are used to produce multiple (50–1000) copies of the antisense RNA transcript of the original target (steps 7–8). These copes of the antisense transcript are immediately converted to T7 promoter-containing, double-stranded cDNA copies (steps 9–12) and used again as transcription templates. This process continues in a self-sustained, cyclic fashion under isothermal conditions (42°C) until enzymes or other components in the reaction become limiting or inactivated. (*Dotted lines*) RNA; (*thin lines*) DNA; (*thick lines*) T7 promoter sequence (see Fig. 2); (*circles*) reverse transcriptase; (*diamonds*) T7 RNA polymerase; (*TCS*) target complementary sequence. (Reprinted from Fahy et al. 1991.)

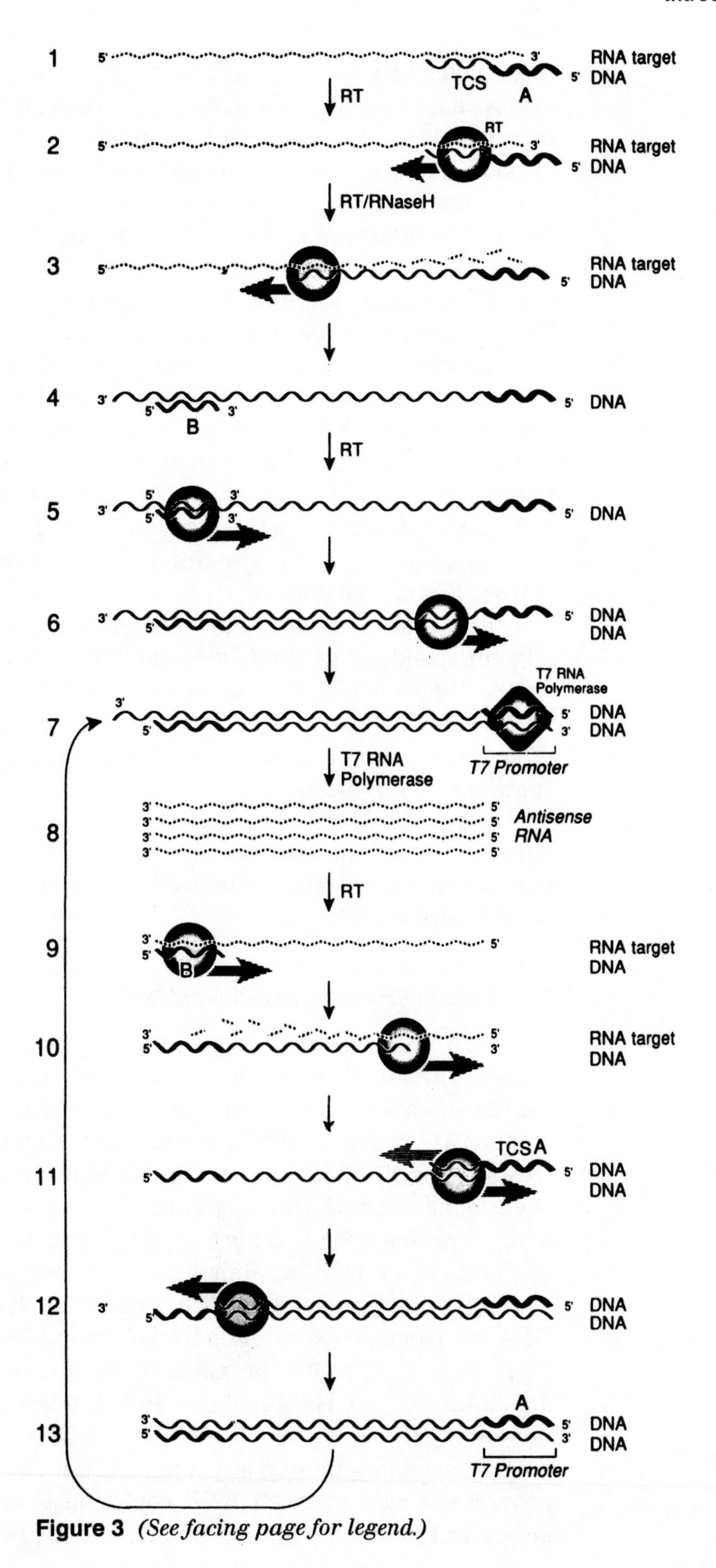

Figure 3 *(See facing page for legend.)*

oligonucleotides (Fig. 1). LCR techniques using either one or two enzymes have been described. Single-enzyme systems utilize only thermostable DNA ligase, which uses NAD^+ as the energy source (Takahashi et al. 1984). Many of the systems based solely on ligation are prone to background arising from ligation events that occur despite the mismatched bases. This background ligation process generates misprimed ligation fragments that are identical in size to the correctly primed product. To eliminate this problem, two-enzyme LCR systems have been developed. These couple the primer extension reaction of a 5′→ 3′ exonuclease-deficient thermostable DNA polymerase such as the Stoffel fragment of *Taq* DNA polymerase with the activity of a thermostable DNA ligase (for review, see Barany 1991b). The two-enzyme system requires deoxynucleotide triphosphates and NAD^+ (Fig. 2). This two-enzyme approach has been developed into pathogen-specific detection kits. Protocols for both single- and dual-enzyme LCR are described in detail in this section in "Ligase Chain Reaction."

Because the product of LCR is essentially twice the size of the input oligonucleotides, product detection has been a challenge. The simplest detection system depends on a ^{32}P signal incorporated into the product followed by acrylamide gel electrophoresis and autoradiography. Other approaches have been implemented for each specific kit using LCR.

Finally, one of the main advantages of LCR when compared with other amplification methods is its general resistance to product contamination. This is partly due to the lesser extent of amplification seen in LCR versus PCR.

Nucleic Acid Sequence-based Amplification

NASBA is an isothermal amplification system that requires three enzymes: reverse transcriptase, RNase H, and T7 RNA polymerase. NASBA takes advantage of the extraordinary activity of T7 RNA polymerase. Since T7 RNA polymerase functions only in conjunction with the specific T7 RNA polymerase-binding site, the NASBA system is designed to attach this sequence to the target nucleic acid first, and then to produce RNA copies of the target. Both DNA and RNA are synthesized in NASBA, thus requiring deoxy- and ribonucleoside triphosphates. Unlike RNA-PCR, where an RT and a thermostable DNA polymerase work in parallel, in NASBA the three enzymes employed must work in concert. In the absence of a physical denaturation step, NASBA relies on the selective enzymatic degradation of one of the intermediates. Once dsDNA containing the T7 RNA polymerase-binding site is produced, the RNA polymerase activity generates single-stranded RNA copies in a template concentration-dependent manner (Fig. 3). Because of this, NASBA is a quantitative amplification technology, providing information about the concentration of the target in the starting material. "Optimization and Characterization of 3SR-based Assays" and "One-tube Quantitative HIV-1 RNA NASBA" describe the conditions for reproducible NASBA and a state-of-the-art quantitation system.

Branched DNA Assay

This amplification system is based solely on multiple nucleic acid hybridization steps followed by only one enzymatic step during the detection phase (Fig. 4). First, capture oligonucleotides attached to the plate hybridize to the target and immobilize it on the plate. The

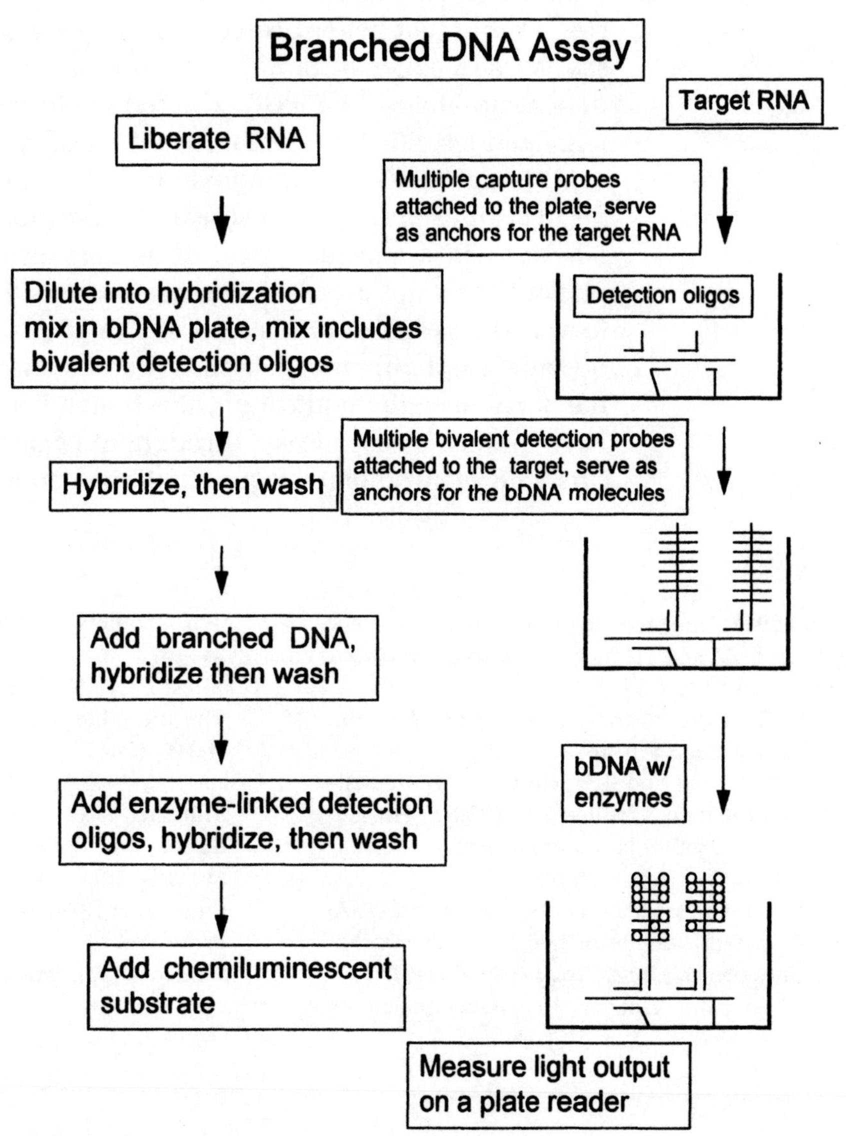

Figure 4 Scheme for branched DNA assay.

bivalent detection oligonucleotides then hybridize with regions of the target RNA and also serve as hybridization substrates for the novel branched DNA molecule. This bDNA then serves to capture oligonucleotides labeled with an enzyme. The chemiluminescent signal of the bDNA assay can achieve detection of 5,000–10,000 specific target molecules per sample (Urdea 1993). Recent advances have improved the lower limit of detection of second-generation bDNA kits to 500 copies per milliliter. Because the detection system is based on the production of light, this assay has a large dynamic range and is quantitative up to 10^8 copies of target per sample. However, its design limits the bDNA assay; it is not useful when exploring possible nucleic acid sequence variations. Since there is no enzymatic amplification of the input target, the bDNA assay has distinct advantages in terms of template preparation. For example, the nucleic acid target can be maintained in buffers that would inhibit enzymes for RNA-PCR or NASBA, but that better protect it from degradation. Also, the bDNA assay is not plagued with either sample-to-sample or product-contamination problems like PCR. Currently there are a number of viral pathogen-specific kits available in the bDNA format.

Although none of the above-described methods will replace PCR for overall usefulness in a research environment, each of these methods may fill a specific research or diagnostic niche. As medical applications in the areas of diagnostics and disease monitoring move forward, it is not at all clear if one or if any of these techniques will become the dominant method. When considering one of these methods for a specific application, the issues listed in Table 1, as well as each method's ease of use, throughput capacity, potential for automation, and requirement for quantitation should be taken into account.

REFERENCES

Barany, F. 1991a. Genetic disease detection and DNA amplification using cloned thermostable ligase. *Proc. Natl. Acad. Sci.* **88:** 189–193.

——. 1991b. The ligase chain reaction in a PCR world. *PCR Methods Appl.* **1:** 5–16.

Fahy, E., D.Y. Kwoh, and T.R. Gingeras. 1991. Self-sustained sequence replication (3SR): An isothermal transcription-based amplification alternative to PCR. *PCR Methods Appl.* **1:** 25–33.

Horn, T. and M. Urdea. 1989. Forks, combs and DNA: The synthesis of branched oligodeoxyribonucleotides. *Nucleic Acids Res.* **17:** 6959–6967.

Miyada, C.G. and R.B. Wallace. 1987. Oligonucleotide hybridization techniques. *Methods Enzymol.* **154:** 94–107.

Takahashi, M., E. Yamaguchi, and T. Uchida. 1984. Thermophilic DNA ligase. *J. Biol. Chem.* **259:** 10041–10047.

Urdea, M. 1993. Synthesis and characterization of branched DNA (bDNA) for direct and quantitative detection of CMV, HBV, HCV and HIV. *Clin. Chem.* **39:** 725–726.

Wetmur, J.G. 1991. DNA probes: Applications of the principles of nucleic acid hybridization. *Crit. Rev. Biochem. Mol. Biol.* **26:** 227–259.

Ligase Chain Reaction

Martin Wiedmann,[1] Francis Barany,[2] and Carl A. Batt[1]

[1]Department of Food Science, Cornell University, Ithaca, New York 14853
[2]Department of Microbiology, Hearst Microbiology Research Center, Cornell University Medical College, New York, New York 10021

INTRODUCTION

From its initial detailed reports, the ligase chain reaction (LCR) has evolved as a very promising diagnostic technique that can be utilized in conjunction with a primary PCR amplification. LCR employs a thermostable ligase and allows the discrimination of DNA sequences differing in only a single base pair (see Fig. 1) (Barany 1991a,b). Although PCR has a greater level of sensitivity, the robust discriminatory power of LCR makes it the method of choice for certain applications. LCR is compatible with other replication-based amplification methods (PCR, 3SR, Qβ-replicase, RT-PCR). Thus, by combining LCR with a primary amplification, one effectively lines up the cross-hairs to distinguish single base-pair differences with pinpoint accuracy.

The intellectual genesis of LCR can be traced back to pioneering work by Whiteley et al. (1989), who described an oligonucleotide probe-based assay using two probes that are ligated together only when they are immediately adjacent to each other. The same concept is the basis for the oligonucleotide ligation assay (OLA) (Landegren et al. 1988; Nickerson et al. 1990). This method was used in conjunction with a primary PCR step to screen for sickle cell anemia, the ΔF508 mutation in cystic fibrosis, and T-cell-receptor polymorphisms. Wu and Wallace (1989) described a similar technique called ligase amplification reaction (LAR), which employs two sets of complementary primers and repeated cycles of denaturation (at 100°C) and ligation (at 30°C) using the mesophilic T4 DNA ligase. Use of mesophilic, i.e., T4 or *Escherichia coli*, ligase has the drawback of requiring the addition of ligase after each denaturation step, as well as the appearance of target-independent ligation products (Wu and Wallace 1989; Bar-

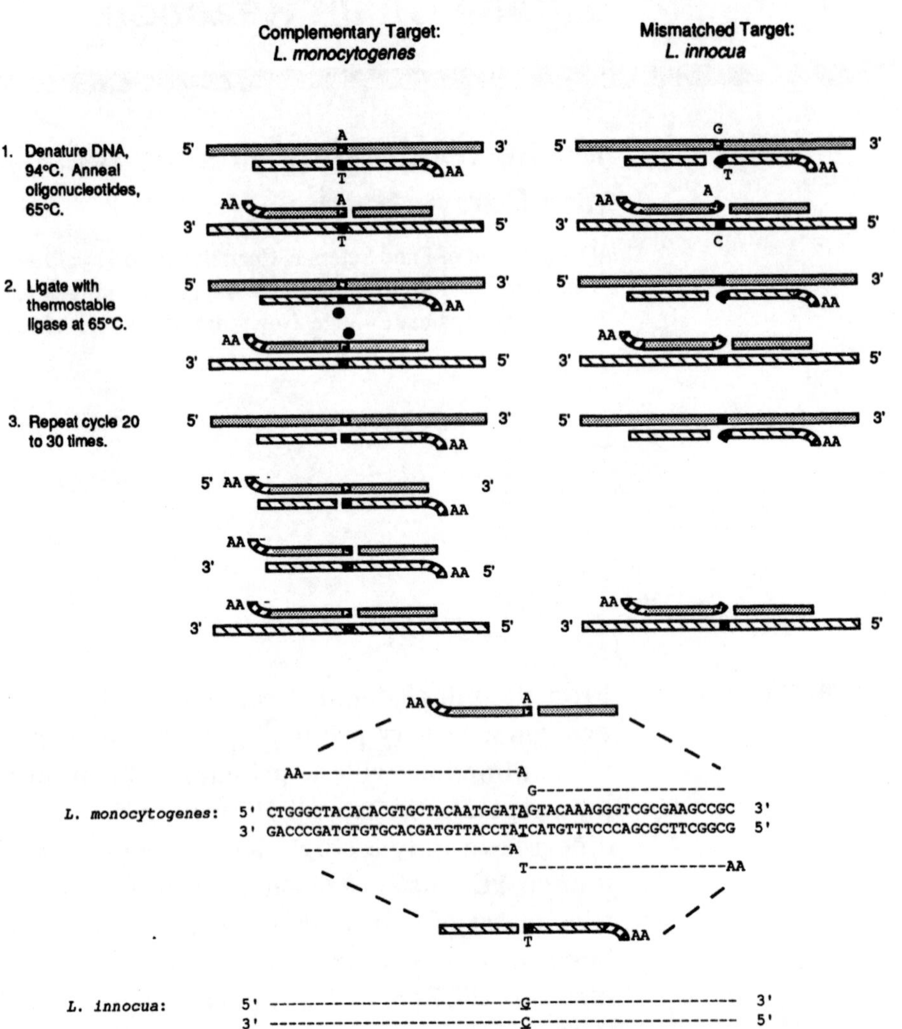

Figure 1 Principle of LCR. (*Top panel*) DNA is denatured at 94°C and the four LCR primers anneal to their complementary strands at 65°C, which is approximately 5°C below their T_m. Thermostable ligase (depicted as a black circle) only ligates primers that are perfectly complementary to their target sequence and hybridize directly adjacent to each other (as shown with *L. monocytogenes*, left). The discriminating bases at the 3′ ends of the upstream primers are depicted as boxes on the target as well as on the primers for clarity. Primers that have at least a single base-pair mismatch at the 3′ end contributing to the junction of the two primers will not ligate (as shown with *L. innocua*, right). The discriminating primers have a 2-base-pair non-complementary AA tail at their 5′ end to avoid ligation of the 3′ ends. (*Bottom panel*) The example shown is an LCR with matched target (*L. monocytogenes*) and mismatched target (*L. innocua*). The pathogenic bacteria *L. monocytogenes* can be distinguished from other closely related *Listeria* species (e.g., *L. innocua*) by a single base-pair difference in the 16S rDNA (Wiedmann et al. 1992). *L. monocytogenes* has an A-T base pair at nucleotide 1258, whereas *L. innocua* has a G-C base pair at this position.

ringer et al. 1990). In contrast, LCR provides a much higher sensitivity and is less susceptible to the formation of false-positive ligation products. Thermostable ligase minimizes target-independent ligation because the reaction is performed at or near the melting temperature (T_m) of the oligonucleotides (Barany 1991b). Furthermore, the use of thermostable ligase avoids the need to add fresh ligase after each denaturation step as required in LAR. Recently, thermostable ligase has become available from a variety of commercial suppliers, which will expand applications of these new ligation-mediated amplification techniques.

Theory of LCR and Similar Amplification Methods

The principle of LCR is based in part on the ligation of two adjacent synthetic oligonucleotide primers that uniquely hybridize to one strand of the target DNA (see Fig. 1). The junction of the two primers is usually positioned such that the nucleotide at the 3′ end of the upstream primer coincides with the targeted single base-pair difference. This single base-pair difference may define two different alleles, species, or other polymorphisms correlated to a given phenotype. If the target nucleotide at that site complements the nucleotide at the 3′ end of the upstream primer, the two adjoining primers can be covalently joined by the ligase. The unique feature of LCR is a second pair of primers, almost entirely complementary to the first pair, which are designed with the nucleotide at the 3′ end of the upstream primer denoting the sequence difference. In a cycling reaction, using a thermostable DNA ligase, both ligated products can then serve as templates for the next reaction cycle leading to an exponential amplification process, analogous to PCR. If there is a mismatch at the primer junction, it will be discriminated against by thermostable ligase, and the primers will not be ligated. The absence of the ligated product therefore indicates at least a single base-pair change in the target sequence (Barany 1991a).

The ligase detection reaction (LDR) is similar to LCR (Barany 1991b). In LDR, one pair of adjacent primers, which hybridize to only one of the target strands, is used to achieve a linear amplification (see Fig. 2). LDR may be used following a primary amplification and has the advantage of accurately quantitating the ratio of two alleles in a target sample (Prchal et al. 1993). LDR coupled to PCR has promise in a multiplex format where several mutations are analyzed in a single amplification (Barany 1991b). This method is currently being applied to the simultaneous detection of multiple mutations in cystic fibrosis (Eggerding et al. 1993; Winn-Deen et al. 1993b) as well as in 21 hydroxylase deficiency (D. Day et al., unpubl.).

pLCR is another ligase-mediated detection method where the 3′

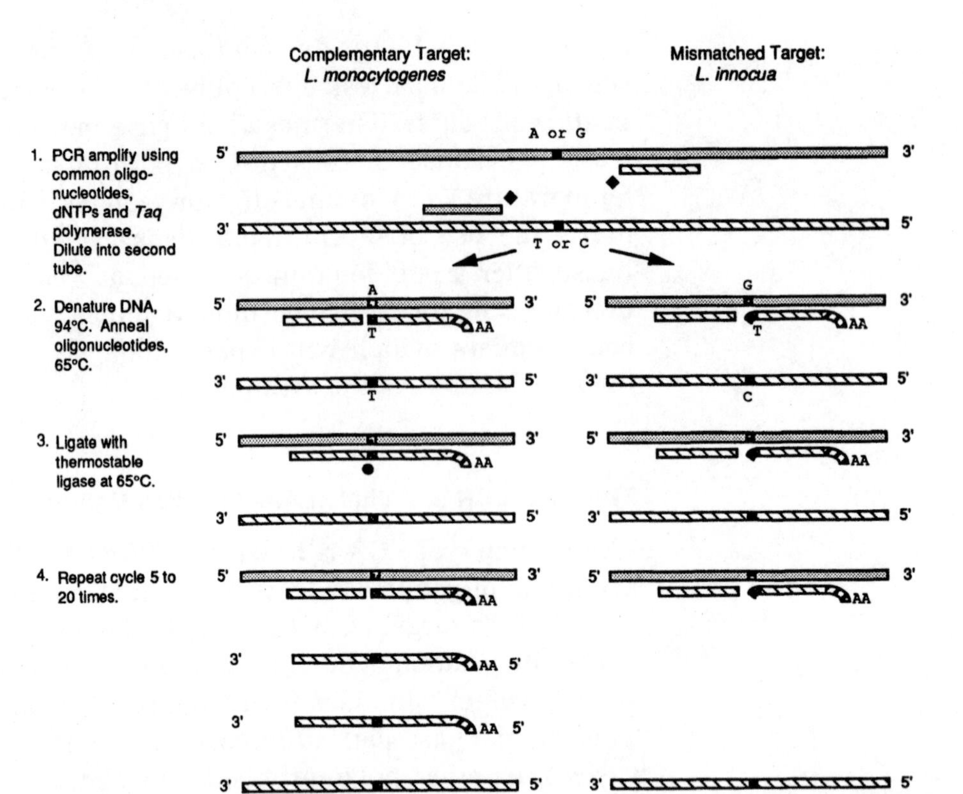

Figure 2 Principle of PCR-coupled LDR. The same two target sequences as in Fig. 1 are used to illustrate the PCR-coupled LDR. The DNA stretch containing the single base-pair difference that distinguishes *L. monocytogenes* from *L. innocua* (see Fig. 1, *bottom*) is PCR-amplified using PCR primers outside the region of the LCR primers. The PCR amplifies both target sequences under standard conditions (for details, see Wiedmann et al. 1992, 1993). *Taq* polymerase is depicted as a black diamond 3′ to each of the two PCR primers. After PCR amplification, *Taq* polymerase is inactivated by incubation at 97°C for 25 minutes (Zebala and Barany 1995). An aliquot of the PCR-amplified DNA (between 1% and 4% of the PCR reaction) is then used in the LDR. The DNA is denatured at 94°C and the two LDR primers anneal to their complementary strands at 65°C, which is approximately 5°C below their T_m. As in LCR (Fig. 1), the thermostable ligase (black circle) only ligates primers that are perfectly complementary to their target sequence and hybridize directly adjacent to each other (as shown with *L. monocytogenes,* left). Primers that have at least a single base-pair mismatch at the 3′ end contributing to the junction of the two primers do not ligate (as shown with *L. innocua,* right). An LDR cycle that consists of a denaturing step for 1 minute at 94°C and an annealing step at 65°C is repeated 5–20 times, so that a linear amplification of ligated LDR primers is achieved with the complementary target (i.e., *L. monocytogenes*).

ends of the discriminating (or "allele-specific") primers coincide with a potential base-pair change. The pLCR primers are designed with a gap between the discriminating and the nondiscriminating primer (see Fig. 3). In this reaction, the gap is filled using *Taq* DNA polymer-

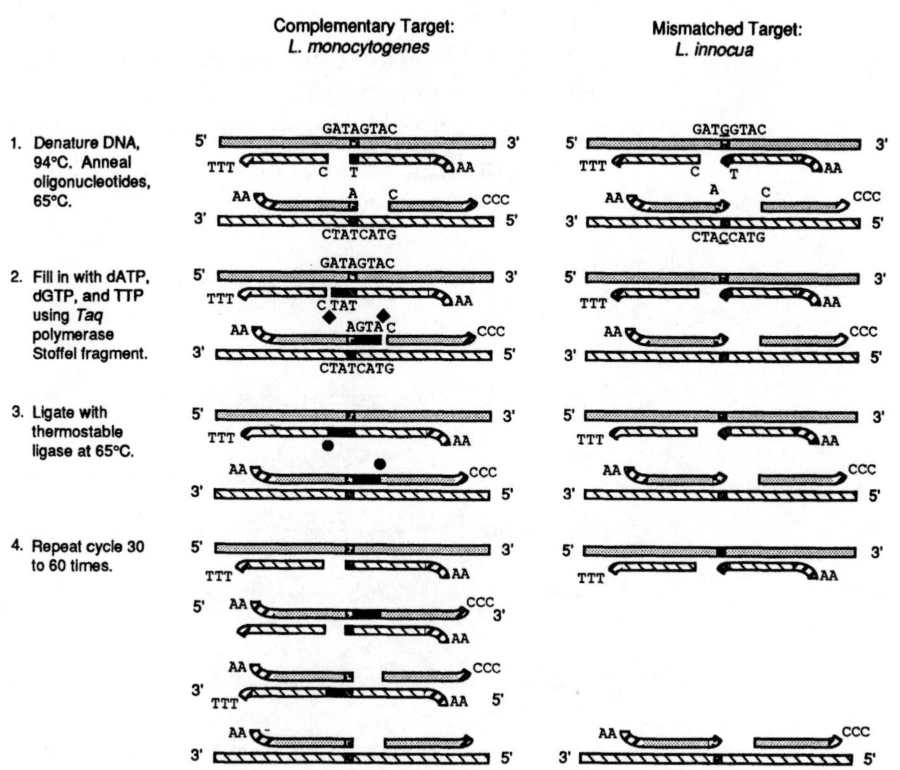

Figure 3 Principle of pLCR. The same target sequences as in Fig. 1 are used to illustrate pLCR. After denaturing of the DNA at 94°C, the four pLCR primers are allowed to anneal at 65°C. These primers anneal so that a 2- or 3-nucleotide gap between the primers of one pair (which anneals to the same strand) is formed. The 3′ end of the discriminating primers (shown with a shaded box at the 3′ end as with the discriminating nucleotides indicated by T and A) can be elongated by *Taq* polymerase Stoffel fragment and the appropriate nucleotides, analogous to the process in PCR. Only the nucleotides needed to fill the 2- or 3-nucleotide gap between the discriminating and the nondiscriminating primers are included in the reaction mix; for the example shown in this figure, dATP, dGTP, and dTTP are needed. After elongation of the discriminating primers by two or three nucleotides, the junction between the elongated discriminating primer and the nondiscriminating primer can be sealed by the thermostable ligase. This cycle is repeated between 30 and 60 times. With a noncomplementary target (right side), no elongation of the discriminating primer is possible, therefore no ligation of the two primers occurs and no pLCR product forms.

ase Stoffel fragment followed by the ligation of the elongated discriminating primer with the nondiscriminating primer. The specificity of this method relies on allele-specific elongation of the discriminating primer by the polymerase. Birkenmeyer and Mushahwar (1991) described another ligase-mediated technique, which was termed gapped LCR (G-LCR). This technique uses four oligonucleotide primers with the two primers of each pair being separated by a gap of one or more consecutive bases that are specific for the target DNA (see Fig. 4). By adding only the missing deoxynucleotides to the reaction, together with a thermostable DNA polymerase and a thermostable DNA ligase, the gap must first be filled in the presence of the matching target before the resulting nick can be sealed by the

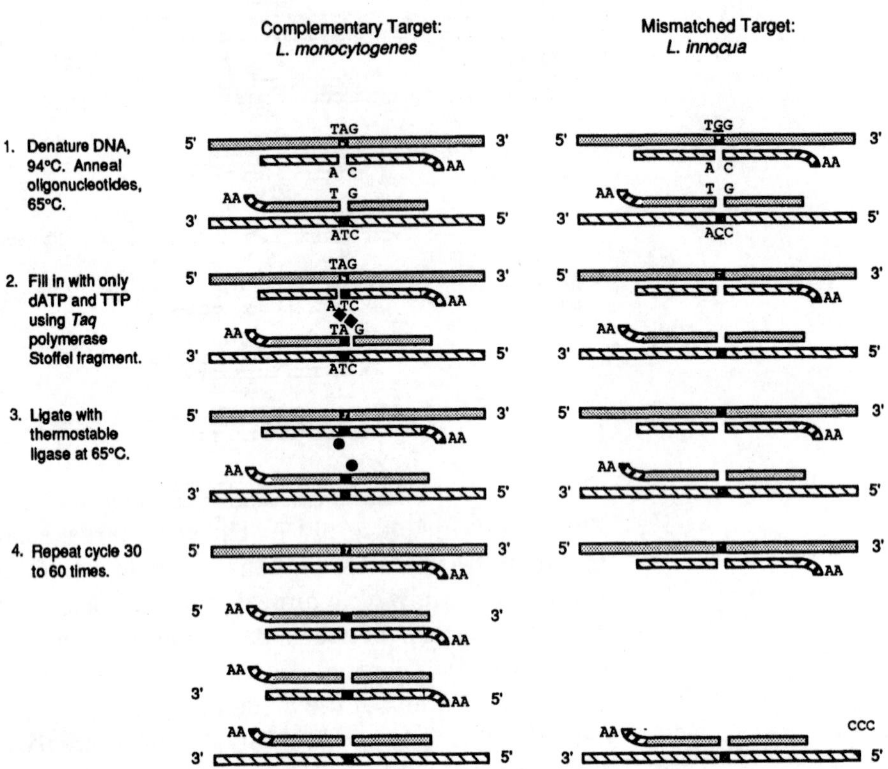

Figure 4 Principle of G-LCR. The same target sequences as in Fig. 1 are used to illustrate G-LCR. After denaturing of the DNA at 94°C, the four G-LCR primers are allowed to anneal at 65°C. These primers anneal so that a 1-nucleotide gap between the primers of one pair (which anneals to the same strand) is formed. This gap is located so that it coincides with the base pair discriminating the two targets (*L. monocytogenes* and *L. innocua* in the example shown) from each other. The 3′ end of the downstream primer can be elongated by *Taq* polymerase Stoffel fragment and the appropriate nucleotides, analogous to the process in PCR. Only the nucleotides needed to fill the 1-nucleotide gap (shown as a shaded box in the target sequence with the discriminating nucleotides indicated by T and A) between the two primers are included in the reaction mix; for the example shown in this figure, only dATP and dTTP are needed. After elongation of the discriminating primers with the appropriate nucleotide, the junction between the elongated downstream primer and the upstream primer can be sealed by the thermostable ligase. This cycle is repeated between 30 and 60 times. With a noncomplementary target (right side), no elongation of the discriminating primer is possible, therefore no ligation of the two primers occurs and no G-LCR product forms.

ligase. This technique limits itself to the detection of base-pair changes from AT/TA to GC/CG or vice versa. For example, G-LCR could not distinguish β^A globin from β^B globin (A→T transversion) because there is no difference in the bases required for filling the gap. A similar principle is also applied in the repair chain reaction (RCR), which has been used for the detection of human papillomavirus (HPV) 16 (Segev 1992).

Comparison of LCR, pLCR, and G-LCR

One performance-linked difference cited among LCR, pLCR, and G-LCR is the relative amount of ligated product formed in the absence of template. Because both pLCR and G-LCR require an initial template-dependent extension, they may be less prone to false positives in the absence of template. To compare LCR, pLCR, and G-LCR, the appropriate primers for the detection of *Listeria monocytogenes* by these three techniques were designed and synthesized (Wiedmann et al. 1992 and unpubl.). Locations of the primers are shown in Figures 1, 3, and 4. To compare the ability of these three methods to differentiate single base-pair differences, the amount of ligation products obtained with a constant amount of PCR-amplified DNA that either perfectly matched (*L. monocytogenes* 16S rDNA) or mismatched in a single base pair (*Listeria innocua* 16S rDNA) with the respective primer sets was determined. The reaction conditions for G-LCR as well as for pLCR were as described in Table 1 for pLCR. In contrast to previous reports for G-LCR, *Taq* DNA polymerase Stoffel fragment was used instead of *Taq* DNA polymerase. The Stoffel fragment lacks 5′→ 3′ exonuclease activity and does not excise bases from the 5′ end of the primer adjacent to the gap. No target-independent ligation products were observed for either LCR, pLCR, or G-LCR. Furthermore, a clear differentiation of *L. monocytogenes* from *L. innocua* based on a single base-pair difference in the 16S rDNA was possible for all three formats. This is the first time that a single base-pair difference was detected using G-LCR. Previous reports described the discrimination of targets with at least two base-pair differences (Birkenmeyer and Armstrong 1992; Dille et al. 1993). LCR, pLCR, and G-LCR can detect single base-pair differences; however, their relative discriminatory ability might depend on the nature and composition of the targets.

REAGENTS

LCR

10× LCR buffer (500 mM Tris-HCl, pH 8.2 at 25°C, 1 M KCl, 100 mM $MgCl_2$, 10 mM EDTA, 100 mM dithiothreitol, 0.1% Triton X-100). This buffer has to be stored in small aliquots at –20°C to avoid multiple freezing and thawing cycles. The inclusion of 0.1% Triton X-100 gives higher yields of ligation product in the LCR compared to a 10× buffer without Triton.

Table 1 Protocols for LCR, pLCR, and G-LCR

	LCR	pLCR	G-LCR
Detailed references	Barany (1991a,b); Feero et al. (1993); Wiedmann et al. (1992, 1993); Winn-Deen and Iovannisci (1991); Wilson et al. (1994)	–	Birkenmeyer and Armstrong (1992); Dille et al. (1993)
Position of discriminating nucleotide	3′ base of both strands (single base 3′ overhang)	3′ base of both strands (one- or two-base overhang)	nucleotides to be filled in
T_m of primers	66–70°C: Barany (1991a); Weidmann et al. (1992) or 60–66°C: Iovannisci and Winn-Deen (1993)	68–70°C	62–76°C
Amount of each primer	1–10 fmole/μl	2 fmole/μl	16.6–20 fmole/μl
Labeling of primers	biotin/digoxigenin; fluorescein; ^{32}P	^{32}P	biotin/fluorescein; unlabeled, used with ^{32}P-labeled nucleotides for fill-in reaction
Reaction volume	10–50 μl	25 μl	25–50 μl
Buffer conditions	20–50 mM Tris-HCl, pH 7.6 100 mM KCl 10 mM $MgCl_2$ 1 mM EDTA 10 mM DTT 1 mM NAD^+ 0.1–0.01% Triton X-100[a]	80 mM KOH/KCl 50 mM EPPS 10 mM $MgCl_2$ 10 mM NH_4Cl 1 mM DTT 10 μg/ml BSA 1 mM NAD^+	80 mM KOH/KCl 50 mM EPPS 10 mM $MgCl_2$ 10 mM NH_4Cl 1 mM DTT 10 μg/ml BSA 0.1 mM NAD^+
Amount of nucleotides for fill-in reaction	–	1 μM	1 μM
Carrier DNA to suppress background	0.4 μg salmon sperm DNA/μl	–	–
Thermostable enzymes/reaction volume	1.5 nick closing units *Taq* ligase/μl; Barany (1991a); Weidmann et al. (1992) or 0.15 units/μl; Kälin et al. (1992)	1.5 nick closing units *Taq* ligase/μl and 0.08 units *Taq* polymerase Stoffel fragment/μl	68 units *Taq* ligase/μl and 0.02 units *Taq* polymerase/μl; Birkenmeyer and Armstrong (1992); Dille et al. (1993)
Cycle conditions	94°C 1 min 65°C 4 min 10–30 cycles[b] or 94°C 1 min 60°C 8 min 30 cycles[c]	97°C 3 min, 1 cycle; 94°C 1 min 65°C 4 min; 50 cycles	100° 3 min 85°C 30 sec 50–60°C 20 sec–1 min 27–60 cycles

[a]Barany (1991a) did not include Triton X-100.
[b]For primers with T_m of 66–72°C (Barany 1991a; Wiedmann et al. 1992)
[c]For primers with T_m of 60–66°C (Iovannisci and Winn-Deen 1993).

NAD^+ (12.5 mM) (β-nicotinamide adenine dinucleotide, grade III; Sigma). Prepare fresh every time used or store frozen in small aliquots.

Salmon sperm DNA (10 mg/ml). Prepare by standard procedures (Sambrook et al. 1989).

Taq DNA ligase (37.5 units/μl) (Barany and Gelfand 1991); currently *Taq* DNA ligase from the same clone is also available from New England Biolabs.

Termination mixture (10 mM EDTA, 0.2% bromophenol blue, 0.2% xylene cyanol in formamide)

Primer Labeling

T4 polynucleotide kinase (10 units/μl)

10x T4 polynucleotide kinase buffer (0.5 M Tris-HCl, pH 7.6, 100 mM $MgCl_2$, 100 mM 2-mercaptoethanol)

[γ-^{32}P]ATP (6000 Ci/mM = 60 pmole ATP)

ATP (20 mM); must be stored at −70°C

TE buffer (10 mM Tris-HCl, pH 8.0 at 25°C, 1 mM EDTA)

Sephadex NAP 5 columns (Pharmacia)

Terminal deoxynucleotidyl transferase (17 units/μl)

5x Terminal deoxynucleotidyl transferase buffer (500 mM sodium cacodylate, pH 7.2, 1 mM 2-mercaptoethanol, 10 mM $CoCl_2$)

Digoxigenin-11-ddUTP solution (1 mM) (Boehringer Mannheim)

Detection of Isotopic LCR Products

TBE buffer

16% Polyacrylamide gel (SequaGEL sequencing system, National Diagnostics)

Kodak X-OMAT AR film

Developer and fixer

Detection of Nonisotopic LCR Products

High-binding EIA 8-well strips, Type I material (Corning Costar)

Streptavidin (Promega)

Anti-digoxigenin Fab fragment, conjugated with alkaline phosphatase (Boehringer Mannheim)

Lumi-Phos 530 (Boehringer Mannheim)

Buffer T (100 mM Tris-HCl, pH 7.5, 150 mM NaCl, 0.05% Tween 20)

Buffer T without Tween (100 mM Tris-HCl, pH 7.5, 150 mM NaCl)

Carbonate buffer (20 mM Na_2CO_3, 30 mM $NaHCO_3$, 1 mM $MgCl_2$); pH should be approximately 9.6; store at 4°C

Plate binding buffer (1 M NaCl, 0.75 M NaOH)

Dry milk (Carnation natural nonfat dry milk, Nestle Food Company)

Salmon sperm DNA (10 mg/ml); prepare by standard procedures (Sambrook et al. 1989)
CAMLIGHT (Camera Luminometer System, Analytical Luminescence Laboratory)
Polaroid Instant Image film type 612 (20,000 ISO)

PROTOCOLS

Accurate results from LCR assays depend on a variety of factors, including primer design and reaction conditions. Based on our experience and that of others over the past years, a discussion of the most important factors to be considered in the development and performance of LCR assays is presented together with detailed experimental protocols. These focus mainly on LCR, because this technique is generally applicable to the differentiation of all potential single base-pair differences. Only occasionally are specifics for pLCR and G-LCR included, with sufficient information for the reader to use these techniques as well.

Design of LCR Primers

To minimize target-independent ligation, LCR primers with a single base-pair overhang, rather than blunt ends, should be used. The importance of single base-pair overhangs is shown by Kälin et al. (1992), who reported a relatively high amount of target-independent ligation using primers with blunt ends. The T_m values of all four LCR primers should be within a narrow temperature range, ideally with an absolute T_m of 70°C ± 2°C. The primer T_m values were estimated by the addition of 2°C for each A or T and 4°C for each G or C base contributing to hybridization to the target DNA. Furthermore, the primers should be designed so that one primer cannot serve as a bridging template for other primers and therefore lead to target-independent ligation. Adding noncomplementary tails of two nucleotides or longer to the nonadjacent 5′ ends of the primers should prevent ligation of these to 3′ ends of other primers. Depending on the discriminated nucleotides, different amounts of ligation product are observed with a mismatched target (Barany 1991a). The expected amounts of false ligation for specific mismatches are shown in Table 2. These data can be used for designing primers with the lowest possible rate of false ligation when a choice between different target sequences exists.

The nature of the base pair at the 3′ end of the primer with the matched target seems to influence the ligation efficiency. Two different sets of LCR primers were used for the detection of their corresponding alleles (D128G) in the bovine CD18 gene (Batt et al. 1994). The discriminating primer set with a G-C primer-template hybridization on the 3′ end gave a more efficient ligation as compared to the second set of primers with an A-T primer-template hybridization. The

Table 2 Noise-to-signal Ratio for Certain Mismatches in the LCR

Oligonucleotide base-target base	Noise-to-signal ratio[a]
A-A, T-T	1.1%
T-T, A-A	<0.2%
G-T, C-A	1.3%
G-A, C-T	<0.2%

[a]Calculated as amount of product with mismatched primers divided by the amount of product with complementary primers. (Adapted from Barany 1991a.)

greater hydrogen bonding of the G-C base-pairing facilitates a more stable hybrid as compared to an A-T base-pairing, thereby allowing a more efficient ligation.

LCR Conditions and Reaction Setup

In the following protocol, standard conditions are described that were successfully used in our laboratories for different LCR assays. Nevertheless, modifications of different reaction conditions should be tested for every newly developed LCR assay to achieve an optimal signal-to-noise ratio for matching targets as compared to targets with a single base-pair difference. The most important modifications are briefly discussed:

1. Reaction cycles are usually 15 seconds to 1 minute at 94°C for denaturation, followed by 4–6 minutes at 60–65°C (ideally 5°C below the lowest T_m values of the primers). Unlike PCR, there is no extension step between annealing and denaturation. In LCR this cycling pattern is repeated 10–30 times, but the number of cycles has to be optimized for each assay. The optimal number of cycles in an isotopic LCR does not always reflect the optimal number of cycles for a nonisotopic LCR with the same primers; usually fewer cycles are required in a nonisotopic LCR. In G-LCR, between 30 and 60 cycles with denaturation at 85°C and annealing at 50–53°C have been used (for details, see Birkenmeyer and Armstrong 1992; Dille et al. 1993).

2. The inclusion of 0.01%–0.1% Triton X-100 in the LCR buffer gives a higher ligation rate but also leads to a slight increase of ligation with a mismatched target (Winn-Deen and Iovannisci 1991; Wiedmann et al. 1992). For different buffers for pLCR and G-LCR, refer to Table 1 and the references therein.

3. Final primer concentrations between 0.5 nM and 5 nM should be tested. The optimal primer concentration in an LCR with isotopically labeled primers does not always give the best signal-to-noise ratio in a nonisotopic LCR with the same primers, and vice versa.

4. A NAD-requiring thermostable ligase (Barany and Gelfand 1991; Lauer et al. 1991) is most often used in ligase-based amplification methods. Recently, another thermostable ligase, which requires ATP as a cofactor, has been cloned and sequenced (Kletzin 1992). However, the use of this enzyme in DNA amplification methods has not yet been explored.

Summaries of protocols, reaction conditions, and detailed references for LCR, as well as pLCR and G-LCR, are outlined in Table 1.

Standard LCR

1. Prepare the LCR mixture at room temperature in a reaction tube suitable for the thermal cycler used. For setup of multiple reactions, prepare a master mix and add 24 μl of it to the target DNA in the reaction tubes. Add reagents in the following order (volumes are for a single reaction):

   1 μl of target DNA
   H_2O up to 25 μl final volume
   2.5 μl of 10x LCR buffer
   2 μl of NAD^+
   1 μl of salmon sperm DNA
   0.5 μl of each unlabeled primer (50 nM) (i.e., 1 nM final concentration)
   1.75 μl of each labeled primer (15 nM) (i.e., 1 nM final concentration)
   1 μl of *Taq* DNA ligase (37.5 units/μl)

   This mixture is overlaid with mineral oil if required for the thermal cycler used.

   If the primers are nonradioactively labeled, use of 100 fmoles of each primer in a 25-μl reaction mixture (i.e., 4 nM final concentration) gave optimal results in our experience.

2. Place tubes in a thermal cycler and perform 25 cycles (for isotopic LCR) or 10 cycles (for nonisotopic LCR) as follows: 1 minute at 94°C, followed by 4 minutes at 65°C (these conditions are suitable for LCR primers with a T_m of approximately 70°C; for primers with a differing T_m, the annealing temperature should be around 5°C lower than the T_m).

3. Stop the reaction by adding 20 μl of termination mixture (only for radioactive detection).

4. The LCR product may be stored at –20°C until used for electrophoresis or for nonisotopic detection.

Detection Methods for LCR Products

Detection of the LCR product was initially accomplished by using a radioactive label, ^{32}P, on the 5′ end of the upstream primer. The separation of LCR products and primers was achieved by denaturing gel electrophoresis, and the LCR product was detected by autoradiography. The level of sensitivity reached with this detection method is on the order of 200 target DNA molecules (Barany 1991a). Winn-Deen and Iovannisci (1991) described a nonisotopic detection method using fluorescence-labeled primers. Detection of the LCR product was accomplished using a fluorescent DNA sequencer in conjunction with a GENESCANNER (Perkin-Elmer, Applied Biosystems Division). One advantage of this method is that it is relatively easy to quantitate the amount of the LCR products. Furthermore, each of the primers can be labeled with a different fluorescent dye, allowing unambiguous assignment of ligation products; incorrect ligation products can be identified by their deviation from the appropriate color combinations (Winn-Deen and Iovannisci 1991). The fluorescent detection system allows multiplexing, with each mutation having a unique LCR primer labeled with a different fluorescent tag or of a different size (Feero et al. 1993). Currently this method is limited by the requirement for sophisticated equipment. An alternative approach for the nonisotopic detection uses one digoxigenin-labeled primer, and the LCR products are detected in a Southern blot format after gel electrophoretic separation (Kälin et al. 1992).

Recently, more convenient methods for the detection of LCR products in microtiter plates have been developed (Wiedmann et al. 1993; Winn-Deen et al. 1993a). In this format, one LCR primer of a pair is labeled with biotin at the 5′ end, and the other primer is labeled with a nonisotopic reporter at the 3′ end. Reporter groups tested so far include a fluorescein dye in blue (FAM, 5-carboxyfluorescein) and digoxigenin. Direct fluorometric detection of FAM-labeled LCR products in solution showed poor sensitivity, and the use of a digoxigenin reporter in conjunction with anti-digoxigenin antibodies coupled to alkaline phosphatase (AP) greatly improved the sensitivity. Subsequent detection of the AP could be achieved using colorimetric, fluorescent, or luminogenic substrates. Winn-Deen et al. (1993a) reported that the luminogenic substrate Lumi-Phos 530 gave the highest sensitivity in a microtiter plate assay. This sensitivity was only tenfold less than with detection methods using radioisotopes or a fluorescent DNA sequencer. Another nonisotopic detection method for LCR products has been reported by Zebala and Barany (1995). They used primer pairs in which one primer was labeled with a poly(dA) tail at the 5′ end while the 3′ end of the other primer was tagged with biotin. The ligated products were captured by hybridization with poly(dT)-coated paramagnetic iron beads and subsequent magnetic separation. Only the LCR products carry a 5′-coupled biotin molecule, which can be detected with a streptavidin-AP conjugate and a colorimetric substrate.

For the detection of the G-LCR products, two different methods have been described. Radioactively labeled nucleotides were used to fill in the gap between the primers, and the G-LCR products were detected by autoradiography (Dille et al. 1993). Alternatively, the primers can be end-labeled with radioisotopes as described for LCR primers (M. Wiedmann et al., unpubl.). Nonisotopic detection of G-LCR products was achieved by using pairs of primers labeled with biotin or fluorescein, respectively. Ligated oligonucleotides were captured on antifluorescein-coated microparticles and detected with an antibiotin-AP conjugate. AP activity was subsequently detected with the fluorescent substrate methylumbelliferone phosphate (Birkenmeyer and Armstrong 1992).

The following protocols describe procedures for the isotopic and nonisotopic labeling of LCR primers, as well as for the detection of the respective LCR products.

Isotopic Labeling of LCR Primers

One set of four different primers is used for the LCR. The upstream primer of each primer pair (designated primer LCR 1 and LCR 2R) has to be labeled isotopically to (1) provide a phosphate group for the ligation and (2) provide a radioactive label of the formed LCR product for later detection.

1. Prepare the labeling mix as follows:

   15 pmoles of primer (primer LCR 1 or LCR 2R)
   2 μl of 10X T4 polynucleotide kinase buffer
   1.5 μl of T4 polynucleotide kinase (10 units/μl)
   4 μl of [γ-^{32}P]ATP

   Make up with H_2O to 20 μl final volume.

2. Incubate for 45 minutes at 37°C.

3. Add 1 μl of unlabeled ATP (20 mM) and continue the incubation for another 2 minutes at 37°C.

4. Add 0.5 μl of 0.5 M EDTA to terminate the reaction.

5. Heat for 10 minutes at 65°C to inactivate the kinase.

6. Add 478.5 μl of TE buffer to bring the final volume to 500 μl.

7. Run the 500 μl of labeling mix through an NAP 5 column, equilibrated in TE buffer. Because the primers are eluted in 1 ml of TE buffer, the final concentration of the labeled primers will be 15 nM. The radioactively labeled primers can be stored at –20°C for up to 2 weeks after they have been kinased.

Nonisotopic Labeling of LCR Primers

The downstream primers of each pair (designated as LCR 3 and LCR 4R) have to be synthesized with a biotin group at the 5′ end using the Biotin-ON (CLONTECH) phosphoramidite method and an ABI 392 DNA synthesizer (Perkin-Elmer, Applied Biosystems Division). These primers should be HPLC-purified after synthesis to give a pure preparation of biotin-labeled primers without contaminating unlabeled primers. Primers LCR 1 and LCR 2R are kinased at the 5′ end and digoxigenin-labeled at the 3′ end using the procedure described below.

1. Prepare the kinasing reaction as follows:

   300 pmoles of primer (LCR 1 or LCR 2R)
   3 μl of 10x T4 polynucleotide kinase buffer
   1 μl of ATP (20 mM)
   6 units of T4 polynucleotide kinase

   Make up with H_2O to 30 μl final volume.

2. Incubate for 45 minutes at 37°C, then heat for 10 minutes at 65°C to inactivate the kinase.

3. Use the kinased primer to set up a digoxigenin labeling reaction:

   50 pmoles of kinased primer (LCR 1 or LCR 2R)
   8 μl of 5x terminal deoxynucleotidyl transferase buffer
   2 μl of digoxigenin-11-ddUTP solution
   85 units of terminal deoxynucleotidyl transferase

   Make up with H_2O to 40 μl final volume.

4. Incubate for 45 minutes at 37°C, then heat for 10 minutes at 65°C.

5. Dilute with H_2O to 100 nM and store at –20°C.

Steps 3–5 can also be used to label a primer with a biotin molecule at the 5′ end with a digoxigenin molecule at the 3′ end to serve as a positive control in the microtiter plate detection procedure.

Analysis of Isotopic LCR Products

1. Prepare a 16% polyacrylamide gel containing 7 M urea in a Mini-PROTEAN II electrophoresis cell (Bio-Rad Laboratories) or a similar apparatus.

2. Allow the gel to polymerize for 30 minutes and then place it into the electrophoresis apparatus with TBE buffer.

3. Heat the LCR samples for 5 minutes at 90°C and load 10-μl aliquots on the gel. It is important to flush the wells of the gel with buffer immediately before loading the sample to remove the urea.

4. Carry out electrophoresis at 175 V constant voltage for about 1 hour (until the front of the bromophenol blue reaches the bottom of the gel).

5. Remove the gel from the apparatus and remove the upper glass plate; then cover the gel with plastic wrap.

6. Place the gel on a Kodak X-OMAT AR film and autoradiograph at -20°C for 12–24 hours. The exposure time should be adjusted to achieve an optimal signal-to-noise ratio.

7. Develop the autoradiogram.

Analysis of Nonisotopic LCR Products in a Microtiter Plate Format

1. Add 60 μl of streptavidin (100 μg/ml in carbonate buffer) to high-binding EIA well strips and incubate for 1 hour at 37°C. Discard the streptavidin solution.

2. Block for 20 minutes at room temperature with 200 μl of buffer T + 0.5% dry milk + 100 μg/ml salmon sperm DNA (prepare this solution fresh every time). Discard the blocking solution.

3. Wash twice with 200 μl of buffer T.

4. Add 5 μl of LCR mixture diluted with 40 μl of buffer T without Tween to one well and add 10 μl of plate binding buffer. Incubate for 30 minutes at room temperature.

5. Wash twice with buffer T.

6. Wash twice with 200 μl of 0.01 M NaOH, 0.05% Tween 20.

7. Wash three times with buffer T.

8. Add 50 μl of anti-digoxigenin-AP, Fab fragment (diluted 1:1000 in buffer T + 0.5% dry milk), and incubate for 30 minutes at room temperature.

9. Wash six times with 200 μl of buffer T.

10. Add 50 μl of Lumi-Phos 530 and hold for 30 minutes at 37°C.

11. Expose to Polaroid type 612 film in CAMLIGHT and develop the film. Vary exposure time between 1 and 5 minutes to determine the exposure time needed for optimal signal-to-noise ratio. Alternatively, the chemiluminescence can be determined in a microtiter plate fluorometer.

The detection of nonisotopic LCR products should always be run in duplicate (one LCR provides enough reaction product for this).

ANALYSIS OF RESULTS AND TROUBLESHOOTING

The development and verification of a new LCR assay is easier using isotopically labeled primers rather than a nonisotopic detection system. Isotopic primers allow easy determination of the kinasing efficiency, because even nonligated primers can be detected after gel electrophoresis if they are radioactively labeled. Potential problems in an isotopic LCR include (1) no formation of LCR product with the matching target or (2) formation of LCR product in the presence of a mismatching target or no target.

The main reason for a failure to form an LCR product is inefficient phosphorylation of the primers. If present, unphosphorylated primers can compete with phosphorylated primers during annealing, but they cannot be ligated to the downstream primer and do not yield LCR products. Proper phosphorylation of primers can be assured by careful preparation of the ATP solution and by avoiding multiple freeze and thaw cycles for the ATP. Another possible cause for observing no amplification products could be defects in the design of the LCR primers, although we never observed such a problem when following the rules outlined above for primer design.

The formation of LCR products in the absence of a matching target could be observed if one of the LCR primers serves as a bridging template even in the presence of no target DNA, leading to template-independent amplification. The performance of each pair of primers in separate LDRs might then allow determination of the responsible primer(s).

Although nonisotopic detection of LCR products is more convenient and amenable to automation, there is also a higher potential for problems during the initial setup of these methods. In our experience, a false-negative nonisotopic LCR is most often caused by either insufficient labeling or phosphorylation of the upstream primers of a pair. As mentioned above, insufficient phosphorylation might be caused by improperly prepared and stored ATP solutions. Troubleshooting with the nonisotopic detection system in a microtiter plate system is complicated because the unligated primers cannot be visualized to verify proper labeling of both the biotin and the digoxigenin-labeled primers. The nonisotopically labeled primers (and LCR products) can be visualized after separation on a 16% polyacrylamide gel, electroblotting onto a membrane, and detection using anti-digoxigenin antibodies for the digoxigenin-labeled primers or streptavidin with suitable reporter enzymes for the biotin-labeled primers. Therefore, it is possible to pinpoint the problem by determining whether (1) the primers are properly labeled and (2) an LCR product is formed.

False-positive results in the microtiter plate detection are usually caused by insufficient blocking of the wells. The best results were obtained in our hands when fresh blocking solution was prepared and when the salmon sperm DNA was boiled and cooled on ice before addition to the blocking solution.

DISCUSSION

LCR assays have been developed for the detection of genetic diseases as well as for the detection of bacteria and viruses. An overview of the current applications of LCR and G-LCR is shown in Table 3. In many of these applications, LCR is preceded by an initial PCR step to achieve a greater sensitivity of the respective assays. PCR products can be used in LCR without prior purification if the removal of any residual polymerase activity is ensured; e.g., by incubation for 25 minutes at 97°C (Zebala and Barany 1995).

With the continuing emergence of sequence data for the human genome as well as the genomes of other species (e.g., bovine, equine), the potential of LCR to detect genetic diseases that result from single base-pair mutations is immense. One inherent advantage of LCR is its potential for automation. The LCR product consists of two covalently joined primers that can be easily detected using different enzyme-linked or direct fluorescent labels. Formatting of multiplex LCR assays will further improve screening samples for an array of different single base-pair changes in a single tube. Automated, multiplex LCR or PCR-coupled LDR/LCR assays have a variety of potential applications (Barany 1991b), such as (1) screening of large populations for monogenic disease polymorphisms; (2) determining HLA haplotypes in tissue typing, e.g., for transplantation; and (3) screening for multiple bacterial species after a generic PCR amplification of 16S rDNA sequences.

Table 3 Current Applications of LCR and G-LCR

Target	Format	Author
Genetic diseases		
β-sickle cell hemoglobinemia	LCR, isotopic	Barany (1991a)
β-sickle cell hemoglobinemia	LCR, fluorescent	Winn-Deen and Iovannisci (1991)
cystic fibrosis	PCR-LDR, fluorescent	Eggerding et al. (1993); Winn-Deen et al. (1993b)
cystic fibrosis	LCR and G-LCR, isotopic	Fang et al. (1992)
Leber's hereditary optic neuropathy	PCR-LCR, nonisotopic	Zebala and Barany (1995)
hyperkalemic periodic paralysis	PCR-LCR, fluorescent	Feero et al. (1993); Wang et al. (1993)
bovine leukocyte adhesion deficiency	PCR-LCR, nonisotopic	Batt et al. (1994)
trinucleotide repeats	repeat extension analysis (RED)	Schalling et al. (1993)
Bacteria		
Borrelia burgdorferi	LCR, nonisotopic	Hu et al. (1991)
Listeria monocytogenes	PCR-LCR, nonisotopic	Wiedmann et al. (1992, 1993)
Neisseria gonorrhoeae	G-LCR, nonisotopic	Birkenmeyer and Armstrong (1992)
Erwinia stewartii	PCR-LCR, nonisotopic	Wilson et al. (1994)
Mycobacterium tuberculosis	LCR, fluorescent	Iovannisci and Winn-Deen (1993)
Chlamydia trachomatis	G-LCR, isotopic	Dille et al. (1993)
Viruses		
human papillomavirus	LCR, nonisotopic	Bond et al. (1990)
herpes simplex virus	LCR, nonisotopic	Rinehardt et al. (1991)
HIV DNA	LCR, nonisotopic	Carrino and Laffler (1991)
Other targets		
Ha-*ras* proto-oncogene	LCR, nonisotopic	Kälin et al. (1992)
Ha-*ras* proto-oncogene	PCR-LCR	Wei et al. (1992)
G-6-PD	RT-PCR-LDR, isotopic	Prchal et al. (1993)
HOXB7	RT-PCR-LCR, isotopic	Chariot et al. (1994)

In the clinical diagnosis of pathogenic bacteria and viruses, the specificity of LCR could be useful in many applications. The detection of single base-pair differences in bacterial pathogens may be valuable to identify antibiotic resistance arising from point mutations, e.g., in some cases of macrolide resistance (Gauthier et al. 1988), or from transformational exchange as occurs in sensitive and resistant strains, e.g., in *Neisseria meningitidis* (Rådström et al. 1992). In viral pathogens, the identification of subpopulations with genetic differences may be important with regard to host range, virulence characteristics, and drug resistance.

Furthermore, the application of LCR and PCR-coupled LCR assays for the detection of specific bacteria based on at least a single base-pair difference in the 16S rDNA gene has great potential (Wiedmann et al. 1992, 1993; Wilson et al. 1994). Such a system circumvents the need to identify species-specific genes, as warranted for PCR or other nucleic-acid-based assays. In this system, a combination of a primary PCR amplification of the target region with a secondary LCR step for the detection of species-specific single base-pair differences allows a combination of the sensitivity of PCR with the specificity of LCR. With emerging interest in poorly characterized bacteria, this method should have great potential as a detection system.

ACKNOWLEDGMENTS Part of the work presented in this review was supported by the Northeast Dairy Foods Research Center (to C.A.B.); Eastern A.I. (to C.A.B.); a grant from the Cornell Center for Advanced Technology (CAT) in Biotechnology, which is sponsored by the New York State Science and Technology Foundation and industrial partners (to C.A.B.); a grant from the Applied Biosystems Division of Perkin-Elmer (to F.B.); and a grant from the National Institutes of Health (GM-41337-03) (to F.B.). M.W. was supported by a stipend of the Gottlieb Daimler- und Carl Benz-Stiftung (2.92.04).

REFERENCES

Barany, F. 1991a. Genetic disease detection and DNA amplification using cloned thermostable ligase. *Proc. Natl. Acad. Sci.* **88:** 189–193.

——. 1991b. The ligase chain reaction in a PCR world. *PCR Methods Appl.* **1:** 5–16.

Barany, F. and D.H. Gelfand. 1991. Cloning, overexpression and nucleotide sequence of a thermostable DNA ligase-encoding gene. *Gene* **109:** 1–11.

Barringer, K., L. Orgel, G. Wahl, and T.R. Gingeras. 1990. Blunt-end and single-stranded ligations by *Escherichia coli* ligase: Influence on an in vitro amplification scheme. *Gene* **89:** 117–122.

Batt, C.A., P. Wagner, M. Wiedmann, J. Luo, and R.O. Gilbert. 1994. Detection of bovine leukocyte adhesion deficiency by nonisotopic ligase chain reaction. *Anim. Genet.* **25:** 95–98.

Birkenmeyer, L. and A.S. Armstrong. 1992. Preliminary evaluation of the ligase chain reaction for specific detection of *Neisseria gonorrhoeae. J. Clin. Microbiol.* **30:** 3089–3094.

Birkenmeyer, L.G. and I.K. Mushahwar. 1991. Mini-review: DNA probe amplification methods. *J. Virol. Methods* **35:** 117–126.

Bond, S., J. Carrino, H. Hampl, K. Hanley, L. Rinehardt, and T. Laffler. 1990. New methods of detection of HPV. In *Serono Symposia* (ed. J. Monsonego), pp. 425–434. Raven Press, Paris.

Carrino, J.J. and T.G. Laffler. 1991. Detection of HIV DNA sequences using the ligase chain reaction (LCR). *Clin. Chem.* **37:** 1059.

Chariot, A., V. Castronovo, G. Senterre, S. Senterre-Lesenfants, M. Kusaka, and M.E. Sobel. 1994. Identification by reverse transcriptase-ligase chain reaction of a HOXB7 stop codon polymorphism in MCF7, a human breast cancer-derived cell line. *Proc. Am. Assoc. Cancer Res. Annu. Meet.* **35:** 597.

Dille, B.J., C.C. Butzen, and L.G. Birkenmeyer. 1993. Amplification of *Chlamydia trachomatis* DNA by ligase chain reaction. *J. Clin. Microbiol.* **31:** 729–731.

Eggerding, F., E. Winn-Deen, W. Giusti, T. Adriano, D. Iovannisci, and E. Brinson. 1993. Detection of mutations in the cystic fibrosis gene by multiplex amplification and oligonucleotide ligation. *Am. J. Hum. Genet.* **53:** 1485.

Fang, P., C. Jou, S. Bouma, and A. Beaudet. 1992. Detection of cystic fibrosis mutations using the ligase chain reaction. *Am. J. Hum. Genet.* **51:** A214.

Feero, W.G., J. Wang, F. Barany, J. Zhou, S.M. Todorovic, R. Conwit, G. Galloway, I. Hausmanowa-Petrusewicz, A. Fidzianska, K. Arahata, H.B. Wessel, C. Wadelius, H.G. Marks, P. Hartlage, H. Hayakawa, and E.P. Hoffman. 1993. Hyperkalemic periodic paralysis: Rapid molecular diagnosis and relationship of genotype to phenotype in 12 families. *Neurology* **43:** 668–673.

Gauthier, A., M. Turmel, and C. Lemieux. 1988. Mapping of chloroplast mutations conferring resistance to antibiotics in *Chlamydomonas*: Evidence for a novel site of streptomycin resistance in the small subunit ribosomal RNA. *Mol. Gen. Genet.* **214:** 192–197.

Hu, H., K. Elmore, I. Facey, and D. Jenderzak. 1991. Detection of *Borrelia burgdorferi* by ligase chain reaction. *Abstr. Gen. Meet. Am. Soc. Microbiol.*, p. 79.

Iovannisci, D.M. and E.S. Winn-Deen. 1993. Ligation amplification and fluorescence detection of *Mycobacterium tuberculosis* DNA. *Mol. Cell. Probes* **7:** 35–43.

Kälin, I., S. Shephard, and U. Candrian. 1992. Evaluation of the ligase chain reaction (LCR) for the detection of point mutations. *Mutat. Res.* **283:** 119–123.

Kletzin, A. 1992. Molecular characterization of a DNA ligase gene of the extremely thermophilic archeon *Desulfurolobus ambivalens* shows close phylogenetic relationship to eukaryotic ligases. *Nucleic Acids Res.* **20:** 5389–5396.

Landegren, U., R. Kaiser, J. Sanders, and L. Hood. 1988. A ligase-mediated gene detection method. *Science* **241:** 1077–1080.

Lauer, G., E.A. Rudd, D.L. McKay, A. Ally, D. Ally, and K.C. Backman. 1991. Cloning, nucleotide sequence, and engineered expression of *Thermus thermophilus* DNA ligase, a homolog of *Escherichia coli* DNA ligase. *J. Bacteriol.* **173:** 5047–5053.

Nickerson, D.A., R. Kaiser, S. Lappin, J. Stewart, L. Hood, and U. Landegren. 1990. Automated DNA diagnostics using an ELISA-based oligonucleotide ligation assay. *Proc. Natl. Acad. Sci.* **87:** 8923–8927.

Prchal, J.T., Y.L. Guan, J.F. Prchal, and F. Barany. 1993. Transcriptional analysis of the active X-chromosome in normal and clonal hematopoiesis. *Blood* **81:** 269–271.

Rådström, P., C. Fermér, B.-E. Kristiansen, A. Jenkins, O. Sköld, and G. Swedeberg. 1992. Transformational exchanges in the dihydropteroate synthase gene of *Neisseria meningitidis:* A novel mechanism for acquisition of sulfonamide resistance. *J. Bacteriol.* **174:** 6386–6393.

Rinehardt, L., H. Hampl, and T.G. Laffler. 1991. Ultrasensitive non-radioactive detection of herpes simplex virus by LCR, the ligase chain reaction. In *20th Annual Meeting of the Keystone Symposia on molecular and cellular biology*, p. 101.

Sambrook, J.E., E.F. Fritsch, and T. Maniatis. 1989. *Molecular cloning: A laboratory manual*, 2nd edition. Cold Spring Harbor Laboratory Press, Cold Spring Harbor, New York.

Schalling, M., T.J. Hudson, K.H. Buetow, and D.E. Housman. 1993. Direct detection of novel expanded trinucleotide repeats in the human genome. *Nature Genet.* **4:** 135–139.

Segev, D. 1992. Amplification of nucleic acid sequences by the repair chain reaction. In *Nonradioactive labeling and detection of biomolecules* (ed. C. Kessler), pp. 212–218. Springer Laboratory, Berlin.

Wang, J., J. Zhou, S.M. Todorovic, W.G. Feero, F. Barany, R. Conwit, I. Hausmanowa-Petrusewicz, A. Fidzianska, K. Arahata, H.B. Wessel, A. Sillen, H.G. Marks, P. Hartlage, G. Galloway, K. Ricker, F. Lehmann-Horn, H. Hayakawa, and E.P. Hoffman. 1993. Molecular genetic and genetic correlations in sodium channelopathies: Lack of funder effect and evidence for a second gene. *Am. J. Hum. Genet.* **52:** 1074–1084.

Wei, Q., F. Barany, and V.L. Wilson. 1992. Oncogenic point mutations detected by combined PCR and LCR techniques. 32nd Annual Meeting of the American Society for Cell Biology. *Mol. Biol. Cell* (suppl.) **3:** 22A.

Whiteley, N.M., M.W. Hunkapiller, and A.N. Glazer. 1989. Detection of specific sequences in nucleic acids. *U.S. Patent no.* 4,883,750.

Wiedmann, M., F. Barany, and C.A. Batt. 1993. Detection of *Listeria monocytogenes* with a nonisotopic polymerase chain reaction-coupled ligase chain reaction assay. *Appl. Environ. Microbiol.* **59:** 2743–2745.

Wiedmann, M., J. Czajka, F. Barany, and C.A. Batt. 1992. Discrimination of *Listeria monocytogenes* from other *Listeria* species by ligase chain reaction. *Appl. Environ. Microbiol.* **58:** 3443–3447.

Wilson, W.J., M. Wiedmann, H.R. Dillard, and C.A. Batt. 1994. Identification of *Erwinia stewartii* by a ligase chain reaction assay. *Appl. Environ. Microbiol.* **60:** 278–284.

Winn-Deen, E.S. and D.M. Iovannisci. 1991. Sensitive fluorescence method for detecting DNA ligation amplification products. *Clin. Chem.* **37:** 1522–1523.

Winn-Deen, E.S., C.A. Batt, and M. Wiedmann. 1993a. Non-radioactive detection of *Mycobacterium tuberculosis* LCR products in a microtitre plate format. *Mol. Cell. Probes* **7:** 179–186.

Winn-Deen, E., P. Grossmann, S. Fung, S. Woo, C. Chang, E. Brinson, and F. Eggerding. 1993b. High density multiplex mutation analysis using the oligonucleotide ligation assay (OLA) and sequence-coded separation. *Am. J. Hum. Gen.* **53:** 1512.

Wu, D.Y. and R.B. Wallace. 1989. The ligation amplification reaction (LAR)–Amplification of specific DNA sequences using sequential rounds of template-dependent ligation. *Genomics* **4:** 560–569.

Zebala, J.A. and F. Barany. 1995. Detection of Leber's hereditary optic neuropathy by non-radioactive LCR. In *PCR Strategies* (ed. D.H. Gelfand et al.). Academic Press, San Diego, California. (In press.)

Optimization and Characterization of 3SR-based Assays

Thomas R. Gingeras,[1] Mathew Biery,[2] Michael Goulden,[2] Soumitea S. Ghosh,[3] and Eoin Fahy[3]

[1]Affymetrix, Santa Clara, California 95051
[2]Life Sciences Research Laboratory, Baxter Diagnostics, San Diego, California 92121
[3]Applied Genetics Inc., San Diego, California 92121

INTRODUCTION

The detection and identification of nucleic acids present at very low levels have been greatly facilitated by the development of in vitro target amplification techniques that exploit the enzyme-mediated processes of DNA replication (Saiki et al. 1985; Mullis and Faloona 1987), DNA ligation (Wu and Wallace 1989; Barany 1991), and RNA transcription. The first reported *t*ranscription-based *a*mplification *s*ystem (TAS) utilized the abilities of avian myeloblastosis virus reverse transcriptase (AMV RT) and T7 RNA polymerase to make multiple RNA copies of target sequences (Kwoh et al. 1989). This strategy, involving a thermal cycling step to separate cDNA strands from their RNA templates, has been applied successfully to the detection of cells infected with human immunodeficiency virus type 1 (HIV-1) (Davis et al. 1990; Gingeras et al. 1990b). More recently, isothermal versions of the TAS protocol, known as the *s*elf-*s*ustained *s*equence *r*eplication (3SR) (Gingeras et al. 1990a; Guatelli et al. 1990) and *n*ucleic *a*cid *s*equence-*b*ased *a*mplification (NASBA) (van Gemen et al. 1993) reactions, have been developed by using the concerted enzymatic activities of *Escherichia coli* RNase H, AMV RT, and T7 RNA polymerase to amplify RNA targets with both high efficiency and specificity. The isothermal characteristic of the 3SR reaction permits the specific amplification of single-stranded RNA target molecules even in the presence of double-stranded DNA molecules containing the same sequence (Gingeras et al. 1990a). This ability to target and to amplify RNA specifically using the 3SR method has led to several clinical applications, including the detection and characterization of nucleoside (Gingeras et al. 1991) and nonnucleoside (Richman et al. 1991) drug-resistant strains of HIV-1 and the detection of HIV-1 in the plasma of pediatric patients (Bush et al. 1992). Importantly, the major reaction product of 3SR reactions is single-stranded RNA, which can be directly cloned and sequenced (Guatelli et al. 1990) and conveniently detected either by heterogeneous isotopic (Fahy et al. 1993), non-isotopic (Bush et al. 1991), or homogeneous (Devlin et al. 1993) hybridization methods.

Each of these areas of clinical application in which the 3SR reaction has been used has evolved to require more absolute rather than relative quantitation. To achieve this requirement, several critical challenges in current 3SR-based assays need to be addressed. The essential aspects in the establishment of a 3SR quantitative assay are (1) determination of reaction conditions that allow for the least variability, (2) control of potential carryover contamination, and (3) characterization of either heterogeneous or homogeneous detection systems used to quantitate the 3SR RNA products in an assay.

3SR Amplification

All aqueous solutions were prepared using deionized H_2O treated with 0.1% (v/v) diethyl pyrocarbonate to suppress nuclease activity. T7 RNA polymerase was obtained from Stratagene. AMV RT and *E. coli* RNase H were purchased from Boehringer Mannheim. Oligonucleotides were synthesized by phosphoramidite chemistry on Applied Biosystems 394 RNA/DNA synthesizer. Potassium glutamate (KGlu) was prepared by adjusting the pH of a 1 M solution of L-glutamic acid to 8.1 with KOH.

REAGENTS

Stock Solutions (keep on ice)

5x Reaction buffer: 200 mM Tris-acetate, pH 8.1, 150 mM Mg $(acetate)_2$
50 mM DTT, 500 mM KGlu
25 mM rNTPs in 50 mM Tris-HCl, pH 8.1
25 mM dNTPs in 50 mM Tris-HCl, pH 8.1

PROTOCOL

1. Add the following to an autoclaved 1.5-ml Eppendorf tube (on ice):

   10 μl of 5x reaction buffer
   2.5 μl of each priming oligonucleotide (0.1 μM each, final)
   2 μl of 25 mM dNTP mix (1 mM final)
   12 μl of 25 mM rNTP mix (6 mM final)
   11 μl of 68.2% sorbitol (15% final)
   5 μl of RNA target (use H_2O for negative control reactions)

It is recommended that a master mix composed of the 5x buffer, NTPs, primers, and sorbitol be made and a 40-μl aliquot of this mix be pipetted into each reaction tube prior to the addition of target.

2. Prepare 3SR enzyme mix on ice. Each reaction requires:

   15 units of AMV reverse transcriptase
   1 unit of *E. coli* RNase H
   50 units of T7 RNA polymerase

   Add 40 mM Tris-HCl, pH 8.1, 10 mM DTT for a final volume of 5 μl per reaction for the enzyme mix.

3. Heat the reaction tubes for 5 minutes at 65°C to denature the RNA target. Transfer the tubes to a heat block at 47°C and incubate for at least 1 minute.

4. Add 5 μl of 3SR enzyme mix to each tube and gently flick several times. Incubate for 90 minutes at 47°C. Store the reactions at –20°C. The reactions are terminated by freezing and thawing.

Control of Carryover Contamination

Both the cDNA and RNA products of 3SR amplification may act as a source of carryover contamination, making the choice of a suitable sterilization method more problematic. Additionally, an effective strategy requires that the modified RNA products be detected efficiently by hybridization. A postamplification approach using the photoactive agent 4′-aminomethyl-4,5-dimethylpsoralen (IP-10) has been found to modify PCR amplification products (Cimino et al. 1991; Isaacs et al. 1991). More recently, the use of IP-10 has been shown to sterilize 3SR amplicons with a 10^6- to 10^8-fold efficiency (Versailles et al. 1993).

REAGENTS

Stock Solutions

4.6 μg/μl of IP-10 (HRI Associates); store in the darkroom at room temperature

1% NuSieve agarose (FMC) in DEPC-treated H_2O; store at room temperature

PROTOCOL

1. Prepare a 0.5% gel containing 1200 μg/ml IP-10. Melt a 1-ml stock of 1% agarose in a microfuge tube in a heat block for 1 minute at 80°C.

2. Add 50 μl of melted gel to 26.1 μl of IP-10 stock and 23.9 μl of DEPC-treated H_2O in an autoclaved microfuge tube to give 100 μl of sterilizing gel. Volumes may be scaled up to 200 μl or more for

use in larger (>10 reactions) amplification experiments. Store the sterilizing gel samples in the darkroom at room temperature.

3. Prior to 3SR amplification, melt the sterilizing gel for 1 minute at 80°C. Assuming a 3SR reaction volume of 50 μl, transfer 10 μl of melted gel to the inside of the reaction-tube caps (the gel solidifies immediately).

4. Cover the tubes with aluminum foil. Proceed as usual for RNA amplification by adding 3SR reagents and target, heating for 1 minute at 65°C, and finally, adding the enzymes. Gently vortex the reaction tubes. It is important not to centrifuge the reaction tubes at this stage to avoid dislodging the gel.

5. After the 3SR reaction has been completed, centrifuge for 5 seconds (to dislodge the agarose from the cap) and heat for 1 minute at 80°C to melt the agarose and inactivate the enzymes.

6. Vortex the tubes and centrifuge for 5 seconds to ensure that the entire solution is at the bottom of the tube prior to irradiation. This procedure gives a final IP-10 concentration of 200 μg/ml.

7. Keep the reaction tubes on ice and UV-irradiate for 20 minutes at 4°C (HRI-100 UV illuminator: HRI Associates).

8. Store the 3SR reactions at –20°C or perform *b*ead-*b*ased *s*andwich *h*ybridization (BBSH) analysis.

ANALYSIS OF RESULTS AND TROUBLESHOOTING

Influence of Buffer Conditions on 3SR Enzymes

Initial efforts to develop the 3SR amplification reaction focused on the issue of increasing the sensitivity of the reaction. The use of organic additives and the alteration of the concentration of ribonucleotide triphosphates resulted in an increase in the productivity of the 3SR reaction typically from 10^6- (Guatelli et al. 1990) to ≥10^9-fold (Fahy et al. 1991; Gingeras and Kwoh 1992). Interestingly, the addition of 15% sorbitol and 10% DMSO enabled the 3SR reaction to proceed in certain amplification reactions without the need for *E. coli* RNase H (Fahy et al. 1991). In these reactions, the additives presumably enhanced the intrinsic RNase activity (specificity or kinetics) of AMV RT to levels otherwise supplied by *E. coli* RNase H. However, in either a two- or three-enzyme protocol, the nature of the 3SR amplification reaction remains a multienzyme, concerted reaction. The consequence of failing to coordinate the concerted enzyme activities is an increase in the variability of the 3SR reactions. Analogously, this has been exemplified for the two-enzyme RT-*Taq* DNA polymerase-mediated PCR assays in which two- to tenfold ranges in reproducibility have been noted (Davis et al. 1990; Oka et al. 1990; Holodniy et al. 1991; Ferre et al. 1992).

Variability in the 3SR reaction can be measured as the lack of either precision or reproducibility. Precision is the measure of agreement in the results of multiple amplification reactions performed repeatedly with a homogeneous target and a consistent pool of amplification reagents (enzymes, buffers, and primers). Reproducibility is the measure of agreement observed in the results of multiple amplification reactions performed repeatedly with heterogeneous target and various preparations of amplification reagents. The lack of agreement in either measurement is expressed as the coefficient of variation (cv = standard deviation x 100/mean).

The precision of the 3SR reaction was measured by targeting a 382-nucleotide region of the third exon of the major immediate early (MIE) mRNA from human cytomegalovirus (HCMV) (see legend to Fig. 1). Using the 3SR reaction conditions described previously (Fahy et al. 1991), a collection of 5 amplification experiments, totaling 50 3SR reactions (10 amplifications per experiment), were performed on the same day using 0.1 attomole (amole) of a pre-quantitated 1066-nucleotide RNA transcript (Versailles et al. 1993) encoding the 382-nucleotide amplification region. Of the 50 reactions performed, no 3SR product could be detected using a BBSH assay in 26 reactions (Table 1, column I). The cv for this collection of experiments was >200%. Removal of the 26 negative reactions from this data set resulted in a cv of 136.7%, still indicating a low degree of precision (Table 1, column II). The precision of the heterogeneous BBSH detection assay itself contributed a cv of only 5.6% (Table 1, column VI) to the variability observed in the 3SR amplification experiments. These results and others like them precipitated a search for the causes of and solutions for such amplification variability.

Effect of Chloride Salts on the Enzymes in the 3SR Reaction

The notion that DNA-protein reactions are highly sensitive to chloride salts has been noted previously in in vitro assays of both *E. coli* RNA polymerase (Leirmo et al. 1987) and restriction endonucleases (Leirmo et al. 1987; McClelland et al. 1988). Specifically, for the T7 RNA polymerase in the 3SR reaction, more than 90% of the activity is lost as the concentration of KCl reaches 150 mM, and at concentrations greater than 200 mM KCl, there is virtually no T7 RNA polymerase activity (Chamberlin and Ring 1973). This effect was determined to be a function specifically of the anion concentration, since both sodium and ammonium salts at the same concentration as the KCl used produced the same inhibitory effects. In contrast, AMV RT and *E. coli* RNase H have broad tolerances for chloride (Berkower

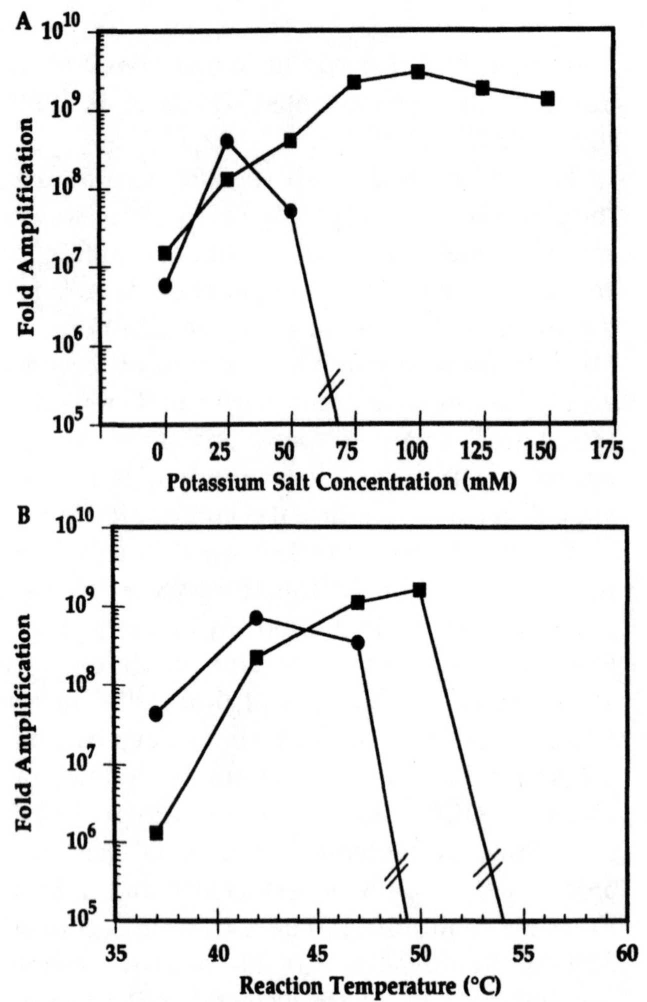

Figure 1 (*A*) The effect of reaction temperature on 3SR amplification efficiency under differing buffer conditions. Plasmid pACYC-HCMV-*Eco*RI-J, based on pACYC184 and containing an *Eco*RI fragment from the genome of the AD 169 strain of cytomegalovirus (CMV) was obtained from Dr. Deborah Spector (University of California, San Diego). A 1066-bp *Bgl*II/*Pvu*II fragment from pACYC-HCMV-*Eco*RI-J was cloned into the transcription vector pSP6/Ta719 and transcribed by T7 RNA polymerase to yield a sense RNA product spanning the MIE region of CMV. A 382-base region of this 1066 transcript (0.1 amole) was amplified by 3SR for 90 minutes with primers 90-635 and 92-135 over a range of temperatures from 37°C to 55°C. Reactions were performed as described in the Protocol (filled squares) or by substituting 40 mM Tris-HCl, pH 8.1, 30 mM $MgCl_2$, and 20 mM KCl for the corresponding acetate and glutamate salts (filled circles). The reaction products were detected by BBSH using Trisacryl OligoBeads 92-155 and ^{32}P-labeled probe 92-132. (*B*) Titration of potassium salts in the 3SR reaction. A 382-base region of a 1066-base transcript (0.1 amole) comprising the MIE region of CMV was amplified by 3SR for 90 minutes with primers 90-635 and 92-135 using a range of KGlu (filled squares) or KCl (filled circles) concentrations. All other reaction conditions were described in Fig. 1. The reaction products were detected by BBSH using TRISACRYL OligoBeads 92-155 and ^{32}P-labeled probe 92-132.

Table 1 Statistical Analysis of Variability of 3SR Amplification of HCMV-MIE RNA

Assay reaction conditions[a]	I[b]	II[c]	III[d]	IV[e]	V[f]	VI[g]
Mean (fmole/μl)	47.0	98.0	3781.2	3781.2	3781.2	4434.0
Standard deviation	104.2	133.9	1618.2	1216.9	860.8	257.7
Samples	50	24	60	20	10	20
Coefficient of variation (%)	221.6	136.7	42.8	35.3	26.0	5.6

[a]Description of MIE reaction conditions as listed in Cimino et al. (1991) and the Protocol section.

[b]CMV-MIE RNA (0.1 amole) was amplified by 3SR in 5 experiments on 1 day, representing a total of 60 reactions. Reaction conditions were as described in Cimino et al. (1991) except that spermidine was omitted and DMSO and sorbitol were present at 5% and 10%, respectively. The 3SR reaction time was 90 minutes.

[c]This column describes a set of the data listed in column I representing the 24 reactions that gave a detectable signal by BBSH.

[d]CMV-MIE RNA (0.1 amole) was amplified by 3SR in two sets of 10 reactions per day for 3 consecutive days. Reaction conditions were as described in the Protocol section. This column describes the analysis of the entire data set (6 reactions).

[e]This column describes a subset of the data in column III representing experiments performed on the same day (two sets of 10 reactions).

[f]This column describes a subset of the data in column III representing an individual experiment (10 reactions).

[g]Twenty identical aliquots of 3SR product generated from amplification of the CMV-MIE region were assayed by BBSH.

et al. 1973; Verma 1977). The 3SR reaction can be expected to operate optimally with low chloride concentration based on the sensitivity of T7 RNA polymerase. The total concentration of chloride in the previously published 3SR reaction conditions (Fahy et al. 1991) is 110 mM and includes contributions from the Tris buffer used to dissolve the nucleotides, primers, target, and enzymes. Inadvertent increase in the chloride concentration can be introduced by errors in reagent pipetting, a change in nucleotide reagent stocks, and carryover from the extraction and precipitations of nucleic acids from clinical samples. Sample-to-sample and experiment-to-experiment variations can also influence the precision in the performance of 3SR reactions. Therefore, the replacement of chloride with an alternative anion in 3SR was explored as a means of increasing the precision of the 3SR reactions. To explore this hypothesis, the Tris-HCl, $MgCl_2$, and KCl components of the previous buffer conditions (Fahy et al. 1991) were replaced with Tris-acetate, magnesium acetate, and KGlu, respectively (see Protocol). The new buffer conditions were not completely devoid of chloride, since the nucleotides, primers, enzymes, and target RNA solutions were all prepared in Tris-HCl buffer (~10 mM final Cl^- concentration). Titrations of KGlu in a 90-minute 3SR amplification of the HCMV-MIE 382-nucleotide region present in the 1066-nucleotide HCMV-MIE RNA transcript were performed. Amplification efficiencies in excess of 10^8-fold were obtained in each reaction utilizing a 25–150 mM range of KGlu concentrations. Optimal efficiencies were observed using 100 mM KGlu (Fig. 1A). This range of KGlu contrasts sharply with the narrow 0–50 mM range noted for 3SR reaction conditions employing chloride buffers.

In addition to increasing the salt tolerance of the 3SR reaction, the substitution of chloride with acetate/glutamate in the reaction buffer permitted a shift in the temperature optimum for the 3SR reaction (Fig. 1B). Using the acetate/glutamate buffer, temperatures as high as 50°C could be employed while maintaining amplification levels in excess of 10^9-fold. In these experiments, the omission of DMSO from the 3SR buffer was found to be essential to obtain amplification above 45°C (data not shown).

Effects of Acetate/Glutamate Substitution on the Precision of 3SR Reaction

When acetate/glutamate buffers are substituted for chloride buffers and the 3SR reactions are conducted at 42°C, the 3SR reactions are again characterized by occasional amplification failures (data not shown). However, performing the 3SR reaction at 47°C under the acetate/glutamate buffer conditions has a dramatic effect on the precision of the reaction.

When the 382-nucleotide HCMV-MIE region was amplified in two sets of 10 reactions per day for 3 consecutive days under conditions described in the Protocol, no amplification failures were noted. All 60 reactions produced $>10^8$-fold amplification. The cv of these experiments was reduced to 42.8% (Table 1, column III). When the experiments performed on a single day (two sets of 10 amplification reactions) were analyzed, the cv was reduced to 35.3% (Table 1, column IV) for both sets of amplifications compared to the 221.6% observed for the single-day experiments performed using the chloride buffers (Table 1, column I). When one set of 10 amplifications was analyzed, a cv of 26% was noted (Table 1, column V). Since 5.6% of this variation is attributable to the BBSH detection method (Table 1, column VI), the net variation compares favorably to that reported recently for RT-PCR amplification assays of HIV-1 RNA performed in competitive (Piatak et al. 1993) and noncompetitive (Mulder et al. 1994) amplification formats.

Although these data refer to the precision of the 3SR reaction, preliminary data on the reproducibility of the 3SR reaction appear to be similar. Reactions using multiple reagent stocks and with different preparations of targets have been performed without amplification failure. In addition, improved precision of the 3SR reaction has been observed using other target templates (in HIV-1 and HCMV). The effect of non-chloride buffer conditions in amplifying longer target molecules also has been studied. Initial results suggest no appreciable improvement in the efficiency of amplifying longer target molecules. However, modifications of nucleotide and enzyme concentrations coupled with the use of non-chloride-containing buffers have not yet been fully explored.

Control of Potential Carryover Contamination

As with the PCR-based assays (Longo et al. 1990), minute amounts of 3SR-amplified products inadvertently transferred to subsequent reactions may be amplified very efficiently, giving rise to false positives. Such an undesirable event can occur through aerosol transfer or contaminated laboratory equipment and clothing. Of course, the prevention of such contamination begins with good laboratory practices of separating pre- and postamplification activities (Dieffenbach and Dveksler 1993). However, the implementation of an added measure of protection with the use of the photoactivatable IP-10 in 3SR applications can be very useful. The results of IP-10 modification of 3SR amplification products in the hybridization reactions have been described previously (Versailles et al. 1993). As observed with PCR products (Cimino et al. 1991; Isaacs et al. 1991), there is a correlation between the number of reactive sites (thymine/uracil residues) and sterilization efficiency of IP-10 for 3SR products (Versailles et al. 1993).

An important technical problem encountered in the application of this photochemical strategy was the inhibition of the 3SR amplification process by IP-10. This problem was overcome by segregating the IP-10 from the 3SR reaction by gel encapsulation, as shown in Figure 2. The IP-10 is mixed with low-melting agarose and the gel is solidified in the caps of Eppendorf reaction tubes. The encapsulated IP-10 remains physically separated from the 3SR solution during the amplification reaction. Following amplification, the tubes are heated to 75–80°C to liquefy the gel to enable delivery of IP-10 to the 3SR reaction. A UV-irradiation step completes the sterilization process.

Homogeneous and Heterogeneous Detection of 3SR Products

The products from 3SR reactions are predominantly single-stranded RNA and are amenable to detection by a variety of hybridization methods. Many 3SR-detection assays have been based on heterogeneous hybridization, where the 3SR product is captured on a solid support and detected with a labeled oligonucleotide probe in a sandwich format (Fig. 3A). To date, probes labeled with ^{32}P (Leirmo et al. 1987; Gingeras et al. 1990, 1991; Guatelli et al. 1990; Fahy et al. 1991, 1993; Bush et al. 1992; Gingeras and Kwoh 1992; Versailles et al. 1993), alkaline phosphatase (Ghosh et al. 1990; Ishii and Ghosh 1993), and lanthanide chelates (Bush et al. 1991) have been employed to

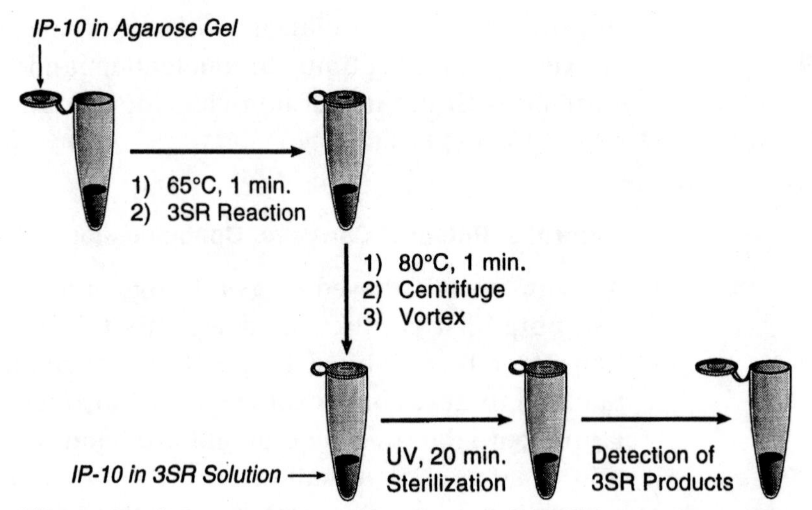

Figure 2 Scheme for sterilization of the 3SR reaction using a gel-based delivery of IP-10.

detect 3SR products by heterogeneous hybridization. For detection with ^{32}P-labeled probes, the macroporous TRISACRYL resin has been the solid support of choice. The combination of a ^{32}P-labeled probe, covalent attachment of the capture oligonucleotides through their 5′ ends, and chemical modification of the surface of the support with anionic groups to reduce nonspecific binding has resulted in detection sensitivities of 0.1–0.5 femtomole (fmole) or 5×10^7 molecules of 3SR-generated RNA products. Typically, 0.5–50 fmoles of analyte is detected when 100 fmoles of ^{32}P-labeled probe is used. This sensitivity is a result of a 40–60% efficiency of the BBSH system. When used in conjunction with 3SR amplification, the BBSH system is easily capable

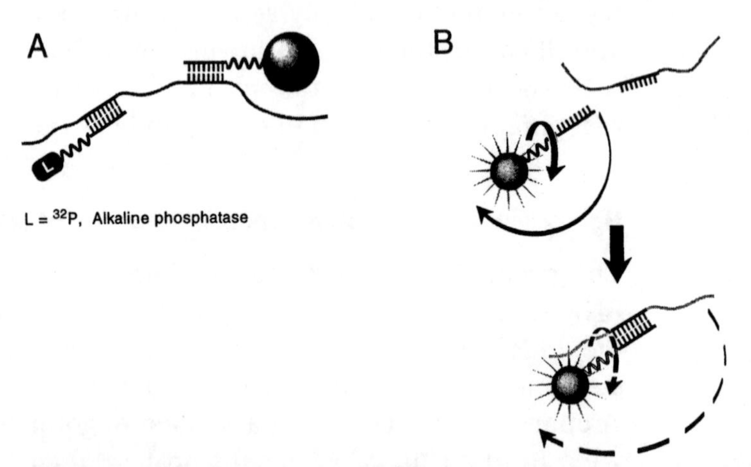

Figure 3 (*A*) Scheme for BBSH using radioisotopic or enzyme labels. (*B*) Scheme for detection of 3SR product using TSPF. Following duplex formation, the rotational freedom of the fluorophor label is restricted, resulting in an increase in fluorescence polarization.

of detecting a single HIV-1-infected cell in a background of 10^6 uninfected cells (Guatelli et al. 1990; Gingeras and Kwoh 1992).

Although heterogeneous assays afford useful sensitivity, they require separation steps to remove unhybridized labeled probe from hybridized probe. Furthermore, hybridization efficiencies may be compromised to some extent by steric interference from the solid support. These disadvantages can be overcome by using a homogeneous format that can discriminate between hybridized and unhybridized probe. One such method is transient-state polarized fluorescence (TSPF), which is particularly well suited to detect single-stranded 3SR RNA products (Devlin et al. 1993). The method is based on the increase in polarized fluorescence resulting from longer rotational relaxation times when a fluorophor-labeled probe hybridizes to its complementary target sequence (Fig. 3B). Fluorescent probes were synthesized by attaching the phthalocyanine dye, La Jolla Blue, to the 5′ end of an oligonucleotide using a flexible alkyl linker. This fluorophor has the advantages of a high molar absorption coefficient and low nonspecific binding characteristics by virtue of its axial polyethylene glycol ligands. Furthermore, its emission maximum in the near-infrared region (705 nm) ensures minimal interference from background fluorescence of biological molecules. When combined with transient-state detection using single photon counting, the TSPF method is capable of detecting nucleic acids with a sensitivity equivalent to that of the ^{32}P-based heterogeneous assay described above. The performance of these detection systems was compared by assaying a 3SR product generated by amplification of a 382-nucleotide region of the *env* gene of HIV-1. The sensitivity of TSPF is demonstrated in Figure 4A, in which the same concentration range of the 3SR product used in the BBSH method (Fig. 4B) was assayed in a simple "mix-and-read" format (Devlin et al. 1993). The high fluorescence intensities (>100,000 photon counts/20 seconds for 0.2 fmoles of probe) obtained in these experiments ensures good precision for the detection measurements. One consideration of the TSPF assay is the relatively poor hybridization efficiency of the 3SR products and the fluorescent probes, requiring 3SR product concentrations considerably higher than that of the probe to drive the reaction to completion. The dependence of the rate of hybrid formation on the concentrations of the RNA target and probe is likely to be the governing factor in determining the ultimate sensitivity of this technique.

DISCUSSION

A 3SR-based assay comprises three component steps: sample preparation, target amplification, and hybridization-detection of the amplified nucleic acid products. This chapter has focused on improvements in the amplification and product-detection steps of the 3SR assay. These improvements in the 3SR reaction provide a means for users to

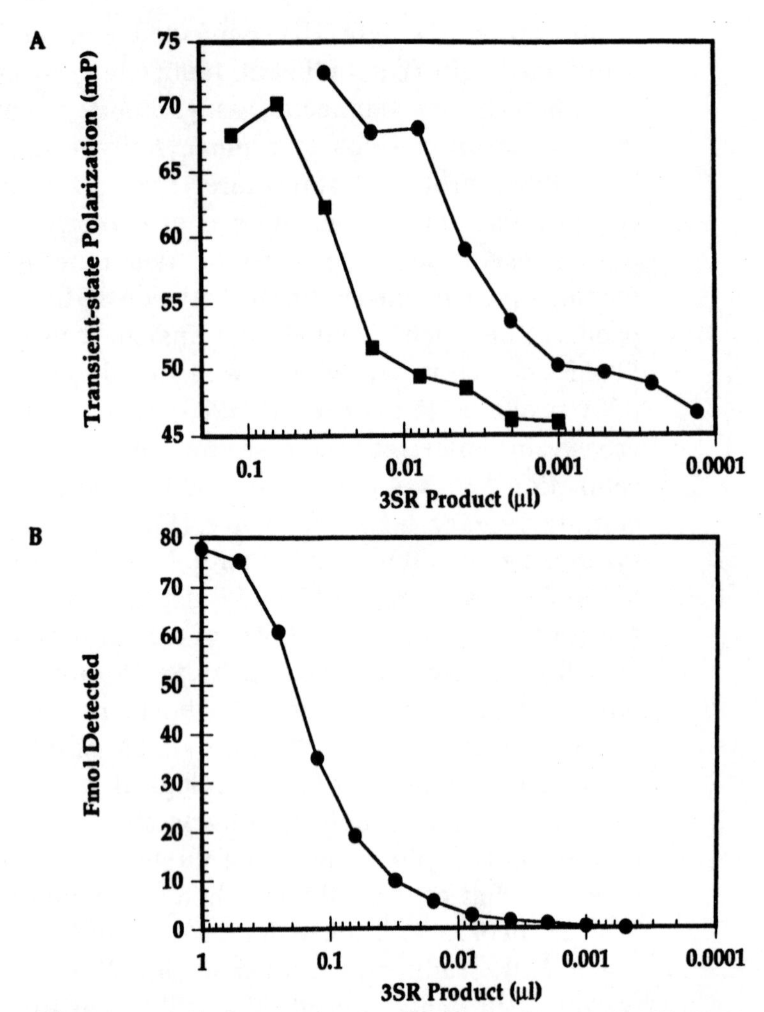

Figure 4 (*A*) Sensitivity for TSPF detection of 3SR RNA product with different La Jolla Blue oligonucleotide 90-422 conjugate probe concentrations. Ten fmoles (filled squares) and 0.2 fmole (filled circles) of La Jolla Blue oligonucleotide 90-422 conjugate probe was used in hybridization reactions with the 3SR-generated RNA target. The 3SR product was obtained by amplifying HIV-1 RNA with the *env*-specific primer pair 89-255/89-263* to give a 382-base antisense RNA product. A stock solution of this 3SR product (1000 fmoles/µl) was serially diluted to deliver varying amounts of target in the hybridization reactions. Ten microliters of these diluted stock solutions was assayed by TSPF as described previously (Devlin et al. 1993). (*B*) ^{32}P-based detection of 3SR product by BBSH. ^{32}P-labeled oligonucleotide 90-422 and TRISACRYL OligoBeads 86-273 were used to detect serial dilutions of the stock solution (1000 fmoles/µl) of 3SR-generated RNA product as described previously (Davis et al. 1990; Fahy et al. 1993).

achieve a measure of robustness not previously associated with this isothermal method. The capability of performing the detection steps of 3SR reaction products using a homogeneous format makes automation of the overall assay more approachable. Continued improvements and optimizations of the 3SR reaction can be expected in the area of substantially more thermostable variants of the enzymes now employed. The capability of performing the 3SR reaction at elevated temperatures is especially attractive, since both target specificity and structure issues are addressed simultaneously. Both of these issues are already addressed in the PCR-based assay.

ACKNOWLEDGMENTS We thank K. Blumeyer, J. Ishii, G. Davis, W. Walker, D. Kwoh, K. Berckhan, and J. Versailles for technical assistance and Diatron Corporation for the use of their TSPF system for the homogeneous detection of nucleic acids.

REFERENCES

Barany, F. 1991. Genetic disease detection and DNA amplification using cloned thermostable ligase. *Proc. Natl. Acad. Sci.* **88:** 189–193.

Berkower, I., J. Leis, and J. Hurwitz. 1973. Isolation and characterization of an endonuclease from *Escherichia coli* specific for ribonucleic acid in ribonucleic acid-deoxyribonucleic acid hybrid structures. *J. Biol. Chem.* **248:** 5914–5921.

Bush, C.E., K.M. VandenBrink, D.G. Sherman, W.R. Peterson, L.A. Beninsig, and J.H. Godsey. 1991. Detection of *Escherichia coli* rRNA using target amplification and time resolved fluorescence detection. *Mol. Cell. Probes* **5:** 467–472.

Bush, C.E., R.M. Donovan, W.R. Peterson, M.B. Jennings, V. Bolton, D.G. Sherman, and J.H. Godsey. 1992. Detection of human immunodeficiency virus type 1 RNA in plasma samples from high-risk pediatric patients by using the self-sustained sequence replication reaction. *J. Clin. Microbiol.* **30:** 281–286.

Chamberlin, M. and J. Ring. 1973. Characterization of T7-specific ribonucleic acid polymerase. *J. Biol. Chem.* **248:** 2245–2250.

Cimino, G.D., K.C. Metchette, J.W. Tessman, J.E. Hearst, and S.T. Isaacs. 1991. Post-PCR sterilization: A method to control carry-over contamination for the polymerase chain reaction. *Nucleic Acids Res.* **19:** 99–107.

Davis, G.R., K. Blumeyer, L.J. DiMichele, K.M. Whitfield, H. Chappelle, N. Riggs, S.S. Ghosh, P.M. Kao, E. Fahy, D.Y. Kwoh, J.C. Guatelli, S.A. Spector, D.D. Richman, and T.R. Gingeras. 1990. Detection of human immunodeficiency virus type 1 in AIDS patients using amplification-mediated hybridization analyses: Reproducibility and quantitative limitations. *J. Infect. Dis.* **162:** 13–20.

Devlin, R., R.M. Studholme, W.B. Dandliker, E. Fahy, K. Blumeyer, and S.S. Ghosh. 1993. Homogeneous detection of nucleic acids by transient state polarized fluorescence. *Clin. Chem.* **39:** 1939–1943.

Dieffenbach, C.W. and G.S. Dveksler. 1993. Setting up a PCR laboratory. *PCR Methods Appl.* (suppl.) **3:** 2–7.

Fahy, E., D.Y. Kwoh, and T.R. Gingeras. 1991. Self-sustained sequence replication (3SR): An isothermal transcription-based amplification system alternative to PCR. *PCR Methods Appl.* **1:** 25–33.

Fahy, E., G.R. Davis, L.J. DiMichele, and S.S. Ghosh. 1993. Design and synthesis of polyacrylamide-based oligonucleotide supports for use in nucleic acid diagnostics. *Nucleic Acids Res.* **21:** 1819–1826.

Ferre, F.A., P.C. Marchese, P.C. Duffy, D.E. Lewis, M.R. Wallace, H.J. Beecham, K.G. Burnett, F.C. Jensen, and D.J. Carlo. 1992. Quantitation of HIV viral burden by PCR in HIV seropositive Navy personnel representing Walter Reed staging 1 to 6. *AIDS Res. Hum. Retroviruses* **8:** 269–275.

Ghosh, S.S., P.M. Kao, A.W. McCue, and H.L. Chappelle. 1990. Use of maleimide-thiol coupling chemistry for efficient synthesis of oligonucleotide-enzyme conjugate hybridization probes. *Bioconjugate Chem.* **1:** 71–76.

Gingeras, T.R. and D.Y. Kwoh. 1992. *In vitro* nucleic acid target amplification: Issues and benefits. In *Biotechnologies Jarhbuch* 4 (ed. P. Präve et al.), pp. 403–429. Carl Hanser, Munich.

Gingeras, T.R., K.M. Whitfield, and D.Y. Kwoh. 1990a. Unique features of the self-sustained sequence replication (3SR) reaction in the *in vitro* amplification of nucleic acids. *Ann. Biol. Clin.* **48:** 498–501.

Gingeras, T.R., G.R. Davis, K.M. Whitfield, H.L. Chappelle, L.J. DiMichele, and D.Y. Kwoh. 1990b. Transcription-based amplification system and the detection of its RNA products by a bead-based sandwich hybridization system. In *PCR protocols: A guide to methods and applications* (ed. M.A. Innis et al.), pp. 245–252. Academic Press, San Diego, California.

Gingeras, T.R., P. Prodanovich, T. Latimer, J.C. Guatelli, D.D. Richman, and K.J. Barringer. 1991. Use of self-sustained sequence replication reaction to analyze and detect mutations in zidovudine-resistant human immunodeficiency virus. *J. Infect. Dis.* **164:** 1066–1074.

Guatelli, J.C., K.M. Whitfield, D.Y. Kwoh, K.J. Barringer, D.D. Richman, and T.R. Gingeras. 1990. Isothermal, *in vitro* amplification of nucleic acids by a multienzyme reaction modeled after retroviral replication. *Proc. Natl. Acad. Sci.* **87:** 1874–1878.

Holodnty, M.D., A. Katzenstein, S. Sengupta, A.M. Wang, C. Casipit, D.H. Schwartz, M. Konrad, E. Groves, and T.C. Merigan. 1991. Detection and quantification of human immunodeficiency virus RNA in patient serum by use of the polymerase chain reaction. *J. Infect. Dis.* **162:** 862–866.

Isaacs, S.T., J.W. Tessman, K.C. Metchette, J.E. Hearst, and G.D. Cimino. 1991. Post-PCR sterilization: Development and application to an HIV-1 diagnostic assay. *Nucleic Acids Res.* **19:** 109–116.

Ishii, J. and S.S. Ghosh. 1993. Bead-based sandwich hybridization characteristics of oligonucleotide-alkaline phophatase conjugates and their potential for quantitating target RNA sequences. *Bioconjugate Chem.* **4:** 34–41.

Kwoh, D.Y., G.R. Davis, K.M. Whitfield, H.L. Chappelle, L.J. DiMichele, and T.R. Gingeras. 1989. Transcription-based amplification system and detection of amplified human immunodeficiency virus type 1 with a bead-based sandwich hybridization format. *Proc. Natl. Acad. Sci.* **86:** 1173–1177.

Leirmo, S., C. Harrison, D.S. Cayley, R.R. Burgess, and T. Record. 1987. Replacement of potassium chloride by potassium glutamate dramatically enchances protein-DNA interactions *in vitro*. *Biochemistry* **26:** 2095–2101.

Longo, M.C., M.S. Berninger, and J.L. Hartley. 1990. Use of uracil DNA glycosylase to control carry over contamination in polymerase chain reactions. *Gene* **93:** 125–128.

McClelland, M., J. Hanish, M. Nelson, and Y. Patel. 1988. KGB: A single buffer for all restriction endonucleases. *Nucleic Acids Res.* **16:** 364.

Mulder, J., N. McKinney, C. Christopherson, J. Sninsky, L. Greenfield, and S. Kwok. 1994. Rapid and simple PCR assay for quantitation of human immunodeficiency virus type 1 RNA in plasma: Application to acute retroviral infection. *J. Clin. Microbiol.* **32:** 292–330.

Mullis, K.B. and F.A. Faloona. 1987. Specific synthesis of DNA *in vitro* via polymerase-catalyzed chain reaction. *Methods Enzymol.* **155:** 335–350.

Oka, S., K. Urayama, Y. Hirabayashi, K. Ohnishi, H. Goto, Y. Mitamura, S. Kimura, and K. Shimada. 1990. Quantitative analysis of human immunodeficiency virus type 1 DNA in asymptomatic carriers using the polymerase chain reaction. *Biochem. Biophys. Res. Commun.* **167:** 1–8.

Piatak, M., M.S. Saag, L.C. Yang, S.J. Clark, J.C. Kappes, K.-C. Luk, B.H. Hahn, G.M. Shaw, and J.D. Lits. 1993. High levels of HIV-1 in plasma during all stages of infection by competitive PCR. *Science* **259:** 1749–1754.

Richman, D.D., C.-K. Shih, I. Lowy, J. Rose, P. Prodanovich, S. Geoff, and J. Griffin. 1991. Human immunodeficiency virus type 1 mutants resistant to nonnucleoside inhibitors of reverse transcriptase arise in tissue culture. *Proc. Natl. Acad. Sci.* **88:** 11241–11245.

Saiki, R.K., S. Scharf, F.A. Faloona, K.B. Mullis, C.T. Horn, H.A. Erlich, and N. Arnheim. 1985. Enzymatic amplification of β-globin genomic sequences and restriction site analysis for diagnosis of sickle cell anemia. *Science* **230:** 1350–1354.

van Gemen, B., T. Kievits, R. Schukkink, D. van Strijp, L.T. Malek, R. Sooknanan, H.G. Huisman, and P. Lens. 1993. Quantitation of HIV-1 RNA in plasma using NASBA during primary infection. *J. Virol. Methods* **43:** 177–188.

Verma, I. 1977. The reverse transcriptase. *Biochem. Biophys. Acta* **473:** 1–38.

Versailles, J., K. Berckhan, S.S. Ghosh, and E. Fahy. 1993. Photochemical sterilization of 3SR reactions. *PCR Methods Appl.* **3:** 151–158.

Wu, D.Y. and R.B. Wallace. 1989. The ligation amplification reaction (LAR): Amplification of specific DNA sequences using sequential rounds of template-dependent ligation. *Genomics* **4:** 560–569.

One-tube Quantitative HIV-1 RNA NASBA

Bob van Gemen,[1] Paul van de Wiel,[3]
Rini van Beuningen,[1] Peter Sillekens,[1]
Suzanne Jurriaans,[2] Carine Dries,[3]
Ron Schoones,[3] and T. Kievits[1]

[1]Organon Teknika, Boxtel, The Netherlands
[2]AMC, Department of Virology, Amsterdam, The Netherlands
[3]Organon Teknika, Turnhout, Belgium

INTRODUCTION

Over the past few years, the quantitation of nucleic acids with target sequence amplification methods has been accomplished in many ways. Theoretically, the co-amplification of internal standard nucleic acid sequences, either DNA or RNA, with wild-type nucleic acid is superior to other quantitation methods, provided that the amplification efficiencies of the wild-type nucleic acid and the internal standard nucleic acid are equal (Nedelman et al. 1992a,b). This method has been applied successfully for PCR (Jurriaans et al. 1992), RT-PCR (Scadden et al. 1992; Piatak et al. 1993), and nucleic acid sequence-based amplification (NASBA) (van Gemen et al. 1993b).

NASBA is an isothermal nucleic acid amplification method that, like 3SR (Gingeras et al. 1990; Guatelli et al. 1990), evolved from TAS (Kwoh et al. 1989), the first RNA transcription-based amplification method described. Nucleic acid amplification in NASBA is accomplished by the concerted enzymatic activities of AMV reverse transcriptase, RNase H, and T7 RNA polymerase, resulting in the accumulation of mainly single-stranded RNA that can readily be used for detection by hybridization methods. The application of an internal RNA standard to NASBA resulted in a quantitative nucleic acid detection method with a dynamic range of 4 logs, but which needed 6 amplification reactions per quantitation (van Gemen et al. 1993b). Recently, this method was improved dramatically by the application of multiple, distinguishable, internal RNA standards added in different amounts and by the use of electrochemiluminesence (ECL) detection technology (van Gemen et al. 1994). This one-tube method, called Q-NASBA, uses only one amplification per quantitation and enables the addition of the internal standards to the clinical sample in a lysis buffer prior to the actual isolation of the nucleic acid (van Gemen et al. 1994). This approach has the advantage that nucleic acid isolation efficiency has no influence on the outcome of the quantitation, which is in contrast to methods in which the internal standards are mixed with the wild-type nucleic acid after its isolation from the clinical sample.

The specific amplification of single-stranded RNA in the presence of double-stranded DNA due to the isothermal nature of NASBA has advanced clinical applications in which the detection of RNA is essential, such as HIV-1 RNA in viral particles (van Gemen et al. 1993b, 1994; Jurriaans et al. 1994), RNA in *Mycobacteria* (van der Vliet et al. 1993, 1994), chronic myelogenous leukemia mRNA in blood lymphocytes (Sooknanan et al. 1994), and hepatitis C virus RNA in viral particles (Sillekens et al. 1994). The first application of the one-tube Q-NASBA was developed for the quantitation of HIV-1 RNA. However, for clinical diagnostic applications of the one-tube HIV-1 Q-NASBA, one should have a good assessment of the precision and accuracy of the method in the dynamic range that is covered by the internal standards. Only then can firm conclusions be reached concerning changes in HIV-1 viral RNA load, which are not attributable to variation in the quantitative assay itself, be made.

In this chapter we describe the precision and accuracy of the one-tube HIV-1 Q-NASBA. In this instance, precision is defined as the variation (standard deviation) between quantitative results obtained on the same sample, and accuracy is defined as the difference between the quantitative result and the actual input.

This precision and accuracy can be achieved by the use of a "gold standard," for which an in vitro-cultured HIV-1 viral stock solution was obtained. The amount of viral particles in this stock solution was determined using electron microscopy (Layne et al. 1992; van Gemen et al. 1993a; Jurriaans et al. 1994). Another important application of the gold standard lies in the comparison of quantitative HIV-1 RNA results obtained using different methods at different laboratories. As more and more quantitative HIV-1 RNA detection methods emerge, the widespread use of a gold standard can help to resolve the discrepancies found in viral RNA load at different laboratories (van Gemen et al. 1993a).

REAGENTS

All one-tube HIV-1 Q-NASBA assays were performed using the NASBA HIV-1 RNA QT kit from Organon Teknika, which contains all necessary reagents for the nucleic acid isolation, amplification, and detection described below. A flowchart describing the procedure is shown in Figure 1.

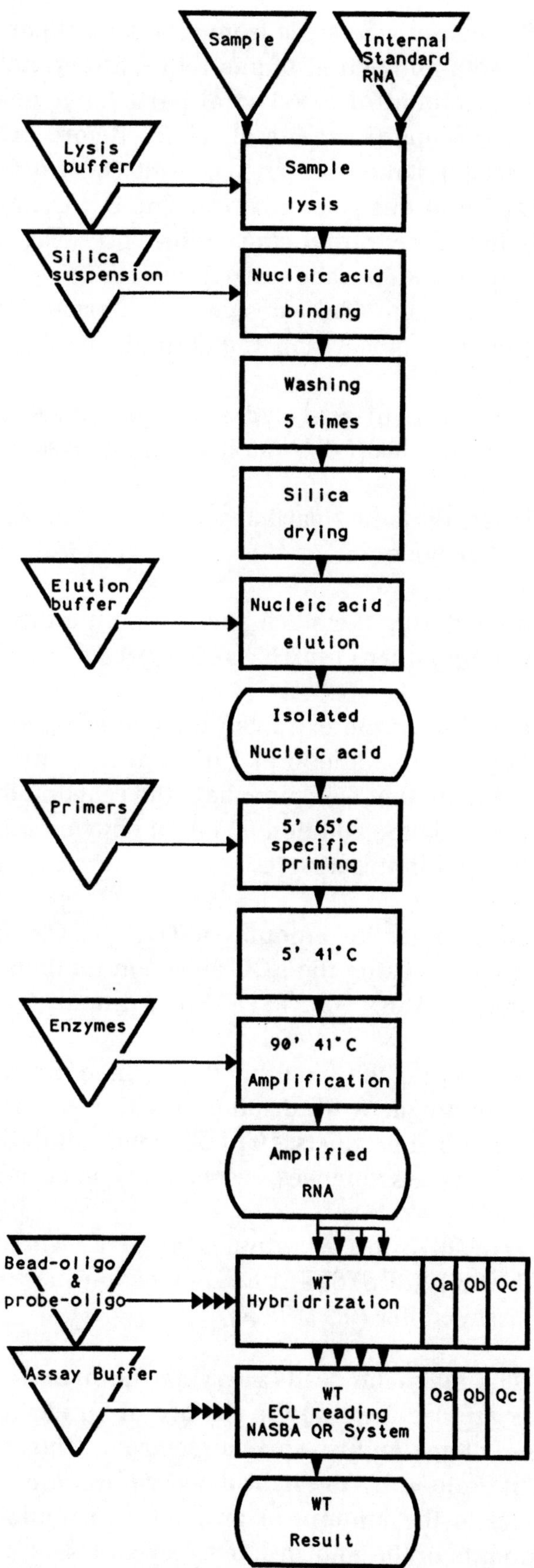

Figure 1 Experimental procedure of the HIV-1 RNA NASBA QT kit.

PROTOCOL

1. Extract nucleic acids from HIV-1 viral particles added to 100 μl of plasma and 900 μl of guanidine thiocyanate (GuSCN) lysis buffer. The amount of HIV-1 viral particles is determined using electron microscopy (Layne et al. 1992). Before extraction, add three RNA internal standards (Q_A, Q_B, and Q_C) from the NASBA HIV-1 RNA QT kit to this lysis mixture. The Q_A, Q_B, and Q_C internal-standard RNAs differ from the wild-type RNA and each other by a 20-nucleotide randomized sequence with the same nucleotide composition. This design of internal standards ensures equal efficiency of isolation and amplification.

2. Add 50 μl of activated silica particle solution to bind all of the nucleic acids (DNA and RNA) in the lysate (Boom et al. 1990).

3. Wash the silica particles with GuSCN wash buffer, 70% ethanol, and acetone.

4. After drying the silica particles, elute the nucleic acids in 50 μl of elution buffer (1 mM Tris-HCl, pH 8.5).

5. Use 5 μl of the extracted nucleic acids as the input for amplification by the addition of 5 μl of primer mix, followed by a 5-minute incubation at 65°C. Incubate the reaction for 5 minutes at 41°C, followed by the addition of 5 μl of enzyme mix and further incubation for 90 minutes at 41°C.

6. Determine the amount of Q_A, Q_B, Q_C, and wild-type amplified products using the ECL detection method (van Gemen et al. 1994) in the NASBA QR SYSTEM instrument.

7. Add 5 μl of 20-fold diluted amplified products to four tubes. Add 10 μl of magnetic bead solution with a generic probe for binding of all amplified products, 10 μl of probe solution, specific Q_A, Q_B, Q_C, and wild-type sequences, respectively, to each tube.

8. Hybridize for 30 minutes at 41°C. Then place the tubes in the NASBA QR SYSTEM ECL detection instrument, and add 300 μl of assay buffer (0.1 M TPA).

The magnetic beads carrying the hybridized products/probe complex are captured on the surface of an electrode by means of a magnet. Voltage applied to this electrode triggers the ECL reaction. The light emitted by the hybridized ruthenium-labeled probes is proportional to the amount of product. Calculations based on the relative amounts of the four products reveal the original amount of wild-type HIV-1 RNA in the sample.

In our laboratory, we have physically separated our NASBA activities (nucleic acid isolation, amplification setup, and detection of amplified products) to prevent carryover contamination. Physical separation was accomplished by the use of different laboratories or the use of fume hoods and flow cabinets. Every designated area is supplied with its own set of laboratory equipment (pipettes); therefore, movement between the areas is minimized.

During this study we analyzed 35 negative controls that were placed among positive samples over a period of weeks. Only once did we observe a false-positive result, which, upon retesting of the amplified nucleic acid, became negative, indicating a carryover contamination event during the detection setup. During this study we did not observe any cross-contamination during nucleic acid isolation or amplification setup.

DISCUSSION

Dynamic Range

The dynamic range of the one-tube Q-NASBA is defined by the concentrations of the internal standards (i.e., Q_A, Q_B, and Q_C RNA). Previously (van Gemen et al. 1994), we have shown that accurate quantitation is possible between a factor 10 below the lowest internal standard (Q_C) and a factor 10 above the highest internal standard (Q_A). The amounts used in this study are 10^6, 10^5, and 10^4 RNA molecules of Q_A, Q_B, and Q_C, respectively. This enables quantitation of the wild-type HIV-1 RNA between 10^3 and 10^7 molecules. The dynamic range of the one-tube Q-NASBA was determined using a dilution series of HIV-1 virus in plasma (100-µl aliquots) containing 1.86, 2.86, 3.86, 4.86, 5.86, and 6.56 log wild-type RNA molecules, respectively. As shown in Figure 2, the quantitation of the lowest wild-type HIV-1 RNA input level (i.e., 1.86 log) revealed an inaccurate result that was not used for linear regression of the data points. The lowest wild-type RNA input level that was quantitated accurately in this study was 2.86 log (725 RNA molecules). In other studies, wild-type RNA input levels of 2.6 log (400 RNA molecules) could be quantitated accurately (data not shown).

Precision and Accuracy

For the clinical diagnostic application of quantitative HIV-1 RNA assays, it is important to know the precision and accuracy of the assay. Only then can one discriminate between fluctuations in HIV-1 viral load results that are attributable to variation in the assay or to differences in HIV-1 viral load as a consequence of factors such as disease stage or antiviral therapy. In this matter, precision is defined as fluctuations in quantitative results obtained on the same sample of

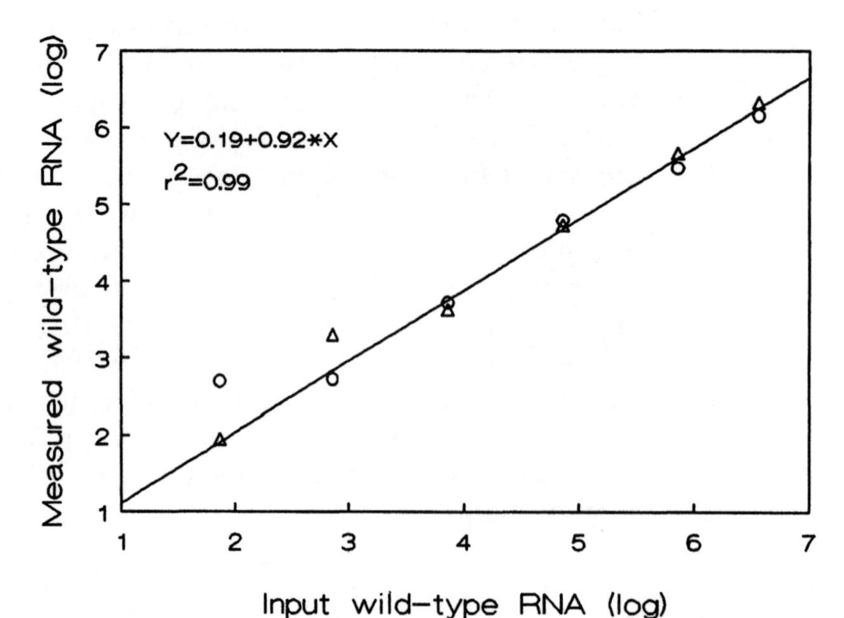

Figure 2 Dynamic range of the one-tube Q-NASBA using 10^6, 10^5, and 10^4 RNA molecules of Q_A, Q_B, and Q_C, respectively. The internal standards were mixed with 1.86, 2.86, 3.86, 4.86, 5.86, and 6.56 log wild-type HIV-1 RNA molecules (i.e., dilution series of HIV-1 virus stock solution) in 100 µl of plasma and 900 µl of lysis buffer. All quantitations were performed in duplicate. For curve fitting (i.e., linear regression), the wild-type RNA quantitation of 1.86 log was not taken into account. Circles and triangles represent duplicate measurements.

HIV-1 RNA input (i.e., the standard deviation). Accuracy is defined as the difference between the outcome of the quantitation and the actual input of HIV-1 RNA. To determine both, we used three concentrations of the HIV-1 virus stock solution (i.e., the gold standard) as input in 100 µl of plasma and 900 µl of lysis buffer and quantitated each concentration 20 times. The results are summarized in Table 1.

The precision was found to increase with increasing wild-type HIV-1 RNA input concentration. A precision of 0.23 log was observed when quantitating 3.34 log wild-type HIV-1 RNA input, whereas the

Table 1 Precision and Accuracy of the One-tube Q-NASBA

	Input wild-type HIV-1 RNA molecules (log)		
	3.34	4.86	6.34
Number	19[a]	20	20
Mean	3.23	4.84	6.15
Precision (= std)	0.23	0.22	0.13
Accuracy	0.11	0.02	0.19

[a]One data point was found to be an outlier by the Grubbs test.

quantitation of 6.34 log wild-type HIV-1 RNA input was accomplished with a precision of 0.13 log. These results are in good agreement with the results we obtained previously (van Gemen et al. 1994). The accuracy of the one-tube Q-NASBA ranges from 0.02 to 0.19 log, indicating the correct calibration of the internal standards (Q_A, Q_B, and Q_C RNA) used in the assay.

Using a standard deviation of 0.23 (Table 1) for the one-tube Q-NASBA, differences found in HIV-1 viral load that are ≥1 log are certainly caused by external factors, like disease stage or antiviral therapy, rather than fluctuations in the one-tube Q-NASBA (p = 0.0035). Differences in the HIV-1 viral load of 0.5 log found as a result of single measurements can be ascribed to external factors with p = 0.13. However, when differences of 0.5 log are found as a result of duplicate measurements, these can be ascribed to external factors with more certainty ($p = 0.03$).

Influencing Factors

The one-tube Q-NASBA assay results might be influenced by external factors related to the clinical sample used for analysis. For this reason, the gold standard (input 4.85 log wild-type HIV-1 RNA) was quantitated with the addition of 100 μl of citrate plasma, heparin plasma, EDTA plasma, or serum. The results are depicted in Figure 3. There appears to be no difference in quantitation of the gold standard in these four sample types. However, the influence of the coagulation

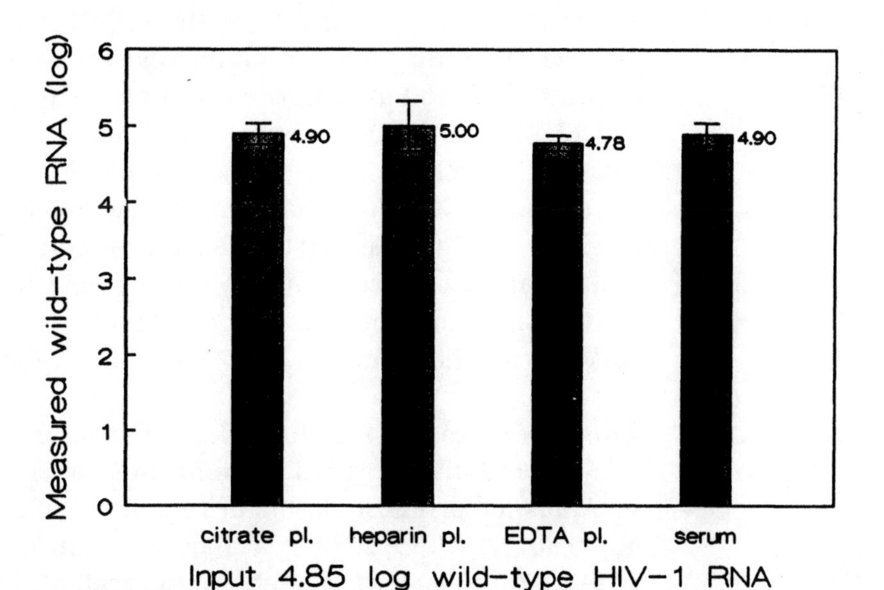

Figure 3 Wild-type HIV-1 RNA (4.85 log) was quantitated five times in citrate plasma, heparin plasma, EDTA plasma, and serum, respectively. The mean result of the quantitations is given above each bar.

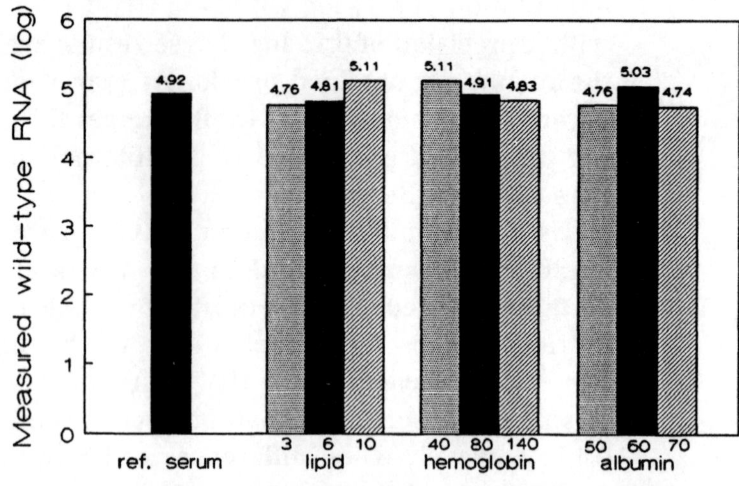

Figure 4 Wild-type HIV-1 RNA (4.85 log) was quantitated in normal human serum (reference serum) with the addition of lipid (3%, 6%, and 10%), hemoglobin (40, 80, and 140 μmoles) or albumin (50, 60, and 70 g/liter). The results of the quantitations are stated above each bar.

process on the quantitation of HIV-1 viral RNA was not investigated in this experiment. Possibly, HIV-1 viral particles are trapped during the coagulation for obtaining serum, resulting in lower viral load results compared with quantitations performed on plasma samples. To address this problem, comparisons should be made between the levels of HIV-1 RNA in plasma and serum samples of HIV-1-infected individuals.

Three components of plasma and serum that may influence the quantitative outcome of the one-tube Q-NASBA are lipids, hemoglobin, and albumin. These components were added in increasing concentrations to 100 μl of normal human serum (reference serum) and 900 μl of lysis buffer in which 4.85 log wild-type HIV-1 RNA was spiked. Neither lipid (3–10%), hemoglobin (40–140 μmoles), nor albumin (50–70 g/liter) influenced the outcome of the quantitation (Fig. 4). During the nucleic acid isolation procedure, these compounds are separated from the nucleic acid in such a way that concentrations remaining in the eluate are too low to influence the NASBA amplification.

APPLICATIONS

The advantages NASBA has over other amplification methods (PCR, LCR, strand displacement amplification, etc.) for the amplification of RNA has led primarily to the development of NASBA assays for those applications where RNA is a diagnostic target. The determination of HIV-1 viral RNA load is such an application. Both as a research tool and as a diagnostic assay, one can appreciate the value of the one-tube Q-NASBA. The simplicity of the assay, combined with its technical advantages when compared with other quantitative assays, allows

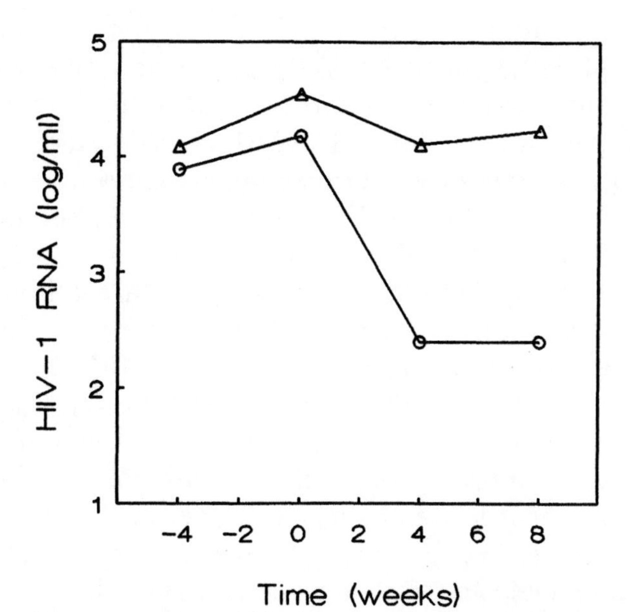

Figure 5 HIV-1 RNA load in plasma of two asymptomatic HIV-1-infected individuals treated with AZT. The AZT therapy was started at week 0. (*Open circle*) Patient 8301; (*open triangle*) patient 8306.

quantitative results to be obtained on a large scale. For HIV-1 it is obvious that viral RNA will be an important marker in the research for vaccines and antiviral drugs. In a recent study, the one-tube Q-NASBA was used to assess the efficacy of AZT therapy in asymptomatic HIV-1-infected individuals (S. Jurriaans, pers. comm.). The results of two representative individuals, a responder, patient 8301, and a non-responder, patient 8306, are depicted in Figure 5. Differences in HIV-1 viral load ≥1 log cannot be explained as effects of the AZT therapy. These results illustrate the useful application of this assay for the rapid assessment of antiviral therapy. In this way, the one-tube Q-NASBA can serve as an important tool in the search for new antiviral drugs.

CONCLUSIONS

A quantitative nucleic acid detection assay generally comprises three components: nucleic acid extraction, amplification, and detection. The quality of the quantitative result is highly dependent on the efficiency and reproducibility of each of the components. In the one-tube Q-NASBA assay, the addition of multiple, distinguishable internal-standard RNAs to the clinical sample prior to nucleic acid isolation abolishes the effect of nucleic acid isolation efficiency on the quantitation outcome. The ratio of wild-type RNA to internal RNA standards remains constant independent of the overall extraction efficiency (van Gemen et al. 1994). This ratio determines the quantitative outcome of the assay. Furthermore, because the amplification efficiencies of wild-type and internal-standard RNAs are equal, the ratio of these RNAs before amplification is reflected in the amplification products. This implies that detection of the amplified products needs to be absolutely quantitative to determine accurately the ratio of wild-type RNA to internal RNA standards. The ECL detection technology has this property and can measure hybridization signals with a dynamic range of 5 orders of magnitude (Blackburn et al. 1991; Kenten et al. 1992).

The one-tube Q-NASBA consisting of Boom nucleic acid isolation technology (Kenten et al. 1991), NASBA amplification (Kievits et al. 1991), and ECL detection (Blackburn et al. 1991; Kenten et al. 1992) has a dynamic range of 4×10^2 to 4×10^7 RNA molecules when using 10^6, 10^5, and 10^4 RNA molecules of the Q_A, Q_B, and Q_C internal standards, respectively (Fig. 1). Within this dynamic range, the precision and accuracy of the assay enables measurement of the changes in HIV-1 RNA load of ≥1 log, and, very probably, changes of ≥0.5 log can be subscribed to factors other than variation in the assay. The use of the nucleic acid isolation method (Boom et al. 1990) makes the assay suitable for almost any kind of clinical sample (plasma, serum, whole blood, cells). Because of the inclusion of efficient washing steps of nucleic acid bound to silica, possible NASBA-inhibiting factors are removed.

As shown in this chapter, the concern about contamination that exists for all amplification methods can be dealt with simply by physical separation of the different NASBA activities (see Protocol). Together with the simplicity of the one-tube Q-NASBA assay, this enables the widespread use of this assay for large-scale quantitation of HIV-1 RNA in both research and clinical settings. With the introduction of a gold standard, enabling comparisons of different quantitation methods, the way will be opened for the application of these assays as a widespread tool in vaccine development and antiviral drug research.

REFERENCES

Blackburn, G.F., H.P. Shah, J.H. Kenten, J. Leland, R.A. Kamin, J. Link, J. Peterman, M.J. Powell, A. Shah, and D.B. Talley. 1991. Electrochemiluminescence detection for development of immunoassays and DNA probe assays for clinical diagnostics. *Clin. Chem.* **37:** 1534–1539.

Boom, R., C.J.A. Sol, M.M.M. Salimans, C.L. Jansen, P.M.E. Wertheim-van Dillen, and J. van der Noordaa. 1990. A rapid and simple method for purification of nucleic acids. *J. Clin. Microbiol.* **28:** 495–503.

Gingeras, T.R., K.M. Whitfield, and D.Y. Kwoh. 1990. Unique features of the self-sustained sequence replication (3SR) reaction in the in vitro amplification of nucleic acids. *Ann. Biol. Clin.* **48:** 498–501.

Guatelli, J.C., K.M. Whitfield, D.Y. Kwoh, K.J. Barringer, D.D. Richman, and T.R. Gingeras. 1990. Isothermal, in vitro amplification of nucleic acids by a multienzyme reaction modeled after retroviral replication. *Proc. Natl. Acad. Sci.* **87:** 7797.

Jurriaans, S., J.T. Dekker, and A. de Ronde. 1992. HIV-1 viral DNA load in peripheral blood mononuclear cells from seroconverters and long-term infected individuals. *AIDS* **6:** 635–641.

Jurriaans, S., B. van Gemen, G.J. Weverling, D. Van Strijp, P.L. Nara, R. Coutinho, M. Koot, H. Schuitemaker, and J. Goudsmit. 1994. The natural history of HIV-1 infection: Virus load and virus phenotype independent determinants of clinical course? *Virology* **204:** 223–233.

Kenten, J.H., S. Gudibande, J. Link, J.J. Willey, B. Curfman, E.O. Major, and R.J. Massey. 1992. Improved electrochemiluminescent label for DNA probe assays: Rapid quantitative assays of HIV-1 polymerase chain reaction products. *Clin. Chem.* **38:** 873–879.

Kievits, T., B. van Gemen, D. van Strijp, R. Schukkink, M. Dircks, H. Adriaanse, L. Malek, R. Sooknanan, and P. Lens. 1991. NASBA TM isothermal enzymatic in vitro nucleic acid amplification optimized for the diagnosis of HIV-1 infection. *J. Virol. Methods* **35:** 273–286.

Kwoh, D.Y., G.R. Davis, K.M. Whitfield, H.L. Chappelle, L.J. DiMichelle, and T.R. Gingeras. 1989. Transcription-based amplification system and detection of amplified human immunodeficiency virus type 1 with a bead-based sandwich hybridization format. *Proc. Natl. Acad. Sci.* **86:** 1173–1177.

Layne, S.P., M.J. Merges, M. Dembo, J.L. Spouge, S.R. Conley, J.P. Moore, J.L. Raina, H. Renz, H.R. Gelderblom, and P.L. Nara. 1992. Factors underlying spontaneous inactivation and susceptibility to neutralization of human immunodeficiency virus. *Virology* **189:** 695–714.

Nedelman, J., P. Heagerty, and C. Lawrence. 1992a. Quantitative PCR: Procedures and precisions. *Bull. Math. Biol.* **54:** 477–502.

——. 1992b. Quantitative PCR with internal controls. *Comput. Appl. Biosci.* **8:** 65–70.

Piatak, M.J., M.S. Saag, L.C. Yang, S.J. Clark, J.C. Kappes, K.C. Luk, B.H. Hahn, G.M. Shaw, and J.D. Lifson. 1993. High levels of HIV-1 in plasma during all stages of infection determined by competitive PCR. *Science* **259:** 1749–1754.

Scadden, D.T., Z.Y. Wang, and J.E. Groopman. 1992. Quantitation of plasma human immunodeficiency virus type-1 RNA by competitive polymerase chain reaction. *J. Infect. Dis.* **165:** 1119–1123.

Sillekens, P., W. Kok, B. van Gemen, P. Lens, H. Huisman, T. Cuypers, and T. Kievits. 1994. Specific detection of HCV RNA using NASBA as a diagnostic tool. In *Hepatitis C virus: New diagnostic tools* (ed. Groupe Francais d'Etudes Moleculaire des Hepatitis), pp. 71–82. John Libbey Eurotext, Paris, France.

Sooknanan, R., L. Malek, W.-I. Wang, T. Siebert, and A. Keating. 1994. Detection and direct sequence identification of BCR-ABL mRNA in Ph+ chronic myeloid leukemia. *Exp. Hematol.* **21:** 1719–1724.

van der Vliet, G.M.E., P. Schepers, R.A.F. Schukkink, B. van Gemen, and P.R. Klatser. 1994. Rapid and sensitive assessment of mycobacterial viability through RNA amplification. *Antimicrob. Agents Chemotherapy* **38:** 1959–1965.

van der Vliet, G.M.E., R.A.F. Schukkink, B. van Gemen, P. Schepers, and P.R. Klatser. 1993. Nucleic acid sequence based amplification (NASBA) for the identification of *Mycobacteria*. *J. Gen. Microbiol.* **139:** 2423–2429.

van Gemen, B., T. Kievits, P. Nara, H. Huisman, S. Jurriaans, J. Goudsmit, and P. Lens. 1993a. Qualitative and quantitative detection of HIV-1 RNA by nucleic acid sequence based amplification. *AIDS* **7:** S107–S110.

van Gemen, B., R. van Beuningen, A. Nabbe, D. van Strijp, S. Jurriaans, P. Lens, and T. Kievits. 1994. A one-tube quantitative HIV-1 RNA NASBA nucleic acid amplification assay using electrochemiluminescent (ecl) labelled probes. *J. Virol. Methods* **49:** 157–168.

van Gemen, B., T. Kievits, R. Schukkink, D. van Strijp, L.T. Malek, R. Sooknanan, H.G. Huisman, and P. Lens. 1993b. Quantification of HIV-1 RNA in plasma using NASBA TM during a HIV-1 primary infection. *J. Virol. Methods* **43:** 177–188.

Appendices

Computer Software for Selecting Primers

PRIMER v. 1.4 (DOS)

Part of Busch & Lucas utilities for DNA sequence analysis. Published in *CABIOS 7:* 525–529 (1991).

Worldwide Distribution
Busch & Lucas Wissenschaftliche Software
Krautgaerten 1A
D-79112 Freiburg, Germany
Phone: 49 7664 1334
Fax: 49 7664 5483
E-mail: busch@mm11.ukl.uni_freiburg.de

PINCERS (Macintosh)

Includes codon usage tables. Helps to select primers from protein sequences. Published in *BioTechniques 10:* 782–784 (1991).

Office of Technology Licensing
2150 Shattuck Avenue, Suite 510
Berkeley, CA 94720-1620
Phone: (510) 643-7201
Fax: (510) 642-4566
E-mail: domino@garnet.berkeley.edu

Oligonucleotide Selection Program (Macintosh, DOS, Digital VAX/VMS, Sun SPARC-based Workstations)

For research purposes only. Published in *PCR Methods and Applications 1:* 124–128 (1991).

Dr. Philip Green
Molecular and Biotechnology Department
University of Washington, FJ-20
Fluke Hall, Mason Road
Seattle, WA 98195
Phone: (206) 685-4341
Fax: (206) 685-7344
E-mail: phg@u.washington.edu

RightPrimer: Primer Design Utility (Macintosh, DOS available 1995)

BioDisk Software
P.O. Box 26447
San Francisco, CA 94126
Phone: (800) 664-3465 *or* (415) 951-9031
Fax: (415) 951-0102
E-mail: 72610.2445@compuserve.com

Gene Runner 3.0

Hastings Software, Inc.
P.O. Box 567
Hastings-on-Hudson, NY 10706
Phone: (800) 834-8574 *or* (914) 969-0855
Fax: 800 834-8799

Oligo 5.0 (DOS)
Oligo 4.0 (Macintosh, 5.0 to be released Fall 1995)

Primer analysis software. Based on *Nucleic Acids Research 17:* 8543–8551 (1989).

National Biosciences, Inc.
3560 Annapolis Lane North, #140
Plymouth, MN 55447-5434
Phone: (800) 747-4362 *or* (612) 550-2012
Fax: (800) 369-5118
E-mail: nbi@biotechnet.com

Europe
MedProbe AS
Postboks 2640
St. Hanshaugen
N-0131 Oslo, Norway
Phone: 47 22 20 0137
Fax: 47 22 20 0189

Japan
Takara Shuzo Co., Ltd.
Bio Products Development Center
2257 Sunaike, Noji-Cho
Kusatsu, Shiga-Ken, Japan 525
Phone: 81 0775 43 7231
Fax: 81 0775 43 9254

DNASIS 2.0 (Windows)
MacDNASIS (Macintosh)

This is a complete DNA sequence analysis package; primer selection is one program within the software.

Americas and Asia
Hitachi Software
1111 Bayhill Drive, Suite #395
San Bruno, CA 94066
Phone: (800) 624-6176 *or* (415) 615-6176
Fax: (415) 615-7699
E-mail: dnasis@biotechnet.com

Europe
Hitachi Software Engineering Europe, S.A.
Parede Limere Zone Industrielle
45160 Ardon, France
Phone: 33 38 69 8693
Fax: 33 38 69 8699

GeneWorks (Macintosh)

This is a complete DNA sequence analysis package; primer selection is one program within the software.

IntelliGenetics, Inc.
700 East El Camino Real, Suite 300
Mountain View, CA 94040
Phone: (415) 962-7300
Fax: (415) 962-7302

Europe
Phone: 32 3 219 5332
Fax: 32 3 219 5334

Japan
Phone: 81 45 661 3414
Fax: 81 45 661 3382

Lasergene (DOS, Windows, Macintosh)

This is a complete DNA sequence analysis package; primer selection is one program within the software.

DNASTAR
1228 South Park Street
Madison, WI 53715
Phone: (608) 258-7420
Fax: (608) 258-7439
E-mail: support@dnastar.com

Europe
Abacus House
West Ealing
London W13 OAS, England
Phone: 44 81 566 8282
Fax: 44 81 566 9555

EuGene™ (DOS)

Daniben Systems, Inc.
1776 Mentor Avenue, Suite 340
Cincinnati, OH 45212
Phone: (513) 531-4219
Fax: (513) 351-0610
E-mail: info@daniben.cincinnati.oh.us

Same address, E-mail, and phone for distribution in Europe, Japan, and Australia

GeneJockey (Macintosh)

United States
Biosoft
P.O. Box 10938
Ferguson, MO 63135
Phone: (314) 524-8029
Fax: (314) 524-8129
E-mail: mpion@artsci.wustl.edu

Europe and Rest of World
Biosoft
49 Bateman Street
Cambridge, CB2 1LR, England
Phone: 44 1223 686 22
Fax: 44 1223 312 873
E-mail: ab47@cityscape.co.uk

Wisconsin Sequence Analysis Package (Digital VAX/VMS, IBM RS6000, Sun SPARC-based Workstations, Silicon Graphics Workstation)

Genetics Computer Group
University Research Park
575 Science Drive
Madison, WI 53711
Phone: (608) 231-5200
Fax: (608) 231-5202
E-mail: help@gcg.com

Japan
Mitsui Knowledge Industry Co., Ltd.
New Business Development Div.
#-7-4 Kojimachi, 3-chome
Chiyoda-Ku, Tokyo 102, Japan
Phone: 81 3 3237 6052
Fax: 81 3 3237 1853
E-mail: akr@mitusui_knowledge.co.tp

MacVector (Macintosh)

Includes AssemblyLIGN (a fragment assembly package)

Scientific Imaging Systems
P.O. Box 9558
New Haven, CT 06535
Phone: (800) 243-2555
Fax: (203) 786-5694
E-mail: macvector@ksis.com

Worldwide distribution available through a number of distributors.

PRIMER PREMIER (Windows, DOS, PowerMac)

PREMIER Biosoft International
3786 Corina Way
Palo Alto, CA 94303-4504
Phone: (415) 856-2703
Fax: (415) 856-7844
E-mail: 74637,3500@compuserve.com

International sales are handled directly by Premier Biosoft at the address above.

DesignerPCR (Windows)

Research Genetics, Inc.
2130 Memorial Parkway S.W.
Huntsville, AL 35801
Phone: (800) 533-4363
Fax: (205) 536-9016
E-mail: info@regen.com
WWW URL: http://www.resgen.com

Vector NTI (Windows, Macintosh)

InforMax, Inc.
444 North Fredrick Avenue, Suite 308
Gaithersburg, MD 20877
Phone: (301) 216-0586
Fax: (301) 216-0586
E-mail: alext@access.digex.net

Primer Designer (Windows, DOS)

Scientific and Educational Software
P.O. Box 440
State Line, PA 17263-0440
Phone: (717) 597-5307
Fax: (717) 597-5108
E-mail: sciedsoft@biotechnet.com

Modifications of Oligonucleotides for Use in PCR

The solid phase synthesis of oligodeoxyribonucleotides, specifically using the 5′-dimethoxytrityl, β-cyanoethyl, diisopropyl phosphoramidite chemistry (Beaucage and Caruthers 1981; Sinha et al. 1984), is presently the most common method utilized for producing high-quality primers for PCR. From this core chemistry, numerous modifications can be made either during the actual synthesis or after the oligonucleotide has been prepared for use. These tables show some of the most commonly used modifications and what is necessary to produce that particular modification. By no means should the tables be construed as the only PCR-relevant manipulations available in the rapidly changing field of nucleic acid solid-phase-synthesis chemistry.

Table 1 Additions to the 5′ End

Additions	Purpose	Potential use(s) in PCR	Necessary for incorporation
^{32}P-^{33}P	radioactively labeled oligo	detection	γ-^{32}P or ^{33}P-labeled ATP with T4 polynucleotide kinase (Richardson 1965)
Biotin	detection with avidin, multiple enzyme conjugates possible	detection, single-strand purification after PCR	amine linker on 5′ end of oligo and NHS-biotin (Chollet and Kawashima 1985) or can be directly added during synthesis
Digoxigenin	detection with anti-Dig Fab, multiple enzyme conjugates possible	detection	amine linker on 5′ end of oligo (Connolly 1987) and NHS-digoxigenin
Fluorescein Texas Red rhodamine	fluorescently labeled oligo, detection by fluorometer or UV light	in situ PCR, detection	amine linker on 5′ end of oligo and NHS-conjugated dye or, in the case of fluorescein, can be added during synthesis
6-FAM HEX TET	fluorescently labeled oligo, detection by fluorometer or charged-coupled device (such as the one in the ABI DNA sequencers)	in situ PCR, detection, quantitative PCR	all three dyes can be directly added during synthesis or added to a 5′ amine linker in their NHS form (Smith et al. 1985)
Ruthenium (TBR)	detection by a specific electrochemiluminescence reaction	quantitative PCR	can be directly added during synthesis (Telser et al. 1989)

5′-End additions, in general, do not block the subsequent use of the oligonucleotide in a PCR. For use in PCR, the 3′ hydroxyl must remain unmodified. (6-FAM) 6-Carboxy-fluorescein; (HEX) hexachloro-6-carboxy-fluorescein; (TET) tetrachloro-6-carboxy-fluorescein; (Ruthenium [TBR]) tris (2,2′-bipyridine) ruthenium(II) chelate; (Dig Fab) fragment of an immunoglobulin specific for digoxigenin; (ABI) Applied Biosystems; (NHS) *N*-hydroxysuccinimide ester.

Table 2 Additions to the 3′ End

Additions	Purpose	Potential use(s) in PCR	Necessary for incorporation
^{32}P-^{33}P	radioactively labeled oligo	detection	TdT, α ^{32}P-^{33}P (usually dC or dA) (Collins and Hunsaker 1985)
^{35}S	radioactively labeled oligo	detection	TdT, dATP-^{35}S
Biotin	detection with avidin, multiple enzyme conjugates possible	detection, single-strand purification after PCR	usually added during synthesis (CPG derivatized) or TdT tailed with dUTP-biotin (Kumar et al. 1988)
Digoxigenin	detection with anti-Dig Fab, multiple enzyme conjugates possible	detection	TdT tailed with dUTP-digoxigenin (BMB)
Fluorescein tetraethyl-rhodamine	fluorescently labeled oligo, detection by a fluorometer or UV light	in situ PCR, detection	uually added during synthesis (CPG derivatized) or TdT tailed (Trainor and Jensen 1988)
TAMRA	fluorescently labeled oligo, additionally used as a quencher in the TaqMan System, detection by a fluorometer or similar luminescence	quantitative PCR (when used with 5′-end-labeled HEX, TET, or 6-FAM), multiplex PCR, detection	NHS-TAMRA, LAN (linker arm nucleotide)
Dideoxy-nucleotide triphosphate (ddNTP)	multiple oligo species, detection by numerous methods (fluorescence, autoradiography)	detection, DNA sequencing	usually added during synthesis
Phosphate	oligo does not serve as primer in PCR	detection (i.e., TaqMan System), inhibition of extension	usually added during synthesis (CPG derivatized) but can be added enzymatically

3′-End additions, in general, prevent further use of the oligonucleotide in PCR. The 3′ hydroxyl is either blocked or missing. (TAMRA) 6-Carboxy-tetramethyl-rhodamine; (Dig Fab) fragment of an immunoglobulin specific for digoxigenin; (HEX) hexachloro-6-carboxy-fluorescein; (TET) tetrachloro-6-carboxy-fluorescein; (6-FAM) 6-carboxy-fluorescein; (CPG) controlled pore glass; (BMB) Boehringer Mannheim.

Table 3 Internal Additions

Additions	Purpose	Potential use(s) in PCR	Necessary for incorporation
Biotin	multi-labeled oligo, detection by avidin, multiple enzyme conjugates possible	detection	biotin phosphoramidite (Misiura et al. 1990) or modified base (Haralambidis et al. 1987)
Digoxigenin	multi-labeled oligo, detection by anti-Dig Fab, multiple enzyme conjugates possible	detection	internal amine phosphoramidite (like amine-VCN from CLONTECH) plus NHS-digoxigenin (BMB)
Fluorescein Texas Red rhodamine	multi-labeled oligo, detection by fluorometer or UV light	in situ PCR, detection	fluorescein phosphoramidite or an internal amine linker with an NHS form of the dye
Degenerative (wobble), inosine	multiple oligo species, detection by DNA sequencing	detection, PCR for related sequences	can be easily added during synthesis
Thioate bond	a more nuclease-resistant oligo than the standard phosphodiester oligo linkage	probe preparation via PCR or probing in general, especially where nucleases are present (i.e., in situ)	TETD (ABI) or similar sulfurizing reagent (Hollway et al. 1993); easily performed on a DNA synthesizer

Internal labeling may, or may not, prevent further use in a PCR. This is dependent on many factors: the number inserted, type of insertion, position of the insertions, and how many base pairs are interrupted. (Dig Fab) Fragment of an immunoglobulin specific for digoxigenin; (NHS) *N*-hydroxy-succinimide ester; (BMB) Boehringer Mannheim; (ABI) Applied Biosystems.

REFERENCES

Beaucage, S.L. and M.H. Caruthers. 1981. Deoxynucleoside phosphoramidites–A new class of key intermediates for deoxypolynucleotide synthesis. *Tetrahedron Lett.* **22:** 1859–1862.

Chollet, A. and E.H. Kawashima. 1985. Biotin-labeled synthetic oligodeoxyribonucleotides: Chemical synthesis and uses as hybridization probes. *Nucleic Acids Res.* **13:** 1529–1541.

Collins, M.L. and W.R. Hunsaker. 1985. Improved hybridization assay employing tailed oligonucleotide probes: A direct comparison with 5′-end-labeled oligonucleotide probes and nick-translated plasmid probes. *Anal. Biochem.* **151:** 211-224.

Connolly, B.A. 1987. The synthesis of oligonucleotides containing a primary amino group at the 5′-terminus. *Nucleic Acids Res.* **15:** 3131–3139.

Haralambidis, J., M. Chai, and A. Chollet. 1987. Preparation of base-modified nucleosides suitable for nonradioactive label attachment and their incorporation into synthetic oligodeoxyribonucleotides. *Nucleic Acids Res.* **15:** 4857–4876.

Hollway, B., D.D. Erdman, E.L. Durigon, and J.J. Murtagh, Jr. 1993. An exonuclease-amplification coupled capture technique improves detection of PCR product. *Nucleic Acids Res.* **21:** 3905–3906.

Kumar, A., P. Tchen, F. Roullet, and J. Cohen. 1988. Non-radioactive labelling of synthetic oligonucleotide probes with terminal deoxynucleotidyl transferase. *Anal. Biochem.* **169:** 376–382.

Misiura, K., I. Durrant, M.R. Evans, and M.J. Gait. 1990. Biotinyl and phosphotyrosinyl phosphoramidite derivatives useful in the incorporation of multiple reporter groups on synthetic oligonucleotides. *Nucleic Acids Res.* **18:** 4345–4354.

Richardson, C.C. 1965. Phosphorylation of nucleic acid by an enzyme from T4 bacteriophage-infected *Escherichia coli. Proc. Natl. Acad. Sci.* **54:** 158–165.

Sinha, N.D., J. Biernat, J. McManus, and H. Köster. 1984. Polymer support oligonucleotide synthesis XVIII: Use of β-cyanoethyl-N,N-dialkylamino-/N-morpholino phosphoramidite of deoxynucleosides for the synthesis of DNA fragments simplifying deprotection and isolation of the final product. *Nucleic Acids Res.* **12:** 4539–4557.

Smith, L.M., S. Fung, M.W. Hunkapillar, T.J. Hunkapillar, and L.E. Hood. 1985. The synthesis of oligonucleotides containing an aliphatic amino group at the 5′ terminus: Synthesis of fluorescent DNA primers for use in DNA sequence analysis. *Nucleic Acids Res.* **13:** 2399–2419.

Telser, J., K.A. Cruickshank, K.S. Schanze, and T.L. Netzel. 1989. DNA oligomers and duplexes containing a covalently attached derivative of tris-(2,2′-bipyridine) ruthenium(II): Synthesis and characterization by thermodynamic and optical spectroscopic measurements. *J. Am. Chem. Soc.* **111:** 7221–7226.

Trainor, G.L. and M.A. Jensen. 1988. A procedure for the preparation of fluorescence-labeled DNA with terminal deoxynucleotidyl transferase. *Nucleic Acids Res.* **16:** 11846.

Suppliers

With the exception of the suppliers listed on the following pages, all suppliers mentioned in this manual can be found in the Cold Spring Harbor Laboratory Press *Lab Manual Source Book.* If you did not receive a copy of the *Source Book* with your *PCR Primer,* you can order your FREE copy by any of the following methods:

Call: (800) 843-4388 *or* (516) 367-8325

Fax: (516) 367-8432

E-mail: cshpress@cshl.org
World Wide Web Site http://www.cshl.org/

Write: Cold Spring Harbor Laboratory Press, 10 Skyline Drive, Plainview, NY 11803-2500

American Ultraviolet
562 Central Avenue
Murray Hill, NJ 07974
Phone: (908) 665-2211
Fax: (908) 665-9523

Collaborative Research, Inc.
2 Oak Park
Bedford, MA 01730
Phone: (617) 275-0004
Fax: (617) 275-0043

Enzo Diagnostics
60 Executive Blvd.
Farmingdale, NY 11735
Phone: (516) 496-8080 *or* (800) 221-7705
Fax: (516) 694-7501

Finnzymes OY
P.O. Box 148
02201 Espoo
Finland
Phone: 358 0 420 8077
Fax: 358 0 420 8653

GenHunter Co.
50 Boylston Street
Brookline, MA 02146
Phone: (617) 739-6771
Fax: (617) 734-5482

HRI Associates
2341 Stanwell Drive
Concord, CA 94520
Phone: (510) 687-1386
Fax: (510) 687-1388

Hybaid Omnigene
111-113 Waldegrave Road
Teddington
Middelsex Twill 8LL
England
Phone: +44 (0) 181 614 1000
Fax: +44 (0) 181 977 0170

Hyperion Inc.
14100 S.W. 136th Street
Miami, FL 33186
Phone: (305) 238-3020
Fax: (305) 232-7375

KREATECH Diagnostics
P.O. Box 12756
1100 AT Amsterdam
The Netherlands
Phone: 31 20 691 9181
Fax: 31 20 696 3531

M.H. Rhodes, Inc.
101 Thompson Road
Avon, CT 06001
Phone: (203) 673-3281

Novabiochem Ltd.
3 Heathcoat Building
Highfields Science Park
University Blvd.
Nottingham NG7 2QJ
England
Phone: 44 602 430840
Fax: 44 602 430951

Nunc, Inc.
2000 North Aurora Road
Naperville, IL 60563-1796
Phone: (708) 983-5700 *or* (800) 288-6862
Fax: (708) 416-2556

Phenix Research Products
3540 Arden Road
Hayward, CA 94545
Phone: (800) 767-0665
Fax: (510) 264-2030

PPG Company
3938 Porett Drive
Gurnee, IL 60031
Phone: (708) 244-3410
Fax: (708) 249-9716

Polyscientific Corp.
70 Cleveland Avenue
Bay Shore, NY 11706
Phone: (516) 586-0400
Fax: (516) 254-0618

Sarstedt, Inc.
P.O. Box 468
Newton, NC 28658-0468
Phone: (704) 465-4000 *or* (800) 257-5101
Fax: (704) 465-4003

SIBIA, Inc.
505 Coast Boulevard South
La Jolla, CA 92037
Phone: (619) 459-4101
Fax: (619) 452-9279

Tera Biotechnology Corp.
11099 N. Torrey Pines Road, Suite 230
La Jolla, CA 92037
Phone: (619) 535-5479
Fax: (619) 535-5472

TaKaRa Biomedicals
Takara Shuzo Co., Ltd.
Biomedical Group
Otsu, Shiga
Japan
Phone: 81 775 43 7247
Fax: 81 775 43 9254

UNELKO Corp.
7428 East Karen Drive
Scottsdale, AZ 85260
Phone: (800) 528-3149
Fax: (602) 483-7674

Trademarks

The following trademarks and registered trademarks are accurate to the best of our knowledge at the time of printing. Please consult individual manufacturers and other resources for specific information.

AMBIS™	CSPI - Scanalytics
AmeriClear™	Stephens Scientific
Amine-VN™	CLONTECH Laboratories, Inc.
AmpErase™	Roche Molecular Systems, Inc.
5′-AmpliFINDER™	CLONTECH Laboratories, Inc.
AmpliSensor®	Biotronics Corp.
AmpliTaq®	Roche Molecular Systems, Inc.
AmpliWax®PCR GEMS®	Roche Molecular Systems, Inc.
A.S.A.P.™	Boehringer Mannheim Corp.
BioMarker®	BioVentures, Inc.
Biomek®	Beckman Instruments, Inc.
Bio-Spin®10 column	Bio-Rad Laboratories
Biotin-ON™	CLONTECH Laboratories, Inc.
Bluescript®	Stratagene
CAMLIGHT™	Analytical Luminescence Laboratory
Carnation® nonfat dry milk	Nestlé Food Co.
ClearCut™	Stratagene
CLONEAMP™	Life Technologies, Inc.
Cyclist™	Stratagene
DeepVent®	New England Biolabs, Inc.
Duralon-UV™	Stratagene

Dynabeads®	DYNAL, Inc.
Econo-Column®	Bio-Rad Laboratories
EluQuik®	Schleicher & Schuell, Inc.
ExAssist™	Stratagene
Excel®	Microsoft Corp.
FALCON®	Becton Dickinson
FastTrack®Kit	Invitrogen Corp.
GELase™	Epicentre Technologies
GEL-MIX™	Life Technologies, Inc.
GeneAmp®	Roche Molecular Systems, Inc.
GENECLEAN®	BIO 101, Inc.
GeneReleaser™	BioVentures, Inc.
Gene Sweep™	Hoefer-Pharmacia Biotech
Gigapack®	Stratagene
Histochoice™	AMESCO
Histopaque®	Sigma
Hot Tub™	Amersham Life Science, Inc.
HYBOND™	Amersham Life Science, Inc.
HydroLink®	FMC BioProducts
ImageQuant™ Software	Molecular Dynamics, Inc.
Intelligenetics®Suite	Intelligenetics, Inc.
Kimwipes®	Kimberly-Clark Corp.
LIPOFECTACE™	Life Technologies, Inc.
Lumi-Phos®530	Lumigen, Inc.
Macintosh™	Apple Computers, Inc.
Macol®LA-12	PPG Industries, Specialty Chemicals Division
MAX EFFICIENCY®	Life Technologies, Inc.
MDE™	FMC BioProducts
MEGAscript™	Ambion
MERMAID®	BIO 101, Inc.
MicroAmp®	The Perkin-Elmer Corp.
Micro-FastTrack™ Kit	Invitrogen Corp.
MICROCON™	Amicon, Inc.
Micronic®tube rack	Micronic
MPC®-E	DYNAL, Inc.
MS-DOS®	Microsoft Corp.
NALGENE®	Nalge Co.
NAP™	Pharmacia Biotech AB
NASBA®HIV-1 RNA QT kit	Organon Teknika Corp.
NASBA®QR SYSTEM	Organon Teknika Corp.
NuSieve®	FMC BioProducts
Nytran®	Schleicher & Schuell, Inc.
OligoBeads™	Organon Teknika Corp.

OmniFix™	American Histology Reagent Co.
OPC™	The Perkin-Elmer Corp.
Opti-MEM™	Life Technologies, Inc.
Opti-Prime™	Stratagene
pCR™3	Invitrogen Corp.
pCR™II	Invitrogen Corp.
PCR Optimizer™ Kit	Invitrogen Corp.
pCR-Script™	Stratagene
pCR-Script™ Amp	Stratagene
pCR-Script™ Cam	Stratagene
pCR-Script™ Direct	Stratagene
pDIRECT™	CLONTECH Laboratories, Inc.
Permount™	Fisher Scientific
pGEM®	Promega Corp.
PhosphorImager™	Molecular Dynamics
Plexiglas®	Rohm and Hass Co.
Polaroid® Instant Image film	Polaroid Corp.
Polytron®	Kinematica AG
Prep-A-Gene®	Bio-Rad Laboratories
PRO/PETTE®	The Perkin-Elmer Corp.
PROTEAN®	Bio-Rad Laboratories
ProtoGEL®	National Diagnostics, Inc.
pSurfscript™	Stratagene
Pyrex®	Corning, Inc.
QIAGEN® Spin-20	Qiagen, Inc.
QIAquick™	Qiagen, Inc.
Retic Lysate IVT™	Ambion, Inc.
RNasin®	Promega Corp.
RNAzol™ B	Tel-Test, Inc.
S+S®903™ Specimen Colec. Paper	Schleicher & Schuell, Inc.
ScreenTest™	Stratagene
SeaKem®	FMC BioProducts
Sephacryl®	Pharmacia Biotech AB
Sephadex®	Pharmacia Biotech AB
Sepharose®	Pharmacia Biotech AB
SequaGEL®	National Diagnostics, Inc.
Sequenase™	Amersham Life Science, Inc.
SHARP Signal™	Digene Diagnostics, Inc.
SpeedVac®	Savant Instruments, Inc.
Spin-X® filter unit	Corning Costar Corp.
StrataClean™	Stratagene
SUPERSCRIPT™	Life Technologies, Inc.
SurfZAP™ λvector	Stratagene

Trademark	Owner
Taq Extender™	Stratagene
TaqMan™	Roche Molecular Systems, Inc.
TaqStart™	CLONTECH Laboratories, Inc.
TNT®	Promega Corp.
TRISACRYL®	BioSepra, Inc.
TRIzol™	Life Technologies, Inc.
Tropix™	Tropix, Inc.
UlTma™ DNA polymerase	Roche Molecular Systems, Inc.
Ultrafree™ MC filter	Millipore Corp.
UNIX™	Unix System Lab
Vent®	New England Biolabs, Inc.
Vortex Genie 2™	Scientific Industries
Whatman® 3MM paper	Whatman Ltd.
Wizard™ DNA Purification System	Promega Corp.
X-Omat™ film	Eastman Kodak Co.
YEASTMAKER™	CLONTECH Laboratories, Inc.
YEAST SPHEROPLAST™	BIO 101, Inc.
ZapCap®	Schleicher & Schuell, Inc.
ZAP-cDNA®	Stratagene
ZAP Express™	Stratagene
ZEPTO™	Molecular Biology Resources, Inc.
Zeta-Probe®	Bio-Rad Laboratories
ZINCOV™	Calbiochem-Novabiochem International, Inc.
ZWITTERGENT®	Calbiochem-Novabiochem International, Inc.
Zymolyase®	Seikaguka America, Inc.
Zysorbin™	Zymed Laboratories, Inc.

Index